T0215439

Lecture Notes in Computer Science　　11560

Commenced Publication in 1973
Founding and Former Series Editors:
Gerhard Goos, Juris Hartmanis, and Jan van Leeuwen

More information about this series at http://www.springer.com/series/7407

Carsten Lutz · Uli Sattler ·
Cesare Tinelli · Anni-Yasmin Turhan ·
Frank Wolter (Eds.)

Description Logic, Theory Combination, and All That

Essays Dedicated to Franz Baader
on the Occasion of His 60th Birthday

 Springer

Editors
Carsten Lutz
University of Bremen
Bremen, Germany

Uli Sattler
University of Manchester
Manchester, UK

Cesare Tinelli
University of Iowa
Iowa City, IA, USA

Anni-Yasmin Turhan
TU Dresden
Dresden, Germany

Frank Wolter
University of Liverpool
Liverpool, UK

ISSN 0302-9743   ISSN 1611-3349   (electronic)
Lecture Notes in Computer Science
ISBN 978-3-030-22101-0   ISBN 978-3-030-22102-7   (eBook)
https://doi.org/10.1007/978-3-030-22102-7

LNCS Sublibrary: SL1 – Theoretical Computer Science and General Issues

Cover illustration: Based on an idea by Anni-Yasmin Turhan, the cover illustration was specifically created for this volume by Stefan Borgwardt.

This Springer imprint is published by the registered company Springer Nature Switzerland AG
The registered company address is: Gewerbestrasse 11, 6330 Cham, Switzerland

Franz Baader

Preface

This Festschrift has been put together on the occasion of Franz Baader's 60th birthday to celebrate his scientific contributions. It was initiated by Anni-Yasmin Turhan, who brought in the other four editors. We contacted Franz's friends and colleagues, asking for their contributions, and the response was enthusiastic.

The result is a volume containing 30 articles from contributors all over the world, starting with an introductory article that provides our personal accounts of Franz's career and achievements and covering many of the several scientific areas Franz has worked on: description logics, unification and matching, term rewriting, and the combination of decision procedures. Although this volume does not come close to covering all of the work that Franz has done, we hope that the reader will gain some insights into the remarkable breadth and depth of his research over the past 30+ years.

We thank all contributors for their great work, for delivering high-quality manuscripts on time, and also for serving as reviewers for submissions by others. This volume would not have been possible without their exceptional effort.

April 2019

Carsten Lutz
Uli Sattler
Cesare Tinelli
Anni-Yasmin Turhan
Frank Wolter

Reviewers

Baumgartner, Peter
Bienvenu, Meghyn
Bonacina, Maria Paola
Borgida, Alex
Brewka, Gerhard
Britz, Arina
Casini, Giovanni
Claßen, Jens
Cuenca Grau, Bernardo
David Santos, Yuri
De Giacomo, Giuseppe
Ecke, Andreas
Eiter, Thomas
Finger, Marcelo
Ghilardi, Silvio
Grädel, Erich
Horrocks, Ian
Hölldobler, Steffen
Krötzsch, Markus
Lakemeyer, Gerhard
Lembo, Domenico
Lenzerini, Maurizio
Meyer, Tommie

Montali, Marco
Möller, Ralf
Narendran, Paliath
Nebel, Bernhard
Özcep, Özgür Lütfü
Peñaloza, Rafael
Ravishankar, Veena
Ringeissen, Christophe
Rosati, Riccardo
Rudolph, Sebastian
Schaub, Torsten
Schmidt, Renate
Schmidt-Schauß, Manfred
Tena Cucala, David
Thielscher, Michael
Toman, David
Ulbricht, Markus
Varzinczak, Ivan
Waldmann, Uwe
Wassermann, Renata
Weddell, Grant
Zakharyaschev, Michael

Contents

A Tour of Franz Baader's Contributions to Knowledge Representation and Automated Deduction

Carsten Lutz[1], Uli Sattler[2], Cesare Tinelli[3], Anni-Yasmin Turhan[4(✉)], and Frank Wolter[5]

[1] Fachbereich 03, Universität Bremen, Bremen, Germany
`clu@uni-bremen.de`
[2] School of Computer Science, University of Manchester, Manchester, UK
`uli.sattler@manchester.ac.uk`
[3] Department of Computer Science, The University of Iowa, Iowa City, USA
`cesare-tinelli@uiowa.edu`
[4] Institute for Theoretical Computer Science, Dresden University of Technology, Dresden, Germany
`Anni-Yasmin.Turhan@tu-dresden.de`
[5] Department of Computer Science, University of Liverpool, Liverpool, UK
`wolter@liverpool.ac.uk`

1 Introduction

This article provides an introduction to the Festschrift that has been put together on the occasion of Franz Baader's 60th birthday to celebrate his fundamental and highly influential scientific contributions. We start with a brief and personal overview of Franz's career, listing some important collaborators, places, and scientific milestones, and then provide first person accounts of how each one of us came in contact with Franz and how we benefitted from his collaboration and mentoring. Our selection is not intended to be complete and it is in fact strongly biased by our own personal experience and preferences. Many of Franz's contributions that we had to leave out are discussed in later chapters of this volume.

Franz was born in 1959 in Spalt, Germany, a small village known for its hop growing and its interesting Bavarian/Franconian accent—which seems to manifest especially after the consumption of hopped beverages. After high school and military service he studied computer science (*Informatik*) in nearby Erlangen. This included programming using punch cards and usage of the vi editor. Rumour has it that Franz still enjoys using some of this technology today. He continued with his Ph.D. on unification and rewriting, under the supervision of Klaus Leeb, an unconventional academic who clearly strengthened Franz's ability for independent research and critical thought. As a Ph.D. student, Franz also worked as a teaching assistant, with unusually high levels of responsibility.

In 1989, Franz completed his Ph.D. *Unifikation und Reduktionssysteme für Halbgruppenvarietäten* and moved to the German Research Center for Artificial

© Springer Nature Switzerland AG 2019
C. Lutz et al. (Eds.): Baader Festschrift, LNCS 11560, pp. 1–14, 2019.
https://doi.org/10.1007/978-3-030-22102-7_1

Intelligence (DFKI) in Saarbrücken as a post-doctoral researcher. It is there that he encountered description logic—then known as "terminological knowledge representation systems" or "concept languages"—met collaborators such as Bernhard Nebel, Enrico Franconi, Phillip Hanschke, Bernhard Hollunder, Werner Nutt, Jörg Siekman, and Gerd Smolka, and added another dimension to his multi-faceted research profile.

After 4 years at DFKI, in 1993, Franz successfully applied for his first professorship at RWTH Aachen and shortly thereafter for his first DFG project on *"Combination of special deduction procedures"*, which allowed him to hire his first externally funded Ph.D. student, Jörn Richts. Within a year of working in Aachen, he assembled his first research group, consisting of 4 Ph.D. students: Can Adam Albayrak, Jörn Richts, and Uli Sattler, together with Diego Calvanese visiting for a year from Rome. This group continued to grow substantially over the next decade, supported by various other DFG and EU-funded research projects as well as a DFG research training group.

In 2002, Franz applied for the Chair of Automata Theory at TU Dresden. Unsurprisingly, his sterling research track record made him the front runner for this post. As a consequence, a group of impressive size moved from Aachen to Dresden, including Franz and his family of three, Sebastian Brandt, Jan Hladik, Carsten Lutz, Anni-Yasmin Turhan, and Uli Sattler (Fig. 1).

Fig. 1. Most of the people who moved with Franz to Dresden: from left to right: Carsten, Anni, Franz, Uli, Sebastian; Jan took the photo, 2001.

They all settled quickly in the beautiful city on the river Elbe, to then experience a millennial flood in their first year. The bottom foot and a half of Franz's new family home in Dresden was flooded—but the damage was fortunately manageable thanks to the house's unusual design and their quick thinking which led them to move all items to the house's higher level. In the following years, Franz continued to grow his group even further, attracting many more students and grants, including, most notably, the QuantLA research training group on Quantitative Logics and Automata which he started and has led from its beginnings to today. He also became an ECCAI Fellow in 2004 and was Dean of the Faculty of Computer Science at TU Dresden from 2012 to 2015.

Throughout his career, Franz has supervised 26 Ph.D. students, five of whom successfully went on to receive their habilitation. He co-authored a textbook on Term Rewriting and one on Description Logic, and co-edited the Description Logic Handbook, all of which have become standard references in their respecting fields. At the time of this writing, according to Google Scholar, his publications have been cited more than 29,000 times. With more than 11,000 citations, the Description Logic Handbook [BCM+03] is his most cited work. The Term Rewriting textbook [BN98] is second with more than 3,000 citation while his research paper on tractable extensions of the description logic $\mathcal{EL}$ [BBL05] takes an impressive 3rd place with more than 1,000 citations. All this provides an excellent example of the high impact that Franz's work has had across several research areas.

2   Contributions

The following subsections provide a first person account of how each one of us came in contact with Franz and ended up enjoying a fruitful collaboration that has spanned many years. Nerdy as we are, we proceed in the order of earliest joint paper with him.

2.1   Uli Sattler: Classification and Subsumption in Expressive Description Logics

In 1993, Franz started his first professorship at RWTH Aachen University, where I joined his young research group in 1994 as one of his first Ph.D. students. I had never heard of description logics but relied on recommendations from former colleagues of Franz from Erlangen who assured me that Franz was a rising star and would make a great Ph.D. supervisor. My first task was to catch up on the already huge body of work that various people, including Franz, had established around description logics.

In the early 90s, Franz worked with Bernhard Hollunder in Saarbruecken on KRIS [BH91], a *terminological knowledge representation system* that implemented classification, realisation, retrieval for extensions of $\mathcal{ALCN}$ (e.g., with feature (dis-)agreement) with respect to acyclic TBoxes and ABoxes. This was a rather brave endeavour at the time, especially after the recent (1989), surprising results by Manfred Schmidt-Schauß that KLONE was undecidable and accumulating evidence (by Bernard Nebel, Klaus Schild, and others) that reasoning in all description logics is intractable once general TBoxes are included. The more common reaction to these insights was to severely restrict the expressive power or to move to even more expressive, undecidable description logics. Franz and Bernhard, however, went for a decidable yet intractable logic, where

*[...] the price one has to pay is that the worst case complexity of the algo-
rithms is worse than NP. But it is not clear whether the behaviour for
"typical" knowledge bases is also that bad.* [BH91]

The reasoner implemented in KRIS was based on a "completion algorithm"
developed by Schmidt-Schauß, Smolka, Nutt, Hollunder which would later be
called *tableau-based*. In [BHN+92], the authors introduce a first *Enhanced
Traversal* method: a crucial optimisation method that reduces the number of
subsumption test from n^2 to $n \log n$ and has been used and further enhanced in
all tableau-based reasoners for expressive DLs. Another highly relevant optimisation method first employed in KRIS is *lazy unfolding*, which enables *early clash
detection* and, again, both have been successfully employed and refined in other
reasoners.

Franz has also developed significant extensions to these first tableau-based
algorithms which required the design of novel, sophisticated techniques: for
example, in [Baa91a] a tableau algorithm for $\mathcal{ALC}$ extended with regular expressions on roles was described. This required not only a suitable cycle detection
mechanism (now known as *blocking*) but also the distinction between good and
bad cycles. Moreover, in this line of work, internalisation of TBoxes was first
described, a technique that can be used to reduce reasoning problems w.r.t. a
general TBox to pure concept reasoning and that turned out to be a powerful
tool to assess the computational complexity of a description logic.

Another significant extension relates to qualifying number restrictions
[HB91]: together with Bernhard Hollunder, Franz discovered the *yo-yo* problem and solved it by introducing explicit (in)equalities on individuals in completion system, thus avoiding non-termination. They also introduced the first
choose rule to avoid incorrectness caused by some tricky behaviour of qualifying
number restrictions.

I have mentioned these technical contributions here to illustrate the kind of
research that was going on at the time, and the many, significant contributions
Franz was involved in developing already at this early stage of his career. In addition, I also want to point out the ways in which Franz influenced the description
logic community, its methodologies, and its value system: as mentioned above,
he was an early advocate of understanding computational complexity beyond
the usual worst case. Moreover, he has always been an amazing explainer and
campaigner. He spent a lot of energy on discussions with colleagues and students
about the trio of soundness, completeness, and termination—and why it matters in description logic reasoners and related knowledge representation systems.
And he developed very clear proof techniques to show that a subsumption or
satisfiability algorithm is indeed sound, complete, and terminating. More generally, we appreciate Franz as a strong supporter of clarity (in proof, definitions,
descriptions, etc.) and as somebody who quickly recognises the murky "then a
miracle occurs" part in a proof or finds an elegant way to improve a definition.
On the occasion of his 60th birthday, I would like to say "Happy birthday, Franz,
and thank you for the ~~fish~~ clarity!".

2.2   Cesare Tinelli: Unification Theory, Term Rewriting and Combination of Decision Procedures

It is easy to argue about the significance of Franz's body of work and its long-lasting impact in several areas of knowledge representation and automated reasoning. Given my expertise, I could comment on the importance of his work in (term) unification theory where he has produced several results [Baa89, Baa91b, BS92, Baa93, BS95, BN96, Baa98, BM10, BBBM16, BBM16] and written authoritative compendiums [BS94, BS98c, BS01] on the topic. I could talk about his contributions to term rewriting, which include both research advances [Baa97] and the publication of a widely used textbook on "term rewriting and all that" [BN98]. I could say how his interest in the general problem of combining formal systems has led him to produce a large number of results on the combination of decision procedures or solvers for various problems [BS92, BS95, Baa97, BS98b, BT02a, BT02b, BGT04, BG05] and create a conference focused on combination, FroCoS [BS96], which is now at its 12-th edition beside being one of the founding member conferences of IJCAR, the biennial Joint Conference on Automated Reasoning. These topics are covered by the contributions in this volume by Peter Baumgartner and Uwe Waldmann, Maria Paola Bonacina et al., Veena Ravishankar et al., Christophe Ringeissen, Manfred Schmidt-Schauss, and Yoni Zohar et al. So, instead, I prefer to focus on more personal anecdotes which nevertheless illustrate why we are celebrating the man and his work with this volume.

At the start of my Ph.D. studies in the early 1990s I became interested in constraint logic programming and automated deduction. I was attracted early on by the problem of combining specialised decision procedures modularly and integrating them into general-purpose proof procedures. However, I found the foundational literature on general-purpose theorem proving and related areas such as term rewriting somewhat unappealing for what I thought was an excessive reliance on syntactical methods for proving correctness properties of the various proof calculi and systems. This was in contrast with much of the foundational work in (constrained) logic programming which was based on elegant and more intuitive algebraic and model-theoretic arguments. I also struggled to understand the literature on the combination decision procedures which I found wanting in clarity and precision.

This was the background when, while searching for related work, I came across a paper by some Franz Baader and Klaus Schulz on combining unification procedures for disjoint equational theories [BS92]. The paper presented a new combination method that, in contrast to previous ones, could be used to combine both decision procedures for unification and procedures for computing complete sets of unifiers. The method neatly extended a previous one by Manfred Schmidt-Schauß and was the start of a series of combination results by Franz and Klaus with increasing generality and formal elegance of the combination method [BS92, BS95, BS98b]. This line of work was significant also for often relying on algebraic arguments to prove the main theoretical results, for instance by exploiting the fact that certain free models of an equational theory are canonical for unification problems, or that computing unifiers in a combined theory can be reduced to solving equations in a model of the theory that is a specific amalgamation of the free models of the component theories.

Those papers, together with their associated technical reports, which included more details and full proofs, had a great impact on me. They showed how one could push the state of the art in automated reasoning with new theoretical results based on solid mathematical foundations while keeping a keen eye on practical implementation concerns. They were remarkable examples of how to write extremely rigorous theoretical material that was nonetheless quite understandable because the authors had clearly put great care in: explicitly highlighting the technical contribution and relating it to previous work; explaining the intuitive functioning of the new method; formatting the description of the method so that it was pleasing to the eye and easy to follow; explaining how the method, described at an abstract level, could be instantiated to concrete and efficient implementations; providing extensive proofs of the theoretical results that clearly explained all the intermediate steps.

Based on the early example of [BS92], I set to write a modern treatment of the well-known combination procedure by Nelson and Oppen [NO79] along the lines of Franz's paper while trying to achieve similar levels of quality. Once I made enough progress, I contacted Franz by email telling him about my attempts and asking for advice of how to address some challenges in my correctness proof. To my surprise and delight, he promptly replied to this email by an unknown Ph.D. student at the University of Illinois, and went on to provide his advice over the course of a long email exchange. I wrote the paper mostly as an exercise; as a way for me to understand the Nelson-Oppen method and explain it well to other novices like me. When I finished it and sent it to Franz for feedback he encouraged me to submit it to the very first edition of a new workshop on combining systems he had started with Klaus Schulz, FroCoS 1996. Not only was the paper accepted [TH96], it also became a widely cited reference in the field that later came to be known as Satisfiability Modulo Theories or SMT. As several people told me in person, the popularity of that paper is in large part due to its clarity and precision both in the description of the method and in the correctness proof, again something that I took from Franz's papers.

After we met in person at FroCoS 1996, Franz proposed to work on combination problems together, an opportunity I immediately accepted. That was the start of a long-lasting collaboration on the combination of decision procedures for the word problem [BT97,BT99,BT00,BT02b] and other more general problems [BT02a,BGT06]. That collaboration gave me the opportunity to appreciate Franz's vast knowledge and prodigious intellect. More important, it also gave me precious insights on how to develop abstract formal frameworks to describe automated reasoning methods, with the goal of understanding and proving their properties. It taught me how to develop soundness, completeness and termination arguments and turn them into reader-friendly mathematical proofs. I learned from him how to constantly keep the reader in mind when writing a technical paper, for instance by using consistent and intuitive notation, defining everything precisely while avoiding verbosity, using redundancy judiciously to remind the reader of crucial points or notions, and so on.

While I have eventually learned a lot also from other outstanding researchers and collaborators, my early exposure to Franz's work and my collaboration with him have profoundly affected the way I do research and write technical papers. I have actively tried over the years to pass on to my students and junior collaborators the principles and the deep appreciation of good, meticulous writing that I have learned from Franz.

Thank you, Franz, for being a collaborator and a model. It has been an honour and a pleasure. Happy 60th birthday, with many more to follow!

2.3 Carsten Lutz: Concrete Domains and the $\mathcal{EL}$ Family

When I was a student of computer science at the university of Hamburg, I became interested in the topic of description logics and I decided that I would like to do a Ph.D. in that area. At the time, I was particularly fascinated by concrete domains, the extension of DLs with concrete qualities such as numbers and strings as well as operations on them. As in Professor $\sqcap \exists$age.$=_{60}$. Franz was the definite authority on DLs, he seemed to have written at least half of all important papers and, what was especially spectacular for me, this guy had actually *invented* concrete domains (together with Hanschke [BHs91]). I was thus very happy when I was accepted as a Ph.D. student in his group at RWTH Aachen. Under Franz's supervision, I continued to study concrete domains and eventually wrote my Ph.D. thesis on the subject. I learned a lot during that time and I feel that I have especially benefitted from Franz's uncompromising formal rigor and from his ability to identify interesting research problems and to ask the right questions (even if, many times, I had no answer). Concrete domains are a good example. He identified the integration of concrete qualities into DLs as the important question that it is and came up with a formalization that was completely to the point and has never been questioned since. In fact, essentially the same setup has later been used in other areas such as XML, constraint LTL, and data words; it would be interesting to reconsider concrete domains today, from the perspective of the substantial developments in those areas. Over the years, Franz has continued to make interesting contributions to concrete domains, for example by adding aggregation (with Sattler [BS98a]) and rather recently by bringing into the picture uncertainty in the form of probability distributions over numerical values (with Koopmann and Turhan [BKT17]).

Another great line of research that Franz has pursued and that I had the pleasure to be involved in concerns lightweight DLs, in particular those of the $\mathcal{EL}$ family. In the early 2000s, there was a strong trend towards identifying more and more expressive DLs that would still be decidable. However, Franz also always maintained an interest in DLs with limited expressive power and better complexity of reasoning. The traditional family of inexpressive DLs was the $\mathcal{FL}$ family, but there even very basic reasoning problems are coNP-complete. In 2003, Franz wrote two IJCAI papers in which he considered $\mathcal{EL}$, which was unusual at the time, showing (among other things) that subsumption can be decided in polynomial time even in the presence of terminological cycles [Baa03a,Baa03b]. A bit later, this positive result was extended to general concept inclusions (GCIs) in joint work with Brandt and myself [Bra04,BBL05]. What followed was a success story. In joint work with Meng Suntisrivaraporn, we implemented the first $\mathcal{EL}$ reasoner called CEL which demonstrated that reasoning in $\mathcal{EL}$ is not only in polynomial time, but also very efficient and robust in practice. Many other reasoners have followed, the most prominent one today being ELK. We also explored the limits of polynomial time reasoning in the $\mathcal{EL}$ family [BLB08] and this resulted in a member of the $\mathcal{EL}$ family of DLs to be standardized as a profile of the W3C's OWL 2 ontology language. Nowadays, $\mathcal{EL}$ is one of the most standard families of DLs, widely used in many applications and also studied in several chapters of this volume, including the ones by Marcelo Finger, by Rafael Peñaloza, by Loris Bozzato et al. and by Ana Ozaki et al. Already in our initial work on $\mathcal{EL}$ with GCIs, we invented a particular kind of polynomial time reasoning procedure, the one that was also implemented in CEL. This type of procedure is now known as consequence-based reasoning and has found applications also far beyond the $\mathcal{EL}$ family of DLs. In fact, a survey on 15 years of consequence-based reasoning is presented in this volume in the chapter by David Tena Cucala, Bernardo Cuenca Grau and Ian Horrocks. It was a tremendous pleasure and privilege to have been involved in all this, together with you, Franz, and building on your prior work. Happy birthday!

2.4 Frank Wolter: Modal, Temporal, and Action Logics

I first met Franz in the summer of 1997 at ESSLLI in Aix-en-Provence, where I (jointly with Michael Zakharyaschev) organised a workshop on combining logics and, I believe, Franz gave a course introducing description logics. I am not entirely sure about the course as I very clearly recall our conversations about description logics but not at all any description logic lectures. At that point, after having failed to sell modal logic to mathematicians, I was looking for new ways of applying modal logic in computing and/or AI. And there the applications were right in front of me! As Franz quickly explained, description logic is nothing but modal logic, but much more relevant and with many exciting new applications and open problems. As long as one does not try to axiomatize description logics, there would be a huge interest in the description logic community in using techniques from modal logic and also in combining modal and description logics. So that is what I did over the next 22 years. The most obvious way to do description logic as a modal logician was to carefully read the papers on modal description logics that Franz et al. had just published [BL95,BO95], ask Franz what he regarded as interesting problems that were left open, and try to solve as many as possible of them (fortunately, Franz has the amazing ability to pose many more open problems than one could ever hope to solve). But Franz did not only pose open problems! He continued to work himself on temporal description logics with Ghilardi and Lutz [BGL12] and, more recently, with Stefan Borgwardt and Marcel Lippmann on combining temporal and description logics to design temporal query languages for monitoring purposes [BL14, BBL15], a topic on which Franz gave an invited keynote address at the Vienna Summer of Logic in 2014.

Other exciting collaborations developed over the years: when working together on combining description logics (with themselves) [BLSW02], I learned a lot from Franz's work on combining equational theories and on combining computational systems in general. Working together on the connection between tableaux and automata [BHLW03] was an excellent opportunity to let Franz explain to me what a tree automaton actually is. We only briefly worked together on extending description logics by action formalisms [BLM+05], but later Franz, together with Gerd Lakemeyer and many others, developed an amazing theory combining GOLOG and description logics, details of this collaboration are given in the article "Situation Calculus Meets Description Logics" by Jens Claßen, Gerhard Lakemeyer, and Benjamin Zarrieß in this volume. So, Franz, many thanks for both the problems and the solutions. It has been a great pleasure to work with you over so many years. Happy Birthday!

2.5   Anni-Yasmin Turhan: Non-standard Inferences in Description Logics

When I studied computer science at the University of Hamburg a project on knowledge representation had sparked my interest in description logics. I found the formal properties, the simplicity and elegance of these logics immediately appealing. As I soon noticed, most of the fundamental research results on description logics were achieved by Franz and his collaborators and, so after completing my studies, I was keen to join Franz's group in Aachen to start my Ph.D. studies. There I started to work in his research project on non-standard inferences in description logics together with Sebastian Brandt.

Non-standard inferences are a collection of various reasoning services for description logic knowledge bases that complement the traditional reasoning problems such as subsumption or satisfiability. The idea is that they assist users in developing, maintaining and integrating knowledge bases. In order to build and augment knowledge bases, inferences that generalize knowledge can be important. Franz had, together with Ralf Küsters and Ralf Molitor, investigated the *most specific concept* that can generalize knowledge about an object into a concept description and the *least common subsumer* that generalizes a set of concepts into a single one. Together these two inferences give rise to example-driven learning of new concepts. Their initial results were achieved for $\mathcal{EL}$ concepts without a general TBox [BKM99]. At that time it was quite a bit non-standard to work on inexpressive, light-weight description logics as many research efforts were dedicated to satisfiability of highly expressive logics. The overall approach to generalization is to ensure the instance-of or the subsumption relationship of the resulting concept by an embedding into the normalized input. This method was explored by us in several settings [BST07]. Franz lifted this also to general TBoxes [Baa03a, Baa03b], which had then lead to the famed polynomial time reasoning algorithms for $\mathcal{EL}$. These are based on canonical models and simulations that are widely used today and have, in turn, fueled further research on new non-standard inferences such as conservative extensions and computation of modules that were investigated in great detail by Carsten Lutz and Frank Wolter.

Besides generalization, Franz also introduced and investigated other non-standard inferences that compute "leaner" representations of concepts. One instantiation of this idea is to compute syntactically minimal, but equivalent rewritings of concepts and another is to compute "(upper) approximations" of concepts written in an expressive description logic in a less expressive one [BK06]. Franz combined his research interests and great expertise in unification and knowledge representation in a strand of work on matching in description logics. This inference is mainly used to detect redundancies in knowledge bases. Here Franz achieved many contributions for sub-Boolean description logics—in the last years predominantly in collaboration with Morawska and Borgwardt [BM10, BBM12, BBM16]. Franz's many contributions on versatile non-standard inferences demonstrates that his research topics are in overwhelming majority driven by a clear motivation that is often drawn from practical applications. Furthermore, they often times establish connections between several sub-areas of knowledge representation and theoretical computer science.

Once a knowledge base is built, explaining and removing unwanted consequences might become necessary as well. This can be done by computing *justifications*, i.e. the minimal axiom sets that are "responsible" for the consequence. This inference was and still is intensively investigated by Franz and his group—especially in collaboration with Rafael Peñaloza. They have mapped out the connection between computing justifications and weighted automata and are studying gentle, i.e. more fine-grained repairs that detect responsible parts of axioms [BKNP18]. Rafael tells the full story about it in "Explaining Axiom Pinpointing" in this volume. Their contributions on justifications are fruitfully applied also in other areas of knowledge representation, such as inconsistency-tolerant reasoning or nonmonotonic reasoning. This is underlined by Gerhard Brewka and Markus Ulbricht in their article on "Strong Explanations for Non-monotonic Reasoning" and also by Vinícius Bitencourt Matos et al. in their article on "Pseudo-contractions as gentle repairs" presented in this volume.

So, what started out as non-standard inferences in description logics—I seem to remember that Franz even coined that term—has become a well-established part of the wider research field. This is certainly due to Franz's ability to identify clear motivation for research questions, his passion for clear explanations and his relentless pursuit of excellence. It has been truly fascinating for me to see this versatile research area grow and evolve over the many years that I have worked with him.

Franz, I thank you for the many chances being given for example-driven learning and the gentle explanations. Happy birthday!

3   Final Words

Although this article and volume are by no means a complete overview, we hope that the reader will gain some insight into the remarkable breadth and depth of the research contributions that Franz Baader has made in the last 30+ years. What is more, he has achieved all this while keeping up his favourite pastimes such as skiing and cycling, while being a proud and loving father to his three children—and without ever cutting off his pony tail, see Fig. 2.

Fig. 2. Franz in a typical pose, giving a clear explanation of a technically complex point.

We hope that Franz will enjoy reading about our views of his research record and our experience in working with him, as well as the many articles in this Franzschrift. We thank him for his advice, guidance, and friendship, and wish him—again and all together now

a very happy birthday and many happy returns!

References

[Baa89]   Baader, F.: Characterizations of unification type zero. In: Dershowitz, N. (ed.) RTA 1989. LNCS, vol. 355, pp. 2–14. Springer, Heidelberg (1989). https://doi.org/10.1007/3-540-51081-8_96

[Baa91a]   Baader, F.: Augmenting concept languages by transitive closure of roles: an alternative to terminological cycles. In: Proceedings of IJCAI (1991)

[Baa91b] Baader, F.: Unification, weak unification, upper bound, lower bound, and generalization problems. In: Book, R.V. (ed.) RTA 1991. LNCS, vol. 488, pp. 86–97. Springer, Heidelberg (1991). https://doi.org/10.1007/3-540-53904-2_88

[Baa93] Baader, F.: Unification in commutative theories, Hilbert's basis theorem and Gröbner bases. J. ACM **40**(3), 477–503 (1993)

[Baa97] Baader, F.: Combination of compatible reduction orderings that are total on ground terms. In: Proceedings of LICS, pp. 2–13 (1997)

[Baa98] Baader, F.: On the complexity of Boolean unification. Inf. Process. Lett. **67**(4), 215–220 (1998)

[Baa03a] Baader, F.: Least common subsumers and most specific concepts in a description logic with existential restrictions and terminological cycles. In: Proceedings of IJCAI, pp. 319–324 (2003)

[Baa03b] Baader, F.: Terminological cycles in a description logic with existential restrictions. In: Proceedings of IJCAI, pp. 325–330 (2003)

[BBBM16] Baader, F., Binh, N.T., Borgwardt, S., Morawska, B.: Deciding unifiability and computing local unifiers in the description logic $\mathcal{EL}$ without top constructor. Notre Dame J. Formal Logic **57**(4), 443–476 (2016)

[BBL05] Baader, F., Brandt, S., Lutz, C.: Pushing the $\mathcal{EL}$ envelope. In: Proceedings of IJCAI, pp. 364–369 (2005)

[BBL15] Baader, F., Borgwardt, S., Lippmann, M.: Temporal query entailment in the description logic $\mathcal{SHQ}$. J. Web Semant. **33**, 71–93 (2015)

[BBM12] Baader, F., Borgwardt, S., Morawska, B.: Extending unification in $\mathcal{EL}$ towards general TBoxes. In: Proceedings of KR, pp. 568–572 (2012)

[BBM16] Baader, F., Borgwardt, S., Morawska, B.: Extending unification in $\mathcal{EL}$ to disunification: the case of dismatching and local disunification. Log. Methods Comput. Sci. **12**(4:1), 1–28 (2016)

[BCM+03] Baader, F., Calvanese, D., McGuinness, D., Nardi, D., Patel-Schneider, P. (eds.): The Description Logic Handbook: Theory, Implementation, and Applications. Cambridge University Press, Cambridge (2003)

[BG05] Baader, F., Ghilardi, S.: Connecting many-sorted theories. In: Nieuwenhuis, R. (ed.) CADE 2005. LNCS, vol. 3632, pp. 278–294. Springer, Heidelberg (2005). https://doi.org/10.1007/11532231_21

[BGL12] Baader, F., Ghilardi, S., Lutz, C.: LTL over description logic axioms. ACM Trans. Comput. Logic **13**(3), 21:1–21:32 (2012)

[BGT04] Baader, F., Ghilardi, S., Tinelli, C.: A new combination procedure for the word problem that generalizes fusion decidability results in modal logics. In: Basin, D., Rusinowitch, M. (eds.) IJCAR 2004. LNCS, vol. 3097, pp. 183–197. Springer, Heidelberg (2004). https://doi.org/10.1007/978-3-540-25984-8_11

[BGT06] Baader, F., Ghilardi, S., Tinelli, C.: A new combination procedure for the word problem that generalizes fusion decidability results in modal logics. Inf. Comput. **10**, 1413–1452 (2006)

[BH91] Baader, F., Hollunder, B.: A terminological knowledge representation system with complete inference algorithms. In: Boley, H., Richter, M.M. (eds.) PDK 1991. LNCS, vol. 567, pp. 67–86. Springer, Heidelberg (1991). https://doi.org/10.1007/BFb0013522

[BHLW03] Baader, F., Hladik, J., Lutz, C., Wolter, F.: From tableaux to automata for description logics. Fundamenta Informatica **57**(2–4), 247–279 (2003)

[BHN+92] Baader, F., Hollunder, B., Nebel, B., Profitlich, H.-J., Franconi, E.: An empirical analysis of optimization techniques for terminological representation systems. In: Proceedings of KR, pp. 270–281 (1992)

[BHs91] Baader, F., Hanschke, P.: A schema for integrating concrete domains into concept languages. In: Proceedings of IJCAI, pp. 452–457 (1991)

[BK06] Baader, F., Küsters, R.: Nonstandard inferences in description logics: the story so far. In: Gabbay, D.M., Goncharov, S.S., Zakharyaschev, M. (eds.) Mathematical Problems from Applied Logic I. IMAT, vol. 4, pp. 1–75. Springer, New York (2006). https://doi.org/10.1007/0-387-31072-X_1

[BKM99] Baader, F., Küsters, R., Molitor, R.: Computing least common subsumers in description logics with existential restrictions. In: Proceedings of IJCAI, pp. 96–101 (1999)

[BKNP18] Baader, F., Kriegel, F., Nuradiansyah, A., Peñaloza, R.: Making repairs in description logics more gentle. In: Proceedings of KR, pp. 319–328 (2018)

[BKT17] Baader, F., Koopmann, P., Turhan, A.-Y.: Using ontologies to query probabilistic numerical data. In: Dixon, C., Finger, M. (eds.) FroCoS 2017. LNCS, vol. 10483, pp. 77–94. Springer, Cham (2017). https://doi.org/10.1007/978-3-319-66167-4_5

[BL95] Baader, F., Laux, A.: Terminological logics with modal operators. In: Proceedings of IJCAI, pp. 808–815 (1995)

[BL14] Baader, F., Lippmann, M.: Runtime verification using the temporal description logic $\mathcal{ALC}$-LTL revisited. J. Appl. Logic **12**(4), 584–613 (2014)

[BLB08] Baader, F., Lutz, C., Brandt, S.: Pushing the $\mathcal{EL}$ envelope further. In: Proceedings of OWLED (2008)

[BLM+05] Baader, F., Lutz, C., Milicic, M., Sattler, U., Wolter, F.: Integrating description logics and action formalisms: first results. In: Proceedings of IJCAI, pp. 572–577 (2005)

[BLSW02] Baader, F., Lutz, C., Sturm, H., Wolter, F.: Fusions of description logics and abstract description systems. J. Artif. Intell. Res. **16**, 1–58 (2002)

[BM10] Baader, F., Morawska, B.: SAT encoding of unification in $\mathcal{EL}$. In: Fermüller, C.G., Voronkov, A. (eds.) LPAR 2010. LNCS, vol. 6397, pp. 97–111. Springer, Heidelberg (2010). https://doi.org/10.1007/978-3-642-16242-8_8

[BN96] Baader, F., Nutt, W.: Combination problems for commutative/monoidal theories: how algebra can help in equational reasoning. J. Appl. Algebra Eng. Commun. Comput. **7**(4), 309–337 (1996)

[BN98] Baader, F., Nipkow, T.: Term Rewriting and All That. Cambridge University Press, Cambridge (1998)

[BO95] Baader, F., Ohlbach, H.J.: A multi-dimensional terminological knowledge representation language. J. Appl. Non-Class. Logics **5**(2), 153–197 (1995)

[Bra04] Brandt, S.: Polynomial time reasoning in a description logic with existential restrictions, GCI axioms, and - what else? In: Proceedings of ECAI, pp. 298–302 (2004)

[BS92] Baader, F., Schulz, K.U.: Unification in the union of disjoint equational theories: combining decision procedures. In: Kapur, D. (ed.) CADE 1992. LNCS, vol. 607, pp. 50–65. Springer, Heidelberg (1992). https://doi.org/10.1007/3-540-55602-8_155

[BS94] Baader, F., Siekmann, J.: Unification theory. In: Gabbay, D.M., Hogger, C.J., Robinson, J.A. (eds.) Handbook of Logic in Artificial Intelligence and Logic Programming, pp. 41–125. Oxford University Press, Oxford (1994)

[BS95]   Baader, F., Schulz, K.: Combination techniques and decision problems for disunification. Theoret. Comput. Sci. **142**, 229–255 (1995)

[BS96]   Baader, F., Schulz, K. (eds.): Frontiers of Combining Systems. Proceedings of First International Workshop. Kluwer Academic Publishers, Dordrecht (1996)

[BS98a] Baader, F., Sattler, U.: Description logics with aggregates and concrete domains. In: Proceedings of ECAI (1998)

[BS98b] Baader, F., Schulz, K.: Combination of constraint solvers for free and quasi-free structures. Theoret. Comput. Sci. **192**, 107–161 (1998)

[BS98c] Baader, F., Schulz, K.: Unification theory. In: Bibel, W., Schmidt, P.H. (eds.) Automated Deduction - A Basis for Applications, Vol. I: Foundations - Calculi and Methods. Applied Logic Series, vol. 8, pp. 225–263. Kluwer Academic Publishers, Dordrecht (1998)

[BS01]   Baader, F., Snyder, W.: Unification theory. In: Robinson, J.A., Voronkov, A. (eds.) Handbook of Automated Reasoning, vol. I, pp. 447–533. Elsevier Science Publishers, Amsterdam (2001)

[BST07] Baader, F., Sertkaya, B., Turhan, A.-Y.: Computing the least common subsumer w.r.t. a background terminology. J. Appl. Logic **5**(3), 392–420 (2007)

[BT97]   Baader, F., Tinelli, C.: A new approach for combining decision procedures for the word problem, and its connection to the Nelson-Oppen combination method. In: McCune, W. (ed.) CADE 1997. LNCS, vol. 1249, pp. 19–33. Springer, Heidelberg (1997). https://doi.org/10.1007/3-540-63104-6_3

[BT99]   Baader, F., Tinelli, C.: Deciding the word problem in the union of equational theories sharing constructors. In: Narendran, P., Rusinowitch, M. (eds.) RTA 1999. LNCS, vol. 1631, pp. 175–189. Springer, Heidelberg (1999). https://doi.org/10.1007/3-540-48685-2_14

[BT00]   Baader, F., Tinelli, C.: Combining equational theories sharing non-collapse-free constructors. In: Kirchner, H., Ringeissen, C. (eds.) FroCoS 2000. LNCS, vol. 1794, pp. 260–274. Springer, Heidelberg (2000). https://doi.org/10.1007/10720084_17

[BT02a] Baader, F., Tinelli, C.: Combining decision procedures for positive theories sharing constructors. In: Tison, S. (ed.) RTA 2002. LNCS, vol. 2378, pp. 352–366. Springer, Heidelberg (2002). https://doi.org/10.1007/3-540-45610-4_25

[BT02b] Baader, F., Tinelli, C.: Deciding the word problem in the union of equational theories. Inf. Comput. **178**(2), 346–390 (2002)

[HB91]   Hollunder, B., Baader, F.: Qualifying number restrictions in concept languages. In: Proceedings of KR, pp. 335–346 (1991)

[NO79]   Nelson, G., Oppen, D.C.: Simplification by cooperating decision procedures. ACM Trans. Program. Lang. Syst. **1**(2), 245–257 (1979)

[TH96]   Tinelli, C., Harandi, M.: A new correctness proof of the Nelson-Oppen combination procedure. In: Baader, F., Schulz, K.U. (eds.) Frontiers of Combining Systems. ALS, vol. 3, pp. 103–119. Springer, Dordrecht (1996). https://doi.org/10.1007/978-94-009-0349-4_5

Hierarchic Superposition Revisited

Peter Baumgartner[1(✉)] and Uwe Waldmann[2(✉)]

[1] Data61/CSIRO, ANU Computer Science and Information Technology (CSIT),
Acton, Australia
`Peter.Baumgartner@data61.csiro.au`
[2] Max-Planck-Institut für Informatik, Saarbrücken, Germany
`uwe@mpi-inf.mpg.de`

Abstract. Many applications of automated deduction require reasoning
in first-order logic modulo background theories, in particular some form
of integer arithmetic. A major unsolved research challenge is to design
theorem provers that are "reasonably complete" even in the presence of
free function symbols ranging into a background theory sort. The hierarchic superposition calculus of Bachmair, Ganzinger, and Waldmann
already supports such symbols, but, as we demonstrate, not optimally.
This paper aims to rectify the situation by introducing a novel form of
clause abstraction, a core component in the hierarchic superposition calculus for transforming clauses into a form needed for internal operation.
We argue for the benefits of the resulting calculus and provide two new
completeness results: one for the fragment where all background-sorted
terms are ground and another one for a special case of linear (integer or
rational) arithmetic as a background theory.

Keywords: Automated deduction · Superposition calculus ·
Combinations of theories

1 Introduction

Many applications of automated deduction require reasoning with respect to a
combination of a background theory, say integer arithmetic, and a foreground
theory that extends the background theory by new sorts such as *list*, new operators, such as *cons* : *int* × *list* → *list* and *length* : *list* → *int*, and first-order
axioms. Developing corresponding automated reasoning systems that are also
able to deal with quantified formulas has recently been an active area of research.
One major line of research is concerned with extending (SMT-based) solvers [24]
for the quantifier-free case by instantiation heuristics for quantifiers [17,18, e. g.].
Another line of research is concerned with adding black-box reasoners for specific background theories to first-order automated reasoning methods (resolution [1,5,19], sequent calculi [26], instantiation methods [8,9,16], etc). In both
cases, a major unsolved research challenge is to provide reasoning support that is
"reasonably complete" in practice, so that the systems can be used more reliably
for both proving theorems and finding counterexamples.

© Springer Nature Switzerland AG 2019
C. Lutz et al. (Eds.): Baader Festschrift, LNCS 11560, pp. 15–56, 2019.
https://doi.org/10.1007/978-3-030-22102-7_2

In [5], Bachmair, Ganzinger, and Waldmann introduced the hierarchical superposition calculus as a generalization of the superposition calculus for black-box style theory reasoning. Their calculus works in a framework of hierarchic specifications. It tries to prove the unsatisfiability of a set of clauses with respect to interpretations that extend a background model such as the integers with linear arithmetic conservatively, that is, without identifying distinct elements of old sorts ("confusion") and without adding new elements to old sorts ("junk"). While confusion can be detected by first-order theorem proving techniques, junk can not – in fact, the set of logical consequences of a hierarchic specifications is usually not recursively enumerable. Refutational completeness can therefore only be guaranteed if one restricts oneself to sets of formulas where junk can be excluded a priori. The property introduced by Bachmair, Ganzinger, and Waldmann for this purpose is called "sufficient completeness with respect to simple instances". Given this property, their calculus is refutationally complete for clause sets that are fully abstracted (i.e., where no literal contains both foreground and background symbols). Unfortunately their full abstraction rule may destroy sufficient completeness with respect to simple instances. We show that this problem can be avoided by using a new form of clause abstraction and a suitably modified hierarchical superposition calculus. Since the new calculus is still refutationally complete and the new abstraction rule is guaranteed to preserve sufficient completeness with respect to simple instances, the new combination is strictly more powerful than the old one.

In practice, sufficient completeness is a rather restrictive property. While there are application areas where one knows in advance that every input is sufficiently complete, in most cases this does not hold. As a user of an automated theorem prover, one would like to see a best effort behavior: The prover might for instance try to *make* the input sufficiently complete by adding further theory axioms. In the calculus from [5], this does not help at all: The restriction to a particular kind of instantiations ("simple instances") renders theory axioms essentially unusable in refutations. We show that this can be prevented by introducing two kinds of variables of the background theory sorts, that can be instantiated in different ways, making our calculus significantly "more complete" in practice. We also include a definition rule in the calculus that can be used to establish sufficient completeness by linking foreground terms to background parameters, thus allowing the background prover to reason about these terms.

The following trivial example demonstrates the problem. Consider the clause set $N = \{C\}$ where $C = f(1) < f(1)$. Assume that the background theory is integer arithmetic and that f is an integer-sorted operator from the foreground (free) signature. Intuitively, one would expect N to be unsatisfiable. However, N is not sufficiently complete, and it admits models in which $f(1)$ is interpreted as some junk element ϕ, an element of the domain of the integer sort that is not a numeric constant. So both the calculus in [5] and ours are excused to not find a refutation. To fix that, one could add an instance $C' = \neg(f(1) < f(1))$ of the irreflexivity axiom $\neg(x < x)$. The resulting set $N' = \{C, C'\}$ is (trivially) sufficiently complete as it has no models at all. However, the calculus in [5] is not helped by adding C', since the abstracted version of N' is again not sufficiently complete and admits a model that interprets $f(1)$ as $\not\in$. Our abstraction mechanism always preserves sufficient completeness and our calculus will find a refutation.

With this example one could think that replacing the abstraction mechanism in [5] with ours gives all the advantages of our calculus. But this is not the case. Let $N'' = \{C, \neg(x < x)\}$ be obtained by adding the more realistic axiom $\neg(x < x)$. The set N'' is still sufficiently complete with our approach thanks to having two kinds of variables at disposal, but it is not sufficiently complete in the sense of [5]. Indeed, in that calculus adding background theory axioms *never* helps to gain sufficient completeness, as variables there have only one kind.

Another alternative to make N sufficiently complete is by adding a clause that forces $f(1)$ to be equal to some background domain element. For instance, one can add a "definition" for $f(1)$, that is, a clause $f(1) \approx \alpha$, where α is a fresh symbolic constant belonging to the background signature (a "parameter"). The set $N''' = \{C, f(1) \approx \alpha\}$ is sufficiently complete and it admits refutations with both calculi. The definition rule in our calculus mentioned above will generate this definition automatically. Moreover, the set N belongs to a syntactic fragment for which we can guarantee not only sufficient completeness (by means of the definition rule) but also refutational completeness.

We present the new calculus in detail and provide a general completeness result, modulo compactness of the background theory, and two specific completeness results for clause sets that do not require compactness – one for the fragment where all background-sorted terms are ground and another one for a special case of linear (integer or rational) arithmetic as a background theory.

We also report on experiments with a prototypical implementation on the TPTP problem library [27].

Sections 1–7, 9–10, and 12 of this paper are a substantially expanded and revised version of [11]. A preliminary version of Sect. 11 has appeared in [10]. However, we omit from this paper some proofs that are not essential for the understanding of the main ideas. They can be found in a slightly extended version of this paper at http://arxiv.org/abs/1904.03776 [12].

Related Work. The relation with the predecessor calculus in [5] is discussed above and also further below. What we say there also applies to other developments rooted in that calculus, [1, e. g.]. The specialized version of hierarchic superposition in [22] will be discussed in Sect. 9 below. The resolution calculus in [19] has built-in inference rules for linear (rational) arithmetic, but is complete only under restrictions that effectively prevent quantification over rationals. Earlier work on integrating theory reasoning into model evolution [8,9] lacks the treatment of background-sorted foreground function symbols. The same applies to the sequent calculus in [26], which treats linear arithmetic with built-in rules for quantifier elimination. The instantiation method in [16] requires an answer-complete solver for the background theory to enumerate concrete solutions of background constraints, not just a decision procedure. All these approaches have in common that they integrate specialized reasoning for background theories into a general first-order reasoning method. A conceptually different approach consists in using first-order theorem provers as (semi-)decision procedures for specific theories in DPLL(T)(-like) architectures [2,13,14]. Notice that in this context the theorem provers do not need to reason modulo background theories themselves, and indeed they don't. The calculus and system in [14], for instance, integrates superposition and DPLL(T). From DPLL(T) it inherits splitting of ground non-unit clauses into their unit components, which determines a (backtrackable) model candidate M. The superposition inference rules are applied to elements from M and a current clause set F. The superposition component guarantees refutational completeness for pure first-order clause logic. Beyond that, for clauses containing background-sorted variables, (heuristic) instantiation is needed. Instantiation is done with ground terms that are provably equal w.r.t. the equations in M to some ground term in M in order to advance the derivation. The limits of that method can be illustrated with an (artificial but simple) example. Consider the unsatisfiable clause set $\{i \leq j \lor \mathsf{P}(i+1, x) \lor \mathsf{P}(j+2, x),\ i \leq j \lor \neg\mathsf{P}(i+3, x) \lor \neg\mathsf{P}(j+4, x)\}$ where i and j are integer-sorted variables and x is a foreground-sorted variable. Neither splitting into unit clauses, superposition calculus rules, nor instantiation applies, and so the derivation gets stuck with an inconclusive result. By contrast, the clause set belongs to a fragment that entails sufficient completeness ("no background-sorted foreground function symbols") and hence is refutable by our calculus. On the other hand, heuristic instantiation does have a place in our calculus, but we leave that for future work.

2   Signatures, Clauses, and Interpretations

We work in the context of standard many-sorted logic with first-order signatures comprised of sorts and operator (or function) symbols of given arities over these sorts. A *signature* is a pair $\Sigma = (\Xi, \Omega)$, where Ξ is a set of *sorts* and Ω is a set of *operator symbols* over Ξ. If $\mathcal{X}$ is a set of sorted variables with sorts in Ξ, then the set of well-sorted terms over $\Sigma = (\Xi, \Omega)$ and $\mathcal{X}$ is denoted by $\mathrm{T}_\Sigma(\mathcal{X})$; T_Σ is short for $\mathrm{T}_\Sigma(\emptyset)$. We require that Σ is a *sensible* signature, i.e., that T_Σ has no empty sorts. As usual, we write $t[u]$ to indicate that the term u is a (not necessarily proper) subterm of the term t. The position of u in t is left implicit.

A Σ-*equation* is an unordered pair (s, t), usually written $s \approx t$, where s and t are terms from $\mathrm{T}_\Sigma(\mathcal{X})$ of the same sort. For simplicity, we use equality as the only predicate in our language. Other predicates can always be encoded as a function into a set with one distinguished element, so that a non-equational atom is turned into an equation $\mathsf{P}(t_1, \ldots, t_n) \approx true_\mathsf{P}$; this is usually abbreviated by $\mathsf{P}(t_1, \ldots, t_n)$.[1] A *literal* is an equation $s \approx t$ or a negated equation $\neg(s \approx t)$, also written as $s \not\approx t$. A *clause* is a multiset of literals, usually written as a disjunction; the empty clause, denoted by $\square$ is a contradiction. If F is a term, equation, literal or clause, we denote by $\mathrm{vars}(F)$ the set of variables that occur in F. We say F is *ground* if $\mathrm{vars}(F) = \emptyset$.

A *substitution* σ is a mapping from variables to terms that is sort respecting, that is, maps each variable $x \in \mathcal{X}$ to a term of the same sort. Substitutions are

---

[1] Without loss of generality we assume that there exists a distinct sort for every predicate.

homomorphically extended to terms as usual. We write substitution application in postfix form. A term s is an *instance* of a term t if there is a substitution σ such that $t\sigma = s$. All these notions carry over to equations, literals and clauses in the obvious way. The composition $\sigma\tau$ of the substitutions σ and τ is the substitution that maps every variable x to $(x\sigma)\tau$.

The *domain* of a substitution σ is the set $\mathrm{dom}(\sigma) = \{x \mid x \neq x\sigma\}$. We use only substitutions with finite domains, written as $\sigma = [x_1 \mapsto t_1, \ldots, x_n \mapsto t_n]$ where $\mathrm{dom}(\sigma) = \{x_1, \ldots, x_n\}$. A *ground substitution* is a substitution that maps every variable in its domain to a ground term. A *ground instance* of F is obtained by applying some ground substitution with domain (at least) $\mathrm{vars}(F)$ to it.

A Σ-*interpretation* I consists of a Ξ-sorted family of carrier sets $\{I_\xi\}_{\xi \in \Xi}$ and of a function $I_f : I_{\xi_1} \times \cdots \times I_{\xi_n} \to I_{\xi_0}$ for every $f : \xi_1 \ldots \xi_n \to \xi_0$ in Ω. The *interpretation* t^I of a ground term t is defined recursively by $f(t_1, \ldots, t_n)^I = I_f(t_1^I, \ldots, t_n^I)$ for $n \geq 0$. An interpretation I is called *term-generated*, if every element of an I_ξ is the interpretation of some ground term of sort ξ. An interpretation I is said to *satisfy* a ground equation $s \approx t$, if s and t have the same interpretation in I; it is said to *satisfy* a negated ground equation $s \not\approx t$, if s and t do not have the same interpretation in I. The interpretation I *satisfies* a ground clause C if at least one of the literals of C is satisfied by I. We also say that a ground clause C is *true in* I, if I satisfies C; and that C is *false in* I, otherwise. A term-generated interpretation I is said to *satisfy* a non-ground clause C if it satisfies all ground instances $C\sigma$; it is called a *model* of a set N of clauses, if it satisfies all clauses of N.[2] We abbreviate the fact that I is a model of N by $I \models N$; $I \models C$ is short for $I \models \{C\}$. We say that N *entails* N', and write $N \models N'$, if every model of N is a model of N'; $N \models C$ is short for $N \models \{C\}$. We say that N and N' are *equivalent*, if $N \models N'$ and $N' \models N$.

If $\mathcal{J}$ is a class of Σ-interpretations, a Σ-clause or clause set is called $\mathcal{J}$-*satisfiable* if at least one $I \in \mathcal{J}$ satisfies the clause or clause set; otherwise it is called $\mathcal{J}$-*unsatisfiable*.

A *specification* is a pair $SP = (\Sigma, \mathcal{J})$, where Σ is a signature and $\mathcal{J}$ is a class of term-generated Σ-interpretations called *models* of the specification SP. We assume that $\mathcal{J}$ is closed under isomorphisms.

We say that a class of Σ-interpretations $\mathcal{J}$ or a specification $(\Sigma, \mathcal{J})$ is *compact*, if every infinite set of Σ-clauses that is $\mathcal{J}$-unsatisfiable has a finite subset that is also $\mathcal{J}$-unsatisfiable.

3 Hierarchic Theorem Proving

In hierarchic theorem proving, we consider a scenario in which a general-purpose foreground theorem prover and a specialized background prover cooperate to

---

[2] This restriction to term-generated interpretations as models is possible since we are only concerned with refutational theorem proving, i. e., with the derivation of a contradiction.

derive a contradiction from a set of clauses. In the sequel, we will usually abbreviate "foreground" and "background" by "FG" and "BG".

The BG prover accepts as input sets of clauses over a *BG signature* $\Sigma_B = (\Xi_B, \Omega_B)$. Elements of Ξ_B and Ω_B are called *BG sorts* and *BG operators*, respectively. We fix an infinite set $\mathcal{X}_B$ of *BG variables* of sorts in Ξ_B. Every BG variable has (is labeled with) a *kind*, which is either *"abstraction"* or *"ordinary"*. Terms over Σ_B and $\mathcal{X}_B$ are called *BG terms*. A BG term is called *pure*, if it does not contain ordinary variables; otherwise it is *impure*. These notions apply analogously to equations, literals, clauses, and clause sets.

The BG prover decides the satisfiability of Σ_B-clause sets with respect to a *BG specification* $(\Sigma_B, \mathcal{B})$, where $\mathcal{B}$ is a class of term-generated Σ_B-interpretations called *BG models*. We assume that $\mathcal{B}$ is closed under isomorphisms.

In most applications of hierarchic theorem proving, the set of BG operators Ω_B contains a set of distinguished constant symbols $\Omega_B^D \subseteq \Omega_B$ that has the property that $d_1^I \neq d_2^I$ for any two distinct $d_1, d_2 \in \Omega_B^D$ and every BG model $I \in \mathcal{B}$. We refer to these constant symbols as *(BG) domain elements*.

While we permit arbitrary classes of BG models, in practice the following three cases are most relevant:

(1) $\mathcal{B}$ consists of exactly one Σ_B-interpretation (up to isomorphism), say, the integer numbers over a signature containing all integer constants as domain elements and $\leq, <, +, -$ with the expected arities. In this case, $\mathcal{B}$ is trivially compact; in fact, a set N of Σ_B-clauses is $\mathcal{B}$-unsatisfiable if and only if some clause of N is $\mathcal{B}$-unsatisfiable.

(2) Σ_B is extended by an infinite number of *parameters*, that is, additional constant symbols. While all interpretations in $\mathcal{B}$ share the same carrier sets $\{I_\xi\}_{\xi \in \Xi_B}$ and interpretations of non-parameter symbols, parameters may be interpreted freely by arbitrary elements of the appropriate I_ξ. The class $\mathcal{B}$ obtained in this way is in general not compact; for instance the infinite set of clauses $\{n \leq \beta \mid n \in \mathbb{N}\}$, where β is a parameter, is unsatisfiable in the integers, but every finite subset is satisfiable.

(3) Σ_B is again extended by parameters, however, $\mathcal{B}$ is now the class of all interpretations that satisfy some first-order theory, say, the first-order theory of linear integer arithmetic.[3] Since $\mathcal{B}$ corresponds to a first-order theory, compactness is recovered. It should be noted, however, that $\mathcal{B}$ contains non-standard models, so that for instance the clause set $\{n \leq \beta \mid n \in \mathbb{N}\}$ is now satisfiable (e. g., $\mathbb{Q} \times \mathbb{Z}$ with a lexicographic ordering is a model).

The FG theorem prover accepts as inputs clauses over a signature $\Sigma = (\Xi, \Omega)$, where $\Xi_B \subseteq \Xi$ and $\Omega_B \subseteq \Omega$. The sorts in $\Xi_F = \Xi \setminus \Xi_B$ and the operator symbols in $\Omega_F = \Omega \setminus \Omega_B$ are called *FG sorts* and *FG operators*. Again we fix an

---

[3] To satisfy the technical requirement that all interpretations in $\mathcal{B}$ are term-generated, we assume that in this case Σ_B is suitably extended by an infinite set of constants (or by one constant and one unary function symbol) that are not used in any input formula or theory axiom.

infinite set $\mathcal{X}_F$ of *FG variables* of sorts in Ξ_F. All FG variables have the kind "ordinary". We define $\mathcal{X} = \mathcal{X}_B \cup \mathcal{X}_F$.

In examples we will use $\{0, 1, 2, \dots\}$ to denote BG domain elements, $\{+, -, <, \leq\}$ to denote (non-parameter) BG operators, and the possibly subscripted letters $\{x, y\}$, $\{X, Y\}$, $\{\alpha, \beta\}$, and $\{\mathsf{a}, \mathsf{b}, \mathsf{c}, \mathsf{f}, \mathsf{g}\}$ to denote ordinary variables, abstraction variables, parameters, and FG operators, respectively. The letter ζ denotes an ordinary variable or an abstraction variable.

We call a term in $T_\Sigma(\mathcal{X})$ a *FG term*, if it is not a BG term, that is, if it contains at least one FG operator or FG variable (and analogously for literals or clauses). We emphasize that for a FG operator $\mathsf{f} : \xi_1 \dots \xi_n \to \xi_0$ in Ω_F any of the ξ_i may be a BG sort, and that consequently FG terms may have BG sorts.

If I is a Σ-interpretation, the *restriction* of I to Σ_B, written $I|_{\Sigma_B}$, is the Σ_B-interpretation that is obtained from I by removing all carrier sets I_ξ for $\xi \in \Xi_F$ and all functions I_f for $f \in \Omega_F$. Note that $I|_{\Sigma_B}$ is not necessarily term-generated even if I is term-generated. In hierarchic theorem proving, we are only interested in Σ-interpretations that extend some model in $\mathcal{B}$ and neither collapse any of its sorts nor add new elements to them, that is, in Σ-interpretations I for which $I|_{\Sigma_B} \in \mathcal{B}$. We call such a Σ-interpretation a $\mathcal{B}$-*interpretation*.

Let N and N' be two sets of Σ-clauses. We say that N *entails* N' *relative to* $\mathcal{B}$ (and write $N \models_{\mathcal{B}} N'$), if every model of N whose restriction to Σ_B is in $\mathcal{B}$ is a model of N'. Note that $N \models_{\mathcal{B}} N'$ follows from $N \models N'$. If $N \models_{\mathcal{B}} \square$, we call N $\mathcal{B}$-*unsatisfiable*; otherwise, we call it $\mathcal{B}$-*satisfiable*.[4]

Our goal in refutational hierarchic theorem proving is to check whether a given set of Σ-clauses N is false in all $\mathcal{B}$-interpretations, or equivalently, whether N is $\mathcal{B}$-unsatisfiable.

We say that a substitution σ is *simple* if $X\sigma$ is a pure BG term for every abstraction variable $X \in \mathrm{dom}(\sigma)$. For example, $[x \mapsto 1 + Y + \alpha]$, $[X \mapsto 1 + Y + \alpha]$ and $[x \mapsto \mathsf{f}(1)]$ all are simple, whereas $[X \mapsto 1 + y + \alpha]$ and $[X \mapsto \mathsf{f}(1)]$ are not. Let F be a clause or (possibly infinite) clause set. By $\mathrm{sgi}(F)$ we denote the set of simple ground instances of F, that is, the set of all ground instances of (all clauses in) F obtained by simple ground substitutions.

For a BG specification $(\Sigma_B, \mathcal{B})$, we define $\mathrm{GndTh}(\mathcal{B})$ as the set of all ground Σ_B-formulas that are satisfied by every $I \in \mathcal{B}$.

Definition 3.1 (Sufficient completeness). *A Σ-clause set N is called* sufficiently complete w.r.t. simple instances *if for every Σ-model J of $\mathrm{sgi}(N) \cup \mathrm{GndTh}(\mathcal{B})$[5] and every ground BG-sorted FG term s there is a ground BG term t such that $J \models s \approx t$.*[6] □

---

[4] If $\Sigma = \Sigma_B$, this definition coincides with the definition of satisfiability w.r.t. a class of interpretations that was given in Sect. 2. A set N of BG clauses is $\mathcal{B}$-satisfiable if and only if some interpretation of $\mathcal{B}$ is a model of N.

[5] In contrast to [5], we include $\mathrm{GndTh}(\mathcal{B})$ in the definition of sufficient completeness. (This is independent of the abstraction method; it would also have been useful in [5].).

[6] Note that J need *not* be a $\mathcal{B}$-interpretation.

For brevity, we will from now on omit the phrase "w.r.t. simple instances" and speak only of "sufficient completeness". It should be noted, though, that our definition differs from the classical definition of sufficient completeness in the literature on algebraic specifications.

4   Orderings

A *hierarchic reduction ordering* is a strict, well-founded ordering on terms that is compatible with contexts, i.e., $s \succ t$ implies $u[s] \succ u[t]$, and stable under simple substitutions, i.e., $s \succ t$ implies $s\sigma \succ t\sigma$ for every simple σ. In the rest of this paper we assume such a hierarchic reduction ordering $\succ$ that satisfies all of the following: (i) $\succ$ is total on ground terms, (ii) $s \succ d$ for every domain element d and every ground term s that is not a domain element, and (iii) $s \succ t$ for every ground FG term s and every ground BG term t. These conditions are easily satisfied by an LPO with an operator precedence in which FG operators are larger than BG operators and domain elements are minimal with, for example, $\cdots \succ -2 \succ 2 \succ -1 \succ 1 \succ 0$ to achieve well-foundedness.

Condition (iii) and stability under *simple* substitutions together justify to always order $s \succ X$ where s is a non-variable FG term and X is an abstraction variable. By contrast, $s \succ x$ can only hold if $x \in \text{vars}(s)$. Intuitively, the combination of hierarchic reduction orderings and abstraction variables affords ordering more terms.

The ordering $\succ$ is extended to literals over terms by identifying a positive literal $s \approx t$ with the multiset $\{s, t\}$, a negative literal $s \not\approx t$ with $\{s, s, t, t\}$, and using the multiset extension of $\succ$. Clauses are compared by the multiset extension of $\succ$, also denoted by $\succ$.

The non-strict orderings $\succeq$ are defined as $s \succeq t$ if and only if $s \succ t$ or $s = t$ (the latter is multiset equality in case of literals and clauses). A literal L is *maximal* (*strictly maximal*) in a clause $L \vee C$ if there is no $K \in C$ with $K \succ L$ ($K \succeq L$).

5   Weak Abstraction

To refute an input set of Σ-clauses, hierarchic superposition calculi derive BG clauses from them and pass the latter to a BG prover. In order to do this, some separation of the FG and BG vocabulary in a clause is necessary. The technique used for this separation is known as *abstraction*: One (repeatedly) replaces some term q in a clause by a new variable and adds a disequations to the clause, so that $C[q]$ is converted into the equivalent clause $\zeta \not\approx q \vee C[\zeta]$, where ζ is a new (abstraction or ordinary) variable.

The calculus by Bachmair, Ganzinger, and Waldmann [5] works on "fully abstracted" clauses: Background terms occurring below a FG operator or in an equation between a BG and a FG term or vice versa are abstracted out until one arrives at a clause in which no literal contains both FG and BG operators.

A problematic aspect of any kind of abstraction is that it tends to increase the number of incomparable terms in a clause, which leads to an undesirable growth of the search space of a theorem prover. For instance, if we abstract out the subterms t and t' in a ground clause $f(t) \approx g(t')$, we get $x \not\approx t \vee y \not\approx t' \vee f(x) \approx g(y)$, and the two new terms $f(x)$ and $g(y)$ are incomparable in any reduction ordering. In [5] this problem is mitigated by considering only instances where BG-sorted variables are mapped to BG terms: In the terminology of the current paper, all BG-sorted variables in [5] have the kind "abstraction". This means that, in the example above, we obtain the two terms $f(X)$ and $g(Y)$. If we use an LPO with a precedence in which f is larger than g and g is larger than every BG operator, then for every simple ground substitution τ, $f(X)\tau$ is strictly larger that $g(Y)\tau$, so we can still consider $f(X)$ as the only maximal term in the literal.

The advantage of full abstraction is that this clause structure is preserved by all inference rules. There is a serious drawback, however: Consider the clause set $N = \{\, 1 + c \not\approx 1 + c \,\}$. Since N is ground, we have $sgi(N) = N$, and since $sgi(N)$ is unsatisfiable, N is trivially sufficiently complete. Full abstraction turns N into $N' = \{\, X \not\approx c \vee 1 + X \not\approx 1 + X \,\}$. In the simple ground instances of N', X is mapped to all pure BG terms. However, there are Σ-interpretations of $sgi(N')$ in which c is interpreted differently from any pure BG term, so $sgi(N') \cup GndTh(\mathcal{B})$ does have a Σ-model and N' is no longer sufficiently complete. In other words, the calculus of [5] is refutationally complete for clause sets that are fully abstracted and sufficiently complete, but full abstraction may destroy sufficient completeness. (In fact, the calculus is not able to refute N'.)

The problem that we have seen is caused by the fact that full abstraction replaces FG terms by abstraction variables, which may not be mapped to FG terms later on. The obvious fix would be to use ordinary variables instead of abstraction variables whenever the term to be abstracted out is not a pure BG term, but as we have seen above, this would increase the number of incomparable terms and it would therefore be detrimental to the performance of the prover.

Full abstraction is a property that is stronger than actually necessary for the completeness proof of [5]. In fact, it was claimed in a footnote in [5] that the calculus could be optimized by abstracting out only non-variable BG terms that occur below a FG operator. This is incorrect, however: Using this abstraction rule, neither our calculus nor the calculus of [5] would be able to refute $\{\, 1 + 1 \approx 2, (1 + 1) + c \not\approx 2 + c \,\}$, even though this set is unsatisfiable and trivially sufficiently complete. We need a slightly different abstraction rule to avoid this problem:

Definition 5.1. *A BG term q is a* target term *in a clause C if q is neither a domain element nor a variable and if C has the form $C[f(s_1, \ldots, q, \ldots, s_n)]$, where f is a FG operator or at least one of the s_i is a FG or impure BG term.*[7]
    A clause is called weakly abstracted *if it does not have any target terms.*
    The weakly abstracted version *of a clause is the clause that is obtained by exhaustively replacing $C[q]$ by*

- *$C[X] \vee X \not\approx q$, where X is a new abstraction variable, if q is a pure target term in C,*
- *$C[y] \vee y \not\approx q$, where y is a new ordinary variable, if q is an impure target term in C.*

The weakly abstracted version of a clause C is denoted by $\mathrm{abstr}(C)$; *if N is a set of clauses then* $\mathrm{abstr}(N) = \{\, \mathrm{abstr}(C) \mid C \in N \,\}$. □

For example, weak abstraction of the clause $\mathsf{g}(1, \alpha, \mathsf{f}(1) + (\alpha + 1), z) \approx \beta$ yields $\mathsf{g}(1, X, \mathsf{f}(1) + Y, z) \approx \beta \vee X \not\approx \alpha \vee Y \not\approx \alpha + 1$. Note that the terms 1, $\mathsf{f}(1) + (\alpha + 1)$, z, and β are not abstracted out: 1 is a domain element; $\mathsf{f}(1) + (\alpha + 1)$ has a BG sort, but it is not a BG term; z is a variable; and β is not a proper subterm of any other term. The clause $\mathsf{write}(\mathsf{a}, 2, \mathsf{read}(\mathsf{a}, 1) + 1) \approx \mathsf{b}$ is already weakly abstracted. Every pure BG clause is trivially weakly abstracted.

Nested abstraction is only necessary for certain impure BG terms. For instance, the clause $\mathsf{f}(z + \alpha) \approx 1$ has two target terms, namely α (since z is an impure BG term) and $z + \alpha$ (since f is a FG operator). If we abstract out α, we obtain $\mathsf{f}(z + X) \approx 1 \vee X \not\approx \alpha$. The new term $z + X$ is still a target term, so one more abstraction step yields $\mathsf{f}(y) \approx 1 \vee X \not\approx \alpha \vee y \not\approx z + X$. (Alternatively, we can first abstract out $z + \alpha$, yielding $\mathsf{f}(y) \approx 1 \vee y \not\approx z + \alpha$, and then α. The final result is the same.)

It is easy to see that the abstraction process described in Definition 5.1 terminates by comparing the multisets of the numbers of non-variable occurrences in the left and right-hand sides of all literals before and after an abstraction step.

Proposition 5.2. *If N is a set of clauses and N' is obtained from N by replacing one or more clauses by their weakly abstracted versions, then* $\mathrm{sgi}(N)$ *and* $\mathrm{sgi}(N')$ *are equivalent and N' is sufficiently complete whenever N is.*

Proof. Let us first consider the case of a single abstraction step applied to a single clause. Let $C[q]$ be a clause with a target term q and let $D = C[\zeta] \vee \zeta \not\approx q$ be the result of abstracting out q (where ζ is a new abstraction variable, if q is pure, and a new ordinary variable, if q is impure). We will show that $\mathrm{sgi}(C)$ and $\mathrm{sgi}(D)$ have the same models.

---

[7] Target terms are terms that need to be abstracted out; so for efficiency reasons, it is advantageous to keep the number of target terms as small as possible. We will show in Sect. 7 why domain elements may be treated differently from other non-variable terms. On the other hand, all the results in the following sections continue to hold if the restriction that q is not a domain element is dropped (i.e., if domain elements are abstracted out as well). We will make use of this fact in Sect. 11.

In one direction let I be an arbitrary model of sgi(C). We have to show that I is also a model of every simple ground instance $D\tau$ of D. If I satisfies the disequation $\zeta\tau \not\approx q\tau$ then this is trivial. Otherwise, $\zeta\tau$ and $q\tau$ have the same interpretation in I. Since dom(τ) $\supseteq$ vars(D) = vars(C) $\cup$ $\{\zeta\}$, $C\tau$ is a simple ground instance of C, so I is a model of $C\tau = C\tau[q\tau]$. By congruence, we conclude that I is also a model of $C\tau[\zeta\tau]$, hence it is a model of $D\tau = C\tau[\zeta\tau] \vee \zeta\tau \not\approx q\tau$.

In the other direction let I be an arbitrary model of sgi(D). We have to show that I is also a model of every simple ground instance $C\tau$ of C. Without loss of generality assume that $\zeta \notin$ dom(τ). If ζ is an abstraction variable, then q is a pure BG term, and since τ is a simple substitution, $q\tau$ is a pure BG term as well. Consequently, the substitutions $[\zeta \mapsto q\tau]$ and $\tau' = \tau[\zeta \mapsto q\tau]$ are again simple substitutions and $D\tau'$ is a simple ground instance of D. This implies that I is a model of $D\tau'$. The clause $D\tau'$ has the form $D\tau' = C\tau'[\zeta\tau'] \vee \zeta\tau' \not\approx q\tau'$; since $\zeta\tau' = q\tau$, $C\tau' = C\tau$ and $q\tau' = q\tau$, this is equal to $C\tau[q\tau] \vee q\tau \not\approx q\tau$. Obviously, the literal $q\tau \not\approx q\tau$ must be false in I, so I must be a model of $C\tau[q\tau] = C[q]\tau = C\tau$.

By induction over the number of abstraction steps we conclude that for any clause C, sgi(C) and sgi(abstr(C)) are equivalent. The extension to clause sets N and N' follows then from the fact that I is a model of sgi(N) if and only if it is a model of sgi(C) for all $C \in N$. Moreover, the equivalence of sgi(N) and sgi(N') implies obviously that N' is sufficiently complete whenever N is. $\square$

In contrast to full abstraction, the weak abstraction rule does not require abstraction of FG terms (which can destroy sufficient completeness if done using abstraction variables, and which is detrimental to the performance of a prover if done using ordinary variables). BG terms are usually abstracted out using abstraction variables. The exception are BG terms that are impure, i. e., that contain ordinary variables themselves. In this case, we cannot avoid to use ordinary variables for abstraction, otherwise, we might again destroy sufficient completeness. For example, the clause set $\{\mathsf{P}(1 + y),\ \neg\mathsf{P}(1 + \mathsf{c})\}$ is sufficiently complete. If we used an abstraction variable instead of an ordinary variable to abstract out the impure subterm $1 + y$, we would get $\{\mathsf{P}(X) \vee X \not\approx 1 + y,\ \neg\mathsf{P}(1 + \mathsf{c})\}$, which is no longer sufficiently complete.

In input clauses (that is, before abstraction), BG-sorted variables may be declared as "ordinary" or "abstraction". As we have seen above, using abstraction variables can reduce the search space; on the other hand, abstraction variables may be detrimental to sufficient completeness. Consider the following example: The set of clauses $N = \{\neg\mathsf{f}(x) > \mathsf{g}(x) \vee \mathsf{h}(x) \approx 1,\ \neg\mathsf{f}(x) \leq \mathsf{g}(x) \vee \mathsf{h}(x) \approx 2,\ \neg\mathsf{h}(x) > 0\}$ is unsatisfiable w.r.t. linear integer arithmetic, but since it is not sufficiently complete, the hierarchic superposition calculus does not detect the unsatisfiability. Adding the clause $X > Y \vee X \leq Y$ to N does not help: Since the abstraction variables X and Y may not be mapped to the FG terms $\mathsf{f}(x)$ and $\mathsf{g}(x)$ in a simple ground instance, the resulting set is still not sufficiently complete. However, if we add the clause $x > y \vee x \leq y$, the set of clauses becomes (vacuously) sufficiently complete and its unsatisfiability is detected.

One might wonder whether it is also possible to gain anything if the abstraction process is performed using ordinary variables instead of abstraction variables. The following proposition shows that this is not the case:

Proposition 5.3. *Let N be a set of clauses, let N' be the result of weak abstraction of N as defined above, and let N'' be the result of weak abstraction of N where all newly introduced variables are ordinary variables. Then $\mathrm{sgi}(N')$ and $\mathrm{sgi}(N'')$ are equivalent and $\mathrm{sgi}(N')$ is sufficiently complete if and only if $\mathrm{sgi}(N'')$ is.*

Proof. By Proposition 5.2, we know already that $\mathrm{sgi}(N)$ and $\mathrm{sgi}(N')$ are equivalent. Moreover, it is easy to check the proof of Proposition 5.2 is still valid if we assume that the newly introduced variable ζ is always an ordinary variable. (Note that the proof requires that abstraction variables are mapped only to pure BG terms, but it does not require that a variable that is mapped to a pure BG term must be an abstraction variable.) So we can conclude in the same way that $\mathrm{sgi}(N)$ and $\mathrm{sgi}(N'')$ are equivalent, and hence, that $\mathrm{sgi}(N')$ and $\mathrm{sgi}(N'')$ are equivalent. From this, we can conclude that N' is sufficiently complete whenever N'' is. □

6   Base Inference System

An *inference system* $\mathcal{I}$ is a set of inference rules. By an $\mathcal{I}$ *inference* we mean an instance of an inference rule from $\mathcal{I}$ such that all conditions are satisfied.

The *base inference system* $\mathrm{HSP}_{\mathrm{Base}}$ of the hierarchic superposition calculus consists of the inference rules Equality resolution, Negative superposition, Positive superposition, Equality factoring, and Close defined below. The calculus is parameterized by a hierarchic reduction ordering $\succ$ and by a "selection function" that assigns to every clause a (possibly empty) subset of its negative FG literals. All inference rules are applicable only to weakly abstracted premise clauses.

$$\text{Equality resolution} \quad \frac{s \not\approx t \vee C}{\mathrm{abstr}(C\sigma)}$$

if (i) σ is a simple mgu of s and t, (ii) $s\sigma$ is not a pure BG term, and (iii) if the premise has selected literals, then $s \not\approx t$ is selected in the premise, otherwise $(s \not\approx t)\sigma$ is maximal in $(s \not\approx t \vee C)\sigma$.[8]

For example, Equality resolution is applicable to $1 + \mathsf{c} \not\approx 1 + x$ with the simple mgu $[x \mapsto \mathsf{c}]$, but it is not applicable to $1 + \alpha \not\approx 1 + x$, since $1 + \alpha$ is a pure BG term.

---

[8] As in [5], it is possible to strengthen the maximality condition by requiring that there exists some simple ground substitution ψ such that $(s \not\approx t)\sigma\psi$ is maximal in $(s \not\approx t \vee C)\sigma\psi$ (and analogously for the other inference rules).

$$\text{Negative superposition} \quad \frac{l \approx r \vee C \qquad s[u] \not\approx t \vee D}{\text{abstr}((s[r] \not\approx t \vee C \vee D)\sigma)}$$

if (i) u is not a variable, (ii) σ is a simple mgu of l and u, (iii) $l\sigma$ is not a pure BG term, (iv) $r\sigma \not\preceq l\sigma$, (v) $(l \approx r)\sigma$ is strictly maximal in $(l \approx r \vee C)\sigma$, (vi) the first premise does not have selected literals, (vii) $t\sigma \not\preceq s\sigma$, and (viii) if the second premise has selected literals, then $s \not\approx t$ is selected in the second premise, otherwise $(s \not\approx t)\sigma$ is maximal in $(s \not\approx t \vee D)\sigma$.

$$\text{Positive superposition} \quad \frac{l \approx r \vee C \qquad s[u] \approx t \vee D}{\text{abstr}((s[r] \approx t \vee C \vee D)\sigma)}$$

if (i) u is not a variable, (ii) σ is a simple mgu of l and u, (iii) $l\sigma$ is not a pure BG term, (iv) $r\sigma \not\preceq l\sigma$, (v) $(l \approx r)\sigma$ is strictly maximal in $(l \approx r \vee C)\sigma$, (vi) $t\sigma \not\preceq s\sigma$, (vii) $(s \not\approx t)\sigma$ is strictly maximal in $(s \approx t \vee D)\sigma$, and (viii) none of the premises has selected literals.

$$\text{Equality factoring} \quad \frac{s \approx t \vee l \approx r \vee C}{\text{abstr}((l \approx r \vee t \not\approx r \vee C)\sigma)}$$

where (i) σ is a simple mgu of s and l, (ii) $s\sigma$ is not a pure BG term, (iii) $(s \approx t)\sigma$ is maximal in $(s \approx t \vee l \approx r \vee C)\sigma$, (iv) $t\sigma \not\preceq s\sigma$, (v) $l\sigma \not\preceq r\sigma$, and (vi) the premise does not have selected literals.

$$\text{Close} \quad \frac{C_1 \quad \cdots \quad C_n}{\square}$$

if $C_1, \ldots, C_n$ are BG clauses and $\{C_1, \ldots, C_n\}$ is $\mathcal{B}$-unsatisfiable, i.e., no interpretation in $\mathcal{B}$ is a Σ_{B}-model of $\{C_1, \ldots, C_n\}$.

Notice that Close is not restricted to take *pure* BG clauses only. The reason is that also impure BG clauses admit simple ground instances that are pure.

Theorem 6.1. *The inference rules of* HSP_{Base} *are sound w.r.t.* $\models_{\mathcal{B}}$, *i.e., for every inference with premises in* N *and conclusion* C, *we have* $N \models_{\mathcal{B}} C$.

Proof. Equality resolution, Negative superposition, Positive superposition, and Equality factoring are clearly sound w.r.t. $\models$, and therefore also sound w.r.t. $\models_{\mathcal{B}}$. For Close, soundness w.r.t. $\models_{\mathcal{B}}$ follows immediately from the definition. $\square$

All inference rules of HSP_{Base} involve (simple) mgus. Because of the two kinds of variables, abstraction and ordinary ones, the practical question arises if standard unification algorithms can be used without or only little modification. For example, the terms Z and $(x + y)$ admit a simple mgu $\sigma = [x \mapsto X, y \mapsto Y, Z \mapsto X + Y]$. This prompts for the use of weakening substitutions as in many-sorted logics with subsorts [28]. A closer inspection of the inference rules shows, however, that such substitutions never need to be considered: All unifiers computed in the inference rules have the property that abstraction variables are only mapped to abstraction variables or domain elements; apart from this additional restriction, we can use a standard unification algorithm.

In contrast to [5], the inference rules above include an explicit weak abstraction in their conclusion. Without it, conclusions would not be weakly abstracted in general. For example Positive superposition applied to the weakly abstracted clauses $f(X) \approx 1 \vee X \not\approx \alpha$ and $P(f(1) + 1)$ would then yield $P(1 + 1) \vee 1 \not\approx \alpha$, whose P-literal is not weakly abstracted. Additionally, the side conditions of our rules differ somewhat from the corresponding rules of [5], this is due on the one hand to the presence of impure BG terms (which must sometimes be treated like FG terms), and on the other hand to the fact that, after weak abstraction, literals may still contain both FG and BG operators.

The inference rules are supplemented by a redundancy criterion, that is, a mapping $\mathcal{R}_{Cl}$ from sets of formulae to sets of formulae and a mapping $\mathcal{R}_{Inf}$ from sets of formulae to sets of inferences that are meant to specify formulae that may be removed from N and inferences that need not be computed. ($\mathcal{R}_{Cl}(N)$ need not be a subset of N and $\mathcal{R}_{Inf}(N)$ will usually also contain inferences whose premises are not in N.)

Definition 6.2. *A pair $\mathcal{R} = (\mathcal{R}_{Inf}, \mathcal{R}_{Cl})$ is called a* redundancy criterion *(with respect to an inference system $\mathcal{I}$ and a consequence relation $\models$), if the following conditions are satisfied for all sets of formulae N and N':*

(i) $N \setminus \mathcal{R}_{Cl}(N) \models \mathcal{R}_{Cl}(N)$.
(ii) *If $N \subseteq N'$, then $\mathcal{R}_{Cl}(N) \subseteq \mathcal{R}_{Cl}(N')$ and $\mathcal{R}_{Inf}(N) \subseteq \mathcal{R}_{Inf}(N')$.*
(iii) *If ι is an inference and its conclusion is in N, then $\iota \in \mathcal{R}_{Inf}(N)$.*
(iv) *If $N' \subseteq \mathcal{R}_{Cl}(N)$, then $\mathcal{R}_{Cl}(N) \subseteq \mathcal{R}_{Cl}(N \setminus N')$ and $\mathcal{R}_{Inf}(N) \subseteq \mathcal{R}_{Inf}(N \setminus N')$.*

The inferences in $\mathcal{R}_{Inf}(N)$ and the formulae in $\mathcal{R}_{Cl}(N)$ are said to be redundant *with respect to N.* □

Let SSP be the ground standard superposition calculus using the inference rules equality resolution, negative superposition, positive superposition, and equality factoring (Bachmair and Ganzinger [3], Nieuwenhuis [23], Nieuwenhuis and Rubio [25]). To define a redundancy criterion for HSP_{Base} and to prove the refutational completeness of the calculus, we use the same approach as in [5] and relate HSP_{Base} inferences to the corresponding SSP inferences.

For a set of ground clauses N, we define $\mathcal{R}_{Cl}^{\mathcal{S}}(N)$ to be the set of all clauses C such that there exist clauses $C_1, \ldots, C_n \in N$ that are smaller than C with respect to $\succ$ and $C_1, \ldots, C_n \models C$. We define $\mathcal{R}_{Inf}^{\mathcal{S}}(N)$ to be the set of all ground SSP inferences ι such that either a premise of ι is in $\mathcal{R}_{Cl}^{\mathcal{S}}(N)$ or else C_0 is the conclusion of ι and there exist clauses $C_1, \ldots, C_n \in N$ that are smaller with respect to $\succ^c$ than the maximal premise of ι and $C_1, \ldots, C_n \models C_0$.

The following results can be found in [3] and [23]:

Theorem 6.3. *The (ground) standard superposition calculus SSP and $\mathcal{R}^{\mathcal{S}} = (\mathcal{R}_{Inf}^{\mathcal{S}}, \mathcal{R}_{Cl}^{\mathcal{S}})$ satisfy the following properties:*

(i) $\mathcal{R}^{\mathcal{S}}$ is a redundancy criterion with respect to $\models$.

(ii) SSP together with $\mathcal{R}^{\mathcal{S}}$ is refutationally complete.

Let ι be an $\mathrm{HSP}_{\mathrm{Base}}$ inference with premises $C_1, \ldots, C_n$ and conclusion $\mathrm{abstr}(C)$, where the clauses $C_1, \ldots, C_n$ have no variables in common. Let ι' be a ground SSP inference with premises $C_1', \ldots, C_n'$ and conclusion C'. If σ is a simple substitution such that $C' = C\sigma$ and $C_i' = C_i\sigma$ for all i, and if none of the C_i' is a BG clause, then ι' is called a *simple ground instance* of ι. The set of all simple ground instances of an inference ι is denoted by $\mathrm{sgi}(\iota)$.

Definition 6.4. Let N be a set of weakly abstracted clauses. We define $\mathcal{R}_{\mathrm{Inf}}^{\mathcal{H}}(N)$ to be the set of all inferences ι such that either ι is not a Close inference and $\mathrm{sgi}(\iota) \subseteq \mathcal{R}_{\mathrm{Inf}}^{\mathcal{S}}(\mathrm{sgi}(N) \cup \mathrm{GndTh}(\mathcal{B}))$, or else ι is a Close inference and $\square \in N$. We define $\mathcal{R}_{\mathrm{Cl}}^{\mathcal{H}}(N)$ to be the set of all weakly abstracted clauses C such that $\mathrm{sgi}(C) \subseteq \mathcal{R}_{\mathrm{Cl}}^{\mathcal{S}}(\mathrm{sgi}(N) \cup \mathrm{GndTh}(\mathcal{B})) \cup \mathrm{GndTh}(\mathcal{B})$.[9]   □

7 Refutational Completeness

To prove that $\mathrm{HSP}_{\mathrm{Base}}$ and $\mathcal{R}^{\mathcal{H}} = (\mathcal{R}_{\mathrm{Inf}}^{\mathcal{H}}, \mathcal{R}_{\mathrm{Cl}}^{\mathcal{H}})$ are refutationally complete for sets of weakly abstracted Σ-clauses and compact BG specifications $(\Sigma_{\mathrm{B}}, \mathcal{B})$, we use the same technique as in [5]:

First we show that $\mathcal{R}^{\mathcal{H}}$ is a redundancy criterion with respect to $\models_{\mathcal{B}}$, and that a set of clauses remains sufficiently complete if new clauses are added or if redundant clauses are deleted. The proofs are rather technical and can be found in [12]. They are similar to the corresponding ones in [5]; the differences are due, on the one hand, to the fact that we include $\mathrm{GndTh}(\mathcal{B})$ in the redundancy criterion and in the definition of sufficient completeness and, on the other hand, to the explicit abstraction steps in our inference rules.

Lemma 7.1. If $\mathrm{sgi}(N) \cup \mathrm{GndTh}(\mathcal{B}) \models \mathrm{sgi}(C)$, then $N \models_{\mathcal{B}} C$.

Proof. Suppose that $\mathrm{sgi}(N) \cup \mathrm{GndTh}(\mathcal{B}) \models \mathrm{sgi}(C)$. Let I' be a Σ-model of N whose restriction to Σ_{B} is contained in $\mathcal{B}$. Clearly, I' is also a model of $\mathrm{GndTh}(\mathcal{B})$. Since I' does not add new elements to the sorts of $I = I'|_{\Sigma_{\mathrm{B}}}$ and I is a term-generated Σ_{B}-interpretation, we know that for every ground Σ-term t' of a BG sort there is a ground BG term t such that t and t' have the same interpretation in I'. Therefore, for every ground substitution σ' there is an equivalent simple ground substitution σ; since $C\sigma$ is valid in I', $C\sigma'$ is also valid.   □

We call the simple most general unifier σ that is computed during an inference ι and applied to the conclusion the pivotal substitution of ι. (For ground inferences, the pivotal substitution is the identity mapping.) If L is the literal $[\neg]\, s \approx t$ or $[\neg]\, s[u] \approx t$ of the second or only premise that is eliminated in ι, we call $L\sigma$ the pivotal literal of ι, and we call $s\sigma$ or $s[u]\sigma$ the pivotal term of ι.

---

[9] In contrast to [5], we include $\mathrm{GndTh}(\mathcal{B})$ in the redundancy criterion. (This is independent of the abstraction method used; it would also have been useful in [5].).

Lemma 7.2. *Let ι be an* $\mathrm{HSP}_{\mathrm{Base}}$ *inference*

$$\frac{C_1}{\mathrm{abstr}(C_0\sigma)} \quad or \quad \frac{C_2 \quad C_1}{\mathrm{abstr}(C_0\sigma)}$$

from weakly abstracted premises with pivotal substitution σ. Let ι' be a simple ground instance of ι of the form

$$\frac{C_1\tau}{C_0\sigma\tau} \quad or \quad \frac{C_2\tau \quad C_1\tau}{C_0\sigma\tau}$$

Then there is a simple ground instance of $\mathrm{abstr}(C_0\sigma)$ *that has the form $C_0\sigma\tau \vee E$, where E is a (possibly empty) disjunction of literals $s \not\approx s$, and each literal of E is smaller than the pivotal literal of ι'.*

As $M \subseteq M'$ implies $\mathcal{R}^{\mathcal{S}}_{\mathrm{Inf}}(M) \subseteq \mathcal{R}^{\mathcal{S}}_{\mathrm{Inf}}(M')$, we obtain $\mathcal{R}^{\mathcal{S}}_{\mathrm{Inf}}(\mathrm{sgi}(N)\backslash\mathrm{sgi}(N')) \subseteq \mathcal{R}^{\mathcal{S}}_{\mathrm{Inf}}(\mathrm{sgi}(N\backslash N'))$. Furthermore, it is fairly easy to see that $\mathrm{sgi}(N)\backslash(\mathcal{R}^{\mathcal{S}}_{\mathrm{Cl}}(\mathrm{sgi}(N) \cup \mathrm{GndTh}(\mathcal{B})) \cup \mathrm{GndTh}(\mathcal{B})) \subseteq \mathrm{sgi}(N \setminus \mathcal{R}^{\mathcal{H}}_{\mathrm{Cl}}(N))$. Using these two results we can prove the following lemmas:

Lemma 7.3. $\mathcal{R}^{\mathcal{H}} = (\mathcal{R}^{\mathcal{H}}_{\mathrm{Inf}}, \mathcal{R}^{\mathcal{H}}_{\mathrm{Cl}})$ *is a redundancy criterion with respect to* $\models_{\mathcal{B}}$.

Lemma 7.4. *Let N, N' and M be sets of weakly abstracted clauses such that $N' \subseteq \mathcal{R}^{\mathcal{H}}_{\mathrm{Cl}}(N)$. If N is sufficiently complete, then so are $N \cup M$ and $N \setminus N'$.*

We now encode arbitrary term-generated Σ_{B}-interpretation by sets of unit ground clauses in the following way: Let $I \in \mathcal{B}$ be a term-generated Σ_{B}-interpretation. For every Σ_{B}-ground term t let $m(t)$ be the smallest ground term of the congruence class of t in I. We define a rewrite system E'_I by $\mathrm{E}'_I = \{t \rightarrow m(t) \mid t \in \mathrm{T}_\Sigma, t \neq m(t)\}$. Obviously, E'_I is right-reduced; since all rewrite rules are contained in $\succ$, E'_I is terminating; and since every ground term t has $m(t)$ as its only normal form, E'_I is also confluent. Now let E_I be the set of all rules $l \rightarrow r$ in E'_I such that l is not reducible by $\mathrm{E}'_I \setminus \{l \rightarrow r\}$. Clearly every term that is reducible by E_I is also reducible by E'_I; conversely every term that is reducible by E'_I has a minimal subterm that is reducible by E'_I and the rule in E'_I that is used to rewrite this minimal subterm is necessarily contained in E_I. Therefore E'_I and E_I define the same set of normal forms, and from this we can conclude that E_I and E'_I induce the same equality relation on ground Σ_{B}-terms. We identify E_I with the set of clauses $\{t \approx t' \mid t \rightarrow t' \in \mathrm{E}_I\}$. Let D_I be the set of all clauses $t \not\approx t'$, such that t and t' are distinct ground Σ_{B}-terms in normal form with respect to E_I.[10]

Lemma 7.5. *Let $I \in \mathcal{B}$ be a term-generated Σ_{B}-interpretation and let C be a ground BG clause. Then C is true in I if and only if there exist clauses $C_1, \ldots, C_n$ in $\mathrm{E}_I \cup \mathrm{D}_I$ such that $C_1, \ldots, C_n \models C$ and $C \succeq C_i$ for $1 \leq i \leq n$.*

---

[10] Typically, E_I contains two kinds of clauses, namely clauses that evaluate non-constant BG terms, such as $2 + 3 \approx 5$, and clauses that map parameters to domain elements, such as $\alpha \approx 4$.

Proof. The "if" part follows immediately from the fact that $I \models E_I \cup D_I$. For the "only if" part assume that the ground BG clause C is true in I. Consequently, there is some literal $s \approx t$ or $s \not\approx t$ of C that is true in I. Then this literal follows from (i) the rewrite rules in E_I that are used to normalize s to its normal form s', (ii) the rewrite rules in E_I that are used to normalize t to its normal form t', and, in the case of a negated literal $s \not\approx t$, (iii) the clause $s' \not\approx t' \in D_I$. It is routine to show that all these clauses are smaller than or equal to $s \approx t$ or $s \not\approx t$, respectively, and hence smaller than or equal to C. $\qquad\square$

Corollary 7.6. *Let $I \in \mathcal{B}$ be a term-generated Σ_B-interpretation. Then $E_I \cup D_I \models \mathrm{GndTh}(\mathcal{B})$.*

Proof. Since $I \in \mathcal{B}$, we have $I \models \mathrm{GndTh}(\mathcal{B})$, hence $E_I \cup D_I \models \mathrm{GndTh}(\mathcal{B})$ by Lemma 7.5. $\qquad\square$

Let N be a set of weakly abstracted clauses and $I \in \mathcal{B}$ be a term-generated Σ_B-interpretation, then N_I denotes the set $E_I \cup D_I \cup \{ C\sigma \mid \sigma \text{ simple, reduced with respect to } E_I, C \in N, C\sigma \text{ ground} \}$.

Lemma 7.7. *If N is a set of weakly abstracted clauses, then $\mathcal{R}^{\mathcal{S}}_{\mathrm{Inf}}(\mathrm{sgi}(N) \cup \mathrm{GndTh}(\mathcal{B})) \subseteq \mathcal{R}^{\mathcal{S}}_{\mathrm{Inf}}(N_I)$.*

Proof. By part (i) of Theorem 6.3 we have obviously $\mathcal{R}^{\mathcal{S}}_{\mathrm{Inf}}(\mathrm{sgi}(N)) \subseteq \mathcal{R}^{\mathcal{S}}_{\mathrm{Inf}}(E_I \cup D_I \cup \mathrm{sgi}(N) \cup \mathrm{GndTh}(\mathcal{B}))$. Let C be a clause in $E_I \cup D_I \cup \mathrm{sgi}(N) \cup \mathrm{GndTh}(\mathcal{B})$ and not in N_I. If $C \in \mathrm{GndTh}(\mathcal{B})$, then it is true in I, so by Lemma 7.5 it is either contained in $E_I \cup D_I \subseteq N_I$ or it follows from smaller clauses in $E_I \cup D_I$ and is therefore in $\mathcal{R}^{\mathcal{S}}_{\mathrm{Cl}}(E_I \cup D_I \cup \mathrm{sgi}(N))$. If $C \notin \mathrm{GndTh}(\mathcal{B})$, then $C = C'\sigma$ for some $C' \in N$, so it follows from $C'\rho$ and $E_I \cup D_I$, where ρ is the substitution that maps every variable ζ to the E_I-normal form of $\zeta\sigma$. Since C follows from smaller clauses in $E_I \cup D_I \cup \mathrm{sgi}(N)$, it is in $\mathcal{R}^{\mathcal{S}}_{\mathrm{Cl}}(E_I \cup D_I \cup \mathrm{sgi}(N))$. Hence $\mathcal{R}^{\mathcal{S}}_{\mathrm{Inf}}(E_I \cup D_I \cup \mathrm{sgi}(N) \cup \mathrm{GndTh}(\mathcal{B})) \subseteq \mathcal{R}^{\mathcal{S}}_{\mathrm{Inf}}(N_I)$. $\qquad\square$

A clause set N is called *saturated (with respect to an inference system $\mathcal{I}$ and a redundancy criterion $\mathcal{R}$)* if $\iota \in \mathcal{R}_{\mathrm{Inf}}(N)$ for every inference ι with premises in N.

Theorem 7.8. *Let $I \in \mathcal{B}$ be a term-generated Σ_B-interpretation and let N be a set of weakly abstracted Σ-clauses. If I satisfies all BG clauses in $\mathrm{sgi}(N)$ and N is saturated with respect to $\mathrm{HSP}_{\mathrm{Base}}$ and $\mathcal{R}^{\mathcal{H}}$, then N_I is saturated with respect to SSP and $\mathcal{R}^{\mathcal{S}}$.*

Proof. We have to show that every SSP-inference from clauses of N_I is contained in $\mathcal{R}^{\mathcal{S}}_{\mathrm{Inf}}(N_I)$. We demonstrate this in detail for the equality resolution and the negative superposition rule. The analysis of the other rules is similar. Note that by Lemma 7.5 every BG clause that is true in I and is not contained in $E_I \cup D_I$ follows from smaller clauses in $E_I \cup D_I$, thus it is in $\mathcal{R}^{\mathcal{S}}_{\mathrm{Cl}}(N_I)$; every inference involving such a clause is in $\mathcal{R}^{\mathcal{S}}_{\mathrm{Inf}}(N_I)$.

The equality resolution rule is obviously not applicable to clauses from $E_I \cup D_I$. Suppose that ι is an equality resolution inference with a premise $C\sigma$, where

$C \in N$ and σ is a simple substitution and reduced with respect to E_I. If $C\sigma$ is a BG clause, then ι is in $\mathcal{R}_{\mathrm{Inf}}^{\mathcal{S}}(N_I)$. If the pivotal term of ι is a pure BG term then the pivotal literal is pure BG as well. Because the pivotal literal is maximal in $C\sigma$ it follows from properties of the ordering that $C\sigma$ is a BG clause. Because we have already considered this case we can assume from now on that the pivotal term of ι is not pure BG and that $C\sigma$ is an FG clause. It follows that ι is a simple ground instance of a hierarchic inference ι' from C. Since ι' is in $\mathcal{R}_{\mathrm{Inf}}^{\mathcal{H}}(N)$, ι is in $\mathcal{R}_{\mathrm{Inf}}^{\mathcal{S}}(\mathrm{sgi}(N) \cup \mathrm{GndTh}(\mathcal{B}))$, by Lemma 7.7, this implies again $\iota \in \mathcal{R}_{\mathrm{Inf}}^{\mathcal{S}}(N_I)$.

Obviously a clause from D_I cannot be the first premise of a negative superposition inference. Suppose that the first premise is a clause from E_I. The second premise cannot be a FG clause, since the maximal sides of maximal literals in a FG clause are reduced; as it is a BG clause, the inference is redundant. Now suppose that ι is a negative superposition inference with a first premise $C\sigma$, where $C \in N$ and σ is a simple substitution and reduced with respect to E_I. If $C\sigma$ is a BG clause, then ι is in $\mathcal{R}_{\mathrm{Inf}}^{\mathcal{S}}(N_I)$. Otherwise, with the same arguments as for the equality resolution case above, the pivotal term is not pure BG and $C\sigma$ is a FG clause. Hence we can conclude that the second premise can be written as $C'\sigma$, where $C' \in N$ is a FG clause (without loss of generality, C and C' do not have common variables). If the overlap takes place below a variable occurrence, the conclusion of the inference follows from $C\sigma$ and some instance $C'\rho$, which are both smaller than $C'\sigma$. Otherwise, ι is a simple ground instance of a hierarchic inference ι' from C. In both cases, ι is contained in $\mathcal{R}_{\mathrm{Inf}}^{\mathcal{S}}(N_I)$. $\square$

The crucial property of abstracted clauses that is needed in the proof of this theorem is that there are no superposition inferences between clauses in E_I and FG ground instances $C\sigma$ in N_I, or in other words, that all FG terms occurring in ground instances $C\sigma$ are reduced w.r.t. E_I. This motivates the definition of target terms in Definition 5.1: Recall that two different domain elements must always be interpreted differently in I and that a domain element is smaller in the term ordering than any ground term that is not a domain element. Consequently, any domain element is the smallest term in its congruence class, so it is reduced by E_I. Furthermore, by the definition of N_I, $\zeta\sigma$ is reduced by E_I for every variable ζ. So variables and domain elements never need to be abstracted out. Other BG terms (such as parameters α or non-constant terms $\zeta_1 + \zeta_2$) have to be abstracted out if they occur below a FG operator, or if one of their sibling terms is a FG term or an impure BG term (since σ can map the latter to a FG term). On the other hand, abstracting out FG terms as in [5] is never necessary to ensure that FG terms are reduced w.r.t. E_I.

If N is saturated with respect to $\mathrm{HSP}_{\mathrm{Base}}$ and $\mathcal{R}^{\mathcal{H}}$ and does not contain the empty clause, then Close cannot be applicable to N. If $(\Sigma_B, \mathcal{B})$ is compact, this implies that there is some term-generated Σ_B-interpretation $I \in \mathcal{B}$ that satisfies all BG clauses in $\mathrm{sgi}(N)$. Hence, by Theorem 7.8, the set of *reduced simple* ground instances of N has a model that also satisfies $E_I \cup D_I$. Sufficient completeness allows us to show that this is in fact a model of *all* ground instances of clauses in N and that I is its restriction to Σ_B:

Theorem 7.9. *If the BG specification* $(\Sigma_\mathrm{B}, \mathcal{B})$ *is compact, then* $\mathrm{HSP}_\mathrm{Base}$ *and* $\mathcal{R}^\mathcal{H}$ *are statically refutationally complete for all sufficiently complete sets of clauses, i. e., if a set of clauses* N *is sufficiently complete and saturated w.r.t.* $\mathrm{HSP}_\mathrm{Base}$ *and* $\mathcal{R}^\mathcal{H}$, *and* $N \models_\mathcal{B} \square$, *then* $\square \in N$.

Proof. Let N be a set of weakly abstracted clauses that is sufficiently complete, and saturated w.r.t. the hierarchic superposition calculus and $\mathcal{R}^\mathcal{H}$ and does not contain $\square$. Consequently, the Close rule is not applicable to N. By compactness, this means that the set of all Σ_B-clauses in sgi(N) is satisfied by some term-generated Σ_B-interpretation $I \in \mathcal{B}$. By Theorem 7.8, N_I is saturated with respect to the standard superposition calculus. Since $\square \notin N_I$, the refutational completeness of standard superposition implies that there is a Σ-model I' of N_I. Since N is sufficiently complete, we know that for every ground term t' of a BG sort there exists a BG term t such that $t' \approx t$ is true in I'. Consequently, for every ground instance of a clause in N there exists an equivalent simple ground instance, thus I' is also a model of all ground instances of clauses in N. To see that the restriction of I' to Σ_B is isomorphic to I and thus in $\mathcal{B}$, note that I' satisfies $\mathrm{E}_I \cup \mathrm{D}_I$, preventing confusion, and that N is sufficiently complete, preventing junk. Since I' satisfies N and $I'|_{\Sigma_\mathrm{B}} \in \mathcal{B}$, we have $N \not\models_\mathcal{B} \square$ $\qquad\square$

A theorem proving derivation $\mathcal{D}$ is a finite or infinite sequence of weakly abstracted clause sets $N_0, N_1, \ldots$, such that N_i and N_{i+1} are equisatisfiable w.r.t. $\models_\mathcal{B}$ and $N_i \setminus N_{i+1} \subseteq \mathcal{R}^\mathcal{H}_\mathrm{Inf}(N_{i+1})$ for all indices i. The set $N_\infty = \bigcup_{i \geq 0} \bigcap_{j \geq i} N_j$ is called the limit of $\mathcal{D}$; the set $N^\infty = \bigcup_{i \geq 0} N_i$ is called the union of $\mathcal{D}$. It is easy to show that every clause in N^∞ is either contained in N_∞ or redundant w.r.t. N_∞. The derivation $\mathcal{D}$ is said to be fair, if every $\mathrm{HSP}_\mathrm{Base}$-inference with (non-redundant) premises in N_∞ becomes redundant at some point of the derivation. The limit of a fair derivation is saturated [4]; this is the key result that allows us to deduce dynamic refutational completeness from static refutational completeness:

Theorem 7.10. *If the BG specification* $(\Sigma_\mathrm{B}, \mathcal{B})$ *is compact, then* $\mathrm{HSP}_\mathrm{Base}$ *and* $\mathcal{R}^\mathcal{H}$ *are dynamically refutationally complete for all sufficiently complete sets of clauses, i. e., if* $N \models_\mathcal{B} \square$, *then every fair derivation starting from* abstr(N) *eventually generates* $\square$.

In the rest of the paper, we consider only theorem proving derivations where each set N_{i+1} results from from N_i by either adding the conclusions of inferences from N_i, or by deleting clauses that are redundant w.r.t. N_{i+1}, or by applying the following generic simplification rule for clause sets:

$$\mathsf{Simp} \ \frac{N \cup \{C\}}{N \cup \{D_1, \ldots, D_n\}}$$

if $n \geq 0$ and (i) D_i is weakly abstracted, for all $i = 1, \ldots, n$, (ii) $N \cup \{C\} \models_\mathcal{B} D_i$, and (iii) $C \in \mathcal{R}^\mathcal{H}_\mathrm{Cl}(N \cup \{D_1, \ldots, D_n\})$.

Condition (ii) is needed for soundness, and condition (iii) is needed for completeness. The Simp rule covers the usual simplification rules of the standard superposition calculus, such as demodulation by unit clauses and deletion of tautologies and (properly) subsumed clauses. It also covers simplification of arithmetic terms, e. g., replacing a subterm $(2 + 3) + \alpha$ by $5 + \alpha$ and deleting an unsatisfiable BG literal $5 + \alpha < 4 + \alpha$ from a clause. Any clause of the form $C \vee \zeta \not\approx d$ where d is domain element can be simplified to $C[\zeta \mapsto d]$. Notice, though, that impure BG terms or FG terms can in general not be simplified by BG tautologies. Although, e. g., $f(X) + 1 \not\approx y + 1$ is larger than $1 + f(X) \not\approx y + 1$ (with a LPO), such a "simplification" is not justified by the redundancy criterion. Indeed, in the example it destroys sufficient completeness.

We have to point out a limitation of the calculus described above. The standard superposition calculus SSP exists in two variants: either using the Equality factoring rule, or using the Factoring and Merging paramodulation rules. Only the first of these variants works together with weak abstraction. Consider the following example. Let $N = \{\, \alpha + \beta \approx \alpha,\ c \not\approx \beta \vee c \not\approx 0,\ c \approx \beta \vee c \approx 0 \,\}$. All clauses in N are weakly abstracted. Since the first clause entails $\beta \approx 0$ relative to linear arithmetic, the second and the third clause are obviously contradictory. The $\mathrm{HSP}_{\mathrm{Base}}$ calculus as defined above is able to detect this by first applying Equality factoring to the third clause, yielding $c \approx 0 \vee \beta \not\approx 0$, followed by two Negative superposition steps and Close. If Equality factoring is replaced by Factoring and Merging paramodulation, however, the refutational completeness of $\mathrm{HSP}_{\mathrm{Base}}$ is lost. The only inference that remains possible is a Negative superposition inference between the third and the second clause. But since the conclusion of this inference is a tautology, the inference is redundant, so the clause set is saturated. (Note that the clause $\beta \approx 0$ is entailed by N, but it is not explicitly present, so there is no way to perform a Merging paramodulation inference with the smaller side of the maximal literal of the third clause.)

8   Local Sufficient Completeness

The definition of sufficient completeness w.r.t. simple instances that was given in Sect. 3 requires that *every* ground BG-sorted FG term s is equal to some ground BG term t in every Σ-model J of $\mathrm{sgi}(N) \cup \mathrm{GndTh}(\mathcal{B})$. It is rather evident, however, that this condition is sometimes stronger than needed. For instance, if the set of input clauses N is ground, then we only have to consider the ground BG-sorted FG terms that actually occur in N [22] (analogously to the Nelson-Oppen combination procedure). A relaxation of sufficient completeness that is also useful for non-ground clauses and that still ensures refutational completeness was given by Kruglov [21]:

Definition 8.1 (Smooth ground instance). *We say that a substitution σ is smooth if for every variable $\zeta \in \mathrm{dom}(\sigma)$ all BG-sorted (proper or non-proper) subterms of $\zeta\sigma$ are pure BG terms. If $F\sigma$ is a ground instance of a term or clause F and σ is smooth, $F\sigma$ is called a smooth ground instance. (Recall that* *every ground BG term is necessarily pure.) If N is a set of clauses,* $\mathrm{smgi}(N)$ *denotes the set of all smooth ground instances of clauses in N.*   □

Every smooth substitution is a simple substitution, but not vice versa. For instance, if x is a FG-sorted variable and y is an ordinary BG-sorted variable, then $\sigma_1 = [x \mapsto \mathsf{cons}(\mathsf{f}(1) + 2, \mathsf{empty})]$ and $\sigma_2 = [y \mapsto \mathsf{f}(1)]$ are simple substitutions, but neither of them is smooth, since $x\sigma_1$ and $y\sigma_2$ contain the BG-sorted FG subterm $\mathsf{f}(1)$.

Definition 8.2 (Local sufficient completeness). *Let N be a Σ-clause set. We say that N is* locally sufficiently complete *w.r.t. smooth instances if for every Σ_{B}-interpretation $I \in \mathcal{B}$, every Σ-model J of $\mathrm{sgi}(N) \cup \mathrm{E}_I \cup \mathrm{D}_I$, and every BG-sorted FG term s occurring in $\mathrm{smgi}(N) \setminus \mathcal{R}^{\mathcal{S}}_{\mathrm{Cl}}(\mathrm{smgi}(N) \cup \mathrm{E}_I \cup \mathrm{D}_I)$ there is a ground BG term t such that $J \models s \approx t$. (Again, we will from now on omit the phrase "w.r.t. smooth instances" for brevity.)*   □

Example 8.3. The clause set $N = \{\, X \not\approx \alpha \lor \mathsf{f}(X) \approx \beta \,\}$ is locally sufficiently complete: The smooth ground instances have the form $s' \not\approx \alpha \lor \mathsf{f}(s') \approx \beta$, where s' is a pure BG term. We have to show that $\mathsf{f}(s')$ equals some ground BG term t whenever the smooth ground instance is not redundant. Let $I \in \mathcal{B}$ be a Σ_{B}-interpretation and J be a Σ-model of $\mathrm{sgi}(N) \cup \mathrm{E}_I \cup \mathrm{D}_I$. If $I \models s' \not\approx \alpha$, then $s' \not\approx \alpha$ follows from some clauses in $\mathrm{E}_I \cup \mathrm{D}_I$, so $s' \not\approx \alpha \lor \mathsf{f}(s') \approx \beta$ is contained in $\mathcal{R}^{\mathcal{S}}_{\mathrm{Cl}}(\mathrm{smgi}(N) \cup \mathrm{E}_I \cup \mathrm{D}_I)$ and $\mathsf{f}(s')$ need not be considered. Otherwise $I \models s' \approx \alpha$, then $\mathsf{f}(s')$ occurs in a non-redundant smooth ground instance of a clause in N and $J \models f(s') \approx \beta$, so $t := \beta$ has the desired property. On the other hand, N is clearly not sufficiently complete, since there are models of $\mathrm{sgi}(N) \cup \mathrm{GndTh}(\mathcal{B})$ in which $\mathsf{f}(\beta)$ is interpreted by some junk element that is different from the interpretation of any ground BG term.

The example demonstrates that local sufficient completeness is significantly more powerful than sufficient completeness, but this comes at a price. For instance, as shown by the next example, local sufficient completeness is not preserved by abstraction:

Example 8.4. Suppose that the BG specification is linear integer arithmetic (including parameters α, β, γ), the FG operators are $\mathsf{f} : int \to int$, $\mathsf{g} : int \to data$, $\mathsf{a} : \to data$, the term ordering is an LPO with precedence $\mathsf{g} > \mathsf{f} > \mathsf{a} > \gamma > \beta > \alpha > 3 > 2 > 1$, and the clause set N is given by

$$\gamma \approx 1 \quad (1) \qquad \mathsf{f}(2) \approx 2 \quad (4) \qquad \mathsf{g}(\mathsf{f}(\alpha)) \approx \mathsf{a} \lor \mathsf{g}(\mathsf{f}(\beta)) \approx \mathsf{a} \quad (6)$$
$$\beta \approx 2 \quad (2) \qquad \mathsf{f}(3) \approx 3 \quad (5) \qquad \mathsf{g}(\mathsf{f}(\alpha)) \not\approx \mathsf{a} \lor \mathsf{g}(\mathsf{f}(\beta)) \approx \mathsf{a} \quad (7)$$
$$\alpha \approx 3 \quad (3) \qquad\qquad\qquad\qquad \mathsf{g}(\mathsf{f}(\gamma)) \approx \mathsf{a} \lor \mathsf{g}(\mathsf{f}(\beta)) \approx \mathsf{a} \quad (8)$$

Since all clauses in N are ground, $\mathrm{smgi}(N) = \mathrm{sgi}(N) = N$. Clause (8) is redundant w.r.t. $\mathrm{smgi}(N) \cup \mathrm{E}_I \cup \mathrm{D}_I$ (for any I): it follows from clauses (6) and (7), and both are smaller than (8). The BG-sorted FG terms in non-redundant clauses are $\mathsf{f}(2)$, $\mathsf{f}(3)$, $\mathsf{f}(\alpha)$, and $\mathsf{f}(\beta)$, and in any Σ-model J of $\mathrm{sgi}(N) \cup \mathrm{E}_I \cup \mathrm{D}_I$, these are necessarily equal to the BG terms 2 or 3, respectively, so N is locally sufficiently complete.

Let $N' = \operatorname{abstr}(N)$, let I be a BG-model such that E_I contains $\alpha \approx 3$, $\beta \approx 2$, and $\gamma \approx 1$ (among others), D_I contains $1 \not\approx 2$, $1 \not\approx 3$, and $2 \not\approx 3$ (among others), and let J be a Σ-model of $\operatorname{sgi}(N') \cup \mathrm{E}_I \cup \mathrm{D}_I$ in which $\mathsf{f}(1)$ is interpreted by some junk element. The set N' contains the clause $\mathsf{g}(\mathsf{f}(X)) \approx \mathsf{a} \lor \mathsf{g}(\mathsf{f}(Y)) \approx \mathsf{a} \lor \gamma \not\approx X \lor \beta \not\approx Y$ obtained by abstraction of (8). Its smooth ground instance $C = \mathsf{g}(\mathsf{f}(1)) \approx \mathsf{a} \lor \mathsf{g}(\mathsf{f}(2)) \approx \mathsf{a} \lor \gamma \not\approx 1 \lor \beta \not\approx 2$ is not redundant: it follows from other clauses in $\operatorname{smgi}(N') \cup \mathrm{E}_I \cup \mathrm{D}_I$, namely

$$\alpha \approx 3 \tag{3}$$

$$\mathsf{g}(\mathsf{f}(3)) \approx \mathsf{a} \lor \mathsf{g}(\mathsf{f}(2)) \approx \mathsf{a} \lor \alpha \not\approx 3 \lor \beta \not\approx 2 \tag{6'}$$

$$\mathsf{g}(\mathsf{f}(3)) \not\approx \mathsf{a} \lor \mathsf{g}(\mathsf{f}(2)) \approx \mathsf{a} \lor \alpha \not\approx 3 \lor \beta \not\approx 2 \tag{7'}$$

but the ground instances $(6')$ and $(7')$ that are needed here are larger than C. Since C contains the BG-sorted FG term $\mathsf{f}(1)$ which is interpreted differently from any BG term in J, N' is not locally sufficiently complete.

Local sufficient completeness of a clause set suffices to ensure refutational completeness. Kruglov's proof [21] works also if one uses weak abstraction instead of strong abstraction and ordinary as well as abstraction variables, but it relies on an additional restriction on the term ordering.[11] We give an alternative proof that works without this restriction.

The proof is based on a transformation on Σ-interpretations. Let J be an arbitrary Σ-interpretation. We transform J into a term-generated Σ-interpretation $\operatorname{nojunk}(J)$ without junk in two steps. In the first step, we define a Σ-interpretation J' as follows:

- For every FG sort ξ, define $J'_\xi = J_\xi$.
- For every BG sort ξ, define $J'_\xi = \{ t^J \mid t \text{ is a ground BG term of sort } \xi \}$.
- For every $f : \xi_1 \ldots \xi_n \to \xi_0$ the function $J'_f : J'_{\xi_1} \times \cdots \times J'_{\xi_n} \to J'_{\xi_0}$ maps $(a_1, \ldots, a_n)$ to $J_f(a_1, \ldots, a_n)$, if $J_f(a_1, \ldots, a_n) \in J'_{\xi_0}$, and to an arbitrary element of J'_{ξ_0} otherwise.

That is, we obtain J' from J be deleting all junk elements from J_ξ if ξ is a BG sort, and by redefining the interpretation of f arbitrarily whenever $J_f(a_1, \ldots, a_n)$ is a junk element.

In the second step, we define the Σ-interpretation $\operatorname{nojunk}(J) = J''$ as the term-generated subinterpretation of J', that is,

- For every sort ξ, $J''_\xi = \{ t^{J'} \mid t \text{ is a ground term of sort } \xi \}$,
- For every $f : \xi_1 \ldots \xi_n \to \xi_0$, the function $J''_f : J''_{\xi_1} \times \cdots \times J''_{\xi_n} \to J''_{\xi_0}$ satisfies $J''_f(a_1, \ldots, a_n) = J'_f(a_1, \ldots, a_n)$.

---

[11] In [21], it is required that every ground term that contains a (proper or improper) BG-sorted FG subterm must be larger than any (BG or FG) ground term that does not contain such a subterm.

Lemma 8.5. *Let J, J', and* $\mathrm{nojunk}(J) = J''$ *be given as above. Then the following properties hold:*

(i) $t^{J''} = t^{J'}$ *for every ground term t.*

(ii) $J''_\xi = J'_\xi$ *for every BG sort ξ.*

(iii) J'' *is a term-generated Σ-interpretation and $J''|_{\Sigma_B}$ is a term-generated Σ_B-interpretation.*

(iv) *If $t = f(t_1, \ldots, t_n)$ is ground, $t_i^{J'} = t_i^J$ for all i, and $t^J \in J'_\xi$, then $t^{J'} = t^J$.*

(v) *If t is ground and all BG-sorted subterms of t are BG terms, then $t^{J'} = t^J$.*

(vi) *If C is a ground BG clause, then $J \models C$ if and only if $J'' \models C$ if and only if $J''|_{\Sigma_B} \models C$.*

Proof. Properties (i)–(iv) follow directly from the definition of J' and J''. Property (v) follows from (iv) and the definition of J' by induction over the term structure. By (i) and (v), every ground BG term is interpreted in the same way in J and J'', moreover it is obvious that every ground BG term is interpreted in the same way in J'' and $J''|_{\Sigma_B}$; this implies (vi). ◻

Lemma 8.6. *If J is a Σ-interpretation and $I = \mathrm{nojunk}(J)$, then for every ground term s there exists a ground term t such that $s^I = t^I$ and all BG-sorted (proper or non-proper) subterms of t are BG terms.*

Proof. If s has a BG sort ξ, then this follows directly from the fact that $s^I \in I_\xi$ and that every element of I_ξ equals t^I for some ground BG term t of sort ξ. If s has a FG sort, let $s_1, \ldots, s_n$ be the maximal BG-sorted subterms of $s = s[s_1, \ldots, s_n]$. Since for every s_i there is a ground BG term t_i with $s_i^I = t_i^I$, we obtain $s^I = (s[s_1, \ldots, s_n])^I = (s[t_1, \ldots, t_n])^I$. Set $t := s[t_1, \ldots, t_n]$. ◻

Corollary 8.7. *Let J be a Σ-interpretation and $I = \mathrm{nojunk}(J)$. Let $C\sigma$ by a ground instance of a clause C. Then there is a smooth ground instance $C\tau$ of C such that $(t\sigma)^I = (t\tau)^I$ for every term occurring in C and such that $I \models C\sigma$ if and only if $I \models C\tau$.*

Proof. Using the previous lemma, we define τ such that for every variable ζ occurring in C, $(\zeta\tau)^I = (\zeta\sigma)^I$ and all BG-sorted (proper or non-proper) subterms of $\zeta\tau$ are BG terms. Clearly τ is smooth. The other properties follow immediately by induction over the term or clause structure. ◻

Lemma 8.8. *Let N be a set of Σ-clauses that is locally sufficiently complete. Let $I \in \mathcal{B}$ be a Σ_B-interpretation, let J be a Σ-model of $\mathrm{sgi}(N) \cup \mathrm{E}_I \cup \mathrm{D}_I$, and let $J'' = \mathrm{nojunk}(J)$. Let $C \in N$ and let $C\tau$ by a smooth ground instance in $\mathrm{smgi}(N) \setminus \mathcal{R}^{\mathcal{S}}_{\mathrm{Cl}}(\mathrm{smgi}(N) \cup \mathrm{E}_I \cup \mathrm{D}_I)$. Then $(t\tau)^J = (t\tau)^{J''}$ for every term t occurring in C and $J \models C\tau$ if and only if $J'' \models C\tau$.*

Proof. Let J' be defined as above, then $(t\tau)^{J'} = (t\tau)^{J''}$ for any term t occurring in C by Lemma 8.5-(i). We prove that $(t\tau)^J = (t\tau)^{J'}$ by induction over the term structure: If t is a variable, then by smoothness all BG-sorted subterms of $t\tau$ are BG terms, hence $(t\tau)^{J'} = (t\tau)^J$ by Lemma 8.5-(v). Otherwise let

$t = f(t_1, \ldots, t_n)$. If $t\tau$ is a BG term, then again $(t\tau)^{J'} = (t\tau)^J$ by Lemma 8.5-(v). If $t\tau$ is a FG term of sort ξ, then t must be a FG term of sort ξ as well. By the induction hypothesis, $(t_i\tau)^J = (t_i\tau)^{J'}$ for every i. If ξ is a FG sort, then trivially $(t\tau)^J = J_f((t_1\tau)^J, \ldots, (t_n\tau)^J)$ is contained in J'_ξ, so $(t\tau)^{J'} = (t\tau)^J$ by Lemma 8.5-(iv). Otherwise, $t\tau$ is a BG-sorted FG term occurring in $\mathrm{smgi}(N) \setminus \mathcal{R}^{\mathcal{S}}_{\mathrm{Cl}}(\mathrm{smgi}(N) \cup \mathrm{E}_I \cup \mathrm{D}_I)$. By local sufficient completeness, there exists a ground BG term s such that $s^J = (t\tau)^J$, hence $(t\tau)^J \in J'_\xi$. Again, Lemma 8.5-(iv) yields $(t\tau)^{J'} = (t\tau)^J$.

Since all left and right-hand sides of equations in $C\tau$ are evaluated in the same way in J'' and J, it follows that $J \models C\tau$ if and only if $J'' \models C\tau$. □

Lemma 8.9. *Let N be a set of Σ-clauses that is locally sufficiently complete. Let $I \in \mathcal{B}$ be a Σ_B-interpretation, let J be a Σ-model of $\mathrm{sgi}(N) \cup \mathrm{E}_I \cup \mathrm{D}_I$, and let $J'' = \mathrm{nojunk}(J)$. Then J'' is a model of N.*

Proof. The proof proceeds in three steps. In the first step we show that J'' is a model of $\mathrm{smgi}(N) \setminus \mathcal{R}^{\mathcal{S}}_{\mathrm{Cl}}(\mathrm{smgi}(N) \cup \mathrm{E}_I \cup \mathrm{D}_I)$: Let $C \in N$ and let $C\tau$ be a smooth ground instance in $\mathrm{smgi}(N) \setminus \mathcal{R}^{\mathcal{S}}_{\mathrm{Cl}}(\mathrm{smgi}(N) \cup \mathrm{E}_I \cup \mathrm{D}_I)$. Since every smooth ground instance is a simple ground instance and J is a Σ-model of $\mathrm{sgi}(N)$, we know that $J \models C\tau$. By Lemma 8.8, this implies $J'' \models C\tau$.

In the second step we show that J'' is a model of $\mathrm{smgi}(N)$. Since we already know that J'' is a model of $\mathrm{smgi}(N) \setminus \mathcal{R}^{\mathcal{S}}_{\mathrm{Cl}}(\mathrm{smgi}(N) \cup \mathrm{E}_I \cup \mathrm{D}_I)$, it is clearly sufficient to show that J'' is a model of $\mathcal{R}^{\mathcal{S}}_{\mathrm{Cl}}(\mathrm{smgi}(N) \cup \mathrm{E}_I \cup \mathrm{D}_I)$: First we observe that by Lemma 8.5 $J'' \models \mathrm{E}_I \cup \mathrm{D}_I$. Using the result of the first step, this implies that J'' is a model of $(\mathrm{smgi}(N) \setminus \mathcal{R}^{\mathcal{S}}_{\mathrm{Cl}}(\mathrm{smgi}(N) \cup \mathrm{E}_I \cup \mathrm{D}_I)) \cup \mathrm{E}_I \cup \mathrm{D}_I$, and since this set is a superset of $(\mathrm{smgi}(N) \cup \mathrm{E}_I \cup \mathrm{D}_I) \setminus \mathcal{R}^{\mathcal{S}}_{\mathrm{Cl}}(\mathrm{smgi}(N) \cup \mathrm{E}_I \cup \mathrm{D}_I)$, J'' is also a model of the latter. By Definition 6.2-(i), $(\mathrm{smgi}(N) \cup \mathrm{E}_I \cup \mathrm{D}_I) \setminus \mathcal{R}^{\mathcal{S}}_{\mathrm{Cl}}(\mathrm{smgi}(N) \cup \mathrm{E}_I \cup \mathrm{D}_I) \models \mathcal{R}^{\mathcal{S}}_{\mathrm{Cl}}(\mathrm{smgi}(N) \cup \mathrm{E}_I \cup \mathrm{D}_I)$. So J'' is a model of $\mathcal{R}^{\mathcal{S}}_{\mathrm{Cl}}(\mathrm{smgi}(N) \cup \mathrm{E}_I \cup \mathrm{D}_I)$.

We can now show the main statement: We know that J'' is a term-generated Σ-interpretation, so $J'' \models N$ holds if and only if J'' is a model of all ground instances of clauses in N. Let $C\sigma$ be an arbitrary ground instance of $C \in N$. By Corollary 8.7, there is a smooth ground instance $C\tau$ such that $J'' \models C\sigma$ if and only if $J'' \models C\tau$. As the latter has been shown in the second step, the result follows. □

Theorem 8.10. *If the BG specification $(\Sigma_B, \mathcal{B})$ is compact and if the clause set N is locally sufficiently complete, then $\mathrm{HSP}_{\mathrm{Base}}$ and $\mathcal{R}^{\mathcal{H}}$ are dynamically refutationally complete for $\mathrm{abstr}(N)$, i. e., if $N \models_{\mathcal{B}} \square$, then every fair derivation starting from $\mathrm{abstr}(N)$ eventually generates $\square$.*

Proof. Let $\mathcal{D} = (N_i)_{i \geq 0}$ be a fair derivation starting from $N_0 = \mathrm{abstr}(N)$, and let N_∞ be the limit of $\mathcal{D}$. By fairness, N_∞ is saturated w.r.t. $\mathrm{HSP}_{\mathrm{Base}}$ and $\mathcal{R}^{\mathcal{H}}$. If $\square \notin N_\infty$, then the Close rule is not applicable to N_∞. Since $(\Sigma_B, \mathcal{B})$ is compact, this means that the set of all Σ_B-clauses in $\mathrm{sgi}(N_\infty)$ is satisfied by some term-generated Σ_B-interpretation $I \in \mathcal{B}$. By Theorem 7.8, $(N_\infty)_I$ is saturated with respect to the standard superposition calculus. Since $\Box \notin (N_\infty)_I$, the refutational completeness of standard superposition implies that there is a Σ-model J of $(N_\infty)_I$, and since $E_I \cup D_I \subseteq (N_\infty)_I$, J is also a Σ-model of $\mathrm{sgi}(N_\infty) \cup E_I \cup D_I$. Since every clause in N_0 is either contained in N_∞ or redundant w.r.t. N_∞, every simple ground instance of a clause in N_0 is a simple ground instance of a clause in N_∞ or contained in $\mathrm{GndTh}(\mathcal{B})$ or redundant w.r.t. $\mathrm{sgi}(N_\infty) \cup \mathrm{GndTh}(\mathcal{B})$. We conclude that J is a Σ-model of $\mathrm{sgi}(N_0)$, and since $\mathrm{sgi}(N_0)$ and $\mathrm{sgi}(N)$ are equivalent, J is a Σ-model of $\mathrm{sgi}(N)$. Now define $J'' = \mathrm{nojunk}(J)$. By Lemma 8.5, J'' is a term-generated Σ-interpretation, $J''|_{\Sigma_B}$ is a term-generated Σ_B-interpretation, and $J''|_{\Sigma_B}$ satisfies $E_I \cup D_I$. Consequently, $J''|_{\Sigma_B}$ is isomorphic to I and thus contained in $\mathcal{B}$. Finally, J'' is a model of N by Lemma 8.9. $\qquad\Box$

If all BG-sorted FG terms in a set N of clauses are ground, local sufficient completeness can be established automatically by adding a "definition" of the form $t \approx \alpha$, where t is a ground BG-sorted FG term and α is a parameter. The following section explains this idea in a more general way.

9   Local Sufficient Completeness by Define

The $\mathrm{HSP}_{\mathrm{Base}}$ inference system will derive a contradiction if the input clause set is inconsistent and (locally) sufficiently complete (cf. Sect. 8). In this section we extend this functionality by adding an inference rule, Define, which can turn input clause sets that are not sufficiently complete into locally sufficiently complete ones. Technically, the Define rule derives "definitions" of the form $t \approx \alpha$, where t is a ground BG-sorted FG term and α is a parameter of the proper sort. For economy of reasoning, definitions are introduced only on a by-need basis, when t appears in a current clause, and $t \approx \alpha$ is used to simplify that clause immediately.

We need one more preliminary definition before introducing Define formally.

Definition 9.1 (Unabstracted clause). *A clause is* unabstracted *if it does not contain any disequation $\zeta \not\approx t$ between a variable ζ and a term t unless $t \neq \zeta$ and $\zeta \in \mathrm{vars}(t)$.* $\qquad\Box$

Any clause can be unabstracted by repeatedly replacing $C \vee \zeta \not\approx t$ by $C[\zeta \mapsto t]$ whenever $t = \zeta$ or $\zeta \notin \mathrm{vars}(t)$. Let $\mathrm{unabstr}(C)$ denote an unabstracted version of C obtained this way. If $t = t[\zeta_1, \ldots, \zeta_n]$ is a term in C and ζ_i is finally instantiated to t_i, we denote its unabstracted version $t[t_1, \ldots, t_n]$ by $\mathrm{unabstr}(t[\zeta_1, \ldots, \zeta_n], C)$. For a clause set N let $\mathrm{unabstr}(N) = \{\mathrm{unabstr}(C) \mid C \in N\}$.

The *full inference system* HSP of the hierarchic superposition calculus consists of the inference rules of $\mathrm{HSP}_{\mathrm{Base}}$ and the following Define inference rule. As for the other inference rules we suppose that all premises are weakly abstracted.

$$\text{Define } \frac{N \cup \{L[t[\zeta_1, \ldots, \zeta_n]] \vee D\}}{N \cup \text{abstr}(\{t[t_1, \ldots, t_n] \approx \alpha_{t[t_1, \ldots, t_n]}, \ L[\alpha_{t[t_1, \ldots, t_n]}] \vee D\}}$$

if

(i) $t[\zeta_1, \ldots, \zeta_n]$ is a minimal BG-sorted non-variable term with a toplevel FG operator,

(ii) $t[t_1, \ldots, t_n] = \text{unabstr}(\{t[\zeta_1, \ldots, \zeta_n], L[t[\zeta_1, \ldots, \zeta_n]] \vee D\})$,

(iii) $t[t_1, \ldots, t_n]$ is ground,

(iv) $\alpha_{t[t_1, \ldots, t_n]}$ is a parameter, uniquely determined by $t[t_1, \ldots, t_n]$, and

(v) $L[t[\zeta_1, \ldots, \zeta_n]] \vee D \in \mathcal{R}_{\text{Cl}}^{\mathcal{H}}(N \cup \text{abstr}(\{t[t_1, \ldots, t_n] \approx \alpha_{t[t_1, \ldots, t_n]}, L[\alpha_{t[t_1, \ldots, t_n]}] \vee D\}))$.

In (i), by minimality we mean that no proper subterm of $t[\zeta_1, \ldots, \zeta_n]$ is a BG-sorted non-variable term with a toplevel FG operator. In effect, the Define rule eliminates such terms inside-out. Conditions (iii) and (iv) are needed for soundness. Condition (v) is needed to guarantee that Define is a simplifying inference rule, much like the Simp rule in Sect. 7.[12] In particular, it makes sure that Define cannot be applied to definitions themselves.

Theorem 9.2. *The inference rules of HSP are satisfiability-preserving w.r.t. $\models_{\mathcal{B}}$, i. e., for every inference with premise N and conclusion N' we have $N \models_{\mathcal{B}} \Box$ if and only if $N' \models_{\mathcal{B}} \Box$. Moreover, $N' \models_{\mathcal{B}} N$.*

Proof. For the inference rules of HSP_{Base}, the result follows from Theorem 6.1.

For Define, we observe first that condition (ii) implies that $L[t[\zeta_1, \ldots, \zeta_n]] \vee D$ and $L[t[t_1, \ldots, t_n]] \vee D$ are equivalent. If $N \cup \{L[t[t_1, \ldots, t_n]] \vee D\}$ is $\mathcal{B}$-satisfiable, let I be a Σ-model of all ground instances of $N \cup \{L[t[t_1, \ldots, t_n]] \vee D\}$ such that $I|_{\Sigma_{\text{B}}}$ is in $\mathcal{B}$. By condition (iii), $t[t_1, \ldots, t_n]$ is ground. Let J be the Σ-interpretation obtained from J by redefining the interpretation of $\alpha_{t[t_1, \ldots, t_n]}$ in such a way that $\alpha_{t[t_1, \ldots, t_n]}^{J} = t[t_1, \ldots, t_n]^{I}$, then J is a Σ-model of every ground instance of N, $t[t_1, \ldots, t_n] \approx \alpha_{t[t_1, \ldots, t_n]}$ and $L[\alpha_{t[t_1, \ldots, t_n]}] \vee D$, and hence also a model of the abstractions of these clauses. Conversely, every model of $t[t_1, \ldots, t_n] \approx \alpha_{t[t_1, \ldots, t_n]}$ and $L[\alpha_{t[t_1, \ldots, t_n]}] \vee D$ is a model of $L[t[t_1, \ldots, t_n]] \vee D$. $\Box$

Example 9.3. Let $C = \mathsf{g}(\mathsf{f}(x, y) + 1, x, y) \approx 1 \vee x \not\approx 1 + \beta \vee y \not\approx \mathsf{c}$ be the premise of a Define inference. We get $\text{unabstr}(C) = \mathsf{g}(\mathsf{f}(1 + \beta, \mathsf{c}) + 1, 1 + \beta, \mathsf{c}) \approx 1$. The (unabstracted) conclusions are the definition $\mathsf{f}(1 + \beta, \mathsf{c}) \approx \alpha_{\mathsf{f}(1 + \beta, \mathsf{c})}$ and the clause $\mathsf{g}(\alpha_{\mathsf{f}(1 + \beta, \mathsf{c})} + 1, x, y) \approx 1 \vee x \not\approx 1 + \beta \vee y \not\approx \mathsf{c}$. Abstraction yields $\mathsf{f}(X, \mathsf{c}) \approx \alpha_{\mathsf{f}(1 + \beta, \mathsf{c})} \vee X \not\approx 1 + \beta$ and $\mathsf{g}(Z, x, y) \approx 1 \vee x \not\approx 1 + \beta \vee y \not\approx \mathsf{c} \vee Z \not\approx \alpha_{\mathsf{f}(1 + \beta, \mathsf{c})} + 1$.

One might be tempted to first unabstract the premise C before applying Define. However, unabstraction may eliminate FG terms (c in the example) which is not undone by abstraction. This may lead to incompleteness. $\Box$

---

[12] Condition (i) of Simp is obviously satisfied and condition (iii) there is condition (v) of Define. Instead of condition (ii), Define inferences are only $\mathcal{B}$-satisfiability preserving, which however does not endanger soundness.

Example 9.4. The following clause set demonstrates the need for condition (v) in Define. Let $N = \{f(c) \approx 1\}$ and suppose condition (v) is absent. Then we obtain $N' = \{f(c) \approx \alpha_{f(c)},\ \alpha_{f(c)} \approx 1\}$. By demodulating the first clause with the second clause we get $N'' = \{f(c) \approx 1,\ \alpha_{f(c)} \approx 1\}$. Now we can continue with N'' as with N. The problem is, of course, that the new definition $f(c) \approx \alpha_{f(c)}$ is greater w.r.t. the term ordering than the parent clause, in violation of condition (v). □

Example 9.5. Consider the weakly abstracted clauses $P(0)$, $f(x) > 0 \vee \neg P(x)$, $Q(f(x))$, $\neg Q(x) \vee 0 > x$. Suppose $\neg P(x)$ is maximal in the second clause. By superposition between the first two clauses we derive $f(0) > 0$. With Define we obtain $f(0) \approx \alpha_{f(0)}$ and $\alpha_{f(0)} > 0$, the latter replacing $f(0) > 0$. From the third clause and $f(0) \approx \alpha_{f(0)}$ we obtain $Q(\alpha_{f(0)})$, and with the fourth clause $0 > \alpha_{f(0)}$. Finally we apply Close to $\{\alpha_{f(0)} > 0,\ 0 > \alpha_{f(0)}\}$. □

It is easy to generalize Theorem 8.10 to the case that local sufficient completeness does not hold initially, but is only established with the help of Define inferences:

Theorem 9.6. *Let $\mathcal{D} = (N_i)_{i \geq 0}$ be a fair HSP derivation starting from $N_0 = \mathrm{abstr}(N)$, let $k \geq 0$, such that $N_k = \mathrm{abstr}(N')$ and N' is locally sufficiently complete. If the BG specification $(\Sigma_B, \mathcal{B})$ is compact, then the limit of $\mathcal{D}$ contains* □ *if and only if N is $\mathcal{B}$-unsatisfiable.*

Proof. Since every derivation step in an HSP derivation is satisfiability-preserving, the "only if" part is again obvious.

For the "if" part, we assume that N_∞, the limit of $\mathcal{D}$, does not contain □. By fairness, N_∞ is saturated w.r.t. HSP and $\mathcal{R}^{\mathcal{H}}$. We start by considering the subderivation $(N_i)_{i \geq k}$ starting with $N_k = \mathrm{abstr}(N')$. Like in the proof of Theorem 8.10, we can show that N' is $\mathcal{B}$-satisfiable, that is, there exists a model J of N' that is a term-generated Σ-interpretation, and whose restriction $J|_{\Sigma_B}$ is contained in $\mathcal{B}$. From Lemma 7.1 and Proposition 5.2 we see that $N' \models_\mathcal{B} N_k$, and similarly $N_0 \models_\mathcal{B} N$. Furthermore, since every clause in $N_0 \setminus N_k$ must be redundant w.r.t. N_k, we have $N_k \models_\mathcal{B} N_0$. Combining these three entailments, we conclude that $N' \models_\mathcal{B} N$, so N is $\mathcal{B}$-satisfiable and J is a model of N. □

Condition (v) of the Define rule requires that the clause that is deleted during a Define inference must be redundant with respect to the remaining clauses. This condition is needed to preserve refutational completeness. There are cases, however, where condition (v) prevents us from introducing a definition for a subterm. Consider the clause set $N = \{C\}$ where $C = f(c) \approx 1 \vee c \approx d$, the constants c and d are FG-sorted, f is a BG-sorted FG operator, and $c \succ d \succ 1$. The literal $f(c) \approx 1$ is maximal in C. The clause set $N = \mathrm{abstr}(N)$ is not locally sufficient complete (the BG-sorted FG-term $f(c)$ may be interpreted differently from all BG terms in a Σ-model). Moreover, it cannot be made locally sufficient complete using the Define rule, since the definition $f(c) \approx \alpha_{f(c)}$ is larger w.r.t. the clause ordering than C, in violation of condition (v) of Define.

However, at the beginning of a derivation, we may be a bit more permissive. Let us define the *reckless* Define inference rule in the same way as the Define rule except that the applicability condition (v) is dropped. Clearly, in the example above, the reckless Define rule allows us to derive the locally sufficiently complete clause set $N' = \{\alpha_{f(c)} \approx 1 \vee c \approx d, f(c) \approx \alpha_{f(c)}\}$ as desired. In fact, we can show that this is always possible if N is a finite clause set in which all BG-sorted FG terms are ground.

Definition 9.7 (Pre-derivation). *Let N_0 be a weakly abstracted clause set. A* pre-derivation *(of a clause set N^{pre}) is a derivation of the form $N_0, N_1,$ $\ldots, (N_k = N^{\mathrm{pre}})$, for some $k \geq 0$, with the inference rule reckless Define only, and such that each clause $C \in N^{\mathrm{pre}}$ either does not contain any BG-sorted FG operator or $C = \mathrm{abstr}(C')$ and C' is a definition, i.e., a ground positive unit clause of the form $f(t_1, \ldots, t_n) \approx t$ where f is a BG-sorted FG operator, $t_1, \ldots, t_n$ do not contain BG-sorted FG operators, and t is a background term.* $\square$

Lemma 9.8. *Let N be a finite clause set in which all BG-sorted FG terms are ground. Then there is a pre-derivation starting from $N_0 = \mathrm{abstr}(N)$ such that N^{pre} is locally sufficiently complete.*

Proof. Since every term headed by a BG-sorted FG operator in $\mathrm{unabstr}(N_0)$ is ground, we can incrementally eliminate all occurrences of terms headed by BG-sorted FG operators from N_0, except those in abstractions of definitions. Let $N_0, N_1, \ldots, (N_k = N^{\mathrm{pre}})$ be the sequence of sets of clauses obtained in this way. We will show that N^{pre} is locally sufficiently complete.

Let $I \in \mathcal{B}$ be a Σ_{B}-interpretation, let J be a Σ-model of $\mathrm{sgi}(N^{\mathrm{pre}}) \cup \mathrm{E}_I \cup \mathrm{D}_I$ and let $C\theta$ be a smooth ground instance in $\mathrm{smgi}(N) \setminus \mathcal{R}^{\mathcal{S}}_{\mathrm{Cl}}(\mathrm{smgi}(N) \cup \mathrm{E}_I \cup \mathrm{D}_I)$. We have to show that for every BG-sorted FG term s occurring in $C\theta$ there is a ground BG term t such that $J \models s \approx t$.

If C does not contain any BG-sorted FG operator, then there are no BG-sorted FG terms in $C\theta$, so the property is vacuously true. Otherwise $C = \mathrm{abstr}(C')$ and C' is a definition $f(t_1, \ldots, t_n) \approx t$ where f is a BG-sorted FG operator, $t_1, \ldots, t_n$ do not contain BG-sorted FG operators, and t is a background term. In this case, C must have the form $f(u_1, \ldots, u_n) \approx u \vee E$, such that E is a BG clause, $u_1, \ldots, u_n$ do not contain BG-sorted FG operators, and u is a BG term. The only BG-sorted FG term in the smooth instance $C\theta$ is therefore $f(u_1\theta, \ldots, u_n\theta)$. If any literal of $E\theta$ were true in J, then it would follow from $\mathrm{E}_I \cup \mathrm{D}_I$, therefore $C\theta \in \mathcal{R}^{\mathcal{S}}_{\mathrm{Cl}}(\mathrm{smgi}(N) \cup \mathrm{E}_I \cup \mathrm{D}_I)$, contradicting the assumption. Hence $J \models f(u_1\theta, \ldots, u_n\theta) \approx u\theta$, and since $u\theta$ is a ground BG term, the requirement is satisfied. $\square$

Lemma 9.8 will be needed to prove a completeness result for the fragment defined in the next section.

10   The Ground BG-Sorted Term Fragment

According to Theorem 8.10, the $\mathrm{HSP}_{\mathrm{Base}}$ calculus is refutationally complete provided that the clause set is locally sufficiently complete and the BG specification is compact. We have seen in the previous section that the (reckless) Define rule can help to establish local sufficient completeness by introducing new parameters. In fact, finite clause sets in which all BG-sorted FG terms are ground can always be converted into locally sufficiently complete clause sets (cf. Lemma 9.8). On the other hand, as noticed in Sect. 3, the introduction of parameters can destroy the compactness of the BG specification. In this and the following section, we will identify two cases where we can not only establish local sufficient completeness, but where we can also guarantee that compactness poses no problems. The *ground BG-sorted term fragment (GBT fragment)* is one such case:

Definition 10.1 (GBT fragment). *A clause C is a GBT clause if all BG-sorted terms in C are ground. A finite clause set N belongs to the GBT fragment if all clauses in N are GBT clauses.*   □

Clearly, by Lemma 9.8 for every clause set N that belongs to the GBT fragment there is a pre-derivation that converts $\mathrm{abstr}(N)$ into a locally sufficiently complete clause set. Moreover, pre-derivations also preserve the GBT property:

Lemma 10.2. *If $\mathrm{unabstr}(N)$ belongs to the GBT fragment and N' is obtained from N by a reckless Define inference, then $\mathrm{unabstr}(N')$ also belongs to the GBT fragment.*

The proof can be found in [12].

As we have seen, N^{pre} is locally sufficiently complete. At this stage this suggests to exploit the completeness result for locally sufficiently complete clause sets, Theorem 8.10. However, Theorem 8.10 requires compact BG specifications, and the question is if we can avoid this. We can indeed get a complete calculus under rather mild assumptions on the Simp rule:

Definition 10.3 (Suitable Simp inference). *Let $\succ_{\mathrm{fin}}$ be a strict partial term ordering such that for every ground BG term s only finitely many ground BG terms t with $s \succ_{\mathrm{fin}} t$ exist.*[13] *We say that a Simp inference with premise $N \cup \{C\}$ and conclusion $N \cup \{D\}$ is* suitable *(for the GBT fragment) if*

(i) *for every BG term t occurring in $\mathrm{unabstr}(D)$ there is a BG term $s \in \mathrm{unabstr}(C)$ such that $s \succeq_{\mathrm{fin}} t$,*
(ii) *every occurrence of a BG-sorted FG operator f in $\mathrm{unabstr}(D)$ is of the form $\mathsf{f}(t_1, \ldots, t_n) \approx t$ where t is a ground BG term,*
(iii) *every BG term in D is pure, and*
(iv) *if every BG term in $\mathrm{unabstr}(C)$ is ground then every BG term in $\mathrm{unabstr}(D)$ is ground.*

We say the Simp inference rule is suitable *if every Simp inference is.*   □

Expected simplification techniques like demodulation, subsumption deletion and evaluation of BG subterms are all covered as suitable Simp inferences. Also, evaluation of BG subterms is possible, because simplifications are not only decreasing

---

[13] A KBO with appropriate weights can be used for $\succ_{\mathrm{fin}}$.

w.r.t. $\succ$ but *additionally* also decreasing w.r.t. $\succeq_{\text{fin}}$, as expressed in condition (i). Without it, e.g., the clause $P(1 + 1, 0)$ would admit infinitely many simplified versions $P(2, 0)$, $P(2, 0 + 0)$, $P(2, 0 + (0 + 0))$, etc.

The HSP_{Base} inferences do in general not preserve the shape of the clauses in $\text{unabstr}(N^{\text{pre}})$; they do preserve a somewhat weaker property – cleanness – which is sufficient for our purposes.

Definition 10.4 (Clean clause). *A weakly abstracted clause C is* clean *if*

(i) *every BG term in C is pure,*
(ii) *every BG term in $\text{unabstr}(C)$ is ground, and*
(iii) *every occurrence of a BG-sorted FG operator f in $\text{unabstr}(C)$ is in a positive literal of the form $\mathsf{f}(t_1, \ldots, t_n) \approx t$ where t is a ground BG term.*

For example, if c is FG-sorted, then $P(\mathsf{f}(\mathsf{c}) + 1)$ is not clean, while $\mathsf{f}(x) \approx 1 + \alpha \vee P(x)$ is. A clause set is called *clean* if every clause in N is. Notice that N^{pre} is clean.

Lemma 10.5. *Let $C_1, \ldots, C_n$ be clean clauses. Assume a HSP_{Base} inference with premises $C_1, \ldots, C_n$ and conclusion C. Then C is clean and every BG term occurring in $\text{unabstr}(C)$ also occurs in some clause $\text{unabstr}(C_1), \ldots, \text{unabstr}(C_n)$.*

The proof can be found in [12].

Thanks to conditions (ii)–(iv) in Definition 10.3, suitable Simp inferences preserves cleanness:

Lemma 10.6. *Let $N \cup \{C\}$ be a set of clean clauses. If $N \cup \{D\}$ is obtained from $N \cup \{C\}$ by a suitable Simp inference then D is clean.*

Proof. Suppose $N \cup \{D\}$ is obtained from $N \cup \{C\}$ by a suitable Simp inference. We need to show properties (i)–(iii) of cleanness for D. That every BG term in D is pure follows from Definition 10.3-(iii). That every BG term in $\text{unabstr}(D)$ is ground follows from Definition 10.3-(iv) and cleanness of C. Finally, property (iii) follows from Definition 10.3-(ii). $\square$

With the above lemmas we can prove our main result:

Theorem 10.7. *The HSP calculus with a suitable Simp inference rule is dynamically refutationally complete for the ground BG-sorted term fragment. More precisely, let N be a finite set of GBT clauses and $\mathcal{D} = (N_i)_{i \geq 0}$ a fair HSP derivation such that reckless Define is applied only in a pre-derivation ($N_0 = \text{abstr}(N)$), $\ldots$, ($N_k = N^{\text{pre}}$), for some $k \geq 0$. Then the limit of $\mathcal{D}$ contains $\square$ if and only if N is $\mathcal{B}$-unsatisfiable.*

Notice that Theorem 10.7 does not appeal to compactness of BG specifications.

Proof. Our goal is to apply Theorem 9.6 and its proof, in a slightly modified way. For that, we first need to know that $N^{\text{pre}} = \text{abstr}(N')$ for some clause set N' that is locally sufficiently complete.

We are given that N is a set of GBT clauses. Recall that weak abstraction (recursively) extracts BG subterms by substituting fresh variables and adding disequations. Unabstraction reverses this process (and possibly eliminates additional disequations). It follows that with N being a set of GBT clauses, so is unabstr(abstr(N)) = unabstr(N_0). From Lemma 10.2 it follows that unabstr(N^{pre}) is also a GBT clause set.

Now chose N' as the clause set that is obtained from N^{pre} by replacing every clause $C \in N^{\mathrm{pre}}$ such that unabstr(C) is a definition by unabstr(C). By construction of definitions, unabstraction reverses weak abstraction of definitions. It follows $N^{\mathrm{pre}} = \mathrm{abstr}(N')$. By definition of pre-derivations, all BG-sorted FG terms occurring in unabstr(N^{pre}) occur in definitions. Hence, with unabstr(N^{pre}) being a set of GBT clauses so is N'. It follows easily that N' is locally sufficiently complete, as desired.

We cannot apply Theorem 9.6 directly now because it requires compactness of the BG specification, which cannot be assumed. However, we can use the following argumentation instead.

Let $N^{\infty} = \bigcup_{i \geq 0} N_i$ be the union of $\mathcal{D}$. We next show that unabstr(N^{∞}) contains only finitely many different BG terms and each of them is ground. Recall that unabstr(N^{pre}) is a GBT clause set, and so every BG term in unabstr(N^{pre}) is ground. Because Define is disabled in $\mathcal{D}$, only HSP$_{\mathrm{Base}}$ and (suitable) Simp inferences need to be analysed. Notice that N^{pre} is clean and both the HSP$_{\mathrm{Base}}$ and Simp inferences preserve cleanness, as per Lemmas 10.5-(1) and 10.6, respectively.

With respect to HSP$_{\mathrm{Base}}$ inferences, together with Definition 10.4-(ii) it follows that every BG term t in the unabstracted version unabstr(C) of the inference conclusion C is ground. Moreover, t also occurs in the unabstracted version of some premise clause by Lemma 10.5-(2). In other words, HSP$_{\mathrm{Base}}$ inferences do not grow the set of BG terms w.r.t. unabstracted premises and conclusions.

With respect to Simp inferences, unabstr(N^{pre}) provide an upper bound w.r.t. the term ordering $\succ_{\mathrm{fin}}$ for all BG terms generated in Simp inferences. There can be only finitely many such terms, and each of them is ground, which follows from Definition 10.3-(i).

Because every BG term occurring in unabstr(N^{∞}) is ground, every BG clause in unabstr(N^{∞}) is a multiset of literals of the form $s \approx t$ or $s \not\approx t$, where s and t are ground BG terms. With only finitely many BG terms available, there are only finitely many BG clauses in unabstr(N^{∞}), modulo equivalence. Because unabstraction is an equivalence transformation, there are only finitely many BG clauses in N^{∞} as well, modulo equivalence.

Let $N_{\infty} = \bigcup_{i \geq 0} \bigcap_{j \geq i} N_j$ be the limit clause set of the derivation $\mathcal{D}$, which is saturated w.r.t. the hierarchic superposition calculus and $\mathcal{R}^{\mathcal{H}}$. Because $\mathcal{D}$ is not a refutation, it does not contain $\square$. Consequently the Close rule is not applicable to N_{∞}. The set N^{∞}, and hence also $N_{\infty} \subseteq N^{\infty}$, contains only finitely many BG clauses, modulo equivalence. This entails that the set of all Σ_{B}-clauses in sgi(N_{∞}) is satisfied by some term-generated Σ_{B}-interpretation $I \in \mathcal{B}$. Now, the rest of the proof is literally the same as in the proof of Theorem 9.6. $\square$

Because unabstraction can also be applied to fully abstracted clauses, it is possible to equip the hierarchic superposition calculus of [5] with a correspondingly modified Define rule and get Theorem 10.7 in that context as well.

Kruglov and Weidenbach [22] have shown how to use hierarchic superposition as a decision procedure for ground clause sets (and for Horn clause sets with constants and variables as the only FG terms). Their method preprocesses the given clause set by "basification", a process that removes BG-sorted FG terms similarly to our reckless Define rule. The resulting clauses then are fully abstracted and hierarchic superposition is applied. Some modifications of the inference rules make sure derivations always terminate. Simplification is restricted to subsumption deletion. The fragment of [22] is a further restriction of the GBT fragment. We expect we can get decidability results for that fragment with similar techniques.

11   Linear Arithmetic

For the special cases of linear integer arithmetic (LIA) and linear rational arithmetic as BG specifications, the result of the previous section can be extended significantly: In addition to ground BG-sorted terms, we can also permit BG-sorted variables and, in certain positions, even variables with offsets.

Recall that we have assumed that equality is the only predicate symbol in our language, so that a non-equational atom, say $s < t$, is to be taken as a shorthand for the equation $(s < t) \approx true$. We refer to the terms that result from this encoding of atoms as *atom terms*; other terms are called *proper terms*.

Theorem 11.1. *Let N be a set of clauses over the signature of linear integer arithmetic (with parameters α, β, etc.), such that every proper term in these clauses is either (i) ground, or (ii) a variable, or (iii) a sum $\zeta + k$ of a variable ζ and a number $k \geq 0$ that occurs on the right-hand side of a positive literal $s < \zeta + k$. If the set of ground terms occurring in N is finite, then N is satisfiable in LIA over $\mathbb{Z}$ if and only if N is satisfiable w.r.t. the first-order theory of LIA.*

Proof. Let N be a set of clauses with the required properties, and let T be the finite set of ground terms occurring in N. We will show that N is equivalent to some *finite* set of clauses over the signature of linear integer arithmetic, which implies that it is satisfiable in the integer numbers if and only if it is satisfiable in the first-order theory of LIA.

In a first step, we replace every negative ordering literal $\neg s < t$ or $\neg s \leq t$ by the equivalent positive ordering literal $t \leq s$ or $t < s$. All literals of clauses in the resulting set N_0 have the form $s \approx t$, $s \not\approx t$, $s < t$, $s \leq t$, or $s < \zeta + k$, where s and t are either variables or elements of T and $k \in \mathbb{N}$. Note that the number of variables in clauses in N_0 may be unbounded.

In order to handle the various inequality literals in a more uniform way, we introduce new binary relation symbols $<_k$ (for $k \in \mathbb{N}$) that are defined by $a <_k b$ if and only if $a < b + k$. Observe that $s <_k t$ entails $s <_n t$ whenever $k \leq n$.

Obviously, we may replace every literal $s < t$ by $s <_0 t$, every literal $s \leq t$ by $s <_1 t$, and every literal $s < \zeta + k$ by $s <_k \zeta$. Let N_1 be the resulting clause set.

We will now transform N_1 into an equivalent set N_2 of ground clauses. We start by eliminating all equality literals that contain variables by exhaustively applying the following transformation rules:

$$
\begin{aligned}
N \cup \{\, C \vee \zeta \not\approx \zeta \,\} &\;\rightarrow\; N \cup \{\, C \,\} \\
N \cup \{\, C \vee \zeta \not\approx t \,\} &\;\rightarrow\; N \cup \{\, C[\zeta \mapsto t] \,\} && \text{if } t \neq \zeta \\
N \cup \{\, C \vee \zeta \approx \zeta \,\} &\;\rightarrow\; N \\
N \cup \{\, C \vee \zeta \approx t \,\} &\;\rightarrow\; N \cup \{\, C \vee \zeta <_1 t,\ C \vee t <_1 \zeta \,\} && \text{if } t \neq \zeta
\end{aligned}
$$

All variables in inequality literals are then eliminated in a Fourier-Motzkin-like manner by exhaustively applying the transformation rule

$$
N \cup \Big\{\, C \vee \bigvee_{i \in I} \zeta <_{k_i} s_i \vee \bigvee_{j \in J} t_j <_{n_j} \zeta \,\Big\} \;\rightarrow\; N \cup \Big\{\, C \vee \bigvee_{i \in I} \bigvee_{j \in J} t_j <_{k_i + n_j} s_i \,\Big\}
$$

where ζ does not occur in C and one of the index sets I and J may be empty.

The clauses in N_2 are constructed over the finite set T of proper ground terms, but the length of the clauses in N_2 is potentially unbounded. In the next step, we will transform the clauses in such a way that any pair of terms s, t from T is related by at most one literal in any clause: We apply one of the following transformation rules as long as two terms s and t occur in more than one literal:

$$
\begin{aligned}
N \cup \{\, C \vee s <_k t \vee s \approx t \,\} &\;\rightarrow\; N \cup \{\, C \vee s <_k t \,\} && \text{if } k \geq 1 \\
N \cup \{\, C \vee s <_0 t \vee s \approx t \,\} &\;\rightarrow\; N \cup \{\, C \vee s <_1 t \,\} \\
N \cup \{\, C \vee s <_k t \vee s \not\approx t \,\} &\;\rightarrow\; N && \text{if } k \geq 1 \\
N \cup \{\, C \vee s <_0 t \vee s \not\approx t \,\} &\;\rightarrow\; N \cup \{\, C \vee s \not\approx t \,\} \\
N \cup \{\, C \vee s <_k t \vee s <_n t \,\} &\;\rightarrow\; N \cup \{\, C \vee s <_n t \,\} && \text{if } k \leq n \\
N \cup \{\, C \vee s <_k t \vee t <_n s \,\} &\;\rightarrow\; N && \text{if } k + n \geq 1 \\
N \cup \{\, C \vee s <_0 t \vee t <_0 s \,\} &\;\rightarrow\; N \cup \{\, C \vee s \not\approx t \,\} \\
N \cup \{\, C \vee L \vee L \,\} &\;\rightarrow\; N \cup \{\, C \vee L \,\} && \text{for any literal } L \\
N \cup \{\, C \vee s \approx t \vee s \not\approx t \,\} &\;\rightarrow\; N
\end{aligned}
$$

The length of the clauses in the resulting set N_3 is now bounded by $\frac{1}{2} m(m + 1)$, where m is the cardinality of T. Still, due to the indices of the relation symbols $<_k$, N_3 may be infinite. We introduce an equivalence relation $\sim$ on clauses in N_3 as follows: Define $C \sim C'$ if for all $s, t \in T$ (i) $s \approx t \in C$ if and only if $s \approx t \in C'$, (ii) $s \not\approx t \in C$ if and only if $s \not\approx t \in C'$, and (iii) $s <_k t \in C$ for some k if and only if $s <_n t \in C'$ for some n. This relation splits N_3 into at most $(\frac{1}{2} m(m + 1))^5$ equivalence classes.[14]

We will now show that each equivalence class is logically equivalent to a finite subset of itself. Let M be some equivalence class. Since any two clauses from

---

[14] Any pair of terms s, t is related in all clauses of an equivalence class by either a literal $s \approx t$, or $s \not\approx t$, or $s <_n t$ for some n, or $t <_n s$ for some n, or no literal at all, so there are five possibilities per unordered pair of terms.

M differ at most in the indices of their $<_k$-literals, we can write every clause $C_i \in M$ in the form

$$C_i \;=\; C \vee \bigvee_{1 \le l \le n} s_l <_{k_{il}} t_l$$

where C and the s_l and t_l are the same for all clauses in M. As we have mentioned above, $s_l <_{k_{il}} t_l$ entails $s_l <_{k_{jl}} t_l$ whenever $k_{il} \le k_{jl}$; so a clause $C_i \in M$ entails $C_j \in M$ whenever the n-tuple $(k_{i1}, \ldots, k_{in})$ is pointwise smaller or equal to the n-tuple $(k_{j1}, \ldots, k_{jn})$ (that is, $k_{il} \le k_{jl}$ for all $1 \le l \le n$).

Let Q be the set of n-tuples of natural numbers corresponding to the clauses in M. By Dickson's lemma [15], for every set of tuples in $\mathbb{N}^n$ the subset of minimal tuples (w.r.t.the pointwise extension of $\le$ to tuples) is finite. Let Q' be the subset of minimal tuples in Q, and let M' be the set of clauses in M that correspond to the tuples in Q'. Since for every tuple in $Q \setminus Q'$ there is a smaller tuple in Q', we know that every clause in $M \setminus M'$ is entailed by some clause in M'. So the equivalence class M is logically equivalent to its finite subset M'. Since the number of equivalence classes is also finite and all transformation rules are sound, this proves our claim.   □

In order to apply this theorem to hierarchic superposition, we must again impose some restrictions on the calculus. Most important, we have to change the definition of weak abstraction slightly: We drop the requirement that target terms are not domain elements from Definition 5.1, i. e., we abstract out a non-variable BG term q occurring in a clause $C[f(s_1, \ldots, q, \ldots, s_n)]$, where f is a FG operator or at least one of the s_i is a FG or impure BG term, even if q is a domain element. As we mentioned, all results obtained so far hold also for the modified definition of weak abstraction. In addition, we must again restrict to *suitable* Simp inferences (Definition 10.3). With these restrictions, we can prove our main result:

Theorem 11.2. *The hierarchic superposition calculus is dynamically refutationally complete w.r.t. LIA over $\mathbb{Z}$ for finite sets of Σ-clauses in which every proper BG-sorted term is either (i) ground, or (ii) a variable, or (iii) a sum $\zeta + k$ of a variable ζ and a number $k \ge 0$ that occurs on the right-hand side of a positive literal $s < \zeta + k$.*

Proof. Let N be a finite set of Σ-clauses with the required properties. By Lemma 9.8, a pre-derivation starting from $N_0 = \text{abstr}(N)$ yields a locally sufficiently complete finite set N_0 of abstracted clauses.

Now we run the hierarchic superposition calculus on N_0 (with the same restrictions on simplifications as in Sect. 10). Let N_1 be the (possibly infinite) set of BG clauses generated during the run. By unabstracting these clauses, we obtain an equivalent set N_2 of clauses that satisfy the conditions of Theorem 11.1, so N_2 is satisfiable in LIA over $\mathbb{Z}$ if and only if N is satisfiable w.r.t.the first-order theory of LIA. Since the hierarchic superposition calculus is dynamically refutationally complete w.r.t. the first-order theory of LIA, the result follows. □

Analogous results hold for linear rational arithmetic. Let n be the least common divisor of all numerical constants in the original clause set; then we define $a <_{2i} b$ by $a < b + \frac{i}{n}$ and $a <_{2i+1} b$ by $a \leq b + \frac{i}{n}$ for $i \in \mathbb{N}$ and express every inequation literal in terms of $<_k$. The Fourier-Motzkin transformation rule is replaced by

$$N \cup \{ C \vee \bigvee_{i \in I} \zeta <_{k_i} s_i \vee \bigvee_{j \in J} t_j <_{n_j} \zeta \} \quad \rightarrow \quad N \cup \{ C \vee \bigvee_{i \in I} \bigvee_{j \in J} t_j <_{k_i \bullet n_j} s_i \}$$

where ζ does not occur in C, one of the index sets I and J may be empty, and $k \bullet n$ is defined as $k + n - 1$ if both k and n are odd, and $k + n$ otherwise. The rest of the proof proceeds in the same way as before.

12   Experiments

We implemented the HSP calculus in the theorem prover *Beagle*.[15] *Beagle* is a testbed for rapidly trying out theoretical ideas but it is not a high-performance prover (in particular it lacks indexing of any form). The perhaps most significant calculus feature not yet implemented is the improvement for linear integer and rational arithmetic of Sect. 11.

Beagle's proof procedure and background reasoning, in particular for linear integer arithmetic, and experimental results have been described in [7]. Here we only provide an update on the experiments and report on complementary aspects not discussed in [7]. More specifically, our new experiments are based on a more recent version of the TPTP problem library [27] (by four years), and we discuss in more detail the impact of the various calculus variants introduced in this paper. We also compare *Beagle*'s performance to that of other provers.

We tested *Beagle* on the first-order problems from the TPTP library, version 7.2.0,[16] that involve some form of arithmetic, including non-linear, rational and real arithmetics. The problems in the TPTP are organized in categories, and the results for some of them are quickly dealt with: none of the HWV-problems in the TPTP library was solvable within the time limit and we ignore these below. We ignore also the SYN category as its sole problem is merely a syntax test, and the GEG category as all problems are zero-rated and easily solved by *Beagle*.

The experiments were run on a MacBook Pro with a 2.3 GHz Intel i7 processor and 16 GB main memory. The CPU time limit was 120 s (a higher time limit does not help much solving more problems). Tables 1 and 2 summarize the results for the problems with a known "theorem" or "unsatisfiable" status with non-zero rating. *Beagle* can also solve some satisfiable problems, but most of them are rather easy and can be solved by the BG solver alone. Unfortunately, the TPTP does not contain reasonably difficult satisfiable problems from the GBT-fragment, which would be interesting for exploiting the completeness result of Sect. 10.

---

[15] *Beagle* is available at https://bitbucket.org/peba123/beagle. The distribution includes the (Scala) source code and a ready-to-run Java jar-file.

[16] http://tptp.org.

Table 1. Number of TPTP version 7.2.0 problems solved, of all non-zero rated "theorem" or "unsatisfiable" problems involving any form of arithmetic. The flag settings giving the best result are in typeset in bold. The CPU time limit was 120 s. The column "Any" is the number of problems solved in the union of the four setting to its left. For the "Auto" column see the description of auto-mode in the main text further below. For auto-mode only, the CPU time limit was increased to 300 s.

Category	#Problems	Ordinary variables		Abstraction variables		Any	Auto
		BG simp cautious	BG simp aggressive	BG simp cautious	BG simp aggressive		
ARI	444	356	**357**	353	355	362	355
DAT	23	9	**12**	6	7	13	12
MSC	3	3	3	3	3	3	3
NUM	36	30	29	**34**	**34**	34	34
PUZ	1	1	1	1	1	1	1
SEV	2	0	0	0	0	0	0
SWV	1	1	1	1	1	1	1
SWW	244	91	88	**92**	89	97	95
SYO	1	0	0	0	0	0	0
Total	755	419	471	**490**	**490**	511	501

Table 1 is a breakdown of *Beagle*'s performance by TPTP problem categories and four flag settings. *Beagle* features a host of flags for controlling its search, but in Table 1 we varied only the two most influential ones: one that controls whether input arithmetic variables are taken as ordinary variables or as abstraction variables. (Sect. 5 discusses the trade-off between these two kinds of variables.) The other controls whether simplification of BG terms is done cautiously or aggressively.

To explain, the cautious simplification rules comprise evaluation of arithmetic terms, e.g. $3 \cdot 5$, $3 < 5$, $\alpha + 1 < \alpha + 1$ (equal lhs and rhs terms in inequations), and rules for TPTP-operators, e.g., $\mathsf{to\_rat}(5)$, $\mathsf{is\_int}(3.5)$. For aggressive simplification, integer sorted subterms are brought into a polynomial-like form and are evaluated as much as possible. For example, the term $5 \cdot \alpha + \mathsf{f}(3 + 6, \alpha \cdot 4) - \alpha \cdot 3$ becomes $2 \cdot \alpha + \mathsf{f}(9, 4 \cdot \alpha)$. These conversions exploit the associativity and commutativity laws for $+$ and $\cdot$. We refer the reader to [7] for additional aggressive simplification rules, but we note here that aggressive simplification does not always preserve sufficient completeness. For example, in the clause set $N = \{\mathsf{P}(1 + (2 + \mathsf{f}(X))), \neg\mathsf{P}(1 + (X + \mathsf{f}(X)))\}$ the first clause is aggressively simplified, giving $N' = \{\mathsf{P}(3 + \mathsf{f}(X)), \neg\mathsf{P}(1 + (X + \mathsf{f}(X)))\}$. Both N and N' are LIA-unsatisfiable, $\mathrm{sgi}(N) \cup \mathrm{GndTh}(\mathrm{LIA})$ is unsatisfiable, but $\mathrm{sgi}(N') \cup \mathrm{GndTh}(\mathrm{LIA})$ is satisfiable. Thus, N is (trivially) sufficiently complete while N' is not.

These two flag settings, in four combinations in total, span a range from "most complete but larger search space" by using ordinary variables and cautious simplification, to "most incomplete but smaller search space" by using abstraction variables and aggressive simplification. As the results in Table 1 show, the flag setting "abstraction variables" solves more problems than "ordinary variables", but not uniformly so. Indeed, as indicated by the "Any" column in Table 1, there are problems that are solved only with either ordinary or abstraction variables.

Some more specific comments, by problem categories:

ARI. Of the 362 solved problems, 14 are not solved in every setting. Of these, four problems require cautious simplification, and five problems require aggressive simplification. This is independent from whether abstraction or ordinary variables are used.

DAT. The DAT category benefits significantly from using ordinary variables. There is only one problem, DAT075=1.p, that is not solved with ordinary variables. Two problems, DAT072=1.p and DAT086=1.p are solvable only with ordinary variables and aggressive simplification.

Many problems in the DAT category, including DAT086=1.p, state *existentially quantified* theorems about data structures such as arrays and lists. If they are of an arithmetic sort, these existentially quantified variables must be taken as ordinary variables. This way, they can be unified with BG-sorted FG terms such as head(cons(x, y)) (which appear in the list axioms) which might be necessary for getting a refutation at all.

A trivial example for this phenomenon is the entailment $\{P(f(1))\} \models \exists x\ P(x)$, where f is BG-sorted, which is provable only with ordinary variables.

NUM. This category requires abstraction variables. With it, four of the problems can be solved in the NUM category (NUM859=1.p, NUM860=1.p, NUM861=1.p, NUM862=1.p), as the search space with ordinary variables is too big.

SWW. By and large, cautious BG simplification fares slightly better on the SWW problems. Of the 97 problems solved, 16 are not solved in every setting, and the settings that do solve it do not follow an obvious pattern.

We were also interested in *Beagle*'s performance, on the same problems, broken down by the calculus features introduced in this paper. Table 2 summarizes our findings for five configurations ①–⑤ obtained by progressively enabling these features. In order to assess the usefulness of the features we filtered the results by problem rating. The column "≥0.75", for instance, lists the number of solved problems, of all 80 known "theorem" or "unsatisfiable" problems with a rating 0.75 or higher and that involve some form of arithmetic.

The predecessor calculus of [5] uses an exhaustive abstraction mechanism that turns every side of an equation into either a pure BG or pure FG term. All BG variables are always abstraction variables. Configuration ① implements this calculus, with the only deviation of an added splitting rule. The splitting rule [29] breaks apart a clause into variable-disjoint parts and leads to a branching search space for finding corresponding sub-proofs. See again [7] for more details.

Table 2. Number of "theorem" or "unsatisfiable" problems solved, by calculus features and problem rating, excluding the HWV-problems.

	Abstraction	Feature	Rating, # Problems			
			≥ 0.1	≥ 0.5	≥ 0.75	≥ 0.88
			756	187	80	55
①	Standard	N/A	355	30	5	1
②		+Define	493	38	5	1
③	Weak	+Define	490	40	5	1
④		+Define +Ordinary vars	500	44	5	1
⑤		+Define +Ordinary vars +BG simp aggressive	511	45	5	1

In our experiments splitting is always enabled, in particular also for configuration ① for better comparability of result. Cautious BG simplification is enabled for configuration ① and the subsequent configurations ②–④.

Configuration ② differs from configuration ① only by an additional Define rule. (As said earlier, the Define rule can be added without problems to the previous calculus.) By comparing the results for ① and ② it becomes obvious that adding Define improves performance dramatically. This applies to the new calculus as well. The Define rule stands out and should always be enabled.

Configuration ③ replaces the standard abstraction mechanism of [5] by the new weak abstraction mechanism of Sect. 5. Weak abstraction seems more effective than standard abstraction for problems with a higher rating, but the data set supporting this conclusion is very small.

There are five problems, all from the SWW category[17] that re solved *only* with configuration ②, and there is one problem, SWW607=2.p, that is solved only by configurations ① and ②.

There are four solvable problems with rating 0.75. These are ARI595=1.p – ARI598=1.p, which are "simple" problems involving a free predicate symbols over the integer background theory. The problem ARI595=1.p, for instance, is to prove the validity of the formula $(\forall z : \mathbb{Z}\ a \leq z \wedge z \leq a + 2 \rightarrow p(z)) \rightarrow \exists x : \mathbb{Z}\ p(3 \cdot x)$.[18] The calculus and implementation techniques needed for solving such problems are rather different to those needed for solving combinatory problems involving trivial arithmetics only, like, e.g., the HWV-problems.

---

[17] SWW583=2.p, SWW594=2.p, SWW607=2.p, SWW626=2.p, SWW653=2.p and SWW657=2.p.

[18] At the time of this writing, there are only four provers (including *Beagle*) registered with the TPTP web infrastructure that can solve these problems. Hence the rating 0.75.

Configuration ④ is the same as ③ except that it includes the results for general variables instead of abstraction variables. Similarly, configuration ⑤ is the same as ④ except that it includes the results for aggressive BG simplification. It is the union of all results in Table 1.

For comparison with other implemented theorem provers for first-order logic with arithmetics, we ran *Beagle* on the problem set used in the 2018 edition of the CADE ATP system competition (CASC-J9).[19]. The competing systems were CVC4 [6], Princess [26], and two versions of Vampire [20].

In the competition, the systems were given 200 problems from the TPTP problem library, 125 problems over the integers as the background theory (TFI category), and 75 over the reals (TFE category). The system that solves the most problems in the union of the TFI and TFE categories within a CPU time limit of 300 s wins. We applied *Beagle* in an "auto" mode, which time-slices (at most) three parameter settings. These differ mainly in their use of abstraction variables or ordinary variables, and the addition of certain arithmetic lemmas.

Table 3. CADE ATP system competition results 2018 and *Beagle*'s performance on the same problem sets.

	Vampire 4.3	Vampire 4.1	CVC4 1.6pre	Princess 170717	Beagle 0.9.51
#Solved TFI (of 125)	93	98	85	62	36
#Solved TFE (of 75)	70	64	72	43	44
#Solved TFA (of 200)	163	162	157	105	70

The results are summarized in Table 3. We note that *Beagle* was run on different hardware but the same timeout of 300 s. The results are thus only indicative of *Beagle*'s performance, but we do not expect significantly different result had it participated. In the TFI category, of the 36 problems solved, 5 require the use of ordinary variables. In the TFE category, 16 problems involve the ceiling or floor function, which is currently not implemented, and hence cannot be attempted.

In general, many problems used in the competition are rather large in size or search space and would require a more sophisticated implementation of *Beagle*.

13   Conclusions

The main theoretical contribution of this paper is an improved variant of the hierarchic superposition calculus. One improvement over its predecessor [5] is a different form of "abstracted" clauses, the clauses the calculus works with internally. Because of that, a modified completeness proof is required. We have

---

[19] http://tptp.cs.miami.edu/~tptp/CASC/J9/.

argued informally for the benefits over the old calculus in [5]. They concern making the calculus "more complete" in practice. It is hard to quantify that exactly in a general way, as completeness is impossible to achieve in presence of background-sorted foreground function symbols (e. g., "head" of integer-sorted lists). To compensate for that to some degree, we have reported on initial experiments with a prototypical implementation on the TPTP problem library. These experiments clearly indicate the benefits of our concepts, in particular the definition rule and the use of ordinary variables. There is no problem that is solved only by the old calculus only. Certainly more experimentation and an improved implementation is needed to also solve bigger-sized problems with a larger combinatorial search space.

We have also obtained two new completeness results for certain clause logic fragments that do not require compactness of the background specification, cf. Sects. 10 and 11. The former is loosely related to the decidability results in [22], as discussed in Sect. 9. It is also loosely related to results in SMT-based theorem proving. For instance, the method in [18] deals with the case that variables appear only as arguments of, in our words, foreground operators. It works by ground-instantiating all variables in order to being able to use an SMT-solver for the quantifier-free fragment. Under certain conditions, finite ground instantiation is possible and the method is complete, otherwise it is complete only modulo compactness of the background theory (as expected). Treating different fragments, the theoretical results are mutually non-subsuming with ours. Yet, on the fragment they consider we could adopt their technique of finite ground instantiation before applying Theorem 10.7 (when it applies). However, according to Theorem 10.7 our calculus needs instantiation of *background-sorted variables only*, this way keeping reasoning with foreground-sorted terms on the first-order level, as usual with superposition.

References

1. Althaus, E., Kruglov, E., Weidenbach, C.: Superposition modulo linear arithmetic SUP(LA). In: Ghilardi, S., Sebastiani, R. (eds.) FroCoS 2009. LNCS (LNAI), vol. 5749, pp. 84–99. Springer, Heidelberg (2009). https://doi.org/10.1007/978-3-642-04222-5_5

2. Armando, A., Bonacina, M.P., Ranise, S., Schulz, S.: New results on rewrite-based satisfiability procedures. ACM Trans. Comput. Log. **10**(1), 4 (2009)

3. Bachmair, L., Ganzinger, H.: Rewrite-based equational theorem proving with selection and simplification. J. Logic Comput. **4**(3), 217–247 (1994)

4. Bachmair, L., Ganzinger, H.: Resolution theorem proving. In: Handbook of Automated Reasoning. North Holland (2001)

5. Bachmair, L., Ganzinger, H., Waldmann, U.: Refutational theorem proving for hierarchic first-order theories. Appl. Algebra Eng. Commun. Comput **5**, 193–212 (1994)

6. Barrett, C., et al.: CVC4. In: Gopalakrishnan, G., Qadeer, S. (eds.) CAV 2011. LNCS, vol. 6806, pp. 171–177. Springer, Heidelberg (2011). https://doi.org/10.1007/978-3-642-22110-1_14

7. Baumgartner, P., Bax, J., Waldmann, U.: Beagle – a hierarchic superposition theorem prover. In: Felty, A.P., Middeldorp, A. (eds.) CADE 2015. LNCS (LNAI), vol. 9195, pp. 367–377. Springer, Cham (2015). https://doi.org/10.1007/978-3-319-21401-6_25

8. Baumgartner, P., Fuchs, A., Tinelli, C.: $\mathcal{ME}$(LIA) - model evolution with linear integer arithmetic constraints. In: Cervesato, I., Veith, H., Voronkov, A. (eds.) LPAR 2008. LNCS (LNAI), vol. 5330, pp. 258–273. Springer, Heidelberg (2008). https://doi.org/10.1007/978-3-540-89439-1_19

9. Baumgartner, P., Tinelli, C.: Model evolution with equality modulo built-in theories. In: Bjørner, N., Sofronie-Stokkermans, V. (eds.) CADE 2011. LNCS (LNAI), vol. 6803, pp. 85–100. Springer, Heidelberg (2011). https://doi.org/10.1007/978-3-642-22438-6_9

10. Baumgartner, P., Waldmann, U.: Hierarchic superposition: completeness without compactness. In: Košta, M., Sturm, T. (eds.), Fifth International Conference on Mathematical Aspects of Computer and Information Sciences, MACIS 2013, pp. 8–12, Nanning, China (2013)

11. Baumgartner, P., Waldmann, U.: Hierarchic superposition with weak abstraction. In: Bonacina, M.P. (ed.) CADE 2013. LNCS (LNAI), vol. 7898, pp. 39–57. Springer, Heidelberg (2013). https://doi.org/10.1007/978-3-642-38574-2_3

12. Baumgartner, P., Waldmann, U.: Hierarchic superposition revisited (2019). http://arxiv.org/abs/1904.03776

13. Bonacina, M.P., Lynch, C., de Moura, L.M.: On deciding satisfiability by theorem proving with speculative inferences. J. Autom. Reason. **47**(2), 161–189 (2011)

14. de Moura, L., Bjørner, N.: Engineering DPLL(T) + saturation. In: Armando, A., Baumgartner, P., Dowek, G. (eds.) IJCAR 2008. LNCS (LNAI), vol. 5195, pp. 475–490. Springer, Heidelberg (2008). https://doi.org/10.1007/978-3-540-71070-7_40

15. Dickson, L.E.: Finiteness of the odd perfect and primitive abundant numbers with n distinct prime factors. Am. J. Math. **35**(4), 413–422 (1913)

16. Ganzinger, H., Korovin, K.: Theory instantiation. In: Hermann, M., Voronkov, A. (eds.) LPAR 2006. LNCS (LNAI), vol. 4246, pp. 497–511. Springer, Heidelberg (2006). https://doi.org/10.1007/11916277_34

17. Ge, Y., Barrett, C., Tinelli, C.: Solving quantified verification conditions using satisfiability modulo theories. In: Pfenning, F. (ed.) CADE 2007. LNCS (LNAI), vol. 4603, pp. 167–182. Springer, Heidelberg (2007). https://doi.org/10.1007/978-3-540-73595-3_12

18. Ge, Y., de Moura, L.: Complete instantiation for quantified formulas in satisfiabiliby modulo theories. In: Bouajjani, A., Maler, O. (eds.) CAV 2009. LNCS, vol. 5643, pp. 306–320. Springer, Heidelberg (2009). https://doi.org/10.1007/978-3-642-02658-4_25

19. Korovin, K., Voronkov, A.: Integrating linear arithmetic into superposition calculus. In: Duparc, J., Henzinger, T.A. (eds.) CSL 2007. LNCS, vol. 4646, pp. 223–237. Springer, Heidelberg (2007). https://doi.org/10.1007/978-3-540-74915-8_19

20. Kovács, L., Voronkov, A.: First-order theorem proving and VAMPIRE. In: Sharygina, N., Veith, H. (eds.) CAV 2013. LNCS, vol. 8044, pp. 1–35. Springer, Heidelberg (2013). https://doi.org/10.1007/978-3-642-39799-8_1

21. Kruglov, E.: Superposition modulo theory. Doctoral dissertation, Universität des Saarlandes, Saarbrücken, October 2013

22. Kruglov, E., Weidenbach, C.: Superposition decides the first-order logic fragment over ground theories. Math. Comput. Sci. **6**, 427–456 (2012)

23. Nieuwenhuis, R.: First-order completion techniques. Technical report, Universidad Politécnica de Cataluña, Dept. Lenguajes y Sistemas Informáticos (1991)
24. Nieuwenhuis, R., Oliveras, A., Tinelli, C.: Solving SAT and SAT modulo theories: from an abstract Davis-Putnam-Logemann-Loveland procedure to DPLL(T). J. ACM 53(6), 937–977 (2006)
25. Nieuwenhuis, R., Rubio, A.: Paramodulation-based theorem proving. In: Handbook of Automated Reasoning, pp. 371–443. Elsevier and MIT Press (2001)
26. Rümmer, P.: A constraint sequent calculus for first-order logic with linear integer arithmetic. In: Cervesato, I., Veith, H., Voronkov, A. (eds.) LPAR 2008. LNCS (LNAI), vol. 5330, pp. 274–289. Springer, Heidelberg (2008). https://doi.org/10.1007/978-3-540-89439-1_20
27. Sutcliffe, G.: The TPTP problem library and associated infrastructure. From CNF to TH0, TPTP v6.4.0. J. Autom. Reason. 59(4), 483–502 (2017)
28. Walther, C.: Many-sorted unification. J. ACM 35(1), 1–17 (1988)
29. Weidenbach, C., Dimova, D., Fietzke, A., Kumar, R., Suda, M., Wischnewski, P.: SPASS version 3.5. In: Schmidt, R.A. (ed.) CADE 2009. LNCS (LNAI), vol. 5663, pp. 140–145. Springer, Heidelberg (2009). https://doi.org/10.1007/978-3-642-02959-2_10

Theory Combination: Beyond Equality Sharing

Maria Paola Bonacina[1]([✉]), Pascal Fontaine[2], Christophe Ringeissen[2], and Cesare Tinelli[3]

[1] Università degli Studi di Verona, Verona, Italy
mariapaola.bonacina@univr.it
[2] Université de Lorraine and Inria & LORIA, Nancy, France
[3] The University of Iowa, Iowa City, USA

Dedicated to
Franz Baader
friend and colleague.

Abstract. Satisfiability is the problem of deciding whether a formula has a model. Although it is not even semidecidable in first-order logic, it is decidable in some first-order theories or fragments thereof (e.g., the quantifier-free fragment). *Satisfiability modulo a theory* is the problem of determining whether a quantifier-free formula admits a model that is a model of a given theory. If the formula mixes theories, the considered theory is their union, and *combination of theories* is the problem of combining decision procedures for the individual theories to get one for their union. A standard solution is the *equality-sharing method* by Nelson and Oppen, which requires the theories to be *disjoint* and *stably infinite*. This paper surveys selected approaches to the problem of reasoning in the union of disjoint theories, that aim at going beyond equality sharing, including: *asymmetric* extensions of equality sharing, where some theories are unrestricted, while others must satisfy stronger requirements than stable infiniteness; *superposition-based* decision procedures; and current work on *conflict-driven satisfiability* (CDSAT).

1 Introduction

Since the early 1980s, it was understood that combination of theories is of paramount importance for software verification [97–99,105], because program checking requires inferences about diverse domains such as arithmetic, data structures, and free predicate and function symbols [20,51,116]. Reasoning about disjunction is just as basic, since the different paths that a program execution may take are logically connected by disjunction. The problem known as *satisfiability modulo theories* (SMT) refers to the problem of determining the satisfiability of an arbitrary (usually quantifier-free) formula modulo a union of theories [12,15,115]. Several solvers for SMT have been developed in the last 20 years or so that support a combination of two or more theories. These include

© Springer Nature Switzerland AG 2019
C. Lutz et al. (Eds.): Baader Festschrift, LNCS 11560, pp. 57–89, 2019.
https://doi.org/10.1007/978-3-030-22102-7_3

(in alphabetical order) Alt-Ergo [46], Boolector [41], CVC4 [13], MathSAT [45], OpenSMT [42], Simplify [56], veriT [37], Yices [57], and Z3 [50]. Because of their power and efficiency in practice, SMT solvers have been interfaced with or integrated in a large number of tools including theorem provers [28,36,109], proofs assistants [1,17], and tools for the analysis, verification, and synthesis of software (see [52] for a survey).

Deductive problems in combination of theories can be formulated as *modularity problems*, or how to get a decision procedure for a problem in a union of theories, given decision procedures for that problem in the component theories. Franz Baader was a pioneer in the study of modularity problems [5–9], propounding the importance of theory combination in automated reasoning.

For the problem of determining the *satisfiability of sets of literals* in a combined theory, an answer to the quest for modularity is offered by the popular *equality sharing method*, by Nelson and Oppen, also known as the *Nelson-Oppen scheme* [99,110,119]. This method combines decision procedures for theories that are *disjoint* and *stably infinite*. In an unsorted setting, this means that the theories' signatures share only the equality symbol and free constants and every quantifier-free formula satisfiable in one of the theories is satisfiable in a model with a countably infinite domain. The Nelson-Oppen scheme *separates* terms that mix function or predicate symbols from different theories by allowing each theory to view maximal *alien* subterms as free constants. The component decision procedures cooperate by agreeing on an *arrangement* of their shared free constants, that is, a complete and consistent set of equalities and disequalities between those constants. For SMT, an equality-sharing decision procedure for a union $\mathcal{T}$ of theories is integrated with a propositional satisfiability (SAT) solver, based on the DPLL/CDCL[1] procedure [49,93,94], according to the DPLL($\mathcal{T}$) framework and its extensions [14,36,38,85,103].

For the problem of *unification modulo theories*, a milestone on the road towards modularity is the *Baader-Schulz combination method* [7]. Unification modulo theories considers *equational theories*, which are presented by sets of universally quantified equalities. The union of two equational theories is the theory presented by the union of the component theories' presentations. Unification is a satisfiability problem that restricts formulas to conjunctions of equations, and models to *Herbrand interpretations*, that is, interpretations where the domain is the universe of terms, and constant and function symbols are interpreted as themselves. In unification *modulo* a theory $\mathcal{T}$, the universe of terms is partitioned into congruence classes induced by the equalities in the presentation of $\mathcal{T}$.

The Baader-Schulz scheme combines decision procedures that allow the addition of free symbols and cooperate by sharing information including an *identification* of free constants, that is, a set of equalities telling which free constants are equal. Both the Baader-Schulz and the Nelson-Oppen schemes require the theories to be disjoint, and feature a separation phase; also, variable identifica-

---

[1] DPLL stands for Davis-Putnam-Logemann-Loveland and CDCL stands for Conflict-Driven Clause Learning. The CDCL procedure is an extension and improvement of the DPLL procedure.

tion is a form of arrangement. On the other hand, the Baader-Schulz method does not deal with inequalities, and does not need stable infiniteness.

The *word problem* in an equational theory $\mathcal{T}$ asks whether a universally quantified equality is valid in the theory, that is, satisfied in every model of $\mathcal{T}$. The *Baader-Tinelli combination method* establishes a modularity result for the word problem [9].

For theories with a finite presentation, these problems can be treated also as *refutational theorem-proving* problems, applying a *superposition-based strategy* (e.g., [10, 22, 34, 71–74, 82]) to the union of the presentation and the negation of the conjecture. For the word problem, a conjecture $\forall \bar{x}.\ s \simeq t$, where $\bar{x}$ contains all variables occurring in $s \simeq t$, is negated into $\exists \bar{x}.\ s \not\simeq t$, whose Skolemization yields a ground target inequality $\hat{s} \not\simeq \hat{t}$, where the hat means that all variables are replaced by Skolem constants. A refutation is reached if $\hat{s}$ and $\hat{t}$ get rewritten to the same term. For the unification problem, a conjecture $\exists \bar{x}.\ s \simeq t$, is negated into $\forall \bar{x}.\ s \not\simeq t$, yielding a non-ground target inequality $s \not\simeq t$ with all variables implicitly universally quantified. Superposition also applies into the target inequality (an inference also known as *narrowing*), and a refutation is reached if s and t get reduced to syntatically unifiable terms, leading to *completion-based approaches to unification modulo theories* (e.g., [62, 89, 102]). Between the word problem and the general validity problem lies the *clausal validity problem*, which queries whether a clause φ, that is, a universally quantified disjunction of literals $\forall \bar{x}.\ l_1 \vee \dots \vee l_k$, is valid in a theory. The conjecture φ is negated into $\exists \bar{x}.\ \neg l_1 \wedge \dots \wedge \neg l_k$, whose Skolemization yields a set of ground literals $Q = \{\neg \hat{l}_1, \dots, \neg \hat{l}_k\}$: φ is valid in the theory if and only if Q is unsatisfiable in the theory. Superposition-based strategies terminate and therefore are *decision procedures* for the satisfiability of sets of ground literals in several theories [2–4, 26].

In this paper we survey selected approaches to go beyond the standard represented by equality sharing for SMT in unions of theories. We begin with methods that generalize equality sharing to *asymmetric combinations*, where some theories are not stably infinite, provided the others are either *shiny* [122], *gentle* [60], or *polite* [76, 108], which means that they are more flexible than stably-infinite theories cardinality-wise. Then we consider superposition, whose application to unions of theories is also formulated as a *modularity problem*, namely *modularity of termination*: knowing that superposition terminates on satisfiability problems in each component theory, show that it terminates on satisfiability problems in their union [2, 3]. This modularity results also allows one to understand the relation between equality sharing and superposition [3, 29]. We conclude with a brief description of a new paradigm for SMT in unions of theories, named CDSAT, for *Conflict-Driven SATisfiability*, which generalizes equality sharing in several ways, including lifting stable infiniteness [31–33]. The interested reader may find additional material in complementary sources (e.g., [64, 66, 67, 92]).

The paper is organized as follows. After providing basic definitions and notations in Sect. 2, we present the equality-sharing method in Sect. 3, including a result showing that the decidability of unions of disjoint decidable theories depends on cardinality requirements. The next three Sects. (4, 5 and 6) are dedicated to shiny, gentle, and polite theories, respectively. Section 7 surveys the application of superposition-based strategies to SMT. Section 8 gives an overview of CDSAT, and Sect. 9 closes the paper with a discussion.

2   Background Definitions

A *first-order language*, or *signature*, is a tuple $\mathcal{L} = \langle \mathcal{S}, \mathcal{F}, \mathcal{P}, \mathcal{V} \rangle$, where $\mathcal{S}$ is a finite set of disjoint sorts, $\mathcal{F}$ is a finite set of function symbols, $\mathcal{P}$ is a finite set of predicate symbols, including an equality symbol for each sort, and $\mathcal{V}$ is a set that contains a denumerable amount of variables for each sort. Every variable, predicate, and function symbol is assigned a sort in $\mathcal{S}$, $\mathcal{S}^n$, and $\mathcal{S}^{n+1}$ respectively, where n is the arity of the symbol. *Propositions* are nullary predicate symbols, and *constants* are nullary function symbols. $\mathcal{L}$ is *one-sorted* if $\mathcal{S}$ is a singleton, *many-sorted* otherwise. As decision procedures may extend the given language $\mathcal{L}$ with a finite set C of *new* constant symbols, $\mathcal{L}^C$ denotes the extended language $\langle \mathcal{S}, \mathcal{F} \cup C, \mathcal{P}, \mathcal{V} \rangle$. Terms over $\mathcal{L}$, or $\mathcal{L}$-*terms*, are defined as usual. A *compound term* contains at least an occurrence of a function symbol. A *context* is a term with a hole: the notation $t[l]$ represents a term where l appears as subterm in context t.

An atomic formula, or *atom*, is a predicate symbol applied to as many terms as its arity. A *literal* is either an atom or the negation of an atom. *Formulas* are built as usual from atoms, connectives ($\neg$, $\wedge$, $\vee$, $\Rightarrow$, $\equiv$), and quantifiers ($\forall$, $\exists$). A *sentence* is a formula where all variables are quantified. A *clause* is a disjunction of literals, where all variables are implicitly universally quantified. A quantifier-free formula is in *conjunctive normal form* (CNF), if it is a conjunction, or a set, of disjunctions of literals; in *disjunctive normal form* (DNF) if it is a disjunction of conjunctions, or sets, of literals. Through Skolemization, every formula can be reduced to an equisatisfiable conjunction, or set, of clauses (*clausal form*). A term is *ground* if it does not contain variables, and the same applies to literals, clauses, and formulas. The set of variables occurring in a term t is denoted by $Var(t)$. Atoms, literals, sentences, and formulas over $\mathcal{L}$ are called $\mathcal{L}$-*atoms*, $\mathcal{L}$-*literals*, $\mathcal{L}$-*sentences*, and $\mathcal{L}$-*formulas*, respectively.

An *interpretation* $\mathcal{M}$ of $\mathcal{L}$, or $\mathcal{L}$-*interpretation*, defines non-empty pairwise disjoint domains $\mathcal{M}[s]$ for all $s \in \mathcal{S}$, a sort- and arity-matching total function $\mathcal{M}[f]$ for all $f \in \mathcal{F}$, a sort- and arity-matching relation $\mathcal{M}[p]$ for all $p \in \mathcal{P}$, and an element $\mathcal{M}[x] \in \mathcal{M}[s]$ for all $x \in \mathcal{V}$ of sort s. $\mathcal{M}$ designates a value in $\mathcal{M}[s]$ for every term of sort s, and a truth value for every formula, with equality interpreted as identity in each domain. An $\mathcal{L}$-*structure* is an $\mathcal{L}$-interpretation over an empty set of variables. A *model* of an $\mathcal{L}$-formula φ is an $\mathcal{L}$-interpretation where φ is true, written $\mathcal{M} \models \varphi$, and read $\mathcal{M}$ satisfies φ. A formula is *satisfiable* if it has a model, *unsatisfiable* otherwise. A model is finite if for all $s \in \mathcal{S}$ the cardinality $|\mathcal{M}[s]|$ is finite.

In this paper we adopt the syntactic definition of a theory. Given a first-order language $\mathcal{L}$, an $\mathcal{L}$-*theory* $\mathcal{T}$ is a set of $\mathcal{L}$-sentences, called *axioms* or $\mathcal{T}$-*axioms*. One can write $\mathcal{T}$-atom, $\mathcal{T}$-literal, or $\mathcal{T}$-formula in place of $\mathcal{L}$-atom, $\mathcal{L}$-literal, or $\mathcal{L}$-formula. Symbols that do not appear in $\mathcal{T}$-axioms are called *free* or *uninterpreted*.

An $\mathcal{L}$-theory $\mathcal{T}$ determines the set $Mod(\mathcal{T})$ of its models, or $\mathcal{T}$-*models*, that is, those $\mathcal{L}$-structures $\mathcal{M}$ such that $\mathcal{M} \models \mathcal{T}$, which means that $\mathcal{M} \models \varphi$ for all φ in $\mathcal{T}$. In turn, a set $\mathcal{C}$ of $\mathcal{L}$-structures determines the set $Th(\mathcal{C})$ of its theorems, that is, those $\mathcal{L}$-sentences that are true in all structures in $\mathcal{C}$. If $\mathcal{C} = Mod(\mathcal{T})$, one can write $Th(\mathcal{T})$, in place of $Th(\mathcal{C})$, for the set of theorems of $\mathcal{T}$ or $\mathcal{T}$-*theorems*. With respect to $Mod(\mathcal{T})$ or $Th(\mathcal{T})$, $\mathcal{T}$ is called *axiomatization* or *presentation*. If $\mathcal{T}$ is finite, the theory is said to be *finitely axiomatized*. Given an $\mathcal{L}_1$-theory $\mathcal{T}_1$ and an $\mathcal{L}_2$-theory $\mathcal{T}_2$, their *union* is the $\mathcal{L}_1 \cup \mathcal{L}_2$-theory $\mathcal{T}_1 \cup \mathcal{T}_2$. Two languages are *disjoint* if their sets of non-nullary functions and predicates are pairwise disjoint, and two theories are *disjoint* if their languages are.

Whenever equational reasoning is built into an inference system or algorithm, the axioms of equality are omitted from $\mathcal{T}$. If all axioms are built into the inference system or algorithm, a finite axiomatization $\mathcal{T}$ may not be given, and an $\mathcal{L}$-theory, for $\mathcal{L} = \langle \mathcal{S}, \mathcal{F}, \mathcal{P}, \mathcal{V} \rangle$, may be characterized by a set $\mathcal{C}$ of $\mathcal{L}$-structures. The implicit, and usually infinite, axiomatization $\mathcal{T}$ of the theory is given by $\mathcal{T} = Th(\mathcal{C})$, so that the structures in $\mathcal{C}$ are still called $\mathcal{T}$-models and $\mathcal{T}$ is still used as the name of the theory. Given another language $\mathcal{L}' = \langle \mathcal{S}', \mathcal{F}', \mathcal{P}', \mathcal{V}' \rangle$ such that $\mathcal{S} \subseteq \mathcal{S}'$, $\mathcal{F} \subseteq \mathcal{F}'$, and $\mathcal{P} \subseteq \mathcal{P}'$, an $\mathcal{L}'$-structure $\mathcal{M}'$ is a $\mathcal{T}$-*model over* $\mathcal{L}'$, if the $\mathcal{L}$-structure $\mathcal{M}$ defined by the $\mathcal{M}'$-interpretation of $\mathcal{L}$-symbols is a $\mathcal{T}$-model, that is, if $\mathcal{M} \in \mathcal{C}$, or, equivalently, $\mathcal{M} \models \mathcal{T}$. Given an $\mathcal{L}_1$-theory $\mathcal{T}_1$ and an $\mathcal{L}_2$-theory $\mathcal{T}_2$ characterized in this style, their *union* $\mathcal{T}_1 \cup \mathcal{T}_2$ is the $\mathcal{L}_1 \cup \mathcal{L}_2$-theory characterized by the class of $\mathcal{L}_1 \cup \mathcal{L}_2$-structures $\mathcal{M}$ that are simultaneously $\mathcal{T}_1$-models over $\mathcal{L}_1 \cup \mathcal{L}_2$ and $\mathcal{T}_2$-models over $\mathcal{L}_1 \cup \mathcal{L}_2$.

A formula φ is $\mathcal{T}$-*satisfiable* if it has a $\mathcal{T}$-model, $\mathcal{T}$-*unsatisfiable* otherwise. A formula φ is $\mathcal{T}$-*valid* if it is true in all $\mathcal{T}$-models, $\mathcal{T}$-invalid otherwise. Since no interpretation satisfies both φ and $\neg\varphi$, a formula φ is $\mathcal{T}$-valid if and only if $\neg\varphi$ is $\mathcal{T}$-unsatisfiable, and φ is $\mathcal{T}$-satisfiable if and only if $\neg\varphi$ is $\mathcal{T}$-invalid. Thus, $\mathcal{T}$-satisfiability is decidable if and only if $\mathcal{T}$-validity is decidable. The $\mathcal{T}$-validity of φ is approached refutationally by proving that $\neg\varphi$ is $\mathcal{T}$-unsatisfiable: for this purpose, $\neg\varphi$ is typically reduced to clausal form. If φ is a clause (*clausal validity problem*), the clausal form of $\neg\varphi$ is a set of ground unit clauses or, equivalently, ground literals. For many theories of interest only the quantifier-free fragment is decidable (e.g., [39], Chap. 3). Let φ be a quantifier-free formula and $\bar{x}$ the tuple of its free variables: φ is $\mathcal{T}$-satisfiable if and only if its existential closure $\exists \bar{x}. \varphi$ is $\mathcal{T}$-satisfiable; φ is $\mathcal{T}$-valid if and only if its universal closure $\forall \bar{x}. \varphi$ is $\mathcal{T}$-valid if and only if $\exists \bar{x}. \neg\varphi$ is $\mathcal{T}$-unsatisfiable. Through Skolemization, both problems reduce to the $\mathcal{T}$-satisfiability of a ground formula, which can be reduced to either CNF, yielding a set of ground clauses, or DNF. CNF is generally preferred, especially if the original problem is a $\mathcal{T}$-validity problem, as most refutational calculi work with clausal form. If the original problem is a $\mathcal{T}$-satisfiability problem, DNF offers the advantage that a DNF formula is satisfiable if and only if at least one of its sets of literals is. In summary, a theory $\mathcal{T}$ is $\exists$-*decidable*, if the $\mathcal{T}$-satisfiability of finite sets of ground literals is decidable, and $\exists_\infty$-*decidable*, if it is $\exists$-decidable and the satisfiability of finite sets of ground literals in infinite $\mathcal{T}$-models is also decidable.

3   The Equality Sharing Method

SMT solvers are generally built around a decision procedure for quantifier-free formulas, whose Boolean structure is handled by the underlying SAT-solver. As a consequence, theory reasoning is only concerned with conjunctions or (finite) sets of literals. In most SMT solvers, the theory reasoners handle a union of theories. The equality-sharing method by Nelson and Oppen [97–99, 105, 110, 119] is a means to build a decision procedure for satisfiability of sets of literals in a union of disjoint theories from decision procedures for satisfiability of sets of literals in each theory. For example, consider the set of literals

$$Q = \{a \leq b, \ b \leq (a + f(a)), \ P(h(a) - h(b)), \ \neg P(0), \ f(a) \simeq 0\}.$$

The first step is to identify the involved theories. Suppose a specification tells us that $\leq$, $+$, $-$, and 0 are to be interpreted over the integers, while the symbols P, f, h, a, and b are free. Since there is no occurrence of product, for the integers it suffices to consider *linear integer arithmetic* (LIA). For the free symbols, the relevant theory is the *theory of (equality with) uninterpreted function symbols* (abbreviated as EUF or UF). Thus, the problem requires the union of LIA and UF sharing the sort int of the integers. However, UF only has equality as predicate, and therefore the problem gets rewritten in the equisatifiable form

$$Q = \{a \leq b, \ b \leq (a + f(a)), \ f_P(h(a) - h(b)) \simeq \bullet, \ f_P(0) \not\simeq \bullet, \ f(a) \simeq 0\}.$$

Let prop be the sort interpreted by all interpretations as the set $\{\mathsf{true}, \mathsf{false}\}$ of the propositional, or Boolean, values. Then, the language of UF has sorts int and prop, function symbols $f, h\colon \mathsf{int} \to \mathsf{int}$ and $f_P\colon \mathsf{int} \to \mathsf{prop}$, and constant symbols a and b of sort int, and $\bullet$ of sort prop. The language of LIA has sorts int and prop, predicate symbol $\leq$ of sort int $\times$ int for the ordering, function symbols $+, -\colon \mathsf{int} \times \mathsf{int} \to \mathsf{int}$ for addition and subtraction, and the constant 0 of sort int. Let $\mathcal{T}_1$ be LIA and $\mathcal{T}_2$ be UF. For the separation phase of the equality-sharing method, Q is *separated* into a set of $\mathcal{T}_1$-literals Q_1 and a set of $\mathcal{T}_2$-literals Q_2 by introducing fresh free constants[2] to produce the equisatisfiable problem $Q_1 \cup Q_2$:

$$Q_1 = \{a \leq b, \ b \leq (a + v_1), \ v_2 \simeq v_3 - v_4, \ v_5 \simeq 0, \ v_1 \simeq 0\}$$
$$Q_2 = \{v_1 \simeq f(a), \ f_P(v_2) \simeq \bullet, \ v_3 \simeq h(a), \ v_4 \simeq h(b), \ f_P(v_5) \not\simeq \bullet\}.$$

Q_1 and Q_2 only share equality between terms of sort int and the free constants in the set $C = \{a, b, v_1, v_2, v_3, v_4, v_5\}$. It is not difficult to see that Q is $\mathcal{T}_1 \cup \mathcal{T}_2$-unsatisfiable whereas Q_1 is $\mathcal{T}_1$-satisfiable and Q_2 is $\mathcal{T}_2$-satisfiable. This means that, in general, it is not sufficient for $\mathcal{T}_1 \cup \mathcal{T}_2$-satisfiability to let the decision procedures for $\mathcal{T}_1$ and $\mathcal{T}_2$ examine only the satisfiability of their subproblem. The decision procedures need to exchange information about their individual sets of literals. A first key idea in equality sharing is that the decision procedures need to *agree* on an *arrangement* of the shared constants.

---

[2] Traditionally combination schemes use free variables for this role (e.g., [39]). Since quantified formulas appear in this paper, we choose to use free constants.

Definition 1. *An* arrangement α *of a set of constant symbols C is a satisfiable set of sorted equalities and inequalities between elements of C such that $a \simeq b \in \alpha$ or $a \not\simeq b \in \alpha$ for all a, b of the same sort in C.*

A second key ingredient is that the decision procedures need to *agree* on the *cardinalities* of the shared sorts. The following theorem states these two requirements for completeness (cf. [60,61,120–122] for equivalent formulations).

Theorem 1. *Assume T_1 and T_2 are theories over disjoint languages $\mathcal{L}_1$ and $\mathcal{L}_2$, and Q_i ($i = 1, 2$) is a set of $\mathcal{L}_i^C$-literals. Then $Q_1 \cup Q_2$ is $T_1 \cup T_2$-satisfiable if and only if there exist an arrangement α of C and a T_i-model $\mathcal{M}_i$ such that $\mathcal{M}_i \models \alpha \cup Q_i$ for $i = 1, 2$ and $|\mathcal{M}_1[s]| = |\mathcal{M}_2[s]|$ for all sorts s common to both languages $\mathcal{L}_1$ and $\mathcal{L}_2$.*

This theorem can be strengthened by restricting the arrangement α to the constants of C that occur in both Q_1 and Q_2. The ($\Rightarrow$) case of the proof is straightforward, and the ($\Leftarrow$) case is proved by building from the T_1-model and the T_2-model a $T_1 \cup T_2$-model (e.g., Theorem 1, [60]). This is possible thanks to the shared arrangement, and because the T_1-model and the T_2-model have the same cardinality for each common sort. Checking the existence of a model is the task of the decision procedures for the component theories. The issue is how to ensure that there are models that agree on the cardinalities of shared sorts.

A theory T is *stably infinite* if every T-satisfiable quantifier-free T-formula has a T-model such that for all sorts other than prop the domain has cardinality $\aleph_0$, the cardinality of the set $\mathbb{N}$ of the natural numbers. Combining only stably infinite theories is a radical solution to the cardinality requirement of Theorem 1: the cardinality is $\aleph_0$ for all shared theories other than prop. Since both LIA and UF are stably infinite, the set Q of our example is $T_1 \cup T_2$-satisfiable if and only if there exists an arrangement α of the free constants in C such that $\alpha \cup Q_i$ is T_i-satisfiable for $i=1, 2$. As no such arrangement exists, Q is $T_1 \cup T_2$-unsatisfiable. In order to include non-stably-infinite theories, combination schemes can rely on the notion of *spectrum* [60].

Definition 2. *The* spectrum *of a one-sorted theory T is the set of the cardinalities of the T-models.*[3]

This notion can be generalized to the many-sorted case by considering tuples of cardinalities, one cardinality for each sort. Using this definition and Theorem 1, it is possible to state completeness requirements for a combination scheme for disjoint theories that are not necessarily stably infinite (cf. Corollary 1, [60]).

Corollary 1. *Given theories T_1 and T_2 over disjoint languages $\mathcal{L}_1$ and $\mathcal{L}_2$, $T_1 \cup T_2$ is $\exists$-decidable if, for all sets of $\mathcal{L}_1^C$-literals Q_1 and $\mathcal{L}_2^C$-literals Q_2, it is possible to determine whether the intersection of the spectra of $T_1 \cup Q_1$ and $T_2 \cup Q_2$ is non-empty for each sort common to both languages $\mathcal{L}_1$ and $\mathcal{L}_2$.*

---

[3] The spectrum of a theory is usually defined as the set of the *finite* cardinalities of its models. We extend the definition slightly for convenience.

For stably infinite theories, if $T_1 \cup Q_1$ and $T_2 \cup Q_2$ are satisfiable, the intersection of their spectra contains $\aleph_0$ for each shared sort, and therefore the following classic combination lemma does not need to mention cardinalities.

Combination Lemma 1 (Stably Infinite Theories). *Assume two stably infinite disjoint theories T_1 and T_2 over languages $\mathcal{L}_1$ and $\mathcal{L}_2$, and let Q_i be a set of $\mathcal{L}_i^C$-literals ($i = 1, 2$). Then $Q_1 \cup Q_2$ is $T_1 \cup T_2$-satisfiable if and only if there exists an arrangement α of C such that $\alpha \cup Q_i$ is T_i-satisfiable for $i = 1, 2$.*

Since the union of disjoint stably infinite theories is stably infinite, the class of ∃-decidable stably infinite theories is closed under disjoint union.

Theorem 2. *The union of disjoint, stably infinite, ∃-decidable theories is stably infinite and ∃-decidable.*

A key result is that some cardinality requirement is necessary for decidability. This finding is a consequence of the following theorem (cf. Proposition 4.1, [29]).

Theorem 3. *There exist an ∃-decidable theory that is not $\exists_\infty$-decidable.*

The proof exhibits a theory TM_∞ with language $\mathcal{L}_{TM_\infty}$, including infinitely many nullary predicates $P_{(e,n)}$ for all $e \in \mathbb{N}$ and $n \in \mathbb{N}$. The meaning of $P_{(e,n)}$ is that e is the index of a Turing machine, and n is an input for the Turing machine of index e. The axioms of TM_∞ involve a kind of clauses called *at-most cardinality constraints*. An *at-most cardinality constraint* is a clause containing only non-trivial (i.e., other than $x \simeq x$) equalities between variables. For example, $\forall x, y, z.\ y \simeq x \vee y \simeq z$ is the at-most-2 cardinality constraint: a model of this clause can have at most 2 elements since the clause says that out of 3 variables at least 2 must be equal. In general,

$$\forall x_0, \ldots, x_m. \bigvee_{0 \leq i \neq k \leq m} x_i \simeq x_k$$

is the *at-most-m cardinality constraint*: a model of this clause can have at most m elements. The axioms of TM_∞ are all the formulas saying that $P_{(e,n)}$ implies the at-most-m cardinality constraint, if Turing Machine e halts on input n in fewer than m steps. The property of being an axiom of TM_∞ is decidable, because it suffices to run the Turing machine and see whether it halts in fewer than m steps. The TM_∞-satisfiability of a finite set Q of ground $\mathcal{L}_{TM_\infty}^C$-literals is also decidable: intuitively, as Q involves finitely many constants, their arrangement dictates the minimum cardinality m of a candidate model; if Q contains both $P_{(e,n)}$ and $\neg P_{(e,n)}$, it is unsatisfiable; otherwise, for each $P_{(e,n)} \in Q$ it suffices to test that Turing Machine e runs on input n for at least m steps. On the other hand, satisfiability in infinite TM_∞-models is undecidable: $Q = \{P_{(e,n)}\}$ is satisfiable in an infinite TM_∞-model if and only if Turing Machine e does not halt on input n, which is undecidable for being the complement of the Halting Problem. While TM_∞ has an infinite language, it is possible to exhibit a theory with the same decidability properties of TM_∞ and a finite language [30]. It follows as a corollary that the union of this theory with any disjoint ∃-decidable theory with only infinite models is not ∃-decidable (cf. Theorem 4.1, [29]).

Theorem 4. *There exist $\exists$-decidable disjoint theories whose union is not $\exists$-decidable.*

This theorem implies that if we want to lift the stable infiniteness requirement on one or more component theories, while maintaining decidability of theory unions, we still need to impose some restrictions on the cardinality of these theories' models. Such restrictions must allow the theories to agree on the cardinality of a joint model, so that Corollary 1 can apply. In the next three sections, we survey combination schemes that achieve precisely such a synchronization of the theories on models' cardinalities without imposing stable infiniteness.

4 Shiny Theories

The theory of uninterpreted symbols (UF) is one of the most useful theories. It is convenient to model arrays, generic functions, or other data structures described by a custom set of axiom handled separately by the reasoner through instantiation or other inferences (e.g., see Sect. 7). This theory is *shiny* [122], a much stronger property than being stably infinite.

Definition 3 (Shiny Theory). *A theory T over a one-sorted language $\mathcal{L}$ is shiny if, for all sets Q of $\mathcal{L}^C$-literals, either the spectrum of $T \cup Q$ is empty or it is the set of all cardinalities greater than or equal to a finite cardinality $mincard_T(Q)$ computable from Q.*

For UF, if Q is unsatisfiable, the spectrum is empty. If Q is satisfiable, let s be its sort. An arrangement of the finitely many constant symbols that appear in Q determines a finite cardinality for the domain $\mathcal{M}[s]$ of a model $\mathcal{M}$. A model $\mathcal{M}'$ such that $|\mathcal{M}'[s]| > |\mathcal{M}[s]|$ is obtained by taking a non-empty set A disjoint from $\mathcal{M}[s]$, and letting $\mathcal{M}'[s]$ be $\mathcal{M}[s] \cup A$. $\mathcal{M}'$ interprets equality as identity on every pair of elements of A, and is otherwise identical to $\mathcal{M}$. This is the argument showing that UF is stably infinite (e.g., see Example 10.3, [39]) plus the observation that an arrangement yields a finite cardinality.

The spectrum of a shiny theory T is *upward closed from $mincard_T(Q)$* or *upward closed* for short: if Q has a T-model, it has a T-model for every larger cardinality. A theory with such a spectrum is called *smooth*. A shiny theory also has the *finite-model property*: if a set Q of $\mathcal{L}^C$-literals is T-satisfiable, it is satisfiable in a finite model of T. Consider the union of a shiny theory T_1 and an arbitrary theory T_2 such that T_1 and T_2 are disjoint and share the one sort s of T_1. For T_1 and T_2 to agree on the cardinality of s it suffices that there is a T_2-model that interprets s with a domain of sufficiently large cardinality. An *at-least cardinality constraint* expresses this requirement. An at-least cardinality constraint is the negation of an at-most cardinality constraint (see Sect. 3), hence it is a conjunction of non-trivial inequalities between variables, all existentially quantified. Through Skolemization, the clausal form of an at-least cardinality constraint is a set of inequalities between constants, known as an *all-different constraint*.

Definition 4. *Given a positive integer m and a sort s, an all-different constraint $\delta_s(m)$ for sort s is a set of literals $\{c_i \not\approx c_j \mid 1 \leq i \neq j \leq n\}$, where $c_1, \ldots, c_m$ are distinct fresh free constants of sort s.*

For a theory $\mathcal{T}$, whose language has a sort s, and a set Q of $\mathcal{T}$-literals, Q has a $\mathcal{T}$-model that interprets s with a domain of cardinality at least m if and only if the set of literals $Q \cup \delta_s(m)$ is $\mathcal{T}$-satisfiable. Thus, the union of a shiny theory $\mathcal{T}_1$ and an arbitrary theory $\mathcal{T}_2$, sharing $\mathcal{T}_1$'s only sort s, is handled by testing in this manner that $\mathcal{T}_2$ has a model $\mathcal{M}$ such that $|\mathcal{M}[s]| \geq m$, with m determined by applying $mincard_{\mathcal{T}_1}$ to the set of $\mathcal{T}_1$-literals and the arrangement.

Combination Lemma 2 (Shiny Theory). *Let $\mathcal{L}_1$ and $\mathcal{L}_2$ be disjoint languages such that $\mathcal{L}_1$ has only one sort s shared with $\mathcal{L}_2$. Assume a shiny $\mathcal{L}_1$-theory $\mathcal{T}_1$ and an arbitrary $\mathcal{L}_2$-theory $\mathcal{T}_2$, and let Q_i be a set of $\mathcal{L}_i^C$-literals ($i = 1, 2$). Then $Q_1 \cup Q_2$ is $\mathcal{T}_1 \cup \mathcal{T}_2$-satisfiable if and only if there exists an arrangement α of C such that $\alpha \cup Q_1$ is $\mathcal{T}_1$-satisfiable and $\alpha \cup Q_2 \cup \delta_s(mincard_{\mathcal{T}_1}(\alpha \cup Q_1))$ is $\mathcal{T}_2$-satisfiable.*

Since one theory is shiny and the other is arbitrary, the combination scheme is *asymmetric*.

Theorem 5. *The union of a shiny $\exists$-decidable theory with an arbitrary $\exists$-decidable theory that shares the single sort of the shiny theory is $\exists$-decidable.*

The generalization of shininess to many-sorted theories leads to *politeness* (see Sect. 6). Several other theories have a spectrum that is not upward closed, but satisfies other useful properties for asymmetric combination schemes, as captured by the concept of *gentleness* in the next section.

5   Gentle Theories

By weakening the requirements on the spectrum of theories, more theories can take part in asymmetric combinations. *Gentleness* is weaker than shininess and captures several other interesting $\exists$-decidable theories [60].

Definition 5. *A theory $\mathcal{T}$ over a one-sorted language $\mathcal{L}$ is gentle if, for all sets Q of $\mathcal{L}^C$-literals, the spectrum of $\mathcal{T} \cup Q$ is computable and is equal to*

1. *Either a finite set of finite cardinalities, or*
2. *A co-finite set of cardinalities given by the union of a finite set of finite cardinalities and the set of all (finite and infinite) cardinalities greater than a computable finite cardinality.*

A gentle theory $\mathcal{T}$ is not necessarily stably infinite, since a $\mathcal{T}$-satisfiable set of literals may have only finite models by Case (1) of Definition 5. A shiny theory is gentle, since it satisfies Case (2) of Definition 5 with an empty set of finite cardinalities. Thus, UF is gentle. Conversely, the spectrum of a gentle theory is like the spectrum of a shiny theory with the addition of finitely many finite cardinalities. Gentle theories are $\exists$-decidable, and theories in the *Bernays-Schönfinkel-Ramsey class* (axioms of the form $\exists^*\forall^*\varphi$, where φ is quantifier-free and without occurrences of non-nullary function symbols), *Löwenheim class* (axioms in first-order relational monadic logic, i.e., no non-nullary functions and only unary predicates), and FO^2 *class* (axioms with only two variables and no non-nullary functions) are gentle [60].

Theorem 6. *The union of disjoint gentle theories is gentle.*

The proof (see [60]) rests on the observation that the intersection of the spectra of gentle theories is a spectrum that satisfies Definition 5.

Let $mincard_\mathcal{T}$ be the partial function from sets of $\mathcal{T}$-literals to cardinal numbers defined by $mincard_\mathcal{T}(Q)=k$, if k is the smallest non-zero finite cardinality such that the spectrum of $\mathcal{T} \cup Q$ is upward closed from k. If Q is $\mathcal{T}$-unsatisfiable, or the spectrum of $\mathcal{T} \cup Q$ is bounded and only contains a finite number of finite cardinalities, then $mincard_\mathcal{T}(Q)$ is undefined. Let $fincard_\mathcal{T}$ be the function that maps a set Q of $\mathcal{T}$-literals to a finite, possibly empty, set of finite cardinalities of $\mathcal{T}$-models of Q as follows: (i) if the spectrum of $\mathcal{T} \cup Q$ is empty, $fincard_\mathcal{T}(Q)$ is empty; (ii) if the spectrum of $\mathcal{T} \cup Q$ is finite, $fincard_\mathcal{T}(Q)$ is the spectrum of $\mathcal{T} \cup Q$; (iii) otherwise, $fincard_\mathcal{T}(Q)$ is the set of the cardinalities in the spectrum of $\mathcal{T} \cup Q$ that are strictly smaller than $mincard_\mathcal{T}(Q)$.

Combination Lemma 3 (Gentle Theory). *Let $\mathcal{L}_1$ and $\mathcal{L}_2$ be disjoint languages such that $\mathcal{L}_1$ has only one sort s shared with $\mathcal{L}_2$. Assume a gentle $\mathcal{L}_1$-theory $\mathcal{T}_1$ and an arbitrary $\mathcal{L}_2$-theory $\mathcal{T}_2$, and let Q_i be a set of $\mathcal{L}_i^C$-literals $(i = 1, 2)$. Then $Q_1 \cup Q_2$ is $\mathcal{T}_1 \cup \mathcal{T}_2$-satisfiable if and only if there exists an arrangement α of C such that $\alpha \cup Q_1$ is $\mathcal{T}_1$-satisfiable and*

1. *Either $mincard_{\mathcal{T}_1}(\alpha \cup Q_1)$ is defined and $\alpha \cup Q_2 \cup \delta_s(mincard_{\mathcal{T}_1}(\alpha \cup Q_1))$ is $\mathcal{T}_2$-satisfiable,*
2. *Or there exists a cardinality $k \in fincard_{\mathcal{T}_1}(Q_1 \cup \alpha)$ such that $\alpha \cup Q_2$ is $\mathcal{T}_2$-satisfiable in a $\mathcal{T}_2$-model $\mathcal{M}$ such that $|\mathcal{M}[s]| = k$.*

Condition (1) follows the pattern of Combination Lemma 2. For Condition (2), note that for finitely axiomatized one-sorted theories, it is decidable whether there is a model of a given finite cardinality k. Indeed, there are finitely many (up to isomorphism) interpretations of cardinality k for a finite language, and it takes a finite amount of time to check whether such an interpretation satisfies the axioms. Several widely used theories are not gentle, but they can be combined with gentle theories [60].

Theorem 7. *Given disjoint one-sorted theories $\mathcal{T}_1$ and $\mathcal{T}_2$, where $\mathcal{T}_1$ is gentle, their union $\mathcal{T}_1 \cup \mathcal{T}_2$ is $\exists$-decidable, provided that:*

1. *$\mathcal{T}_2$ also is gentle, or*
2. *$\mathcal{T}_2$ is an $\exists$-decidable finitely axiomatized theory, or*

3. T_2 is an ∃-decidable theory that only admits a fixed finite (possibly empty) known set of finite cardinalities for its models, and possibly infinite models.

The proof shows that in each case of Theorem 7, either Condition (1) or Condition (2) of Combination Lemma 3 applies, possibly after some preliminary work. For example, if both T_1 and T_2 are gentle, the intersection of their spectra is computed first. In Case (3), either the intersection of the finite sets of finite cardinalities admitted by the two theories is non-empty, or else the Löwenheim-Skolem theorem for first-order logic (if a theory has an infinite model, it has models for every infinite cardinality) is invoked to imply that the spectrum of T_2 is upward closed from some infinite cardinality. For example, the theory of arrays (e.g., [95] and Chap. 3 [39]) belongs to Case (2) of Theorem 7, while LIA and the theory of real closed fields (RCF) fit in Case (3).

    Gentleness can be extended to many-sorted theories as done for $\mathcal{P}$-*gentleness*, a generalization of gentleness introduced to handle unions of non-disjoint theories sharing only unary predicates [43].

6   Polite Theories

Politeness can be considered as a many-sorted extension of shininess [76,108]. The concept of politeness is instrumental to combine data-structure theories with an arbitrary theory of elements, as illustrated below for *arrays*, *records*, *sets*, and *multisets*. In this section, we work directly with quantifier-free formulas, instead of sets of literals. A theory is *polite* with respect to a given set $\mathcal{S}$ of sorts, if it is *smooth* and *finitely witnessable* with respect to $\mathcal{S}$. As seen in Sect. 4 for the one-sorted case, *smooth* means that it is possible to enlarge arbitrarily the $\mathcal{S}$-sorted domains of a model to get another model with the desired cardinalities.

Definition 6 (Smooth Theory). *An $\mathcal{L}$-theory T is smooth with respect to a set $\mathcal{S} = \{s_1, \ldots, s_n\}$ of sorts of its many-sorted language $\mathcal{L}$, if:*

- *For all ground formulas φ in $\mathcal{L}^C$,*
- *For all T-models $\mathcal{M}$ of φ,*
- *For all cardinal numbers $k_1, \ldots, k_n$ such that $k_i \geq |\mathcal{M}[s_i]|$, for $i = 1, \ldots, n$,*

there exists a T-model $\mathcal{N}$ of φ such that $|\mathcal{N}[s_i]| = k_i$ for $i = 1, \ldots, n$.

    Finite witnessability complements smoothness by establishing a starting point for the upward movement. The starting point is given by a finite T-model obtained from formulas called *finite witnesses* (see Sects. 6.1 and 6.2 for examples).

Definition 7 (Finite Witness). *Let $\mathcal{S}$ be a set of sorts of a many-sorted language $\mathcal{L}$ and T an $\mathcal{L}$-theory. Given a ground $\mathcal{L}^C$-formula φ, a ground $\mathcal{L}^D$-formula ψ, where $D \supseteq C$, is a finite witness of φ in T with respect to $\mathcal{S}$ if:*

1. For all T-models $\mathcal{M}$ over $\mathcal{L}^D$, $\mathcal{M} \models (\varphi \equiv \psi)$, and

2. *For all arrangements α of all the S-sorted constants in D, if $\{\psi\} \cup \alpha$ is T-satisfiable then there exists a T-model $\mathcal{M}^*$ of $\{\psi\} \cup \alpha$ such that $\mathcal{M}^*[s] = \{\mathcal{M}^*[d] \mid d \in D,\ d$ of sort $s\}$, for all $s \in \mathcal{S}$.*

Thanks to Property (2) in this definition, finite witnesses provide the finite T-model $\mathcal{M}^*$ that is the starting point for the upward movement made possible by smoothness: for all $s \in \mathcal{S}$, the domain $\mathcal{M}^*[s]$ comprises precisely the elements used to interpret the constant symbols occuring in the finite witness.

Definition 8 (Finitely Witnessable Theory). *An $\mathcal{L}$-theory T is finitely witnessable with respect to a set $\mathcal{S}$ of sorts of $\mathcal{L}$, if there exists a computable function* witness *such that, for all ground formulas φ in $\mathcal{L}^C$,* witness(φ) *is a finite witness of φ in T with respect to $\mathcal{S}$.*

Definition 9 (Polite Theory). *An $\mathcal{L}$-theory T is polite with respect to a set of sorts $\mathcal{S}$ of $\mathcal{L}$, if it is smooth and finitely witnessable with respect to $\mathcal{S}$.*

In the classical Nelson-Oppen procedure for disjoint stably infinite theories, it suffices to compute an arrangement of shared constants (see Sect. 3). Polite theories allow us to extend the equality-sharing scheme to non-stably-infinite theories [108], provided the procedure computes an arrangement of a larger set of constants, that includes the new ones introduced by witnesses [76].

Combination Lemma 4 (Polite Theory). *Let $\mathcal{L}_1$ and $\mathcal{L}_2$ be disjoint languages sharing a set $\mathcal{S}$ of sorts. Assume an $\mathcal{L}_1$-theory T_1 polite with respect to $\mathcal{S}$ and an arbitrary $\mathcal{L}_2$-theory T_2. Let φ_i be a ground $\mathcal{L}_i^C$-formula ($i = 1, 2$), and* witness(φ_1) *be a ground formula in $\mathcal{L}_1^D$, $D \supseteq C$, that is a finite witness of φ_1 in T_1 with respect to $\mathcal{S}$. Then $\varphi_1 \wedge \varphi_2$ is $T_1 \cup T_2$-satisfiable if and only if there exists an arrangement α of all $\mathcal{S}$-sorted constants in D such that $\alpha \wedge$ witness(φ_1) is T_1-satisfiable and $\alpha \wedge \varphi_2$ is T_2-satisfiable.*

The ($\Rightarrow$) case of the proof is straightforward. For the ($\Leftarrow$) case the reasoning goes as follows. First, by Property (1) of Definition 7, witness(φ_1) is equivalent to φ_1 in T_1. Second, by Property (2) of Definition 7, witness(φ_1) determines finite cardinalities for the shared sorts. Third, by smoothness of T_1, it is possible to scale up these cardinalities to meet those required by the T_2-model. As there is agreement on both shared constants, as provided by the arrangement, and cardinalities of shared sorts, the result follows by Theorem 1. Since one theory is polite and the other is arbitrary, the combination scheme is *asymmetric*.

Theorem 8. *The union of two $\exists$-decidable disjoint theories is $\exists$-decidable, if one of them is polite with respect to the set of sorts shared by the two theories.*

The main difficulty in applying this theorem is to show that a given theory is polite, and especially to show that finite witnesses can be computed for all input formulas. All known polite theories are theories of data structures. We distinguish two classes of polite data-structure theories. The first one comprises theories characterized by sets of *standard interpretations*, and is covered in Sect. 6.1: the theories of *arrays*, *records*, *sets*, and *multisets* belong to this class. All these theories feature an *extensionality axiom* whereby two data-structures are equal if and only if their corresponding elements are. For all these theories, a witness function can be built by using a common principle based on the translation of an inequality of data structures into some constraint on their elements. The second class includes extensions of UF with axioms such as *projection*, *injectivity*, or *acyclicity*, leading to axiomatizations of *recursive data structures*, also known as *algebraic data types*, and is covered in Sect. 6.2. For this second class of theories, a witness function can be defined using the saturated set of clauses computed by a *superposition-based decision procedure* (see Sect. 7).

6.1 Arrays, Records, Sets, and Multisets

The *theory of arrays* $\mathcal{T}_{\text{array}}$ is an especially important example of polite theory. This theory is widely studied and usually presented as an axiomatized theory (e.g., [2–4,39,40,95]). Its language $\mathcal{L}_{\text{array}}$ has a sort elem for elements, a sort index for indices, a sort array for arrays, and the function symbols read: $\text{array} \times \text{index} \rightarrow \text{elem}$ and write: $\text{array} \times \text{index} \times \text{elem} \rightarrow \text{array}$. Semantically, an array is seen as a function $a\colon I \rightarrow E$ from some set of *indices* to some set of *elements*. We use the notation E^I for the set of functions from I to E. Consider an $\mathcal{L}_{\text{array}}$-interpretation $\mathcal{M}$ with domains $\mathcal{M}[\text{elem}] = E$ and $\mathcal{M}[\text{index}] = I$. $\mathcal{M}$ interprets a term of sort array as a function $a\colon I \rightarrow E$. However, arrays can be updated, and $\mathcal{L}_{\text{array}}$ employs the function write to represent such an update. For the interpretation of write, a function $a_{i \mapsto e}\colon I \rightarrow E$ is defined as follows: $a_{i \mapsto e}(i) = e$ and $a_{i \mapsto e}(j) = a(j)$, for $j \neq i$. Then, a *standard* $\mathcal{L}_{\text{array}}$-*interpretation* $\mathcal{M}$ is an $\mathcal{L}_{\text{array}}$-interpretation such that: $\mathcal{M}[\text{array}] = (\mathcal{M}[\text{elem}])^{\mathcal{M}[\text{index}]}$, $\mathcal{M}[\text{read}](a, i) = a(i)$ for all $a \in \mathcal{M}[\text{array}]$ and $i \in \mathcal{M}[\text{index}]$, and $\mathcal{M}[\text{write}](a, i, e) = a_{i \mapsto e}$ for all $a \in \mathcal{M}[\text{array}]$, $i \in \mathcal{M}[\text{index}]$, and $e \in \mathcal{M}[\text{elem}]$. $\mathcal{T}_{\text{array}}$ is the $\mathcal{L}_{\text{array}}$-theory characterized by the *class of all standard* array-*structures*. The following *extensionality axiom* is $\mathcal{T}_{\text{array}}$-valid:

$$\forall x, y\colon \text{array. } (x \simeq y) \equiv (\forall z\colon \text{index. } \text{read}(x, z) \simeq \text{read}(y, z)).$$

The clausal form of its ($\Leftarrow$) direction is $x \simeq y \vee \text{read}(x, sk(x, y)) \not\simeq \text{read}(y, sk(x, y))$, where the Skolem term $sk(x, y)$ represents the index where x and y differ, the "witness" that the two arrays are different. A finite witness of a set Q of $\mathcal{T}_{\text{array}}$-literals is constructed by replacing each array-sorted inequality $l \not\simeq r$ in Q with $\text{read}(l, i) \not\simeq \text{read}(r, i)$, where i is a new constant symbol of sort index. This transformation originated as a reduction of the satisfiability of sets of ground literals in $\mathcal{T}_{\text{array}}$ with extensionality to the satisfiability of sets of ground literals in $\mathcal{T}_{\text{array}}$ without extensionality [4]: using a Skolem constant i in place of the compound Skolem term $sk(x, y)$ preserves equisatisfiability. Intuitively, adding constants to name the positions where arrays differ suffices to glean from the number of constants occurring in the set of literals the minimum cardinalities for sorts elem and index, leading to the following politeness result [108].

Theorem 9. $\mathcal{T}_{\text{array}}$ *is polite with respect to* {elem, index}.

Basically, $\mathcal{T}_{\text{array}}$ inherits smoothness with respect to $\{\text{elem}, \text{index}\}$ from the shininess of UF. Once extensionality has been eliminated, the reasoning focuses solely on equality of indices and elements. The remaining occurrences of write can be eliminated by a *case analysis* with respect to the read-over-write axioms, on whether the index-sorted argument of read is equal or different from that of the nested write (see Sect. 9.5, [39]). Once all occurrences of write have been eliminated, $\mathcal{T}_{\text{array}}$ essentially reduces to UF, as a term $\text{read}(l, i)$ can be written as $f_a(l)$, by introducing a free function symbol f_a for every array-term l (e.g., Chap. 9, [39]).

Records aggregate attribute-value pairs and resemble arrays if attributes are considered as indices [2, 3, 108]): if there are n attributes, the set of "indices" has cardinality n. The *theory of records* $\mathcal{T}_{\text{rec}}$ has a language $\mathcal{L}_{\text{rec}}$ with a sort rec for records, a sort elém for values, and a pair of read and write function symbols for each attribute: $\text{read}_i \colon \text{rec} \to \text{elem}$ and $\text{write}_i \colon \text{rec} \times \text{elem} \to \text{rec}$, for $i = 1, \ldots, n$. A *standard* rec-*interpretation* $\mathcal{M}$ is an $\mathcal{L}_{\text{rec}}$-interpretation such that: $\mathcal{M}[\text{rec}] = (\mathcal{M}[\text{elem}])^n$, $\mathcal{M}[\text{read}_i](a) = a(i)$ for all $a \in \mathcal{M}[\text{rec}]$, for $i = 1, \ldots, n$, and $\mathcal{M}[\text{write}](a, i, e) = a_{i \mapsto e}$ for all $a \in \mathcal{M}[\text{rec}]$ and $e \in \mathcal{M}[\text{elem}]$, for $i = 1, \ldots, n$. $\mathcal{T}_{\text{rec}}$ is the $\mathcal{L}_{\text{rec}}$-theory characterized by the *class of all standard* rec-*structures*. The $\mathcal{T}_{\text{rec}}$-valid *extensionality axiom* has the following form:

$$\forall x, y \colon \text{rec}. \ (x \simeq y) \equiv (\text{read}_1(x) \simeq \text{read}_1(y) \wedge \cdots \wedge \text{read}_n(x) \simeq \text{read}_n(y)).$$

Sets also resemble arrays if a set X, $X \subseteq I$, is viewed as its *characteristic function* $X \colon I \to \{0, 1\}$ [4]. The language $\mathcal{L}_{\text{set}}$ of the *theory of sets* $\mathcal{T}_{\text{set}}$ has a sort elem for set elements, a sort set for sets, and the predicate symbol $\in \colon \text{elem} \times \text{set} \to \text{prop}$. A *standard* set-*interpretation* $\mathcal{M}$ is an $\mathcal{L}_{\text{set}}$-interpretation such that $\mathcal{M}[\text{set}] = 2^{\mathcal{M}[\text{elem}]}$ and $\mathcal{M}[\in](e, x) = \text{true}$ if and only if $x(e) = 1$ for all $x \in \mathcal{M}[\text{set}]$ and $e \in \mathcal{M}[\text{elem}]$. $\mathcal{T}_{\text{set}}$ is the $\mathcal{L}_{\text{set}}$-theory characterized by the *class of all standard* set-*structures*. The *extensionality axiom* for $\mathcal{T}_{\text{set}}$ says that two sets are equal if and only if they contain the same elements:

$$\forall x, y \colon \text{set}. \ (x \simeq y) \equiv (\forall e \colon \text{elem}. \ (e \in x) \equiv (e \in y)).$$

Similarly, a multiset X with elements in I is viewed as its *multiplicity function* $X \colon I \to \mathbb{N}$. The *theory of multisets* $\mathcal{T}_{\text{bag}}$ requires a language $\mathcal{L}_{\text{bag}}$ with sorts int^+ for the non-negative integers, elem for multiset elements, bag for multisets, and the function symbol $\text{count} \colon \text{elem} \times \text{bag} \to \text{int}^+$. A *standard* $\mathcal{L}_{\text{bag}}$-*interpretation* $\mathcal{M}$ is an $\mathcal{L}_{\text{bag}}$-interpretation such that: $\mathcal{M}[\text{int}^+] = \mathbb{N}$; $\mathcal{M}[\text{bag}] = \mathbb{N}^{\mathcal{M}[\text{elem}]}$; and $\mathcal{M}[\text{count}](e, x) = x(e)$ for all $e \in \mathcal{M}[\text{elem}]$ and $x \in \mathcal{M}[\text{bag}]$. $\mathcal{T}_{\text{bag}}$ is the $\mathcal{L}_{\text{bag}}$-theory characterized by the *class of all standard* bag-*structures*, with *extensionality axiom*

$$\forall x, y \colon \text{bag}. \ (x \simeq y) \equiv (\forall e \colon \text{elem}. \ \text{count}(e, x) \simeq \text{count}(e, y)).$$

The politeness of these three theories is a corollary [108] of Theorem 9.

Corollary 2. *The theories* $\mathcal{T}_{\text{rec}}$, $\mathcal{T}_{\text{set}}$ *and* $\mathcal{T}_{\text{bag}}$ *are polite with respect to* $\{\text{elem}\}$.

In order to construct finite witnesses for sets of ground literals in these theories, the ($\Leftarrow$) direction of their extensionality axiom is used to translate every inequality between terms of the data-structure sort into some constraint on their elements. For records, a rec-sorted inequality $l \not\simeq r$ is replaced with $\text{read}_i(l) \not\simeq \text{read}_i(r)$ for some attribute i, $1 \le i \le n$: the attribute is the "witness" that the records differ. For sets, an inequality $l \not\simeq r$ between terms of sort set is translated into the literal sets $\{e \in l,\ e \notin r\}$ or $\{e \notin l,\ e \in r\}$, where e is a new constant symbol of sort elem denoting an element that differentiates the sets. For multisets, a bag-sorted inequality $l \not\simeq r$ yields $\text{count}(e, l) \not\simeq \text{count}(e, r)$: the new constant symbol e of sort elem denotes an element that occurs with different multiplicities in the two multisets.

6.2   Recursive Data Structures

Theories of *recursive data structures* (RDS) [26,39], also known as theories of *inductive data types* [16], or *algebraic data types*, are convenient to describe several types of data structures commonly used in programming languages. Classical examples are *lists* and *trees*. These theories adopt a language $\mathcal{L}_{\text{rds}}$ with a sort struct for the data structures, and a sort elem for their elements. The set of function symbols of $\mathcal{L}_{\text{rds}}$ is the disjoint union of a set $\mathcal{F}_c$ of *constructors* and a set $\mathcal{F}_{\text{sel}}$ of *selectors*. A constructor symbol $c \in \mathcal{F}_c$ has the sort $c: s_1, \ldots, s_n \to \text{struct}$, for $s_1, \ldots, s_n \in \{\text{elem}, \text{struct}\}$, as a constructor takes elements and structures to build more structures. For example, in theories of lists [3,4,26,39,100,117] the constructor cons takes an element and a list, and returns the list with the given element as the head and the given list as the tail. For every constructor $c \in \mathcal{F}_c$ of sort $c: s_1, \ldots, s_n \to \text{struct}$, $\mathcal{F}_{\text{sel}}$ contains selector symbols $\text{sel}_i^c: \text{struct} \to s_i$ for $i = 1, \ldots, n$. For example, in theories of lists the selectors associated to cons are named car and cdr: the first one applies to a cons-term to return the head; the second one returns the tail.

The axiomatizations of these theories may contain the following sets of axioms, where all variables are implicitly universally quantified:

- *Projection axioms:* $Proj = \{\text{sel}_i^c(c(x_1, \ldots, x_n)) \simeq x_i \mid \text{sel}_i^c \in \mathcal{F}_{\text{sel}}\}$, that show how selectors operate as *projection operators* over selectors;
- *Distinctiveness axioms:* $Dis = \{c(x_1, \ldots, x_{n_c}) \not\simeq d(y_1, \ldots, y_{n_d}) \mid c, d \in \mathcal{F}_c, c \ne d\}$, where n_c and n_d are the arities of constructors c and d, respectively: these axioms state that *distinct* constructors build *distinct* data structures, so that a term whose root symbol is a constructor cannot be equal to a term whose root symbol is a distinct constructor;
- *Acyclicity axioms:* $Acyc = \{x \not\simeq t[x] \mid t \text{ is an } \mathcal{F}_c\text{-context}\}$, where an $\mathcal{F}_c$-context is a context made only of symbols in $\mathcal{F}_c$, that is, constructors; these axioms ensures that constructors do not build cyclic structures;
- *Injectivity axioms:* $Inj = \{c(x_1, \ldots, x_{n_c}) \simeq c(y_1, \ldots, y_{n_c}) \Rightarrow \bigwedge_{i=1}^{n_c} x_i \simeq y_i \mid c \in \mathcal{F}_c\}$, where n_c is the arity of constructor c; these axioms stipulate that constructors are to be interpreted as *injective* functions.

The following general politeness theorem holds for these theories [16,118]:

Theorem 10. *For all theories* $\mathcal{T}$ *included in Inj* $\cup$ *Dis, the theories* $\mathcal{T} \cup Proj$ *and* $\mathcal{T} \cup Acyc \cup Proj$ *are polite with respect to* {elem}.

For all theories mentioned in Theorem 10 a *superposition-based strategy* (see Sect. 7) is a decision procedure for problems of the form $\mathcal{T} \cup Q$, where $\mathcal{T}$ is the (finite) axiomatization of the theory and Q is a (finite) set of $\mathcal{T}$-literals. If $\mathcal{T} \cup Q$ is satisfiable, the superposition-based strategy returns a set of clauses that is *saturated* (no irredundant inference applies), and contains equalities useful to construct a finite witness of Q. Dedicated tableaux-style decision procedures based on a combination of congruence closure and unification steps also exist [16].

All these theories can be extended with additional axioms defining a *bridging function* [118], such as the *length* of lists or the *size* of trees. Such a function represents a bridge between two theories: for example, a length function for lists is a bridge between a theory of lists and a theory of the integers, since the length of a list is a non-negative integer. Theories of lists and trees with bridging functions are polite theories [44]. These results rely on the characterization of the theory as a set of standard structures satisfying an extensionality axiom (cf. Sect. 6.1). For example, the *theory of possibly empty lists*, with constructors nil and cons, selectors car and cdr, extensionality axiom

$$\forall x : \text{list. } x \not\simeq \text{nil} \Rightarrow x \simeq \text{cons}(\text{car}(x), \text{cdr}(x)),$$

and bridging function length is polite, and the bridging function helps the construction of finite witnesses [44].

7   Superposition-Based Decision Procedures

From the perspective of reasoning in a theory $\mathcal{T}$, theorem proving is the problem of $\mathcal{T}$-validity, approached refutationally by proving $\mathcal{T}$-unsatisfiability of the negation of the conjecture. A complete theorem-proving strategy for first-order logic is a *semidecision procedure* for $\mathcal{T}$-validity for all finitely axiomatized first-order theories $\mathcal{T}$: termination with a proof is guaranteed for all unsatisfiable inputs $\mathcal{T} \cup Q$, where $\mathcal{T}$ contains the axioms of the theory in clausal form, and Q is the clausal form of the negation of the conjecture. On the other hand, termination on satisfiable inputs is a challenge. In this section we survey *termination* results that allow one to apply *superposition-based theorem-proving strategies* to decide SMT problems for some theories [2,3,26,27], including most of the *polite theories* described in Sect. 6. The central result covered in this section is a *modularity theorem for termination* of superposition [2,3], that opened the way to understanding the relationship between superposition and equality sharing [29], and to designing integrations of SMT-solving and theorem proving [24,28,35,36], yielding more decision procedures.

A *theorem-proving strategy* is given by an *inference system*, which is a set of *inference rules*, and a *search plan*, which is an algorithm that controls the application of the inference rules. We consider a class of theorem-proving strategies known in the literature under various names:

- *Resolution-based* or *superposition-based*, to emphasize the main *expansion inference rules*,
- *Rewrite-based* or *simplification-based* or *ordering-based*, to highlight the removal of redundant clauses by *contraction inference rules* based on well-founded orderings, and
- *Completion-based* or *saturation-based*, to convey the overall process of expanding and contracting the existing set of clauses until either a contradiction arises or no more irredundant inferences apply.

In this paper we adopt the name *superposition-based*, because the surveyed results depend mostly on the inference system, and superposition plays the main role, as equality is the only shared symbol among the theories.

Superposition $\dfrac{C \vee l[s'] \bowtie r \quad D \vee s \simeq t}{(C \vee D \vee l[t] \bowtie r)\sigma}$ $(i), (ii), (iii), (iv)$

Reflection $\dfrac{C \vee s' \not\simeq s}{C\sigma}$ $\forall l \in C : (s' \not\simeq s)\sigma \not\prec l\sigma$

Equational Factoring $\dfrac{C \vee s \simeq t \vee s' \simeq t'}{(C \vee t \not\simeq t' \vee s \simeq t')\sigma}$ $(i), \forall l \in \{s' \simeq t'\} \cup C : (s \simeq t)\sigma \not\prec l\sigma$

where $\bowtie$ stands for either $\simeq$ or $\not\simeq$, σ is the most general unifier (mgu) of s and s', in superposition s' is not a variable, and the following abbreviations hold:

(i) is $s\sigma \not\preceq t\sigma$,
(ii) is $\forall m \in D : (s \simeq t)\sigma \not\preceq m\sigma$,
(iii) is $l[s']\sigma \not\preceq r\sigma$, and
(iv) is $\forall m \in C : (l[s'] \bowtie r)\sigma \not\preceq m\sigma$.

Simplification $\dfrac{C[l] \qquad s \simeq t}{C[t\sigma] \qquad s \simeq t}$ $l = s\sigma, \quad s\sigma \succ t\sigma, \quad C[l] \succ (s \simeq t)\sigma$

Strict Subsumption $\dfrac{C \quad D}{C}$ $D \stackrel{.}{\succ} C$

Deletion $\dfrac{C \vee t \simeq t}{}$

where $D \stackrel{.}{\succ} C$ if $D \stackrel{.}{\succeq} C$ and $C \not\stackrel{.}{\succeq} D$; and $D \stackrel{.}{\succeq} C$ if $C\sigma \subseteq D$ (as multisets) for some substitution σ. Theorem provers also apply subsumption of variants (if $D \stackrel{.}{\succeq} C$ and $C \stackrel{.}{\succeq} D$, the oldest clause is retained) and tautology deletion (that removes clauses such as $C \vee s \simeq t \vee s \not\simeq t$).

Fig. 1. SP: a standard superposition-based inference system.

A standard superposition-based inference system, named SP from superposition (cf. Fig. 1–2, [3], Fig. 1, [27], and Fig. 1, [36]), is reported in Fig. 1: expansion rules add what is below the single inference line to what is above; contraction rules replace what is above the double inference line by what is below. SP is parametric with respect to a *complete simplification ordering (CSO)* $\succ$ on terms, extended to literals and clauses by multiset extension. A simplification ordering is *stable* ($l \succ r$ implies $l\sigma \succ r\sigma$ for all terms l and r and substitutions σ), *monotonic* ($l \succ r$ implies $t[l] \succ t[r]$ for all terms l and r and contexts t), and has the *subterm property* (i.e., it contains the *strict subterm ordering* $\rhd$: $l \rhd r$ implies $l \succ r$ for all terms l and r). An ordering with these properties is *well-founded*. A CSO is also *total* on ground terms. Definitions, results, and references on orderings are accessible in several surveys (e.g., [54,55]).

The main expansion rule in SP is *superposition*, where $l[s'] \bowtie r$ ($C \vee l[s'] \bowtie r$) is the literal (clause) *superposed into*, and $s \simeq t$ ($D \vee s \simeq t$) is the literal (clause) *superposed from*, where $\bowtie$ stands for either $\simeq$ or $\not\simeq$. Depending on whether s is a variable, a constant, or a compound term, superposition is from a variable, a constant, or a compound term. *Reflection* captures ordered resolution with the reflexivity axiom $\forall x.\ x \simeq x$. *Equational factoring* allows the inference system to impose all four ordering-based restrictions listed for superposition [10]. The main contraction rule in SP is *simplification*, that performs rewriting by an equality. *Subsumption* eliminates a clause that is less general than another clause according to the subsumption ordering $\geqslant$ defined in the caption of Fig. 1. *Deletion* removes clauses containing trivial equalities.

Superposition dates back to the late 1960's [82,112]; inference systems of this kind appear in many papers (e.g., [10,34,72,113]); several general treatments or surveys with additional references and historic background are available (e.g., [19,21,22,54,55,87,104,106,107]). Superposition-based strategies yield decision procedures for several fragments of first-order logic (e.g., [59,63] and [58] for a survey), and are implemented in many theorem-provers including, in alphabetical order, E [114], SPASS [124], Vampire [84], WALDMEISTER [70], and Zipperposition [48]. An *SP-derivation* is a series

$$Q_0 \underset{\text{SP}}{\vdash} Q_1 \underset{\text{SP}}{\vdash} \ldots Q_i \underset{\text{SP}}{\vdash} Q_{i+1} \underset{\text{SP}}{\vdash} \ldots,$$

where for all i, $i \geq 0$, the set of clauses Q_{i+1} is derived from Q_i by applying an SP-inference rule. A derivation is characterized by its *limit* $Q_\infty = \bigcup_{j \geq 0} \bigcap_{i \geq j} Q_i$, that is the set of *persistent* clauses, those that are either input or generated at some stage, and never deleted afterwards. A derivation is a *refutation* if there exists an i such that $\Box \in Q_i$, where $\Box$ is the empty clause, the contradiction in clausal form. SP is *refutationally complete*: whenever the input set Q_0 is unsatisfiable, then there exist SP-refutations from Q_0. Inference systems are nondeterministic: multiple SP-derivations are possible from a given input set Q_0. The pairing of SP with a search plan yields an *SP-strategy*, and then the SP-derivation generated from Q_0 by a given SP-strategy is unique.

Refutational completeness of the inference system is not sufficient for the completeness of a strategy: the complementary requirement on the search plan is *fairness*. A derivation is *fair* if it is guaranteed to be a refutation whenever the input set is unsatisfiable. A search plan is *fair* if it generates a fair derivation for all inputs. A strategy is *fair* if its search plan is. A strategy is *complete* if its inference system is refutationally complete and its search plan is fair. In practice, a derivation that considers eventually all irredundant inferences is fair, and its limit is *saturated*. An inference is *redundant* if it uses or generates redundant clauses, and *irredundant* otherwise. Definitions of redundancy and sufficient conditions for fairness can be given based on well-founded orderings on either clauses [10] or proofs [22,34]. In the sequel SP-strategy stands for *complete SP-strategy*.

In order to prove that an SP-strategy is a *decision procedure* for a certain class of problems, one needs to show that it is guaranteed to *terminate* on inputs in that class. If one shows that SP only generates *finitely many* clauses from such inputs, termination is guaranteed. We begin with ∃-*decidability*, that is, we consider T-satisfiability problems $T \cup Q$, where T contains the axioms of the theory in clausal form and Q is a set of ground T-literals. SP generates finitely many clauses from such problems in the theories of:

- *Equality*, for which T is empty [4,11,86],
- *Non-empty possibly cyclic lists* [4] and *possibly empty possibly cyclic lists* [3],
- *Arrays* with or without extensionality [2–4],
- Finite *sets* with or without extensionality [4],
- *Records* with or without extensionality [2,3],
- *Integer offsets* and *integer offsets modulo*, a theory useful to describe data structures such as *circular queues* [2,3], and
- *Recursive data structures* with a constructor and k selectors [26], of which integer offsets and *acyclic non-empty lists* are special cases for $k=1$ and $k=2$, respectively.

Therefore, these theories are ∃-*SP-decidable* [27], meaning that an SP-strategy is a *decision procedure* for ∃-decidability in these theories.

For each theory T the proof of termination rests on an analysis of the possible SP-inferences from an input of the form $T \cup Q$, showing that there are only finitely many. This analysis assumes that a preprocessing phase *flattens* all literals in Q, by introducing new constant symbols and equalities, in such a way that every positive literal contains at most one occurrence of function symbol, and every negative literal contains no occurrence of function symbols. For example, the literal $f(a) \not\simeq f(b)$ yields the set of flat literals $\{f(a) \simeq a', \ f(b) \simeq b', \ a' \not\simeq b'\}$ by introducing fresh constants a' and b'.

The preprocessing phase may involve some other simple mechanical transformation, called T-*reduction*: for example, for the theories of *arrays with extensionality* [2–4] and *records with extensionality* [2,3], T-reduction replaces array-sorted, and rec-sorted, inequalities, via the introduction of "witnesses," as already described in Sect. 6.1, so that the extensionality axiom can be removed. For both theories, T-reduction preserves equisatisfiability also in the presence of free function symbols, which is relevant for their union with the theory of equality, provided the sorts array and rec do not appear in the sorts of the free function symbols [3]. Since the presentation of recursive data structures includes *infinitely many acyclicity axioms* (cf. Sect. 6.2), T-reduction in this case transforms the problem to an equisatisfiable problem with finitely many acyclicity axioms [3,26].

For some theories, the $\mathcal{T}$-reduction is empty and preprocessing consists only of flattening. Also, the CSO $\succ$ employed by SP is required to be *good*, meaning that $t \succ c$ for all ground compound terms t and constants c [3,25,27].

Once termination is established, the *complexity* of the resulting superposition-based decision procedure may be characterized abstractly in terms of *meta-saturation* [88,91]. More concretely, the superposition-based decision procedures for the above mentioned theories are exponential, except for *records without extensionality* and *integer offsets modulo* [3,26]. For the theory of arrays this is unavoidable, because the already mentioned case analysis over whether two indices are equal (see Sect. 6.1) means that $\mathcal{T}_{\mathsf{array}}$-satisfiability is as hard as SAT, and therefore has an exponential lower bound. For the theories of *records with extensionality* and *integer offsets* the superposition-based decision procedures were improved to be *polynomial* [27], showing that there is a big difference between records and arrays complexity-wise.

The *modularity problem* for $\exists$-SP-decidability is the problem of showing that if $\mathcal{T}_1$ and $\mathcal{T}_2$ are $\exists$-SP-decidable, then $\mathcal{T} = \mathcal{T}_1 \cup \mathcal{T}_2$ also is $\exists$-SP-decidable. Since $\mathcal{T}_i$-reduction applies separately for each theory, and flattening is harmless, the modularity of $\exists$-SP-decidability reduces to that of *termination*. The problem is to show that if SP is guaranteed to generate finitely many clauses from inputs of the form $\mathcal{T}_i \cup Q_i$ $(i = 1, 2)$, where Q_i is a set of ground $\mathcal{T}_i$-literals, then SP is guaranteed to generate finitely many clauses from inputs of the form $\mathcal{T} \cup Q$, where Q is a set of ground $\mathcal{T}$-literals. The issue is to find sufficient conditions for this result. Two conditions are easy: $\mathcal{T}_1$ and $\mathcal{T}_2$ should not share non-nullary function symbols, which is implied by their being disjoint, and the CSO $\succ$ should be good for both theories. The key condition is that $\mathcal{T}_1$ and $\mathcal{T}_2$ are *variable-inactive* [2,3], which prevents *superposition from variables* across theories.

Definition 10. *A clause φ is* variable-inactive *if no $\succ$-maximal literal in φ is an equality $t \simeq x$ where $x \notin Var(t)$. A set of clauses is* variable-inactive *if all its clauses are.*

Variable-inactivity is concerned only with equalities $t \simeq x$ where $x \notin Var(t)$, because if $x \in Var(t)$, the ordering-based restrictions on superposition suffice to bar superposition from x, as $t \succ x$ in any ordering $\succ$ with the subterm property.

Definition 11. *A theory $\mathcal{T}$ is* variable-inactive, *if for all $\mathcal{T}$-satisfiability problems $\mathcal{T} \cup Q$ the limit of every fair SP-derivation from $\mathcal{T} \cup Q$ is variable-inactive.*

The absence of shared non-nullary function symbols prevents superposition from compound terms across theories. Thus, the only superpositions across theories are superpositions from constants into constants, and because there are finitely many constant symbols in the problem, the modularity of termination follows (cf. Theorem 5, [2]; Theorem 4.1 and Corollary 3, [3]).

Theorem 11. *If theories $\mathcal{T}_1$ and $\mathcal{T}_2$ are disjoint, variable-inactive, and $\exists$-SP-decidable, then their union $\mathcal{T}_1 \cup \mathcal{T}_2$ is $\exists$-SP-decidable.*

All the theories above satisfy the hypotheses of this theorem and therefore their unions are ∃-SP-decidable (cf. Corollary 1, [2] and Theorem 4.6, [3]).

Superpositions from constants into constants across theories are superpositions from shared constants into shared constants. Also, the proof of the modularity result (see the proof of Theorem 4.1, [3]) shows that the equalities superposed from are equalities between constants. Thus, the only equalities that are active across variable-inactive theories in superposition are *equalities between shared constants*. This suggests an analogy with equality sharing, where the decision procedures can build an arrangement for stably infinite theories by propagating only clauses made of *equalities between shared constants* (e.g., Sect. 10.3, [39]). Theorem-proving strategies do not require stable infiniteness upfront. However, *variable-inactivity implies stable-infiniteness*, a finding that reinforces the analogy between superposition and equality sharing, and the intuition that they capture the same essential features of reasoning in a union of theories. The discovery that variable-inactivity implies stable-infiniteness descends from a lemma showing that superposition has the power of revealing the lack of infinite models by generating at-most cardinality constraints (cf. Lemma 5.2, [29]).

Lemma 1. *A finite satisfiable set of clauses Q admits no infinite models if and only if the limit of every fair SP-derivation from Q contains an at-most cardinality constraint.*

Since an at-most cardinality constraint is *not* variable-inactive, the result that variable-inactivity implies stable-infiniteness follows (cf. Theorem 4.5, [3]).

Theorem 12. *If a theory T is variable-inactive, then it is stably infinite.*

Indeed, if T is not stably-infinite, there is a quantifier-free T-satisfiable T-formula φ with no infinite T-model. By the above lemma, the limit of every fair SP-derivation from the clausal form of $T \cup \{\varphi\}$ contains an at-most cardinality constraint, and T is not variable-inactive.

We consider next T-satisfiability problems $T \cup Q$ where T contains the axioms of the theory in clausal form and Q is a set of ground T-clauses. If SP is guaranteed to generate finitely many clauses from these problems, an SP-strategy is a decision procedure for the T-satisfiability of quantifier-free formulas and theory T is *SP-decidable*. Results of this kind are obtained by replacing variable inactivity with a stronger property named *subterm inactivity*: the theories of *equality, arrays with or without extensionality*, possibly augmented with an *injectivity* predicate, a *swap* predicate, or both, *finite sets with or without extensionality, recursive data structures*, and their unions, are shown to be SP-decidable in this manner [23, 25]. Other variable-inactive theories, such as *possibly empty lists, records*, and *integer offsets modulo*, are not subterm-inactive. By a simpler approach [27] it is possible to show that variable-inactivity alone suffices for SP-decidability (cf. Theorem 3.5, [27]).

Theorem 13. *If a theory T is variable-inactive and ∃-SP-decidable, then it is SP-decidable.*

Furthermore, for *arrays with or without extensionality, records with or without extensionality, integer offsets*, and their unions, superposition can be a preprocessor for an SMT-solver: the problem is decomposed in such a way that superposition is applied only to the axiomatization $\mathcal{T}$ and ground unit $\mathcal{T}$-clauses, realizing an inference-based reduction of $\mathcal{T}$ to the theory of equality [24, 28].

Although the results surveyed in this section were obtained for languages where equality is the only predicate, it is simple to generalize them to languages with more predicate symbols (see [36], Sect. 3). The discovery that variable-inactivity implies stable infiniteness is rich in implications. Variable-inactivity and meta-saturation [88, 91] were used to test for stable infiniteness [81, 90]. A superposition-based decision procedure for a variable-inactive theory can be a component of an equality-sharing combination: this is a theoretical underpinning for a method named DPLL($\Gamma + \mathcal{T}$) [35, 36] which integrates a superposition-based inference system Γ into the DPLL($\mathcal{T}$) framework for SMT [14, 85, 103]. DPLL($\Gamma + \mathcal{T}$) handles axiomatized theories by superposition and built-in theories by DPLL($\mathcal{T}$). Superposition offers complete reasoning about quantifiers, decision procedures for some axiomatized theories, and it can detect the lack of infinite models. DPLL($\Gamma + \mathcal{T}$) also enriches DPLL($\mathcal{T}$) with *speculative inferences* to yield decision procedures for more theories and unions of theories.

8   CDSAT: An Overview

The philosophy of equality sharing is to combine decision procedures as *black-boxes*. Clearly, black-box combination has advantages: existing procedures can be combined without modifying them, and their communication can be minimized. In equality sharing, communication is limited to the propagation of disjunctions of equalities between shared constants (e.g., Sect. 10.3, [39]). The DPLL($\mathcal{T}$) framework for SMT [14, 85, 103] extends this philosophy to the interaction between the *conflict-driven clause learning* (CDCL) procedure for SAT-solving [49, 93, 94] and the equality-sharing-based $\mathcal{T}$-decision procedure, where $\mathcal{T}$ is a union of theories. An *abstraction function* maps $\mathcal{T}$-atoms to propositional atoms and its inverse performs the opposite translation. Every disjunction propagated by the $\mathcal{T}$-decision procedure is handled by the CDCL procedure, which searches for a propositional model of the set of clauses. The current candidate model is represented as an assignment, called *trail*:

$$\Gamma = u_1 \leftarrow \mathfrak{b}_1, \ldots, u_m \leftarrow \mathfrak{b}_m,$$

where, for all i, $1 \leq i \leq m$, u_i is a propositional atom and $\mathfrak{b}_i$ is either true or false. The $\mathcal{T}$-decision procedure contributes by signalling that a subset of these Boolean assignments implies in $\mathcal{T}$ either a contradiction ($\mathcal{T}$-*conflict*), or another Boolean assignment to an existing $\mathcal{T}$-atom, which is thus added to the trail ($\mathcal{T}$-*propagation*).

Some decision procedures for fragments of arithmetic are *conflict-driven theory procedures*, in the sense that they generalize features of CDCL to theory reasoning [47, 69, 78–80, 83, 96, 123, 126]. They assign values to first-order variables, like CDCL assign truth values to atoms, and they *explain* theory conflicts by theory inferences, like CDCL explains Boolean conflicts by resolution. A significant difference is that propositional resolution generates resolvents made of input atoms, whereas theory inferences may generate *new* (i.e., non-input) T-atoms. If such a conflict-driven procedure for a single theory is integrated as a black-box, the search for a T-model cannot take direct advantage of its guesses and inferences, and the conflict-driven reasoning remains propositional. The MCSAT method, where MCSAT stands for *Model-Constructing SATisfiability*, shows how to integrate a conflict-driven procedure for one theory [53,68,75,127], or for a specific union of theories [18,77], with CDCL, allowing them to cooperate on a single trail that contains assignments to both Boolean and first-order variables. CDSAT, which stands for *Conflict-Driven SATisfiability* [31–33], generalizes MCSAT to *generic unions of disjoint theories*, and generalizes equality sharing to accommodate *both* black-box and conflict-driven theory procedures.

The idea of CDSAT is to open the boxes and let *theory modules*, one for each theory in the union T, cooperate in the search for a T-model. Propositional logic is considered as one of the theories, and the CDCL procedure is its theory module. All theory modules access the same trail

$$\Gamma = u_1 \leftarrow c_1, \ldots, u_m \leftarrow c_m,$$

where, for all i, $1 \leq i \leq m$, u_i is a T-term and c_i is a concrete value of the appropriate sort for u_i. If u_i is a Boolean term, that is, a formula, c_i is either true or false. If u_i is a term of sort int, for example, c_i is an integer value, as in $x \leftarrow 3$ or $(y+1) \leftarrow -3$. The notion of *theory extension* is used to make sure that values can be named with fresh constant symbols, which, however, remain separate from the original language and do not occur in terms.

Once first-order (i.e., non-Boolean) assignments are allowed on the trail, there is no reason for barring them from appearing in the input problem, which is also viewed as an assignment. Therefore, CDSAT solves a generalization of SMT dubbed SMA for *satisfiability modulo assignments*. An SMT problem is written as $\{u_1 \leftarrow \text{true}, \ldots, u_m \leftarrow \text{true}\}$, where, for all i, $1 \leq i \leq m$, u_i is a quantifier-free T-formula. An SMA problem is written as $\{u_1 \leftarrow \text{true}, \ldots, u_m \leftarrow \text{true}, u_{m+1} \leftarrow c_1, \ldots, u_{m+j} \leftarrow c_j\}$, where $\{u_1 \leftarrow \text{true}, \ldots, u_m \leftarrow \text{true}\}$ is an SMT problem, and, for all i, $m+1 \leq i \leq m+j$, u_i is a first-order term, typically a variable occurring in some of the input formulas. Either way, the trail is initialized with the input problem, and CDSAT works to determine whether it is T-satisfiable.

Since the assignment on the trail mixes symbols and values from the different theories, each theory has its *view* of the trail. Suppose that T is the union of theories $T_1, \ldots, T_n$. The T_k-view ($1 \leq k \leq n$) includes the pairs $t \leftarrow c$, where c comes from the extension of T_k, and those equalities and disequalities determined by first-order assignments to terms of a T_k-sort. For example, if the trail contains $\{x \leftarrow 3, y \leftarrow 3, z \leftarrow 2\}$, the theory view of every theory with sort int contains $x \simeq y$, $x \not\simeq z$, and $y \not\simeq z$. Indeed, in the presence of first-order assignments, the truth value of an equality can be determined in two ways: either by assigning it true or false, or by assigning the same or different values to its sides. CDSAT employs a notion of *relevance* of a term to a theory to determine which theory uses one way or the other. A T_k-model $\mathcal{M}_k$ *satisfies* an assignment if for all pairs $t \leftarrow c$ in the T_k-view of the assignment $\mathcal{M}_k$ interprets t and c as the same element. An assignment is T-*satisfiable* if there is a T-model that endorses the T-view, or *global view*, which contains everything. Otherwise, the assignment is T-*unsatisfiable*. A T-unsatisfiable subset of the trail represents a *conflict*.

Since the conflict-driven search is provided centrally for all theories by CDSAT, theory modules are *theory inference systems*. Thus, the reasoning in the union of the theories is conflict-driven, and combination of theories becomes *conflict-driven combination of theory inference systems*. A theory decision procedure that is *not* conflict-driven is still incorporated as a black-box, by viewing it as an inference system whose only inference rule invokes the procedure to detect the T_k-unsatisfiability of the T_k-view of the trail.

CDSAT is defined as a *transition system* with *trail rules* and *conflict-state rules*. The trail rules transform the trail with *decisions* or *deductions*, and detect conflicts. A *decision* is a guess of a value for a term: a T_k-module is allowed to decide a value for a term t that is relevant to T_k. CDSAT has a notion of *acceptable* assignment to exclude decisions that are obviously bad, because repetitious or causing trivial conflicts. With a *deduction*, a T_k-module posts on the trail a Boolean assignment inferred from assignments on the trail. As a deduction may bring to the trail a *new* term, all deduced terms must come from a *finite global basis* to avoid jeopardizing termination. The inferred assignment is a *justified assignment*, whose *justification* is the set of premises from which it was inferred. This mechanism encompasses both T_k-*propagations* and *explanations* of T_k-conflicts. All assignments on the trail that are not decisions are justified assignments, including input assignments, that have empty justifications.

The *conflict-state rules* intervene after a conflict has been detected, so that they work on the trail and the conflict, until either the conflict is solved, or the input problem is recognized as T-unsatisfiable. CDSAT applies a form of *resolution* to unfold the conflict, by replacing a justified assignment in the conflict with its justification. This process continues until CDSAT identifies either a first-order decision that needs to be undone, or a Boolean assignment that needs to be flipped: the flipped assignment is also a justified assignment inheriting its justification from the process of unfolding the conflict.

CDSAT is *sound*, *terminating*, and *complete*, under suitable hypotheses on T_k-modules and global basis [31,33]. CDSAT requires the T_k's to be disjoint, but *not stably infinite*, provided there is a *leading theory* that knows all sorts in T. For completeness, every T_k-module must be *leading-theory-complete*, which ensures that when no trail rule applies to the trail, there is a T_k-model that satisfies the T_k-view of the trail, and agrees with a model of the leading theory on *arrangement of shared terms* and *cardinalities of shared sorts*. Whenever CDSAT terminates without reporting unsatsfiability, the T_k-models can be combined in a T-model satisfying the trail, hence the input assignment. CDSAT is a nondeterministic system, as there is nondeterminism in the CDSAT transition system and in each theory module. A CDSAT procedure is obtained by adding a search plan, that establishes priorities among CDSAT transition rules, theories, and inference rules within each theory module.

9 Discussion

Reasoning in a union of theories can be approached in several ways: the equality-sharing method and its extensions combine theory decision procedures; theorem-proving strategies unite theory presentations and reason about them; and CDSAT combines in a conflict-driven manner theory inference systems. With respect to lifting stable infiniteness, extensions of equality sharing based on shiny, gentle, or polite theories are asymmetric; the superposition-based methodology is symmetric, as it treats all theories evenly, and it handles cardinality issues seamlessly. CDSAT is also asymmetric, as the leading theory knows more than the other theories, including the cardinalities of the shared sorts. Another way to go beyond equality sharing is to admit combinations of *non-disjoint* theories (e.g., [65,67,101,118,125]). Work on this direction has begun for methods based on gentleness and politeness [43,44], as well as for superposition-based decision procedures [111], while it is a direction of future work for CDSAT.

Acknowledgments. The authors thank the co-authors of their papers covered in this survey. Part of this work was done when the first author was visiting LORIA Nancy and the Computer Science Laboratory of SRI International: the support of both institutions is greatly appreciated. This work was funded in part by grant "Ricerca di base 2017" of the Università degli Studi di Verona.

References

1. Armand, M., Faure, G., Grégoire, B., Keller, C., Théry, L., Werner, B.: A modular integration of SAT/SMT solvers to Coq through proof witnesses. In: Jouannaud, J.-P., Shao, Z. (eds.) CPP 2011. LNCS, vol. 7086, pp. 135–150. Springer, Heidelberg (2011). https://doi.org/10.1007/978-3-642-25379-9_12
2. Armando, A., Bonacina, M.P., Ranise, S., Schulz, S.: On a rewriting approach to satisfiability procedures: extension, combination of theories and an experimental appraisal. In: Gramlich, B. (ed.) FroCoS 2005. LNCS (LNAI), vol. 3717, pp. 65–80. Springer, Heidelberg (2005). https://doi.org/10.1007/11559306_4
3. Armando, A., Bonacina, M.P., Ranise, S., Schulz, S.: New results on rewrite-based satisfiability procedures. ACM TOCL **10**(1), 129–179 (2009)
4. Armando, A., Ranise, S., Rusinowitch, M.: A rewriting approach to satisfiability procedures. Inf. Comput. **183**(2), 140–164 (2003)
5. Baader, F., Ghilardi, S.: Connecting many-sorted structures and theories through adjoint functions. In: Gramlich, B. (ed.) FroCoS 2005. LNCS (LNAI), vol. 3717, pp. 31–47. Springer, Heidelberg (2005). https://doi.org/10.1007/11559306_2
6. Baader, F., Schulz, K.U.: Combination techniques and decision problems for disunification. Theor. Comput. Sci. **142**(2), 229–255 (1995)
7. Baader, F., Schulz, K.U.: Unification in the union of disjoint equational theories: combining decision procedures. J. Symb. Comput. **21**(2), 211–243 (1996)

8. Baader, F., Schulz, K.U.: Combination of constraint solvers for free and quasi-free structures. Theor. Comput. Sci. **192**(1), 107–161 (1998)
9. Baader, F., Tinelli, C.: Deciding the word problem in the union of equational theories. Inf. Comput. **178**(2), 346–390 (2002)
10. Bachmair, L., Ganzinger, H.: Rewrite-based equational theorem proving with selection and simplification. J. Logic Comput. **4**(3), 217–247 (1994)
11. Bachmair, L., Tiwari, A., Vigneron, L.: Abstract congruence closure. J. Autom. Reasoning **31**(2), 129–168 (2003)
12. Barrett, C.W., Tinelli, C.: Satisfiability modulo theories. Handbook of Model Checking, pp. 305–343. Springer, Cham (2018). https://doi.org/10.1007/978-3-319-10575-8_11
13. Barrett, C.W., et al.: CVC4. In: Gopalakrishnan, G., Qadeer, S. (eds.) CAV 2011. LNCS, vol. 6806, pp. 171–177. Springer, Heidelberg (2011). https://doi.org/10.1007/978-3-642-22110-1_14
14. Barrett, C.W., Nieuwenhuis, R., Oliveras, A., Tinelli, C.: Splitting on demand in SAT modulo theories. In: Hermann, M., Voronkov, A. (eds.) LPAR 2006. LNCS (LNAI), vol. 4246, pp. 512–526. Springer, Heidelberg (2006). https://doi.org/10.1007/11916277_35
15. Barrett, C.W., Sebastiani, R., Seshia, S.A., Tinelli, C.: Satisfiability modulo theories. In: Biere, A., Heule, M., Maaren, H.V., Walsh, T. (eds.) Handbook of Satisfiability, Chap. 26, pp. 825–886. IOS Press (2009)
16. Barrett, C.W., Shikanian, I., Tinelli, C.: An abstract decision procedure for satisfiability in the theory of inductive data types. J. Satisfiability Bool. Model. Comput. **3**, 21–46 (2007)
17. Blanchette, J.C., Böhme, S., Paulson, L.C.: Extending Sledgehammer with SMT solvers. In: Bjørner, N., Sofronie-Stokkermans, V. (eds.) CADE 2011. LNCS (LNAI), vol. 6803, pp. 116–130. Springer, Heidelberg (2011). https://doi.org/10.1007/978-3-642-22438-6_11
18. Bobot, F., Graham-Lengrand, S., Marre, B., Bury, G.: Centralizing equality reasoning in MCSAT. In: Proceedings of SMT-16 (2018)
19. Bonacina, M.P.: A taxonomy of theorem-proving strategies. In: Wooldridge, M.J., Veloso, M. (eds.) Artificial Intelligence Today. LNCS (LNAI), vol. 1600, pp. 43–84. Springer, Heidelberg (1999). https://doi.org/10.1007/3-540-48317-9_3
20. Bonacina, M.P.: On theorem proving for program checking - Historical perspective and recent developments. In: Fernàndez, M. (ed.) Proceedings of PPDP-12, pp. 1–11. ACM Press (2010)
21. Bonacina, M.P.: Parallel theorem proving. In: Hamadi, Y., Sais, L. (eds.) Handbook of Parallel Constraint Reasoning, pp. 179–235. Springer, Cham (2018). https://doi.org/10.1007/978-3-319-63516-3_6
22. Bonacina, M.P., Dershowitz, N.: Abstract canonical inference. ACM TOCL **8**(1), 180–208 (2007)
23. Bonacina, M.P., Echenim, M.: Generic theorem proving for decision procedures. TR 41/2006, Univ. degli Studi di Verona (2006). http://profs.sci.univr.it/~bonacina/rewsat.html. Revised March 2007
24. Bonacina, M.P., Echenim, M.: $\mathcal{T}$-decision by decomposition. In: Pfenning, F. (ed.) CADE 2007. LNCS (LNAI), vol. 4603, pp. 199–214. Springer, Heidelberg (2007). https://doi.org/10.1007/978-3-540-73595-3_14
25. Bonacina, M.P., Echenim, M.: Rewrite-based decision procedures. In: Archer, M., Boy de la Tour, T., Muñoz, C. (eds.) Proceedings of STRATEGIES-6, volume 174(11) of ENTCS, pp. 27–45. Elsevier (2007)

26. Bonacina, M.P., Echenim, M.: Rewrite-based satisfiability procedures for recursive data structures. In: Cook, B., Sebastiani, R. (eds.) Proceedings of PDPAR-4, volume 174(8) of ENTCS, pp. 55–70. Elsevier (2007)

27. Bonacina, M.P., Echenim, M.: On variable-inactivity and polynomial $\mathcal{T}$-satisfiability procedures. J. Logic Comput. **18**(1), 77–96 (2008)

28. Bonacina, M.P., Echenim, M.: Theory decision by decomposition. J. Symb. Comput. **45**(2), 229–260 (2010)

29. Bonacina, M.P., Ghilardi, S., Nicolini, E., Ranise, S., Zucchelli, D.: Decidability and undecidability results for Nelson-Oppen and rewrite-based decision procedures. In: Furbach, U., Shankar, N. (eds.) IJCAR 2006. LNCS (LNAI), vol. 4130, pp. 513–527. Springer, Heidelberg (2006). https://doi.org/10.1007/11814771_42

30. Bonacina, M.P., Ghilardi, S., Nicolini, E., Ranise, S., Zucchelli, D.: Decidability and undecidability results for Nelson-Oppen and rewrite-based decision procedures. TR 308–06, Univ. degli Studi di Milano (2006). http://profs.sci.univr.it/~bonacina/rewsat.html

31. Bonacina, M.P., Graham-Lengrand, S., Shankar, N.: Satisfiability modulo theories and assignments. In: de Moura, L. (ed.) CADE 2017. LNCS (LNAI), vol. 10395, pp. 42–59. Springer, Cham (2017). https://doi.org/10.1007/978-3-319-63046-5_4

32. Bonacina, M.P., Graham-Lengrand, S., Shankar, N.: Proofs in conflict-driven theory combination. In: Andronick, J., Felty, A. (eds.) Proceedings of CPP-7, pp. 186–200. ACM Press (2018)

33. Bonacina, M.P., Graham-Lengrand, S., Shankar, N.: Conflict-driven satisfiability for theory combination: transition system and completeness. J. Autom. Reasoning, 1–31 (2019, in press). https://doi.org/10.1007/s10817-018-09510-y

34. Bonacina, M.P., Hsiang, J.: Towards a foundation of completion procedures as semidecision procedures. Theor. Comput. Sci. **146**, 199–242 (1995)

35. Bonacina, M.P., Lynch, C.A., de Moura, L.: On deciding satisfiability by DPLL($\Gamma + \mathcal{T}$) and unsound theorem proving. In: Schmidt, R.A. (ed.) CADE 2009. LNCS (LNAI), vol. 5663, pp. 35–50. Springer, Heidelberg (2009). https://doi.org/10.1007/978-3-642-02959-2_3

36. Bonacina, M.P., Lynch, C.A., de Moura, L.: On deciding satisfiability by theorem proving with speculative inferences. J. Autom. Reasoning **47**(2), 161–189 (2011)

37. Bouton, T., Caminha B. de Oliveira, D., Déharbe, D., Fontaine, P.: veriT: an open, trustable and efficient SMT-solver. In: Schmidt, R.A. (ed.) CADE 2009. LNCS (LNAI), vol. 5663, pp. 151–156. Springer, Heidelberg (2009). https://doi.org/10.1007/978-3-642-02959-2_12

38. Bozzano, M., et al.: Efficient satisfiability modulo theories via delayed theory combination. In: Etessami, K., Rajamani, S.K. (eds.) CAV 2005. LNCS, vol. 3576, pp. 335–349. Springer, Heidelberg (2005). https://doi.org/10.1007/11513988_34

39. Bradley, A.R., Manna, Z.: The Calculus of Computation - Decision Procedures with Applications to Verification. Springer, Heidelberg (2007). https://doi.org/10.1007/978-3-540-74113-8

40. Bradley, A.R., Manna, Z., Sipma, H.B.: What's decidable about arrays? In: Emerson, E.A., Namjoshi, K.S. (eds.) VMCAI 2006. LNCS, vol. 3855, pp. 427–442. Springer, Heidelberg (2005). https://doi.org/10.1007/11609773_28

41. Brummayer, R., Biere, A.: Boolector: an efficient SMT solver for bit-vectors and arrays. In: Kowalewski, S., Philippou, A. (eds.) TACAS 2009. LNCS, vol. 5505, pp. 174–177. Springer, Heidelberg (2009). https://doi.org/10.1007/978-3-642-00768-2_16

42. Bruttomesso, R., Pek, E., Sharygina, N., Tsitovich, A.: The OpenSMT solver. In: Esparza, J., Majumdar, R. (eds.) TACAS 2010. LNCS, vol. 6015, pp. 150–153. Springer, Heidelberg (2010). https://doi.org/10.1007/978-3-642-12002-2_12

43. Chocron, P., Fontaine, P., Ringeissen, C.: A gentle non-disjoint combination of satisfiability procedures. In: Demri, S., Kapur, D., Weidenbach, C. (eds.) IJCAR 2014. LNCS (LNAI), vol. 8562, pp. 122–136. Springer, Cham (2014). https://doi.org/10.1007/978-3-319-08587-6_9

44. Chocron, P., Fontaine, P., Ringeissen, C.: Politeness and combination methods for theories with bridging functions. J. Automat. Reasoning (2019). https://doi.org/10.1007/s10817-019-09512-4

45. Cimatti, A., Griggio, A., Schaafsma, B.J., Sebastiani, R.: The MathSAT5 SMT solver. In: Piterman, N., Smolka, S.A. (eds.) TACAS 2013. LNCS, vol. 7795, pp. 93–107. Springer, Heidelberg (2013). https://doi.org/10.1007/978-3-642-36742-7_7

46. Conchon, S., Contejean, E., Iguernelala, M.: Canonized rewriting and ground AC completion modulo Shostak theories: design and implementation. Logical Methods Comput. Sci. 8(3), 1–29 (2012)

47. Cotton, S.: Natural domain SMT: a preliminary assessment. In: Chatterjee, K., Henzinger, T.A. (eds.) FORMATS 2010. LNCS, vol. 6246, pp. 77–91. Springer, Heidelberg (2010). https://doi.org/10.1007/978-3-642-15297-9_8

48. Cruanes, S.: Extending superposition with integer arithmetic, structural induction, and beyond. Ph.D. thesis, École Polytechnique, Univ. Paris-Saclay (2015)

49. Davis, M., Logemann, G., Loveland, D.: A machine program for theorem-proving. Commun. ACM 5(7), 394–397 (1962)

50. de Moura, L., Bjørner, N.: Z3: an efficient SMT solver. In: Ramakrishnan, C.R., Rehof, J. (eds.) TACAS 2008. LNCS, vol. 4963, pp. 337–340. Springer, Heidelberg (2008). https://doi.org/10.1007/978-3-540-78800-3_24

51. de Moura, L., Bjørner, N.: Bugs, moles and skeletons: symbolic reasoning for software development. In: Giesl, J., Hähnle, R. (eds.) IJCAR 2010. LNCS (LNAI), vol. 6173, pp. 400–411. Springer, Heidelberg (2010). https://doi.org/10.1007/978-3-642-14203-1_34

52. de Moura, L., Bjørner, N.: Satisfiability modulo theories: introduction and applications. Commun. ACM 54(9), 69–77 (2011)

53. de Moura, L., Jovanović, D.: A model-constructing satisfiability calculus. In: Giacobazzi, R., Berdine, J., Mastroeni, I. (eds.) VMCAI 2013. LNCS, vol. 7737, pp. 1–12. Springer, Heidelberg (2013). https://doi.org/10.1007/978-3-642-35873-9_1

54. Dershowitz, N., Jouannaud, J.-P.: Rewrite systems. In: van Leeuwen, J. (ed.) Handbook of Theoretical Computer Science, vol. B, pp. 243–320. Elsevier (1990)

55. Dershowitz, N., Plaisted, D.A.: Rewriting. In: Robinson, J.A., Voronkov, A. (eds.) Handbook of Automated Reasoning, vol. 1, Chap. 9, pp. 535–610. Elsevier (2001)

56. Detlefs, D.L., Nelson, G., Saxe, J.B.: Simplify: a theorem prover for program checking. J. ACM 52(3), 365–473 (2005)

57. Dutertre, B.: Yices 2.2. In: Biere, A., Bloem, R. (eds.) CAV 2014. LNCS, vol. 8559, pp. 737–744. Springer, Cham (2014). https://doi.org/10.1007/978-3-319-08867-9_49

58. Fermüller, C., Leitsch, A., Hustadt, U., Tammet, T.: Resolution decision procedures. In: Robinson, J.A., Voronkov, A. (eds.) Handbook of Automated Reasoning, vol. II, Chap. 25, pp. 1793–1849. Elsevier (2001)

59. Fietzke, A., Weidenbach, C.: Superposition as a decision procedure for timed automata. Math. Comput. Sci. 6(4), 409–425 (2012)

60. Fontaine, P.: Combinations of theories for decidable fragments of first-order logic. In: Ghilardi, S., Sebastiani, R. (eds.) FroCoS 2009. LNCS (LNAI), vol. 5749, pp. 263–278. Springer, Heidelberg (2009). https://doi.org/10.1007/978-3-642-04222-5_16

61. Fontaine, P., Gribomont, E.P.: Combining non-stably infinite, non-first order theories. In: Ahrendt, W., Baumgartner, P., de Nivelle, H., Ranise, S., Tinelli, C. (eds.) Proceedings of PDPAR-2, volume 125 of ENTCS, pp. 37–51. Elsevier (2005)

62. Gallier, J.H., Snyder, W.: Designing unification procedures using transformations: a survey. Bull. EATCS **40**, 273–326 (1990)

63. Ganzinger, H., de Nivelle, H.: A superposition decision procedure for the guarded fragment with equality. In: Proceedings of LICS-14. IEEE Computer Society (1999)

64. Ganzinger, H., Rueß, H., Shankar, N.: Modularity and refinement in inference systems. Technical report, CSL-SRI-04-02, SRI International (2004)

65. Ghilardi, S., Nicolini, E., Zucchelli, D.: A comprehensive framework for combined decision procedures. In: Gramlich, B. (ed.) FroCoS 2005. LNCS (LNAI), vol. 3717, pp. 1–30. Springer, Heidelberg (2005). https://doi.org/10.1007/11559306_1

66. Ghilardi, S., Nicolini, E., Zucchelli, D.: Recent advances in combined decision problems. In: Ballo, E., Franchella, M. (eds.) Logic and Philosophy in Italy: Trends and Perspectives, pp. 87–104. Polimetrica (2006)

67. Ghilardi, S., Nicolini, E., Zucchelli, D.: A comprehensive combination framework. ACM TOCL **9**(2), 1–54 (2008)

68. Graham-Lengrand, S., Jovanović, D.: An MCSAT treatment of bit-vectors. In: Brain, M., Hadarean, L. (eds.) Proceedings of SMT-15 (2017)

69. Haller, L., Griggio, A., Brain, M., Kroening, D.: Deciding floating-point logic with systematic abstraction. In: Cabodi, G., Singh, S. (eds.) Proceedings of FMCAD-12. ACM/IEEE (2012)

70. Hillenbrand, T.: Citius, altius, fortius: lessons learned from the theorem prover WALDMEISTER. In: Dahn, I., Vigneron, L. (eds.) Proceedings of FTP-4, volume 86 of ENTCS. Elsevier (2003)

71. Hsiang, J., Rusinowitch, M.: On word problems in equational theories. In: Ottmann, T. (ed.) ICALP 1987. LNCS, vol. 267, pp. 54–71. Springer, Heidelberg (1987). https://doi.org/10.1007/3-540-18088-5_6

72. Hsiang, J., Rusinowitch, M.: Proving refutational completeness of theorem proving strategies: the transfinite semantic tree method. J. ACM **38**(3), 559–587 (1991)

73. Huet, G.: A complete proof of correctness of the Knuth-Bendix completion algorithm. J. Comput. Syst. Sci. **23**(1), 11–21 (1981)

74. Jouannaud, J.-P., Kirchner, H.: Completion of a set of rules modulo a set of equations. SIAM J. Comput. **15**(4), 1155–1194 (1986)

75. Jovanović, D.: Solving nonlinear integer arithmetic with MCSAT. In: Bouajjani, A., Monniaux, D. (eds.) VMCAI 2017. LNCS, vol. 10145, pp. 330–346. Springer, Cham (2017). https://doi.org/10.1007/978-3-319-52234-0_18

76. Jovanović, D., Barrett, C.W.: Polite theories revisited. In: Fermüller, C.G., Voronkov, A. (eds.) LPAR 2010. LNCS, vol. 6397, pp. 402–416. Springer, Heidelberg (2010). https://doi.org/10.1007/978-3-642-16242-8_29

77. Jovanović, D., Barrett, C.W., de Moura, L.: The design and implementation of the model-constructing satisfiability calculus. In: Jobstman, B., Ray, S. (eds.) Proceedings of FMCAD-13. ACM/IEEE (2013)

78. Jovanović, D., de Moura, L.: Cutting to the chase solving linear integer arithmetic. In: Bjørner, N., Sofronie-Stokkermans, V. (eds.) CADE 2011. LNCS (LNAI), vol. 6803, pp. 338–353. Springer, Heidelberg (2011). https://doi.org/10.1007/978-3-642-22438-6_26

79. Jovanović, D., de Moura, L.: Solving non-linear arithmetic. In: Gramlich, B., Miller, D., Sattler, U. (eds.) IJCAR 2012. LNCS (LNAI), vol. 7364, pp. 339–354. Springer, Heidelberg (2012). https://doi.org/10.1007/978-3-642-31365-3_27

80. Jovanović, D., de Moura, L.: Cutting to the chase: solving linear integer arithmetic. J. Autom. Reasoning **51**, 79–108 (2013)

81. Kirchner, H., Ranise, S., Ringeissen, C., Tran, D.-K.: Automatic combinability of rewriting-based satisfiability procedures. In: Hermann, M., Voronkov, A. (eds.) LPAR 2006. LNCS (LNAI), vol. 4246, pp. 542–556. Springer, Heidelberg (2006). https://doi.org/10.1007/11916277_37

82. Knuth, D.E., Bendix, P.B.: Simple word problems in universal algebras. In: Leech, J. (ed.) Proceedings of Computational Problems in Abstract Algebras, pp. 263–298. Pergamon Press (1970)

83. Korovin, K., Tsiskaridze, N., Voronkov, A.: Conflict resolution. In: Gent, I.P. (ed.) CP 2009. LNCS, vol. 5732, pp. 509–523. Springer, Heidelberg (2009). https://doi.org/10.1007/978-3-642-04244-7_41

84. Kovács, L., Voronkov, A.: First-order theorem proving and VAMPIRE. In: Sharygina, N., Veith, H. (eds.) CAV 2013. LNCS, vol. 8044, pp. 1–35. Springer, Heidelberg (2013). https://doi.org/10.1007/978-3-642-39799-8_1

85. Krstić, S., Goel, A.: Architecting solvers for SAT modulo theories: Nelson-Oppen with DPLL. In: Konev, B., Wolter, F. (eds.) FroCoS 2007. LNCS (LNAI), vol. 4720, pp. 1–27. Springer, Heidelberg (2007). https://doi.org/10.1007/978-3-540-74621-8_1

86. Lankford, D.S.: Canonical inference. Memo ATP-32, Automatic Theorem Proving Project, Univ. of Texas at Austin (1975)

87. Lifschitz, V., Morgenstern, L., Plaisted, D.A.: Knowledge representation and classical logic. In: van Harmelen, F., Lifschitz, V., Porter, B. (eds.) Handbook of Knowledge Representation, vol. 1, pp. 3–88. Elsevier (2008)

88. Lynch, C.A., Morawska, B.: Automatic decidability. In: Plotkin, G. (ed.) Proceedings of LICS-17. IEEE Computer Society (2002)

89. Lynch, C.A., Morawska, B.: Basic syntactic mutation. In: Voronkov, A. (ed.) CADE 2002. LNCS (LNAI), vol. 2392, pp. 471–485. Springer, Heidelberg (2002). https://doi.org/10.1007/3-540-45620-1_37

90. Lynch, C.A., Ranise, S., Ringeissen, C., Tran, D.: Automatic decidability and combinability. Inf. Comput. **209**(7), 1026–1047 (2011)

91. Lynch, C.A., Tran, D.-K.: Automatic decidability and combinability revisited. In: Pfenning, F. (ed.) CADE 2007. LNCS (LNAI), vol. 4603, pp. 328–344. Springer, Heidelberg (2007). https://doi.org/10.1007/978-3-540-73595-3_22

92. Manna, Z., Zarba, C.G.: Combining decision procedures. In: Aichernig, B.K., Maibaum, T. (eds.) Formal Methods at the Crossroads. From Panacea to Foundational Support. LNCS, vol. 2757, pp. 381–422. Springer, Heidelberg (2003). https://doi.org/10.1007/978-3-540-40007-3_24

93. Marques Silva, J., Sakallah, K.A.: GRASP: a search algorithm for propositional satisfiability. IEEE Trans. Comput. **48**(5), 506–521 (1999)

94. Marques Silva, J.P., Lynce, I., Malik, S.: Conflict-driven clause learning SAT solvers. In: Biere, A., Heule, M., Van Maaren, H., Walsh, T. (eds.) Handbook of Satisfiability, volume 185 of Frontiers in Artificial Intelligence and Applications, pp. 131–153. IOS Press (2009)

88   M. P. Bonacina et al.

95. McCarthy, J.W.: Towards a mathematical science of computation. In: Popplewell, C.M. (ed.) Proceedings of IFIP 1962, North Holland, pp. 21–28 (1963)
96. McMillan, K.L., Kuehlmann, A., Sagiv, M.: Generalizing DPLL to richer logics. In: Bouajjani, A., Maler, O. (eds.) CAV 2009. LNCS, vol. 5643, pp. 462–476. Springer, Heidelberg (2009). https://doi.org/10.1007/978-3-642-02658-4_35
97. Nelson, G.: Techniques for program verification. Technical report, CSL-81-10, Xerox, Palo Alto Research Center (1981)
98. Nelson, G.: Combining satisfiability procedures by equality sharing. In: Bledsoe, W.W., Loveland, D.W. (eds.) Automatic Theorem Proving: After 25 Years, pp. 201–211. American Mathematical Society (1983)
99. Nelson, G., Oppen, D.C.: Simplification by cooperating decision procedures. ACM TOPLAS 1(2), 245–257 (1979)
100. Nelson, G., Oppen, D.C.: Fast decision procedures based on congruence closure. J. ACM 27(2), 356–364 (1980)
101. Nicolini, E., Ringeissen, C., Rusinowitch, M.: Data structures with arithmetic constraints: a non-disjoint combination. In: Ghilardi, S., Sebastiani, R. (eds.) FroCoS 2009. LNCS (LNAI), vol. 5749, pp. 319–334. Springer, Heidelberg (2009). https://doi.org/10.1007/978-3-642-04222-5_20
102. Nieuwenhuis, R.: Decidability and complexity analysis by basic paramodulation. Inf. Comput. 147(1), 1–21 (1998)
103. Nieuwenhuis, R., Oliveras, A., Tinelli, C.: Solving SAT and SAT modulo theories: from an abstract Davis-Putnam-Logemann-Loveland procedure to DPLL(T). J. ACM 53(6), 937–977 (2006)
104. Nieuwenhuis, R., Rubio, A.: Paramodulation-based theorem proving. In: Robinson, J.A., Voronkov, A. (eds.) Handbook of Automated Reasoning, vol. 1, Chap. 7, pp. 371–443. Elsevier (2001)
105. Oppen, D.C.: Complexity, convexity and combinations of theories. Theor. Comput. Sci. 12, 291–302 (1980)
106. Plaisted, D.A.: Equational reasoning and term rewriting systems. In: Gabbay, D.M., Hogger, C.J., Robinson, J.A. (eds.) Handbook of Logic in Artificial Intelligence and Logic Programming, volume I: Logical Foundations, pp. 273–364. Oxford University Press (1993)
107. Plaisted, D.A.: Automated theorem proving. Wiley Interdisc. Rev. Cogn. Sci. 5(2), 115–128 (2014)
108. Ranise, S., Ringeissen, C., Zarba, C.G.: Combining data structures with nonstably infinite theories using many-sorted logic. In: Gramlich, B. (ed.) FroCoS 2005. LNCS (LNAI), vol. 3717, pp. 48–64. Springer, Heidelberg (2005). https://doi.org/10.1007/11559306_3
109. Reger, G., Suda, M., Voronkov, A.: Playing with AVATAR. In: Felty, A.P., Middeldorp, A. (eds.) CADE 2015. LNCS (LNAI), vol. 9195, pp. 399–415. Springer, Cham (2015). https://doi.org/10.1007/978-3-319-21401-6_28
110. Ringeissen, C.: Cooperation of decision procedures for the satisfiability problem. In: Baader, F., Schulz, K.U. (eds.) Proceedings of FroCoS-1, Applied Logic, Kluwer, pp. 121–140 (1996)
111. Ringeissen, C., Senni, V.: Modular termination and combinability for superposition modulo counter arithmetic. In: Tinelli, C., Sofronie-Stokkermans, V. (eds.) FroCoS 2011. LNCS (LNAI), vol. 6989, pp. 211–226. Springer, Heidelberg (2011). https://doi.org/10.1007/978-3-642-24364-6_15
112. Robinson, G.A., Wos, L.: Paramodulation and theorem-proving in first-order theories with equality. In: Michie, D., Meltzer, B. (eds.) Machine Intelligence, vol. 4, pp. 135–150. Edinburgh University Press (1969)

113. Rusinowitch, M.: Theorem-proving with resolution and superposition. J. Symb. Comput. **11**(1 & 2), 21–50 (1991)
114. Schulz, S.: System description: E 1.8. In: McMillan, K., Middeldorp, A., Voronkov, A. (eds.) LPAR 2013. LNCS, vol. 8312, pp. 735–743. Springer, Heidelberg (2013). https://doi.org/10.1007/978-3-642-45221-5_49
115. Sebastiani, R.: Lazy satisfiability modulo theories. J. Satisfiability Bool. Model. and Comput. **3**, 141–224 (2007)
116. Shankar, N.: Automated deduction for verification. ACM Comput. Surv. **41**(4), 40–96 (2009)
117. Shostak, R.E.: An algorithm for reasoning about equality. Commun. ACM **21**(7), 583–585 (1978)
118. Sofronie-Stokkermans, V.: Locality results for certain extensions of theories with bridging functions. In: Schmidt, R.A. (ed.) CADE 2009. LNCS (LNAI), vol. 5663, pp. 67–83. Springer, Heidelberg (2009). https://doi.org/10.1007/978-3-642-02959-2_5
119. Tinelli, C., Harandi, M.T.: A new correctness proof of the Nelson-Oppen combination procedure. In: Baader, F., Schulz, K.U. (eds.) Proceedings of FroCoS-1, Applied Logic, pp. 103–120. Kluwer (1996)
120. Tinelli, C., Ringeissen, C.: Unions of non-disjoint theories and combinations of satisfiability procedures. Theor. Comput. Sci. **290**(1), 291–353 (2003)
121. Tinelli, C., Zarba, C.G.: Combining decision procedures for sorted theories. In: Alferes, J.J., Leite, J. (eds.) JELIA 2004. LNCS (LNAI), vol. 3229, pp. 641–653. Springer, Heidelberg (2004). https://doi.org/10.1007/978-3-540-30227-8_53
122. Tinelli, C., Zarba, C.G.: Combining non-stably infinite theories. J. Autom. Reasoning **34**(3), 209–238 (2005)
123. Wang, C., Ivančić, F., Ganai, M., Gupta, A.: Deciding separation logic formulae by SAT and incremental negative cycle elimination. In: Sutcliffe, G., Voronkov, A. (eds.) LPAR 2005. LNCS (LNAI), vol. 3835, pp. 322–336. Springer, Heidelberg (2005). https://doi.org/10.1007/11591191_23
124. Weidenbach, C., Dimova, D., Fietzke, A., Kumar, R., Suda, M., Wischnewski, P.: SPASS Version 3.5. In: Schmidt, R.A. (ed.) CADE 2009. LNCS (LNAI), vol. 5663, pp. 140–145. Springer, Heidelberg (2009). https://doi.org/10.1007/978-3-642-02959-2_10
125. Wies, T., Piskac, R., Kuncak, V.: Combining theories with shared set operations. In: Ghilardi, S., Sebastiani, R. (eds.) FroCoS 2009. LNCS (LNAI), vol. 5749, pp. 366–382. Springer, Heidelberg (2009). https://doi.org/10.1007/978-3-642-04222-5_23
126. Wolfman, S.A., Weld, D.S.: The LPSAT engine and its application to resource planning. In: Dean, T. (ed.) Proceedings of IJCAI-16, vol. 1, pp. 310–316. Morgan Kaufmann (1999)
127. Zeljić, A., Wintersteiger, C.M., Rümmer, P.: Deciding bit-vector formulas with mcSAT. In: Creignou, N., Le Berre, D. (eds.) SAT 2016. LNCS, vol. 9710, pp. 249–266. Springer, Cham (2016). https://doi.org/10.1007/978-3-319-40970-2_16

Initial Steps Towards a Family of Regular-Like Plan Description Logics

Alexander Borgida$^{(\boxtimes)}$

Department of Computer Science, Rutgers University, New Brunswick, NJ, USA
borgida@cs.rutgers.edu

Abstract. A wide range of ordinary Description Logics (DLs) have been explored by considering collections of concept/role constructors, and types of terminologies, yielding an array of complexity results. Representation and reasoning with plans is a very important topic in AI, yet there has been very little work on finding and studying DL constructors for plan concepts.

We start to remedy this problem here by considering Plan DLs where concept instances are sequences of action instances, and hence plan concepts can be viewed as analogues of formal languages, describing sets of strings. Inspired by the CLASP system, we consider using regular-like expressions, obtaining a rich variety of Plan DLs based on combinations of regular-like expression constructors, including sequence (concatenation), alternation (union, disjunction), looping (Kleene star), conjunction (intersection), and complement. To model the important notion of concurrency, we also consider interleaving.

We present results from the formal language literature which have immediate bearing on the complexity of DL-like reasoning tasks. However, we also focus on succinctness of representation, and on expressive power, issues first studied by Franz Baader for ordinary DLs.

1 Introduction

Ordinary Description Logics (DLs) are families of knowledge representation and reasoning formalisms based on concepts (unary predicates) and roles (binary predicates), which allow complex concept expressions to be built using *concept/role constructors*. These are used to build knowledge bases (KBs) consisting of subsumption axioms relating concepts, called TBoxes (terminologies), and sometimes additional axioms asserting information about individuals, including membership in concepts, called ABoxes. Research over the past decades has explored a wide range of DL variants by considering alternative collections of concept and role constructors, and mapping out the decidability and complexity of algorithms for operations such as concept consistency, subsumption checking, and instance membership testing.[1]

---

[1] See the table at http://www.cs.man.ac.uk/~ezolin/dl/, for example.

© Springer Nature Switzerland AG 2019
C. Lutz et al. (Eds.): Baader Festschrift, LNCS 11560, pp. 90–109, 2019.
https://doi.org/10.1007/978-3-030-22102-7_4

Brachman and Levesque [11] introduced the so-called *expressiveness vs. complexity trade off* by showing two DLs, $\mathcal{FL}^-$ and a more expressive extension $\mathcal{FL}$, which differ only by one inoffensive-looking role constructor, yet which have $O(n^2)$ vs. co-NP complete subsumption problems. In turn, Nebel [31] showed that the presence of certain kinds of TBoxes also causes the complexity of subsumption to become intractable, even for a sublanguage $\mathcal{FL}_0$ of $\mathcal{FL}^-$. Finally, pioneering work by Baader [2, 3] gave the first formal definition for the notion of *"expressiveness"*, and also pointed out, echoing Woods [40], that *succinctness/compactness* is another important dimension of KR formalisms.

In modern applications, one needs to represent not just static but also dynamic aspects of a domain. In this chapter we will concentrate on the description of complex plans made up of property-less atomic actions. Gil's survey paper [18] details the following applications of Plan DLs, esp. taxonomies: (i) the organization of plan classes; (ii) the retrieval of plan types and instances with description-based queries; (iii) the validation of plans based on descriptions of valid classes of plans; and (iv) the recognition of plan executions/instances. Weida [39] also lists many advantages for using DL formalisms for the planning domain.

In general, we are interested in DLs that allow the representation of "control flow" in concepts, so that one can describe that making a telephone call consists of dialing a number, followed by some phone rings, and then either talking or hanging up; plus *the ability to reason* that (i) placing a call, followed by either talking or hanging up, is logically equivalent to (ii) placing a call followed by talking, or placing a call followed by hanging up. A literature review, starting with [18], shows that there have been only a few DLs with specific features for representing plans, as opposed to describing actions in DLs, and then using a separate kind of formalism to represent plans composed of these actions.

In this chapter we explore variations on DL-like formalisms for describing *plan concepts having as instances sequences of action instances*. We will not consider the representation of atomic actions in DLs, which have been widely studied. Instead, following the CLASP system [15], we will draw inspiration from regular-like expressions. We provide almost no new technical results but instead gather relevant results from the vast literature on formal languages[2]

As mentioned, each standard DL is characterized by its set of "concept constructors", and the different kinds of axioms one is allowed to keep in TBoxes. We continue this tradition here, and for each selection of plan concept constructors we consider issues such as expressiveness (adapting Baader's definition), computational complexity of various tasks, and importantly, succinctness/descriptive complexity. For the latter, we give examples of plan concepts that can be described more succinctly.

The potentially interesting aspects (at least to this author) of this chapter include:

---

[2] We assume the reader is only familiar with basic properties of regular expressions and finite automata, as taught in undergraduate CS courses.

- A formal semantics is provided for CLASP [15] (but with no action classes), which is based on regular expressions.
- CLASP also provided some features whose complexity has not been analyzed. For example, the ability to define and use sub-plans, **subplan($\langle$plan name$\rangle$)**, may lead to exponentially shorter descriptions, and correspondingly increase complexity. The same is true of the constructor used to describe counted loops **repeat($\langle intCounter \rangle, \langle plan\ concept \rangle$)**, even when $intCounter = 2$.
- In addition to the standard regular expression (RE) constructors, we also consider plan concept conjunction/intersection and complementation (so-called regular-*like* expressions). This is motivated in part by the desire to extend ordinary DLs to allow plan concepts as role restrictions. Even if there is no explicit conjunction of plan concepts, in most DLs the implementation requires that reasoning with $\forall f.C \sqcap \forall f.D$ result in reasoning with $C \sqcap D$. This is true not just for all DLs that are extensions of the basic boolean DL $\mathcal{ALC}$, but also weaker DLs such as CLASSIC [10] and $\mathcal{E}^{++}$ [4].
- Concurrent execution is a natural feature of everyday plans. The natural corresponding formal language construct is "interleaving/shuffle", so we examine the effect of adding the corresponding plan constructor.
- Throughout, we explore in detail the notion of *succinctness/descriptive complexity* provided by different collections of concept constructors.

2   Description Logics

We assume the reader is familiar with ordinary DLs, but give a few definitions here because we plan to give a parallel development for Plan DLs.

As an example, the following DL concept can be interpreted as describing books authored by Canadians: *Book $\sqcap \forall authoredBy.Canadian$*, as it consists of the intersection of *Book*s and objects related by *authoredBy* only to instances of *Canadian*. We have argued [9] that although such concepts can be translated to FOL, the distinguishing feature of most DLs is that concepts can be described by variable free terms, as in most early implemented DLs. So, the above example can also be written as

$$\textbf{and}(Book, \textbf{all}(authoredBy, Canadian))$$

In this chapter, the purpose of this notation will be to make sure that a certain formalism is indeed "DL-like".[3]

Terminologies provide facilities to state a variety of axioms about concepts:

1. A *necessary condition* for members of an atomic concept A, of the form $A \sqsubseteq C$ where C is a general concept and A is an atomic name, asserts that C *subsumes* A: every instance of A is an instance of C. A specialized form of such axioms only allows atomic concept names for C, in which case the TBox is called a *taxonomy*.

---

[3] Some mathematical formalisms such as quantifiers over variables in temporal DLs (e.g., [35]) do not appear to have an obvious representation in such a notation.

2. A *definition* of a concept, $A \doteq C$, expresses both the necessary and sufficient conditions for membership in A in terms of concept C, with the usual constraint that A can be defined at most once in a terminology.
3. Terminologies can also be distinguished based on the "directly depends on" relationship they give rise to: in $A \sqsubseteq C$ and $A \doteq C$, the identifiers in C directly depend on A. Concepts that do not depend on any others are called *primitive/base*.

    – A TBox is said to be *cyclic* if the transitive closure of "directly depends on" contains (B, B) for some identifier B. Otherwise the TBox is said to be *acyclic*.

    – A TBox is said to be *unfolded* if all concepts appearing on the right hand side of definitions are primitive, and there are no necessary conditions.

2.1 Formal Semantics and Reasoning

The semantics of DL concepts is provided by an *interpretation* $\mathcal{I} = (\Delta^{\mathcal{I}}, \cdot^{\mathcal{I}})$, which consists of a non-empty, potentially infinite domain $\Delta^{\mathcal{I}}$, and a function $(\cdot)^{\mathcal{I}}$ which behave as follows for the standard DL:

Name of constructor	Syntax	Term notation	Semantics
Concept name A	A	A	$A^{\mathcal{I}} \subseteq \Delta^{\mathcal{I}}$
Role name r	r	r	$r^{\mathcal{I}} \subseteq \Delta^{\mathcal{I}} \times Dom$
Individual name b	b	b	$b^{\mathcal{I}} \ (\in \Delta^{\mathcal{I}})$
Top-concept	$\top$	**Top**	$\Delta^{\mathcal{I}}$
Bottom-concept	$\bot$	**Bottom**	$\emptyset$
Conjunction	$C_1 \sqcap C_2$	**and**(C_1,C_2)	$C_1^{\mathcal{I}} \cap C_2^{\mathcal{I}}$
Disjunction	$C_1 \sqcup C_2$	**or**(C_1,C_2)	$C_1^{\mathcal{I}} \cup C_2^{\mathcal{I}}$
Complement	$\neg C$	**not**(C)	$\Delta^{\mathcal{I}} - C^{\mathcal{I}}$
Existential	$\exists r$	**some**(r)	$\{b \in \Delta^{\mathcal{I}} \mid \exists y.\ (b, y) \in r^{\mathcal{I}}\}$
value restriction	$\forall r.C$	**all**(r,C)	$\{b \in \Delta^{\mathcal{I}} \mid \forall y.\ (b, y) \in r^{\mathcal{I}} \Rightarrow y \in C^{\mathcal{I}}\}$

Based on this semantics, we can formally define the following predicates/formulas involving concepts C, D and individual b:

Axiom type	Syntax	Semantic truth condition
Subsumption of C by D	$C \sqsubseteq D$	$C^{\mathcal{I}} \subseteq D^{\mathcal{I}}$
Equivalence of C and D	$C \equiv D$	$C^{\mathcal{I}} = D^{\mathcal{I}}$
Definition of atomic A as D	$A \doteq D$	$A^{\mathcal{I}} = D^{\mathcal{I}}$
Inconsistency of C	$C \equiv \bot$	$C^{\mathcal{I}} = \emptyset$
Membership of b in C	$b : C$	$b^{\mathcal{I}} \in C^{\mathcal{I}}$

In the presence of a TBox $\mathbf{T}$, one can of course infer more such relationships by considering only those interpretations $\mathcal{I}$ which satisfy the axioms in $\mathbf{T}$. In this case we write $\mathbf{T} \models C \sqsubseteq D$, etc. for each of the non-definitional axioms above.

The semantics of *cyclic terminologies*, such as $Person \doteq \forall parents.$ $PERSON$ are somewhat complex. We refer the reader to [5] for a discussion, and simply summarize the observation that there are three possible approaches normally considered: (i) descriptive semantics yields the interpretations of the FOL translation (e.g., those of $\forall x.PERSON(x) \Leftrightarrow (\forall y.parents(x,y) \Rightarrow Person(y)))$; (ii) least fixed point (lfp) and (iii) greatest fixed point (gfp) semantics are based on the definition of a monotone function over the lattice of interpretations. In this example, the lfp would yield $Person^{\mathcal{I}} = \emptyset$, $parents^{\mathcal{I}} = \emptyset$, and the gfp would yield $Person^{\mathcal{I}} = \Delta^{\mathcal{I}}$, $parents^{\mathcal{I}} = \Delta^{\mathcal{I}} \times \Delta^{\mathcal{I}}$.

2.2   Complexity Theory Fundamentals

As usual [36], we start from the following basic complexity classes of time- and space-bounded Turing Machine computations, parametrized by function f: DTIME(f(n)), NTIME(f(n)), DSPACE(f(n)), NSPACE(f(n)), where prefixes D and N stand for "deterministic" and "non-deterministic" respectively. In addition, we need NC^j – the class of problems solvable in polylogarithmic time on a *parallel computer* with a $O(n^j)$ number of processors.

The following are some specific classes we will encounter, with containment relationships:

$$
\begin{array}{lll}
NC^1 & \subseteq & \\
LOGSPACE \overset{\text{def}}{=} DSPACE(log\ n) & \subseteq NLOGSPACE \overset{\text{def}}{=} NSPACE(log\ n) & \subseteq \\
LogCFL_{(\text{described later})} & \subseteq NC^2 & \subseteq \\
P \overset{\text{def}}{=} DTIME(n^{O(1)}) & \subseteq NP \overset{\text{def}}{=} NTIME(n^{O(1)}) & \subseteq \\
PSPACE \overset{\text{def}}{=} DSPACE(n^{O(1)}) & = NSPACE \overset{\text{def}}{=} NSPACE(n^{O(1)}) & \subseteq \\
EXPTIME \overset{\text{def}}{=} DTIME(2^{n^{O(1)}}) & \subseteq NEXPTIME \overset{\text{def}}{=} NTIME(2^{n^{O(1)}}) & \subseteq \\
EXPSPACE \overset{\text{def}}{=} DSPACE(n^{2^{O(1)}}) & = NEXPSPACE \overset{\text{def}}{=} NSPACE(n^{2^{O(1)}}) &
\end{array}
$$

$$DLINSPACE \overset{\text{def}}{=} DSPACE(O(1)) \subseteq NLINSPACE \overset{\text{def}}{=} DSPACE(O(1))$$

All inclusions $\subseteq$ above are suspected to be strict.

The above definitions require a distinction between the familiar polynomial reductions (p-reductions) and log-space reductions; the latter are necessary to distinguish complexity classes contained in P, and require the reduction to be accomplished using only logarithmic space for computation. For even sharper distinctions, one uses log-lin reductions, which can be computed in LOGSPACE but have the additional property that the size of the output is linearly bounded by the size of the input.

2.3   Expressive Power, Computational Complexity, and Succinctness

Baader [3] examined the question of when adding new constructors or extending the type of TBox axioms allowed in standard DLs leads to the ability to "say more things"—increase the *expressiveness* of a language.

Clearly, the addition of a new constructor does not necessarily lead to an increase in expressive power. For example, in the presence of concept conjunction and complementation, adding intersection does not increase expressive power because it can easily be simulated using de Morgan's laws.

Baader approaches the problem of expressiveness by considering the interpretations that satisfy a KB (called models) expressed in a language. Intuitively, KR language $\mathcal{L}_2$ is intended to be considered to be as expressive as language $\mathcal{L}_1$ if for every KB Γ_1 in $\mathcal{L}_1$ there is a KB Γ_2 in $\mathcal{L}_2$ which has the same models.

However, the following example [3] shows that there is a need for more subtlety: one can eliminate necessary conditions in a TBox **T** by replacing axioms of the form $A \sqsubseteq C$ by definitions of the form $A \doteq C \sqcap A'$, where A' is a new, primitive name. It is easy to see that the subsumptions entailed **T** and **T'** are identical when restricted to concepts with unprimed identifiers; but the models of **T'** are different since they also involve the new, primed identifiers. To account for this. The formal definition of "model equality" ignores new identifiers:

Definition 1 (Baader). *Let $\mathcal{L}_1$ and $\mathcal{L}_2$ be description logics. Let Γ_1 and Γ_2 be TBoxes expressed in $\mathcal{L}_1$ and $\mathcal{L}_2$ respectively. Also, let $\mathcal{I}_1 \in Int(\Gamma_1)$ and $\mathcal{I}_2 \in Int(\Gamma_2)$ be interpretations satisfying Γ_1 and Γ_2, and let f be a function mapping names in Γ_1 to names in Γ_2. Then*

- *$\mathcal{I}_1$ is embedded in $\mathcal{I}_2$ by f ($\mathcal{I}_1 \subseteq_f \mathcal{I}_2$) if for all names Q occurring in Γ_1, $Q^{\mathcal{I}_1} = Q^{\mathcal{I}_2}$.*
- *We write $Int(\Gamma_1) =_f Int(\Gamma_2)$ if for all $\mathcal{I}_1 \in Int(\Gamma_1)$ there is $\mathcal{I}_2 \in Int(\Gamma_2)$ such that $\mathcal{I}_1 \subseteq_f \mathcal{I}_2$, **and** for all $\mathcal{I}_2 \in Int(\Gamma_2)$ there is $\mathcal{I}_1 \in Int(\Gamma_1)$ such that $\mathcal{I}_1 \subseteq_f \mathcal{I}_2$.*
- *$\mathcal{L}_1$ can be expressed by $\mathcal{L}_2$ through embeddings ($\mathcal{L}_1 \leq_e \mathcal{L}_2$) if for all $\Gamma_1 \in \mathcal{L}_1$ there exists $\Gamma_2 \in \mathcal{L}_2$ and embedding f such that $Int(\Gamma_1) =_f Int(\Gamma_2)$ and Γ_1 is expressed by Γ_2. Two languages have the same expressive power if they can be expressed by each other through embeddings.*

This definition allows one to prove that indeed $\mathcal{FL}_0$ with only definitions in TBoxes has the same expressive power as $\mathcal{FL}_0$ with both necessary conditions and definitions in TBoxes, since the sets of models are identical when ignoring the newly introduced identifiers. Moreover the sizes of the corresponding terminologies are within a constant factor of each other. So we have a situation where expressive power, computational complexity and descriptive complexity remain unchanged.

On the other hand, $\mathcal{FL}_0$ theories with *acyclic definitional terminologies*, whose reasoning is co-NP complete, can be turned into $\mathcal{FL}_0$ theories with *unfolded terminologies*, whose subsumption complexity is in P, by repeatedly replacing each defined concept name on the right hand side by its definition.

In this case, again, the expressive power remains the same, but the complexity is different, because definitional TBoxes can provide exponentially greater succinctness (lower descriptive complexity).

Finally, Baader shows that there is a way to efficiently reduce reasoning in $\mathcal{FL}^-$ to $\mathcal{FL}_0$, by replacing each concept expression of the form $\exists r$ by a new atomic concept A_r. So the two languages have essentially the same complexity, but not the same expressive power, because role names are not properly restricted in the $\mathcal{FL}_0$ models. Moreover, over the set of concepts definable in both $\mathcal{FL}_0$ and $\mathcal{FL}^-$ (i.e., without $\exists r$), there is no difference in succinctness.

The above examples illustrate how expressive power, computational complexity and descriptive complexity can vary, though when $\mathcal{L}_1$ is significantly more succinct than $\mathcal{L}_2$, the former will have higher complexity since the size of the input is smaller.

3   Plan Libraries and DL Reasoning

This chapter is concerned with representing and reasoning about plans and processes using DL-like languages, with plan-specific concept constructors. Among others, researchers are interested in exploiting precomputed libraries of plans, where a plan is composed of subplans and primitive actions. As mentioned earlier, Gil [18] surveys the following applications of reasoning about plans: organization of plan classes; retrieval of plan types and instances with description-based queries; validation of plans based on descriptions of valid classes of plans; and recognition of plan executions/instances. The reader is referred to that survey paper for considerably more details.

Among the systems reasoning with plans reviewed in [18], only one, CLASP[15], provides explicit plan-oriented concept constructors.

3.1   The CLASP System

The CLASP system was developed to help reason about telephonics software projects, and as such it had to handle information such as the fact that a phone call consists of picking up the phone, getting a dial tone, dialing, getting a ring tone, etc. For this reason, CLASP provides a language for describing plan concepts (whose instances are called *scenarios*), and algorithms for computing subsumption between these, as well as recognizing scenarios as concept instances. The representation is built on top of atomic STRIPS-like actions/operators such as *Ring*, with add- and delete-lists, etc. where specific states, such as *phone1-is-ringing*, are instances of atomic concepts, such as *PhoneRinging*. All of this information is represented in the CLASSIC DL.

We are not interested here in the process of planning itself, and hence we will treat atomic actions as primitive objects, without the usual aspects of planning operators such as parameters, preconditions, etc.

The key novelty of CLASP is the ability to describe plan concepts, whose EBNF syntax is given as follows, based on the analysis in [9]:

⟨plan-concept-expression⟩ ::=

  act(⟨action-concept-name⟩) |

  seq(⟨plan-concept-expression⟩$^{+}$) |

  loop(⟨plan-concept-expression⟩) |

  or(⟨plan-concept-expression⟩$^{+}$) |

  repeat(⟨integer⟩ , ⟨plan-concept-expression⟩) |

  subplan(⟨plan-concept-name⟩)

Using these, one might then describe the concept *MakingAPhoneCall* as

$$\mathbf{seq}(\mathbf{act}(Dial), \mathbf{loop}(\mathbf{act}(Ring)), \mathbf{or}(\mathbf{act}(Talk), \mathbf{act}(HangUp)))$$

An instance scenario of this plan might be
[1234dials1212at6am, 1212ringsAt6am, 1212ringsAt6:01am, 1234hangsUpAt6:02am].

The implementation of plan reasoning is based on the observation that {**seq**, **or**, **loop**} correspond to regular expression constructors {∘, ∪, *}, when the set *Actions* of action concept names is viewed as the alphabet Σ used in regular expressions.[4] For example, in order to check the subsumption $P1 \sqsubseteq P2$, Devanbu and Litman construct a product automaton from the automata for $P1$ and the complement of the deterministic automaton for $P2$ (with a potential single exponential explosion when eliminating non-determinism), and then check it for emptiness.

4   Formalizing Plan Concepts as Sets of Strings

As noted above, a CLASP scenario is a sequence of instances of concepts in the finite set *Actions*. Since in this paper we will not consider information about individual actions other than their type, nor action taxonomies, we will not distinguish different instances of the same action concept, and assume that each action class a has a single instance, "a". By abuse of notation, and following formal language tradition, we will often drop the quotes on strings, and use a to represent both the action class and its instance.

The simplest language for describing classes of scenarios therefore starts with a finite set, *Actions*, of atomic concept names for actions, a disjoint set of identifiers N for plan concepts, and plan constructors **act**, for single action plans, and **seq**, for sequence concatenation. The semantics of plan concepts is provided by an interpretation $\mathcal{I} = (Actions^{\mathcal{I}}, \cdot^{\mathcal{I}})$ where $Actions^{\mathcal{I}}$ is, in our simplified case, just a set isomorphic to *Actions*, and $\cdot^{\mathcal{I}}$ maps plan names to subsets of the set of strings/sequences over $Actions^{\mathcal{I}}$, written here for clarity as $Sequences(Actions^{\mathcal{I}})$. $\mathcal{I}$ is then extended in the natural way to some constants and the constructors in the manner shown in Fig. 1.

---

[4] CLASP actually does more, because it takes into account action concept taxonomies and the structure of actions.

name of constructor	Syntax	Term notation	Semantics
atomic action a $a \in Actions$	a	$\mathbf{act}(a)$	$\{ \text{"a"} \}$
plan concept name A $A \in N$	A	$\mathbf{subplan}(A)$	$A^{\mathcal{I}} \subseteq Sequences(Actions^{\mathcal{I}})$
no-action plan	NULL	NULL	$\{ \lambda \}$
empty plan	$\perp_P$	$\mathbf{Bottom}_{Plan}$	$\emptyset$
top-plan concept	$\top_P$	$\mathbf{Top}_{Plan}$	$Sequences(Actions^{\mathcal{I}})$
any one action	$Actions$	$\mathbf{Actions}$	$Actions^{\mathcal{I}}$
sequence	$P_1 \circ P_2$	$\mathbf{seq}(P_1, P_2)$	$\{ uw \mid u \in P_1^{\mathcal{I}}, w \in P_2^{\mathcal{I}} \}$
alternative	$P_1 \sqcup P_2$	$\mathbf{or}(P_1, P_2)$	$P_1^{\mathcal{I}} \cup P_2^{\mathcal{I}}$
repetition (base)	P^0	$\mathbf{repeat}(0, P)$	$\equiv$ NULL
repetition (ind'n)	P^{k+1}	$\mathbf{repeat}(k+1, P)$	$\equiv P \circ P^k$
loop	P^*	$\mathbf{loop}(P)$	$\equiv \bigcup_{i>0} P^i$

Fig. 1. Syntax and semantics of CLASP Plan DL

Note that unlike traditional DLs, in the absence of a terminology, every plan concept has a unique interpretation, and in fact in such situations we will drop the **act** constructor, assuming that names refer to action concepts.

We will eventually also consider terminologies of plan concepts with definitions of the form $\langle PlanName \; in \; N \rangle \doteq \langle PlanExpression \rangle$, with the standard semantics. In CLASP, TBoxes must be acyclic, and there are no ABoxes. We will also not study ABoxes in this chapter.

We are now ready to consider a variety of Plan DLs. Not surprisingly, they are inspired by formal language theory, which has dealt with the description of sets of strings.[5] In each case we will consider a subset of the following issues:

Expressive Power: For this purpose we adopt Baader's notion of expressive power, with the only difference being that interpretations now assign sets of strings to plan concepts. This works just as desired, because in showing that certain grammatical formalisms can simulate others, one often introduces new non-terminals, and Definition 1 allows one to ignore their effect. As a result, we obtain the standard notion of expressiveness for various techniques used in formal language theory, such as grammars and automata.

Computational Complexity: We will be interested in the complexity of the standard questions usually associated with DLs: *concept inconsistency/emptiness, subsumption,* and *membership.* Concerning subsumption, formal language theorists usually study two simpler variants of this: (i) non-equality with the set of all strings ("notTopPlan") and (ii) non-equality of languages ("nonEqual"). The general reason to consider the "non" variants of the above problems is that these are easily checked using non-determinism, and thus avoid

---

[5] For succinctness, we will frequently refer to *Actions* and *Actions** by their more usual formal language symbols Σ and Σ^*.

the need for co-$\mathcal{C}$ complexity results for complexity classes $\mathcal{C}$ not closed under complement.[6] We report the actual results in the literature.

In some cases, we will also give ordinary algorithms that can be implemented on (serial) computers, which allow random access to memory locations without the cost of traversing Turing Machine tape. We'll call such algorithms RAM (for Random Access Machine).

Succinctness/Descriptive Complexity: Especially in cases of Plan DLs with equal expressive power, it will be interesting to see when one allows for more succinct descriptions than another. For example, Non-deterministic Finite Automata (NFAs) are well-known to be exponentially more succinct than Deterministic Finite Automata (DFAs), because one can exhibit a family of languages L_n accepted by NFAs having $O(n)$ states, but for which every DFA requires $O(2^n)$ states. Here, L_n is the set of strings whose n'th last digit is a 1.

5 Plan DLs Based on Regular-Like Expressions

As we saw above, CLASP can be viewed as a Plan DL based on regular expressions, so let us explore the properties of Plan DLs based on so called "regular-like expressions". We will use the notation $RegExp(\{S\})$ to refer to the set of all regular-*like* expressions (over an implicit alphabet Σ) built using constructors in $\{\mathcal{S}\}$. Thus, $RegExp(\{\sqcup, \circ, ^*\})$ is just the set of ordinary standard REs, while $RegExp(\{\sqcup, \circ\})$ defines plan concepts built using only **seq** and **or**, but no looping. Unless otherwise stated, all formal results in this section can be found in the survey paper by Holzer and Kutrib [20], where references to the original papers are given.

5.1 Ordinary Regular Expressions

A variant of an earlier example

$$\mathbf{seq}(Dial, \mathbf{loop}(Ring), \mathbf{or}(HangUp, \mathbf{seq}(Talk, HangUp)))$$

has corresponding infix math notation

$$Dial \circ Ring^* \circ (HangUp \sqcup (Talk \circ HangUp))$$

The way one usually reasons about regular expressions (REs) is by converting them to NFA. The standard approach constructs an NFA with λ-transitions, which is of size linear in the size of the RE. Interestingly, this construction can be done in log-space, with output size linear in the input.

Emptiness: The usual way to test for this is to search if the corresponding NFA has a path from its start state to the end state. This is the prototypical NLOGSPACE-complete problem. A CLASP regular expression can have empty

---

[6] Recall that many space complexity classes are known to be closed under complement.

interpretation only if it contains **Bottom**$_{Plan}$, which is not likely to occur in practice. However, once we consider the conjunction or complement of REs, these may be empty even without explicit **Bottom**, and hence can become inconsistent/empty in a natural way.

In addition, in CLASP, the plan concept $b \circ c$ could be inconsistent in the case when the post-condition of activity b was incompatible with the pre-condition of activity c. In our current approach, such information cannot be encoded directly since we treat actions as atomic letters. Instead we would want to know the emptiness of the language obtained by intersecting our plan concept with complements of REs of the form $Actions^* \circ b \circ c \circ Actions^*$ for all pairs of such incompatible b and c. The complexity of emptiness in this case needs to take into account the presence of top-level complement and intersection – see below.

Subsumption: The problem of deciding if the language described by RE R_1 is not equal to that of RE R_2 is log space complete in NLINSPACE. In fact, the simpler problem of deciding if the complement of an RE R is non-empty (i.e., R does not represent Σ^*) is already NLINSPACE complete.

Membership: Membership of a string in the language of an ordinary RE (written as $Member(\{\sqcup, \circ, ^*\})$) is log-complete in NLOGSPACE [23].

5.2   Adding Plan Conjunction/Intersection

The syntax and semantics of plan conjunction, **and**, is naturally defined as $(R_1 \sqcap R_2)^{\mathcal{I}} = \{w \mid w \in R_1^{\mathcal{I}}, w \in R_2^{\mathcal{I}}\}$.

The utility of **and** arises especially in situations where one uses plan concepts as restrictions in ordinary DLs. For example, if we want to consider people who talked at least once before hanging up, we could conjoin the restriction

$$\exists\, callsMade.(Actions^* \circ Talk \circ Actions^*)$$

to the description of $Person$, which contains, among others,

$$\forall\, callsMade.(Dial \circ Ring^* \circ (HangUp \sqcup (Talk \circ HangUp)))$$

Expressiveness: It is known that regular languages are closed under intersection (one can construct the deterministic product automaton to recognize it) so this does not increase the expressive power of the plan language.

Emptiness: If we do not limit the number of intersections in extended regular expression as its size grows, then $notEmpty(\{\sqcup, \circ, ^*, \sqcap\})$ is complete in PSPACE [21], even when we do not nest intersections ([25], Proof of Lemma 3.2.3).

On a RAM, we can determine the emptiness of only $R_1 \sqcap R_2$, where R_1 and R_2 are ordinary regular expressions, by constructing in linear time/log space the NFA F_1 and F_2 recognizing R_1 and R_2, and then in quadratic time the cross-product NFA recognizing $R_1 \sqcap R_2$. Now test the emptiness of the final resulting NFA, which is a path problem solvable on a RAM in quadratic time.

Subsumption: It is known that $notTopPlan(\{\sqcup, \circ, {}^*, \sqcap\})$ and $nonEqual$ $(\{\sqcup, \circ, {}^*, \sqcap\})$ are log-lin complete for EXPSPACE [16].

Membership: The recognition problem for context-free languages can be reduced to the membership problem for $RegExp(\{\sqcup, \circ, {}^*, \sqcap\})$ [33] using a log-lin reduction, and hence $Member(\{\sqcup, \circ, {}^*, \sqcap\})$ is log-lin complete in LOGCFL, where the class LOGCFL consists of all those decision problems that are log-space reducible to a context-free language.

Succinctness: When constructing a regular expression defining the intersection of a fixed (resp. an arbitrary number of regular expressions), an exponential (resp. a double exponential) size increase, cannot be avoided in the worst-case [17]. The naive approach of constructing it does in fact achieve this bound.

5.3   Adding plan complement

Define $\neg R$ over $Actions^*$ as having interpretation $\{w \in Actions^*, w \notin R^{\mathcal{I}}\}$. So $\mathbf{not}(\mathbf{seq}(\mathbf{loop}(Actions), \text{Talk}, \mathbf{loop}(Actions)))$ denotes all plan instances where there is no talking involved.

Expressiveness: It is known that the complement of a regular language is regular (switch final and non-final states of the DFA for it), so this does not increase expressive power.

Emptiness: The language emptiness problem for "star-free regular expressions" $RegExp(\{\sqcup, \circ, \neg\})$, which don't even involve looping, has been shown to be non-elementary [37], which means that there is no function constructed from the fixed finite composition of: arithmetic operations, exponentials, logarithms, constants, and solutions of algebraic equations that bounds its complexity.

Subsumption: The complexity of checking subsumption in the presence of complementation is extremely high, even in the absence of looping. Define the tower function tow recursively as $tow(0, j) = j$, $tow(k + 1, j) = 2^{tow(k,j)}$. It was proved in [37] that although $nonEqual(\{\sqcup, \circ, {}^*, \neg\})$ is in NSPACE$(tow(n, 0))$, every problem in NSPACE$(tow(\lceil log_b(n) \rceil, 0))$ can be polynomially reduced to one in $notTopPlan(\{\sqcup, \circ, \neg\})$, and in addition $notTopPlan(\{\sqcup, \circ, \neg\})$ is itself not in NSPACE$(tow(\lceil log_b(n) \rceil, 0))$.

Membership: $Member(\{\sqcup, \circ, {}^*, \neg\})$ is log-complete in P, and RAM algorithms running in time complexity $O(n^3)$ are known.

Succinctness: When constructing a regular expression defining the complement of a given regular expression, a double exponential size increase cannot be avoided, even if the alphabet is restricted to 4 letters [17].

5.4   Adding both Intersection and Complement

The non-emptiness and equivalence problems for $RegExp(\{\sqcup, \circ, {}^*, \sqcap, \neg\})$ are non-elementary, and such extended regular expressions can be non-elementarily more succinct than classical ones [38].

5.5   Counted Iteration

As mentioned, CLASP proposed a constructor **repeat**, which can be used as in the following plan concept

$$\textbf{repeat}(10, \text{DialOneDigit}))$$

This construct bears an interesting relationship to a widely studied formal language construct called "squaring", where the extended RE R^2 is simply a short form for $R \circ R$ or **repeat**$(2, R)$. This allows us to immediately provide complexity lower bounds for some reasoning problems involving **repeat**.

Expressiveness: Clearly **repeat** does not extend the expressiveness of REs since one can use **seq**(R, ..., R) to replace **repeat**(n,R). If looping is omitted, the resulting language can only represent finite sets so it is less expressive.

Subsumption: It is known that $notTopPlan(\{\sqcup, \circ, *, ^2\})$ and $nonEqual$ $(\{\sqcup, \circ, *, ^2\})$ is log-lin complete EXPSPACE.

Succinctness: Consider the sequence of expressions $E_1 = a^2, ..., E_{n+1} = E_n^2, ...$ used to describe strings of a's of length 2^n, n=1,2,... Any NFA requires at least 2^n states to recognize E_n, and hence no ordinary RE can describe E_n in less than exponential size since the conversion from REs to NFAs is linear in size.

5.6   Omitting Looping

If we eliminate looping/star from extended regular expressions we get a variety of ways to represent finite plan concepts, as well as some infinite ones when using complement.

Expressiveness: Without complement, these are less expressive than regular expressions, since they represent only finite sets. For complement, it is known that $RegExp(\{\sqcup, \circ, \neg\})$ cannot represent $(00)^*$.

Emptiness: The emptiness problem for $RegExp(\{\sqcup, \circ\})$ is in P using polynomial reduction, but for $RegExp(\{\sqcup, \circ, \sqcap\})$ it is NP-complete [21].

Subsumption: The following results shed some light on the problem of determining subsumption: $nonEqual(\{\sqcup, \circ\})$ is log-complete in NP according to [30], while $nonEqual(\{\sqcup, \circ, \sqcap\})$ is p-complete in NP according to [21]. And $nonEqual(\{\sqcup, \circ, ^2\})$ is log-lin complete in NEXPTIME according to [30].
As noted above, even $notTopPlan(\{\sqcup, \circ, \neg\}) \notin$ NSPACE($tow(\lceil log_b\ n \rceil, 0)$).

Membership: It is known that $Member(\{\sqcup, \circ, \neg\})$ is complete in P and also $Member(\{\sqcup, \circ, \neg, ^2\})$ is log complete in P ([20] citing [32]). In fact $Member(\{\sqcup, \circ, \sqcap, ^2\})$ is log-lin-complete in LOGCFL [33], where LOGCFL is the complexity class that contains all decision problems that can be reduced in logarithmic space to a context-free language.

5.7 Adding Concurrency

The notion of concurrency in the style of programming has been studied in the formal language literature under the notion of interleaving/shuffle. Formally,

Definition 2. *Given alphabet Σ, and symbols $a, b \in \Sigma$, the shuffle of two strings, $x \# y$, is defined recursively as follows:*

$$a \# \lambda = \lambda \# a = a$$
$$(a \circ s) \# (b \circ t) = a \circ (s \# b \circ t) \sqcup b \circ (a \circ s \#)\ for\ s, t \in \Sigma^*$$

Shuffle is extended to languages, in the natural way: $\mathcal{L}_1 \# \mathcal{L}_2 = \{u \# w \mid u \in \mathcal{L}_1, w \in \mathcal{L}_2\}$. Finally, the shuffle closure $L^\#$ *of a language L is defined in a manner similar to Kleene closure: $L^\# = \bigcup L^{\#(i)}$, where $L^{\#(0)} = \{\lambda\}$, and $L^{\#(i+1)} = L \# L^{\#(i)}$.*

The shuffle constructor can be very useful in recognizing the interleaving of the actions of different plans (e.g., making a phone call while walking).

The following results are from [29].

Expressiveness: Regular languages are closed under shuffle So $\#$ does not increase expressive power. The same is not true of shuffle closure. For example, if we take $E_1 = (abc)^\#$ and $E_2 = a^* b^* c^*$, then the intersection of their languages is $\{a^n b^n c^n \mid n \geq 0\}$, which is non-regular. Since regular languages are closed under intersection, this means that E_1 cannot represent a regular language, and shuffle closure does increase expressive power.

Emptiness: In [8], a "concurrent finite state automaton" CFSA is defined, and it is shown that the emptiness problem for CFSA is decidable in polynomial time. This is relevant because in the same paper it is shown that the subclass of "acyclic CFSA" recognize $RegExp(\{\sqcup, \circ, ^*, \#, \cdot^\#\})$, and hence the polynomial bound applies to this class.

Subsumption: Mayer and Stockmeyer [29] have studied the complexity of various extensions of regular expressions with shuffle, and proved that $notTopPlan(\{\circ, \sqcup, ^*, \#\})$ is complete in EXPSPACE and $nonEqual(\{\circ, \sqcup, ^*, \#\})$ is complete in EXPSPACE.

Even without looping, $nonEqual(\{\circ, \sqcup, \#\})$ is complete in Σ_2^p. In fact this is one of the few natural problems known to be complete for Σ_2^p.

Membership: Maier and Stockmeyer also show that $Member(\{\sqcup, \circ, ^*, \sqcap, \#\})$ is NP-complete. Interestingly, the membership problem remains NP-hard even if (1) only $\{\sqcup, \#\}$ are used in expressions, (2) only $\{^*, \#\}$ are used, or (3) $\{\sqcup, \circ, \sqcap, \#\}$ are used, and $\#$ appears only once. This means that the membership problems in all these cases is NP-complete.

The data complexity of membership with shuffle and shuffle closure has been shown to be in one-way-NSPACE(log n), and hence in P [22].

Succinctness Gruber and Holzer [19] show that any ordinary regular expression defining the language $(a_1 \circ b_1)^* \# (a_2 \circ b_2)^* \# \ldots \# (a_n \circ b_n)^*$ must be of size at least *double exponential* in n.

Table 1. Summary of some reasoning complexity results from literature

Problem	Constructors	Reduction	Class [Ref.]
Regular Plan DLs			
notEmpty	$\{\sqcup, \circ, ^*\}$	log-lin complete	NLOGSPACE [23]
notTopPlan	$\{\sqcup, \circ, ^*\}$	log-lin complete	NLINSPACE [30, 37]
nonEqual	$\{\sqcup, \circ, ^*\}$	log-lin complete	NLINSPACE [38]
member	$\{\sqcup, \circ, ^*\}$	log complete	NLOGSPACE [23] (citing [24])
Regular Plan DLs + Conjunction			
notEmpty	$\{\sqcup, \circ, ^*, \sqcap\}$	complete	PSPACE [33] (citing [21])
notEmpty	$\{\sqcup, \circ, ^*, \text{only top } \sqcap\}$	complete	PSPACE [33] (citing [25])
nonEqual	$\{\sqcup, \circ, ^*, \sqcap\}$	complete	EXPSPACE [16]
member	$\{\sqcup, \circ, ^*, \sqcap\}$	log-lin complete	LOGCFL [33]
Regular Plan DLs + Complement			
notEmpty	$\{\sqcup, \circ, ^*, \neg\}$		Not bded by elementary fn [37]
nonEqual	$\{\sqcup, \circ, ^*, \neg\}$		$\in$ NSPACE$(tow(n, 0))$ [37]
nonEqual	$\{\sqcup, \circ, \neg\}$		$\notin$ NSPACE$(tow(log_b(n), 0))$ [37]
member	$\{\sqcup, \circ, ^*, \neg\}$	log complete	P [20] (citing [32])
Regular Plan DLs + Conjunction, Complement			
inconsistent	$\{\sqcup, \circ, \neg, \sqcap, ^*\}$		non-elementary
equivalence	$\{\sqcup, \circ, \neg, \sqcap, ^*\}$		non-elementary
Loop-less Plan DLs			
notEmpty	$\{\sqcup, \circ, \sqcap, \neg\}$	p-complete	NP [37]
nonEqual	$\{\sqcup, \circ\}$	log complete	NP [38]
nonEqual	$\{\sqcup, \circ, \neg\}$		$\notin$ NSPACE$(tow(\lceil log_b n \rceil), 0))$ [37]
nonEqual	$\{\sqcup, \circ, \text{k-nested } \neg\}$	hard for	NSPACE$(tow(k-3, c\sqrt{n}))$ [37]
member	$\{\sqcup, \circ, \neg\}$	complete	P [20] (citing [32])
Effect of Adding Squaring			
nonEqual	$\{\sqcup, \circ, ^2\}$	log-lin complete	NEXPTIME [30]
nonEqual	$\{\sqcup, \circ, ^*, ^2\}$	log-lin complete	EXPSPACE [30]
member	$\{\sqcup, \circ, \sqcap, ^2\}$	log-lin complete	LOGCFL [33]
member	$\{\sqcup, \circ, \neg, ^2\}$	log complete	P [20] (citing [32])
Plan DLs with Concurrency			
notEmpty	$\{\sqcup, \circ, ^*, \#\}$		$\in$ P [8]
nonEqual	$\{\sqcup, \circ, \#\}$	complete	Σ_2^p [29]
nonEqual	$\{\sqcup, \circ, ^*, \#\}$	complete	EXPSPACE [29]
member	$\{\sqcup, \circ, ^*, \sqcap, \#\}$	complete	NP [29]

5.8   Regular Plan DLs with Acyclic TBoxes

The ability to give names to sub-plans (without recursion) was already in the original CLASP Plan DL, as the **subplan** constructor. For example, *Dial* itself, in *MakeAPhoneCall*, could have been defined as

$$Dial \doteq \mathbf{seq}(PickUpReceiver, ListenForTone, \mathbf{repeat}(10, DialOneNumber))$$

Expressiveness: This does not increase the expressive power of ordinary regular expression plans, because one can unfold the definitions by substituting the definiens, as in the case of $\mathcal{FL}_0$ in Sect. 2.3.

Emptiness: For unfolded TBoxes, given a definition $N \doteq R$, one can determine the emptiness of N by checking the emptiness of R.

However, in the more general case of acyclic plan TBoxes the naive computation in cases like $\mathbf{T}_n = \{N_{i+1} \doteq N_i \circ N_i , i = n, ..., 1\}$ could lead to an exponential number of tests. Instead, one has to topologically sort the names in the TBox, and then compute in reverse order, and cache, the result of emptiness tests, resulting in a linear time RAM algorithm.

Subsumption: A formal complexity lower bound for subsumption comes from the study of regular expressions with the squaring constructor $(\cdot)^2$ introduced earlier. The point is that squaring can be replaced by TBox definitions: in an RE, repeatedly replace in an inside-out manner every subexpression R^2 by a new name $\hat{R}$, and add to the TBox the definition $\hat{R} \doteq R \circ R$. Note that the result is an acyclic TBox.

It is then known that $notTopPlan(\{\sqcup, \circ, {}^*, {}^2\})$ is log-lin complete in EXPSPACE, and $nonEqual(\{\sqcup, \circ, {}^*, {}^2\})$ is log-lin complete in EXPSPACE [30]. Hence subsumption with acyclic TBoxes is EXPSPACE-hard.

Membership: We found no sub-polynomial lower bounds in the literature for the recognition problem with squaring. A general PTIME upper bound on RAMs is obtained by translating the regular expressions and the TBox definitions into context free grammar (CFG) rules, and then using any of the CFG parsing algorithms, which run in time $\Omega(n^3)$.

Succinctness: The same argument used for the exponentially improved descriptive complexity of squaring applies here.

Table 1 summarizes some of the computational complexity results mentioned above concerning regular-like expressions.

5.9 Regular Plan DLs with Cyclic TBox

Cyclic definitions lead to interesting results in this case too. The extension of the concept D in the following TBox

$$D \doteq \mathbf{or}(NULL, \mathbf{seq}(a, D, b))$$

can be shown to consist of the language $\{a^n b^n \mid n \geq 0\}$ if one takes a least-fixed point semantics, for example. So one can express non-regular context-free languages, and therefore adding cyclic TBoxes leads to an increase in expressive power. We leave to a future paper the exploration of CFG-based Plan DLs.

6 Summary, Related and Future Work

Motivated by the use of DLs in planning, especially plan hierarchies and plan recognition [18,39], we have started to explore the space of Plan Description Logics, which describe sequences of action instances. In particular, inspired by the

CLASP system [15] and our earlier analysis of it [9] as part of an extensible architecture, we viewed plan concepts as ways of describing sets of valid sequences – strings, which are therefore formal languages. We explored subsets of plan constructors for "regular-like" expressions: {sequencing, alternation/disjunction, squaring, looping, conjunction, complementation}. We also considered interleaving as a way to model concurrent execution of plans. In each case, we sought out results in the formal language literature which shed light on the expressive power, DL reasoning complexity, and descriptive complexity of the resulting subset. The issues of expressive power and conciseness for ordinary DLs were raised already in 1990 by Franz Baader [2], and they extend naturally to Plan DLs.

Clearly, the next step in this research is to complete the results in Table 1 to deal with the three reasoning tasks of interest for DLs (e.g., replace notTopPlan results with ones about subsumption).

There are additional formalisms for describing plans and processes which we are exploring. Some of these are based on other language formalisms, such as finite state machines and various kinds of grammars. For most of these, descriptive complexity is improved, and hence reasoning complexity increases or even becomes undecidable, so one would have to look for restrictions.

There are two relevant strands of DL research, which could be used to represent plan concepts, and hence plan hierarchies.

First, dynamic logics deal directly with programs, which can be viewed as plans to perform primitive actions. For example, Propositional Dynamic Logic (PDL) allows programs to be described in a manner very similar to CLASP plans, with $\{;,',\cup,^*\}$ corresponding to $\{\circ,\ \sqcup,\ ^*\}$. In this regard, of relevance are the works of Schild [34], and De Giacomo & Lenzerini [12,13], which use extended sets of PDL program constructors as complex DL *role constructors*. More recent work by De Giacomo & Vardi, starting from [14], deals with linear temporal logic and PDL over finite traces, so is also very relevant. Future work is intended to clarify the precise relationship between Plan DLs investigated in this chapter, and the ones based on PDL.

Second, temporal DLs have been shown to be able to represent actions and their relationships in plans [1], by relating state descriptions holding at different times. By using modal variants of temporal DL [28] to capture some of the temporal constraints without the use of variables, one could have a language for describing plan concepts that has no variables, only term constructors. Once again, we leave for future work establishing the relationship between such temporal descriptions and our regular-expression based plan concepts.

Finally, it is fitting to end by mentioning a sizable collection of work by Franz Baader and (former) students, concerning reasoning about actions described using DLs [6,7,26,27]. This work considers, for example, what conditions hold at the end of simple sequences of actions (the "projection problem").

Acknowledgement. I am very grateful to my colleague, Eric Allender for his patient guidance through the landscape of modern complexity theory, and various kinds of reductions. Grant Weddell and David Toman provided useful comments and probing questions about the goal of the entire enterprise.

References

1. Artale, A., Franconi, E.: A temporal description logic for reasoning about actions and plans. J. Artif. Intell. Res. **9**, 463–506 (1998)
2. Baader, F.: A formal definition for the expressive power of knowledge representation languages. In: Proceedings of the ECAI, pp. 53–58 (1990)
3. Baader, F.: A formal definition for the expressive power of terminological knowledge representation languages. J. Log. Comput. **6**(1), 33–54 (1996)
4. Baader, F., Brandt, S., Lutz, C.: Pushing the EL envelope. In: IJCAI 2005, Proceedings of the Nineteenth International Joint Conference on Artificial Intelligence, Edinburgh, Scotland, UK, 30 July–5 August 2005, pp. 364–369 (2005)
5. Baader, F., Calvanese, D., McGuinness, D., Patel-Schneider, P., Nardi, D.: The Description Logic Handbook: Theory, Implementation and Applications. Cambridge University Press (2003)
6. Baader, F., Liu, H., ul Mehdi, A.: Verifying properties of infinite sequences of description logic actions. In: ECAI, Frontiers in Artificial Intelligence and Applications, vol. 215, pp. 53–58. IOS Press (2010)
7. Baader, F., Lutz, C., Milicic, M., Sattler, U., Wolter, F.: Integrating description logics and action formalisms: first results. In: AAAI, pp. 572–577. AAAI Press (2005)
8. Berglund, M., Björklund, H., Björklund, J.: Shuffled languages–representation and recognition. Theor. Comput. Sci. **489**, 1–20 (2013)
9. Borgida, A.: Towards the systematic development of description logic reasoners: CLASP reconstructed. In: Proceedings of the KR 1992, Cambridge, MA, USA, pp. 259–269 (1992)
10. Borgida, A., Brachman, R.J., McGuinness, D.L., Resnick, L.A.: CLASSIC: a structural data model for objects. In: Proceedings of SIGMOD 1989, pp. 58–67 (1989)
11. Brachman, R.J., Levesque, H.J.: The tractability of subsumption in frame-based description languages. In: AAAI, vol. 84, pp. 34–37 (1984)
12. De Giacomo, G., Lenzerini, M.: Boosting the correspondence between description logics and propositional dynamic logics. In: Proceedings of the AAAI 1994, pp. 205–212 (1994)
13. De Giacomo, G., Lenzerini, M.: Tbox and Abox reasoning in expressive description logics. In: Proceedings of the AAAI, pp. 37–48. AAAI Press (1996)
14. De Giacomo, G., Vardi, M.Y.: Linear temporal logic and linear dynamic logic on finite traces. In: IJCAI 2013, pp. 854–860. IJCAI/AAAI (2013)
15. Devanbu, P.T., Litman, D.J.: Taxonomic plan reasoning. Artif. Intell. **84**(1–2), 1–35 (1996)
16. Fürer, M.: The complexity of the inequivalence problem for regular expressions with intersection. In: de Bakker, J., van Leeuwen, J. (eds.) ICALP 1980. LNCS, vol. 85, pp. 234–245. Springer, Heidelberg (1980). https://doi.org/10.1007/3-540-10003-2_74
17. Gelade, W., Neven, F.: Succinctness of the complement and intersection of regular expressions. ACM Trans. Comput. Logic **4**(1), 1–19 (2012)
18. Gil, Y.: Description logics and planning. AI Mag. **26**(2), 73–84 (2005)
19. Gruber, H., Holzer, M.: Tight bounds on the descriptional complexity of regular expressions. In: Diekert, V., Nowotka, D. (eds.) DLT 2009. LNCS, vol. 5583, pp. 276–287. Springer, Heidelberg (2009). https://doi.org/10.1007/978-3-642-02737-6_22

20. Holzer, M., Kutrib, M.: The complexity of regular(-like) expressions. In: Gao, Y., Lu, H., Seki, S., Yu, S. (eds.) DLT 2010. LNCS, vol. 6224, pp. 16–30. Springer, Heidelberg (2010). https://doi.org/10.1007/978-3-642-14455-4_3

21. Hunt III, H.B.: The equivalence problem for regular expressions with intersections is not polynomial in tape. Technical report, pp. 73–161, Department of Computer Science, Cornell University, Ithaca, New York (1973)

22. Jędrzejowicz, J., Szepietowski, A.: Shuffle languages are in P. Theor. Comput. Sci. **250**(1–2), 31–53 (2001)

23. Jiang, T., Ravikumar, B.: A note on the space complexity of some decision problems for finite automata. Inf. Process. Lett. **40**(1), 25–31 (1991)

24. Jones, N.D.: Space-bounded reducibility among combinatorial problems. J. Comput. Syst. Sci. **11**(1), 68–85 (1975)

25. Kozen, D.: Lower bounds for natural proof systems. In: 18th Annual Symposium on Foundations of Computer Science, Providence, Rhode Island, USA, 31 October–1 November 1977, pp. 254–266 (1977)

26. Liu, H., Lutz, C., Milicic, M., Wolter, F.: DL actions with GCIs: a pragmatic approach. In: CEUR Workshop Proceedings of Description Logics, vol. 189. CEUR-WS.org (2006)

27. Liu, H., Lutz, C., Miličić, M., Wolter, F.: Reasoning about actions using description logics with general TBoxes. In: Fisher, M., van der Hoek, W., Konev, B., Lisitsa, A. (eds.) JELIA 2006. LNCS (LNAI), vol. 4160, pp. 266–279. Springer, Heidelberg (2006). https://doi.org/10.1007/11853886_23

28. Lutz, C., Wolter, F., Zakharyaschev, M.: Temporal description logics: a survey. In Proceedings of Temporal Representation and Reasoning, pp. 3–14. IEEE (2008)

29. Mayer, A.J., Stockmeyer, L.J.: The complexity of word problems-this time with interleaving. Inf. Comput. **115**(2), 293–311 (1994)

30. Meyer, A.R., Stockmeyer, L.J.: The equivalence problem for regular expressions with squaring requires exponential space. In: 13th Annual Symposium on Switching and Automata Theory, College Park, Maryland, USA, 25–27 October 1972, pp. 125–129 (1972)

31. Nebel, B.: Terminological reasoning is inherently intractable. Artif. Intell. **43**(2), 235–249 (1990)

32. Petersen, H.: Decision problems for generalized regular expressions. In: Descriptional Complexity of Automata, Grammars and Related Structures, Proceedings, DCAGRS 2000, pp. 22–29 (2000)

33. Petersen, H.: The membership problem for regular expressions with intersection is complete in LOGCFL. In: Alt, H., Ferreira, A. (eds.) STACS 2002. LNCS, vol. 2285, pp. 513–522. Springer, Heidelberg (2002). https://doi.org/10.1007/3-540-45841-7_42

34. Schild, K.: A correspondence theory for terminological logics: preliminary report. In: IJCAI, pp. 466–471. Morgan Kaufmann (1991)

35. Schmiedel, A.: Temporal terminological logic. In: Proceedings of AAAI 1990, pp. 640–645 (1990)

36. Sipser, M.: Introduction to the Theory of Computation. PWS Publishing Company (1997)

37. Stockmeyer, L.J.: The complexity of decision problems in automata theory and logic. Ph.D. thesis, Massachusetts Institute of Technology, Cambridge, Massachusetts (1974)

38. Stockmeyer, L.J., Meyer, A.R.: Word problems requiring exponential time. In: Symposium on Theory of Computing (STOC 1973), pp. 1–9 (1973)
39. Weida, R.: Knowledge representation for plan recognition. In: IJCAI 1995 Workshop on the Next Generation of Plan Recognition Systems (1995)
40. Woods, W.A.: What's important about knowledge representation. IEEE Comput. **16**(10), 22–26 (1983)

Reasoning with Justifiable Exceptions in $\mathcal{EL}_\perp$ Contextualized Knowledge Repositories

Loris Bozzato[1]($\boxtimes$)(iD), Thomas Eiter[2](iD), and Luciano Serafini[1](iD)

[1] Fondazione Bruno Kessler, Via Sommarive 18, 38123 Trento, Italy
{bozzato,serafini}@fbk.eu
[2] Institute of Logic and Computation, Technische Universität Wien,
Favoritenstraße 9-11, 1040 Vienna, Austria
eiter@kr.tuwien.ac.at

Abstract. The Contextualized Knowledge Repository (CKR) framework has been proposed as a description logics-based approach for contextualization of knowledge, a well-known area of study in AI. The CKR knowledge bases are structured in two layers: a global context contains context-independent knowledge and contextual structure, while a set of local contexts hold specific knowledge bases. In practical uses of CKR, it is often desirable that global knowledge can be "overridden" at the local level, that is to recognize local pieces of knowledge that do not need to satisfy the general axiom. By targeting this need, in our recent works we presented an extension of CKR with global defeasible axioms, which apply local instances unless an exception for overriding exists; such an exception, however, requires that justification is provable from the knowledge base. In this paper we apply this framework to the basic description logic $\mathcal{EL}_\perp$. We provide a formalization of $\mathcal{EL}_\perp$ CKRs with global defeasible axioms and study their semantic and computational properties. Moreover, we present a translation of CKRs to datalog programs under the answer set semantics for instance checking.

1 Introduction

The problem of reasoning over context dependent knowledge is a well-known area of study in Knowledge Representation and Reasoning: proposals for its formalization date back to the works of McCarthy [28], Lenat [27], and Giunchiglia et al. [22]. As such, the interest in representing and reasoning with contexts has also been recognized in the field of Description Logics (DLs) and led to the proposal of different approaches for introducing a notion of context in DL knowledge bases e.g. [24,25,33].

In this regard, the *Contextualized Knowledge Repository (CKR) framework* [11–13,33] is one of the most recent DL based formalisms for the representation of contextualized knowledge. A CKR knowledge base is a two-layer structure composed of a *global context* and a set of *local contexts*. The global context

© Springer Nature Switzerland AG 2019
C. Lutz et al. (Eds.): Baader Festschrift, LNCS 11560, pp. 110–134, 2019.
https://doi.org/10.1007/978-3-030-22102-7_5

contains two types of knowledge: (i) properties and structure of local contexts (*meta-knowledge*) and (ii) facts about the domain of discourse, visible by all the local contexts (*global object knowledge*). Local contexts contain *local object knowledge* that holds under specific situations (e.g. during a certain period of time, region in space) and thus they represent different partial and perspective views of the domain. The context independent knowledge from the global context is propagated to the local contexts: in other words, the axioms in the global object knowledge need to hold in the situation described by the (more specific) local contexts. In practical uses of CKR, however, it is often desirable that global axioms can be "overridden" at the local level: since they represent local specializations, local instances might need to violate the general rule established by a global axiom for some "exceptional" individual.

By targeting this need, in [10] we presented an extension of CKR by introducing a notion of *justifiable exceptions*: axioms in the global context may be specified as *defeasible* and they apply to local instances unless an exception for overriding exists. Such an exception, however, requires that a *justification* is provable from the knowledge base: an axiom can be locally "overridden" on some exceptional instance, if we can prove that they would cause a local contradiction. In [10] we presented the general syntax and semantics of CKR extended with defeasible axioms and we studied a datalog encoding for reasoning over CKRs based on $\mathcal{SROIQ}$-RL, which is a DL related to the OWL 2 RL profile of the Web Ontology Language (OWL) standard recommended by the W3C [29]. Like its siblings OWL 2 QL and OWL 2 EL, this profile offers tractable reasoning, but has the peculiarity that for these tasks only individuals that occur in the knowledge base matter; that is, the profile (and the underlying DL $\mathcal{SROIQ}$-RL) can not manage unnamed individuals.

The OWL 2 EL profile is rooted in the seminal work of Baader [2]. In search for a tractable yet expressive fragment of description logic (that is, of the DL $\mathcal{ALC}$), he proposed the $\mathcal{EL}$ language, in which only existential restriction and conjunction are available to form new concepts, but cyclic relationships among concepts are allowed; as he argued, these means are sufficient for a number of interesting use cases. In successive work [3,5], the language has been extended while keeping tractability, and the growing family of $\mathcal{EL}$ DLs has become one of the most important ones in the whole field.

In this paper, we apply the framework developed in [10] to the description logic $\mathcal{EL}_\perp$, which extends $\mathcal{EL}$ with the $\perp$ concept (in $\mathcal{EL}$, exceptions are futile as no inconsistency can arise). By adopting $\mathcal{EL}_\perp$ as the base logic, we need to take unnamed individuals introduced by existential formulas into account, and in particular for the justifications of exceptions. In case of conflict, making different exceptions can lead to different models; as deciding satisfiability of $\mathcal{EL}_\perp$ knowledge bases is PTime-complete, we may expect that exceptions on top of it would make reasoning using justified models intractable. In a benign behaviour, the complexity would not go (much) beyond NP respectively co-NP, and ideally stay within this bound. This in fact holds true for the version of CKR over $\mathcal{EL}_\perp$ that we develop here.

The main contributions of this paper are summarized as follows:

- We instantiate the CKR framework to the DL $\mathcal{EL}_\perp$: while the CKR definitions provided in [10] are mostly independent from the DL chosen as base language, we must pay further attention to the management of the semantics, due to the relevance of unnamed individuals for defeasible axioms.
- We investigate the effects of the adaptation to semantic properties and computational complexity of major reasoning tasks. It turns out that several of the properties carry over to this setting, and that the complexity of reasoning tasks is similar.
- We revise the translation to datalog (with negation under answer set semantics) proposed in [9,10] for reasoning over CKRs with defeasible axioms in $\mathcal{EL}_\perp$. We can prove that this procedure provides a sound and complete materialization calculus [26] for instance checking over such CKRs.

The rest of the paper is organized as follows. After a brief summary of the preliminaries in the next section, we provide in Sect. 3 the syntax and semantics for CKRs with defeasible axioms over $\mathcal{EL}_\perp$. In Sect. 4 we then consider semantic properties and complexity of reasoning, after which we adapt and extend in Sect. 5 the datalog translation for instance checking on CKRs with defeasible axioms in $\mathcal{SROIQ}$-RL to $\mathcal{EL}_\perp$. Finally, in Sect. 6 we briefly discuss our results, where we shall also address another important contribution of Franz Baader – namely, terminological default logic [4] – and we conclude with possible directions for future work.

2   Preliminaries

2.1   Description Logics and $\mathcal{EL}_\perp$ Language

In this work, we assume the common definitions of description logics [1] and the definition of the logic $\mathcal{EL}$ [2]: for reference, we summarize in the following the basic definitions used in our work.

A *DL vocabulary* Σ consists of the mutually disjoint countably infinite sets NC of *atomic concepts*, NR of *atomic roles*, and NI of *individual constants*. Complex *concepts* are then recursively defined as the smallest set containing all concepts that can be inductively constructed using the constructors of the considered DL language. In this paper we consider the DL $\mathcal{EL}_\perp$, which extends the basic definition of $\mathcal{EL}$ provided in [2] with the $\perp$ concept constructor. Thus, in $\mathcal{EL}_\perp$ a concept C can take the form defined by the following grammar:

$$C := A \mid \top \mid \perp \mid C_1 \sqcap C_2 \mid \exists R.C_1 \tag{1}$$

where A is a concept name and R is role name. An $\mathcal{EL}_\perp$ *knowledge base* $\mathcal{K} = \langle \mathcal{T}, \mathcal{A} \rangle$ consists of a TBox $\mathcal{T}$ containing *general concept inclusion (GCI)* axioms $C \sqsubseteq D$, where C, D are concepts, and an ABox $\mathcal{A}$ composed of assertions of the forms $C(a)$, $R(a,b)$, with C a concept, $R \in$ NR and $a, b \in$ NI.

A *DL interpretation* is a pair $\mathcal{I} = \langle \Delta^{\mathcal{I}}, \cdot^{\mathcal{I}} \rangle$ where $\Delta^{\mathcal{I}}$ is a non-empty set called *domain* and $\cdot^{\mathcal{I}}$ is the *interpretation function* which assigns denotations for language elements: $a^{\mathcal{I}} \in \Delta^{\mathcal{I}}$, for $a \in \mathrm{NI}$; $A^{\mathcal{I}} \subseteq \Delta^{\mathcal{I}}$, for $A \in \mathrm{NC}$; $R^{\mathcal{I}} \subseteq \Delta^{\mathcal{I}} \times \Delta^{\mathcal{I}}$, for $R \in \mathrm{NR}$. The interpretation of non-atomic concepts and roles is defined by the evaluation of their description logic operators (see [2] for $\mathcal{EL}$).

An interpretation $\mathcal{I}$ *satisfies* an axiom ϕ, denoted $\mathcal{I} \models_{\mathrm{DL}} \phi$, if it verifies the respective semantic condition, in particular: for $\phi = D(a)$, $a^{\mathcal{I}} \in D^{\mathcal{I}}$; for $\phi = R(a,b)$, $\langle a^{\mathcal{I}}, b^{\mathcal{I}} \rangle \in R^{\mathcal{I}}$; for $\phi = C \sqsubseteq D$, $C^{\mathcal{I}} \subseteq D^{\mathcal{I}}$. $\mathcal{I}$ is a *model* of $\mathcal{K}$, denoted $\mathcal{I} \models_{\mathrm{DL}} \mathcal{K}$, if it satisfies all axioms of $\mathcal{K}$.

Without loss of generality, we adopt the *standard name assumption (SNA)* in the DL context [17]. We consider an infinite subset $\mathrm{NI}_S \subseteq \mathrm{NI}$ of individual constants, called *standard names* s.t. in every interpretation $\mathcal{I}$ we have (i) $\Delta^{\mathcal{I}} = \mathrm{NI}_S^{\mathcal{I}} = \{ c^{\mathcal{I}} \mid c \in \mathrm{NI}_S \}$; (ii) $c^{\mathcal{I}} \neq d^{\mathcal{I}}$, for every distinct $c, d \in \mathrm{NI}_S$. Thus, we may assume that $\Delta^I = \mathrm{NI}_S$ and $c^{\mathcal{I}} = c$ for each $c \in \mathrm{NI}_S$. The *unique name assumption (UNA)* can be enforced by assertions $c \neq d$ for all constants in $\mathrm{NI} \setminus \mathrm{NI}_S$ resp. occurring in the knowledge base.

Moreover, we restrict to the case in which no axioms or defeasible axioms of the form $\top \sqsubseteq C$ occur: we will discuss in Sect. 6 how the presented results could be extended to the general case.

Note that negative assertions of the kind $\neg A(a)$ are not allowed by the syntax of $\mathcal{EL}_\perp$: however, these can be easily simulated by the axioms $C_a \sqcap A \sqsubseteq \perp, C_a(a)$ (with C_a a new atomic concept for the individual a). In the following, for simplicity of presentation, we can use such negative assertions in our definitions (in particular to identify negative information relative to an individual).

2.2   Datalog Programs and Answer Sets

As in the case of [10], we express our rules in *datalog* with *(default) negation* under answer sets semantics [18]: in particular, *(default) negation* not and its interpretation under answer set semantics is needed for the representation of defeasibility.

A *signature* is a tuple $\langle \mathbf{C}, \mathbf{P} \rangle$ of a finite set $\mathbf{C}$ of *constants* and a finite set $\mathbf{P}$ of *predicates*. We assume a set $\mathbf{V}$ of *variables*; the elements of $\mathbf{C} \cup \mathbf{V}$ are *terms*. An *atom* is of the form $p(t_1, \ldots, t_n)$ where $p \in \mathbf{P}$ and $t_1, \ldots, t_n$, are terms.

A *(datalog) rule* r is an expression of the form

$$a \leftarrow b_1, \ldots, b_k, \mathrm{not}\, b_{k+1}, \ldots, \mathrm{not}\, b_m. \tag{2}$$

where $a, b_1, \ldots, b_m$ are atoms and not is the negation as failure symbol (NAF). We denote with $Head(r)$ the head a of rule r and with $Body(r) = \{ b_1, \ldots, b_k, \mathrm{not}\, b_{k+1}, \ldots, \mathrm{not}\, b_m \}$ the body of r, respectively. A *(datalog) program* P is a finite set of rules.

An atom (rule etc.) is *ground*, if no variables occur in it. A *ground substitution* σ for $\langle \mathbf{C}, \mathbf{P} \rangle$ is any function $\sigma : \mathbf{V} \to \mathbf{C}$; the *ground instance* of an atom (rule, etc.) χ from σ, denoted $\chi\sigma$, is obtained by replacing in χ each occurrence of variable $v \in \mathbf{V}$ with $\sigma(v)$. A *fact* H is a ground rule r with empty body. The *grounding* of a rule r, $grnd(r)$, is the set of all ground instances of r, and the *grounding* of a program P is $grnd(P) = \bigcup_{r \in P} grnd(r)$.

Given a program P, the *(Herbrand) universe* U_P of P is the set of all constants occurring in P and the *(Herbrand) base* B_P of P is the set of all the ground atoms constructable from the predicates in P and the constants in U_P. An *interpretation* $I \subseteq B_P$ is any subset of B_P. An atom l is *true* in I, denoted $I \models l$, if $l \in I$.

Given a rule $r \in grnd(P)$, we say that $Body(r)$ is true in I, denoted $I \models Body(r)$, if (i) $I \models b$ for each atom $b \in Body(r)$ and (ii) $I \not\models b$ for each atom $\text{not } b \in Body(r)$. A rule r is *satisfied* in I, denoted $I \models r$, if either $I \models Head(r)$ or $I \not\models Body(r)$. An interpretation I is a *model* of P, denoted $I \models P$, if $I \models r$ for each $r \in grnd(P)$; moreover, I is *minimal*, if $I' \not\models P$ for each subset $I' \subset I$.

Given an interpretation I for P, the (Gelfond-Lifschitz) *reduct* of P w.r.t. I, denoted by $G_I(P)$ [18], is the set of rules obtained from $grnd(P)$ by (i) removing every rule r such that $I \models l$ for some $\text{not } l \in Body(r)$; and (ii) removing the NAF part from the bodies of the remaining rules. Then I is an *answer set* of P, if I is a minimal model of $G_I(P)$; the minimal model is unique and exists iff $G_I(P)$ has some model. Moreover, if I is an answer set for P, then I is a minimal model of P. We say that an atom $a \in B_P$ is a *consequence* of P and we write $P \models a$ iff for every answer set I of P we have that $I \models a$.

3   CKR Knowledge Bases with Defeasible Axioms on $\mathcal{EL}_\bot$

In this section, we review the definition for the syntax and semantics of CKR introduced in [9,10] by considering their application to the $\mathcal{EL}_\bot$ logic. While the definition of the syntax remains basically unchanged, further considerations are needed in the revision of the semantics due to the interpretation of existential axioms in presence of defeasible axioms.

3.1   Syntax

The CKR framework is defined as a two layered structure: the upper layer is a DL knowledge base $\mathfrak{G}$ representing the contextual structure (*meta-knowledge*) and (possibly defeasible) globally valid axioms (*global knowledge*); the lower layer contains a set of local contexts representing axioms and facts that are locally valid. To facilitate knowledge reuse, the local knowledge is organized as a set of knowledge modules K_m, sets of DL axioms that can be associated to one or more contexts: the associations between contexts and modules is specified in the meta-knowledge by means of the role mod. To specify the meta-knowledge of a CKR, we establish a DL vocabulary defining the elements of the contextual structure:

Definition 1 (meta-vocabulary). *A* meta-vocabulary *is a DL vocabulary* $\Gamma = \mathrm{NC}_\Gamma \uplus \mathrm{NR}_\Gamma \uplus \mathrm{NI}_\Gamma$ *of mutually disjoint sets* NC_Γ *of atomic concepts,* NR_Γ *of atomic roles, and* NI_Γ *of individual constants containing the following sets of symbols[1]:*

1. $\mathsf{N} \subseteq \mathrm{NI}_\Gamma$ *of context names;*
2. $\mathsf{M} \subseteq \mathrm{NI}_\Gamma$ *of module names;*
3. $\mathcal{C} \subseteq \mathrm{NC}_\Gamma$ *of context classes, including* Ctx
4. $\mathcal{R} \subseteq \mathrm{NR}_\Gamma$ *of contextual relations.*

Context classes in $\mathcal{C}$ are used to specify properties on types of contexts, while $\mathcal{R}$ contains relations that can be specified across contexts. The role mod $\in \mathrm{NR}_\Gamma$ defined on $\mathsf{N} \times \mathsf{M}$ defines associations between contexts and modules in the meta-knowledge.

Definition 2 (meta-language). *The* meta-language $\mathcal{L}_\Gamma$ *of a CKR is a DL language over* Γ *where axioms of the kind* $\mathsf{A} \sqsubseteq \exists \mathsf{R}.\mathsf{C}$ *for* $\mathsf{C} \in \mathcal{C}$ *are disallowed.*

Note that, differently from [10], in $\mathcal{EL}_\perp$ we can not directly assign modules to a context class with $\mathsf{C} = \exists \mathsf{mod}.\{\mathsf{m}\}$ and $\mathsf{m} \in \mathsf{M}$.

The knowledge in contexts of a CKR is expressed via a DL language called *object-language* $\mathcal{L}_\Sigma$ over an object-vocabulary $\Sigma = \mathrm{NC}_\Sigma \cup \mathrm{NR}_\Sigma \cup \mathrm{NI}_\Sigma$ akin to Γ. Intuitively, the expressions in $\mathcal{L}_\Sigma$ are evaluated locally in a (local) interpretation of the context of reference: we extend the language to access the interpretation of expressions inside other contexts as follows.

Definition 3 (object language with eval). *The language* $\mathcal{L}_\Sigma^e$ *extends* $\mathcal{L}_\Sigma$ *with* eval *expressions*

$$\mathsf{eval}(X, \mathsf{C}), \tag{3}$$

where X *is a concept or role expression of* $\mathcal{L}_\Sigma$ *and* C *is a concept expression of* $\mathcal{L}_\Gamma$ *(with* $\mathsf{C} \sqsubseteq \mathsf{Ctx}$*).*

The DL language $\mathcal{L}_\Sigma^e$ extends $\mathcal{L}_\Sigma$ with the set of eval-expressions in $\mathcal{L}_\Sigma$.

In the global context we allow for axioms of the object language $\mathcal{L}_\Sigma$ which are intended to be valid for all local contexts. We specify the global object axioms that are to be treated as defeasible, i.e. that allow for exceptions in the local contexts, as follows:

Definition 4 (defeasible axiom). *A* defeasible axiom *is any expression of the form* $\mathrm{D}(\alpha)$*, where* $\alpha \in \mathcal{L}_\Sigma$*.*

Definition 5 (object language with defeasible axioms). *The DL language* $\mathcal{L}_\Sigma^D$ *extends* $\mathcal{L}_\Sigma$ *with the set of defeasible axioms in* $\mathcal{L}_\Sigma$*.*

Using these language definitions, we are ready to provide a definition of contextualized repository.

Definition 6 (contextualized knowledge repository, CKR). *A* contextualized knowledge repository (CKR) *over a meta-vocabulary* Γ *and an object vocabulary* Σ *is a structure* $\mathfrak{K} = \langle \mathfrak{G}, \{\mathsf{K}_\mathsf{m}\}_{\mathsf{m} \in \mathsf{M}} \rangle$*, where:*

---

[1] Intuitively, Ctx will be used to denote the class of all contexts.

- $\mathfrak{G}$ is a DL knowledge base over $\mathcal{L}_\Gamma \cup \mathcal{L}_\Sigma^D$, and
- every K_m is a DL knowledge base over $\mathcal{L}_\Sigma^e$, for each module name $m \in \mathbf{M}$.

Furthermore, $\mathfrak{K}$ is an $\mathcal{EL}_\perp$ CKR, if $\mathfrak{G}$ and all K_m are knowledge bases over the extended language of $\mathcal{EL}_\perp$ where eval-expressions can occur only in left-concepts.

In the following, we tacitly focus on $\mathcal{EL}_\perp$ CKRs.

Example 1. We introduce a simple example showing the definition and interpretation of a defeasible existential axiom. In the organization of a university, we want to specify that "in general" members of a research department need also to have at least one course which they are teaching. However, in the context of the Computer Science department, we want to specify that PhD students, while recognized as members of the department, are not allowed to be holder of a course. We can represent this scenario as a CKR $\mathfrak{K}_{dept} = \langle \mathfrak{G}, \{K_{csd\_m}\} \rangle$ where:

$$\mathfrak{G} : \left\{ \begin{array}{l} D(DepartmentMember \sqsubseteq \exists hasCourse.Course), \\ \mathsf{mod}(\mathsf{cs\_dept}, \mathsf{csd\_m}) \end{array} \right\}$$

$$K_{\mathsf{csd\_m}} : \left\{ \begin{array}{l} Professor \sqsubseteq DepartmentMember, \\ PhDStudent \sqsubseteq DepartmentMember, \\ PhDStudent \sqcap \exists hasCourse.Course \sqsubseteq \perp, \\ Professor(alice), PhDStudent(bob) \end{array} \right\}$$

Intuitively, at the local context cs_dept, given the local definition, we want to override the fact that there exists some course assigned to the PhD student *bob*. However, for the individual *alice* no overriding should happen and the global axiom can be applied.   ◇

3.2 Semantics

We can now present the model-based interpretation of CKR: in particular, with respect to [10], in the instantiation of the framework with the logic $\mathcal{EL}_\perp$ we need further attention to the interpretation of exceptions to defeasible axioms. This is due to existential axioms of the form $A \sqsubseteq \exists R.B$ which, intuitively, allow one to make reference to "unnamed" elements of the domain. Thus, we need to formulate our semantics definition in a way that it preserves the properties from [10], where we can concentrate on exceptions only on the "named" elements of the domain. We first provide the definition of CKR interpretations.

Definition 7 (CKR interpretation). *A CKR interpretation for $\langle \Gamma, \Sigma \rangle$ is a structure $\mathfrak{I} = \langle \mathcal{M}, \mathcal{I} \rangle$ where*

(i) *$\mathcal{M}$ is a DL interpretation of $\Gamma \cup \Sigma$ such that $c^\mathcal{M} \in \mathsf{Ctx}^\mathcal{M}$, for every $c \in \mathbf{N}$, and $C^\mathcal{M} \subseteq \mathsf{Ctx}^\mathcal{M}$, for every $C \in \mathcal{C}$;*

(ii) *for every $x \in \mathsf{Ctx}^\mathcal{M}$, $\mathcal{I}(x)$ is a DL interpretation over Σ s.t., $\Delta^{\mathcal{I}(x)} = \Delta^\mathcal{M}$ and $a^{\mathcal{I}(x)} = a^\mathcal{M}$, for every $a \in \mathrm{NI}_\Sigma$.*

The interpretation of ordinary DL expressions on $\mathcal{M}$ and $\mathcal{I}(x)$ in $\mathfrak{I} = \langle \mathcal{M}, \mathcal{I} \rangle$ is as usual; *eval* expressions are interpreted as follows: for every $x \in \mathsf{Ctx}^\mathcal{M}$,

$$eval(X, \mathsf{C})^{\mathcal{I}(x)} = \bigcup_{e \in \mathsf{C}^\mathcal{M}} X^{\mathcal{I}(e)}$$

As we have shown in the case of $\mathcal{SROIQ}$-RL [10], we can express $\mathcal{EL}_\perp$ knowledge bases in first-order (FO) logic, where every axiom $\alpha \in \mathcal{L}_\Sigma$ is translated into an equivalent FO-sentence $\forall \boldsymbol{x}.\phi_\alpha(\boldsymbol{x})$ where $\boldsymbol{x}$ contains all free variables of ϕ_α depending on the type of the axiom. In particular given an existential axiom of the kind $\alpha = A \sqsubseteq \exists R.B$, its FO-translation $\phi_\alpha(\boldsymbol{x})$ is defined as:

$$A(x_1) \rightarrow R(x_1, f_\alpha(x_1)) \wedge B(f_\alpha(x_1)).$$

The cases for other axiom types can be defined analogously to the FO-translation presented in [10]. Note that in order to treat right existential formulas we need to introduce a Skolem function $f_\alpha(x_1)$ which provides new "existential" individuals (depending on the existential axiom α and instance x_1). Formally, for every right existential axiom $\alpha \in \mathcal{L}_\Sigma \cup \mathcal{L}_\Gamma$, we define a Skolem function $f_\alpha : \mathrm{NI} \mapsto \mathcal{E}$ where $\mathcal{E}$ is a set of new individual constants not appearing in NI. In particular, for a set of individual names $N \subseteq \mathrm{NI}$, we will write $sk(N)$ to denote the extension of N with the set of Skolem constants for elements in N.

After this transformation, as in the case of $\mathcal{SROIQ}$-RL, the resulting formulas $\phi_\alpha(\boldsymbol{x})$ amount semantically to Horn formulas, since left-side concepts C can be expressed by an existential positive FO-formula, and right-side concepts D by a conjunction of Horn clauses. As in [10], axioms are contextualized by extending the translation with a further argument x_c for the context, such that the formula $\forall \boldsymbol{x}.\phi_\alpha(\boldsymbol{x}, x_c)$ expresses the axiom α within context x_c. Furthermore, this translation can be extended to $\mathcal{L}_\Sigma^e$ such that the Horn property is maintained for $\mathcal{EL}_\perp$, due to the restrictions on the occurrence of *eval* expressions.

The following property from [10] is then preserved for such translation:

Lemma 1. *For any DL knowledge base $\mathcal{K}$ over $\mathcal{L}_\Gamma$ (resp. $\mathcal{L}_\Sigma^e$), its FO-translation (resp. its contextualized FO-translation)*

$$\phi_\mathcal{K} := \bigwedge_{\alpha \in \mathcal{K}} \forall \boldsymbol{x}\phi_\alpha(\boldsymbol{x}) \quad (\textit{resp.} \quad \phi_{\mathcal{K}, x_c} := \bigwedge_{\alpha \in K} \forall \boldsymbol{x}\phi_\alpha(\boldsymbol{x}, x_c)) \qquad (4)$$

is semantically equivalent to a conjunction of universal Horn clauses.

It is important to note that the elements of the kind $f_\alpha(a)$ do not identify a single named domain element (i.e. that can be uniquely mapped to an individual name in NI in every model of the axiom), but can be interpreted differently in different interpretations satisfying α. In the following, we may write such elements in ABox assertions to denote these abstract (i.e. existentially quantified) individuals: namely, we can say that $\mathcal{I} \models \exists R.B(a)$ iff $\mathcal{I} \models R(a, f_\alpha(a))$ and $\mathcal{I} \models B(f_\alpha(a))$. With these considerations on the definition of the FO-translation, we can now revise our definition of axiom instantiation and clashing assumptions:

Definition 8 (axiom instantiation). *Given an axiom* $\alpha \in \mathcal{L}_\Sigma$ *with FO-translation* $\forall \boldsymbol{x}.\phi_\alpha(\boldsymbol{x})$, *the instantiation of* α *with a tuple* $\mathbf{e}$ *of individuals in* NI_Σ, *written* $\alpha(\mathbf{e})$, *is the specialization of* α *to* $\mathbf{e}$, *i.e.,* $\phi_\alpha(\mathbf{e})$, *depending on the type of* α.

Note that, since we are assuming standard names, this basically means that we can express instantiations (and exceptions) to any element of the domain (identified by a standard name in NI_Σ). We next introduce clashing assumptions and clashing sets.

Definition 9 (clashing assumptions and sets). *A* clashing assumption *is a pair* $\langle \alpha, \mathbf{e} \rangle$ *such that* $\alpha(\mathbf{e})$ *is an axiom instantiation for an axiom* $\alpha \in \mathcal{L}_\Sigma$. *A* clashing set *for a clashing assumption* $\langle \alpha, \mathbf{e} \rangle$ *is a satisfiable set* S *that consists of ABox assertions over* $\mathcal{L}_\Sigma$ *and negated ABox assertions of the forms* $\neg C(a)$ *and* $\neg R(a, b)$, *where in* C *no conjunction* $\sqcap$ *and no qualified role restriction occurs (i.e. $R.B$ implies $B = \top$), such that* $S \cup \{\alpha(\mathbf{e})\}$ *is unsatisfiable.*

A clashing assumption $\langle \alpha, \mathbf{e} \rangle$ represents (the assumption) that $\alpha(\mathbf{e})$ is not satisfiable, while a clashing set S provides an assertional "justification" for the assumption of local overriding of α on $\mathbf{e}$. We allow for negated ABox assertions in order to make violations of axioms such as $A \sqsubseteq B$ for a possible; here a clashing set would be $\{A(a), \neg B(a)\}$. The restricted form of $\neg C(a)$ excludes disjunction, so that we have true factual evidence of the form $\neg A(a)$ or $\neg \exists R.\top(a)$.

Then, we extend the notion of CKR interpretation with a set of clashing assumptions for each local context:

Definition 10 (CAS-interpretation). *A CAS-interpretation is a structure* $\mathfrak{I}_{CAS} = \langle \mathcal{M}, \mathcal{I}, \chi \rangle$ *where* $\mathfrak{I} = \langle \mathcal{M}, \mathcal{I} \rangle$ *is a CKR interpretation and* χ *maps every* $x \in \Delta^{\mathcal{M}}$ *to a set* $\chi(x)$ *of clashing assumptions for* x.

By extending the notion of satisfaction with respect to CAS-interpretations, we can disregard the application of defeasible axioms to the exceptional elements in the sets of clashing assumptions. For convenience, we call two DL interpretations $\mathcal{I}_1$ and $\mathcal{I}_2$ (resp. CAS-interpretations $\mathfrak{I}^i_{CAS} = \langle \mathcal{M}_i, \mathcal{I}_i, \chi_i \rangle$, $i \in \{1, 2\}$) NI-*congruent*, if $c^{\mathcal{I}_1} = c^{\mathcal{I}_2}$ (resp. $c^{\mathcal{M}_1} = c^{\mathcal{M}_2}$) holds for every $c \in \mathrm{NI}$.

Definition 11 (CAS-model). *Given a CKR* $\mathfrak{K} = \langle \mathfrak{G}, \{K_m\}_{m \in \mathsf{M}} \rangle$, *a CAS-interpretation* $\mathfrak{I}_{CAS} = \langle \mathcal{M}, \mathcal{I}, \chi \rangle$ *is a CAS-model for* $\mathfrak{K}$ *(denoted* $\mathfrak{I}_{CAS} \models \mathfrak{K}$*), if the following holds:*

(i). for every $\alpha \in \mathcal{L}_\Sigma \cup \mathcal{L}_\Gamma$ *in* $\mathfrak{G}$, $\mathcal{M} \models \alpha$;
(ii). for every $\mathrm{D}(\alpha) \in \mathfrak{G}$ *(where* $\alpha \in \mathcal{L}_\Sigma$*),* $\mathcal{M} \models \alpha$;
(iii). for every $\langle x, y \rangle \in \mathsf{mod}^{\mathcal{M}}$ *such that* $y = m^{\mathcal{M}}$, $\mathcal{I}(x) \models K_m$;
(iv). for every $\alpha \in \mathfrak{G} \cap \mathcal{L}_\Sigma$ *and* $x \in \mathsf{Ctx}^{\mathcal{M}}$, $\mathcal{I}(x) \models \alpha$, *and*
(v). for every $\mathrm{D}(\alpha) \in \mathfrak{G}$ *(where* $\alpha \in \mathcal{L}_\Sigma$*),* $x \in \mathsf{Ctx}^{\mathcal{M}}$, *and* $|\boldsymbol{x}|$-*tuple* $\boldsymbol{d}$ *of elements in* NI_Σ *such that* $\boldsymbol{d} \notin \{\mathbf{e} \mid \langle \alpha, \mathbf{e} \rangle \in \chi(x)\}$, *we have* $\mathcal{I}(x) \models \phi_\alpha(\boldsymbol{d})$.

We want to have only the assumptions on exceptional elements that have a provable evidence from the contents of the initial CKR: thus, we are interested in models in which all assumptions are *justified*. Formally, we say that a clashing assumption $\langle \alpha, \mathbf{e} \rangle \in \chi(x)$ is *justified* for a *CAS* model $\mathfrak{I}_{CAS} = \langle \mathcal{M}, \mathcal{I}, \chi \rangle$, if some clashing set $S = S_{\langle \alpha, \mathbf{e} \rangle, x}$ for $\langle \alpha, \mathbf{e} \rangle$ and context x exists such that, for every CAS-model $\mathfrak{I}'_{CAS} = \langle \mathcal{M}', \mathcal{I}', \chi \rangle$ of $\mathfrak{K}$ that is NI-congruent with $\mathfrak{I}_{CAS}$, it holds that $\mathcal{I}'(x) \models S_{\langle \alpha, \mathbf{e} \rangle, x}$. CKR models are then only the CAS-models in which all of the clashing assumptions are justified.

Definition 12 (justified CAS model and CKR model). *A CAS model* $\mathfrak{I}_{CAS} = \langle \mathcal{M}, \mathcal{I}, \chi \rangle$ *of a CKR* $\mathfrak{K}$ *is* justified, *if every* $\langle \alpha, \mathbf{e} \rangle \in \bigcup_{x \in \mathsf{Ctx}^{\mathcal{M}}} \chi(x)$ *is justified. An interpretation* $\mathfrak{I} = \langle \mathcal{M}, \mathcal{I} \rangle$ *is a* CKR model *of* $\mathfrak{K}$ *(in symbols,* $\mathfrak{I} \models \mathfrak{K}$*), if* $\mathfrak{K}$ *has some justified CAS model* $\mathfrak{I}_{CAS} = \langle \mathcal{M}, \mathcal{I}, \chi \rangle$.

Example 2. We can now show an example of CKR model satisfying the CKR $\mathfrak{K}_{dept}$ from Example 1. Considering the contents of the CKR, a model providing the intended interpretation of defeasible axioms is $\mathfrak{I}_{CAS_{dept}} = \langle \mathcal{M}, \mathcal{I}, \chi_{dept} \rangle$, where:

$$\chi_{dept}(\mathsf{cs\_dept}^{\mathcal{M}}) = \{ \langle \alpha, bob \rangle \}$$

with $\alpha = DepartmentMember \sqsubseteq \exists hasCourse.Course$. The fact that this model is justified is easily verifiable considering the clashing set S:

$$\mathcal{I}(\mathsf{cs\_dept}^{\mathcal{M}}) \models S \text{ for } S = \{ DepartmentMember(bob), \neg\exists hasCourse.\top(bob) \}$$

On the other hand, note that a similar clashing assumption for the individual *alice* is not justifiable: it is not possible from the contents of $\mathfrak{K}_{dept}$ to derive a clashing set S' such that $S' \cup \{\alpha(alice)\}$ is unsatisfiable. Indeed, by Definition 11, this allows us to apply the defeasible axiom to this individual as expected and thus $\mathcal{I}(\mathsf{cs\_dept}^{\mathcal{M}}) \models \exists hasCourse.Course(alice)$. $\Diamond$

4 Properties

In this section, we consider some properties of the notion of justified CAS-models as defined above, where we look at both semantic and computational properties.

4.1 Semantic Properties

As for semantic properties, we recall first properties of CAS-models and justified CAS-models for $\mathcal{SROIQ}$-RL from [10]:

P1: Irrelevance of Syntax. Suppose $\mathfrak{K} = \langle \mathfrak{G}, \{K_m\}_{m \in M} \rangle$ has in $\mathfrak{G}$ a defeasible axiom $D(\alpha)$. If $\beta \in \mathcal{L}_\Sigma$ satisfies $\phi_\alpha(\boldsymbol{x}) \equiv \phi_\beta(\boldsymbol{x})$ (i.e., β is of the same genus and logically equivalent to α), then $\mathfrak{K}$ and $\mathfrak{K}' = \langle (\mathfrak{G} \setminus \alpha) \cup \{\beta\}, \{K_m\}_{m \in M} \rangle$ have the same CKR-models.

P2: Non-Monotonicity. Suppose $\mathfrak{I}_{CAS} = \langle \mathcal{M}, \mathcal{I}, \chi \rangle$ is a justified CAS-model of a CKR $\mathfrak{K}'$. Then $\mathfrak{I}_{CAS}$ is not necessarily a justified CAS-model of every $\mathfrak{K} \subseteq \mathfrak{K}'$.

P3: Context Focus. Suppose $\mathfrak{I}_{CAS} = \langle \mathcal{M}, \mathcal{I}, \chi \rangle \models \mathfrak{K}$ for a CAS-interpretation of a CKR $\mathfrak{K}$ and that χ' coincides with χ on $\mathsf{Ctx}^{\mathcal{M}}$. Then $\mathfrak{I}'_{CAS} = \langle \mathcal{M}, \mathcal{I}, \chi' \rangle$ $\models \mathfrak{K}$. Furthermore, if $\mathfrak{I}_{CAS}$ is justified, then also $\mathfrak{I}'_{CAS}$ is justified.

P4: Minimality of Justification. Suppose that $\mathfrak{I}_{CAS} = \langle \mathcal{M}, \mathcal{I}, \chi \rangle$ and $\mathfrak{I}'_{CAS} = \langle \mathcal{M}', \mathcal{I}', \chi' \rangle$ are justified CAS-models of a CKR $\mathfrak{K}$ that are NI-congruent. Then, $\mathsf{Ctx}^{\mathcal{M}} = \mathsf{Ctx}^{\mathcal{M}'}$ and $\chi'(x) \subseteq \chi(x)$ for every $x \in \mathsf{Ctx}^{\mathcal{M}}$ implies $\chi = \chi'$.

P5: Intersection Property. Let $\mathfrak{I}^i_{CAS} = \langle \mathcal{M}_i, \mathcal{I}_i, \chi \rangle, i \in \{1,2\}$ be NI-congruent CAS-models of a CKR $\mathfrak{K}$. Then $\mathfrak{I}_{CAS} = \langle \mathcal{M}, \mathcal{I}, \chi \rangle$ where $\mathcal{M} = \mathcal{M}_1 \cap \mathcal{M}_2$ and $\mathcal{I} = \mathcal{I}_1 \cap \mathcal{I}_2$ is the intersection of the $\mathcal{M}_i$ resp. $\mathcal{I}_i$, is also a CAS-model of $\mathfrak{K}$. Furthermore, $\mathfrak{I}_{CAS}$ is justified if some $\mathfrak{I}^i_{CAS}$ is justified, $i \in \{1, 2\}$.

P6: Least Model Property. Let a *name assignment* be a function $\nu : \mathrm{NI} \to \Delta$ respecting SNA. In particular, the name assignment of a CAS-interpretation $\mathfrak{I}_{CAS} = \langle \mathcal{M}, \mathcal{I}, \chi \rangle$ is the one induced by $\mathrm{NI}^{\mathcal{M}}$. We say that χ for a CKR $\mathfrak{K}$ is *satisfiable* (resp., *justified*) for a name assignment ν, if $\mathfrak{K}$ has some CAS-model (resp., justified CAS-model) $\mathfrak{I}_{CAS}$ on χ with name assignment ν. Then, if a clashing assumption map χ for a CKR $\mathfrak{K}$ is satisfiable for name assignment ν, $\mathfrak{K}$ has a least (unique minimal) CAS-model $\hat{\mathfrak{I}}_{\mathfrak{K}}(\chi, \nu) = \langle \hat{\mathcal{M}}, \hat{\mathcal{I}}, \chi \rangle$ w.r.t. inclusion $\mathcal{M}' \subseteq \mathcal{M}$ and $\mathcal{I}' \subseteq \mathcal{I}$ for ν. Furthermore, $\hat{\mathfrak{I}}_{\mathfrak{K}}(\chi, \nu)$ is justified if χ is justified.

Examining these properties, we find that P1–P4 hold for $\mathcal{EL}_\perp$ as well; this can be shown with arguments similar to those in [10]. On the other hand, property P5 does not hold since the intersection of models (where $A^{\mathcal{I}} = A^{\mathcal{I}_1} \cap A^{\mathcal{I}_2}$, $R^{\mathcal{I}} = R^{\mathcal{I}_1} \cap R^{\mathcal{I}_2}$) may not yield a model due to existential axioms $A \sqsubseteq \exists R.B$. However, under skolemization, where the Skolem functions $f_\alpha(\cdot)$ are part of the signature, property P5 holds if they coincide, that is $f_\alpha^{\mathcal{I}_1} = f_\alpha^{\mathcal{I}_2}$. Similarly P6 does not hold in general but when the Skolem functions coincide; i.e., relative to an interpretation of the individuals and the witnesses proving existential axioms, a least model does exist.

We next consider the restriction of models to their named part, where we use the skolemized form. Given a set $N \subseteq \mathrm{NI} \setminus \mathrm{NI}_S$ of individual names and an DL interpretation $\mathcal{I}$, denote by $\mathcal{I}^{sk(N)}$ the restriction of $\mathcal{I}$ to the elements of $sk(N)$; that is, all unnamed elements are (virtually) removed from $\mathcal{I}$.

Lemma 2. *Suppose $\mathcal{I}$ is a model of a $\mathcal{EL}_\perp$ knowledge base $\mathcal{K}$ and $N \subseteq \mathrm{NI} \setminus \mathrm{NI}_S$ includes all individuals occurring in $\mathcal{K}$. Then the Skolem-restriction $\mathcal{I}^{sk(N)}$ is named w.r.t. $sk(N)$ and a model of $\mathcal{K}$.*

Armed with this, we find that we can, similar as in the case of $\mathcal{SROIQ}$-RL-knowledge bases, restrict CAS-models and CKR models to their $sk(N)$-part. Let $N_{\mathfrak{K}}$ denote the set of all individuals that occur in a CKR $\mathfrak{K}$, then:

Theorem 1 (Skolem named model focus). *Let $\mathfrak{I}_{CAS}$ be a CAS-model of $\mathfrak{K}$ and suppose $N_{\mathfrak{K}} \subseteq N \subseteq \mathrm{NI} \setminus \mathrm{NI}_S$. Then, also $\mathfrak{I}_{CAS}^{sk(N)}$, and in particular $\mathfrak{I}_{CAS}^{sk(N_{\mathfrak{K}})}$, is a CAS-model for $\mathfrak{K}$. Furthermore, $\mathfrak{I}_{CAS}^{sk(N)}$ is justified if $\mathfrak{I}_{CAS}$ is justified, and every clashing assumption $\langle \alpha, \mathbf{e} \rangle$ in $\mathfrak{I}_{CAS}^{sk(N)}$ is justified by some clashing set S formulated with terms from $sk(N)$.*

In fact, it turns out that for justified CAS-models an even stronger version of Theorem 1 is possible. This is because no exceptions for unnamed individuals can be made.

Proposition 1. *Suppose* $\mathfrak{I}_{CAS} = \langle \mathcal{M}, \mathcal{I}, \chi \rangle$ *is a justified CAS-model of a CKR* $\mathfrak{K}$. *Then, for every clashing assumption* $\langle \alpha, \mathbf{e} \rangle \in \chi(c)$ *with* $c \in \mathsf{Ctx}^{\mathcal{M}}$, $\mathbf{e}$ *is named, i.e. for all elements* $e \in \mathbf{e}$, *there exists an individual name* $a \in \mathrm{NI}_\Sigma$ *such that* $e = a^{\mathcal{I}(c)}$.

This allows us to reason only on the named part of the model when dealing with clashing assumptions.

An interesting consequence of this property is that it is possible to define nominals $\{a\}$ by the use of defeasible axioms under the unique names assumption. To this end, we can introduce a fresh concept name and make assertions $C_{\{a\}}(a)$ and $\neg C_{\{a\}}(b)$ for every individual b different from a in the ABox; and we add the $D(C_{\{a\}} \sqsubseteq \perp)$. Since no exceptions on unnamed individuals are possible, in any justified CAS-model $C_{\{a\}}$ must amount to $\{a\}$.

4.2 Complexity

As regards the satisfiability problem, clearly a CKR $\mathfrak{K}$ has some CAS-model if the CKR $\mathfrak{K}'$ that results from dropping all defeasible axioms from $\mathfrak{K}$ (i.e., make each defeasible axiom instance an exception) has some CAS model. As satisfiability of an $\mathcal{EL}_\perp$ knowledge base is decidable in polynomial time (and in fact PTime-complete), the following can be established.

Theorem 2. *Given a CKR* $\mathfrak{K} = \langle \mathfrak{G}, \{K_m\}_{m \in \mathbf{M}} \rangle$, *deciding whether* $\mathfrak{K}$ *has some CAS-model is* PTime-*complete*.

The membership in PTime can be argued from the datalog translation that we provide in Sect. 5, as it can be used to map $\mathfrak{K}$ into a datalog program that is evaluable in polynomial time.

On the other hand, satisfiability under CKR-models is more complex, as exceptions must be justified. In case of inconsistent axioms, this leads to alternatives: e.g., given $D(A(a))$ and $D(\neg A(a))$, we have the option to either make an exception for $A(a)$ or for $\neg A(a)$. This choice mechanism results in intractability.

Theorem 3. *Given a CKR* $\mathfrak{K} = \langle \mathfrak{G}, \{K_m\}_{m \in \mathbf{M}} \rangle$, *deciding whether* $\mathfrak{K}$ *has some justified CAS-model resp. some CKR-model is* NP-*complete*.

Proof (Sketch). As for membership in NP, if $\mathfrak{K}$ has some CKR model, then a justified CAS-model $\mathfrak{I}_{CAS} = \langle \mathcal{M}, \mathcal{I}, \chi \rangle$ of $\mathfrak{K}$ named relative to $sk(N)$ exists, for $N = N_{\mathfrak{K}} \subseteq \mathrm{NI} \setminus \mathrm{NI}_S$. The clashing assumptions χ can be guessed (they are over $N_{\mathfrak{K}}$), along with clashing sets $S_{\alpha(\mathbf{e})}$ for each clashing assumption $\langle \alpha, \mathbf{e} \rangle$, as well as a partial interpretation over N. One then can check whether each clashing set $S_{\alpha(\mathbf{e})}$ is derivable and whether the partial interpretation can be extended to a model of $\mathfrak{K}$ relative to $sk(N)$ using the chase procedure (respectively, the materialization calculus). Each such test is feasible in polynomial time, and there is a polynomial number of such tests. Thus, we overall obtain membership in NP. Notably, the datalog translation in Sect. 5 implements this membership test.

The NP-hardness part is shown by a reduction from not-all-equal 3SAT (NAE3SAT), and in fact for a fixed set of inclusion axioms. Let $E = \bigwedge_{i=1}^{m} \gamma_i$ be an instance of NAE3SAT over propositional atoms $X = \{x_1, \ldots, x_n\}$, i.e. a CNF where each clause contains three literals. An assignment σ to X NAE-satisfies E, if in each clause σ evaluates some literal to true and some literal to false.

Without loss of generality, each clause γ_i in E is positive. We then construct $\mathfrak{K}$ as follows, where V, F, T, A are concepts, R is a role, and $x_1, \ldots, x_n, c_1, \ldots, c_m$ are individual constants.

- the global knowledge $\mathfrak{G}$ contains defeasible axioms $D(V \sqsubseteq T)$, $D(V \sqsubseteq F)$, $D(T \sqsubseteq \bot)$, $D(F \sqsubseteq \bot)$ and a module association $\mathsf{mod}(\mathsf{m}, \mathsf{c})$;
- a single module K_m that contains the following inclusion axioms:
    - $T \sqcap F \sqsubseteq \bot$,
    - $A \sqsubseteq \exists R.T$, and $A \sqsubseteq \exists R.F$.
  K_m contains the following assertions
    - $V(x_i)$, $i = 1, \ldots, n$,
    - $A(c_j)$, $\neg T(c_j)$, $\neg F(c_j)$, for $j = 1, \ldots, m$, and
    - $R(c_i, x_{i_j})$ for $i = 1, \ldots, m$ and $j = 1, 2, 3$, such that the clause γ_i is of form $x_{i_1} \vee x_{i_2} \vee x_{i_3}$.

Intuitively, at context c we must make for each atom x_h either an exception to $V \sqsubseteq F$ (then x_h is true) or to $V \sqsubseteq T$ (then x_h is false); the respective minimal clashing set is $\{V(x_h), \neg F(x_h)\}$ resp. $\{V(x_h), \neg T(x_h)\}$. To justify $\neg F(x_h)$ (resp. $\neg T(x_h)$), we can keep the axiom $F \sqsubseteq \bot$ (resp. $T \sqsubseteq \bot$) for x_h. Note that we can make an exception to both $V \sqsubseteq F$ or $V \sqsubseteq T$; this means that x_h is unassigned. On the other hand, since for no unnamed individual defeasible axioms can have an exception, each unnamed individual e must belong to $\neg F$ and $\neg T$.

To have a (justified) model, the axiom $A \sqsubseteq \exists R.T$ must be satisfied for each c_i: as T is false at all elements except for $x_1, \ldots, x_n$, we have that for some j it holds that x_{i_j} belongs to T; analogously, for $A \sqsubseteq \exists R.F$ we have that some x'_{i_j} belongs to F, Thus, if a justified CAS-model exists, then the formula E is NAE-satisfiable. Conversely, from an NAE-satisfying assignment of E, we can construct a justified CAS model in the intuitive way. In conclusion, NP-hardness under data complexity is established.

Both Theorems 2 and 3 hold also for data complexity, i.e., if the module structure in $\mathfrak{K}$ is fixed and only the assertions in the modules K_m vary; for CAS-satisfiability, the hardness is inherited from $\mathcal{EL}_\bot$.

As for reasoning, we recall the entailment problem for CKR from [10].

Definition 13 (c-entailment, global entailment). *Given a CKR $\mathfrak{K}$ over $\langle \Gamma, \Sigma \rangle$, an axiom $\alpha \in \mathcal{L}_\Sigma^e$ is (i) is c-entailed by $\mathfrak{K}$ for $\mathsf{c} \in N_\mathfrak{K}$ (denoted $\mathfrak{K} \models \mathsf{c} : \alpha$) if $\mathcal{I}(\mathsf{c}^\mathcal{M}) \models \alpha$ for every CKR-model $\mathfrak{I} = \langle \mathcal{M}, \mathcal{I} \rangle$ of $\mathfrak{K}$, and (ii) is (globally) entailed by $\mathfrak{K}$ (denoted $\mathfrak{K} \models \alpha$) if $\mathfrak{K} \models \mathsf{c} : \alpha$ for every $\mathsf{c} \in \mathbf{N}$. Furthermore, an axiom $\alpha \in \mathcal{L}_\Gamma$ is entailed by $\mathfrak{K}$ if $\mathcal{M} \models \alpha$ for every CKR-model $\mathfrak{I} = \langle \mathcal{M}, \mathcal{I} \rangle$ of $\mathfrak{K}$.*

The complexity of entailment checking is dual to the one of satisfiability.

Theorem 4. *Given a CKR $\mathfrak{K} = \langle \mathfrak{G}, \{K_m\}_{m\in M}\rangle$, a context name c and an axiom α, deciding whether $\mathfrak{K}$ c-entails (resp. globally entails) α is co-NP-complete.*

Proof (Sketch). In order to refute $\mathfrak{K} \models c : \alpha$, it follows from Theorem 1 that a justified CAS-model $\mathfrak{I}_{CAS} = \langle \mathcal{M}, \mathcal{I}, \chi\rangle$ of $\mathfrak{K}$ named relative to $sk(N)$ exists, with $N_{\mathfrak{K}} \subseteq N \subseteq \mathrm{NI} \setminus \mathrm{NI}_S$, such that $\mathcal{I}(c^{\mathcal{I}}) \not\models \alpha$, where N includes fresh individual names such that $\mathcal{I}(c^{\mathcal{I}})$ violates the instance of α for some elements e over $sk(N)$. Similarly as for satisfiability testing, one can guess χ, clashing assumptions $S_{\alpha(e)}$ and a partial interpretation over N, and check derivability of all $S_{\alpha(e)}$ and that the interpretation extends to a model of $\mathfrak{K}$ relative to $sk(N)$ in polynomial time. Thus, we overall obtain membership of c-entailment in co-NP.

The co-NP-hardness is immediate from Theorem 3, since $\mathfrak{K}$ in c entails $\perp(a)$ iff $\mathfrak{K}$ has no CKR-model. For global entailment, the proof is similar.   □

As above, the co-NP-hardness holds also under data complexity and assertional queries α. In conclusion, we obtain for $\mathcal{EL}_\perp$ similar complexity characteristics as for $\mathcal{SROIQ}$-RL on the above reasoning tasks. We omit here conjunctive query answering, but also this problem is expected to have the same complexity, and thus to be Π_2^p-complete.

5   Datalog Translation

Following the line of work in [10], we present a translation of reasoning from $\mathcal{EL}_\perp$ CKRs with defeasible axioms to datalog. The translation provides a reasoning method for positive instance queries under c-entailment (resp. global entailment).

As in the case of $\mathcal{SROIQ}$-RL, we will limit ourselves to the fragment of $\mathcal{EL}_\perp$ in which $D \sqcap D$ can not appear as a right-side concept.[2] For the interpretation of right-hand side existential axioms, we follow the original approach of [26]: for every existential axiom of the kind $\alpha = A \sqsubseteq \exists R.B$, an auxiliary abstract individual aux^α is introduced in the translation to represent the class of all R-successors introduced by α.

We introduce a normal form for axioms of $\mathcal{EL}_\perp$ CKRs, so that we can represent contents of the CKRs as datalog facts.

Definition 14. *A CKR $\mathfrak{K} = \langle \mathfrak{G}, \{K_m\}_{m\in M}\rangle$ is in* normal form, *if every non-defeasible axiom in $\mathfrak{G}$ and K_m matches a form in Table 1, and every defeasible axiom in $\mathfrak{G}$ is of the form $D(\alpha)$ where α is of the form (I) in Table 1.*

As in [10], we can easily provide a set of rules to transform any $\mathcal{EL}_\perp$ CKR into normal form and show that the rewritten CKR is "equivalent" to the original.

---

[2] This restriction allows us to simplify the characterization of the datalog encoding: this is demonstrated as an example in [10].

Table 1. Normal form for $\mathfrak{G}$ axioms from $\mathcal{L}_\Gamma \cup \mathcal{L}_\Sigma$ (I) and for K$_m$ axioms from $\mathcal{L}_\Sigma$ (I) and $\mathcal{L}_\Sigma^e \setminus \mathcal{L}_\Sigma$ (II)

(I) for $A, B, C \in \mathcal{C}$ (resp., $\in$ NC$_\Sigma$), $R \in \mathcal{R}$ (resp., $\in$ NR$_\Sigma$), $a, b \in \mathbf{N}$ (resp., $\in$ NI$_\Sigma$):
$\quad A(a) \qquad R(a,b) \qquad A \sqsubseteq B \qquad A \sqcap B \sqsubseteq C \qquad \exists R.A \sqsubseteq B \qquad A \sqsubseteq \exists R.B$

(II) for $A, B \in$ NC$_\Sigma$ and $C \in \mathcal{C}$:
$$eval(A, C) \sqsubseteq B$$

5.1 Translation Rules Overview

We can now present the components of our datalog translation for $\mathcal{EL}_\perp$ based CKRs. The encoding builds on the translation of CKR with defeasible axioms into datalog introduced in [9,10], which extended the encoding without defeasibility proposed in [13]. These translations were an adaptation to the DL $\mathcal{SROIQ}$-RL and the structure of CKR of the techniques used by Krötzsch [26] in defining the materialization calculus K_{inst} for instance checking in the description logic $\mathcal{SROEL}(\sqcap, \times)$.

As noted in [10], the extension of the translation to defeasible axioms is non-trivial, as it requires us to deal with strong negation, as provable falsity of assertions is needed for clashing sets. The extension of the materialization calculus to conclude negative literals requires us to deal with negative disjunctive information: however, encoding this reasoning in form of disjunctive datalog rules it is not only undesirable, as it may generate a large number of models, but also in general not complete, i.e. it is not sufficient to derive all negative consequences needed to prove justifications. As a solution, we encode inference of negative literals as individual proofs by contradiction: these "tests" on negative literals will be indicated by presence of the atom $\mathsf{unsat}(\cdots)$ for the literal in the answer set and, from its the absence in the model, we can conclude that the literal is not derivable. In this new translation for $\mathcal{EL}_\perp$, as we show in the following, this mechanism has to be adapted to the case of proofs by contradiction on the negative information that appears in clashing sets for existential axioms.

In the following, for each component, we describe the newly introduced rules for managing the interpretation of $\mathcal{EL}_\perp$ constructs.

(i). $\mathcal{EL}_\perp$ input translation: the set of rules in I_{el} translate to datalog facts the $\mathcal{EL}_\perp$ axioms and signature of each context from the input CKR. The rules of I_{el} are listed in Table 2. In the case of existential axioms, these are translated with the rule $A \sqsubseteq \exists R.B \mapsto \{\mathsf{supEx}(A, R, B, aux^\alpha, c)\}$: note that this rule, in the spirit of [26], introduces an auxiliary element aux^α, which intuitively represents the class of all new R-successors generated by the axiom α.

(ii). $\mathcal{EL}_\perp$ deduction rules: the set of rules in P_{el} provide the deduction rules for the translated $\mathcal{EL}_\perp$ axioms. In the case of existential axioms, the rule (pel-supex1) introduces a new relation to the auxiliary individual as follows:

Table 2. $\mathcal{EL}_\perp$ input and deduction rules

$\mathcal{EL}_\perp$ input translation $I_{el}(S, c)$

(iel-nom)	$a \in \text{NI} \mapsto \{\texttt{nom}(a, c)\}$	(iel-top)	$\top(a) \mapsto \{\texttt{insta}(a, \texttt{top}, c, \texttt{main})\}$
(iel-cls)	$A \in \text{NC} \mapsto \{\texttt{cls}(A, c)\}$	(iel-bot)	$\perp(a) \mapsto \{\texttt{insta}(a, \texttt{bot}, c, \texttt{main})\}$
(iel-rol)	$R \in \text{NR} \mapsto \{\texttt{rol}(R, c)\}$	(iel-subc)	$A \sqsubseteq B \mapsto \{\texttt{subClass}(A, B, c)\}$
(iel-inst)	$A(a) \mapsto \{\texttt{insta}(a, A, c, \texttt{main})\}$	(iel-subcnj)	$A_1 \sqcap A_2 \sqsubseteq B \mapsto \{\texttt{subConj}(A_1, A_2, B, c)\}$
(iel-triple)	$R(a, b) \mapsto \{\texttt{triplea}(a, R, b, c, \texttt{main})\}$	(iel-subex)	$\exists R.A \sqsubseteq B \mapsto \{\texttt{subEx}(R, A, B, c)\}$
		(iel-supex)	$A \sqsubseteq \exists R.B \mapsto \{\texttt{supEx}(A, R, B, aux^\alpha, c)\}$

$\mathcal{EL}_\perp$ deduction rules P_{el}

(pel-instd)	$\texttt{instd}(x, z, c, t) \leftarrow \texttt{insta}(x, z, c, t).$
(pel-tripled)	$\texttt{tripled}(x, r, y, c, t) \leftarrow \texttt{triplea}(x, r, y, c, t).$
(pel-top)	$\texttt{instd}(x, \texttt{top}, c, \texttt{main}) \leftarrow \texttt{nom}(x, c).$
(pel-bot)	$\texttt{unsat}(t) \leftarrow \texttt{instd}(x, \texttt{bot}, c, t).$
(pel-subc)	$\texttt{instd}(x, z, c, t) \leftarrow \texttt{subClass}(y, z, c), \texttt{instd}(x, y, c, t).$
(pel-subcnj)	$\texttt{instd}(x, z, c, t) \leftarrow \texttt{subConj}(y_1, y_2, z, c), \texttt{instd}(x, y_1, c, t), \texttt{instd}(x, y_2, c, t).$
(pel-subex)	$\texttt{instd}(x, z, c, t) \leftarrow \texttt{subEx}(v, y, z, c), \texttt{tripled}(x, v, x', c, t), \texttt{instd}(x', y, c, t).$
(pel-supex1)	$\texttt{tripled}(x, r, x', c, t) \leftarrow \texttt{supEx}(y, r, z, x', c), \texttt{instd}(x, y, c, t).$
(pel-supex2)	$\texttt{instd}(x', z, c, t) \leftarrow \texttt{supEx}(y, r, z, x', c), \texttt{instd}(x, y, c, t).$
(pel-sat)	$\leftarrow \texttt{unsat}(\texttt{main}).$

$$\texttt{tripled}(x, r, x', c, t) \leftarrow \texttt{supEx}(y, r, z, x', c), \texttt{instd}(x, y, c, t).$$

Namely, for a local existential axiom $\alpha = A \sqsubseteq \exists R.B$ in context c ($\texttt{supEx}(y, r, z, x', c)$, where $y = A, r = R, z = B, x' = aux^\alpha$ by the input translation) and a local instance $x = e$ with $A(e)$ ($\texttt{instd}(x, y, c, t)$), then the rule adds $R(e, aux^\alpha)$ to the local context ($\texttt{tripled}(x, r, x', c, t)$). Similarly, (pel-supex2) classifies the new individual as member of the concept z. The rules of P_{el} are listed in Table 2.

(iii). Global and local translations: Input global rules in I_{glob} encode the interpretation of the contextual and module structure. In a similar way, input and deduction local rules in I_{loc} and P_{loc} define the translation for axioms of the local object language, in particular for the interpretation of axioms that make use of *eval* expressions. Rules for global and local translation are shown in Table 3.

(iv). Defeasible axioms input translations: the set of input rules I_D (shown in Table 4) provides the translation of defeasible axioms $D(\alpha) \in \mathfrak{G}$: in particular, they are used to specify that the axiom α need to be considered as defeasible. For example, $D(A \sqsubseteq \exists R.B)$ is translated to $\texttt{def\_supex}(A, R, B, aux^\alpha)$.

(v). Overriding rules: conditions for overriding of defeasible axioms are encoded in the overriding rules in P_D, shown in Table 5. These rules define when a defeasible axiom has to be locally overridden: intuitively, they correspond to the proof of existence of a clashing set for an instance and axiom at hand. For example, for axioms of the form $D(A \sqsubseteq \exists R.B)$, the translation introduces the rule:

$$\texttt{ovr}(\texttt{supEx}, x, y, r, z, w, c) \leftarrow \texttt{def\_supex}(y, r, z, w), \texttt{prec}(c, g),$$
$$\texttt{instd}(x, y, c, \texttt{main}),$$
$$\texttt{not test\_fails}(\texttt{nex}(x, r, z, w, c)).$$

Table 3. Global, local and output rules

Global input rules $I_{glob}(\mathfrak{G})$

(igl-subctx1) $C \in \mathcal{C} \mapsto \{\texttt{subClass}(C, \mathsf{Ctx}, \mathsf{gm})\}$

(igl-subctx2) $c \in \mathbf{N} \mapsto \{\texttt{insta}(c, \mathsf{Ctx}, \mathsf{gm}, \mathsf{main})\}$

Local input rules $I_{loc}(\mathsf{K_m}, \mathsf{c})$

(ilc-subevalat) $eval(A, \mathsf{C}) \sqsubseteq B \mapsto \{\texttt{subEval}(A, \mathsf{C}, B, \mathsf{c})\}$

Local deduction rules P_{loc}

(plc-subevalat) $\texttt{instd}(x, b, c, t) \leftarrow \texttt{subEval}(a, c_1, b, c), \texttt{instd}(c', c_1, \mathsf{gm}, t), \texttt{instd}(x, a, c', t).$

Output translation $O(\alpha, \mathsf{c})$

(o-concept) $A(a) \mapsto \{\texttt{instd}(a, A, \mathsf{c}, \mathsf{main})\}$

(o-role) $R(a, b) \mapsto \{\texttt{tripled}(a, R, b, \mathsf{c}, \mathsf{main})\}$

Table 4. Input rules $I_D(S)$ for defeasible axioms

(id-inst)	$D(A(a)) \mapsto \{\,\texttt{def\_insta}(A, a).\,\}$
(id-triple)	$D(R(a, b)) \mapsto \{\,\texttt{def\_triplea}(R, a, b).\,\}$
(id-subc)	$D(A \sqsubseteq B) \mapsto \{\,\texttt{def\_subclass}(A, B).\,\}$
(id-subcnj)	$D(A_1 \sqcap A_2 \sqsubseteq B) \mapsto \{\,\texttt{def\_subcnj}(A_1, A_2, B).\,\}$
(id-subex)	$D(\exists R.A \sqsubseteq B) \mapsto \{\,\texttt{def\_subex}(R, A, B).\,\}$
(id-supex)	$D(A \sqsubseteq \exists R.B) \mapsto \{\,\texttt{def\_supex}(A, R, B, aux^\alpha).\,\}$

Namely, this rule states that if there exists some global defeasible axiom $\alpha = D(A \sqsubseteq \exists R.B)$ ($\texttt{def\_supex}(y, r, z, w)$, where $y = A, r = R, z = B, w = aux^\alpha$ by the input translation) and in a context c we can prove for $x = e$ that $A(e)$ (i.e. $\texttt{instd}(x, y, c, \mathsf{main})$) but $\neg \exists R.B(e)$ ($\texttt{not test\_fails}(\texttt{nex}(x, r, z, w, c))$), then there is an overriding for this axiom with respect to e in context c ($\texttt{ovr}(\texttt{supEx}, x, y, r, w, c)$). As discussed above, as in [10] the derivation of the negative part of the clashing set of the axiom (in this case $\neg \exists R.B(e)$) is encoded by a proof by contradiction (namely, by proving that inconsistency can be derived by adding $R(e, d), B(d)$, with d a new constant, to the current context).

(vi). Inheritance rules: the set of rules P_D then includes the rules for the inheritance of (possibly defeasible) axioms from the global context. For example, the rule (prop-supex1) propagates an existential axiom $\alpha = A \sqsubseteq \exists R.B$:

$$\texttt{tripled}(x, r, x', c, t) \leftarrow \texttt{supEx}(y, r, z, x', g), \texttt{instd}(x, y, c, t),$$
$$\texttt{prec}(c, g), \texttt{not ovr}(\texttt{supEx}, x, y, r, z, x', c).$$

As in the rule (pel-supex1) above, this applies α to a local instance $e = x$ with $A(e)$ and adds $R(e, aux^\alpha)$: the rule is applied only if such instance e is in a local context ($\texttt{prec}(c, g)$) and no overriding can be proved on e ($\texttt{not ovr}(\texttt{supEx}, x, y, r, z, x', c)$). Similarly, (prop-supex2) propagates the qualification of the related element to the concept in z. If no overriding is recognized,

Table 5. Deduction rules P_D for defeasible axioms: overriding rules

(ovr-inst)	$\mathtt{ovr}(\mathtt{insta}, x, y, c) \leftarrow \mathtt{def\_insta}(x, y), \mathtt{prec}(c, g), \mathtt{not\,test\_fails}(\mathtt{nlit}(x, y, c)).$
(ovr-triple)	$\mathtt{ovr}(\mathtt{triplea}, x, r, y, c) \leftarrow \mathtt{def\_triplea}(x, r, y), \mathtt{prec}(c, g), \mathtt{not\,test\_fails}(\mathtt{nrel}(x, r, y, c)).$
(ovr-subc)	$\mathtt{ovr}(\mathtt{subClass}, x, y, z, c) \leftarrow \mathtt{def\_subclass}(y, z), \mathtt{prec}(c, g), \mathtt{instd}(x, y, c, \mathtt{main}),$ $\mathtt{not\,test\_fails}(\mathtt{nlit}(x, z, c)).$
(ovr-cnj)	$\mathtt{ovr}(\mathtt{subConj}, x, y_1, y_2, z, c) \leftarrow \mathtt{def\_subcnj}(y_1, y_2, z), \mathtt{prec}(c, g), \mathtt{instd}(x, y_1, c, \mathtt{main}),$ $\mathtt{instd}(x, y_2, c, \mathtt{main}), \mathtt{not\,test\_fails}(\mathtt{nlit}(x, z, c)).$
(ovr-subex)	$\mathtt{ovr}(\mathtt{subEx}, x, r, y, z, c) \leftarrow \mathtt{def\_subex}(r, y, z), \mathtt{prec}(c, g), \mathtt{tripled}(x, r, w, c, \mathtt{main}),$ $\mathtt{instd}(w, y, c, \mathtt{main}), \mathtt{not\,test\_fails}(\mathtt{nlit}(x, z, c)).$
(ovr-supex)	$\mathtt{ovr}(\mathtt{supEx}, x, y, r, z, w, c) \leftarrow \mathtt{def\_supex}(y, r, z, w), \mathtt{prec}(c, g),$ $\mathtt{instd}(x, y, c, \mathtt{main}), \mathtt{not\,test\_fails}(\mathtt{nex}(x, r, z, w, c)).$

Table 6. Deduction rules P_D for defeasible axioms: inheritance rules

(prop-inst)	$\mathtt{instd}(x, z, c, t) \leftarrow \mathtt{insta}(x, z, g, t), \mathtt{prec}(c, g), \mathtt{not\,ovr}(\mathtt{insta}, x, z, c).$
(prop-triple)	$\mathtt{tripled}(x, r, y, c, t) \leftarrow \mathtt{triplea}(x, r, y, g, t), \mathtt{prec}(c, g), \mathtt{not\,ovr}(\mathtt{triplea}, x, r, y, c).$
(prop-subc)	$\mathtt{instd}(x, z, c, t) \leftarrow \mathtt{subClass}(y, z, g), \mathtt{instd}(x, y, c, t),$ $\mathtt{prec}(c, g), \mathtt{not\,ovr}(\mathtt{subClass}, x, y, z, c).$
(prop-cnj)	$\mathtt{instd}(x, z, c, t) \leftarrow \mathtt{subConj}(y_1, y_2, z, g), \mathtt{instd}(x, y_1, c, t), \mathtt{instd}(x, y_2, c, t),$ $\mathtt{prec}(c, g), \mathtt{not\,ovr}(\mathtt{subConj}, x, y_1, y_2, z, c).$
(prop-subex)	$\mathtt{instd}(x, z, c, t) \leftarrow \mathtt{subEx}(v, y, z, g), \mathtt{tripled}(x, v, x', c, t), \mathtt{instd}(x', y, c, t),$ $\mathtt{prec}(c, g), \mathtt{not\,ovr}(\mathtt{subEx}, x, v, y, z, c).$
(prop-supex1)	$\mathtt{tripled}(x, r, x', c, t) \leftarrow \mathtt{supEx}(y, r, z, x', g), \mathtt{instd}(x, y, c, t),$ $\mathtt{prec}(c, g), \mathtt{not\,ovr}(\mathtt{supEx}, x, y, r, z, x', c).$
(prop-supex2)	$\mathtt{instd}(x', z, c, t) \leftarrow \mathtt{supEx}(y, r, z, x', g), \mathtt{instd}(x, y, c, t),$ $\mathtt{prec}(c, g), \mathtt{not\,ovr}(\mathtt{supEx}, x, y, r, z, x', c).$

then the global axiom is applied to each of its local instances. Note that this rule propagates also to local instances of strict axioms, since their overriding is never verified (Table 6).

(vii). Test rules: the last set of rules in P_D is the set of test rules: following the translation in [10], these rules are used to instantiate and define the "test environments" for the proofs by contradiction used in the overriding rules. The rules are presented in Table 7. Notably, in the translation for $\mathcal{EL}_\perp$, we introduce rules for the case of tests on "existential" negative ABox assertions (**nex**) with the rule (test-supex). The rule (test-add3) and (test-add4) are used to introduce in the test environment the positive instance of the kind $R(e, d), B(d)$ (with d a new element represented by aux^α) of the negative literal to be verified. If a contradiction is not proved, then the failure of the test is recognized by the rule (test-fails3). Note that, in order to be consistent with the semantics (i.e. we can not have exceptions on unnamed individuals), the rules (test-fails4) and (test-fails5) are needed to exclude the possibility to instantiate tests on auxiliary individuals.

(viii). Output rules: rules in O are used to translate (atomic) ABox assertions to be verified to hold in a given context by the final program. The rules in O are listed in Table 3.

Table 7. Deduction rules P_{D} for defeasible axioms: test rules

(test-inst)	$\mathtt{test(nlit}(x,y,c)) \leftarrow \mathbf{def\_insta}(x,y), \mathtt{prec}(c,g).$
(constr-inst)	$\leftarrow \mathtt{test\_fails(nlit}(x,y,c)), \mathtt{ovr(insta}, x,y,c).$
(test-triple)	$\mathtt{test(nrel}(x,r,y,c)) \leftarrow \mathbf{def\_triplea}(x,r,y), \mathtt{prec}(c,g).$
(constr-triple)	$\leftarrow \mathtt{test\_fails(nrel}(x,r,y,c)), \mathtt{ovr(triplea}, x,r,y,c).$
(test-subc)	$\mathtt{test(nlit}(x,z,c)) \leftarrow \mathbf{def\_subclass}(y,z), \mathtt{instd}(x,y,c,\mathsf{main}), \mathtt{prec}(c,g).$
(constr-subc)	$\leftarrow \mathtt{test\_fails(nlit}(x,z,c)), \mathtt{ovr(subClass}, x,y,z,c).$
(test-subcnj)	$\mathtt{test(nlit}(x,z,c)) \leftarrow \mathbf{def\_subcnj}(y_1,y_2,z), \mathtt{instd}(x,y_1,c,\mathsf{main}),$
	$\mathtt{instd}(x,y_2,c,\mathsf{main}), \mathtt{prec}(c,g).$
(constr-subcnj)	$\leftarrow \mathtt{test\_fails(nlit}(x,z,c)), \mathtt{ovr(subConj}, x,y_1,y_2,z,c).$
(test-subex)	$\mathtt{test(nlit}(x,z,c)) \leftarrow \mathbf{def\_subex}(r,y,z), \mathtt{tripled}(x,r,w,c,\mathsf{main}),$
	$\mathtt{instd}(w,y,c,\mathsf{main}), \mathtt{prec}(c,g).$
(constr-subex)	$\leftarrow \mathtt{test\_fails(nlit}(x,z,c)), \mathtt{ovr(subEx}, x,r,y,z,c).$
(test-supex)	$\mathtt{test(nex}(x,r,z,w,c)) \leftarrow \mathbf{def\_supex}(y,r,z,w), \mathtt{instd}(x,y,c,\mathsf{main}), \mathtt{prec}(c,g).$
(constr-supex)	$\leftarrow \mathtt{test\_fails(nex}(x,r,z,w,c)), \mathtt{ovr(supEx}, x,r,z,w,c).$
(test-fails1)	$\mathtt{test\_fails(nlit}(x,z,c)) \leftarrow \mathtt{instd}(x,z,c,\mathtt{nlit}(x,z,c)), \mathtt{not\ unsat(nlit}(x,z,c)).$
(test-fails2)	$\mathtt{test\_fails(nrel}(x,r,y,c)) \leftarrow \mathtt{tripled}(x,r,y,c,\mathtt{nrel}(x,r,y,c)), \mathtt{not\ unsat(nrel}(x,r,y,c)).$
(test-fails3)	$\mathtt{test\_fails(nex}(x,r,z,y,c)) \leftarrow \mathtt{tripled}(x,r,y,c,\mathtt{nex}(x,r,z,y,c)), \mathtt{not\ unsat(nex}(x,r,z,y,c)).$
(test-fails4)	$\mathtt{test\_fails(nlit}(x,z,c)) \leftarrow \mathtt{supEx}(y',r',z',x,c'), \mathtt{instd}(x,z,c,\mathtt{nlit}(x,z,c)).$
(test-fails5)	$\mathtt{test\_fails(nex}(x,r,z,y,c)) \leftarrow \mathtt{supEx}(y',r',z',x,c'), \mathtt{tripled}(x,r,y,c,\mathtt{nex}(x,r,z,y,c)).$
(test-add1)	$\mathtt{instd}(x,z,c,\mathtt{nlit}(x,z,c)) \leftarrow \mathtt{test(nlit}(x,z,c)).$
(test-add2)	$\mathtt{tripled}(x,r,y,c,\mathtt{nrel}(x,r,y,c)) \leftarrow \mathtt{test(nrel}(x,r,y,c)).$
(test-add3)	$\mathtt{tripled}(x,r,y,c,\mathtt{nex}(x,r,z,y,c)) \leftarrow \mathtt{test(nex}(x,r,z,y,c)).$
(test-add4)	$\mathtt{instd}(x,z,c,\mathtt{nex}(x,r,z,y,c)) \leftarrow \mathtt{test(nex}(x,r,z,y,c)).$
(test-copy1)	$\mathtt{instd}(x_1,y_1,c,t) \leftarrow \mathtt{instd}(x_1,y_1,c,\mathsf{main}), \mathtt{test}(t).$
(test-copy2)	$\mathtt{tripled}(x_1,r,y_1,c,t) \leftarrow \mathtt{tripled}(x_1,r,y_1,c,\mathsf{main}), \mathtt{test}(t).$

5.2 Translation Process

The translation process that, given a CKR $\mathfrak{K} = \langle \mathfrak{G}, \{\mathrm{K_m}\}_{\mathsf{m}\in\mathsf{M}} \rangle$ in $\mathcal{EL}_\perp$ normal form, produces a program $PK(\mathfrak{K})$ encoding instance checking from the CKR-models of $\mathfrak{K}$ is analogous to the one presented in [10]:

1. the *global program* for $\mathfrak{G}$ is constructed as (where gm, gk are new context names):

$$PG(\mathfrak{G}) = P_{el} \cup I_{glob}(\mathfrak{G}_\Gamma) \cup I_{\mathrm{D}}(\mathfrak{G}_\Sigma) \cup I_{el}(\mathfrak{G}_\Gamma, \mathsf{gm}) \cup I_{el}(\mathfrak{G}_\Sigma \cup \mathfrak{G}_\Sigma^{\mathrm{D}}, \mathsf{gk})$$

where $\mathfrak{G}_\Gamma = \mathfrak{G} \cap \mathcal{L}_\Gamma$, $\mathfrak{G}_\Sigma = \mathfrak{G} \cap \mathcal{L}_\Sigma^{\mathrm{D}}$ and $\mathfrak{G}_\Sigma^{\mathrm{D}} = \{\alpha \in \mathcal{L}_\Sigma \mid \mathrm{D}(\alpha) \in \mathfrak{G}_\Sigma\}$.
2. Let $\mathbf{N}_\mathfrak{G}$ be the set of contexts:

$$\mathbf{N}_\mathfrak{G} = \{\mathsf{c} \in \mathbf{N} \mid PG(\mathfrak{G}) \models \mathtt{instd(c, Ctx, gm, main)}\},$$

and, for every $\mathsf{c} \in \mathbf{N}_\mathfrak{G}$, let its associated knowledge base $\mathrm{K_c}$ be defined as:

$$\mathrm{K_c} = \bigcup \{\mathrm{K_m} \in \mathfrak{K} \mid PG(\mathfrak{G}) \models \mathtt{tripled(c, mod, m, gm, main)}\}.$$

3. For each $\mathsf{c} \in \mathbf{N}_\mathfrak{G}$, we define the *local program* $PC(\mathsf{c}, \mathfrak{K})$ as:

$$PC(\mathsf{c}, \mathfrak{K}) := P_{el} \cup P_{loc} \cup P_{\mathrm{D}} \cup I_{loc}(\mathrm{K_c}, \mathsf{c}) \cup I_{el}(\mathrm{K_c}, \mathsf{c}) \cup \{\mathtt{prec(c, gk)}\};$$

4. The *CKR program* $PK(\mathfrak{K})$, encoding the whole input CKR, is defined as:

$$PK(\mathfrak{K}) = PG(\mathfrak{G}) \cup \bigcup_{\mathsf{c}\in\mathbf{N}_\mathfrak{G}} PC(\mathsf{c}, \mathfrak{K}) \qquad (5)$$

5.3 Correctness

The presented translation procedure provides a sound and complete materialization calculus for instance checking on $\mathcal{EL}_\perp$ CKRs in normal form. The proof for this result can be verified with the same line of reasoning presented in [10] by establishing a correspondence between minimal justified CKR-models of $\mathfrak{K}$ and answer sets of $PK(\mathfrak{K})$: indeed, the non trivial aspect of this proof in the case of $\mathcal{EL}_\perp$ resides in the management of existential axioms, where there is the need to define a correspondence between the auxiliary individuals in the translation and the interpretation of existential axioms in the semantics. In this regard, we basically follow the approach used by Krötzsch in [26]: in building the correspondence with justified models, auxiliary constants aux^α are mapped to the class of R-successors (i.e. Skolem individuals) for existential axiom α.

As in [10], we consider UNA and named models in our translation: thus we can show the correctness result on Herbrand models, that will be denoted $\hat{\mathfrak{I}}(\chi)$.

Let $\mathfrak{I}_{CAS}$ be a justified named CAS-model. We define the set of overriding assumptions as follows:

$$OVR(\mathfrak{I}_{CAS}) = \{\, \mathtt{ovr}(p(\mathbf{e})) \mid \langle \alpha, \mathbf{e} \rangle \in \chi(\mathsf{c}),\ I_{el}(\alpha, \mathsf{c}) = p \,\}.$$

Given a CAS-interpretation $\mathfrak{I}_{CAS} = \langle \mathcal{M}, \mathcal{I}, \chi \rangle$, we can define a corresponding Herbrand interpretation $I(\mathfrak{I}_{CAS})$ for $PK(\mathfrak{K})$ by including the following atoms in it:

(1). all facts of $PK(\mathfrak{K})$;
(2). $\mathtt{instd}(a, A, \mathsf{c}, \mathtt{main})$, if $\mathcal{I}(\mathsf{c}) \models A(a)$;
(3). $\mathtt{tripled}(a, R, b, \mathsf{c}, \mathtt{main})$, if $\mathcal{I}(\mathsf{c}) \models R(a,b)$;
(4). $\mathtt{tripled}(a, R, aux^\alpha, \mathsf{c}, \mathtt{main}), \mathtt{instd}(aux^\alpha, B, \mathsf{c}, \mathtt{main})$, if $\mathcal{I}(\mathsf{c}) \models \exists R.B(a)$;
(5). each ovr-literal from $OVR(\mathfrak{I}_{CAS})$;
(6). each literal l with environment $t \neq \mathtt{main}$, if $\mathtt{test}(t) \in I(\mathfrak{I}_{CAS})$ and l is in the head of a rule $r \in grnd(PK(\mathfrak{K}))$ with $Body(r) \subseteq I(\mathfrak{I}_{CAS})$;
(7). $\mathtt{test}(t)$, if $\mathtt{test\_fails}(t)$ appears in the body of an overriding rule r in $grnd(PK(\mathfrak{K}))$ and the head of r is an ovr literal in $OVR(\mathfrak{I}_{CAS})$;
(8). $\mathtt{unsat}(\mathtt{nlit}(a, A, \mathsf{c}))$, if $\mathcal{I}(\mathsf{c}) \not\models \mathsf{K_c} \cup \{A(a)\}$;
(9). $\mathtt{unsat}(\mathtt{nrel}(a, R, b, \mathsf{c}))$, if $\mathcal{I}(\mathsf{c}) \not\models \mathsf{K_c} \cup \{R(a,b)\}$;
(10). $\mathtt{unsat}(\mathtt{nex}(a, R, B, aux^\alpha, \mathsf{c}))$, if $\mathcal{I}(\mathsf{c}) \not\models \mathsf{K_c} \cup \{\neg \exists R.B(a)\}$ for $\alpha = A \sqsubseteq \exists R.B$;
(11). $\mathtt{test\_fails}(t)$, if $\mathtt{unsat}(t) \notin I(\mathfrak{I}_{CAS})$.

In the following we provide a sketch of the correctness proof by highlighting the newly added aspects for the management of existential axioms.

The next proposition shows that the least Herbrand model of the global context $\mathfrak{G}$ can be represented by the answer set of the global program $PG(\mathfrak{G})$. Let us consider $I(\mathcal{M}_\mathfrak{G})$ as the Herbrand interpretation for $PG(\mathfrak{G})$ defined as $I(\mathfrak{I}_{CAS})$ above for $PK(\mathfrak{K})$.

Proposition 2. *Let $\mathfrak{K} = \langle \mathfrak{G}, \{\mathsf{K_m}\}_{\mathsf{m} \in \mathsf{M}} \rangle$ be a CKR in $\mathcal{EL}_\perp$ normal form. If $\mathfrak{G}$ is satisfiable, then $I(\mathcal{M}_\mathfrak{G})$ is the unique answer set of $PG(\mathfrak{G})$; otherwise, $PG(\mathfrak{G})$ has no answer sets.*

Proof (Sketch). The result is shown by proving, on one side, that $I(\mathcal{M}_{\mathfrak{G}})$ is an answer set for $PG(\mathfrak{G})$ if $\mathfrak{G}$ is satisfiable: the fact that $I(\mathcal{M}_{\mathfrak{G}})$ satisfies rules of the form of (pel-supex1) and (pel-supex2) in $PG(\mathfrak{G})$ is verified by the newly added condition (4) on existential formulas in the construction of the model above.

In the other direction, we need to show that from an answer set M of $PG(\mathfrak{G})$, we can build a model $\mathcal{M} = \langle \Delta^{\mathcal{M}}, \cdot^{\mathcal{M}} \rangle$ for $\mathfrak{G}$. The construction of the model is similar to the original proof, but we need to consider auxiliary individuals in the domain of the model, that is thus defined as: $\Delta^{\mathcal{M}} = \{c \mid c \in \mathrm{NI}_{\Gamma} \cup \mathrm{NI}_{\Sigma}\} \cup \{aux^{\alpha} \mid \alpha = A \sqsubseteq \exists R.B \in \mathfrak{G}\}$. The result can then be proved by considering the effect of the new local rules in P_{el} for existential axioms: auxiliary individuals provide the domain elements in $\mathcal{M}$ needed to verify this kind of axioms. ☐

Then, in the following lemma we can establish the correspondence between the answer sets of the final program $PK(\mathfrak{K})$ and the least justified models of $\mathfrak{K}$.

Lemma 3. *Let $\mathfrak{K}$ be a CKR in $\mathcal{EL}_{\perp}$ normal form. Then:*

(i). for every (named) justified clashing assumption χ, the interpretation $S = I(\hat{\mathfrak{J}}(\chi))$ is an answer set of $PK(\mathfrak{K})$;
(ii). every answer set S of $PK(\mathfrak{K})$ is of the form $S = I(\hat{\mathfrak{J}}(\chi))$ where χ is a (named) justified clashing assumption for $\mathfrak{K}$.

Proof (Sketch). We consider $S = I(\hat{\mathfrak{J}}(\chi))$ built as above and reason over the reduct $G_S(PK(\mathfrak{K}))$ of $PK(\mathfrak{K})$ with respect S: basically, $G_S(PK(\mathfrak{K}))$ contains all ground rules from $PK(\mathfrak{K})$ that are not falsified by some NAF literal in S.

Then, item (i) can be proved by showing that given a justified χ, then S is an answer set for $G_S(PK(\mathfrak{K}))$ (and thus $PK(\mathfrak{K})$): the proof follows the same reasoning of the one in [10], where condition (10) in the construction of $I(\hat{\mathfrak{J}}(\chi))$ is used to show the correctness of the corresponding overriding rule.

For item (ii), we can show that from any answer set S we can build a justified model for $\mathcal{K}$ such that $S = I(\hat{\mathfrak{J}}(\chi))$ holds. The model can be built with the same considerations adopted for the proof of previous Proposition 2 on auxiliary individuals. The justification of the model follows by noting that the newly added tests for negative literals **nex** correctly encode the (negative part of) possible clashing sets for existential formulas. ☐

The correctness result then directly follows from Lemma 3 and properties in Sect. 4.1.

Theorem 5. *Let $\mathfrak{K}$ be a CKR in $\mathcal{EL}_{\perp}$ normal form, and let α and c be such that $O(\alpha, c)$ is defined. Then $\mathfrak{K} \models c : \alpha$ iff $PK(\mathfrak{K}) \models O(\alpha, c)$.*

6 Discussion and Conclusion

6.1 Syntax Restrictions

In our development, we have excluded the use of the $\top$ concept for axioms of the form $\top \sqsubseteq C$. The reason for this limitation stands in the fact that such axioms allow to derive properties for any (named or unnamed) individual, and thus may make in particular defeasible axioms $D(\top \sqsubseteq C)$ applicable to any individual.

A possible solution to include such axioms may consist in introducing a number of assertions of the kind $\top(a_i)$ where each a_i is a fresh individual. Then, intuitively, a_i individuals can be used as "proxy" individuals for making exceptions on unnamed individuals, by relating provability of a clashing set $S_{\alpha(\mathbf{e})}$ for such individuals to the provability for a_i.

The formulation of this solution, however, is not trivial: in general, multiple fresh individuals are needed to represent the different unnamed individuals that would be considered in exceptions (consider, e.g., the case of conflicting defeasible axioms of the kind $D(\top \sqsubseteq A)$, $D(A \sqsubseteq \perp)$). The intuition is that exceptions on unnamed elements should be proved (by using strict axioms) on proxy individuals and then "injected" to unnamed elements: in the case of conflicting axioms but also in order to consider abstract elements arising from different existential axioms, different such injections should be needed to correctly capture all intended exceptions and axiom applications. In order to reason on such exceptions in the datalog translation, then, we would need to replicate this relation by correctly associating the exceptions on the freshly added proxy individuals and the aux^α elements representing the class of unnamed individuals generated by existential axiom α.

Thus, the formulation of this correspondence is currently not in the scope of this paper, but its definition will be a step ahead in the direction of applying our framework to the full expressiveness of $\mathcal{EL}_\perp$.

6.2 Related Work

The relation of our justified exception approach to nonmonotonic description logics has been discussed in [10], where in particular typicality in DLs [20,21], normality [7,8] and overriding [6] were considered in more depth; we point out that our work is distinctive in that it aims at a basic mechanism for a formalism with explicit hierarchical structure, which is usually not reflected in nonmonotonic entailment relations. For these works, $\mathcal{EL}$ extensions and in particular $\mathcal{EL}_\perp$ have been important (for typicality, we mention here [19,20]). A noticeable recent work on a defeasible version of $\mathcal{EL}_\perp$ is [30,31], which aimed at overcoming issues with the approach by [14] on quantified concepts, especially in nested expressions. The approach is to extend classical canonical models in $\mathcal{EL}_\perp$, by adding multiple representatives of concepts and individuals, in order to model higher typicality; inference is then determined from a canonical model of the extended domain. Apparently, this approach is different from ours, which works on all models and uses factual justifications that need to be derived; canonical models are useful for characterization and implementation. However, it would be interesting to see whether ideas from [30,31] could be used to extend our approach. Another recent work on nonmonotonic $\mathcal{EL}_\perp$ is [15]: the work presents a polynomial time subsumption procedure that, notably, can be reduced to classical monotonic $\mathcal{EL}_\perp$ reasoning.

We would not like to close this short discussion without recalling another pioneering work of Franz Baader, which has predated by far the approaches to non-monotonic default logics mentioned above. In the early 1990s, when description logics where in their early rise, he has recognized the need for an extension of DLs with nonmonotonic features, and in particular to allow for defaults as in Reiter's Default Logic [32] of the form $\alpha : \beta_1, \ldots, \beta_n / \gamma$, which intuitively means that if α is derived and each β_i can be consistently assumed, then γ is concluded. Specifically, in terminological default logic (TDL) [4], the formulas in a default are concept terms; so $manager : \exists supervises / \exists supervises$ would intuitively express that by default, managers act as supervisors. In a critical analysis, Baader and Hollunder pointed out issues of Reiter's approach for description logics due to skolemization, and weaknesses of proposals to overcome them; as moreover reasoning turned out to be undecidable, they defined a restricted semantics in which defaults are only applicable to individuals from the ABox.

Our justified exceptions are related to TDL, as we could see a defeasible axiom $D(\alpha)$ as a default $\top : \alpha/\alpha$, which informally is applied whenever possible; however, there are some differences. A minor difference is that α is an axiom and not a concept term. More substantial is that non-applicability of the default (e.g., making an exception) requires in our approach a stronger condition (derivation of a clashing set), and that inapplicability of exceptions for unnamed individuals – if axioms $\top \sqsubseteq C$ are excluded – is an emerging property and not by design. Finally, our approach aims at singling out models, while TDL determines extensions (sets of formulas) in the tradition of Default Logic. It will be interesting to see fragments of $\mathcal{EL}_\perp$ which can be mapped to TDL, such that implementations of the latter (for instance, the one in [16]) can be exploited.

6.3   Summary and Outlook

In this article, we have described a formalization of making exceptions to axioms in $\mathcal{EL}_\perp$, by adapting the justified exception approach in [10]. While the resulting formalism has acceptable complexity and basic reasoning tasks can be translated to datalog with expressive negation, exceptions are effectively limited to individuals known from the knowledge base. This will be insufficient for scenarios which need exceptions on unnamed individuals, in particular due to cyclic axioms of the form $A \sqsubseteq \exists R.A$.

In order to allow for more exceptions, justification of clashing assumptions may be restricted to CAS-models $\mathfrak{I}'_{CAS}$ that are NI-congruent with $\mathfrak{I}_{CAS}$ and coincide on the Skolem functions f_α. Justified CAS-models will then have types of elements with respect to the exceptions made on them. The number of different such types will be finite, yet can get exponential and thus an increase in complexity is to be expected; to consider this semantics is an interesting issue for future work.

Another direction for further development concerns considering other extensions of $\mathcal{EL}$, such as $\mathcal{EL}^{\neg A}$ and $\mathcal{E}^{++}$.

References

1. Baader, F., Calvanese, D., McGuinness, D., Nardi, D., Patel-Schneider, P. (eds.): The Description Logic Handbook. Cambridge University Press, Cambridge (2003)
2. Baader, F.: Terminological cycles in a description logic with existential restrictions. In: Gottlob, G., Walsh, T. (eds.) Proceedings of the Eighteenth International Joint Conference on Artificial Intelligence, IJCAI-03, Acapulco, Mexico, 9–15 August 2003, pp. 325–330. Morgan Kaufmann (2003). http://ijcai.org/Proceedings/03/Papers/048.pdf
3. Baader, F., Brandt, S., Lutz, C.: Pushing the EL envelope. In: Kaelbling, L.P., Saffiotti, A. (eds.) IJCAI-05, Proceedings of the Nineteenth International Joint Conference on Artificial Intelligence, Edinburgh, Scotland, UK, 30 July–August 5 2005, pp. 364–369. Professional Book Center (2005). http://ijcai.org/Proceedings/05/Papers/0372.pdf
4. Baader, F., Hollunder, B.: Embedding defaults into terminological knowledge representation formalisms. J. Autom. Reasoning **14**(1), 149–180 (1995). https://doi.org/10.1007/BF00883932
5. Baader, F., Lutz, C., Brandt, S.: Pushing the EL envelope further. In: Clark, K., Patel-Schneider, P.F. (eds.) Proceedings of the Fourth OWLED Workshop on OWL: Experiences and Directions, CEUR Workshop Proceedings, Washington, DC, USA, 1–2 April 2008, vol. 496. CEUR-WS.org (2008). http://ceur-ws.org/Vol-496/owled2008dc_paper_3.pdf
6. Bonatti, P.A., Faella, M., Petrova, I., Sauro, L.: A new semantics for overriding in description logics. Artif. Intell. **222**, 1–48 (2015). https://doi.org/10.1016/j.artint.2014.12.010
7. Bonatti, P.A., Faella, M., Sauro, L.: Defeasible inclusions in low-complexity DLs. J. Artif. Intell. Res. **42**, 719–764 (2011). https://doi.org/10.1613/jair.3360
8. Bonatti, P.A., Lutz, C., Wolter, F.: Description logics with circumscription. In: 10th International Conference on Principles of Knowledge Representation and Reasoning (KR 2006), pp. 400–410. AAAI Press (2006)
9. Bozzato, L., Eiter, T., Serafini, L.: Contextualized knowledge repositories with justifiable exceptions. In: DL2014. CEUR-WP, vol. 1193, pp. 112–123. CEUR-WS.org (2014)
10. Bozzato, L., Eiter, T., Serafini, L.: Enhancing context knowledge repositories with justifiable exceptions. Artif. Intell. **257**, 72–126 (2018). https://doi.org/10.1016/j.artint.2017.12.005
11. Bozzato, L., Ghidini, C., Serafini, L.: Comparing contextual and flat representations of knowledge: a concrete case about football data. In: Benjamins, V.R., d'Aquin, M., Gordon, A. (eds.) 2013 Proceedings of the 7th International Conference on Knowledge Capture (K-CAP 2013), pp. 9–16. ACM (2013)
12. Bozzato, L., Homola, M., Serafini, L.: Towards more effective tableaux reasoning for CKR. In: Kazakov, Y., Lembo, D., Wolter, F. (eds.) Proceedings of the 2012 International Workshop on Description Logics, DL-2012, CEUR Workshop Proceedings, Rome, Italy, 7–10 June 2012, vol. 846, pp. 114–124. CEUR-WS.org (2012)
13. Bozzato, L., Serafini, L.: Materialization calculus for contexts in the semantic web. In: DL 2013. CEUR-WP, vol. 1014, pp. 552–572. CEUR-WS.org (2013)
14. Casini, G., Straccia, U.: Rational closure for defeasible description logics. In: Janhunen, T., Niemelä, I. (eds.) JELIA 2010. LNCS (LNAI), vol. 6341, pp. 77–90. Springer, Heidelberg (2010). https://doi.org/10.1007/978-3-642-15675-5_9. [23]

15. Casini, G., Straccia, U., Meyer, T.: A polynomial time subsumption algorithm for nominal safe $\mathcal{ELO}_\perp$ under rational closure. Inf. Sci. (2018, in press). https://doi.org/10.1016/j.ins.2018.09.037
16. Dao-Tran, M., Eiter, T., Krennwallner, T.: Realizing default logic over description logic knowledge bases. In: Sossai, C., Chemello, G. (eds.) ECSQARU 2009. LNCS (LNAI), vol. 5590, pp. 602–613. Springer, Heidelberg (2009). https://doi.org/10.1007/978-3-642-02906-6_52
17. Eiter, T., Ianni, G., Lukasiewicz, T., Schindlauer, R., Tompits, H.: Combining answer set programming with description logics for the semantic web. Artif. Intell. **172**(12–13), 1495–1539 (2008). https://doi.org/10.1016/j.artint.2008.04.002
18. Gelfond, M., Lifschitz, V.: Classical negation in logic programs and disjunctive databases. New Gener. Comput. **9**, 365–385 (1991)
19. Giordano, L., Dupré, D.T.: Defeasible reasoning in $\mathcal{SROEL}$: from rational entailment to rational closure. Fundam. Inform. **161**(1–2), 135–161 (2018). https://doi.org/10.3233/FI-2018-1698
20. Giordano, L., Gliozzi, V., Olivetti, N., Pozzato, G.L.: Reasoning about typicality in low complexity DLs: the logics $\mathcal{EL}^\perp T_{min}$ and DL-Lite$_c T_{min}$. In: Walsh, T. (ed.) Proceedingsof the 22nd International Joint Conference on Artificial Intelligence, IJCAI 2011, Barcelona, Catalonia, Spain, 16–22 July 2011, pp. 894–899. IJCAI/AAAI (2011). https://doi.org/10.5591/978-1-57735-516-8/IJCAI11-155
21. Giordano, L., Gliozzi, V., Olivetti, N., Pozzato, G.L.: A non-monotonic description logic for reasoning about typicality. Artif. Intell. **195**, 165–202 (2013)
22. Giunchiglia, F., Serafini, L.: Multilanguage hierarchical logics, or: how we can do without modal logics. Artif. Intell. **65**(1), 29–70 (1994)
23. Janhunen, T., Niemelä, I. (eds.): JELIA 2010. LNCS (LNAI), vol. 6341. Springer, Heidelberg (2010). https://doi.org/10.1007/978-3-642-15675-5
24. Klarman, S.: Reasoning with contexts in description logics. Ph.D. thesis, Free University of Amsterdam (2013)
25. Klarman, S., Gutiérrez-Basulto, V.: Two-dimensional description logics of context. In: DL 2011. CEUR-WP, vol. 745. CEUR-WS.org (2011)
26. Krötzsch, M.: Efficient inferencing for OWL EL. In: Janhunen, T., Niemelä, I. (eds.) JELIA 2010. LNCS (LNAI), vol. 6341, pp. 234–246. Springer, Heidelberg (2010). https://doi.org/10.1007/978-3-642-15675-5_21
27. Lenat, D.: The dimensions of context space. Technical report, CYCorp (1998). http://www.cyc.com/doc/context-space.pdf
28. McCarthy, J.: Notes on formalizing context. In: Bajcsy, R. (ed.) IJCAI 1993, pp. 555–562. Morgan Kaufmann (1993)
29. Motik, B., Fokoue, A., Horrocks, I., Wu, Z., Lutz, C., Grau, B.C.: OWL 2 web ontology language profiles. W3C recommendation, W3C, October 2009. http://www.w3.org/TR/2009/REC-owl2-profiles-20091027/
30. Pensel, M., Turhan, A.-Y.: Including quantification in defeasible reasoning for the description logic $\mathcal{EL}_\perp$. In: Balduccini, M., Janhunen, T. (eds.) LPNMR 2017. LNCS (LNAI), vol. 10377, pp. 78–84. Springer, Cham (2017). https://doi.org/10.1007/978-3-319-61660-5_9
31. Pensel, M., Turhan, A.: Reasoning in the defeasible description logic $\mathcal{EL}_\perp$ - computing standard inferences under rational and relevant semantics. Int. J. Approx. Reasoning **103**, 28–70 (2018). https://doi.org/10.1016/j.ijar.2018.08.005
32. Reiter, R.: A logic for default reasoning. Artif. Intell. **13**(12), 81–132 (1980). https://doi.org/10.1016/0004-3702(80)90014-4
33. Serafini, L., Homola, M.: Contextualized knowledge repositories for the semantic web. J. Web Semant. **12**, 64–87 (2012)

Strong Explanations
for Nonmonotonic Reasoning

Gerhard Brewka$^{(\boxtimes)}$ ⓘ and Markus Ulbricht ⓘ

Department of Informatics, Leipzig University,
Augustusplatz 10, 04109 Leipzig, Germany
brewka@informatik.uni-leipzig.de

Abstract. The ability to generate explanations for inferences drawn from a knowledge base is of utmost importance for intelligent systems. A central notion in this context are minimal subsets of the knowledge base entailing a certain formula. Such subsets are often referred to as justifications, and their identification is called axiom pinpointing. As observed by Franz Baader, this concept of explanations is useful for monotonic logics in which additional information can never invalidate former conclusions. However, for nonmonotonic logics the concept simply makes no sense. In this paper, we introduce a different notion, called strong explanation. Strong explanations coincide with the standard notion for monotonic logics, but also handle the nonmonotonic case adequately.

1 Introduction

Explainable AI is a highly relevant topic of current research, see e.g. the DARPA XAI initiative[1]. It aims to come up with intelligent systems able to provide reasons for decisions made and actions taken. The ultimate goal is to enable human users to understand and to appropriately trust artificially intelligent systems.

Although the generation of convincing and easy to understand explanations is less of a problem for symbolic than for subsymbolic approaches, there are still many issues to be solved, in particular for nonmonotonic formalisms where additional information may lead to the withdrawal of earlier consequences. In approaches based on monotonic logics, like classical logic, description logics, definite logic programs etc., the central notion underlying the definition of explanations are minimal sets of axioms A implying a certain consequence p. Such sets are sometimes also called *justifications* for p, see for instance [16,17,19], where a *justification* for a formula p in a description logic knowledge base K is a minimal subset K' of K such that K' entails p. The identification of adequate subsets K' of a knowledge base K is often referred to as axiom pinpointing, or simply pinpointing (see [3,22,23] for some recent references).

Unfortunately, this notion of explanation, respectively justification[2], is not helpful at all for nonmonotonic logics. As Baader and Peñaloza put it [3]:

---

[1] See www.darpa.mil/program/explainable-artificial-intelligence.

[2] We will use both terms interchangeably in this paper.

C. Lutz et al. (Eds.): Baader Festschrift, LNCS 11560, pp. 135–146, 2019.
https://doi.org/10.1007/978-3-030-22102-7_6

"In fact, for non-monotonic logics, looking at minimal sets of axioms that have a given consequence does not make much sense."

We could not agree more with this statement. The problem, intuitively, is the following: a nonmonotonic knowledge base K may have a minimal subset K' entailing p. However, this does not guarantee that K itself also entails p. There may be information in $K \backslash K'$ that explicitly blocks the derivation of p. This makes the standard notion of explanations useless in the broader context of nonmonotonic reasoning.

In a nutshell, the observation made by Baader and Peñaloza is the starting point of this paper. The question we try to answer is the following: given that minimal sets of axioms entailing a specific formula do not do the job, *what is the adequate notion of explanations that works for both monotonic and nonmonotonic logics?*

The answer we will provide is closely related to strong inconsistency, as thoroughly investigated in [7]. Strong inconsistency is a strengthening of the standard notion of inconsistency which guarantees that the Reiter hitting set duality [25] also applies to nonmonotonic logics. Recall that a hitting set of a set of sets S is a minimal set H such that H has a nonempty intersection with all elements of S. Reiter's duality states that a set of formulas K' is a maximal consistent subset of a knowledge base K iff $K \backslash K'$ is a hitting set of the collection of all minimal inconsistent subsets of K. This duality is particularly relevant for diagnosis and repair of knowledge bases: if a knowledge base needs to be repaired because it is inconsistent, one can simply compute its minimal inconsistent subsets and a hitting set of these sets. Now Reiter's result guarantees that eliminating the hitting set from the original knowledge base not only removes the inconsistency, but also does so without unnecessarily giving up information.

As to be expected, the duality breaks down for nonmonotonic formalisms for obvious reasons: in a nonmonotonic setting, a consistent knowledge base may have inconsistent - and thus minimal inconsistent - subsets. Thus, applying Reiter's result *as is* to a nonmonotonic knowledge base can lead to the elimination of formulas although there is nothing to repair at all.

The solution developed in [7] is based on a stronger notion of inconsistency. $K' \subseteq K$ is strongly inconsistent relative to K iff each K'' such that $K' \subseteq K'' \subseteq K$ is inconsistent. This obviously avoids situations where the inconsistency in K' is resolved by additional information in $K \backslash K'$. One of the main results in [7] states that by replacing inconsistency with strong inconsistency, Reiter's hitting set theorem can indeed be generalized to arbitrary logics, including nonmonotonic ones.

Detecting inconsistencies is a special case of pinpointing whenever there is a formula which is entailed if and only if the knowledge base is inconsistent. In description logics formulas like $\bot(c)$ or $A \sqcap \neg A(c)$ will do, that is K is inconsistent iff K entails $\bot(c)$, or equivalently $A \sqcap \neg A(c)$. Justifications for these assertions then correspond to minimal inconsistent subsets of K. It is thus not overly surprising that the basic ideas underlying strong inconsistency also prove useful for generalizing explanations to the nonmonotonic case. In particular, we will present in this paper a notion of explanations which strengthens the standard notion and at the same time relativizes it to a specific knowledge base: we will define what we mean by a strong explanation relative to a knowledge base K.

The results in [7] cover arbitrary logics. The investigation thus had to be based on a highly abstract notion of a logic. For the sake of readability, we will not discuss arbitrary monotonic and nonmonotonic logics in this paper. Rather, we will focus here on one quite successful nonmonotonic formalism, namely logic programs under answer set semantics. This way it will be a lot easier to illustrate our approach. We emphasize, though, that what we present in this paper can be generalized to arbitrary monotonic and nonmonotonic logics.

The rest of the paper is organized as follows. We start with some background on logic programs under answer set semantics. The need for a stronger notion of explanations for nonmonotonic formalisms is then illustrated in Sect. 3, using logic programs to illustrate the underlying issues. Our new notion of strong explanations is introduced in Sect. 4. Section 5 discusses some repercussions of our analysis to description logics. Finally, Sect. 6 concludes.

2 Background

Logic programs under the answer set semantics [13,14] are a popular non-monotonic formalism for knowledge representation and reasoning which consists of rules possibly containing default-negated literals. In this paper, we consider so-called extended logic programs with two kinds of negation, namely strong negation "¬" and default negation "not", under the answer set semantics [13,14]. Our focus is on propositional programs. Note that this is not a severe restriction as programs with variables are commonly treated as compact representations of their ground instantiations [6].

A (ground) extended logic program P over a set of literals $\mathcal{L}$ is a finite set of rules r of the form

$$l_0 \leftarrow l_1, \ldots, l_m, \text{not } l_{m+1}, \ldots, \text{not } l_n. \tag{1}$$

where $l_0, \ldots, l_n \in \mathcal{L}$ and $0 \leq m \leq n$. In particular, no function symbols or variables occur in r.

For a rule r of the form (1), we write $head(r) = l_0$, $pos(r) = \{l_1, \ldots, l_m\}$, $neg(r) = \{l_{m+1}, \ldots, l_n\}$ and $body(r) = \{l_1, \ldots, l_m, \text{not } l_{m+1}, \ldots, \text{not } l_n\}$. For a set M of literals, let $\mathcal{A}(M)$ be the set of all atoms occurring in M. We let $\mathcal{A}(r)$ and $\mathcal{L}(r)$ be the set of all atoms and literals occurring in r, respectively. Similarly, let $\mathcal{A}(P)$ and $\mathcal{L}(P)$ be the set of all atoms and literals that occur in a program P, respectively. We often write "l_0." instead of "$l_0 \leftarrow .$" for rules with trivial body and call such rules a *fact*.

We now turn to the semantics, i.e. the definition of answer sets. Intuitively speaking, an answer set is a set of literals which is (1) closed under the rules of the program, that is, whenever the body of a rule is satisfied, then its head must be included in an answer set, and (2) grounded, that is, for each literal in an answer set there must be a valid derivation. A derivation is valid if it is non-circular and based only on rules whose default negated literals are not contained in the answer set. The challenge then is to deal adequately with the circularity implicit in the second condition.

There are actually variants of the formal definition in the literature which differ in whether inconsistent answer sets are admitted or not. We follow the original definition in [14] which allows for a single inconsistent answer set, namely $\mathcal{L}$, in cases where a subset of rules without default negation generates an inconsistency.

For a set M of literals and a literal l we say M satisfies l ($M \models l$) iff $l \in M$. If L is a set of literals, then $M \models L$ iff $M \models l$ for all $l \in L$. Now consider a classical rule, i.e. a rule r of the form (1) with $m = n$. We say M satisfies r, denoted by $M \models r$ iff $M \models l_0$ whenever $M \models body(r)$. For a program P we say M satisfies P, denoted $M \models P$, whenever M satisfies r for each $r \in P$. Finally, a set M of literals is called *consistent* if it does not contain both a and $\neg a$ for any atom a.

Now we are ready to define answer sets of a given program.

Definition 1. *Let P be an extended logic program without default negation. A set M of literals is the answer set of P if*

1. *M is consistent, $M \models P$ and there is no $M' \subsetneq M$ with $M' \models P$, or*
2. *there is no consistent set of literals M' such that $M' \models P$ and $M = \mathcal{L}$.*

For an arbitrary program P, M is an answer set of P iff M is an answer set of P^M, where P^M is the reduct of P wrt. M, i.e.

$$P^M = \{head(r) \leftarrow pos(r) \mid r \in P, \, neg(r) \cap M = \emptyset\}.$$

Note that programs without default negation always have a unique answer set. Programs with default negation may have an arbitrary number of answer sets, including zero. If a program has the inconsistent answer set $\mathcal{L}$, then this is its only answer set. A program P is called *consistent* if it has at least one consistent answer set (in which case all of its answer sets are consistent), otherwise it is called *inconsistent*. A program P can be inconsistent because

- it has no answer set, or
- it only has the inconsistent answer set.

The first type of inconsistency is often called incoherence. Answer sets give rise to two different notions of consequence. We say P *skeptically entails* a literal l if l is contained in all answer sets of P. P *credulously entails* l if l is contained in at least one answer set of P.

3   Axiom Pinpointing in Logic Programming

As pointed out in the introduction, axiom pinpointing, or simply pinpointing (see [3,22,23] for recent references), is the identification of minimal subsets of axioms which are responsible for a certain (unintended) conclusion. Following [16,17,19], a *justification* for a formula p in a knowledge base K is a minimal subset K' of K such that K' entails p. Although the references introduce these notions in the context of description logics, they directly apply to logic programs without default negation, which we call classical logic programs. The reason why these notions are applicable, as we will see, is the monotonicity of this class of logic programs.

Classical logic programs always have a single answer set. Credulous and skeptical consequence thus coincide.

Example 1. Consider the classical logic program P_0:

$$peng \leftarrow$$
$$bird \leftarrow peng$$
$$\neg flies \leftarrow peng$$
$$flies \leftarrow bird, \neg peng$$

P_0 has the answer set $\{peng, bird, \neg flies\}$. The justification for $\neg flies$, that is, a minimal subset P_0' of P_0 entailing $\neg flies$, is:

$$peng \leftarrow$$
$$\neg flies \leftarrow peng$$

So far, everything is fine. In the rest of this section, however, we illustrate why the standard notion of explanation fails for nonmonotonic logic programs, that is, logic programs with default negation. As there are two consequence relations for such logic programs, namely skeptical and credulous consequence, justifications come in two forms: let P be a logic program, l a literal. $P' \subseteq P$ is a skeptical (respectively credulous) justification for l iff P' is a minimal subset of P such that l is a skeptical (respectively credulous) consequence of P'.

Example 2. Consider the extended logic program P_1, a straightforward formalization of the famous flying birds example:

$$peng \leftarrow$$
$$bird \leftarrow peng$$
$$\neg flies \leftarrow peng$$
$$flies \leftarrow bird, not \ \neg flies$$

P_1 has a single answer set, namely $\{peng, bird, \neg flies\}$. In case of a single answer set credulous and skeptical consequence coincide, and obviously *flies* is not among the consequences.

Note, however, that there is a "justification" for *flies*, that is, a minimal subset P_1' of P_1 entailing *flies*. In this case P_1' is:

$$peng \leftarrow$$
$$bird \leftarrow peng$$
$$flies \leftarrow bird, not \ \neg flies$$

Like P_1 the program P_1' has a single answer set, in this case $\{peng, bird, flies\}$. Apparently *flies* is both a skeptical and a credulous consequence of P_1'. Moreover, P_1' is a minimal subset of P_1 and thus a justification for *flies*.

This illustrates that the standard notion of justifications may lead to justifications for formulas which are not at all consequences of the original program P. This shows that a different notion of explanations is needed. The example already suggests the problem that needs to be addressed: we have to exclude explanations that cease to exist when additional information included in the original program P is taken into account. In our example, the rule $\neg flies \leftarrow peng$ which is missing from the justification destroys the derivation of *flies*. The example also clarifies why the standard notion works properly for monotonic logics: if reasoning is monotonic, then justifications can never be invalidated by taking additional information into account. It is for nonmonotonic formalisms only that the standard notion of explanations breaks down. We show in the next section how this can be avoided by an adequate strengthening of the notion of explanations.

4   Strong Explanations

In this section we develop a stronger notion of explanations for logic programs. As mentioned earlier, similar notions can be developed for arbitrary nonmonotonic formalisms, but for the sake of clarity we restrict our discussion to logic programming in this paper.

As discussed in the background section, there are two notions of consequence for logic programs under answer set semantics, namely skeptical and credulous consequence. Fortunately, these notions can be handled uniformly in a single definition.

As pointed out at the end of the last section, the reason for the failure of the standard notion of justifications is that it does not account for the possibility of information which is outside the justification but still part of the program and which blocks a derivation from a justification. To fill this gap we introduce a new, strong type of explanations which are useful also in the nonmonotonic case.

Definition 2. *Let P be an extended logic program, l a literal. A set of rules $P' \subseteq P$ is called a strong skeptical (respectively credulous) explanation of l with respect to P iff*

1. *l is a skeptical (respectively credulous) consequence of each P'' satisfying $P' \subseteq P'' \subseteq P$, and*
2. *there is no proper subset of P' satisfying condition 1.*

This definition makes sure that strong explanations cannot be invalidated by any additional information in the program P.

Example 3. Consider again Example 2 discussed in Sect. 4. There the program P_1':

$$peng \leftarrow$$
$$bird \leftarrow peng$$
$$flies \leftarrow bird, \text{not } \neg flies$$

provided a justification for *flies*. It is easy to see that this program is not a strong explanation with respect to $P_1 = P_1' \cup \{\neg flies \leftarrow peng\}$ as P_1 does not entail *flies*. In fact, the only strongly explainable literal besides *peng* and *bird* is $\neg flies$. The strong explanation (both skeptical and credulous) for the latter literal is P_1 itself.

We next discuss an (abstract) example where strong skeptical and strong credulous explanations differ.

Example 4. Consider the program P_2

$$a \leftarrow \text{not } b$$
$$b \leftarrow \text{not } a$$
$$c \leftarrow a$$
$$c \leftarrow b$$

P_2 has two answer sets, namely $\{a, c\}$ and $\{b, c\}$. Literal a has the strong credulous explanation $\{a \leftarrow \text{not } b\}$. Note that a remains a credulous consequence no matter which rules of P_2 are added to the explanation. Similarly, b has the strong credulous explanation $\{b \leftarrow \text{not } a\}$. None of these two literals possesses a strong skeptical explanation. There are two strong skeptical explanation for c, though, namely X_1:

$$a \leftarrow \text{not } b$$
$$c \leftarrow a$$
$$c \leftarrow b$$

and also X_2:

$$b \leftarrow \text{not } a$$
$$c \leftarrow a$$
$$c \leftarrow b$$

Note that both rules with head c must be included in the explanations as otherwise adding $b \leftarrow \text{not } a$ to $X_1 \backslash \{c \leftarrow b\}$, respectively $a \leftarrow \text{not } b$ to $X_2 \backslash \{c \leftarrow a\}$ would destroy the skeptical derivability of c.

Let us now briefly discuss how to characterize strong explanations of a given literal l. We focus on credulous explanations. As already mentioned a notion quite similar to strong explanations is *strong inconsistency* (see [7]). Recall that a subprogram H of P is strongly inconsistent if H' is inconsistent for each $H \subseteq H' \subseteq P$. Let $SI_{min}(P)$ be the set of all minimal strongly subsets of P. In [7] maximal consistent subsets of a program (denoted by $C_{max}(P)$) are characterized via hitting sets of strongly inconsistent sets, more precisely: S is a minimal hitting set of $SI_{min}(P)$ iff $P \backslash S \in C_{max}(P)$; here, S is a hitting set of a set $\mathcal{M}$ of sets iff $S \cap M \neq \emptyset$ for each $M \in \mathcal{M}$. We can give a similar result. For this, let $StrEx(P, l)$ be the set of strong explanations of l.

Proposition 1 (Credulous Explanations). *Let P be a logic program, l a literal. A set S is a minimal hitting set of $StrEx(P, l)$ (with respect to credulous reasoning) if and only if $P \backslash S$ is maximal such that $P \backslash S$ does not possess a consistent answer set that contains l.*

Proof. The duality result from [7] actually applies to arbitrary logics. We may thus define an (artificial) logic which models logic programs such that a program P is considered "inconsistent" if

– at least one answer set M of P is consistent and contains l.

Then, a program P is considered "consistent" if it does not possess a consistent answer set containing l (including the case that P does not possess a consistent answer set at all). Now, the claim is just the duality characterization from [7] applied to this modified logic.

Let us also consider skeptical explanations.

Proposition 2 (Skeptical Explanations). *Let P be a logic program, l a literal. A set S is a minimal hitting set of $StrEx(P, l)$ (with respect to skeptical reasoning) if and only if $P \backslash S$ is maximal such that $P \backslash S$ is either inconsistent or possess a consistent answer set that contains l.*

Proof. As before, we define an (artificial) logic which models logic programs such that a program P is considered "inconsistent" if

– P possesses consistent answer sets and all of them contain l.

Then, a program P is considered "consistent" if

– P does not possess a consistent answer set or
– at least one consistent answer set of P does not contain l.

Again, the claim is the duality characterization from [7] applied to the modified logic.

5   Explanation for Description Logics

Description logics are a family of knowledge representation formalisms which are tailored towards the representation of concepts and relationships between them. Description logic knowledge bases are commonly split into two parts, the so-called TBox which defines the terminology, that is, it contains the definition of concepts and their relationships, and the ABox which uses these concepts to describe a particular domain.

Specific description logics differ in the constructors available to define concepts, and in the way relationships among concepts are defined. Typical concept constructors are intersection ($\sqcap$), union ($\sqcup$) and complement ($\neg$) of concepts, as well as existential restriction and value restriction. For instance, the existential restriction $\exists r.\, C$ represents the objects standing in relation r to some C-object. Relationships between concepts are expressed using axioms of the form $C_1 \sqsubseteq C_2$ expressing that C_1 is subsumed by C_2. The symbol $\bot$ usually represents the empty, unsatisfiable concept. The semantics of description logics is standard first order semantics. We refer the reader to [2] for further details.

In this section we give a brief overview of what has been done in terms of explanations in the area of (monotonic) description logics.

As pointed out in the introduction, axiom pinpointing, or simply pinpointing (see [3,22,23] for recent references), is the identification of minimal subsets of axioms which are responsible for a certain (unintended) conclusion. Following [16,17,19], a *justification* for a formula p in a description logic knowledge base K is a minimal subset K' of K such that K' entails p.

The identification of justifications is particularly relevant for debugging description logic knowledge bases. A typical modelling error one would like to detect (and repair) are unsatisfiable concepts, that is, concepts which cannot have instances and thus are subsumed by $\bot$. The reasons for the unsatisfiability of a concept C can be detected by computing the justifications for $C \sqsubseteq \bot$. Justifications for formulas of this form consisting of TBox elements only are called MUPS (minimal unsatisfiability-preserving sub-TBox) in [26–28], justifications for formulas of the form $C_1 \sqsubseteq C_2$ are called MinA in [3,23]. Note that the presence of an unsatisfiable concept C in K does not mean K is inconsistent. However, as soon as the ABox contains an assertion of the form $C(a)$ for some constant a an inconsistency arises.

The importance of justifications for explaining results, detecting modelling errors and potentially repairing description logic knowledge bases - via Reiter's hitting set duality - is apparent. Corresponding explanation services were first built into the ontology editor Swoop [18] and are by now standard in modern ontology development tools like Protégé[3].

Description logics are monotonic and thus lack means to express what is typically the case and to handle exceptions. For this reason, several nonmonotonic extensions of description logics have been proposed over the last decades, starting from combinations of description logics with Reiter's default logic [1], to circumscriptive description logics [4], autoepistemic description logics [12], description logics with typicality [15], and preferential, respectively rational description logics [8–11,24]. The latter are based on the influential KLM theory of propositional nonmonotonic reasoning [20,21].

---

[3] See https://protege.stanford.edu/.

A detailed analysis of these logics is beyond the scope of this paper. Nevertheless, we would like to illustrate that our analysis of explanations also applies to other nonmonotonic formalisms and use nonmonotonic description logics as an example.

Without going into any formal detail regarding how the nonmonotonic inference relation is defined, let us assume there is – in addition to the standard subsumption relation $\sqsubseteq$ – a defeasible form of subsumption $\sqsubseteq\!\!\!\sim$. Moreover, we assume there is a single set of nonmonotonic consequences such that the distinction between skeptical and credulous consequence is obsolete. In this case a strong explanation of a formula p with respect to a nonmonotonic description logic knowledge base K is a minimal subset X of K such that p is entailed by all K' with $X \subseteq K' \subseteq K$. For monotonic description logics justifications and strong justifications obviously coincide. For nonmonotonic logics the latter notion is the relevant one, as justifications may exist for formulas which are not entailed.

Example 5. Consider the description logic knowledge base[4]

$$K = \{P \sqsubseteq B, P \sqsubseteq \neg F, B \sqsubseteq\!\!\!\sim F, P(t)\}.$$

Here $\sqsubseteq\!\!\!\sim$ represents defeasible subsumption, that is, $B \sqsubseteq\!\!\!\sim F$ means Bs are normally Fs. In all nonmonotonic description logics with defeasible subsumption we are aware of, e.g. [8–11], the assertion $\neg F(t)$ is entailed by K, the justification being $X = \{P(t), P \sqsubseteq \neg F\}$. X is also a strong justification, as adding arbitrary formulas from K to X does not invalidate this entailment. Moreover, there is no proper subset of X entailing $\neg F(t)$.

Now consider $F(t)$. There is an explanation for $F(t)$, namely $\{P(t), P \sqsubseteq B, B \sqsubseteq\!\!\!\sim F\}$. This explanation, however, is obviously not a strong explanation as $F(t)$ is not a consequence of K.

This illustrates that for nonmonotonic description logics (and nonmonotonic logics in general) strong explanation is the right notion to use for explaining results.

6   Conclusions

In this paper we pointed out that the standard notion of justification/explanation fails to adequately cover nonmonotonic formalisms. For illustrative purposes we focused on extended logic programs under answer set semantics and developed a new, stronger type of explanations for such programs. Strong explanations coincide with standard explanations for monotonic logics, but provide an adequate generalization also to the nonmonotonic case. We also illustrated some of the repercussions of our analysis for description logics, in particular for defeasible extensions of such logics.

---

[4] Since Tbox and ABox elements differ syntactically, we do not explicitly distinguish these to sets.

Throughout the paper we repeatedly made what might be considered strong claims about the applicability of our notions to arbitrary monotonic and non-monotonic logics, without clarifying what we mean by such logics. Let us just point out that an abstract notion of a logic, which slightly refines the one used for multi-context systems in [5], was developed in [7]. The framework used there makes little assumptions about logics, except that logics have a syntax defining knowledge bases and a semantics which assigns - potentially multiple - sets of explicit beliefs to each knowledge base. It was shown that strong inconsistency applies to arbitrary logics captured by the abstract framework. Strong explanations work for arbitrary logics for the same reasons strong inconsistency works for them.

Acknowledgements. We thank the reviewers for their comments which helped to significantly improve this paper. The work presented in this paper was supported by QuantLA, the joint Dresden/Leipzig doctoral school on Quantitative Logics and Automata (DFG Research Training Group 1763) which was initiated and run by Franz Baader. The second author was a PhD student in QuantLA from 2015–2018. There was additional support from the DFG Research Unit Hybris (Hybrid Reasoning in Intelligent Systems, FOR 1513) in which the first author had the great pleasure to cooperate with Franz Baader for 6 years. The first author is deeply indebted to Franz for more than three decades of challenge, inspiration, and insight.

References

1. Baader, F., Hollunder, B.: Embedding defaults into terminological knowledge representation formalisms. J. Autom. Reason. **14**(1), 149–180 (1995)
2. Baader, F., Horrocks, I., Lutz, C., Sattler, U.: An Introduction to Description Logic. Cambridge University Press, Cambridge (2017)
3. Baader, F., Peñaloza, R.: Axiom pinpointing in general tableaux. J. Log. Comput. **20**(1), 5–34 (2010)
4. Bonatti, P.A., Lutz, C., Wolter, F.: Description logics with circumscription. In: Proceedings of Tenth International Conference on Principles of Knowledge Representation and Reasoning, Lake District of the United Kingdom, 2–5 June 2006, pp. 400–410 (2006)
5. Brewka, G., Eiter, T.: Equilibria in heterogeneous nonmonotonic multi-context systems. In: Proceedings of the Twenty-Second AAAI Conference on Artificial Intelligence, Vancouver, British Columbia, Canada, 22–26 July 2007, pp. 385–390 (2007)
6. Brewka, G., Eiter, T., Truszczynski, M.: Answer set programming at a glance. Commun. ACM **54**(12), 92–103 (2011)
7. Brewka, G., Thimm, M., Ulbricht, M.: Strong inconsistency. Artif. Intell. **267**, 78–117 (2019)
8. Casini, G., Meyer, T., Moodley, K., Varzinczak, I.J.: Towards practical defeasible reasoning for description logics. In: Informal Proceedings of the 26th International Workshop on Description Logics, Ulm, Germany, 23–26 July 2013, pp. 587–599 (2013)
9. Casini, G., Meyer, T., Varzinczak, I.J., Moodley, K.: Nonmonotonic reasoning in description logics: rational closure for the ABox. In: Informal Proceedings of the 26th International Workshop on Description Logics, Ulm, Germany, 23–26 July 2013, pp. 600–615 (2013)

10. Casini, G., Meyer, T., Moodley, K., Sattler, U., Varzinczak, I.: Introducing defeasibility into OWL ontologies. In: Arenas, M., et al. (eds.) ISWC 2015. LNCS, vol. 9367, pp. 409–426. Springer, Cham (2015). https://doi.org/10.1007/978-3-319-25010-6_27

11. Casini, G., Straccia, U.: Rational closure for defeasible description logics. In: Janhunen, T., Niemelä, I. (eds.) JELIA 2010. LNCS (LNAI), vol. 6341, pp. 77–90. Springer, Heidelberg (2010). https://doi.org/10.1007/978-3-642-15675-5_9

12. Donini, F.M., Nardi, D., Rosati, R.: Autoepistemic description logics. In: Proceedings of the Fifteenth International Joint Conference on Artificial Intelligence, IJCAI 1997, Nagoya, Japan, 23–29 August 1997, vol. 2, pp. 136–141 (1997)

13. Gelfond, M., Leone, N.: Logic programming and knowledge representation - the a-Prolog perspective. Artif. Intell. **138**(1–2), 3–38 (2002)

14. Gelfond, M., Lifschitz, V.: Classical negation in logic programs and disjunctive databases. New Gener. Comput. **9**(3/4), 365–386 (1991)

15. Giordano, L., Gliozzi, V., Olivetti, N., Pozzato, G.L.: A non-monotonic description logic for reasoning about typicality. Artif. Intell. **195**, 165–202 (2013)

16. Horridge, M., Bail, S., Parsia, B., Sattler, U.: Toward cognitive support for OWL justifications. Knowl.-Based Syst. **53**, 66–79 (2013)

17. Horridge, M., Parsia, B., Sattler, U.: Laconic and precise justifications in OWL. In: Sheth, A., et al. (eds.) ISWC 2008. LNCS, vol. 5318, pp. 323–338. Springer, Heidelberg (2008). https://doi.org/10.1007/978-3-540-88564-1_21

18. Kalyanpur, A., Parsia, B., Sirin, E., Grau, B.C., Hendler, J.A.: Swoop: a web ontology editing browser. J. Web Semant. **4**(2), 144–153 (2006)

19. Kalyanpur, A., Parsia, B., Sirin, E., Hendler, J.A.: Debugging unsatisfiable classes in OWL ontologies. J. Web Semant. **3**(4), 268–293 (2005)

20. Kraus, S., Lehmann, D.J., Magidor, M.: Nonmonotonic reasoning, preferential models and cumulative logics. Artif. Intell. **44**(1–2), 167–207 (1990)

21. Lehmann, D.J., Magidor, M.: What does a conditional knowledge base entail? Artif. Intell. **55**(1), 1–60 (1992)

22. Manthey, N., Peñaloza, R., Rudolph, S.: Efficient axiom pinpointing in $\mathcal{EL}$ using SAT technology. In: Proceedings of the 29th International Workshop on Description Logics, Cape Town, South Africa, 22–25 April 2016 (2016). http://ceur-ws.org/Vol-1577/paper_33.pdf

23. Peñaloza, R., Sertkaya, B.: Understanding the complexity of axiom pinpointing in lightweight description logics. Artif. Intell. **250**, 80–104 (2017)

24. Pensel, M., Turhan, A.: Reasoning in the defeasible description logic $\mathcal{EL}_\perp$ - computing standard inferences under rational and relevant semantics. Int. J. Approx. Reason. **103**, 28–70 (2018). https://doi.org/10.1016/j.ijar.2018.08.005

25. Reiter, R.: A theory of diagnosis from first principles. Artif. Intell. **32**(1), 57–95 (1987)

26. Schlobach, S.: Diagnosing terminologies. In: Proceedings of the Twentieth National Conference on Artificial Intelligence and the Seventeenth Innovative Applications of Artificial Intelligence Conference, Pittsburgh, Pennsylvania, USA, 9–13 July 2005, pp. 670–675 (2005)

27. Schlobach, S., Cornet, R.: Non-standard reasoning services for the debugging of description logic terminologies. In: Proceedings of the Eighteenth International Joint Conference on Artificial Intelligence, IJCAI 2003, Acapulco, Mexico, 9–15 August 2003, pp. 355–362 (2003)

28. Schlobach, S., Huang, Z., Cornet, R., van Harmelen, F.: Debugging incoherent terminologies. J. Autom. Reason. **39**(3), 317–349 (2007)

A KLM Perspective on Defeasible Reasoning for Description Logics

Katarina Britz[1], Giovanni Casini[2], Thomas Meyer[3(✉)], and Ivan Varzinczak[1,4]

[1] CAIR, Stellenbosch University, Stellenbosch, South Africa
`abritz@sun.ac.za`
[2] CSC, Université du Luxembourg, Esch-sur-Alzette, Luxembourg
`giovanni.casini@gmail.com`
[3] CAIR, University of Cape Town, Cape Town, South Africa
`tmeyer@cs.uct.ac.za`
[4] CRIL, Université d'Artois & CNRS, Lens, France
`varzinczak@cril.fr`

Abstract. In this paper we present an approach to defeasible reasoning for the description logic $\mathcal{ALC}$. The results discussed here are based on work done by Kraus, Lehmann and Magidor (KLM) on defeasible conditionals in the propositional case. We consider versions of a preferential semantics for two forms of defeasible subsumption, and link these semantic constructions formally to KLM-style syntactic properties via representation results. In addition to showing that the semantics is appropriate, these results pave the way for more effective decision procedures for defeasible reasoning in description logics. With the semantics of the defeasible version of $\mathcal{ALC}$ in place, we turn to the investigation of an appropriate form of *defeasible entailment* for this enriched version of $\mathcal{ALC}$. This investigation includes an algorithm for the computation of a form of defeasible entailment known as *rational closure* in the propositional case. Importantly, the algorithm relies completely on classical entailment checks and shows that the computational complexity of reasoning over defeasible ontologies is no worse than that of the underlying classical $\mathcal{ALC}$. Before concluding, we take a brief tour of some existing work on defeasible extensions of $\mathcal{ALC}$ that go beyond defeasible subsumption.

Keywords: Knowledge representation and reasoning ·
Description logics · Defeasible reasoning · Preferential semantics

1 Introduction

Description logics (DLs) [1] are central to many modern AI and database applications since they provide the logical foundation of formal ontologies. Yet, as classical formalisms, DLs do not allow for the proper representation of and reasoning with defeasible information, as shown up in the following example, adapted from Giordano et al. [39]: Students do not get tax invoices; employed

C. Lutz et al. (Eds.): Baader Festschrift, LNCS 11560, pp. 147–173, 2019.
https://doi.org/10.1007/978-3-030-22102-7_7

students do; employed students who are also parents do not. From a naïve (classical) formalisation of this scenario, one concludes that the notion of employed student is an oxymoron, and consequently the concept of employed student is unsatisfiable. A more nuanced view is to represent such statements as *defeasible*.

Endowing DLs with defeasible reasoning features is therefore a promising endeavour from the point of view of applications of knowledge representation and reasoning. Indeed, the past 25 years have witnessed many attempts to introduce defeasible reasoning capabilities in a DL setting, usually drawing on a well-established body of research on non-monotonic reasoning (NMR). These comprise the so-called preferential approaches [19–21,30,32,40,44,45,57,58,63] and the circumscription-based ones [8,9,60], amongst others [7,36,46–48,53,56,62]. Not surprisingly, Franz was among those who first made a meaningful contribution in this regard [2,3].

Preferential extensions of DLs turn out to be particularly promising, mainly because they are based on an elegant, comprehensive and well-studied framework for non-monotonic reasoning in the propositional case proposed by Kraus, Lehmann and Magidor [49,52], often referred to as the *KLM approach*. Such a framework is valuable for a number of reasons. First, it provides for a thorough analysis of some formal properties that any consequence relation deemed as appropriate in a non-monotonic setting ought to satisfy. Such formal properties play a central role in assessing how intuitive the obtained results are and enable a more comprehensive characterisation of the introduced non-monotonic conditional from a logical point of view. Second, the KLM approach allows for many decision problems to be reduced to classical entailment checking, sometimes without blowing up the computational complexity compared to the underlying classical case. Finally, it has a well-known connection with the AGM approach to belief revision [38,59] and with frameworks for reasoning under uncertainty [6,37]. It is therefore reasonable to expect that most, if not all, of the aforementioned features of the KLM approach should transfer to KLM-based extensions of DLs too.

Following the motivation laid out above, several extensions to the KLM approach to description logics have been proposed recently [19,21,23,24,27,30,32, 39,40,44,45], each of them investigating particular constructions and variants of the preferential approach. Here we provide an overview of the formal foundations of preferential defeasible reasoning in DLs. By that we mean (*i*) providing a general and intuitive *semantics*; (*ii*) showing that the corresponding *representation results* (in the KLM sense of the term) hold, linking the semantic constructions to the KLM-style set of properties, and (*iii*) presenting an appropriate analysis of *entailment* in the context of ontologies with defeasible information with an associated decision procedure that is implementable.

After a brief introduction to the required background on the DL we consider here (in Sect. 2), we introduce the notion of defeasible subsumption along with a set of KLM-inspired properties it ought to satisfy (Sect. 3). In particular, using an intuitive semantics for the idea that "usually, an element of the class C is also an element of the class D", we provide a characterisation (via representation results) of two important classes of defeasible statements, namely preferential and rational subsumption. In Sect. 4, we discuss two obvious candidates for the notion of entailment in the context of defeasible DLs, namely preferential and modular entailment. These turn out *not* to have all properties seen as important in a non-monotonic DL setting, mimicking a similar feature in the propositional case [52]. This is followed in Sect. 5 by the presentation of a version of rational entailment satisfying all the required properties, and which can thus be seen as a suitable candidate for defeasible entailment. In Sect. 6 we discuss aspects of defeasible reasoning going beyond defeasible concept inclusion. We conclude in Sect. 7 with some pointers to research following on from the work presented here, and remarks on future related endeavours.

The overview presented in this paper relies heavily on research conducted by the present authors, et al. [15].

2   Background

Description Logics (DLs) [1] are decidable fragments of first-order logic with interesting properties and a variety of applications. There is a whole family of description logics, an example of which is $\mathcal{ALC}$ and on which we shall focus in the present paper. The (concept) language of $\mathcal{ALC}$ is built upon a finite set of atomic *concept names* C, a finite set of *role names* R and a finite set of *individual names* I such that C, R and I are pairwise disjoint. With $A, B, \ldots$ we denote atomic concepts, with $r, s, \ldots$ role names, and with $a, b, \ldots$ individual names. Complex concepts are denoted with $C, D, \ldots$ and are built according to the following rule:

$$C ::= \top \mid \bot \mid \mathsf{C} \mid \neg C \mid C \sqcap C \mid C \sqcup C \mid \forall r.C \mid \exists r.C$$

With $\mathcal{L}$ we denote the *language* of all $\mathcal{ALC}$ concepts.

The semantics of $\mathcal{L}$ is the standard set theoretic Tarskian semantics. An *interpretation* is a structure $\mathcal{I} =_{\mathrm{def}} \langle \Delta^{\mathcal{I}}, \cdot^{\mathcal{I}} \rangle$, where $\Delta^{\mathcal{I}}$ is a non-empty set called the *domain*, and $\cdot^{\mathcal{I}}$ is an *interpretation function* mapping concept names A to subsets $A^{\mathcal{I}}$ of $\Delta^{\mathcal{I}}$, role names r to binary relations $r^{\mathcal{I}}$ over $\Delta^{\mathcal{I}}$, and individual names a to elements of the domain $\Delta^{\mathcal{I}}$, i.e., $A^{\mathcal{I}} \subseteq \Delta^{\mathcal{I}}$, $r^{\mathcal{I}} \subseteq \Delta^{\mathcal{I}} \times \Delta^{\mathcal{I}}$, $a^{\mathcal{I}} \in \Delta^{\mathcal{I}}$. Define $r^{\mathcal{I}}(x) =_{\mathrm{def}} \{ y \mid (x, y) \in r^{\mathcal{I}} \}$. We extend the interpretation function $\cdot^{\mathcal{I}}$ to interpret complex concepts of $\mathcal{L}$ in the following way:

$$\top^{\mathcal{I}} =_{\mathrm{def}} \Delta^{\mathcal{I}}, \quad \bot^{\mathcal{I}} =_{\mathrm{def}} \emptyset, \quad (\neg C)^{\mathcal{I}} =_{\mathrm{def}} \Delta^{\mathcal{I}} \setminus C^{\mathcal{I}}$$
$$(C \sqcap D)^{\mathcal{I}} =_{\mathrm{def}} C^{\mathcal{I}} \cap D^{\mathcal{I}}, \quad (C \sqcup D)^{\mathcal{I}} =_{\mathrm{def}} C^{\mathcal{I}} \cup D^{\mathcal{I}}$$
$$(\exists r.C)^{\mathcal{I}} =_{\mathrm{def}} \{ x \in \Delta^{\mathcal{I}} \mid r^{\mathcal{I}}(x) \cap C^{\mathcal{I}} \neq \emptyset \}, \quad (\forall r.C)^{\mathcal{I}} =_{\mathrm{def}} \{ x \in \Delta^{\mathcal{I}} \mid r^{\mathcal{I}}(x) \subseteq C^{\mathcal{I}} \}$$

Given $C, D \in \mathcal{L}$, $C \sqsubseteq D$ is called a *subsumption statement*, or *general concept inclusion* (GCI). $C \equiv D$ is an abbreviation for both $C \sqsubseteq D$ and $D \sqsubseteq C$. An $\mathcal{ALC}$ *TBox* $\mathcal{T}$ is a finite set of GCIs. We denote subsumption statements with $\alpha, \beta, \ldots$

An interpretation $\mathcal{I}$ *satisfies* a GCI $C \sqsubseteq D$ (denoted $\mathcal{I} \Vdash C \sqsubseteq D$) if $C^{\mathcal{I}} \subseteq D^{\mathcal{I}}$. An interpretation $\mathcal{I}$ is a *model* of a TBox TB (denoted $\mathcal{I} \Vdash \mathcal{T}$) if $\mathcal{I} \Vdash \alpha$ for every $\alpha \in \mathcal{T}$. A statement α is (classically) *entailed* by $\mathcal{T}$, denoted $\mathcal{T} \models \alpha$, if every model of $\mathcal{T}$ satisfies α.

Given $C \in \mathcal{L}$, $r \in \mathsf{R}$ and $a, b \in \mathsf{I}$, an *assertional statement* (*assertion*, for short) is an expression of the form $a : C$ or $(a, b) : r$. An $\mathcal{ALC}$ *ABox* $\mathcal{A}$ is a finite set of assertions. Given $\mathcal{T}$ and $\mathcal{A}$, with $\mathcal{KB} =_{\mathrm{def}} \mathcal{T} \cup \mathcal{A}$ we denote an $\mathcal{ALC}$ *knowledge base*, a.k.a. an *ontology*. This chapter focuses on defeasibility for description logic TBoxes only, and does not consider the extension to defeasible knowledge bases that include ABox statements. Various solutions for defeasible ABox reasoning have been proposed, that can be associated with the present approach for TBoxes [29, 30, 35, 45].

3   Defeasible Concept Inclusions

In a sense, class subsumption (alias concept inclusion) of the form $C \sqsubseteq D$ is the main notion in DL ontologies. Given its implication-like intuition, subsumption lends itself naturally to defeasibility: "provisionally, if an object falls under C, then it also falls under D", as in "usually, students are tax exempted". In this respect, a defeasible version of concept inclusion is the starting point for an investigation of defeasible reasoning in DL ontologies. We also address defeasibility of the entailment relation in later sections.

Definition 1 (Defeasible Concept Inclusion). *Let* $C, D \in \mathcal{L}$. *A **defeasible concept inclusion axiom** (DCI, for short) is a statement of the form* $C \mathbin{\rotatebox[origin=c]{0}{$\sqsubseteq$}\kern-0.6em\raisebox{-0.3ex}{$\sim$}} D$.

A DCI of the form $C \mathbin{\sqsubseteq\!\!\!\sim} D$ is to be read as "*usually*, an instance of the class C is also an instance of the class D". For instance, the DCI

$$\mathsf{Stud} \mathbin{\sqsubseteq\!\!\!\sim} \neg\exists\mathsf{receives}.\mathsf{TaxInv}$$

formalises the example above. Paraphrasing Lehmann [50], the intuition of $C \mathbin{\sqsubseteq\!\!\!\sim} D$ is that "if C were all the information about an object available to an agent, then D would be a sensible conclusion to draw about such an object". It is worth noting that $\mathbin{\sqsubseteq\!\!\!\sim}$, just as $\sqsubseteq$, is a 'connective' positioned between the concept language (object level) and the meta-language (that of entailment) and it is meant to be the defeasible counterpart of the classical subsumption $\sqsubseteq$.

Definition 2 (Defeasible TBox). *A **defeasible TBox** (dTBox, for short) is a finite set of DCIs.*

Given a TBox $\mathcal{T}$ and a dTBox $\mathcal{D}$, we let $\mathcal{KB} =_{\mathrm{def}} \mathcal{T} \cup \mathcal{D}$ and refer to it as a *defeasible knowledge base* (alias *defeasible ontology*).

Example 1. The following defeasible knowledge base gives a formal specification for our student scenario:

$$\mathcal{T} = \{\mathsf{EmpStud} \sqsubseteq \mathsf{Stud}\}, \quad \mathcal{D} = \left\{ \begin{array}{l} \mathsf{Stud} \mathbin{\sqsubseteq\!\!\!\sim} \neg\exists\mathsf{receives}.\mathsf{TaxInv}, \\ \mathsf{EmpStud} \mathbin{\sqsubseteq\!\!\!\sim} \exists\mathsf{receives}.\mathsf{TaxInv}, \\ \mathsf{EmpStud} \sqcap \mathsf{Parent} \mathbin{\sqsubseteq\!\!\!\sim} \neg\exists\mathsf{receives}.\mathsf{TaxInv} \end{array} \right\}$$

In the semantic construction later on, it will also be useful to be able to refer to infinite sets of concept inclusions. Let $\mathcal{KB}_{\mathrm{inf}}$ therefore denote a *defeasible theory*, defined as a defeasible knowledge base but without the restriction on $\mathcal{T}$ and $\mathcal{D}$ to finite sets.

In order to assess the behaviour of the new connective and check it against both the intuition and the set of properties usually considered in a non-monotonic setting, it is convenient to look at a set of $\sqsubseteq\!\!\!\sim$-statements as a binary relation of the 'antecedent-consequent' kind.

Definition 3 (Defeasible Subsumption Relation). *A **defeasible subsumption relation** is a binary relation* $\sqsubseteq\!\!\!\sim\ \subseteq \mathcal{L} \times \mathcal{L}$.

The idea is to mimic the analysis of defeasible entailment relations carried out by Kraus et al. [49] in the propositional case, where entailment is seen as a binary relation on the set in propositional sentences. Here we adopt the view of subsumption as a binary relation on concepts of our description language.

Sometimes (e.g. in the structural properties below) we write $(C, D) \in\ \sqsubseteq\!\!\!\sim$ in the infix notation, i.e., as $C \sqsubseteq\!\!\!\sim D$. The context will make clear when we will be talking about elements of a relation or statements (DCIs) in a defeasible knowledge base.

Definition 4 (Preferential Subsumption Relation). *A defeasible subsumption relation* $\sqsubseteq\!\!\!\sim$ *is a **preferential subsumption relation** if it satisfies the following set of properties, which we refer to as (the DL versions of the) preferential KLM properties:*

$$(Ref)\ C \sqsubseteq\!\!\!\sim C \qquad (LLE)\ \frac{C \equiv D,\ C \sqsubseteq\!\!\!\sim E}{D \sqsubseteq\!\!\!\sim E} \qquad (And)\ \frac{C \sqsubseteq\!\!\!\sim D,\ C \sqsubseteq\!\!\!\sim E}{C \sqsubseteq\!\!\!\sim D \sqcap E}$$

$$(Or)\ \frac{C \sqsubseteq\!\!\!\sim E,\ D \sqsubseteq\!\!\!\sim E}{C \sqcup D \sqsubseteq\!\!\!\sim E} \qquad (RW)\ \frac{C \sqsubseteq\!\!\!\sim D,\ D \sqsubseteq E}{C \sqsubseteq\!\!\!\sim E} \qquad (CM)\ \frac{C \sqsubseteq\!\!\!\sim D,\ C \sqsubseteq\!\!\!\sim E}{C \sqcap E \sqsubseteq\!\!\!\sim D}$$

The properties in Definition 4 result from a translation of those for preferential consequence relations proposed by Kraus et al. [49] in the propositional setting. They have been discussed at length in the literature for both the propositional and the DL cases [19, 21, 41, 42, 49, 52] and we shall not repeat so here.

If, in addition to the preferential properties above, the relation $\sqsubseteq\!\!\!\sim$ also satisfies rational monotonicity (RM) below, then it is said to be a *rational* subsumption relation:

$$(RM)\ \frac{C \sqsubseteq\!\!\!\sim D,\ C \not\sqsubseteq\!\!\!\sim \neg E}{C \sqcap E \sqsubseteq\!\!\!\sim D}$$

Rational monotonicity is often considered a desirable property to have, one of the reasons stemming from the fact that it is a necessary condition for the satisfaction of the principle of *presumption of typicality* (more on that in Sect. 4).

In what follows, we present a semantics for preferential and rational subsumption by enriching standard DL interpretations $\mathcal{I}$ with an ordering on the elements of the domain $\Delta^{\mathcal{I}}$. The intuition underlying this is simple and natural, and extends similar work in the propositional case by Shoham [61], Kraus et al. [49], Lehmann and Magidor [52] and Booth et al. [10–12] to the case for description logics. This is not the first extension of this kind, as evidenced by the work of Boutilier [14], Baltag and Smets [4,5], Giordano et al. [39,41–45], Britz et al. [17–21] and Britz and Varzinczak [22–27]. The present paper presents a cohesive semantic account of both preferential and rational subsumption, with accompanying representation results and computational characterisation based on the standard semantics for description logics.

Definition 5 (Preferential Interpretation). *A **preferential interpretation** is a tuple* $\mathcal{P} =_{\text{def}} \langle \Delta^{\mathcal{P}}, \cdot^{\mathcal{P}}, \prec^{\mathcal{P}} \rangle$, *where* $\langle \Delta^{\mathcal{P}}, \cdot^{\mathcal{P}} \rangle$ *is a (standard) DL interpretation (which we denote by* $\mathcal{I}_{\mathcal{P}}$ *and refer to as the classical interpretation associated with* $\mathcal{P}$ *), and* $\prec^{\mathcal{P}}$ *is a strict partial order on* $\Delta^{\mathcal{P}}$ *(i.e.,* $\prec^{\mathcal{P}}$ *is irreflexive and transitive) satisfying the smoothness condition (for every* $C \in \mathcal{L}$, *if* $C^{\mathcal{P}} \neq \emptyset$, *then* $\min_{\prec^{\mathcal{P}}} C^{\mathcal{P}} \neq \emptyset$).[1]

Preferential interpretations provide us with a simple and intuitive way to give a semantics to DCIs.

Definition 6 (Satisfaction). *Let* $\mathcal{P}$ *be a preferential interpretation and let* $C, D \in \mathcal{L}$. *The **satisfaction relation** $\Vdash$ is defined as follows:*

- $\mathcal{P} \Vdash C \sqsubseteq D$ *if* $C^{\mathcal{P}} \subseteq D^{\mathcal{P}}$;
- $\mathcal{P} \Vdash C \precsim D$ *if* $\min_{\prec^{\mathcal{P}}} C^{\mathcal{P}} \subseteq D^{\mathcal{P}}$.

If $\mathcal{P} \Vdash \alpha$, *then we say* $\mathcal{P}$ ***satisfies*** α. $\mathcal{P}$ *satisfies a defeasible knowledge base* $\mathcal{KB}$, *written* $\mathcal{P} \Vdash \mathcal{KB}$, *if* $\mathcal{P} \Vdash \alpha$ *for every* $\alpha \in \mathcal{KB}$, *in which case we say* $\mathcal{P}$ *is a **preferential model** of* $\mathcal{KB}$. *We say* $C \in \mathcal{L}$ *is **satisfiable** w.r.t.* $\mathcal{KB}$ *if there is a model* $\mathcal{P}$ *of* $\mathcal{KB}$ *s.t.* $C^{\mathcal{P}} \neq \emptyset$.

It is easy to see that the addition of the $\prec^{\mathcal{P}}$-component preserves the truth of all classical subsumption statements holding in the remaining structure:

Lemma 1. *Let* $\mathcal{P}$ *be a preferential interpretation. For every* $C, D \in \mathcal{L}$, $\mathcal{P} \Vdash C \sqsubseteq D$ *if and only if* $\mathcal{I}_{\mathcal{P}} \Vdash C \sqsubseteq D$.

It is worth noting that, due to the smoothness of $\prec^{\mathcal{P}}$, every (classical) subsumption statement is equivalent, with respect to preferential interpretations, to some DCI.

Lemma 2. *For every preferential interpretation* $\mathcal{P}$, *and every* $C, D \in \mathcal{L}$, $\mathcal{P} \Vdash C \sqsubseteq D$ *if and only if* $\mathcal{P} \Vdash C \sqcap \neg D \precsim \bot$.

An obvious question that can now be raised is: "How do we know our preferential semantics provides an appropriate meaning to the notion of DCI?" The following definition will help us in answering this question:

---

[1] Given $X \subseteq \Delta^{\mathcal{P}}$, with $\min_{\prec^{\mathcal{P}}} X$ we denote the set $\{ x \in X \mid$ for every $y \in X, y \not\prec^{\mathcal{P}} x \}$.

Definition 7 ($\mathcal{P}$-Induced Defeasible Subsumption). *Let $\mathcal{P}$ be a preferential interpretation. Then $\mathrel{\vcenter{\hbox{$\scriptstyle\sqsubset\!\sim$}}}_{\mathcal{P}} =_{\mathrm{def}} \{(C, D) \mid \mathcal{P} \Vdash C \mathrel{\vcenter{\hbox{$\scriptstyle\sqsubset\!\sim$}}} D\}$ is the* **defeasible subsumption relation induced by** $\mathcal{P}$.

The first important result we present here, which also answers the above raised question, shows that there is a full correspondence between the class of preferential subsumption relations and the class of defeasible subsumption relations induced by preferential interpretations. It is the DL analogue of a representation result proved by Kraus et al. for the propositional case [49, Theorem 3].

Theorem 1 (Representation Result for Preferential Subsumption). *A defeasible subsumption relation $\mathrel{\vcenter{\hbox{$\scriptstyle\sqsubset\!\sim$}}} \subseteq \mathcal{L} \times \mathcal{L}$ is preferential if and only if there is a preferential interpretation $\mathcal{P}$ such that $\mathrel{\vcenter{\hbox{$\scriptstyle\sqsubset\!\sim$}}}_{\mathcal{P}} = \mathrel{\vcenter{\hbox{$\scriptstyle\sqsubset\!\sim$}}}$.*

What is perhaps surprising about this result is that no additional properties based on the syntactic structure of the underlying DL are necessary to characterise the defeasible subsumption relations induced by preferential interpretations.

In addition to preferential interpretations, we are also interested in the study of *modular* interpretations, which are preferential interpretations in which the $\prec$-component is a *modular* ordering:

Definition 8 (Modular Order). *Given a set X, $\prec \subseteq X \times X$ is* **modular** *if it is a strict partial order, and its associated incomparability relation $\sim$, defined by $x \sim y$ if neither $x \prec y$ nor $y \prec x$, is transitive.*

Definition 9 (Modular Interpretation). *A* **modular interpretation** *is a preferential interpretation $\mathcal{R} = \langle \Delta^{\mathcal{R}}, \cdot^{\mathcal{R}}, \prec^{\mathcal{R}} \rangle$ such that $\prec^{\mathcal{R}}$ is modular.*

Intuitively, modular interpretations allow us to compare any two objects w.r.t. their plausibility. Those that are incomparable are viewed as being equally plausible. As such, modular interpretations are special cases of the preferential ones, where plausibility can be represented by any smooth strict partial order.

The main reason to consider modular interpretations is that they provide the semantic foundation of rational subsumption relations. This is made precise by our second important result below, which shows that the defeasible subsumption relations induced by modular interpretations are precisely the rational subsumption relations. Again, this is the DL analogue of a representation result proved by Lehmann and Magidor for the propositional case [52, Theorem 5].

Theorem 2 (Representation Result for Rational Subsumption). *A defeasible subsumption relation $\mathrel{\vcenter{\hbox{$\scriptstyle\sqsubset\!\sim$}}} \subseteq \mathcal{L} \times \mathcal{L}$ is rational if and only if there is a modular interpretation $\mathcal{R}$ such that $\mathrel{\vcenter{\hbox{$\scriptstyle\sqsubset\!\sim$}}}_{\mathcal{R}} = \mathrel{\vcenter{\hbox{$\scriptstyle\sqsubset\!\sim$}}}$.*

It is worth pausing for a moment to emphasise the significance of these two results (Theorems 1 and 2). They provide exact semantic characterisations of two important classes of defeasible subsumption relations, namely preferential and rational subsumption, in terms of the classes of preferential and modular interpretations, respectively. As we shall see in Sect. 4, these results form the core of the investigation into an appropriate form of entailment for defeasible DL ontologies.

4   Defeasible Entailment

From the standpoint of knowledge representation and reasoning, a pivotal question is that of deciding which statements are *entailed* by a knowledge base. In the present section we lay out the formal foundations for that.

4.1   Preferential Entailment

In the exploration of a notion of entailment for defeasible ontologies, an obvious starting point is to consider a Tarskian definition of consequence:

Definition 10 (Preferential Entailment). *A statement α is **preferentially entailed** by a defeasible knowledge base $\mathcal{KB}$, written $\mathcal{KB} \models {}_{\mathsf{pref}}\alpha$, if every preferential model of $\mathcal{KB}$ satisfies α.*

As usual, this form of entailment is accompanied by a corresponding notion of closure.

Definition 11 (Preferential Closure). *Let $\mathcal{KB}$ be a defeasible knowledge base. With $\mathcal{KB}^*_{\mathsf{pref}} =_{\mathrm{def}} \{\alpha \mid \mathcal{KB} \models {}_{\mathsf{pref}}\alpha\}$ we denote the **preferential closure** of $\mathcal{KB}$.*

Intuitively, the preferential closure of a defeasible knowledge base $\mathcal{KB}$ corresponds to the 'core' set of statements, classical and defeasible, that should hold given those in $\mathcal{KB}$. Hence, preferential entailment and preferential closure are two sides of the same coin, mimicking an analogous result for preferential reasoning in both the propositional [49] and the DL [16,21] cases.

Recall (cf. the discussion following Definition 2) that a defeasible theory $\mathcal{KB}_{\mathrm{inf}}$ is a defeasible knowledge base without the restriction to finite sets. When assessing how appropriate a notion of entailment for defeasible ontologies is, the following definitions turn out to be useful, as will become clear in the sequel:

Definition 12 ($\mathcal{KB}_{inf}$-Induced Defeasible Subsumption). *Let $\mathcal{KB}_{inf}$ be a defeasible theory. Then $\mathcal{D}_{\mathcal{KB}_{inf}} =_{\mathrm{def}} \{C \mathbin{\sqsubseteq\!\!\!\sim} D \mid C \mathbin{\sqsubseteq\!\!\!\sim} D \in \mathcal{KB}_{inf}\} \cup \{C \sqcap \neg D \mathbin{\sqsubseteq\!\!\!\sim} \bot \mid C \sqsubseteq D \in \mathcal{KB}_{inf}\}$ is the **dTBox induced by $\mathcal{KB}_{inf}$** and $\mathbin{\sqsubseteq\!\!\!\sim}_{\mathcal{KB}_{inf}} =_{\mathrm{def}} \{(C,D) \mid C \mathbin{\sqsubseteq\!\!\!\sim} D \in \mathcal{D}_{\mathcal{KB}_{inf}}\}$ is the **defeasible subsumption relation induced by $\mathcal{KB}_{inf}$**.*

So, the dTBox induced by $\mathcal{KB}_{\mathrm{inf}}$ is the set of defeasible subsumption statements contained in $\mathcal{KB}_{\mathrm{inf}}$, together with the defeasible versions of the classical subsumption statements in $\mathcal{KB}_{\mathrm{inf}}$. The defeasible subsumption relation induced by $\mathcal{KB}_{\mathrm{inf}}$ is simply the defeasible subsumption relation corresponding to $\mathcal{D}_{\mathcal{KB}_{\mathrm{inf}}}$.

Definition 13. *A defeasible theory $\mathcal{KB}_{inf}$ is called **preferential** if the subsumption relation induced by it satisfies the preferential properties in Definition 4.*

It turns out that the defeasible subsumption relation induced by the preferential closure of a defeasible knowledge base $\mathcal{KB}$ is exactly the intersection of the defeasible subsumption relations induced by the preferential defeasible theories containing $\mathcal{KB}$.

Lemma 3. *Let $\mathcal{KB}$ be a defeasible knowledge base. Then*

$$\mathrel{\vcenter{\hbox{$\sim$}}}_{\mathcal{KB}^*_{\mathrm{pref}}} = \bigcap \{ \mathrel{\vcenter{\hbox{$\sim$}}}_{\mathcal{KB}_{inf}} \mid \mathcal{KB} \subseteq \mathcal{KB}_{inf} \text{ and } \mathcal{KB}_{inf} \text{ is preferential} \}.$$

It follows immediately that the preferential closure of a defeasible knowledge base $\mathcal{KB}$ is preferential, and induces the smallest defeasible subsumption relation induced by a preferential defeasible theory containing $\mathcal{KB}$.

Preferential entailment is not always desirable, one of the reasons being that it is monotonic, courtesy of the Tarskian notion of consequence it relies on (see Definition 10). In most cases, as witnessed by the great deal of work in the non-monotonic reasoning community, a move towards rationality is in order. Thanks to the definitions above and the result in Theorem 2, we already know where to start looking for it.

Definition 14 (Modular Entailment). *A statement α is **modularly entailed** by a defeasible knowledge base $\mathcal{KB}$, written $\mathcal{KB} \models _{\mathrm{mod}} \alpha$, if every modular model of $\mathcal{KB}$ satisfies α.*

As is the case for preferential entailment, modular entailment is accompanied by a corresponding notion of closure.

Definition 15 (Modular Closure). *Let $\mathcal{KB}$ be a defeasible knowledge base. With $\mathcal{KB}^*_{\mathrm{mod}} =_{\mathrm{def}} \{ \alpha \mid \mathcal{KB} \models _{\mathrm{mod}} \alpha \}$ we denote the **modular closure** of $\mathcal{KB}$.*

Definition 16. *A defeasible theory $\mathcal{KB}_{inf}$ is called **rational** if it is preferential and $\mathrel{\vcenter{\hbox{$\sim$}}}_{\mathcal{KB}_{inf}}$ is also closed under the rational monotonicity rule (RM).*

For modular closure we get a result similar to Lemma 3.

Lemma 4. *Let $\mathcal{KB}$ be a defeasible knowledge base. Then*

$$\mathrel{\vcenter{\hbox{$\sim$}}}_{\mathcal{KB}^*_{\mathrm{mod}}} = \bigcap \{ \mathrel{\vcenter{\hbox{$\sim$}}}_{\mathcal{KB}_{inf}} \mid \mathcal{KB} \subseteq \mathcal{KB}_{inf} \text{ and } \mathcal{KB}_{inf} \text{ is rational} \}.$$

That is, the modular closure of a defeasible knowledge base $\mathcal{KB}$ induces the smallest defeasible subsumption relation induced by a rational defeasible theory containing $\mathcal{KB}$. However, the modular closure of $\mathcal{KB}$ is not necessarily rational. That is, if one looks at the set of statements (in particular the $\mathrel{\vcenter{\hbox{$\sim$}}}$-ones) modularly entailed by a knowledge base as a defeasible subsumption relation, then it need not satisfy the RM property. This is so because modular entailment coincides with preferential entailment, as the following result, adapted from a well-known similar result in the propositional case [52, Theorem 4.2], shows.

Lemma 5. $\mathcal{KB}^*_{\mathrm{mod}} = \mathcal{KB}^*_{\mathrm{pref}}$.

Hence, modular entailment unfortunately falls short of providing us with an appropriate notion of defeasible entailment. In what follows, we overcome precisely this issue.

4.2 Rational Entailment

We now present a definition of semantic entailment which is appropriate in the light of the discussion above. The constructions we are going to present are inspired by the semantic characterisation of rational closure by Booth and Paris [13] in the propositional case.

We start by focusing our attention on a subclass of modular orders, referred to as *ranked orders*:

Definition 17 (Ranked Order). *Given a set X, the binary relation $\prec \subseteq X \times X$ is a **ranked order** if there is a mapping $h_{\mathcal{R}} : X \longrightarrow \mathbb{N}$ satisfying the following convexity property:*

– *for every $i \in \mathbb{N}$, if for some $x \in X$ $h_{\mathcal{R}}(x) = i$, then, for every j such that $0 \leq j < i$, there is a $y \in X$ for which $h_{\mathcal{R}}(y) = j$,*

and such that for every $x, y \in X$, $x \prec y$ iff $h_{\mathcal{R}}(x) < h_{\mathcal{R}}(y)$.

It is easy to see that a ranked order $\prec$ is also modular: $\prec$ is a strict partial order, and, since two objects x, y are incomparable (i.e., $x \sim y$) if and only if $h_{\mathcal{R}}(x) = h_{\mathcal{R}}(y)$, $\sim$ is a transitive relation. By constraining our preference relations to the ranked orders, we can identify a subset of the modular interpretations we refer to as the *ranked interpretations*.

Definition 18 (Ranked Interpretation). *A **ranked interpretation** is a modular interpretation $\mathcal{R} = \langle \Delta^{\mathcal{R}}, \cdot^{\mathcal{R}}, \prec^{\mathcal{R}} \rangle$ s.t. $\prec^{\mathcal{R}}$ is a ranked order.*

We now provide two basic results about ranked interpretations. First, all finite modular interpretations are ranked interpretations.

Lemma 6. *A modular interpretation $\mathcal{R} = \langle \Delta^{\mathcal{R}}, \cdot^{\mathcal{R}}, \prec^{\mathcal{R}} \rangle$ s.t. $\Delta^{\mathcal{R}}$ is finite is a ranked interpretation.*

Next, for every ranked interpretation, the function $h_{\mathcal{R}}$ is unique.

Proposition 1. *Given a ranked interpretation $\mathcal{R} = \langle \Delta^{\mathcal{R}}, \cdot^{\mathcal{R}}, \prec^{\mathcal{R}} \rangle$, there is only one function $h_{\mathcal{R}} : \Delta^{\mathcal{R}} \longrightarrow \mathbb{N}$ satisfying the convexity property and s.t. for every $x, y \in \Delta^{\mathcal{R}}$, $x \prec y$ iff $h_{\mathcal{R}}(x) < h_{\mathcal{R}}(y)$.*

Proposition 1 allows us to use the function $h_{\mathcal{R}}(\cdot)$ to define the notions of *height* and *layers*.

Definition 19 (Height and Layers). *Let $\mathcal{R} = \langle \Delta^{\mathcal{R}}, \cdot^{\mathcal{R}}, \prec^{\mathcal{R}} \rangle$ be a ranked interpretation with characteristic ranking function $h_{\mathcal{R}}(\cdot)$. Given an object $x \in \Delta^{\mathcal{R}}$, $h_{\mathcal{R}}(x)$ is called the **height** of x in $\mathcal{R}$. For every ranked interpretation $\mathcal{R} = \langle \Delta^{\mathcal{R}}, \cdot^{\mathcal{R}}, \prec^{\mathcal{R}} \rangle$, we can partition the domain $\Delta^{\mathcal{R}}$ into a sequence of **layers** $(L_0, \dots, L_n, \dots)$, where, for every object $x \in \Delta^{\mathcal{R}}$, we have $x \in L_i$ iff $h_{\mathcal{R}}(x) = i$.*

Intuitively, the lower the height of an object in an interpretation $\mathcal{R}$, the more typical (or normal) the object is in $\mathcal{R}$. We can also think of a level of typicality for concepts: the height of a concept $C \in \mathcal{L}$ in $\mathcal{R}$ is the index of the layer to which the restriction of the concept's extension to its $\prec^{\mathcal{R}}$-minimal elements belong, i.e., $h_{\mathcal{R}}(C) = i$ if $\emptyset \subset \min_{\prec^{\mathcal{R}}} C^{\mathcal{R}} \subseteq L_i$.

Given a set of ranked interpretations, we can introduce a new form of model merging, *ranked union*.

Definition 20 (Ranked Union). *Given a countable set of ranked interpretations* $\mathfrak{R} = \{\mathcal{R}_1, \mathcal{R}_2, \ldots\}$, *a ranked interpretation* $\mathcal{R}^{\mathfrak{R}} =_{\text{def}} \langle \Delta^{\mathfrak{R}}, \cdot^{\mathfrak{R}}, \prec^{\mathfrak{R}} \rangle$ *is the* **ranked union** *of* $\mathfrak{R}$ *if the following holds:*

- $\Delta^{\mathfrak{R}} =_{\text{def}} \coprod_{\mathcal{R} \in \mathfrak{R}} \Delta^{\mathcal{R}}$, *i.e., the disjoint union of the domains from* $\mathfrak{R}$, *where each* $\mathcal{R} \in \mathfrak{R}$ *has the elements* $x, y, \ldots$ *of its domain renamed as* $x_{\mathcal{R}}, y_{\mathcal{R}}, \ldots$ *so that they are all distinct in* $\Delta^{\mathfrak{R}}$;
- $x_{\mathcal{R}} \in A^{\mathfrak{R}}$ *iff* $x \in A^{\mathcal{R}}$;
- $(x_{\mathcal{R}}, y_{\mathcal{R}'}) \in r^{\mathfrak{R}}$ *iff* $\mathcal{R} = \mathcal{R}'$ *and* $(x, y) \in r^{\mathcal{R}}$;
- *for every* $x_{\mathcal{R}} \in \Delta^{\mathfrak{R}}$, $h_{\mathfrak{R}}(x_{\mathcal{R}}) = h_{\mathcal{R}}(x)$.

The latter condition corresponds to imposing that $x_{\mathcal{R}} \prec^{\mathfrak{R}} y_{\mathcal{R}'}$ *iff* $h_{\mathcal{R}}(x) < h_{\mathcal{R}'}(y)$.

The following lemma will be useful in what follows.

Lemma 7. *Ranked interpretations are closed under ranked union.*

Let $\mathcal{KB}$ be a defeasible knowledge base and let Δ be a fixed countably infinite set. Define

$$Mod_{\Delta}(\mathcal{KB}) =_{\text{def}} \{\mathcal{R} = \langle \Delta^{\mathcal{R}}, \cdot^{\mathcal{R}}, \prec^{\mathcal{R}} \rangle \mid \mathcal{R} \Vdash \mathcal{KB}, \mathcal{R} \text{ is ranked and } \Delta^{\mathcal{R}} = \Delta\}.$$

The following result shows that the set $Mod_{\Delta}(\mathcal{KB})$ suffices to characterise modular entailment:

Lemma 8. *For every* $\mathcal{KB}$ *and every* $C, D \in \mathcal{L}$, $\mathcal{KB} \models_{\text{mod}} C \mathrel{\raise.17ex\hbox{$\subset\hspace{-0.8em}\sim$}} D$ *iff* $\mathcal{R} \Vdash C \mathrel{\raise.17ex\hbox{$\subset\hspace{-0.8em}\sim$}} D$, *for every* $\mathcal{R} \in Mod_{\Delta}(\mathcal{KB})$.

Therefore, we can use just the set of interpretations in $Mod_{\Delta}(\mathcal{KB})$ to decide the consequences of $\mathcal{KB}$ w.r.t. modular entailment.

We can now use the set $Mod_{\Delta}(\mathcal{KB})$ as a springboard to introduce what will turn out to be a canonical modular interpretation for $\mathcal{KB}$. Using $Mod_{\Delta}(\mathcal{KB})$ and ranked union we can define the following relevant model.

Definition 21 (Big ranked model). *Let* $\mathcal{KB}$ *be a defeasible knowledge base. The* **big ranked model** *of* $\mathcal{KB}$ *is the ranked model* $\mathcal{O} =_{\text{def}} \langle \Delta^{\mathcal{O}}, \cdot^{\mathcal{O}}, \prec^{\mathcal{O}} \rangle$ *that is the ranked union of the models in* $Mod_{\Delta}(\mathcal{KB})$.

Since ranked interpretations are closed under ranked unions (Lemma 7), we can state the following:

Lemma 9. $\mathcal{O}$ *is a ranked model of* $\mathcal{KB}$.

Armed with the definitions and results above, we are now ready to provide an alternative definition of entailment in the context of defeasible ontologies:

Definition 22 (Rational Entailment). *A statement α is **rationally entailed** by a knowledge base $\mathcal{KB}$, written $\mathcal{KB} \models_{\mathsf{rat}} \alpha$, if $\mathcal{O} \Vdash \alpha$.*

That such a notion of entailment indeed deserves its name is witnessed by the following result, a consequence of Lemma 9 and Theorem 2:

Corollary 1. *Let $\mathcal{KB}$ be a defeasible knowledge base and $\mathcal{O}$ its big ranked model. Then $\{C \mathrel{\sqsubseteq\!\!\!\sim} D \mid \mathcal{O} \Vdash C \mathrel{\sqsubseteq\!\!\!\sim} D\}$ is rational.*

We shall see below that this form of entailment corresponds to the DL version of a well-known form of propositional defeasible entailment [52].

In conclusion, rational entailment is a good candidate for the appropriate notion of consequence we have been looking for. Of course, a question that arises is whether a notion of closure, in the spirit of preferential and modular closures, that is equivalent to it can be defined. In the next section, we address precisely this matter.

5   Rational Closure for Defeasible Knowledge Bases

We now turn our attention to the exploration, in a DL setting, of the well-known notion of *rational closure* of a defeasible knowledge base as studied by Lehmann and Magidor [52]. For the most part, we shall base the presentation of the constructions on the work by Casini and Straccia [30,32], amending it wherever necessary. An alternative semantic characterisation of rational closure in DLs has also been proposed by Giordano et al. [44,45]; their characterisation and the one we present here are equivalent [35, Appendix A].

As we shall see, rational closure provides a proof-theoretic characterisation of rational entailment and the complexity of its computation is no higher than that of computing entailment in the underlying classical DL.

5.1   Rational Closure and a Correspondence Result

Rational closure is a form of inferential closure based on modular entailment $\models_{\mathsf{mod}}$, but it extends its inferential power. Such an extension of modular entailment is obtained by formalising the already mentioned principle of *presumption of typicality* [51, Section 3.1]. That is, under possibly incomplete information, we always assume that we are dealing with the most typical possible situation that is compatible with the information at our disposal. We first define what it means for a concept to be *exceptional*, a notion that, as we shall see, is central to the definition of rational closure:

Definition 23 (Exceptionality). *Let $\mathcal{KB}$ be a defeasible knowledge base and $C \in \mathcal{L}$. We say C is **exceptional** in $\mathcal{KB}$ if $\mathcal{KB} \models_{\mathsf{mod}} \top \mathrel{\sqsubseteq\!\!\!\sim} \neg C$. A DCI $C \mathrel{\sqsubseteq\!\!\!\sim} D$ is exceptional in $\mathcal{KB}$ if C is exceptional in $\mathcal{KB}$.*

A concept C is considered exceptional in a knowledge base $\mathcal{KB}$ if it is not possible to have a modular model of $\mathcal{KB}$ in which there is a typical object (i.e., an object at least as typical as all the others) that is in the interpretation of C. Intuitively, a DCI is exceptional if it does not concern the most typical objects, i.e., it is about less normal (or exceptional) ones. This is an intuitive translation of the notion of exceptionality used by Lehmann and Magidor [52] in the propositional framework, and has already been used by Casini and Straccia [30] and Giordano et al. [45] in their investigations into defeasible reasoning for DLs.

Applying the notion of exceptionality iteratively, we associate with every concept C a *rank* in $\mathcal{KB}$, which we denote by $\mathsf{rank}_{\mathcal{KB}}(C)$. We extend this to DCIs and associate with every statement $C \mathrel{\vrule height 1.2ex depth 0pt\joinrel\sqsubset} D$ a rank, denoted $\mathsf{rank}_{\mathcal{KB}}(C \mathrel{\vrule height 1.2ex depth 0pt\joinrel\sqsubset} D)$:

1. Let $\mathsf{rank}_{\mathcal{KB}}(C) = 0$, if C is not exceptional in $\mathcal{KB}$, and let $\mathsf{rank}_{\mathcal{KB}}(C \mathrel{\vrule height 1.2ex depth 0pt\joinrel\sqsubset} D) = 0$ for every DCI having C as antecedent, with $\mathsf{rank}_{\mathcal{KB}}(C) = 0$. The set of DCIs in $\mathcal{D}$ with rank 0 is denoted as $\mathcal{D}_0^{\mathsf{rank}}$.
2. Let $\mathsf{rank}_{\mathcal{KB}}(C) = 1$, if C does not have a rank of 0 and it is not exceptional in the knowledge base $\mathcal{KB}^1$ composed of $\mathcal{T}$ and the exceptional part of $\mathcal{D}$, that is, $\mathcal{KB}^1 = \langle \mathcal{T}, \mathcal{D} \setminus \mathcal{D}_0^{\mathsf{rank}} \rangle$. If $\mathsf{rank}_{\mathcal{KB}}(C) = 1$, then let $\mathsf{rank}_{\mathcal{KB}}(C \mathrel{\vrule height 1.2ex depth 0pt\joinrel\sqsubset} D) = 1$ for every DCI $C \mathrel{\vrule height 1.2ex depth 0pt\joinrel\sqsubset} D$. The set of DCIs in $\mathcal{D}$ with rank 1 is denoted $\mathcal{D}_1^{\mathsf{rank}}$.
3. In general, for $i > 0$, a concept C is assigned a rank of i if it does not have a rank of $i - 1$ and it is not exceptional in $\mathcal{KB}^i = \langle \mathcal{T}, \mathcal{D} \setminus \bigcup_{j=0}^{i-1} \mathcal{D}_j^{\mathsf{rank}} \rangle$. If $\mathsf{rank}_{\mathcal{KB}}(C) = i$, then $\mathsf{rank}_{\mathcal{KB}}(C \mathrel{\vrule height 1.2ex depth 0pt\joinrel\sqsubset} D) = i$, for every DCI $C \mathrel{\vrule height 1.2ex depth 0pt\joinrel\sqsubset} D$. The set of DCIs in $\mathcal{D}$ with rank i is denoted $\mathcal{D}_i^{\mathsf{rank}}$.
4. By iterating the previous steps, we eventually reach a subset $\mathcal{E} \subseteq \mathcal{D}$ such that all the DCIs in $\mathcal{E}$ are exceptional (since $\mathcal{D}$ is finite, we must reach such a point). If $\mathcal{E} \neq \emptyset$, we define the rank of the DCIs in $\mathcal{E}$ as ∞, and the set $\mathcal{E}$ is denoted $\mathcal{D}_\infty^{\mathsf{rank}}$.

The notion of rank can also be extended to GCIs as follows: $\mathsf{rank}_{\mathcal{KB}}(C \sqsubseteq D) = \mathsf{rank}_{\mathcal{KB}}(C)$.

Following on the procedure above, $\mathcal{D}$ is partitioned into a finite sequence $\langle \mathcal{D}_0^{\mathsf{rank}}, \ldots, \mathcal{D}_n^{\mathsf{rank}}, \mathcal{D}_\infty^{\mathsf{rank}} \rangle$ $(n \geq 0)$, where $\mathcal{D}_\infty^{\mathsf{rank}}$ may possibly be empty. So, through this procedure we can assign a rank to every DCI.

It is easy to see that for a concept C to have a rank of ∞ corresponds to not being satisfiable in any model of $\mathcal{KB}$, that is, $\mathcal{KB} \models_{\mathsf{mod}} C \sqsubseteq \bot$.

Lemma 10. $\mathsf{rank}_{\mathcal{KB}}(C) = \infty$ *iff* $\mathcal{KB} \models_{\mathsf{mod}} C \sqsubseteq \bot$.

Example 2. Let $\mathcal{KB} = \mathcal{T} \cup \mathcal{D}$, where $\mathcal{T}$ and $\mathcal{D}$ are as in Example 1, i.e., $\mathcal{T} = \{\mathsf{EmpStud} \sqsubseteq \mathsf{Stud}\}$ and

$$
\mathcal{D} = \left\{
\begin{array}{c}
\mathsf{Stud} \mathrel{\vrule height 1.2ex depth 0pt\joinrel\sqsubset} \neg\exists\mathsf{receives}.\mathsf{TaxInv}, \\
\mathsf{EmpStud} \mathrel{\vrule height 1.2ex depth 0pt\joinrel\sqsubset} \exists\mathsf{receives}.\mathsf{TaxInv}, \\
\mathsf{EmpStud} \sqcap \mathsf{Parent} \mathrel{\vrule height 1.2ex depth 0pt\joinrel\sqsubset} \neg\exists\mathsf{receives}.\mathsf{TaxInv}
\end{array}
\right\}
$$

Examining the concepts on the LHS of each DCI in $\mathcal{KB}$, one can verify that Stud is not exceptional w.r.t. $\mathcal{KB}$. Therefore, $\mathsf{rank}_{\mathcal{KB}}(\mathsf{Stud}) = 0$. We also find that $\mathrm{rank}_{\mathcal{KB}}(\mathsf{EmpStud}) \neq 0$ and $\mathrm{rank}_{\mathcal{KB}}(\mathsf{EmpStud} \sqcap \mathsf{Parent}) \neq 0$ because both concepts are exceptional w.r.t. $\mathcal{KB}$.

$\mathcal{KB}^1$ is composed of $\mathcal{T}$ and $\mathcal{D} \setminus \mathcal{D}_0^{\mathsf{rank}}$, which consists of the DCIs in $\mathcal{D}$ except for $\mathsf{Stud} \mathrel{\underset{\sim}{\sqsubseteq}} \neg\exists\mathsf{receives}.\mathsf{TaxInv}$. We find that $\mathsf{EmpStud}$ is *not* exceptional w.r.t. $\mathcal{KB}^1$ and therefore $\mathrm{rank}_{\mathcal{KB}}(\mathsf{EmpStud}) = 1$. Since $\mathsf{EmpStud} \sqcap \mathsf{Parent}$ is exceptional w.r.t. $\mathcal{KB}^1$, $\mathrm{rank}_{\mathcal{KB}}(\mathsf{EmpStud} \sqcap \mathsf{Parent}) \neq 1$. Similarly, $\mathcal{KB}^2$ is composed of $\mathcal{T}$ and $\{\mathsf{EmpStud} \sqcap \mathsf{Parent} \mathrel{\underset{\sim}{\sqsubseteq}} \neg\exists\mathsf{receives}.\mathsf{TaxInv}\}$. We have that $\mathsf{EmpStud} \sqcap \mathsf{Parent}$ is not exceptional w.r.t. $\mathcal{KB}^2$ and therefore $\mathrm{rank}_{\mathcal{KB}}(\mathsf{EmpStud} \sqcap \mathsf{Parent}) = 2$.

Adapting Lehmann and Magidor's construction for propositional logic [52], the rational closure of a defeasible knowledge base $\mathcal{KB}$ is defined as follows:

Definition 24 (Rational Closure). *Let $\mathcal{KB}$ be a defeasible knowledge base and $C, D \in \mathcal{L}$.*

1. *$C \mathrel{\underset{\sim}{\sqsubseteq}} D$ is in the rational closure of $\mathcal{KB}$ if*

$$\mathrm{rank}_{\mathcal{KB}}(C \sqcap D) < \mathrm{rank}_{\mathcal{KB}}(C \sqcap \neg D) \text{ or } \mathrm{rank}_{\mathcal{KB}}(C) = \infty.$$

2. *$C \sqsubseteq D$ is in the rational closure of $\mathcal{KB}$ if $\mathrm{rank}_{\mathcal{KB}}(C \sqcap \neg D) = \infty$.*

Informally, the definition above says that the DCI $C \mathrel{\underset{\sim}{\sqsubseteq}} D$ is in the rational closure of $\mathcal{KB}$ if the modular models of $\mathcal{KB}$ tell us that some instances of $C \sqcap D$ are more plausible than all instances of $C \sqcap \neg D$, while the GCI $C \sqsubseteq D$ is in the rational closure of $\mathcal{KB}$ if the instances of $C \sqcap \neg D$ are impossible. The attentive reader will note that this definition has some similarity with the *epistemic entrenchment* orderings used in belief revision [38, 59].

Example 2 (continued). Applying the definition above to the knowledge base in Example 2, we can verify that $\mathsf{Stud} \mathrel{\underset{\sim}{\sqsubseteq}} \neg\exists\mathsf{receives}.\mathsf{TaxInv}$ is in the rational closure of $\mathcal{KB}$ because $\mathrm{rank}_{\mathcal{KB}}(\mathsf{Stud} \sqcap \neg\exists\mathsf{receives}.\mathsf{TaxInv}) = 0$ and $\mathrm{rank}_{\mathcal{KB}}(\mathsf{Stud} \sqcap \exists\mathsf{receives}.\mathsf{TaxInv}) > 0$. The latter can be derived from the fact that $\mathsf{Stud} \sqcap \exists\mathsf{receives}.\mathsf{TaxInv}$ is exceptional w.r.t. $\mathcal{KB}$.

Similarly, one can derive that both DCIs $\mathsf{EmpStud} \mathrel{\underset{\sim}{\sqsubseteq}} \exists\mathsf{receives}.\mathsf{TaxInv}$ and $\mathsf{EmpStud} \sqcap \mathsf{Parent} \mathrel{\underset{\sim}{\sqsubseteq}} \neg\exists\mathsf{receives}.\mathsf{TaxInv}$ are in the rational closure of $\mathcal{KB}$ as well.

□

We now state the main result of the present section, which provides an answer to the question raised at the end of Sect. 4.2.

Theorem 3. *Let $\mathcal{KB}$ be a defeasible knowledge base having a modular model. A statement α is in the rational closure of $\mathcal{KB}$ iff $\mathcal{KB} \models_{\mathsf{rat}} \alpha$.*

An easy corollary of this result is that rational closure preserves the equivalence between GCIs (of the form $C \sqsubseteq D$) and their defeasible counterparts ($C \sqcap \neg D \mathrel{\underset{\sim}{\sqsubseteq}} \bot$).

Corollary 2. *$C \sqsubseteq D$ is in the rational closure of a defeasible knowledge base $\mathcal{KB}$ iff $C \sqcap \neg D \mathrel{\underset{\sim}{\sqsubseteq}} \bot$ is the restriction of the closure of $\mathcal{KB}$ under rational entailment to defeasible concept inclusions.*

Rational entailment from a knowledge base can therefore be formulated as membership checking of the rational closure of the knowledge base. Of course, from an application-oriented point of view, this raises the question of how to compute membership of the rational closure of a knowledge base, and what is the complexity thereof. This is precisely the topic of the next section.

5.2   Rational Entailment Checking

We now present an algorithm to effectively check the rational entailment of a DCI from a defeasible knowledge base. Our algorithm is based on the one given by Casini and Straccia [30] for defeasible $\mathcal{ALC}$.

Let $\mathcal{KB} = \mathcal{T} \cup \mathcal{D}$ be a defeasible knowledge base. The first step of the algorithm is to assign a rank to each DCI in $\mathcal{D}$. Central to this step is the exceptionality function $\mathsf{Exceptional}(\cdot)$, which computes the semantic notion of exceptionality of Definition 23. Given a set of DCIs $\mathcal{D}' \subseteq \mathcal{D}$, $\mathsf{Exceptional}(\mathcal{T}, \mathcal{D}')$ returns a subset $\mathcal{E}$ of $\mathcal{D}'$ such that $\mathcal{E}$ is exceptional w.r.t. $\mathcal{T} \cup \mathcal{D}'$.

Function. $\mathsf{Exceptional}(\mathcal{T}, \mathcal{D}')$

Input: $\mathcal{T}$ and $\mathcal{D}' \subseteq \mathcal{D}$
Output: $\mathcal{E} \subseteq \mathcal{D}'$ such that $\mathcal{E}$ is exceptional w.r.t. $\mathcal{T} \cup \mathcal{D}'$

1 $\mathcal{E} \leftarrow \emptyset$
2 **foreach** $C \sqsubset\!\!\sim D \in \mathcal{D}'$ **do**
3 $\quad$ **if** $\mathcal{T} \models \bigsqcap \overline{\mathcal{D}'} \sqsubseteq \neg C$ **then**
4 $\quad\quad$ $\mathcal{E} \leftarrow \mathcal{E} \cup \{C \sqsubset\!\!\sim D\}$

5 **return** $\mathcal{E}$

The function makes use of the notion of *materialisation* to reduce concept exceptionality checking to entailment checking:

Definition 25 (Materialisation). *Let $\mathcal{D}$ be a set of DCIs. With $\overline{\mathcal{D}} =_{\mathrm{def}} \{\neg C \sqcup D \mid C \sqsubset\!\!\sim D \in \mathcal{D}\}$ we denote the **materialisation** of $\mathcal{D}$.*

We can show that, given $\mathcal{KB} = \mathcal{T} \cup \mathcal{D}$ and $\mathcal{D}' \subseteq \mathcal{D}$, if $\mathcal{T} \models \bigsqcap \overline{\mathcal{D}'} \sqsubseteq \neg C$, a DCI $C \sqsubset\!\!\sim D$ is exceptional w.r.t. $\mathcal{T} \cup \mathcal{D}'$, thereby justifying the use of Line 3 of function Exceptional.

Lemma 11. *For $\mathcal{KB} = \mathcal{T} \cup \mathcal{D}$, if $\mathcal{T} \models \bigsqcap \overline{\mathcal{D}} \sqsubseteq \neg C$ then $C \sqsubset\!\!\sim D$ is exceptional w.r.t. $\mathcal{T} \cup \mathcal{D}$.*

While the converse of Lemma 11 does not hold, it follows from Lemma 13 below that this reduction to classical entailment checking, when applied iteratively (lines 4–14 in function ComputeRanking below), fully captures the semantic notion of exceptionality of Definition 23.

Example 2 (continued). If we feed the knowledge base in Example 2 to the function Exceptional($\cdot$), we obtain the output

$$\mathcal{E} = \{\mathsf{EmpStud} \sqsubseteq \exists \mathsf{receives}.\mathsf{TaxInv}, \mathsf{EmpStud} \sqcap \mathsf{Parent} \sqsubseteq \neg \exists \mathsf{receives}.\mathsf{TaxInv}\}.$$

This is because both concepts on the LHS of the DCIs in $\mathcal{D}'$ are exceptional w.r.t. $\mathcal{KB}$ in Example 2.

We now describe the overall ranking algorithm, presented in the function ComputeRanking($\cdot$) below. The algorithm makes a finite sequence of calls to the function Exceptional($\cdot$), starting from the knowledge base $\mathcal{KB} = \mathcal{T} \cup \mathcal{D}$. The algorithm terminates with a partitioning of the axioms in the dTBox, from which a ranking of axioms can easily be obtained.

Function. ComputeRanking($\mathcal{KB}$)

Input: $\mathcal{KB} = \mathcal{T} \cup \mathcal{D}$
Output: $\mathcal{KB}^* = \mathcal{T}^* \cup \mathcal{D}^*$ and a partitioning $R = \{\mathcal{D}_0, \ldots, \mathcal{D}_n\}$ for $\mathcal{D}^*$
1   $\mathcal{T}^* \leftarrow \mathcal{T}$
2   $\mathcal{D}^* \leftarrow \mathcal{D}$
3   $R \leftarrow \emptyset$
4   **repeat**
5      $i \leftarrow 0$
6      $\mathcal{E}_0 \leftarrow \mathcal{D}^*$
7      $\mathcal{E}_1 \leftarrow \mathrm{Exceptional}(\mathcal{T}^*, \mathcal{E}_0)$
8   **while** $\mathcal{E}_{i+1} \neq \mathcal{E}_i$ **do**
9         $i \leftarrow i+1$
10        $\mathcal{E}_{i+1} \leftarrow \mathrm{Exceptional}(\mathcal{T}^*, \mathcal{E}_i)$
11      $\mathcal{D}_\infty^* \leftarrow \mathcal{E}_i$
12      $\mathcal{T}^* \leftarrow \mathcal{T}^* \cup \{C \sqsubseteq D \mid C \sqsubseteq D \in \mathcal{D}_\infty^*\}$
13      $\mathcal{D}^* \leftarrow \mathcal{D}^* \setminus \mathcal{D}_\infty^*$
14 **until** $\mathcal{D}_\infty^* = \emptyset$
15 **for** $j \leftarrow 1$ *to* i **do**
16      $\mathcal{D}_{j-1} \leftarrow \mathcal{E}_{j-1} \setminus \mathcal{E}_j$
17      $R \leftarrow R \cup \{\mathcal{D}_{j-1}\}$
18 **return** $\mathcal{KB}^* = \mathcal{T}^* \cup \mathcal{D}^*, R$

We initialise $\mathcal{T}^*$ to $\mathcal{T}$ and $\mathcal{D}^*$ to $\mathcal{D}$ (Lines 1 and 2 of ComputeRanking). We then repeatedly invoke the function Exceptional to obtain a sequence of sets of DCIs $\mathcal{E}_0, \mathcal{E}_1, \ldots$, where $\mathcal{E}_0 = \mathcal{D}^*$ and each $\mathcal{E}_{i+1}$ is the set of exceptional axioms in $\mathcal{E}_i$ (Lines 4–14 of ComputeRanking($\cdot$)).

Now, let $\mathcal{C}_{\mathcal{D}^*} =_{\mathrm{def}} \{C \mid C \sqsubseteq D \in \mathcal{D}^*\}$, i.e., $\mathcal{C}_{\mathcal{D}^*}$ is the set of all *antecedents* of DCIs in $\mathcal{D}^*$. The exceptionality ranking of the DCIs in $\mathcal{D}^*$ computed by Exceptional($\cdot$) makes use of $\mathcal{T}^*, \overline{\mathcal{D}^*}$, and $\mathcal{C}_{\mathcal{D}^*}$. That is, it checks, for each concept $C \in \mathcal{C}_{\mathcal{D}^*}$, whether $\mathcal{T}^* \models \sqcap \overline{\mathcal{D}^*} \sqsubseteq \neg C$. In case C is exceptional, every DCI $C \sqsubseteq D \in \mathcal{D}^*$ is exceptional w.r.t. $\mathcal{KB}^* = \mathcal{T}^* \cup \mathcal{D}^*$ and is added to the set $\mathcal{E}_1$.

If $\mathcal{E}_1 \neq \mathcal{E}_0$, then we call Exceptional($\cdot$) for $\mathcal{T}^* \cup \mathcal{E}_1$, defining the set $\mathcal{E}_2$, and so on. Hence, given $\mathcal{KB}^* = \mathcal{T}^* \cup \mathcal{D}^*$, we construct a sequence $\mathcal{E}_0, \mathcal{E}_1, \ldots$ in the following way, for $i \geq 0$:

- $\mathcal{E}_0 =_{\mathrm{def}} \mathcal{D}^*$
- $\mathcal{E}_{i+1} =_{\mathrm{def}}$ Exceptional($\mathcal{T}^*, \mathcal{E}_i$)

Example 2 (continued). Using the knowledge base of Example 2, we initialise $\mathcal{T}^*$ as {EmpStud $\sqsubseteq$ Stud} and let

$$\mathcal{D}^* = \left\{ \begin{array}{c} \mathsf{Stud} \mathrel{\text{$\sqsubseteq$\raisebox{-2pt}{$\sim$}}} \neg\exists\mathsf{receives}.\mathsf{TaxInv}, \\ \mathsf{EmpStud} \mathrel{\text{$\sqsubseteq$\raisebox{-2pt}{$\sim$}}} \exists\mathsf{receives}.\mathsf{TaxInv}, \\ \mathsf{EmpStud} \sqcap \mathsf{Parent} \mathrel{\text{$\sqsubseteq$\raisebox{-2pt}{$\sim$}}} \neg\exists\mathsf{receives}.\mathsf{TaxInv} \end{array} \right\}$$

We then obtain the following exceptionality sequence:

$$\mathcal{E}_0 = \left\{ \begin{array}{c} \mathsf{Stud} \mathrel{\text{$\sqsubseteq$\raisebox{-2pt}{$\sim$}}} \neg\exists\mathsf{receives}.\mathsf{TaxInv}, \\ \mathsf{EmpStud} \mathrel{\text{$\sqsubseteq$\raisebox{-2pt}{$\sim$}}} \exists\mathsf{receives}.\mathsf{TaxInv}, \\ \mathsf{EmpStud} \sqcap \mathsf{Parent} \mathrel{\text{$\sqsubseteq$\raisebox{-2pt}{$\sim$}}} \neg\exists\mathsf{receives}.\mathsf{TaxInv} \end{array} \right\}$$

$$\mathcal{E}_1 = \left\{ \begin{array}{c} \mathsf{EmpStud} \mathrel{\text{$\sqsubseteq$\raisebox{-2pt}{$\sim$}}} \exists\mathsf{receives}.\mathsf{TaxInv}, \\ \mathsf{EmpStud} \sqcap \mathsf{Parent} \mathrel{\text{$\sqsubseteq$\raisebox{-2pt}{$\sim$}}} \neg\exists\mathsf{receives}.\mathsf{TaxInv} \end{array} \right\}$$

$$\mathcal{E}_2 = \{\mathsf{EmpStud} \sqcap \mathsf{Parent} \mathrel{\text{$\sqsubseteq$\raisebox{-2pt}{$\sim$}}} \neg\exists\mathsf{receives}.\mathsf{TaxInv}\}$$

Since $\mathcal{D}^*$ is finite, the construction will eventually terminate with a fixed point $\mathcal{E}_{\mathrm{fix}} = $ Exceptional($\mathcal{T}^*, \mathcal{E}_{\mathrm{fix}}$). If this fixed point is non-empty, then the axioms in there are said to have infinite rank. We therefore set $\mathcal{D}^*_\infty$ as $\mathcal{E}_{\mathrm{fix}}$ (Line 11 of ComputeRanking($\cdot$)), and the classical translations of these axioms are moved to the TBox. Hence we redefine the knowledge base in the following way (Lines 12 and 13 of ComputeRanking($\cdot$)):

- $\mathcal{T}^* \leftarrow \mathcal{T}^* \cup \{C \sqsubseteq D \mid C \mathrel{\text{$\sqsubseteq$\raisebox{-2pt}{$\sim$}}} D \in \mathcal{D}^*_\infty\}$;
- $\mathcal{D}^* \leftarrow \mathcal{D}^* \setminus \mathcal{D}^*_\infty$.

Function ComputeRanking($\cdot$) must terminate since $\mathcal{D}$ is finite, and at every iteration, $\mathcal{D}^*$ becomes smaller (hence, we have at most $|\mathcal{D}|$ iterations). In the end, we obtain a knowledge base $\mathcal{KB}^* = \mathcal{T}^* \cup \mathcal{D}^*$ which is modularly equivalent to the original knowledge base $\mathcal{KB} = \mathcal{T} \cup \mathcal{D}$ (see Lemma 12 below), in which $\mathcal{D}^*$ has no DCIs of infinite rank (all the strict knowledge 'hidden' in the dTBox has been moved to the TBox). In the following, we say that such a knowledge base is in *rank normal form*.

Once we have obtained the knowledge base $\mathcal{KB}^* = \mathcal{T}^* \cup \mathcal{D}^*$ and the final sequence $\mathcal{E}_0, \mathcal{E}_1, \ldots, \mathcal{E}_{\mathrm{fix}}$, we partition the set $\mathcal{D}^*$ into the sets $\mathcal{D}_0, \ldots, \mathcal{D}_n$, for some $n \geq 0$ (Lines 15–17 of ComputeRanking($\cdot$)).

Example 2 *(continued)*. For $\mathcal{KB}$ as in Example 2, we obtain the sequence:
$$\mathcal{D}_0 = \{\mathsf{Stud} \mathrel{\subset\!\!\!\sim} \neg\exists\mathsf{receives}.\mathsf{TaxInv}\}$$
$$\mathcal{D}_1 = \{\mathsf{EmpStud} \mathrel{\subset\!\!\!\sim} \exists\mathsf{receives}.\mathsf{TaxInv}\}$$
$$\mathcal{D}_2 = \{\mathsf{EmpStud} \sqcap \mathsf{Parent} \mathrel{\subset\!\!\!\sim} \neg\exists\mathsf{receives}.\mathsf{TaxInv}\}$$

At this stage, we have moved all the classical information possibly 'hidden' inside the dTBox to the TBox, and ranked all the remaining DCIs, where the rank of a DCI is the index of the unique partition to which it belongs, defined as follows:

Definition 26 (Ranking). *For every* $C, D \in \mathcal{L}$:

- $\mathsf{rk}(C) =_{\mathrm{def}} i$, $0 \le i \le n$, *if* $\bigsqcap \overline{\mathcal{E}_i}$ *is the first element in* $(\bigsqcap \overline{\mathcal{E}_0}, \ldots, \bigsqcap \overline{\mathcal{E}_n})$ *s.t.* $\mathcal{T}^* \not\models \bigsqcap \overline{\mathcal{E}_i} \sqcap C \sqsubseteq \bot$;
- $\mathsf{rk}(C) =_{\mathrm{def}} \infty$ *if there is no such* $\bigsqcap \overline{\mathcal{E}_i}$;
- $\mathsf{rk}(C \mathrel{\subset\!\!\!\sim} D) =_{\mathrm{def}} \mathsf{rk}(C)$.

Remark 1. For every $i \le j \le n$, $\models \bigsqcap \overline{\mathcal{E}_j} \sqsubseteq \bigsqcap \overline{\mathcal{E}_i}$.

Remark 2. For every $i < j \le n$, $\mathcal{D}_i \cap \mathcal{D}_j = \emptyset$.

To summarise, we transform our initial knowledge base $\mathcal{KB} = \mathcal{T} \cup \mathcal{D}$, obtaining a modularly equivalent knowledge base $\mathcal{KB}^* = \mathcal{T}^* \cup \mathcal{D}^*$ (see Lemma 12 below) and a ranking of DCIs in the form of a partitioning of $\mathcal{D}^*$. The main difference between $\mathsf{ComputeRanking}(\cdot)$ and the analogous procedure by Casini and Straccia [30] is the reiteration of the ranking procedure until $\mathcal{D}^*_\infty = \emptyset$ (lines 4–14 in $\mathsf{ComputeRanking}(\cdot)$). While the two procedures behave identically in the case where there are no DCIs $C \mathrel{\subset\!\!\!\sim} D$ s.t. $\mathsf{rank}_{\mathcal{KB}}(C \mathrel{\subset\!\!\!\sim} D) = \infty$ in $\mathcal{D}$, the original procedure [30] did not handle all the cases correctly in which there is strict information 'hidden' inside the dTBox.

Given the knowledge base $\mathcal{KB}^* = \mathcal{T}^* \cup \mathcal{D}^*$, we can now define the main algorithm for deciding whether a DCI $C \mathrel{\subset\!\!\!\sim} D$ is in the rational closure of $\mathcal{KB}$. To do that, we use the same approach as in the function $\mathsf{Exceptional}(\cdot)$, that is, given $\mathcal{KB}^* = \mathcal{T}^* \cup \mathcal{D}^*$ and our sequence of sets $\mathcal{E}_0, \ldots, \mathcal{E}_n$, we use the TBox $\mathcal{T}^*$ and the sets of conjunctions of materialisations $\bigsqcap \overline{\mathcal{E}_0}, \ldots, \bigsqcap \overline{\mathcal{E}_n}$.

Definition 27 (Rational Deduction). *Let* $\mathcal{KB} = \mathcal{T} \cup \mathcal{D}$ *and let* $C, D \in \mathcal{L}$. *We say that* $C \mathrel{\subset\!\!\!\sim} D$ *is* **rationally deducible** *from* $\mathcal{KB}$, *denoted* $\mathcal{KB} \vdash_{\mathsf{rat}} C \mathrel{\subset\!\!\!\sim} D$, *if* $\mathcal{T}^* \models \bigsqcap \overline{\mathcal{E}_i} \sqcap C \sqsubseteq D$, *where* $\bigsqcap \overline{\mathcal{E}_i}$ *is the first element of the sequence* $\bigsqcap \overline{\mathcal{E}_0}, \ldots, \bigsqcap \overline{\mathcal{E}_n}$ *s.t.* $\mathcal{T}^* \not\models \bigsqcap \overline{\mathcal{E}_i} \sqsubseteq \neg C$. *If there is no such element,* $\mathcal{KB} \vdash_{\mathsf{rat}} C \mathrel{\subset\!\!\!\sim} D$ *if* $\mathcal{T}^* \models C \sqsubseteq D$.

Observe that $\mathcal{KB} \vdash_{\mathsf{rat}} C \sqsubseteq D$ if and only if $\mathcal{KB} \vdash_{\mathsf{rat}} C \sqcap \neg D \mathrel{\subset\!\!\!\sim} \bot$, i.e., if and only if $\mathcal{KB} \vdash_{\mathsf{rat}} C \sqcap \neg D \mathrel{\subset\!\!\!\sim} \bot$ (that is to say, $\mathcal{T}^* \models C \sqsubseteq D$).

The algorithm corresponding to the steps above is presented in the function $\mathsf{RationalClosure}(\cdot)$ below.

Example 2 *(continued)*. Let $\mathcal{KB}$ be as in Example 2 and assume we want to check whether $\mathsf{EmpStud} \mathrel{\subset\!\!\!\sim} \exists\mathsf{receives}.\mathsf{TaxInv}$ is in the rational closure of $\mathcal{KB}$. Then,

---

Function. RationalClosure($\mathcal{KB}$, α)

Input: $\mathcal{KB} = \mathcal{T} \cup \mathcal{D}$, the corresponding $\mathcal{KB}^* = \mathcal{T}^* \cup \mathcal{D}^*$, the sequence
     $\mathcal{E}_0, \ldots, \mathcal{E}_n$, and a query $\alpha = C \mathrel{\vert\!\approx} D$.
Output: `true` if $\mathcal{KB} \vdash_{\mathsf{rat}} C \mathrel{\vert\!\approx} D$, `false` otherwise

1   $i \leftarrow 0$
2   **while** $\mathcal{T}^* \models \bigsqcap \overline{\mathcal{E}_i} \sqcap C \sqsubseteq \bot$ *and* $i \leq n$ **do**
3     $\lfloor \; i \leftarrow i + 1$
4   **if** $i \leq n$ **then**
5     $\lfloor$ **return** $\mathcal{T}^* \models \bigsqcap \overline{\mathcal{E}_i} \sqcap C \sqsubseteq D$
6   **else**
7     $\lfloor$ **return** $\mathcal{T}^* \models C \sqsubseteq D$

---

the while-loop on Line 2 of function RationalClosure($\cdot$) terminates when $i = 1$. At this stage, $\bigsqcap \overline{\mathcal{E}_i} = (\neg\mathsf{EmpStud} \sqcup \exists\mathsf{receives.TaxInv}) \sqcap (\neg\mathsf{EmpStud} \sqcup \neg\mathsf{Parent} \sqcup \neg\exists\mathsf{receives.TaxInv})$. Given this, one can check that $\mathcal{T}^* \not\models \bigsqcap \overline{\mathcal{E}_i} \sqcap C \sqsubseteq \bot$, i.e., $\{\mathsf{EmpStud} \sqsubseteq \mathsf{Stud}\} \not\models (\neg\mathsf{EmpStud} \sqcup \exists\mathsf{receives.TaxInv}) \sqcap (\neg\mathsf{EmpStud} \sqcup \neg\mathsf{Parent} \sqcup \neg\exists\mathsf{receives.TaxInv}) \sqcap \mathsf{EmpStud} \sqsubseteq \bot$.

Finally, we can confirm that $\mathcal{T}^* \not\models \bigsqcap \overline{\mathcal{E}_i} \sqcap C \sqsubseteq D$, i.e., $\{\mathsf{EmpStud} \sqsubseteq \mathsf{Stud}\} \not\models$ $(\neg\mathsf{EmpStud} \sqcup \exists\mathsf{receives.TaxInv}) \sqcap (\neg\mathsf{EmpStud} \sqcup \neg\mathsf{Parent} \sqcup \neg\exists\mathsf{receives.TaxInv}) \sqcap$ $\mathsf{EmpStud} \sqsubseteq \exists\mathsf{receives.TaxInv}$.

Before we state the main theorem of this section, we need to establish the correspondence between the ranking function $\mathsf{rank}_{\mathcal{KB}}(\cdot)$ presented in Sect. 5.1 in the construction of the rational closure of $\mathcal{KB}$ and linked by Theorem 3 to the definition of rational entailment, and the ranking function $\mathsf{rk}(\cdot)$ of Definition 26 used in the above algorithm. We also need to establish that the normalisation of a knowledge base by our algorithm maintains modular equivalence.

Lemma 12. *Let $\mathcal{KB} = \mathcal{T} \cup \mathcal{D}$ and let $\mathcal{KB}^* = \mathcal{T}^* \cup \mathcal{D}^*$ be obtained from $\mathcal{KB}$ through function* ComputeRanking($\cdot$). *Then $\mathcal{KB}$ and $\mathcal{KB}^*$ are modularly equivalent.*

Lemma 13. *For every defeasible knowledge base $\mathcal{KB} = \mathcal{T} \cup \mathcal{D}$ and every $C \in \mathcal{L}$, $\mathsf{rank}_{\mathcal{KB}}(C) = \mathsf{rk}(C)$.*

Now we can state the main theorem, which links rational entailment to rational deduction via Theorem 3.

Theorem 4. *Let $\mathcal{KB} = \mathcal{T} \cup \mathcal{D}$ and let $C, D \in \mathcal{L}$. Then $\mathcal{KB} \vdash_{\mathsf{rat}} C \mathrel{\vert\!\approx} D$ iff $\mathcal{KB} \models {}_{\mathsf{rat}} C \mathrel{\vert\!\approx} D$.*

As an immediate consequence, we have that the function RationalClosure($\cdot$) is correct w.r.t. the definition of rational closure in Definition 24.

Corollary 3. *Checking rational entailment is* EXPTIME-*complctc.*

Hence entailment checking for defeasible ontologies is just as hard as classical subsumption checking.

We conclude this section by noting that although rational closure is viewed as an appropriate form of defeasible reasoning, it does have its limitations, the first of which is that it does not satisfy the *presumption of independence* [51, Section 3.1]. To consider a well-worn example, suppose we know that birds usually fly and usually have wings, that both penguins and robins are birds, and that penguins usually do not fly. That is, we have the following knowledge base: $\mathcal{KB} = \{\text{Bird} \sqsubseteq_{\sim} \text{Flies}, \text{Bird} \sqsubseteq_{\sim} \text{Wings}, \text{Penguin} \sqsubseteq \text{Bird}, \text{Robin} \sqsubseteq \text{Bird}, \text{Penguin} \sqsubseteq_{\sim} \neg\text{Flies}\}$. Rational closure allows us to conclude that robins usually have wings, since they are viewed as typical birds, thereby satisfying the presumption of typicality. But with penguins being atypical birds, rational closure does not allow us to conclude that penguins usually have wings, thus violating the presumption of independence which, in this context, would require the atypicality of penguins w.r.t. flying to be independent of the typicality of penguins w.r.t. having wings.

This deficiency is well-known, and there are other forms of defeasible reasoning that can overcome this, most notably lexicographic closure [31], relevance closure [33], and inheritance-based closure [32,34]. But note that the presumption of independence is *propositional* in nature. In fact, the DL version of lexicographic closure is essentially a lifting to the DL case of a propositional solution to the problem [51].

What is perhaps of more interest is the inability of rational closure to deal with defeasibility relating to the *non-propositional* aspects of descriptions logics. For example, Pensel and Turhan [54,55] have shown that rational closure across role expressions does not always support defeasible inheritance appropriately.

Suppose we know that bosses are workers, do not have workers as their superiors, and are usually responsible. Furthermore, suppose we know that workers usually have bosses as their superiors. We thus have the knowledge base:

$$\mathcal{KB} = \left\{ \begin{array}{c} \text{Boss} \sqsubseteq \ \text{Worker}, \\ \text{Boss} \sqsubseteq \ \neg\exists\text{hasSuperior.Worker}, \\ \text{Boss} \sqsubseteq_{\sim} \ \text{Responsible}, \\ \text{Worker} \sqsubseteq_{\sim} \ \exists\text{hasSuperior.Boss} \end{array} \right\}$$

Since workers usually have bosses as their superiors, and bosses are usually responsible, one would expect to be able to conclude that workers usually have responsible superiors. But rational closure is unable to do so. From the perspective of the algorithm for rational closure, this can be traced back to the use of materialisation (Definition 25) when computing exceptionality, as Pensel and Turhan [54] show. A more detailed semantic explanation for this inability is still forthcoming, though.

6   Beyond Defeasible Concept Inclusion

Defeasible reasoning in description logics extends beyond defeasible concept inclusion. In this section, we outline two such extensions following on from the work presented here, firstly to account for named individuals in defeasible knowledge bases, and secondly to introduce defeasible class descriptions.

The introduction of defeasible reasoning also for ABox reasoning is a necessary extension of the results we have presented in this chapter. We want to be able to derive assertions of the kind "Presumably, the individual a falls under the concept C", and, in the present framework, the natural way of doing it would be to model the *presumption of typicality* also w.r.t. the individuals named in the ABox, that is, to maximise the amount of defeasible information we associate with each individual: If all we know about Ann is that she is a student, we want to be able to conclude that *presumably* Ann does not get a tax invoice. The main technical problem in the present framework is the possibility of having multiple distinct configurations that maximise the presumption of typicality w.r.t. the individuals [30, Example 7]. Different solutions have been proposed [29,30,35,45,55], but, as mentioned in Sect. 2, we are not going to introduce here the different proposals regarding the introduction of defeasible reasoning for the ABox.

The systems proposed by Giordano and others [39,40,44,45] introduce an operator $\mathbf{T}$ (*typical*) associated to the concepts. This allows extra expressivity in modelling defeasible information: an inclusion like Stud $\sqcap \neg \exists$receives.TaxInv $\sqsubseteq$ $\mathbf{T}$(Stud), indicating that the students that do not receive the a tax invoice must be considered typical students, is not expressible in a language using only defeasible subsumptions. However, in most of the systems they introduce, $\mathbf{T}$ can be used only in expressions of the form $\mathbf{T}(C) \sqsubseteq D$, which is interpreted exactly as an expression $C \mathrel{\raise0.3ex\hbox{$\sqsubset$}\kern-0.5em\lower0.5ex\hbox{$\sim$}} D$. Booth and others [10] have shown that, even at the propositional level, using freely an operator like $\mathbf{T}$ creates the possibility of multiple configurations satisfying the *presumption of typicality*, in a way that, from the formal point of view, is analogous to the problem registered working with the ABoxes.

Given the special status of subsumption in DLs in particular and the historical importance of argument forms and entailment in logic in general, the bulk of the effort in non-monotonic reasoning has quite naturally been spent on the definition of a proper account of defeasible subsumption and the characterisation of appropriate notions of defeasible entailment.

However, given the importance of concept descriptions in DLs, an extension of this work to also represent *defeasible classes* is called for. This includes the ability to represent notions such as plausible value or existential restrictions in complex concept descriptions [17,23,24,27]. There are several ways to accomplish this, and we focus here on one such proposal.

We could, for example, ask whether the constraint that workers usually have bosses as their superior is necessarily correctly captured by the defeasible subsumption: Worker $\mathrel{\raise0.3ex\hbox{$\sqsubset$}\kern-0.5em\lower0.5ex\hbox{$\sim$}} \exists$hasSuperior.Boss. An alternative reading of the phrase is that all workers have some superior, who is usually a boss. It is therefore the class description $\exists$hasSuperior.Boss which is defeasible. rather than the subsumption statement. This can be captured by extending the concept language of $\mathcal{ALC}$ as follows:

$$C ::= \top \mid \bot \mid \mathsf{C} \mid \neg C \mid C \sqcap C \mid C \sqcup C \mid \forall r.C \mid \exists r.C \mid \forall\!\!\!/\, r.C \mid \exists\!\!\!/\, r.C$$

With $\widetilde{\mathcal{L}}$ we denote the extended language of all (possibly defeasible) $\mathcal{ALC}$ concepts.

Definition 28. *Let* $\mathcal{P} = \langle \Delta^{\mathcal{P}}, \cdot^{\mathcal{P}}, \prec_{\mathcal{P}} \rangle$ *be a preferential interpretation. Let* $r \in \mathsf{R}$ *and* $C \in \mathsf{C}$. *The truth conditions for defeasible universal restriction* $\forall\!\!\!/\, r.C$ *and strict existential restriction* $\exists\!\!\!/\, r.C$ *are given by:*

$$(\forall\!\!\!/\, r.C)^{\mathcal{P}} =_{\text{def}} \{x \in \Delta^{\mathcal{P}} \mid \min_{\prec_{\mathcal{P}}} r^{\mathcal{P}}(x) \subseteq C^{\mathcal{P}}\};$$
$$(\exists\!\!\!/\, r.C)^{\mathcal{P}} =_{\text{def}} \{x \in \Delta^{\mathcal{P}} \mid \min_{\prec_{\mathcal{P}}} r^{\mathcal{P}}(x) \cap C^{\mathcal{P}} \neq \emptyset\}.$$

That $\exists\!\!\!/\, r.C$ captures the notion of strict existential restriction follows since, not only does the semantics require that some r-filler be in $C^{\mathcal{P}}$, but it also demands that some most preferred r-filler be in $C^{\mathcal{P}}$. In contrast, defeasible universal (value) restriction relaxes the condition that all r-fillers be in $C^{\mathcal{P}}$, requiring only that all most preferred r-fillers be in $C^{\mathcal{P}}$.

Definition 28 now allows us to state that every worker has some typical superior who is a boss, i.e., Worker $\sqsubseteq \exists\!\!\!/\, $hasSuperior.Boss, or that any superior of a worker is usually a boss, i.e., Worker $\sqsubseteq \forall\!\!\!/\, $hasSuperior.Boss.

The defeasible quantifiers of Definition 28 are based on a single order on objects, but this generalises naturally to a parameterised ordering on either objects or role interpretations [23, 27], the details of which we omit here. The ramifications of extending the language with defeasible quantification have also been investigated for modal logics, where it assumes the form of defeasible modalities [25, 26].

7 Concluding Remarks

In this paper we have provided an overview of a specific approach to defeasible reasoning—one that is based on work initiated by Kraus, Lehmann and Magidor for the propositional case [49, 52]. This approach has a number of attractive characteristics: It has a simple and intuitive semantics for defeasible subsumption in description logics that is general enough to constitute the core framework within which to investigate defeasible extensions to DLs. It also allows for the characterisation of two forms of defeasible subsumption relations—preferential and rational subsumption—providing weight to the claim that the semantic constructions are intuitively appropriate. In addition, it provides the basis for defining an appropriate form of defeasible entailment—a description logic version of what is known as *rational closure* in the propositional case. Moreover, it comes equipped with an algorithm for computing the DL version of rational closure with computational complexity that is no worse than the complexity of entailment checking in $\mathcal{ALC}$. Importantly from a practical perspective, the algorithm can be reduced to a number of classical entailment checks, which means that it can be implemented on top of existing (highly optimised) description logic reasoners. In terms of performance, a relatively naïve version of such an algorithm has already been shown to scale well in practice [28].

Section 6 touched on some ways in which defeasible reasoning for description logics has already been extended beyond defeasible concept inclusion, but all these proposals are only preliminary investigations with much work that still needs to be done. Further topics for future research include the study of role-based defeasible constructors [23,24,27] and the investigation of defeasible versions of query answering [64]. Finally, a somewhat different area for future exploration is one that is aimed at exploiting the well-known connection between belief revision and rational consequence in the propositional case [38]. Given this connection on the propositional level, it seems reasonable to expect that the results presented in this paper can form the basis of a different perspective on belief revision for description logics.

Acknowledgments. Giovanni Casini and Thomas Meyer have received funding from the EU Horizon 2020 research and innovation programme under the Marie Skłodowska-Curie grant agr. No. 690974 (MIREL). The work of Thomas Meyer has been supported in part by the National Research Foundation of South Africa (grant No. UID 98019).

References

1. Baader, F., Calvanese, D., McGuinness, D., Nardi, D., Patel-Schneider, P. (eds.): The Description Logic Handbook: Theory Implementation and Applications, 2nd edn. Cambridge University Press, Cambridge (2007)
2. Baader, F., Hollunder, B.: How to prefer more specific defaults in terminological default logic. In: Bajcsy, R. (ed.) Proceedings of the 13th International Joint Conference on Artificial Intelligence (IJCAI), pp. 669–675. Morgan Kaufmann Publishers (1993)
3. Baader, F., Hollunder, B.: Embedding defaults into terminological knowledge representation formalisms. J. Autom. Reason. **14**(1), 149–180 (1995)
4. Baltag, A., Smets, S.: Dynamic belief revision over multi-agent plausibility models. In: van der Hoek, W., Wooldridge, M. (eds.) Proceedings of LOFT, pp. 11–24. University of Liverpool (2006)
5. Baltag, A., Smets, S.: A qualitative theory of dynamic interactive belief revision. In: Bonanno, G., van der Hoek, W., Wooldridge, M. (eds.) Logic and the Foundations of Game and Decision Theory (LOFT7). Texts in Logic and Games, no. 3, pp. 13–60. Amsterdam University Press (2008)
6. Benferhat, S., Dubois, D., Prade, H.: Possibilistic and standard probabilistic semantics of conditional knowledge bases. J. Log. Comput. **9**(6), 873–895 (1999)
7. Bonatti, P., Faella, M., Petrova, I., Sauro, L.: A new semantics for overriding in description logics. Artif. Intell. **222**, 1–48 (2015)
8. Bonatti, P., Faella, M., Sauro, L.: Defeasible inclusions in low-complexity DLs. J. Artif. Intell. Res. **42**, 719–764 (2011)
9. Bonatti, P., Lutz, C., Wolter, F.: The complexity of circumscription in description logic. J. Artif. Intell. Res. **35**, 717–773 (2009)
10. Booth, R., Casini, G., Meyer, T., Varzinczak, I.: On the entailment problem for a logic of typicality. In: Proceedings of the 24th International Joint Conference on Artificial Intelligence (IJCAI), pp. 2805–2811 (2015)

11. Booth, R., Meyer, T., Varzinczak, I.: PTL: a propositional typicality logic. In: del Cerro, L.F., Herzig, A., Mengin, J. (eds.) JELIA 2012. LNCS (LNAI), vol. 7519, pp. 107–119. Springer, Heidelberg (2012). https://doi.org/10.1007/978-3-642-33353-8_9

12. Booth, R., Meyer, T., Varzinczak, I.: A propositional typicality logic for extending rational consequence. In: Fermé, E., Gabbay, D., Simari, G. (eds.) Trends in Belief Revision and Argumentation Dynamics, Studies in Logic - Logic and Cognitive Systems, vol. 48, pp. 123–154. King's College Publications (2013)

13. Booth, R., Paris, J.: A note on the rational closure of knowledge bases with both positive and negative knowledge. J. Log. Lang. Inf. **7**(2), 165–190 (1998)

14. Boutilier, C.: Conditional logics of normality: a modal approach. Artif. Intell. **68**(1), 87–154 (1994)

15. Britz, K., Casini, G., Meyer, T., Moodley, K., Sattler, U., Varzinczak, I.: Theoretical foundations of defeasible description logics. Technical report. arXiv:1904.07559 [cs.AI], ArXiV (April 2019). http://arxiv.org/abs/1904.07559

16. Britz, K., Casini, G., Meyer, T., Moodley, K., Varzinczak, I.: Ordered interpretations and entailment for defeasible description logics. Technical report, CAIR, CSIR Meraka and UKZN, South Africa (2013). http://tinyurl.com/cydd6yy

17. Britz, K., Casini, G., Meyer, T., Varzinczak, I.: Preferential role restrictions. In: Proceedings of the 26th International Workshop on Description Logics, pp. 93–106 (2013)

18. Britz, K., Heidema, J., Labuschagne, W.: Semantics for dual preferential entailment. J. Philos. Log. **38**, 433–446 (2009)

19. Britz, K., Heidema, J., Meyer, T.: Semantic preferential subsumption. In: Lang, J., Brewka, G. (eds.) Proceedings of the 11th International Conference on Principles of Knowledge Representation and Reasoning (KR), pp. 476–484. AAAI Press/MIT Press (2008)

20. Britz, K., Heidema, J., Meyer, T.: Modelling object typicality in description logics. In: Nicholson, A., Li, X. (eds.) AI 2009. LNCS (LNAI), vol. 5866, pp. 506–516. Springer, Heidelberg (2009). https://doi.org/10.1007/978-3-642-10439-8_51

21. Britz, K., Meyer, T., Varzinczak, I.: Semantic foundation for preferential description logics. In: Wang, D., Reynolds, M. (eds.) AI 2011. LNCS (LNAI), vol. 7106, pp. 491–500. Springer, Heidelberg (2011). https://doi.org/10.1007/978-3-642-25832-9_50

22. Britz, K., Varzinczak, I.: Defeasible modalities. In: Proceedings of the 14th Conference on Theoretical Aspects of Rationality and Knowledge (TARK), pp. 49–60 (2013)

23. Britz, K., Varzinczak, I.: Introducing role defeasibility in description logics. In: Michael, L., Kakas, A. (eds.) JELIA 2016. LNCS (LNAI), vol. 10021, pp. 174–189. Springer, Cham (2016). https://doi.org/10.1007/978-3-319-48758-8_12

24. Britz, K., Varzinczak, I.: Toward defeasible $\mathcal{SROIQ}$. In: Proceedings of the 30th International Workshop on Description Logics (2017)

25. Britz, K., Varzinczak, I.: From KLM-style conditionals to defeasible modalities, and back. J. Appl. Non-Class. Log. (JANCL) **28**(1), 92–121 (2018)

26. Britz, K., Varzinczak, I.: Preferential accessibility and preferred worlds. J. Log. Lang. Inf. (JoLLI) **27**(2), 133–155 (2018)

27. Britz, K., Varzinczak, I.: Rationality and context in defeasible subsumption. In: Ferrarotti, F., Woltran, S. (eds.) FoIKS 2018. LNCS, vol. 10833, pp. 114–132. Springer, Cham (2018). https://doi.org/10.1007/978-3-319-90050-6_7

28. Casini, G., Meyer, T., Moodley, K., Sattler, U., Varzinczak, I.: Introducing defeasibility into OWL ontologies. In: Arenas, M., et al. (eds.) ISWC 2015. LNCS, vol. 9367, pp. 409–426. Springer, Cham (2015). https://doi.org/10.1007/978-3-319-25010-6_27
29. Casini, G., Meyer, T., Moodley, K., Varzinczak, I.: Nonmonotonic reasoning in description logics: rational closure for the ABox. In: Proceedings of the 26th International Workshop on Description Logics, pp. 600–615 (2013)
30. Casini, G., Straccia, U.: Rational closure for defeasible description logics. In: Janhunen, T., Niemelä, I. (eds.) JELIA 2010. LNCS (LNAI), vol. 6341, pp. 77–90. Springer, Heidelberg (2010). https://doi.org/10.1007/978-3-642-15675-5_9
31. Casini, G., Straccia, U.: Lexicographic closure for defeasible description logics. In: Proceedings of the 8th Australasian Ontology Workshop (AOW), vol. 969, pp. 4–15. CEUR Workshop Proceedings (2012)
32. Casini, G., Straccia, U.: Defeasible inheritance-based description logics. J. Artif. Intell. Res. (JAIR) **48**, 415–473 (2013)
33. Casini, G., Meyer, T., Moodley, K., Nortjé, R.: Relevant closure: a new form of defeasible reasoning for description logics. In: Fermé, E., Leite, J. (eds.) JELIA 2014. LNCS (LNAI), vol. 8761, pp. 92–106. Springer, Cham (2014). https://doi.org/10.1007/978-3-319-11558-0_7
34. Casini, G., Straccia, U.: Defeasible inheritance-based description logics. In: Walsh, T. (ed.) Proceedings of the 22nd International Joint Conference on Artificial Intelligence (IJCAI), pp. 813–818 (2011)
35. Casini, G., Straccia, U., Meyer, T.: A polynomial time subsumption algorithm for nominal safe $\mathcal{ELO}_\perp$ under rational closure. Inf. Sci. (2018). https://doi.org/10.1016/j.ins.2018.09.037
36. Donini, F., Nardi, D., Rosati, R.: Description logics of minimal knowledge and negation as failure. ACM Trans. Comput. Log. **3**(2), 177–225 (2002)
37. Dubois, D., Lang, J., Prade, H.: Possibilistic logic. In: Gabbay, D., Hogger, C., Robinson, J. (eds.) Handbook of Logic in Artificial Intelligence and Logic Programming, vol. 3, pp. 439–513. Oxford University Press (1994)
38. Gärdenfors, P., Makinson, D.: Nonmonotonic inference based on expectations. Artif. Intell. **65**(2), 197–245 (1994)
39. Giordano, L., Gliozzi, V., Olivetti, N., Pozzato, G.L.: Preferential description logics. In: Dershowitz, N., Voronkov, A. (eds.) LPAR 2007. LNCS (LNAI), vol. 4790, pp. 257–272. Springer, Heidelberg (2007). https://doi.org/10.1007/978-3-540-75560-9_20
40. Giordano, L., Gliozzi, V., Olivetti, N., Pozzato, G.L.: Reasoning about typicality in preferential description logics. In: Hölldobler, S., Lutz, C., Wansing, H. (eds.) JELIA 2008. LNCS (LNAI), vol. 5293, pp. 192–205. Springer, Heidelberg (2008). https://doi.org/10.1007/978-3-540-87803-2_17
41. Giordano, L., Gliozzi, V., Olivetti, N., Pozzato, G.: Analytic tableaux calculi for KLM logics of nonmonotonic reasoning. ACM Trans. Comput. Log. **10**(3), 18:1–18:47 (2009)
42. Giordano, L., Gliozzi, V., Olivetti, N., Pozzato, G.: $\mathcal{ALC} + T$: a preferential extension of description logics. Fundam. Inform. **96**(3), 341–372 (2009)
43. Giordano, L., Gliozzi, V., Olivetti, N., Pozzato, G.L.: A minimal model semantics for nonmonotonic reasoning. In: del Cerro, L.F., Herzig, A., Mengin, J. (eds.) JELIA 2012. LNCS (LNAI), vol. 7519, pp. 228–241. Springer, Heidelberg (2012). https://doi.org/10.1007/978-3-642-33353-8_18
44. Giordano, L., Gliozzi, V., Olivetti, N., Pozzato, G.: A non-monotonic description logic for reasoning about typicality. Artif. Intell. **195**, 165–202 (2013)

45. Giordano, L., Gliozzi, V., Olivetti, N., Pozzato, G.: Semantic characterization of rational closure: from propositional logic to description logics. Artif. Intell. **226**, 1–33 (2015)
46. Governatori, G.: Defeasible description logics. In: Antoniou, G., Boley, H. (eds.) RuleML 2004. LNCS, vol. 3323, pp. 98–112. Springer, Heidelberg (2004). https://doi.org/10.1007/978-3-540-30504-0_8
47. Grosof, B., Horrocks, I., Volz, R., Decker, S.: Description logic programs: combining logic programs with description logic. In: Proceedings of the 12th International Conference on World Wide Web (WWW), pp. 48–57. ACM (2003)
48. Heymans, S., Vermeir, D.: A defeasible ontology language. In: Meersman, R., Tari, Z. (eds.) OTM 2002. LNCS, vol. 2519, pp. 1033–1046. Springer, Heidelberg (2002). https://doi.org/10.1007/3-540-36124-3_66
49. Kraus, S., Lehmann, D., Magidor, M.: Nonmonotonic reasoning, preferential models and cumulative logics. Artif. Intell. **44**, 167–207 (1990)
50. Lehmann, D.: What does a conditional knowledge base entail? In: Brachman, R., Levesque, H. (eds.) Proceedings of the 1st International Conference on Principles of Knowledge Representation and Reasoning (KR), pp. 212–222 (1989)
51. Lehmann, D.: Another perspective on default reasoning. Ann. Math. Artif. Intell. **15**(1), 61–82 (1995)
52. Lehmann, D., Magidor, M.: What does a conditional knowledge base entail? Artif. Intell. **55**, 1–60 (1992)
53. Padgham, L., Zhang, T.: A terminological logic with defaults: a definition and an application. In: Bajcsy, R. (ed.) Proceedings of the 13th International Joint Conference on Artificial Intelligence (IJCAI), pp. 662–668. Morgan Kaufmann Publishers (1993)
54. Pensel, M., Turhan, A.: Making quantification relevant again - the case of defeasible $el_\perp$. In: Booth, R., Casini, G., Varzinczak, I.J. (eds.) Proceedings of the 4th International Workshop on Defeasible and Ampliative Reasoning (DARe), pp. 44–57 (2017)
55. Pensel, M., Turhan, A.: Reasoning in the defeasible description logic $\mathcal{EL}_\perp$ - computing standard inferences under rational and relevant semantics. Int. J. Approx. Reason. **103**, 28–70 (2018)
56. Qi, G., Pan, J.Z., Ji, Q.: Extending description logics with uncertainty reasoning in possibilistic logic. In: Mellouli, K. (ed.) ECSQARU 2007. LNCS (LNAI), vol. 4724, pp. 828–839. Springer, Heidelberg (2007). https://doi.org/10.1007/978-3-540-75256-1_72
57. Quantz, J., Royer, V.: A preference semantics for defaults in terminological logics. In: Proceedings of the 3rd International Conference on Principles of Knowledge Representation and Reasoning (KR), pp. 294–305 (1992)
58. Quantz, J., Ryan, M.: Preferential default description logics. Technical report, TU Berlin (1993). www.tu-berlin.de/fileadmin/fg53/KIT-Reports/r110.pdf
59. Rott, H.: Change, Choice and Inference: A Study of Belief Revision and Nonmonotonic Reasoning. Oxford University Press, Oxford (2001)
60. Sengupta, K., Krisnadhi, A.A., Hitzler, P.: Local closed world semantics: grounded circumscription for OWL. In: Aroyo, L., et al. (eds.) ISWC 2011. LNCS, vol. 7031, pp. 617–632. Springer, Heidelberg (2011). https://doi.org/10.1007/978-3-642-25073-6_39
61. Shoham, Y.: Reasoning about Change: Time and Causation from the Standpoint of Artificial Intelligence. MIT Press, Cambridge (1988)

62. Straccia, U.: Default inheritance reasoning in hybrid KL-ONE-style logics. In: Bajcsy, R. (ed.) Proceedings of the 13th International Joint Conference on Artificial Intelligence (IJCAI), pp. 676–681. Morgan Kaufmann Publishers (1993)
63. Varzinczak, I.: A note on a description logic of concept and role typicality for defeasible reasoning over ontologies. Logica Universalis **12**(3–4), 297–325 (2018)
64. Xiao, G., et al.: Ontology-based data access: a survey. In: Proceedings of the Twenty-Seventh International Joint Conference on Artificial Intelligence, IJCAI 2018, 13–19 July 2018, Stockholm, Sweden, pp. 5511–5519 (2018). https://doi.org/10.24963/ijcai.2018/777

Temporal Logic Programs with Temporal Description Logic Axioms

Pedro Cabalar[1(✉)] and Torsten Schaub[2]

[1] University of Corunna, Corunna, Spain
cabalar@udc.es
[2] University of Potsdam, Potsdam, Germany
torsten@cs.uni-potsdam.de

Abstract. In this paper we introduce a combination of Answer Set Programming (ASP) and Description Logics (DL) (in particular, $\mathcal{ALC}$) on top of a modal temporal basis using connectives from Linear-time Temporal Logic (LTL). On the one hand, for the temporal extension of $\mathcal{ALC}$, we depart from Baader et al.'s proposal $\mathcal{ALC}$-LTL that restricts the use of temporal operators to occur only in front of DL axioms. On the other hand, for the temporal extension of ASP we use its formalization in terms of Temporal (Quantified) Equilibrium Logic (TEL). This choice is convenient since (non-temporal) Equilibrium Logic has been already used to capture the semantics of hybrid theories, that is, combinations of ASP programs with DL axioms. Our proposal, called $\mathcal{ALC}$-TEL, actually interprets $\mathcal{ALC}$ axioms in terms of their translation into first order sentences, so that the semantics of TEL is eventually used in the background. The resulting formalism conservatively extends TEL, hybrid theories and $\mathcal{ALC}$-LTL as particular cases.

1 Introduction

Due to its versatility, Answer Set Programming (ASP) [1,2] is one of the paradigms for non-monotonic reasoning that has been more frequently extended in the literature (if not the most). Each extension has been motivated by a given type of reasoning problem or family of application domains. For instance, the treatment of dynamic scenarios and transition systems was present from the very beginning of ASP [3] and eventually led to a combination of ASP with modal operators from Linear-time Temporal Logic (LTL) [4,5], giving birth to so-called *Temporal Equilibrium Logic* (TEL) [6]. As another example, the ASP extension of *Hybrid Knowledge Bases* [7] allows for combining non-monotonic logic programs with classical inference about ontologies, in terms of Description Logic (DL) [8]. Both extensions are based on the underlying formalism of *Equilibrium Logic* [9] but work in different directions: a natural question is what happens

This work was partially supported by MINECO, Spain (grant TIC2017-84453-P), Xunta de Galicia, Spain (grant 2016-2019 ED431G/01, CITIC) and DFG, Germany, (grant SCHA 550/9).

© Springer Nature Switzerland AG 2019
C. Lutz et al. (Eds.): Baader Festschrift, LNCS 11560, pp. 174–186, 2019.
https://doi.org/10.1007/978-3-030-22102-7_8

when we try to embrace both features, time and ontologies, in a common ASP extension. In the monotonic case, several approaches considered the introduction of LTL operators in DL at different levels, at the cost of a high complexity, or even undecidability for some reasoning tasks. A simple approach that avoids these inconveniences is $\mathcal{ALC}$-LTL [10], proposed by Baader, Ghilardi and Lutz, that extends $\mathcal{ALC}$ [11] with LTL constructs, but restricts the use of temporal operators to occur only in front of DL axioms.

In this paper, we consider the same temporal extension of DL in $\mathcal{ALC}$-LTL but under the answer set semantics for temporal logic programs provided by TEL, so that temporal $\mathcal{ALC}$ expressions can be combined with temporal logic programs. The resulting formalism, $\mathcal{ALC}$-TEL, conservatively extends TEL, hybrid theories and $\mathcal{ALC}$-LTL as particular cases. This work is a preliminary step to introduce the logic and informally explain its behavior using a simple example.

The rest of the paper is organized as follows. In the next section, we recall the basic definition of $\mathcal{ALC}$ and its translation to First Order Logic. In Sect. 3 we present the first order version of TEL as introduced in [12], but with a slight modification to allow open domains and capture $\mathcal{ALC}$ quantification. Section 4 defines the $\mathcal{ALC}$-LTL syntax whereas Sect. 5 incorporates those constructs into TEL using their first order translation together with some additional axiomatization. Finally, Sect. 6 concludes the paper.

2   Description Logic $\mathcal{ALC}$

The *alphabet* of an $\mathcal{ALC}$ theory [11,13] is a triple $\langle N_C, N_R, N_I \rangle$ of mutually disjoint sets of names referring to *concepts*, *roles* and *individuals*, respectively. As an example, consider the alphabet $N_C = \{\texttt{Disease}, \texttt{Treatment}, \texttt{Vaccine}, \texttt{Medication}\}$, $N_R = \{\texttt{curedBy}\}$, $N_I = \{\texttt{AIDS}, \texttt{Smallpox}\}$.

A *concept (description)* C is an expression that follows the grammar:

$$C ::= \mathsf{c} \mid \neg C \mid C \sqcap C \mid \exists \mathsf{r}.C$$

where $\mathsf{c} \in N_C$ is a concept name and $\mathsf{r} \in N_R$ a role name. We use the following abbreviations for concept descriptions:

$$C \sqcup D \overset{\text{def}}{=} \neg(\neg C \sqcap \neg D)$$
$$\top \overset{\text{def}}{=} \mathsf{c} \sqcup \neg \mathsf{c}$$
$$\bot \overset{\text{def}}{=} \neg \top$$
$$\forall \mathsf{r}.C \overset{\text{def}}{=} \neg \exists \mathsf{r}.(\neg C)$$

for some concept name $\mathsf{c} \in N_C$. A *general concept inclusion* (GCI) axiom is an expression of the form $C \sqsubseteq D$ where C and D are concept descriptions. A *T-Box* is a set of GCI axioms. We sometimes write $C \equiv D$ as an element of a T-Box

Θ to mean that the two axioms $C \sqsubseteq D$ and $D \sqsubseteq C$ are elements of Θ. As an example, consider the T-Box:

$$\text{Vaccine} \sqcup \text{Medication} \sqsubseteq \text{Treatment} \tag{1}$$

$$\exists\text{curedBy.Treatment} \sqsubseteq \text{Disease} \tag{2}$$

meaning that vaccines and medications are treatments, and that anything cured by a treatment must be a disease. An *assertion* (axiom) is a construct of one of the forms:

$$\text{a} : C \qquad (\text{a}, \text{b}) : \text{r}$$

where $\text{a}, \text{b} \in N_I$ are individual names, $\text{r} \in N_R$ is a role name and C is an arbitrary concept description. An *A-Box* is a set of assertions. For instance, the A-Box:

$$\text{Smallpox} : \exists\text{curedBy.Vaccine} \tag{3}$$

$$\text{AIDS} : \text{Disease} \sqcap \neg\exists\text{curedBy.Treatment} \tag{4}$$

tells us that smallpox is cured[1] by a vaccine whereas AIDS is a disease and has no treatment for its cure. The fact that `Smallpox` is a disease can be derived from the previous T-Box since it is cured by some vaccine, and the later is a treatment. A *knowledge base* $\langle \Theta, \Omega \rangle$ consists of a T-Box Θ and an A-Box Ω.

In the rest of the paper, we treat $\mathcal{ALC}$ through its standard First Order Logic translation (see for instance [13]). However, for the sake of completeness, we provide next the standard definition of the $\mathcal{ALC}$ semantics.

Definition 1 ($\mathcal{ALC}$ interpretation). *An $\mathcal{ALC}$ interpretation $\mathcal{I}$ is a pair $(\Delta^{\mathcal{I}}, \cdot^{\mathcal{I}})$ where $\Delta^{\mathcal{I}}$ is a non-empty set called the* domain *(containing individuals) and $\cdot^{\mathcal{I}}$ is a mapping on $N_C \cup N_R \cup N_I$ that assigns: an individual $\text{a}^{\mathcal{I}} \in \Delta^{\mathcal{I}}$ to each individual name $\text{a} \in N_I$; a set of individuals $\text{c}^{\mathcal{I}} \subseteq \Delta^{\mathcal{I}}$ to each concept name $\text{c} \in N_C$; and a set of pairs of individuals $\text{r}^{\mathcal{I}} \subseteq \Delta^{\mathcal{I}} \times \Delta^{\mathcal{I}}$ to each role name $\text{r} \in N_R$.* ☐

Definition 2 (Interpretation of concept descriptions). *Given interpretation $\mathcal{I} = (\Delta^{\mathcal{I}}, \cdot^{\mathcal{I}})$ its extension to concept descriptions follows the recursive rules:*

$$(\neg C)^{\mathcal{I}} \stackrel{\text{def}}{=} \Delta^{\mathcal{I}} \setminus C^{\mathcal{I}}$$

$$(C \sqcap D)^{\mathcal{I}} \stackrel{\text{def}}{=} C^{\mathcal{I}} \cap D^{\mathcal{I}}$$

$$(\exists\text{r}.C)^{\mathcal{I}} \stackrel{\text{def}}{=} \{d \in \Delta^{\mathcal{I}} \mid \text{ there is a } d' \text{ with } (d, d') \in \text{r}^{\mathcal{I}} \text{ such that } d' \in C^{\mathcal{I}}\}$$

The interpretation of derived concepts can be easily deduced:

$$(C \sqcup D)^{\mathcal{I}} = C^{\mathcal{I}} \cup D^{\mathcal{I}}$$

$$\top^{\mathcal{I}} = \Delta^{\mathcal{I}}$$

$$\bot^{\mathcal{I}} = \emptyset$$

$$(\forall\text{r}.C)^{\mathcal{I}} = \{d \in \Delta^{\mathcal{I}} \mid \text{ all } d' \text{ with } (d, d') \in \text{r}^{\mathcal{I}} \text{ satisfy } d' \in C^{\mathcal{I}}\}$$

---

[1] Understanding here `curedBy` as "cured or prevented by.".

As expected, an interpretation $\mathcal{I}$ *satisfies* a GCI axiom $C \sqsubseteq D$, written $\mathcal{I} \models C \sqsubseteq D$, iff $C^{\mathcal{I}} \subseteq D^{\mathcal{I}}$. Similarly, we define satisfaction for assertions as: $\mathcal{I} \models \mathsf{a} : C$ iff $\mathsf{a}^{\mathcal{I}} \in C^{\mathcal{I}}$; and $\mathcal{I} \models (\mathsf{a}, \mathsf{b}) : r$ iff $(\mathsf{a}^{\mathcal{I}}, \mathsf{b}^{\mathcal{I}}) \in r^{\mathcal{I}}$. Interpretation $\mathcal{I}$ is a *model* of a knowledge base $\langle \Theta, \Omega \rangle$ iff it satisfies all GCIs in the T-Box Θ and all assertions in the A-Box Ω.

As said before, we are interested in the translation of $\mathcal{ALC}$ into First Order Logic (FOL) [13]. Given an $\mathcal{ALC}$ alphabet $N_C \cup N_R \cup N_I$ we define the corresponding First Order signature with one unary predicate $\mathsf{c}(x)$ per each $\mathsf{c} \in N_C$, binary predicate $\mathsf{r}(x,y)$ per each $\mathsf{r} \in N_R$ and constant name a per each $\mathsf{a} \in N_I$. The FOL translation of a concept description C with respect to a free variable x is a formula denoted as $t_x(C)$ and recursively defined as follows:

$$
\begin{aligned}
t_x(\mathsf{c}) &\overset{\text{def}}{=} \mathsf{c}(x) \qquad \text{for any concept name} \mathsf{c} \in N_C \\
t_x(\neg C) &\overset{\text{def}}{=} \neg t_x(C) \\
t_x(C \sqcap D) &\overset{\text{def}}{=} t_x(C) \wedge t_x(D) \\
t_x(\exists \mathsf{r}.C) &\overset{\text{def}}{=} \exists y (\ \mathsf{r}(x,y) \wedge t_y(C)\ )
\end{aligned}
$$

Notice that y is a variable name[2] different from x and bound in $\exists y$. It is relatively easy to check that the translation of derived concepts can be captured by the following equivalent FOL formulas:

$$
\begin{aligned}
t_x(C \sqcup D) &\leftrightarrow t_x(C) \vee t_x(D) \\
t_x(\top) &\leftrightarrow \top \\
t_x(\bot) &\leftrightarrow \bot \\
t_x(\forall \mathsf{r}.C) &\leftrightarrow \forall y (\ \mathsf{r}(x,y) \rightarrow t_y(C)\ )
\end{aligned}
$$

The translation of a GCI axiom $C \sqsubseteq D$ is defined as

$$
t(C \sqsubseteq D) \overset{\text{def}}{=} \forall x (t_x(C) \rightarrow t_x(D))
$$

For instance, the translation of (2) corresponds to:

$$
\forall x (\exists y\ (\mathtt{curedBy}(x,y) \wedge \mathtt{Treatment}(y)) \rightarrow \mathtt{Disease}(x))
$$

We also define the translation of assertions as:

$$
t(\mathsf{a} : C) \overset{\text{def}}{=} t_x(C)[x/\mathsf{a}] \qquad t((\mathsf{a}, \mathsf{b}) : \mathsf{r}) \overset{\text{def}}{=} \mathsf{r}(\mathsf{a}, \mathsf{b})
$$

where $[x/\mathsf{a}]$ stands for the substitution of variable x by the individual name a. As an example, the translation of (3) amounts to:

$$
\exists y\ (\mathtt{curedBy}(\mathtt{Smallpox}, y) \wedge \mathtt{Vaccine}(y))
$$

Given a knowledge base $\langle \Theta, \Omega \rangle$, we define its translation as the union $t(\Theta) \cup t(\Omega)$ of the sets of translations of all GCIs in Θ and assertions in Ω, respectively.

Proposition 1. *There is a one-to-one correspondence between $\mathcal{ALC}$ models of $\langle \Theta, \Omega \rangle$ and FOL models of $t(\Theta) \cup t(\Omega)$.*

---

[2] In fact, we can define translation $t_y(C)$ using x as new bound variable, and the whole translation belongs to the 2-variable fragment of FOL.

3   Temporal Quantified Equilibrium Logic

The definition of Temporal Quantified Equilibrium logic we use in the current paper is an extension of a previous version defined in [12] to cope with open domains as in Quantified Equilibrium Logic from [7]. Syntactically, we consider function-free first-order languages $\mathcal{L} = \langle \mathcal{C}, \mathcal{P} \rangle$ built over a set of *constant* symbols, $\mathcal{C}$, and a set of *predicate* symbols, $\mathcal{P}$. Additionally, each $p \in \mathcal{P}$ has an associated arity or number of arguments. An *atom* is any $p(t_1, \ldots, t_n)$ where $p \in \mathcal{P}$ is a predicate with arity $n \geq 0$ and each t_i is a *term*, that is, a constant or a variable in its turn. We assume the existence of a binary equality predicate '=' $\in \mathcal{P}$, written in infix notation. Using $\mathcal{L}$, connectors and variables, an $\mathcal{L}$-*formula* φ is defined by following the grammar:

$$\varphi ::= p(t_1, \ldots, t_n) \mid \bot \mid \varphi_1 \wedge \varphi_2 \mid \varphi_1 \vee \varphi_2 \mid \varphi_1 \rightarrow \varphi_2 \mid$$
$$\bigcirc \varphi \mid \varphi_1 \text{ U } \varphi_2 \mid \varphi_1 \text{ R } \varphi_2 \mid \forall x \; \varphi \mid \exists x \; \varphi \mid (\varphi)$$

where $p(t_1, \ldots, t_n)$ is an atom, x is a variable and $\bigcirc$, U and R respectively stand for "next", "until" and "release." A *theory* is a finite set of formulas. We use the following derived operators:

$$\neg \varphi \overset{\text{def}}{=} \varphi \rightarrow \bot \qquad\qquad \Diamond \varphi \overset{\text{def}}{=} \top \text{ U } \varphi$$
$$\top \overset{\text{def}}{=} \neg \bot \qquad\qquad \Box \varphi \overset{\text{def}}{=} \bot \text{ R } \varphi$$
$$\varphi \leftrightarrow \psi \overset{\text{def}}{=} (\varphi \rightarrow \psi) \wedge (\psi \rightarrow \varphi)$$

for any formulas φ, ψ. Note that $\neg \varphi$ will be used to represent default negation. The application of i consecutive $\bigcirc$'s is denoted as follows: $\bigcirc^i \varphi \overset{\text{def}}{=} \bigcirc(\bigcirc^{i-1}\varphi)$ for $i > 0$ and $\bigcirc^0 \varphi \overset{\text{def}}{=} \varphi$. We say that a term, atom, formula or theory is *ground* if it does not contain variables. A *sentence* or closed-formula is a formula without free-variables (defined as usual). A *theory* Γ is a set of sentences.

A *universe* is a pair $(\mathcal{D}, \sigma)$ where $\mathcal{D}$ is a non-empty set called the *domain* and σ is a mapping $\sigma \colon \mathcal{C} \cup \mathcal{D} \rightarrow \mathcal{D}$ satisfying $\sigma(d) = d$ for every $d \in \mathcal{D}$. We call d an *unnamed* individual if there is no constant $c \in \mathcal{C}$ with $\sigma(c) = d$. Throughout this paper, σ is subject to the *unique names assumption (UNA)* stating that different individual names are mapped to different domain elements, that is, $\sigma(c) \neq \sigma(c')$ if $c \neq c'$ for any $c, c' \in \mathcal{C}$. This is a common assumption both in Description Logics and in Logic Programming. In fact, the latter usually makes a stronger assumption, taking the *Herbrand Universe* $(\mathcal{C}, \sigma)$ where $\mathcal{D} = \mathcal{C}$, and so, $\sigma(c) = c$ for all $c \in \mathcal{C}$. In this paper, however, we adopt an open domain as in [7] to accommodate the use of quantification from Description Logic.

By $At_{\mathcal{D}}(\mathcal{C}, \mathcal{P})$ we denote the set of ground atoms constructible from the language $\mathcal{L}' = \langle \mathcal{C} \cup \mathcal{D}, \mathcal{P} \rangle$. A first-order *LTL-interpretation* for language $\mathcal{L} = \langle \mathcal{C}, \mathcal{P} \rangle$ is a structure $\langle (\mathcal{D}, \sigma), \mathbf{T} \rangle$ where $(\mathcal{D}, \sigma)$ is a universe as above and $\mathbf{T}$ is an infinite sequence of sets, $\mathbf{T} = \{T_i\}_{i \geq 0}$ with $T_i \subseteq At_{\mathcal{D}}(\mathcal{C}, \mathcal{P})$. Intuitively, T_i contains those ground atoms that are true at situation i. For any $\mathbf{T} = \{T_i\}_{i \geq 0}$ and $k \geq 0$, by $\mathbf{T}[k]$ we denote the LTL-interpretation $\mathbf{T} = \{T_i\}_{i \geq k}$ that starts at the k-th position of $\mathbf{T}$. Given two sequences of sets $\mathbf{H}$ and $\mathbf{T}$ we say that $\mathbf{H}$ is *smaller than* $\mathbf{T}$, written $\mathbf{H} \leq \mathbf{T}$, when $H_i \subseteq T_i$ for all $i \geq 0$. As usual, $\mathbf{H} < \mathbf{T}$ stands for: $\mathbf{H} \leq \mathbf{T}$ and $\mathbf{H} \neq \mathbf{T}$.

Definition 3. *A temporal quantified here-and-there (or just TQHT) interpretation is a tuple* $\mathcal{M} = \langle (\mathcal{D}, \sigma), \mathbf{H}, \mathbf{T} \rangle$ *where* $\langle (\mathcal{D}, \sigma), \mathbf{H} \rangle$ *and* $\langle (\mathcal{D}, \sigma), \mathbf{T} \rangle$ *are two LTL-interpretations satisfying* $\mathbf{H} \leq \mathbf{T}$. □

In the definition above, we respectively call $\mathbf{H}$ and $\mathbf{T}$ the "here" and "there" components of $\mathcal{M}$. A TQHT-interpretation of the form $\mathcal{M} = \langle (\mathcal{D}, \sigma), \mathbf{T}, \mathbf{T} \rangle$ is said to be *total*. If $\mathcal{M} = \langle (\mathcal{D}, \sigma), \mathbf{H}, \mathbf{T} \rangle$ we write $\mathcal{M}[k]$ to stand for $\langle (\mathcal{D}, \sigma), \mathbf{H}[k], \mathbf{T}[k] \rangle$. The satisfaction relation for $\mathcal{M} = \langle (\mathcal{D}, \sigma), \mathbf{H}, \mathbf{T} \rangle$ and a formula α, written $\mathcal{M} \models \alpha$, is recursively defined as follows:

$$
\begin{aligned}
&\mathcal{M} \models p(t_1, \ldots, t_n) && \text{iff } p(\sigma(t_1), \ldots, \sigma(t_n)) \in H_0. \\
&\mathcal{M} \models t = s && \text{iff } \sigma(t) = \sigma(s) \\
&\mathcal{M} \not\models \bot && \\
&\mathcal{M} \models \varphi \wedge \psi && \text{iff } \mathcal{M} \models \varphi \text{ and } \mathcal{M} \models \psi. \\
&\mathcal{M} \models \varphi \vee \psi && \text{iff } \mathcal{M} \models \varphi \text{ or } \mathcal{M} \models \psi. \\
&\mathcal{M} \models \varphi \to \psi && \text{iff } \langle (\mathcal{D}, \sigma), w, \mathbf{T} \rangle \not\models \varphi \text{ or } \langle (\mathcal{D}, \sigma), w, \mathbf{T} \rangle \models \psi \\
&&& \text{for all } w \in \{\mathbf{H}, \mathbf{T}\} \\
&\mathcal{M} \models \bigcirc \varphi && \text{iff } \mathcal{M}[1] \models \varphi. \\
&\mathcal{M} \models \varphi \, \mathrm{U} \, \psi && \text{iff } \exists j \geq 0, \ \mathcal{M}[j] \models \psi \\
&&& \text{and } (\mathcal{M}[i] \models \varphi \text{ for all } i, \, 0 \leq i < j). \\
&\mathcal{M} \models \varphi \, \mathrm{R} \, \psi && \text{iff } \forall j \geq 0, \ \mathcal{M}[j] \models \psi \\
&&& \text{or } (\mathcal{M}[i] \models \varphi \text{ for some } i, \, 0 \leq i < j). \\
&\mathcal{M} \models \forall x \, \varphi(x) && \text{iff } \langle (\mathcal{D}, \sigma), w, \mathbf{T} \rangle \models \varphi(d) \\
&&& \text{for every } d \in \mathcal{D} \text{ and every } w \in \{\mathbf{H}, \mathbf{T}\}. \\
&\mathcal{M} \models \exists x \, \varphi(x) && \text{iff } \mathcal{M} \models \varphi(d) \text{ for some } d \in \mathcal{D}.
\end{aligned}
$$

where by $\varphi(d)$ we denote the replacement by d of all free occurrences of x in $\varphi(x)$. An interpretation $\mathcal{M}$ is a model of a theory Γ, written $\mathcal{M} \models \Gamma$, if it satisfies all the sentences in Γ. The resulting logic is called *Temporal Quantified Here-and-There Logic* with equality and static[3] domains, and we simply abbreviate it as TQHT. It is not difficult to see that, if we restrict ourselves to total TQHT-interpretations, $\langle (\mathcal{D}, \sigma), \mathbf{T}, \mathbf{T} \rangle \models \varphi$ iff $\langle (\mathcal{D}, \sigma), \mathbf{T} \rangle \models \varphi$ in first-order LTL. Furthermore, the following properties can be easily checked by structural induction.

Proposition 2. *For any formula* φ, *and interpretation* $\langle (\mathcal{D}, \sigma), \mathbf{H}, \mathbf{T} \rangle$:

(i) *if* $\langle (\mathcal{D}, \sigma), \mathbf{H}, \mathbf{T} \rangle \models \varphi$, *then* $\langle (\mathcal{D}, \sigma), \mathbf{T}, \mathbf{T} \rangle \models \varphi$
(ii) $\langle (\mathcal{D}, \sigma), \mathbf{H}, \mathbf{T} \rangle \models \neg \varphi$ *iff* $\langle (\mathcal{D}, \sigma), \mathbf{T}, \mathbf{T} \rangle \not\models \varphi$

In general, it is clear that the other direction of (i) does not hold: any non-total interpretation contains atoms $\varphi = p(t_1, \ldots, t_n) \in T_i \setminus H_i$ for some $i \geq 0$.

---

[3] The name "static" refers here to the fact that the same domain $\mathcal{D}$ is used both for $\mathbf{H}$ and $\mathbf{T}$.

Without loss of generality, suppose $i = 0$ (we can always take $\mathcal{M}[i]$ instead). Then, for those atoms, $\langle(\mathcal{D}, \sigma), \mathbf{T}, \mathbf{T}\rangle \models \varphi$ but $\langle(\mathcal{D}, \sigma), \mathbf{H}, \mathbf{T}\rangle \not\models \varphi$. Moreover, by (ii), the former also means $\langle(\mathcal{D}, \sigma), \mathbf{H}, \mathbf{T}\rangle \not\models \neg\varphi$, so we conclude that non-total interpretations *falsify* the formula $\varphi \vee \neg\varphi$, a classical tautology known as the *excluded middle axiom*. This axiom is not valid either in intuitionistic logic or in the intermediate logic of Here-and-There [14], where '$\neg$' is weaker than classical negation. It is still possible to add this axiom for some predicates $p \in \mathcal{P}$ by forcing the condition:

$$\Box\, \forall x_1 \ldots \forall x_n\; (p(x_1, \ldots, x_n) \vee \neg p(x_1, \ldots, x_n)) \qquad (\text{EM}_p)$$

The following results explain the effect of including $((\text{EM}_p))$ among the formulas of our theory.

Proposition 3. *An interpretation $\mathcal{M} = \langle(\mathcal{D}, \sigma), \mathbf{H}, \mathbf{T}\rangle$ satisfies (EM$_p$)for some $p \in \mathcal{P}$ iff, for all $i \geq 0$: $p(t_1, \ldots, t_n) \in T_i$ is equivalent to $p(t_1, \ldots, t_n) \in H_i$.*

Corollary 1. *Given language $\mathcal{L} = \langle\mathcal{C}, \mathcal{P}\rangle$, let $\mathcal{P}' \subseteq \mathcal{P}$ be a subset of predicates and let $\mathcal{M} \models (EM_p)$ for all $p \in \mathcal{P}'$. Then, $\langle(\mathcal{D}, \sigma), \mathbf{H}, \mathbf{T}\rangle \models \varphi$ amounts to $\langle(\mathcal{D}, \sigma), \mathbf{T}, \mathbf{T}\rangle \models \varphi$ for any formula φ in the language $\mathcal{L} = \langle\mathcal{C}, \mathcal{P}'\rangle$.*

Corollary 2. *Given language $\mathcal{L} = \langle\mathcal{C}, \mathcal{P}\rangle$, the addition of (EM$_p$)for all $p \in \mathcal{P}$ makes TQHT collapse into LTL.*

As an illustration of TQHT satisfaction, consider the propositional formula:

$$\neg inmune \rightarrow vulnerable \qquad (5)$$

This formula corresponds to the ASP ground rule:

```
vulnerable :- not inmune.
```

Any model $\mathcal{M} = \langle(\mathcal{D}, \sigma), \mathbf{H}, \mathbf{T}\rangle$ of (5) must satisfy that $\langle(\mathcal{D}, \sigma), w, \mathbf{T}\rangle, 0 \not\models \neg inmune$ or $\langle(\mathcal{D}, \sigma), w, \mathbf{T}\rangle, 0 \models vulnerable$ for all $w \in \{\mathbf{H}, \mathbf{T}\}$. By Proposition 2 (ii), the former is equivalent to $\langle(\mathcal{D}, \sigma), \mathbf{T}, \mathbf{T}\rangle, 0 \models inmune$, that is, $inmune \in T_0$, whereas the latter ammounts to $vulnerable \in H_0$ for $w = \mathbf{H}$ and $vulnerable \in T_0$ for $w = \mathbf{T}$. Therefore, models of (5) are such that, if $inmune \notin T_0$ then $vulnerable \in H_0 \subseteq T_0$.

To introduce non-monotonicity, we define a set of selected total TQHT models we will call *temporal equilibrium models*, or just *temporal stable models*, if we consider their corresponding LTL representation.

Definition 4 (Temporal Equilibrium Model). *A temporal equilibrium model of a theory Γ is a total model $\mathcal{M} = \langle(\mathcal{D}, \sigma), \mathbf{T}, \mathbf{T}\rangle$ of Γ such that there is no $\mathbf{H} < \mathbf{T}$ satisfying $\langle(\mathcal{D}, \sigma), \mathbf{H}, \mathbf{T}\rangle \models \Gamma$. When this happens, we further say that the LTL-interpretation $\langle(\mathcal{D}, \sigma), \mathbf{T}\rangle$ is a temporal stable model of Γ.* $\quad\Box$

The logic induced by temporal equilibrium models is called *Temporal Quantified Equilibrium Logic* (TEL, for short). We can identify temporal logic programs with variables as a fragment of first order temporal theories. For a detailed definition of this fragment see [12]. In the previous simple example (5), we can observe that any total interpretation $\mathcal{M} = \langle(\mathcal{D},\sigma),\mathbf{T},\mathbf{T}\rangle$ with $inmune \in T_0$ is a TQHT model, but we can always form another interpretation $\mathcal{M}' = \langle(\mathcal{D},\sigma),\mathbf{H},\mathbf{T}\rangle$ with $H_0 = T_0 \setminus \{inmune, vulnerable\}$ and $H_i = T_i$ for $i > 0$ such that it is also a TQHT model of (5) but $\mathbf{H} < \mathbf{T}$, so $\mathcal{M}$ is not in equilibrium. If, on the contrary, $inmune \notin T_0$, then the satisfaction of (5) requires $vulnerable \in H_0 \subseteq T_0$ for any TQHT model, and there is no way to form a smaller model by removing atoms in T_0. For the rest of situations $i > 0$, any T_i containing at least one atom can always be reduced to $H_i = \emptyset$ while keeping the satisfaction of (5), since this formula only affects to the initial situation. It is not difficult to see that the only temporal equilibrium model of (5) corresponds to $T_0 = \{vulnerable\}$ and $T_i = \emptyset$ for $i > 0$. Let us consider next a more elaborated example.

Example 1. Take the following temporal logic program:

$$Person(x) \land Disease(y) \land \neg Immune(x,y) \rightarrow Vulnerable(x,y) \qquad (6)$$

$$\Box(Immune(x,y) \rightarrow \bigcirc Immune(x,y)) \qquad (7)$$

$$\Box(Vulnerable(x,y) \land \neg \bigcirc Immune(x,y) \rightarrow \bigcirc Vulnerable(x,y)) \qquad (8)$$

$$\Box(Vaccinate(x,y) \rightarrow Immune(x,y)) \qquad (9)$$

$$\Box Person(\text{John}) \land \Box Disease(\text{Smallpox}) \qquad (10)$$

$$\bigcirc^3 Vaccinate(\text{John}, \text{Smallpox}) \qquad (11)$$

where we assume that all free variables in a formula are universally quantified. Formula (6) asserts that, initially, any person x is vulnerable to any desease y, unless we can prove it is immune. As we saw before, the effect of $\neg\varphi$ in TEL is that of *default negation* of φ, that is, $\neg\varphi$ holds when there is no evidence on φ. Formula (7) tells us that once somebody becomes immune to some disease, it remains so forever. A similar expression is (8), saying that someone vulnerable remains so, but this time is under the default condition that there is no evidence of becoming immune. Formulas of the form (8) are called *inertia rules*. The expression (9) means that the effect of vaccinating x against y is becoming immune. Finally, (10) contains some typing information saying that John is (always) a person and Smallpox is (always) a disease, whereas (11) asserts that John has been vaccinated at situation $i = 3$. Program (6)–(11) has a temporal stable model $\langle(\mathcal{D},\sigma),\mathbf{T}\rangle$ where $\mathcal{D} = \mathcal{C} = \{\text{John}, \text{Smallpox}\}$, σ is the identity relation and the only states making $Vulnerable(\text{John}, \text{Smallpox})$ true are $i \in \{0,1,2\}$ whereas $Immune(\text{John}, \text{Smallpox})$ becomes true for all $i \geq 3$. The rest of stable models only vary in the extension of $\mathcal{D}$ (we can have arbitrary unnamed individuals) and the assignment σ, provided that UNA is respected. Suppose we are said now that John has some genetic anomaly that made him immune to Smallpox from the very beginning. If we add the formula $Immune(\text{John}, \text{Smallpox})$ to (6)–(11) then $Vulnerable(\text{John}, \text{Smallpox})$ is never derived and we obtain $\Box Immune(\texttt{John},\texttt{Smallpox})$ as a conclusion. This last variation illustrates the non-monotonic behavior of TEL entailment relation. $\Box$

Without entering into further detail and just as an illustration, Fig. 1 shows an encoding of Example 1 in the language of the temporal ASP solver `telingo` [15]. The correspondence of program rules with the respective formulas (6)–(11) is pretty obvious in most cases. The only difference is that `telingo` uses the previous operator in rules representing transitions between two states, rather than the next operator. Thus, for instance, `'inmune(X,Y)` must be read as "previously, `inmune(X,Y)` was true". On the other hand, the next operator used on facts is represented as `>`, as we can see in the last line.

```
#program initial.
vulnerable(X,Y) :- person(X), disease(Y), not inmune(X,Y).

#program dynamic.
inmune(X,Y) :- 'inmune(X,Y).
vulnerable(X,Y) :- 'vulnerable(X,Y), not inmune(X,Y).

#program always.
inmune(X,Y) :- vaccinate(X,Y).
person(john).
disease(smallpox).

#program initial.
&tel{ > > > vaccinate(john,smallpox) }.
```

Fig. 1. An encoding of Example 1 in the language of the temporal ASP solver `telingo`.

4 $\mathcal{ALC}$-LTL

The combination of description logics with temporal patterns is an important field of knowledge representation that has been widely studied in the literature (see, for instance, the surveys [16–18]). In a cornerstone paper, Baader, Ghilardi and Lutz [10] proposed the temporal extension $\mathcal{ALC}$-LTL where temporal operators are only introduced in front of $\mathcal{ALC}$ axioms, but not as concept constructors. This guaranteed decidability and significantly reduced the complexity of different reasoning tasks (depending on whether rigid roles are considered or not) while keeping enough expressiveness for solving many practical problems. According to [10], an $\mathcal{ALC}$-LTL *formula* φ is defined by the grammar:

$$\alpha \mid \varphi_1 \wedge \varphi_2 \mid \varphi_1 \vee \varphi_2 \mid \neg\varphi \mid \varphi_1 \ U \ \varphi_2 \mid \bigcirc\varphi$$

where α is an $\mathcal{ALC}$ axioms. We assume the same abbreviations for temporal operators seen in Sect. 3. For $\mathcal{ALC}$-LTL formulas, $\varphi \rightarrow \psi$ can be defined as $\neg\varphi \vee \psi$

and $\varphi \, \mathrm{R} \, \psi$ can be defined as $\neg(\neg\varphi \, \mathrm{U} \, \neg\psi)$ (something that, in general, TEL does not satisfy). The semantics for $\mathcal{ALC}$-LTL is provided in [10] by considering an infinite sequence $\{\mathcal{I}_i\}_{i \geq 0}$ of $\mathcal{ALC}$ interpretations $\mathcal{I}_i$. In our case, however, we will be more interested in the first order translation of $\mathcal{ALC}$-LTL. We assume that any $\mathcal{ALC}$ axiom α actually represents the first order formula $t(\alpha)$ as defined in Sect. 2. Then, an $\mathcal{ALC}$-LTL formula may be simply seen as an abbreviation of first order temporal formula. To give an example, the $\mathcal{ALC}$-LTL formula:

$$\Diamond\Box(\mathtt{AIDS} : \exists\mathtt{curedBy.Treatment})$$

expresses the wish that a definitive treatment for AIDS is eventually found and, after applying translation $t(\cdot)$ becomes the first order temporal formula:

$$\Diamond\Box \; \exists y(\mathtt{curedBy}(\mathtt{AIDS}, y) \wedge \mathtt{Treatment}(y))$$

Baader et al. define *rigid concepts and roles* as those whose interpretation does not vary along time (otherwise, they are called *flexible* instead). Using the FOL representation, for any rigid concept $\mathtt{c} \in N_C$ and rigid role $\mathtt{r} \in N_R$ we have:

$$\forall x \; (\mathtt{c}(x) \leftrightarrow \Box\mathtt{c}(x))$$
$$\forall x \forall y \; (\mathtt{r}(x, y) \leftrightarrow \Box\mathtt{r}(x, y))$$

5 $\mathcal{ALC}$-TEL

Following the encoding in [7] to incorporate hybrid theories in Equilibrium Logic, we describe now how $\mathcal{ALC}$ can be easily embodied in TEL. Given language $\mathcal{L} = \langle \mathcal{C}, \mathcal{P} \rangle$ we suppose that $N_C \subseteq \mathcal{P}$ and $N_R \subseteq \mathcal{P}$ become unary and binary predicates, respectively, and that $N_I \subseteq \mathcal{C}$ become constant names. The crucial point in the encoding is the *addition of the excluded middle axiom* $(\mathrm{EM}_p)$for every predicate $p \in N_C \cup N_R$. In this way, the translation of an $\mathcal{ALC}$ axiom is interpreted under classical FOL whereas the translation of any $\mathcal{ALC}$-LTL formula is interpreted under quantified LTL. The final result provides an expressive formalism that allows combining temporal logic programming and terminological knowledge. For instance, we can modify now our running example as follows.

Example 2 (Example 1 continued). We can incorporate axioms (1)–(4) assuming that $\mathtt{Vaccine}$, $\mathtt{Medication}$, $\mathtt{Treatment}$ and $\mathtt{Disease}$ are rigid concepts, whereas $\mathtt{curedBy}$ is flexible. We also include the rigid concept $\mathtt{Person}$ and the constant John. Our logic program can be modified to include $\mathcal{ALC}$ expressions accordingly. For instance, we can keep untouched the formulas (7)–(9) and (11) since they do not refer to terminological knowledge, but we replace now (6) by[4]:

$$(x : \mathtt{Person}) \wedge (y : \mathtt{Disease}) \wedge \neg Immune(x, y) \rightarrow Vulnerable(x, y)$$

---

[4] We allow now logical variables in assertions, but their translation is straightforward, playing the role of generic individual names.

and (10) by the assertions:

$$\texttt{John} : \texttt{Person} \qquad \texttt{Smallpox} : \texttt{Disease}$$

that do not need temporal operators, since these concepts are rigid.   □

An important issue may occur when dealing with flexible concepts or roles. For instance, since `curedBy` is flexible, the fact that smallpox is cured by a vaccine, (3), is not guaranteed to persist throughout the temporal narrative. To do so, we can add a rule for strict persistence like (7) as follows:

$$\Box(\texttt{curedBy}(x,y) \rightarrow \bigcirc\texttt{curedBy}(x,y))$$

which works in this case since we can assume that a curable disease does not cease to be so. However, if we wanted to transform this rule into a general inertia default, it would not be directly possible, since `curedBy` behaves as a classical predicate due to (EM_p). An additional auxiliary predicate could still be used for that purpose. A more ambitious solution would be removing the (EM_p)axiom and allowing concepts and roles to behave as logic programming predicates. This would allow expressing defaults on Description Logic axioms, but would depart from the standard interpretation of $\mathcal{ALC}$.

6   Conclusions

We have defined a logical formalism $\mathcal{ALC}$-TEL that, under a modal temporal basis, combines the Description Logic $\mathcal{ALC}$ [11] with logic programming under Equilibrium Logic semantics. On the one hand, if we disregard the temporal operators, this formalism embeds hybrid theories from [7], allowing the combination of description logics (in our case, $\mathcal{ALC}$) with logic programming. On the other hand, if we add the excluded middle axiom (EM_p) for all the predicates in the language, $\mathcal{ALC}$-TEL collapses into $\mathcal{ALC}$-LTL as defined by Baader et al. in [10]. Moreover, $\mathcal{ALC}$ is encoded in terms of its First Order translation, so that, once $\mathcal{ALC}$ expressions are translated, we simply get Temporal (Quantified) Equilibrium Logic [6,12] as underlying formalism.

The current proposal opens the exploration of many possible directions. A first obvious line of future work is the study of syntactic fragments and the analysis of complexity for their satisfiability problem. An obviously related line has to do with implementation. For instance, model checking techniques have been applied both to $\mathcal{ALC}$-LTL [19] and to TEL [20,21] and their efficient combination could be a interesting topic for future investigation. An adaptation of TEL for practical problem solving in the spirit of ASP has led to a variant [15] defined on finite traces and its corresponding ASP solver, `telingo`. The $\mathcal{ALC}$-TEL formalism may help us to incorporate terminological knowledge in `telingo` in the form of DL knowledge bases. Besides, the use of finite traces on temporal description logics has also been recently proposed in [22]. Another exploratory line could be a more integrated combination of DL and logic programs where defaults were also introduced in DL concepts and roles. Finally, another possible research direction is the use of $\mathcal{ALC}$-TEL in application domains that involve temporal reasoning and rich ontologies, following similar steps as [23] in the medical domain.

Acknowledgments. We dedicate this work to Franz Baader for his inspiring work and fundamental contributions to the field of Knowledge Representation and Reasoning.

References

1. Marek, V.W., Truszczyński, M.: Stable models and an alternative logic programming paradigm. In: Apt, K.R., Marek, V.W., Truszczynski, M., Warren, D.S. (eds.) The Logic Programming Paradigm. AI, pp. 375–398. Springer, Heidelberg (1999). https://doi.org/10.1007/978-3-642-60085-2_17
2. Niemelä, I.: Logic programs with stable model semantics as a constraint programming paradigm. Ann. Math. Artif. Intell. **25**(3–4), 241–273 (1999)
3. Lifschitz, V.: Answer set planning. In: de Schreye, D., (ed.) Proceedings of the International Conference on Logic Programming (ICLP 1999), pp. 23–37. MIT Press (1999)
4. Kamp, H.: Tense logic and the theory of linear order. Ph.D. thesis, UCLA (1968)
5. Pnueli, A.: The temporal logic of programs. In: 18th Annual Symposium on Foundations of Computer Science, pp. 46–57. IEEE Computer Society Press (1977)
6. Aguado, F., Cabalar, P., Diéguez, M., Pérez, G., Vidal, C.: Temporal equilibrium logic: a survey. J. Appl. Non-Class. Log. **23**(1–2), 2–24 (2013)
7. de Bruijn, J., Pearce, D., Polleres, A., Valverde, A.: A semantic framework for hybrid knowledge bases. Knowl. Inf. Syst. **25**(1), 81–104 (2010)
8. Baader, F., Calvanese, D., McGuinness, D.L., Nardi, D., Patel-Schneider, P.F. (eds.): The Description Logic Handbook: Theory, Implementation, and Applications. Cambridge University Press, New York (2003)
9. Pearce, D.: A new logical characterisation of stable models and answer sets. In: Dix, J., Pereira, L.M., Przymusinski, T.C. (eds.) NMELP 1996. LNCS, vol. 1216, pp. 57–70. Springer, Heidelberg (1997). https://doi.org/10.1007/BFb0023801
10. Baader, F., Ghilardi, S., Lutz, C.: LTL over description logic axioms. In: Proceedings of the Eleventh International Conference on Principles of Knowledge Representation and Reasoning, KR 2008, pp. 684–694. AAAI Press (2008)
11. Schmidt-Schauß, M., Smolka, G.: Attributive concept descriptions with complements. Artif. Intell. **48**(1), 1–26 (1991)
12. Aguado, F., Cabalar, P., Pérez, G., Vidal, C., Diéguez, M.: Temporal logic programs with variables. TPLP **17**(2), 226–243 (2017)
13. Baader, F., Horrocks, I., Sattler, U.: Description logics, Chap. 3. In: van Harmelen, F., Lifschitz, V., Porter, B. (eds.) Handbook of Knowledge Representation. Elsevier (2007)
14. Heyting, A.: Die formalen Regeln der intuitionistischen Logik. Sitzungsberichte der Preussischen Akademie der Wissenschaften. Physikalisch-mathematische Klasse (1930)
15. Cabalar, P., Kaminski, R., Schaub, T., Schuhmann, A.: Temporal answer set programming on finite traces. Theory Pract. Log. Program. **18**(3–4), 406–420 (2018)
16. Artale, A., Franconi, E.: A survey of temporal extensions of description logics. Ann. Math. Artif. Intell. **30**(1–4), 171–210 (2000)

17. Artale, A., Franconi, E.: Temporal description logics. In: Handbook of Time and Temporal Reasoning in AI. The MIT Press (2001)
18. Lutz, C., Wolter, F., Zakharyaschev, M.: Temporal description logics: a survey. In: Demri, S., Jensen, C.S. (eds.) 15th International Symposium on Temporal Representation and Reasoning, TIME 2008, Université du Québec à Montréal, Canada, 16–18 June 2008, pp. 3–14. IEEE Computer Society (2008)
19. Baader, F., Lippmann, M.: Runtime verification using the temporal description logic ALC-LTL revisited. J. Appl. Log. **12**(4), 584–613 (2014)
20. Cabalar, P., Diéguez, M.: STELP - a tool for temporal answer set programming. In: Delgrande, J.P., Faber, W. (eds.) LPNMR 2011. LNCS (LNAI), vol. 6645, pp. 370–375. Springer, Heidelberg (2011). https://doi.org/10.1007/978-3-642-20895-9_43
21. Cabalar, P., Diéguez, M.: Strong equivalence of non-monotonic temporal theories. In: Proceedings of the 14th International Conference on Principles of Knowledge Representation and Reasoning (KR 2014), Vienna, Austria (2014)
22. Artale, A., Mazzullo, A., Ozaki, A.: Temporal description logics over finite traces. In: Ortiz, M., Schneider, T. (eds.) Proceedings of the 31st International Workshop on Description Logics. CEUR Workshop Proceedings, vol. 2211. CEUR-WS.org (2018)
23. Baader, F., Borgwardt, S., Forkel, W.: Patient selection for clinical trials using temporalized ontology-mediated query answering. In: Companion Proceedings of the Web Conference, WWW 2018, Republic and Canton of Geneva, Switzerland, pp. 1069–1074. International World Wide Web Conferences Steering Committee (2018)

The What-To-Ask Problem
for Ontology-Based Peers

Diego Calvanese[1]([⊠]) [iD], Giuseppe De Giacomo[2] [iD], Domenico Lembo[2] [iD],
Maurizio Lenzerini[2] [iD], and Riccardo Rosati[2] [iD]

[1] Free University of Bozen-Bolzano, Bolzano, Italy
`calvanese@inf.unibz.it`
[2] Sapienza Università di Roma, Rome, Italy
{`degiacomo,lembo,lenzerini,rosati`}`@diag.uniroma1.it`

Abstract. The issue of cooperation, integration, and coordination between information peers has been addressed over the years both in the context of the Semantic Web and in several other networked environments, including data integration, Peer-to-Peer and Grid computing, service-oriented computing, distributed agent systems, and collaborative data sharing. One of the main problems arising in such contexts is how to exploit the mappings between peers in order to answer queries posed to one peer. We address this issue for peers managing data through ontologies and in particular focus on ontologies specified in logics of the *DL-Lite* family. Our goal is to present some basic, fundamental results on this problem. In particular, we focus on a simplified setting based on just two interoperating peers, and we investigate how to solve the so-called "What-To-Ask" problem: find a way to answer queries posed to a peer by relying only on the query answering service available at the queried peer and at the other peer. We show both a positive and a negative result. Namely, we first prove that a solution to this problem always exists when the ontology is specified in $DL\text{-}Lite_{\mathcal{R}}$, and we provide an algorithm to compute it. Then, we show that for the case of $DL\text{-}Lite_{\mathcal{F}}$ the problem may have no solution. We finally illustrate that a solution to our problem can still be found even for more general networks of peers, and for any language of the *DL-Lite* family, provided that we interpret mappings according to an epistemic semantics, rather than the usual first-order semantics.

1 Introduction

In the era towards a data-driven society, the issue of cooperation, integration, and coordination between data stored in different nodes of a network is of paramount importance. Indeed, recent years have shown the need to deal with networked data in large-scale, distributed settings, and it is not surprising that the abstraction of networked data systems appears in many disciplines, including Web Science and Peer-to-Peer computing [3,8,26], Semantic Web [1,42], Data Management [12,27,31,37,44], and Knowledge Representation [25,30,42,46].

C. Lutz et al. (Eds.): Baader Festschrift, LNCS 11560, pp. 187–211, 2019.
https://doi.org/10.1007/978-3-030-22102-7_9

Put in an abstract way, all these systems are characterized by an architecture constituted by various autonomous nodes (called sites, sources, agents, or, as we call them here, peers) which hold information, and which are linked to other nodes by means of mappings. A mapping is a statement specifying that some relationship exists between pieces of information held by one peer and pieces of information held by another peer. The whole knowledge of the system is fully distributed, without any central entity holding a global view of information, or controlling the overall operation of the system.

The basic problems arising in this architecture include the following:

– how to discover, express, and compose the mappings between peers (see, for instance, [8, 23, 26, 33, 39]),
– how to exchange data between peers based on the specified mappings (see, for instance, [24, 31, 32]),
– how to exploit the mappings in order to answer queries posed to one peer [28, 37, 40].

The latter is the problem studied in this paper. Although several interesting results have been reported in each of the above mentioned contexts, we argue that a deep understanding of the problem of answering queries in a networked environment is still lacking, in particular when the information in each peer is modelled in terms of an ontology.

Our goal is to present some basic, fundamental results on this problem. Given the fundamental nature of our investigation, we consider a simplified setting where the whole system is constituted by only two peers, called local and remote, respectively. Information in the remote peer is related to the information in the local peer by means of suitable mappings (cf. Fig. 1). Interestingly, despite the fact that this setting might look elementary, it will nevertheless allow us to uncover various subtleties of an interoperating ontology-based peer system.

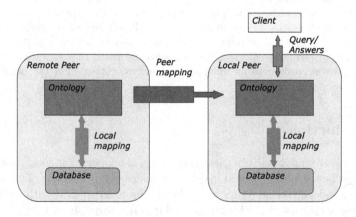

Fig. 1. Ontology-to-ontology: a simple form of interoperation among peer ontologies

In our study, we make several assumptions, that are made explicit here:

– In contrast with most of the papers in peer-to-peer data management, we assume that each peer does not simply store data, but holds a knowledge base. In particular, we explore the context where each peer models its knowledge base by means of an ontology.
– The ontology at each peer specifies both intensional knowledge (general rules) and extensional one (individual facts). Actually, the latter may be managed through a relational DBMS, and therefore represented by a database connected to the ontology via local mappings, as shown in Fig. 1. So, if we have data sources linked to our ontology through mappings, they are seen as internal components within a peer. In other words, each peer can be seen as an Ontology-based data access (OBDA) system [13], and the novelty with respect to the usual notion of OBDA is represented by the fact that mappings connect peers, and not simply data sources to ontologies.
– We concentrate our attention to the issue of answering queries posed to the local peer.
– We assume that each of the two peers provides the service of answering queries expressed over its underlying ontology. Note that answering a query for a peer requires reasoning over the ontology by means of deduction, rather than simply evaluating the query expression over a database.
– We assume that query answering is the only basic service provided by each peer. In other words, while processing a query posed to the local peer, the query answering services provided by each of the two peers are the only basic services that can be relied upon.
– In order to address the problem in the most general way, we assume that the local peer can only collect the answers received by the remote peers, and add them to the answers obtained by accessing its own data. In other words, no computational power is available at the local peer to process the tuples returned by the remote peer, except for just adding them to the result of the whole query.

We believe that the above assumptions faithfully capture the modular structure of a peer-to-peer system, and generalize the existing investigation of peer-to-peer architectures to the case where each peer is seen as an agent holding complex knowledge, instead of simply data.

In this context, the basic problem we address is the following: given a query posed to the local peer, find a way to answer the query by relying only on the two query answering services available at the two peers. Thus, when answering the query posed to the local peer, we have to figure out which queries to send to the remote peer in order for the local peer to be able to return the correct and complete set of answers to the original query. This is why we call this problem the "What-To-Ask" problem (cf. [14]).

Example 1. Consider a music sharing system, and assume that the peer SongUniverse stores its own information about songs, and has a mapping specifying that other songs, in particular live rock songs, can be retrieved from the remote peer RockPlanet. Now, suppose that Carol interacts with the SongUniverse peer, and asks for all live songs of U.K. artists. What this peer can do in order to answer Carol's query at best is to: *(i)* directly provide her with the live songs of U.K. artists that it stores locally, *(ii)* use its general knowledge about music to deduce that also live rock songs suit Carol's needs, *(iii)* use the mapping to reformulate Carol's request in terms of RockPlanet knowledge, in particular asking to the remote peer the right query to retrieve all live rock songs of U.K. artists. ◁

In this paper, we study the What-To-Ask problem in a setting where the two peers hold an ontology expressed in a Description Logic of the *DL-Lite* family [16]. Specifically, we present the following contributions.

1. We formalize the above mentioned two-peer architecture, we define its semantics, and we give a precise characterization of the semantics of query answering (Sect. 3).
2. We provide both the intuition and the formal definition of the "What-To-Ask" problem, taking into account both the semantics of query answering and the fact that, when answering a query posed to the local peer, only the query answering services available at the two peers can be relied upon (Sect. 3).
3. We show that in the case of ontologies specified in *DL-Lite$_R$* there is an algorithm that allows us to solve any instance of "What-To-Ask", i.e., that allows us to compute what we should ask to the remote peer in order to answer a query posed to the local peer. One of the basic ingredients of the algorithm is the ability of reformulating the query on the basis of the local peer ontology and the mappings, so as to deduce the correct queries to send to the remote peer (Sect. 4).
4. We show that in the case of *DL-Lite$_F$*, the "What-To-Ask" problem may not admit any solution. This shows that particular attention should be devoted to the trade-off between the expressive power of the ontology language and the complexity/feasibility of reasoning (Sect. 5).
5. We finally discuss how to overcome the limitation above by making use of mappings that explicitly take into account that ontologies are autonomous agents that provide as query answering service the (independent) generation of certain answers. This calls for the usage of (auto-)epistemic operators (Sect. 6).

To complete the description of the organization of the paper, Sect. 2 illustrates some preliminary notions that will be used in the technical development, and Sect. 7 presents some concluding remarks. Finally, we note that this paper is a revised and extended version of [14].

2 Preliminaries

We introduce now the ontology languages on which we base the technical development in the next sections. Specifically, we rely on Description Logics (DLs) [6], which are logic's that represent the domain of interest in terms of *concepts*, denoting sets of objects, and *roles*, denoting binary relations between objects. Complex concept and role expressions are constructed by applying suitable constructs, starting from a set of atomic concepts and roles.

2.1 The *DL-Lite* Family

We focus here on a family of lightweight DLs, called the *DL-Lite family* [16], and introduce three prominent logics of this family, namely *DL-Lite*, *DL-Lite*$_\mathcal{R}$ and *DL-Lite*$_\mathcal{F}$. In the core language of the family, called *DL-Lite*, (basic) *concepts C* and *roles R* are formed according to the following syntax:

$$C \longrightarrow A \mid \exists R \qquad\qquad R \longrightarrow P \mid P^-$$

where A denotes an atomic concept, P an atomic role, P^- the *inverse* of P, and $\exists R$ an unqualified existential quantification. Intuitively, P^- denotes the inverse of the binary relation denoted by P, while $\exists R$ denotes the domain of (the binary relation denoted by) R, i.e., the projection of R on its first component.

A DL *ontology* $\mathcal{O} = \langle \mathcal{T}, \mathcal{A} \rangle$ encodes the knowledge about the domain of interest in two distinct components: the *TBox* (for terminological box) $\mathcal{T}$ specifies general knowledge about the conceptual elements of the domain, while the *ABox* (for assertional box) $\mathcal{A}$, specifies extensional knowledge about individual elements of the domain.

In *DL-Lite*, a TBox is formed by a finite set of *inclusion* and *disjointness assertions* between concepts, respectively of the form

$$B_1 \sqsubseteq B_2 \qquad\qquad B_1 \sqsubseteq \neg B_2$$

where B_1 and B_2 are basic concepts. The first assertion expresses that every instance of concept B_1 is also an instance of concept B_2, while the second assertion expresses that the two sets of instances are disjoint. An ABox consists of concept and role membership assertions, respectively of the form

$$A(c) \qquad\qquad P(c, c')$$

where A is an atomic concept, P an atomic role, and c, c' two constants. The first assertion expresses that the individual denoted by c is an instance of concept A, while the second assertion expresses that the two individuals denoted by c and c' are in relation P.

In *DL-Lite*$_\mathcal{R}$, a TBox may additionally contain *role inclusion* and *disjointness assertions*, respectively of the form

$$R_1 \sqsubseteq R_2 \qquad\qquad R_1 \sqsubseteq \neg R_2$$

where R_1 and R_2 are arbitrary roles. The meaning of such assertions is analogous to the one for concepts.

Instead, in *DL-Lite*$_\mathcal{F}$, a TBox may contain also *functionality assertions* of the form

$$(\mathsf{funct}\ R)$$

asserting that R is a functional role. Such a role R can relate each object to at most one other object.

The semantics of a DL is given in terms of first-order logic interpretations, where an *interpretation* $\mathcal{I} = (\Delta^\mathcal{I}, \cdot^\mathcal{I})$ consists of a non-empty *interpretation domain* $\Delta^{\mathcal{I}}$ and an *interpretation function* $\cdot^{\mathcal{I}}$ that assigns to each concept C a subset $C^{\mathcal{I}}$ of $\Delta^{\mathcal{I}}$, and to each role R a binary relation $R^{\mathcal{I}}$ over $\Delta^{\mathcal{I}}$, in such a way that the following conditions hold. In particular, for the constructs of *DL-Lite* we have:

$$A^{\mathcal{I}} \subseteq \Delta^{\mathcal{I}} \qquad\qquad P^{\mathcal{I}} \subseteq \Delta^{\mathcal{I}} \times \Delta^{\mathcal{I}}$$
$$(\exists R)^{\mathcal{I}} = \{o \mid \exists o'.\,(o,o') \in R^{\mathcal{I}}\} \qquad (P^-)^{\mathcal{I}} = \{(o_2,o_1) \mid (o_1,o_2) \in P^{\mathcal{I}}\}$$
$$(\neg C)^{\mathcal{I}} = \Delta^{\mathcal{I}} \setminus C^{\mathcal{I}} \qquad\qquad (\neg R)^{\mathcal{I}} = \Delta^{\mathcal{I}} \times \Delta^{\mathcal{I}} \setminus R^{\mathcal{I}}$$

To specify the semantics of membership assertions, we extend interpretations to constants, by assigning to each constant c a *distinct* object $c^{\mathcal{I}} \in \Delta^{\mathcal{I}}$. Note that this implies that we enforce the *unique name assumption* on constants [6]. Then, to assign semantics to an ontology, we first define when an interpretation $\mathcal{I}$ *satisfies an assertion* α, denoted $\mathcal{I} \models \alpha$, as follows:

- $\mathcal{I} \models E_1 \sqsubseteq E_2$, if $E_1^{\mathcal{I}} \subseteq E_2^{\mathcal{I}}$;
- $\mathcal{I} \models E_1 \sqsubseteq \neg E_2$, if $E_1^{\mathcal{I}} \cap E_2^{\mathcal{I}} = \emptyset$;
- $\mathcal{I} \models (\text{funct } R)$, if whenever $\{(o,o_1),(o,o_2) \subseteq R^{\mathcal{I}}$, then $o_1 = o_2$;
- $\mathcal{I} \models A(c)$, if $c^{\mathcal{I}} \in A^{\mathcal{I}}$;
- $\mathcal{I} \models P(c,c')$, if $(c^{\mathcal{I}}, c'^{\mathcal{I}}) \in P^{\mathcal{I}}$.

An interpretation $\mathcal{I}$ that satisfies all assertions of an ontology $\mathcal{O}$ is called a *model* of $\mathcal{O}$, and is denoted as $\mathcal{I} \models \mathcal{O}$. An ontology that admits a model is called *satisfiable*. Finally, we say that an ontology $\mathcal{O}$ *logically implies* an assertion α, denoted $\mathcal{O} \models \alpha$, if every model of $\mathcal{O}$ satisfies α. Analogous definitions hold when we replace the ontology $\mathcal{O}$ with a TBox $\mathcal{T}$ or an ABox $\mathcal{A}$.

We observe that, despite the simplicity of the language, the logics of the *DL-Lite* family are able to capture the main elements of conceptual modeling formalisms used in databases and software engineering (e.g., Entity-Relationship and UML class diagrams), cf. [13]. Furthermore, *DL-Lite* is one of the classes of DLs for which conjunctive query answering is tractable in data complexity. Other DLs showing this property are $\mathcal{EL}$ [4,5], and all Horn DLs [41]. Moreover, query answering remains tractable in the DL $\mathcal{FL}_0$ for instance queries (whereas answering conjunctive queries in this logic is coNP-complete), as shown in [7].

2.2   Queries over a DL Ontology

We start with a general notion of queries in first-order logic, and then we move to the definition of queries over a DL ontology.

In general, a *query* is an open formula of first-order logic with equalities (FOL in the following). We denote a (FOL) query q as follows

$$\{\, x_1, \ldots, x_n \mid \phi(x_1, \ldots, x_n) \,\}$$

where $\phi(x_1, \ldots, x_n)$ is a FOL formula with free variables $x_1, \ldots, x_n$. We call n the *arity* of the query q. Given an interpretation $\mathcal{I}$, $q^{\mathcal{I}}$ is the set of tuples of domain elements that, when assigned to the free variables, make the formula ϕ true in $\mathcal{I}$ [2].

A query over an ontology is a FOL query as above, in which the predicates in ϕ are concepts and roles of the ontology. Among the various queries, we are interested in conjunctive queries, which provide a reasonable trade-off between expressive power and complexity of query processing.

A *conjunctive query* (CQ) q of arity n over an ontology $\mathcal{O}$ is a FOL query of the form

$$\{x_1, \ldots, x_n \mid \exists y_1, \ldots, y_m \cdot \phi(x_1, \ldots, x_n, y_1, \ldots, y_m)\},$$

where $x_1, \ldots, x_n$ are pairwise distinct variables[1], and $\phi(x_1, \ldots, x_n, y_1, \ldots, y_m)$ is a conjunction of atoms whose predicates are concept and roles of $\mathcal{O}$, and whose free variables are the variables in $x_1, \ldots, x_n, y_1, \ldots, y_m$. We call $\exists y_1, \ldots, y_m \cdot \phi(x_1, \ldots, x_n, y_1, \ldots, y_m)$ the *body* of q, $x_1, \ldots, x_n$ the *distinguished variables* of q, and $y_1, \ldots, y_m$ the *non-distinguished variables* of q.

In the following we will not indicate existential variables in queries when not explicitly needed, i.e., we will use $\phi(x_1, \ldots, x_n)$ to indicate $\exists y_1, \ldots, y_m \cdot \phi(x_1, \ldots, x_n, y_1, \ldots, y_m)$.

When a query is posed to an ontology, the ontology should answer the query by returning all tuples of constants from the alphabet Γ that satisfy the query in every interpretation that is a model of the ontology. This is formalized by the following notion of certain answers,

Given a CQ q of arity n over an ontology $\mathcal{O}$, the *certain answers* $cert(q, \mathcal{O})$ to q over $\mathcal{O}$ is the set of tuples of constants:

$$cert(q, \mathcal{O}) = \{\langle c_1, \ldots, c_n \rangle \mid \langle c_1^{\mathcal{I}}, \ldots, c_n^{\mathcal{I}} \rangle \in q^{\mathcal{I}} \text{ for all } \mathcal{I} \text{ such that } \mathcal{I} \models \mathcal{O}\}.$$

3   What-To-Ask

In this section we set up a formal framework for interoperation between ontology-based peers, and we formally define the What-To-Ask problem.

3.1   Ontology-Based Peer Framework

As already said, the extensional level of the ontology can be virtually generated by means of local mappings connecting the intensional level of the ontology to a database. In this case, a peer is actually an autonomous ontology-based data access system, or an ontology-based data integration system in the case where the underlying database is federated [43]. For the sake of simplicity, in this paper we consider a peer ontology of a more plain form, in which both the intensional and the extensional knowledge are represented in a first-order logic theory, and more precisely as a DL ontology. All the results we present in fact apply almost straightforwardly to peers that are ontology-based data integration systems.

---

[1] For simplicity of presentation, we have assumed here that conjunctive queries contain neither constants nor repeated variables among $x_1, \ldots, x_n$, but all our results extend to the case where this restrictions do not apply.

Each peer contains an ontology $\mathcal{O} = \langle \mathcal{T}, \mathcal{A} \rangle$ that it can use to make logical inferences. Agents willing to use the peer, here called *clients*, can *ask* the peer queries specified over the peer ontology (i.e., over its TBox).

Besides using its ontology $\mathcal{O}$ for answering queries, each peer can be connected with other peers by means of *mappings*. Mappings establish the relationship between the concepts represented in the peers. When answering a query, each peer can also *ask* queries to the other peers based on such mappings.

In this paper we focus on a system made up by two interoperating peers. One of them, called local peer, is the one the client interacts with by asking queries. The other peer will be referred to as the remote peer, and the knowledge contained in it can be exploited by the local peer through the mappings, so as to enhance the capability of the local peer to provide answers to queries posed by the client. We further assume that, while the local peer exploits the remote peer through the mappings, the remote peer has no information about the local peer, and thus it cannot use in any way the knowledge of the local peer.

Next, we move to the formalization of the framework. We assume that all peers share the same set of constants, denoted by Γ, and we assume that Γ is part of the alphabet of the ontology in each peer. We also assume that in every interpretation different constants are interpreted with different domain elements, i.e., we adopt the *unique name assumption*. With this assumption in place, we turn our attention to the definition of ontology-based peers.

Definition 1. An *ontology-based peer* (or simply *peer*) is a pair $P = \langle \mathcal{O}, M \rangle$ where:

- $\mathcal{O} = \langle \mathcal{T}, \mathcal{A} \rangle$ is the peer ontology, where $\mathcal{T}$ is a TBox and $\mathcal{A}$ an ABox;
- M is a set of mapping assertions, whose form will be illustrated below.

We also call the pair $\langle \mathcal{T}, M \rangle$ the *specification* of P, denoted by P^S, and call the ABox $\mathcal{A}$ the *instance* of P.   ◁

Queries posed to a peer are specified over its TBox $\mathcal{T}$. The queries that we consider are conjunctive queries (cf. Sect. 2). We concentrate on systems consisting of two peers, namely $P_\ell = \langle \mathcal{O}_\ell, M_\ell \rangle$, called *local peer*, which is the peer to which the client may connect, and $P_r = \langle \mathcal{O}_r, \emptyset \rangle$, called *remote peer*. The alphabets of $\mathcal{O}_\ell$ and $\mathcal{O}_r$ share the set of constants Γ, but contain disjoint sets of relation names. Observe that the remote peer does not contain any mapping assertion. We also assume that both peers may process conjunctive queries posed over them, i.e., they are able to compute certain answers to CQs specified over $\mathcal{O}_\ell$ and $\mathcal{O}_r$, respectively. We say that the class of CQs is *accepted by* P_ℓ and P_r, and we call the pair $\langle P_\ell, P_r \rangle$ an *ontology-to-ontology system*.

The mapping M_ℓ in the local peer is constituted by a finite set of *assertions* of the form

$$q_r \rightsquigarrow \{x \mid C(x)\} \quad \text{or}$$

$$q_r' \rightsquigarrow \{x_1, x_2 \mid R(x_1, x_2)\},$$

where q_r is a CQ of arity 1 and q_r' a CQ of arity 2 over the remote peer, C is a concept and R a role of the local peer, x is a variable, and x_1 and x_2 are distinct variables.

A mapping assertion $q_r \rightsquigarrow \{x \mid C(x)\}$ has an immediate interpretation as an implication in FOL: it states that

$$\forall x. \phi_r(x) \rightarrow C(x),$$

where ϕ_r is the open formula constituting the query q_r. Analogously, the mapping assertion $q'_r \rightsquigarrow \{x_1, x_2 \mid R(x_1, x_2)\}$ is interpreted as

$$\forall x_1, x_2. \phi_r(x_1, x_2) \rightarrow R(x_1, x_2).$$

We note that, in data integration terminology, the mappings we have considered here would correspond to a form of mappings called global-as-view (GAV), where the local ontology corresponds to a global schema of a data integration system, the remote ontology corresponds to a set of data sources, and each concept of the global schema is defined by means of a CQ over the data sources.

3.2 The What-To-Ask Problem

A natural task to consider, given a client's query q specified over the local peer P_ℓ, is to return the answers that can be inferred from all the knowledge in the system, that is, return the certain answers $cert(q, \mathcal{O}_\ell \cup M_\ell \cup \mathcal{O}_r)$[2]. Clearly, such a task is meaningful in the case where the axioms in $\mathcal{O}_\ell$, M_ℓ, and $\mathcal{O}_r$ are known and usable by the query answering algorithm.

Here, however, we consider a different setting, in which we assume that the remote peer can only be used by invoking its query answering service, and the local peer has minimal computational capabilities to perform post-processing of the answers provided by the remote peer. More precisely, we assume that:

- each peer $P = \langle \mathcal{O}, M \rangle$ is able to provide the certain answers $cert(q, \mathcal{O})$ to queries q specified over P itself, and
- each peer does not have additional computation capabilities, and is only able to redirect its own answers and those produced by the other peer to the output[3].

Under these assumptions, computing the certain answers to a query posed to the local peer requires to determine the set of queries to send to the remote peer in such a way that the union of such answers with the certain answers computed locally provides the certain answers to the query. This challenge is formalized in what we call the *What-To-Ask* problem.

Definition 2. Consider a local peer $P_\ell = \langle \mathcal{O}_\ell, M_\ell \rangle$, a remote peer specification $P_r^S = \langle \mathcal{T}_r, \emptyset \rangle$, and a query q specified over P_ℓ. The *What-To-Ask* problem, $WTA(q, P_\ell, P_r^S)$, is defined as follows: *Given as input q, P_ℓ and P_r^S, find a finite set $\{q_r^1, \ldots, q_r^n\}$ of queries, each specified over the remote peer P_r, such that for every instance of the remote peer $\mathcal{A}_r$:*

$$cert(q, \mathcal{O}_\ell \cup M_\ell \cup \mathcal{O}_r) = cert(q, \mathcal{O}_\ell) \cup cert(q_r^1, \mathcal{O}_r) \cup \cdots \cup cert(q_r^n, \mathcal{O}_r).$$

where $\mathcal{O}_r = \langle \mathcal{T}_r, \mathcal{A}_r \rangle$.   ◁

---

[2] Whenever we refer to M_ℓ as part of an ontology, we consider its FOL formulation.
[3] This formally corresponds to computing the *union* of the two sets of answers.

The above definition clearly points out the specific nature of the What-To-Ask problem, where the answers coming from the remote peer are combined using *union only*. In particular, it clarifies the difference with other data interoperability architectures, such as data federation. Indeed, in data federation, the mediator has to decide how to send the query to the various federated databases, and then in principle it can use the whole power of SQL (or relational algebra) to combine the answers returned by the data sources.

Notice that, in general, several solutions to the What-To-Ask problem may exist. However, it is easy to see that all solutions are equivalent from a semantic point of view, i.e., each of them allows us to obtain all certain answers that can be inferred from the knowledge managed by the peer system. Syntactic differences might exist between different solutions that could lead one to prefer one solution to another, e.g., if the set of queries in the former one is contained in the set of queries in the latter one. However, we focus here on solving the What-To-Ask problem, i.e., finding *any* solution that satisfies Definition 2, in the specific setting described in the next section, whereas the problem of characterizing when a solution is "better" than another, or finding the "best" solutions with respect to some criteria, is outside the scope of this paper.

In the following, for simplicity, we consider only systems of peers that are *consistent*, i.e., such that their FOL formalization admits at least one model. We will then briefly come back to the issue of (in)consistency in the conclusions.

4   What-To-Ask Problem: Positive Results

We now consider a particular instantiation of the formal framework described in Sect. 3, i.e., we consider specific choices for both the language in which a peer ontology is expressed, and the queries appearing in the mapping assertions. We then study the What-To-Ask problem in the specialized framework. We first present an algorithm, called computeWTA, for the What-To-Ask problem, and then we both prove its termination and correctness, and establish its computational complexity. We also comment on the relationship between the What-To-Ask problem and the task of computing the answers to queries posed to the local peer.

4.1   *DL-Lite$_R$* Peer Ontologies

We concentrate first on the ontology language in which to express the peer ontology. The language we use for this purpose is *DL-Lite$_R$*.

Example 2. Consider a local peer specification $P_\ell = \langle \mathcal{T}_\ell, M_\ell \rangle$ such that $\mathcal{T}_\ell$ is the following *DL-Lite$_R$* TBox:

$$\exists \text{member} \sqsubseteq \text{Employee} \qquad \exists \text{director} \sqsubseteq \text{Manager}$$
$$\exists \text{member}^- \sqsubseteq \text{Dept} \qquad \exists \text{director}^- \sqsubseteq \text{Dept}$$
$$\text{Employee} \sqsubseteq \exists \text{member} \qquad \text{Dept} \sqsubseteq \exists \text{director}^-$$
$$\text{Manager} \sqsubseteq \text{Employee} \qquad \text{director} \sqsubseteq \text{member}$$

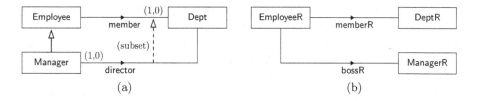

Fig. 2. Intensional component of the local and remote ontologies for Example 2

In this case, such TBox can be directly represented by means of a UML class diagram [45]. Indeed, concepts and roles correspond to UML classes and binary associations, respectively, and role typing assertions are represented in UML by the participation of classes to associations. ISA assertions between concepts correspond to sub-classing, while mandatory participation to roles can be specified in UML by means of multiplicity constraints. Also, ISA assertions between roles can be specified by means of association subsetting. The UML representation of $\mathcal{T}_\ell$ is shown in Fig. 2(a).

Similarly, the following set of $DL\text{-}Lite_\mathcal{R}$ assertions, providing the representation of the TBox $\mathcal{T}_r$ of a remote peer P_r, corresponds to the UML class diagram shown in Fig. 2(b)[4]:

$$\exists \mathsf{memberR} \sqsubseteq \mathsf{EmployeeR} \qquad \exists \mathsf{bossR} \sqsubseteq \mathsf{EmployeeR}$$
$$\exists \mathsf{memberR}^- \sqsubseteq \mathsf{DeptR} \qquad \exists \mathsf{bossR}^- \sqsubseteq \mathsf{ManagerR}$$

Possible ABoxes $\mathcal{A}_\ell$ and $\mathcal{A}_r$ for the TBoxes given above are represented by the $DL\text{-}Lite_\mathcal{R}$ assertions below:

$$\mathsf{Manager}(\mathtt{Mary}) \qquad \mathsf{memberR}(\mathtt{Mary}, \mathtt{D2})$$
$$\mathsf{Dept}(\mathtt{D1}) \qquad \mathsf{DeptR}(\mathtt{D3})$$

Finally, a possible set of assertions for the mapping M_ℓ of the local peer P_ℓ is the following:

$$\{x \mid \mathsf{DeptR}(x)\} \rightsquigarrow \{x \mid \mathsf{Dept}(x)\}$$
$$\{x \mid \mathsf{EmployeeR}(x)\} \rightsquigarrow \{x \mid \mathsf{Employee}(x)\}$$
$$\{x \mid \mathsf{ManagerR}(x)\} \rightsquigarrow \{x \mid \mathsf{Manager}(x)\}$$
$$\{x,y \mid \exists z.\mathsf{bossR}(x,z) \wedge \mathsf{memberR}(z,y)\} \rightsquigarrow \{x,y \mid \mathsf{director}(x,y)\}$$
$$\{x,y \mid \mathsf{memberR}(x,y)\} \rightsquigarrow \{x,y \mid \mathsf{member}(x,y)\} \qquad \triangleleft$$

4.2   The Algorithm ComputeWTA

Consider a local peer $P_\ell = \langle \mathcal{O}_\ell, M_\ell \rangle$ and a remote peer specification $P_r = \langle \mathcal{T}_r, \emptyset \rangle$, and a client's conjunctive query q that is specified over P_ℓ. In a nutshell, our

---

[4] Note that, differently from classical UML semantics, we do not consider as disjoint those classes that in the class diagram do not have a common ancestor.

algorithm first reformulates the client's query q into a set Q of conjunctive queries expressed over $\mathcal{T}_\ell$, in which it compiles the knowledge of the local peer that is relevant for answering q; then the algorithm reformulates the queries of Q into a new set of queries specified over the remote peer P_r.

In the following, given a remote instance $\mathcal{A}_r$, we assume that the theory $\mathcal{O}_\ell \cup M_\ell \cup \mathcal{O}_r$, where $\mathcal{O}_r = \langle \mathcal{T}_r, \mathcal{A}_r \rangle$, is consistent, i.e., there exists at least one first-order interpretation $\mathcal{I}$ such that $\mathcal{I} \models \mathcal{O}_\ell \cup M_\ell \cup \mathcal{O}_r$. Notice that when the theory is inconsistent, the certain answers to a query q of arity n over $\mathcal{O}_\ell \cup M_\ell \cup \mathcal{O}_r$ are all the n-tuples constructible from constants of Γ. Therefore, computing the certain answers to q in this situation does not lead to a meaningful result[5].

Algorithm computeWTA(q, P_ℓ)
Input: CQ q, local peer $P_\ell = \langle \mathcal{O}_\ell, M_\ell \rangle$, where $\mathcal{O}_\ell = \langle \mathcal{T}_\ell, \mathcal{A}_\ell \rangle$ is a *DL-Lite*$_\mathcal{R}$ ontology
Output: set of conjunctive queries
begin
    $Q_{\mathsf{Pref}} \leftarrow \mathsf{PerfectRef}(q, \mathcal{T}_\ell)$;
    $Q \leftarrow \mathsf{Mref}(Q_{\mathsf{Pref}}, M_\ell, \mathcal{O}_\ell)$;
    return Q
end

Fig. 3. Algorithm computeWTA

In Fig. 3, we define the algorithm computeWTA. The algorithm makes use of two main procedures: the first one, called PerfectRef, reformulates the query in accordance with the local TBox $\mathcal{T}_\ell$, whereas the second procedure, called Mref, is concerned with the reformulation based on the mapping.

The algorithm PerfectRef is the query rewriting algorithm for *DL-Lite*$_\mathcal{R}$ defined in [16,18,43]. Intuitively, it compiles the knowledge of the local TBox $\mathcal{T}_\ell$ needed to answer the input query q into a set of conjunctive queries over $\mathcal{T}_\ell$.

Example 3. Continuing Example 2, consider the query

$$q_0 = \{y \mid \exists x.\mathsf{Manager}(x) \wedge \mathsf{member}(x, y)\}$$

that is specified over the local peer P_ℓ, and execute computeWTA(q_0, P_ℓ). Since the first component of the role director is typed by the concept Manager (assertion $\exists \mathsf{director} \sqsubseteq \mathsf{Manager}$ in $\mathcal{T}_\ell$), the algorithm rewrites the first atom of q_0 and produces the query $q_1 = \{y \mid \exists x.\mathsf{director}(x, \_) \wedge \mathsf{member}(x, y)\}$[6]. Since the role director is subsumed by the role member (assertion director $\sqsubseteq$ member in $\mathcal{T}_\ell$), the algorithm rewrites the second atom of q_1 and produces the query $q_2 = \{y \mid \exists x.\mathsf{director}(x, \_) \wedge \mathsf{director}(x, y)\}$. It is not possible to directly rewrite the query

---

[5] For an analysis on the inconsistency problem in the context of database and ontology integration see, for example, [9,11,36,47].

[6] We use the symbol '$\_$' to denote non-shared variables that are existentially quantified.

q_2 by exploiting the TBox assertions. However, the two atoms in q_2 unify, and hence PerfectRef "reduces" q_2, thus producing the query $q_3 = \{y \mid director(\_, y)\}$. Actually, the reduction transforms the bound variable x of q_2 in an unbound variable in q_3. Therefore, the algorithm can now rewrite q_3 by means of the assertion $\mathsf{Dept} \sqsubseteq \exists director^-$, and produces the query $q_4 = \{y \mid \mathsf{Dept}(y)\}$. Then, by the TBox assertion $\exists member^- \sqsubseteq \mathsf{Dept}$, the algorithm produces $q_5 = \{y \mid member(\_, y)\}$. Notice also that due to the role subsumption assertion in $\mathcal{T}_\ell$, from the query q_0, the algorithm produces also the query $q_6 = \{y \mid \exists x.\mathsf{Manager}(x) \wedge director(x, y)\}$. The algorithm does not generate other reformulations.   ◁

Algorithm $\mathsf{Mref}(Q, M_\ell, \mathcal{O}_\ell)$
Input: set of CQs Q, mapping M_ℓ, local ontology $\mathcal{O}_\ell$
Output: set of CQs Q over P_r
begin
  $Q_{aux} \leftarrow \emptyset$; $Q_{ris} \leftarrow \emptyset$;
  for each $q \in Q$ **do**
    $Q_{aux} = Q_{aux} \cup unfold(q, M_\ell)$
  for each $q \in Q_{aux}$ **do**
    if q is a mixed query
    then $Q_{ris} \leftarrow Q_{ris} \cup R_{ref}(q, \mathcal{O}_\ell)$
    else $Q_{ris} \leftarrow Q_{ris} \cup q$
  return Q_{ris}
end

Fig. 4. Algorithm Mref

We now turn our attention to the algorithm Mref, shown in Fig. 4, which reformulates the queries over the local TBox $\mathcal{T}_\ell$ returned by PerfectRef into a new set of queries specified over the remote peer P_r. To this aim, Mref makes use of two operators, *unfold* and R_{ref}. Informally, the former reformulates a query q that is specified over the local TBox $\mathcal{T}_\ell$ by replacing atoms of q with the queries over the remote peer P_r associated to such atoms by the mapping M_ℓ. The latter operator computes a set of queries specified over the remote peer for each query that is specified over both the local and the remote TBox. Notice that queries of this form cannot be directly evaluated in our framework. In the following, we formally describe the two operators.

Definition 3. Let $P = \langle \mathcal{T}, M \rangle$ be a peer, let $R(z_1, z_2)$ be an atom, and let m be a mapping assertion $q_r \rightsquigarrow q_\ell$ in M such that

$$q_\ell = \{x_1, x_2 \mid R(x_1, x_2)\}, \quad \text{and}$$
$$q_r = \{x_1', x_2' \mid \exists y_1, \ldots, y_m.\phi(x_1', x_2', y_1, \ldots, y_m)\}.$$

Then $unfold(R(z_1, z_2), m) = \phi(z_1, z_2, y_1, \ldots, y_m)$. Similarly, let $C(z)$ be an atom, and let m be a mapping assertion $q_r \rightsquigarrow q_\ell$ in M such that

$$q_\ell = \{x \mid C(x)\}, \quad \text{and}$$
$$q_r = \{x' \mid \exists y_1, \ldots, y_m.\phi(x', y_1, \ldots, y_m)\}.$$

Then $unfold(C(z), m) = \phi(z, y_1, \ldots, y_m)$.

If there is no mapping assertion $q_r \leadsto q_\ell$ in M such that $q_\ell = \{x_1, x_2 \mid R(x_1, x_2)\}$ (resp., $q_\ell = \{x \mid C(x)\}$), then $R(z_1, z_2)$ (resp., $C(z)$) is said to be *non unfoldable* in M, otherwise it is said to be *unfoldable* in M.   ◁

The above notion is extended below to unfolding of conjunctive queries. The following definition generalizes the well-known concept of query unfolding [48].

Definition 4. Let $P = \langle T, M \rangle$ be a peer, and let $q = \{z_1, \ldots, z_n \mid \phi(z_1, \ldots, z_n)\}$ be a conjunctive query specified over P. The *unfolding of q w.r.t. M* is the set of conjunctive queries $unfold(q, M)$ defined as follows:

$$unfold(q, M) = \{$$
$$\{z_1, \ldots, z_n \mid unfold(g_1, m_1) \wedge \cdots \wedge unfold(g_h, m_h) \wedge g_{h+1} \wedge \cdots \wedge g_k\} \mid$$
$$m_1, \ldots, m_h \in M, \{g_1, \ldots, g_h\} \text{ is a non-empty subset of the unfoldable}$$
$$\text{atoms of q, and } g_{h+1}, \ldots, g_k \text{ are the remaining atoms of } q\}.   ◁$$

Note that, if no atom in a query q is unfoldable in the mapping M, then $unfold(q, M) = \emptyset$. Therefore, the unfolding operator produces either CQs completely specified over the alphabet of T_r, and hence specified over P_r, or CQs specified over both the alphabets of T_ℓ and T_r. Such queries are called *mixed queries* (we recall that queries specified over P_ℓ are called local queries, whereas queries specified over P_r are called remote queries). It is easy to see that mixed queries are not queries specified over either the local or remote peer, and therefore there is no means in our framework for directly evaluating them. To solve this problem, the algorithm Mref reformulates each mixed query in a set of remote queries, in such a way that the set of answers to the reformulated queries with respect to an instance for the remote peer $\mathcal{A}_r$, computed by the remote peer, coincides with the set of answers that we would have obtained by directly evaluating mixed queries over $\mathcal{O}_\ell \cup \mathcal{O}_r$, where $\mathcal{O}_r = \langle T_r, \mathcal{A}_r \rangle$. Since each tuple in the answer to a mixed query is partially supported by extensional assertions provided by the local ontology, the idea at the basis of such a reformulation is to cast into the new remote queries those constants occurring in $\mathcal{O}_\ell$ that support the answers to the mixed query. Such a mechanism is realized trough the operator R_{ref}, formally described below.

Definition 5. Let $P_\ell = \langle \mathcal{O}_\ell, M_\ell \rangle$ be a local peer, $P_r^S = \langle T_r, \emptyset \rangle$ a remote peer specification, and let

$$q = \{x_1, \ldots, x_n, y_1, \ldots, y_m \mid \exists z_1, \ldots, z_i, w_1, \ldots, w_j . g_\ell^1 \wedge \cdots \wedge g_\ell^h \wedge g_r^1 \wedge \cdots \wedge g_r^k\}$$

be a mixed conjunctive query (i.e., it is such that $h \neq 0$ and $k \neq 0$), where $g_\ell^1, \ldots, g_\ell^h$ are local atoms, $g_r^1, \ldots, g_r^k$ are remote atoms, $x_1, \ldots, x_n$ are the distinguished variables that occur in $g_\ell^1, \ldots, g_\ell^h$ (and possibly also in $g_r^1, \ldots, g_r^k$), $y_1, \ldots, y_m$ are the distinguished variables that occur only in $g_r^1, \ldots, g_r^k$, $z_1, \ldots, z_i$ are the non-distinguished variables that occur both in $g_\ell^1, \ldots, g_\ell^h$ and in $g_r^1, \ldots, g_r^k$, and $w_1, \ldots, w_j$ are the remaining non-distinguished variables of q.

Then, the *remote reformulation of* q w.r.t. $\mathcal{O}_\ell$ is the set $R_{ref}(q, \mathcal{O}_\ell)$ of conjunctive queries specified over P_r defined as follows:

$$R_{ref}(q, \mathcal{O}_\ell) = \{\ \{d_1, \ldots, d_n, y_1, \ldots, y_m \mid \exists w_1, \ldots, w_j.\sigma(g_r^1) \wedge \cdots \wedge \sigma(g_r^k)\} \mid$$
$$\langle d_1, \ldots, d_n, c_1, \ldots, c_i \rangle \in$$
$$cert(\{x_1, \ldots, x_n, z_1, \ldots, z_i \mid \exists w_1, \ldots, w_j.g_\ell^1 \wedge \cdots \wedge g_\ell^h\}, \mathcal{O}_\ell),$$
$$\text{and } \sigma = \{x_1 \to d_1, \ldots, x_n \to d_n, z_1 \to c_1, \ldots, z_i \to c_i\}\}.$$

◁

Roughly speaking, R_{ref} first computes "local answers" to the mixed query q "projected" on its local component, selecting as distinguished variables $z_1, \ldots, z_i$, i.e., the variables that are non-distinguished in q and that also occur in the remote component of the query. Then, for each computed tuple t, R_{ref} constructs a new remote query by projecting the body of q on its remote component, and substituting $z_1, \ldots, z_i$ and $x_1, \ldots, x_n$ with the corresponding constants in t (notice that in such a way the remote peer receives through the reformulated query those extensional information of the local ontology which is needed to answer the mixed query). Obviously, if no local answers to the mixed query exists, the remote reformulation of q is empty.

Example 4. We continue Example 2. The procedure Mref is executed with the set $Q = \{q_0, q_1, q_2, q_3, q_4, q_5, q_6\}$ as input. Let's focus on the query q_0. It is unfolded in the remote query $\{y \mid \exists x.\mathsf{ManagerR}(x) \wedge \mathsf{memberR}(x, y)\}$, and in two mixed queries. Since there are no facts in the local peer for the predicate member, the mixed mentioning this predicate can be ignored. We thus consider the mixed query $q_m = \{y \mid \exists x.\mathsf{Manager}(x) \wedge \mathsf{memberR}(x, y)\}$. Since $cert(\{x \mid \mathsf{Manager}(x)\}, \mathcal{O}_\ell) = \{\mathtt{Mary}\}$, the remote reformulation of q_m produced by R_{ref} is $\{y \mid \mathsf{memberR}(\mathtt{Mary}, y)\}$. We proceed analogously for the other queries in Q^7. The result returned by computeWTA is the following set of queries:

$$\{y \mid \exists x.\mathsf{ManagerR}(x) \wedge \mathsf{memberR}(x, y)\},$$
$$\{y \mid \mathsf{memberR}(\mathtt{Mary}, y)\},$$
$$\{y \mid \exists x, z, w.\mathsf{bossR}(x, z) \wedge \mathsf{memberR}(z, w) \wedge \mathsf{memberR}(x, y)\},$$
$$\{y \mid \exists x, z.\mathsf{bossR}(x, z) \wedge \mathsf{memberR}(z, y)\},$$
$$\{y \mid \mathsf{DeptR}(y)\},$$
$$\{y \mid \exists x, z.\mathsf{ManagerR}(x) \wedge \mathsf{bossR}(x, z) \wedge \mathsf{memberR}(z, y)\},$$
$$\{y \mid \exists z.\mathsf{bossR}(\mathtt{Mary}, z) \wedge \mathsf{memberR}(z, y)\},$$
$$\{y \mid \exists x.\mathsf{memberR}(x, y)\}.$$

The set of certain answers returned by the remote peer is then $\{\mathtt{D3}, \mathtt{D2}\}$. Furthermore, the set of certain answers to q_0 computed by the local peer is $\{\mathtt{D1}\}$. It is easy to see that the union of the above sets is exactly the set that we would have obtained by computing $cert(q_0, \mathcal{O}_\ell \cup M_\ell \cup \mathcal{O}_r)$, i.e., the algorithm computeWTA returned a solution to the What-To-Ask problem for our ongoing example.   ◁

---

[7] We do not reformulate q_2 since it is contained in q_3.

As for the correctness of our technique, it is possible to show that the algorithm computeWTA provides a solution to the What-To-Ask problem in our specialized setting, based on the following properties:

(i) From a client's query q over the local ontology $\mathcal{O}_\ell$, the algorithm PerfectRef is able to compute a set of CQs over $\mathcal{O}_\ell$ that can be evaluated in order to provide the certain answers to q, without taking into account the TBox of $\mathcal{O}_\ell$.

(ii) The unfolding operator used in the algorithm Mref allows us to obtain, from a query q specified over the local peer P_ℓ, a set of CQs over $\mathcal{O}_\ell$ and $\mathcal{O}_r$, which can be evaluated in order to compute the certain answers to q, without taking into account the mapping $\mathcal{M}_\ell$.

(iii) In order to compute the certain answers to a mixed CQ q, i.e., referring to at least one predicate of $\mathcal{O}_\ell$ and one predicate of $\mathcal{O}_r$, we can resort to the remote reformulation of q which produces only queries over the remote ontology $\mathcal{O}_r$.

Theorem 1. *Let $P_\ell = \langle \mathcal{O}_\ell, \mathcal{M}_\ell \rangle$ be a local peer such that $\mathcal{O}_\ell$ is a DL-Lite$_\mathcal{R}$ ontology, let $P_r^S = \langle \mathcal{T}_r, \emptyset \rangle$ be a remote peer specification such that $\mathcal{T}_r$ is a DL-Lite$_\mathcal{R}$ TBox, and let q be a CQ over P_ℓ. Then, computeWTA(q, P_ℓ) returns a solution for WTA(q, P_ℓ, P_r^S).*   ◁

Next, we turn to computational complexity of the algorithm and provide the following result, which follows from the fact that the algorithm PerfectRef runs in polynomial time with respect to the size of the input TBox $\mathcal{T}_\ell$ [16], and from the fact that Mref runs in polynomial time with respect to the size of $\mathcal{O}_\ell$.

Theorem 2. *Let $P_\ell = \langle \mathcal{O}_\ell, \mathcal{M}_\ell \rangle$ be a local peer such that $\mathcal{O}_\ell$ is a DL-Lite$_\mathcal{R}$ ontology, let $P_r^S = \langle \mathcal{T}_r, \emptyset \rangle$ be a remote peer specification such that $\mathcal{T}_r$ is a DL-Lite$_\mathcal{R}$ TBox, and let q be a CQ over P_ℓ. Then, the computational complexity of computeWTA(q, P_ℓ) is polynomial in the size of $\mathcal{O}_\ell$ and $\mathcal{M}_\ell$.*   ◁

We point out that, in general, the size of the set of queries generated by computeWTA may be exponential in the size of the initial query, which obviously implies that the algorithm runs in exponential time in the query size. However, since typically the input query size can be assumed to be small, this exponential blow-up is not likely to be a problem in practice.

5   What-To-Ask Problem: Negative Result

In this section we consider peers equipped with ontologies specified in DL-Lite$_\mathcal{F}$, the other basic language of the DL-Lite family, which does not admit role inclusions, as in DL-Lite$_\mathcal{R}$, but allows for functionalities on roles, without any restriction (cf. Sect. 2). Interestingly, despite the fact that, as in DL-Lite$_\mathcal{R}$, conjunctive query answering in DL-Lite$_\mathcal{F}$ can be solved through query rewriting into a set of conjunctive queries (cf. [16]), the What-To-Ask problem in this case may not admit a solution.

To prove this result, we first provide a complexity lower bound for the problem of instance checking in our framework when both the remote and local peer hosts ontologies specified in *DL-Lite$_{\mathcal{F}}$*.

Theorem 3. *The instance checking (and thus query answering) problem in an ontology-to-ontology system $\langle P_\ell, P_r \rangle$ where the ontologies of both P_ℓ and P_r are expressed in DL-Lite$_{\mathcal{F}}$ is* NLOGSPACE-*hard in data complexity.*   ◁

PROOF. We prove this result by a reduction from reachability in directed graphs.

Let $G = (N, E)$ be a directed graph, where N is the set of its nodes and E is the set of its edges, i.e., pairs (n_i, n_j) such that n_i and n_j belongs to N. We consider the problem of verifying whether a node $d \in N$ is reachable from a node $s \in N$. We define the remote peer $\mathcal{P}_r = \langle \mathcal{O}_r, \emptyset \rangle$, where $\mathcal{O}_r = \langle \mathcal{T}_r, \mathcal{A}_r \rangle$, as follows:

- the alphabet of the predicates of P_r contains the atomic concept A, the atomic role P, and the atomic role $\hat{P}$, and $\mathcal{T}_r$ consists of the inclusion assertions

$$A \sqsubseteq \exists P \qquad \exists P^- \sqsubseteq A$$

- the ABox $\mathcal{A}_r$ is the set of facts

$$\{A(s)\} \cup \{\hat{P}(n_i, n_j) \mid (n_i, n_j) \in E \text{ is an edge of } G\}$$

We then construct the local peer $\mathcal{P}_\ell = \langle \mathcal{O}_\ell, M_\ell \rangle$, with $\mathcal{O}_\ell = \langle \mathcal{T}_\ell, \mathcal{A}_\ell \rangle$, as follows:

- the alphabet of the predicates of P_ℓ consists of the atomic concept C and the atomic role Q, and the TBox $\mathcal{T}_\ell$ contains the assertion

$$(\text{funct } Q)$$

- the ABox $\mathcal{A}_\ell$ is empty;
- the mapping M_ℓ contains the following assertions

$$\{x, y \mid P(x, y)\} \rightsquigarrow \{x, y \mid Q(x, y)\}$$
$$\{x, y \mid \hat{P}(x, y)\} \rightsquigarrow \{x, y \mid Q(x, y)\}$$
$$\{x \mid A(x)\} \rightsquigarrow \{x \mid C(x)\}$$

It is then easy to see that there is a path in G from s to d if and only if $d \in cert(q, \mathcal{O}_\ell \cup M_\ell \cup \mathcal{O}_r)$, where $q = \{x \mid C(x)\}$.   □

From the complexity characterization given above, it follows that peer query answering in the setting considered requires at least the power of linear recursive Datalog (NLOGSPACE). The following result is therefore a straightforward consequence of Theorem 3.

Theorem 4. *There exists a local peer $P_\ell = \langle \mathcal{O}_\ell, M_\ell \rangle$, where $\mathcal{O}_\ell$ is a DL-Lite$_{\mathcal{F}}$ ontology, a remote peer specification $P_r^S = \langle \mathcal{T}_r, \emptyset \rangle$, where $\mathcal{T}_r$ is a DL-Lite$_{\mathcal{F}}$ TBox, and a CQ q (in fact an instance query) specified over P_ℓ such that $WTA(q, P_\ell, P_r^S)$ has no solution.*   ◁

We finally remark that for $DL\text{-}Lite_{\mathcal{F}}$ peers we miss the property that a solution to the What-To-Ask problem exists even if we empower the local peer with the ability of combining the certain answers from the remote peer through FOL rather than simply union, since query answering in this setting requires to go beyond a FOL processing of the data.

6   Towards a Different Semantic Interpretation of Peer Mappings

We have seen above that the What-To-Ask problem admits solutions for two $DL\text{-}Lite_{\mathcal{R}}$ ontology-based peers where the local ontology contains mappings towards the remote ontology, but not vice-versa. In fact, it is immediate to extend this result to any number of remote ontologies as long as this hierarchical topology on the mapping is maintained, i.e., the remote ontologies contain no mappings between them nor towards the local ontology. Instead, if we allow for a network of peers with arbitrary topology of the mappings, even for ontologies with no TBox, peer query answering becomes undecidable [19,26]. On the other hand, we have just shown above that even if we maintain a hierarchical structure of the mapping, but include functionalities, in fact replacing $DL\text{-}Lite_{\mathcal{R}}$ with $DL\text{-}Lite_{\mathcal{F}}$, the What-To-Ask problem becomes unsolvable even if we allow for arbitrary FOL combinations of the certain answers returned by the remote peer.

These results together question the use of first-order mappings, i.e., mappings whose interpretation is an implication between FOL formulas, typically adopted in data peer frameworks [8,19,26].

A radical solution to this is adopting an (auto) epistemic view of the mappings, as suggested in [19]. According to this view each peer is seen as an autonomous agent that interacts with other autonomous agents through peer mappings, and the entire network of peers is not interpreted as a single first-order logic theory, obtained as the disjoint union of the various peer theories, but it is rather considered as a set of different modules, each with its own knowledge about the world and about the other peers in the network. We formalize these ideas below.

6.1   The Logic K

We present a logical formalization of a peer-to-peer network of peer-ontologies based on the use of epistemic logic [10,20,22,29]. In particular, we adopt a *multimodal* epistemic logic, based on the premise that each peer in the system can be seen as a rational agent. More precisely, the formalization we provide is based on $\mathbf{K}$, the multi-modal version of the well-known modal logic of knowledge/belief $K45$ [20] (a.k.a. *weak-S5* [29], see also [38]).

The language $\mathcal{L}(\mathbf{K})$ of $\mathbf{K}$ is obtained from first-order logic by adding a set $\mathbf{K}_1, \ldots, \mathbf{K}_n$ of modal operators, for the forming rule: if ϕ is a (possibly open) formula, then also $\mathbf{K}_i \phi$ is so, for $1 \leq i \leq n$ for a fixed n. In $\mathbf{K}$, each modal operator is used to formalize the epistemic state of a different agent. Informally, the formula $\mathbf{K_i}\phi$ should be read as "ϕ is known to hold by the agent i". The semantics of $\mathbf{K}$ is such that what is known by an agent must hold in the real world: in other words, the agent cannot have inaccurate knowledge of what is true, i.e., believe something to be true although in reality it is false. Moreover, $\mathbf{K}$ states that the agent has complete information on what it knows, i.e., if agent i knows ϕ then it knows of knowing ϕ, and if agent i does not know ϕ, then it knows that it does not know ϕ. In other words, the following assertions hold for every $\mathbf{K}$ formula ϕ:

$$\mathbf{K_i}\phi \rightarrow \phi, \qquad \text{known as the axiom schema T}$$
$$\mathbf{K_i}\phi \rightarrow \mathbf{K_i}(\mathbf{K_i}\phi), \qquad \text{known as the axiom schema 4}$$
$$\neg\mathbf{K_i}\phi \rightarrow \mathbf{K_i}(\neg\mathbf{K_i}\phi), \quad \text{known as the axiom schema 5}$$

To define the semantics of $\mathbf{K}$, we start from first-order interpretations. We restrict our attention to first-order interpretations that share a fixed infinite domain Δ and assume that constants of the set Γ act as standard names for Δ.

Formulas of $\mathbf{K}$ are interpreted over $\mathbf{K}$-structures. A $\mathbf{K}$-*structure* is a Kripke structure E of the form $(W, \{R_1, \ldots R_n\}, V)$, where: W is a set whose elements are called *possible worlds*; V is a function assigning to each $w \in W$ a first-order interpretation $V(w)$; and each R_i, called the *accessibility relation* for the modality $\mathbf{K_i}$, is a binary relation over W, with the following constraints:

if $w \in W$ then $(w, w) \in R_i$, i.e., R_i is reflexive
if $(w_1, w_2) \in R_i$ and $(w_2, w_3) \in R_i$ then $(w_1, w_3) \in R_i$, i.e., R_i is transitive
if $(w_1, w_2) \in R_i$ and $(w_1, w_3) \in R_i$ then $(w_2, w_3) \in R_i$, i.e., R_i is euclidean.

An $\mathbf{K}$-*interpretation* is a pair E, w, where $E = (W, \{R_1, \ldots R_n\}, V)$ is an $\mathbf{K}$-structure, and w is a world in W. We inductively define when a sentence (i.e., a closed formula) ϕ *is true in an interpretation* E, w (or, is true on world $w \in W$ in E), written $E, w \models \phi$, as follows:[8]

$$\begin{array}{lll}
E, w \models P(c_1, \ldots, c_n) & \text{iff} & V(w) \models P(c_1, \ldots, c_n) \\
E, w \models \phi_1 \wedge \phi_2 & \text{iff} & E, w \models \phi_1 \text{ and } E, w \models \phi_2 \\
E, w \models \neg\phi & \text{iff} & E, w \not\models \phi \\
E, w \models \exists x.\psi & \text{iff} & E, w \models \psi_c^x \text{ for some constant } c \\
E, w \models \mathbf{K_i}\phi & \text{iff} & E, w' \models \phi \text{ for every } w' \text{ such that } (w, w') \in R_i
\end{array}$$

We say that a sentence ϕ is *satisfiable* if there exists an $\mathbf{K}$-*model* for ϕ, i.e., an $\mathbf{K}$-interpretation E, w such that $E, w \models \phi$, *unsatisfiable* otherwise. A *model* for a set Σ of sentences is a model for every sentence in Σ. A sentence ϕ is *logically implied* by a set Σ of sentences, written $\Sigma \models_{\mathbf{K}} \phi$, if and only if in every $\mathbf{K}$-model E, w of Σ, we have that $E, w \models \phi$.

Notice that, since each accessibility relation of a $\mathbf{K}$-structure is reflexive, transitive and Euclidean, all instances of axiom schemas T, 4 and 5 are satisfied in every $\mathbf{K}$-interpretation.

---

[8] We use ψ_c^x to denote the formula obtained from ψ by substituting each free occurrence of the variable x with the constant c.

6.2   The What-To-Ask Problem Under the Epistemic Semantics

Due to the characteristics mentioned above, see also [15], $\mathbf{K}$ is well-suited to formalize mappings between peers. We recall that an ontology-based peer ontology P_i has the form $P_i = \langle \mathcal{O}_i, M_i \rangle$, where $\mathcal{O}_i$ is an ontology, and M_i is a set of peer mapping assertions of the form (cf. Sect. 3.1)

$$\{x \mid \exists \boldsymbol{y}.conj(x, \boldsymbol{y})\} \rightsquigarrow \{x \mid C(x)\} \ \text{ or}$$
$$\{x_1, x_2 \mid \exists \boldsymbol{y}.conj(x_1, x_2, \boldsymbol{y})\} \rightsquigarrow \{x_1, x_2 \mid R(x_1, x_2)\},$$

where $conj(x, \boldsymbol{y})$ and $conj(x_1, x_2, \boldsymbol{y})$ are specified over another peer P_j.

For a peer P_i, we define the theory $\mathcal{T}_K(P_i)$ in $\mathbf{K}$ as the union of the following sentences:

- Ontology $\mathcal{O}_i$ of P_i: for each sentence ϕ in $\mathcal{O}_i$, we have

$$\mathbf{K_i}\phi$$

  Observe that ϕ is a first-order sentence expressed in the alphabet of P_i, which is disjoint from the alphabet of all the other peers.
- peer mapping assertions M_i: for each peer mapping assertion from peer P_j to peer P_i in M, we have

$$\forall x.\mathbf{K_j}(\exists \boldsymbol{y}.conj(x, \boldsymbol{y})) \rightarrow \mathbf{K_i}(C(x))$$
$$\forall x_1, x_2.\mathbf{K_j}(\exists \boldsymbol{y}.conj(x_1, x_2, \boldsymbol{y})) \rightarrow \mathbf{K_i}(R(x_1, x_2)).$$

In words, the first sentence specifies the following rule: for each object a, if peer P_j knows the sentence $\exists \boldsymbol{y}.conj(a, \boldsymbol{y})$, then peer P_i knows the assertion $C(a)$. Similarly, the second sentence specifies that for each pair of objects a, b, if peer P_j knows the sentence $\exists \boldsymbol{y}.conj(a, b, \boldsymbol{y})$, then peer P_i knows the assertion $R(a, b)$.

Given a network of peer-ontologies $\mathcal{P} = \{P_1, \ldots, P_n\}$, we denote by $\mathcal{T}_K(\mathcal{P})$ the theory corresponding to the network of peer-ontologies $\mathcal{P}$, i.e., $\mathcal{T}_K(\mathcal{P}) = \bigcup_{i=1,\ldots,n} \mathcal{T}_K(P_i)$.

The semantics of a (conjunctive) query q posed to a peer $P_i = \langle \mathcal{O}_i, M_i \rangle$ of $\mathcal{P}$ is defined as the set of tuples

$$cert_{\mathbf{K}}(q, P_i, \mathcal{P}) = \{\boldsymbol{t} \mid \mathcal{T}_K(\mathcal{P}) \models_{\mathbf{K}} \mathbf{K_i}q(\boldsymbol{t})\}$$

where $q(\boldsymbol{t})$ denotes the sentence obtained from the open formula $q(\boldsymbol{x})$ by replacing all occurrences of the free variables in $\boldsymbol{x}$ with the corresponding constants in $\boldsymbol{t}$.

Let us now turn our attention to ontology-to-ontology systems of the form defined in Sect. 3.1. It is immediate to apply the epistemic-based interpretation given above to systems of this kind, which contain only a remote peer and a local peer. Then, we can rephrase the What-To-ask problem under the epistemic semantics as follows.

Definition 6. Let $P_\ell = \langle \mathcal{O}_\ell, M_\ell \rangle$ be a local peer, $P_r^S = \langle \mathcal{T}_r, \emptyset \rangle$ a remote peer specification, and q a client's query specified over P_ℓ. The *What-To-Ask* problem under the epistemic interpretation of peer mappings, $WTA_e(q, P_\ell, P_r^S)$, is defined as follows: Given as input q, P_ℓ, and P_r^S, find a finite set $\{q_r^1, \ldots, q_r^n\}$ of queries, each specified over the remote peer P_r, such that for every instance $\mathcal{A}_r$ of the remote peer:

$$cert_{\mathbf{K}}(q, P_\ell, \mathcal{P}) = cert(q, \mathcal{O}_\ell) \cup cert(q_r^1, \mathcal{O}_r) \cup \ldots \cup cert(q_r^n, \mathcal{O}_r)$$

where $\mathcal{O}_r = \langle \mathcal{T}_r, \mathcal{A}_r \rangle$ and $\mathcal{P} = \{P_\ell, P_r\}$, with $P_r = \langle \mathcal{O}_r, \emptyset \rangle$. ◁

Notably, it is possible to show that under this interpretation of the system, the What-To-Ask problem admits solutions when ontologies are specified in *DL-Lite$_\mathcal{A}$* [43], which is the logic combining the features of both *DL-Lite$_\mathcal{R}$*, and *DL-Lite$_\mathcal{F}$*, but where the functionality axiom can be asserted only on roles that have no specializations.

Theorem 5. *Let $P_\ell = \langle \mathcal{O}_\ell, M_\ell \rangle$ be a local peer, such that $\mathcal{O}_\ell$ is a DL-Lite$_\mathcal{A}$ ontology, let $P_r^S = \langle \mathcal{T}_r, \emptyset \rangle$ be a remote peer specification, such that $\mathcal{T}_r$ is a DL-Lite$_\mathcal{A}$ TBox, and let q be a CQ specified over P_ℓ. Then,* computeWTA(q, P_ℓ) *returns a solution for $WTA_e(q, P_\ell, P_r)$.* ◁

Finally, we point out that when the ability of the local peer of combining certain answers returned by the remote peer goes beyond the simple union, peer query answering can be solved also through mechanisms that are different from the algorithm computeWTA. For example, when the local peer is able to combine tuples coming from the remote peer with local tuples for computing joins in mixed queries, the procedure Mref in the algorithm computeWTA might be substituted with a more efficient procedure, based for example on the (partial) local materialization of remote data accessible through mapping assertions [34, 35]. Some smart strategies can be adopted in this case to limit materialization only to data relevant for answering the query at hand.

6.3   Epistemic Semantics for Networks of Peer-Ontologies

Interestingly, by virtue of the epistemic interpretation of the peer mappings, techniques for query answering as the one discussed above can be generalized to peer-ontologies networks of arbitrary topology, provided that each peer has the ability of reformulating queries posed over the local ontology in queries to be posed to the other peers in the network (e.g., via the algorithm given in [19] where the external database system can be seen as an autonomous peer in the network). These techniques have been studied in the relational setting in [21].

7   Conclusions

The peer-to-peer paradigm represents an abstraction that captures several types of system studied in different disciplines, such as Multi-agent systems, Semantic

Web, Data Management, Knowledge Representations, and others. In this paper, we have carried out a fundamental study on data-intensive peer-to-peer systems in the case where the whole system is constituted by two peers connected by mappings, and each peer is structured as a knowledge base expressed in a Description Logic of the *DL-Lite* family. In particular, we have addressed the so-called "What-To-Ask" problem, which, given a query q on a local peer P_ℓ, requires to figure out which queries to send to the remote peer in order for P_ℓ to be able to return the correct and complete set of answers to q.

The investigation discussed in this paper can be continued along several interesting directions. In particular, it would be interesting to explore methods for dealing with inconsistencies between peers, a problem that has been ignored by the present paper (see, for instance, [17]). Finally, another relevant problem is to design methods for update propagation between peers, so that all relevant data from the remote peers can be stored in the local peer, thus avoiding asking queries at run time.

References

1. Aberer, K.: Peer-to-Peer Data Management. Synthesis Lectures on Data Management. Morgan & Claypool Publishers, San Rafael (2011). https://doi.org/10.2200/S00338ED1V01Y201104DTM015
2. Abiteboul, S., Hull, R., Vianu, V.: Foundations of Databases. Addison Wesley Publishing Co., Boston (1995)
3. Adjiman, P., Chatalic, P., Goasdoué, F., Rousset, M.C., Simon, L.: Distributed reasoning in a peer-to-peer setting: application to the Semantic Web. J. Artif. Intell. Res. **25**, 269–314 (2006)
4. Baader, F., Brandt, S., Lutz, C.: Pushing the $\mathcal{EL}$ envelope. In: Proceedings of the 19th International Joint Conference on Artificial Intelligence (IJCAI), pp. 364–369 (2005)
5. Baader, F., Brandt, S., Lutz, C.: Pushing the $\mathcal{EL}$ envelope further. In: Clark, K., Patel-Schneider, P.F. (eds.) Proceedings of the 4th International Workshop on OWL: Experiences and Directions (OWLED DC) (2008)
6. Baader, F., Calvanese, D., McGuinness, D., Nardi, D., Patel-Schneider, P.F. (eds.): The Description Logic Handbook: Theory, Implementation and Applications. Cambridge University Press, Cambridge (2003)
7. Baader, F., Marantidis, P., Pensel, M.: The data complexity of answering instance queries in $\mathcal{FL}_0$. In: Proceedings of the 27th International World Wide Web Conferences (WWW), pp. 1603–1607 (2018)
8. Bernstein, P.A., Giunchiglia, F., Kementsietsidis, A., Mylopoulos, J., Serafini, L., Zaihrayeu, I.: Data management for peer-to-peer computing: a vision. In: Proceedings of the 5th International Workshop on the Web and Databases (WebDB) (2002)
9. Bienvenu, M., Bourgaux, C., Goasdoué, F.: Computing and explaining query answers over inconsistent DL-Lite knowledge bases. J. Artif. Intell. Res. **64**, 563–644 (2019). https://doi.org/10.1613/jair.1.11395
10. Blackburn, P., van Benthem, J.F.A.K., Wolter, F.: Handbook of Modal Logic. Elsevier, New York (2006)

11. Bravo, L., Bertossi, L.: Disjunctive deductive databases for computing certain and consistent answers to queries from mediated data integration systems. J. Appl. Logic **3**(2), 329–367 (2005). Special Issue on Logic-based Methods for Information Integration

12. Calvanese, D., Damaggio, E., De Giacomo, G., Lenzerini, M., Rosati, R.: Semantic data integration in P2P systems. In: Aberer, K., Koubarakis, M., Kalogeraki, V. (eds.) DBISP2P 2003. LNCS, vol. 2944, pp. 77–90. Springer, Heidelberg (2004). https://doi.org/10.1007/978-3-540-24629-9_7

13. Calvanese, D., et al.: Ontologies and databases: the *DL-Lite* approach. In: Tessaris, S., et al. (eds.) Reasoning Web 2009. LNCS, vol. 5689, pp. 255–356. Springer, Heidelberg (2009). https://doi.org/10.1007/978-3-642-03754-2_7

14. Calvanese, D., De Giacomo, G., Lembo, D., Lenzerini, M., Rosati, R.: What to ask to a peer: ontology-based query reformulation. In: Proceedings of the 9th International Conference on the Principles of Knowledge Representation and Reasoning (KR), pp. 469–478 (2004)

15. Calvanese, D., De Giacomo, G., Lembo, D., Lenzerini, M., Rosati, R.: EQL-Lite: effective first-order query processing in description logics. In: Proceedings of the 20th International Joint Conference on Artificial Intelligence (IJCAI), pp. 274–279 (2007)

16. Calvanese, D., De Giacomo, G., Lembo, D., Lenzerini, M., Rosati, R.: Tractable reasoning and efficient query answering in description logics: the *DL-Lite* family. J. Autom. Reason. **39**(3), 385–429 (2007)

17. Calvanese, D., De Giacomo, G., Lembo, D., Lenzerini, M., Rosati, R.: Inconsistency tolerance in P2P data integration: an epistemic logic approach. Inf. Syst. **33**(4–5), 360–384 (2008)

18. Calvanese, D., De Giacomo, G., Lembo, D., Lenzerini, M., Rosati, R.: Data complexity of query answering in description logics. Artif. Intell. **195**, 335–360 (2013). https://doi.org/10.1016/j.artint.2012.10.003

19. Calvanese, D., De Giacomo, G., Lenzerini, M., Rosati, R.: Logical foundations of peer-to-peer data integration. In: Proceedings of the 23rd ACM Symposium on Principles of Database Systems (PODS), pp. 241–251 (2004)

20. Chellas, B.F.: Modal Logic: An introduction. Cambridge University Press, Cambridge (1980)

21. De Giacomo, G., Lembo, D., Lenzerini, M., Rosati, R.: On reconciling data exchange, data integration, and peer data management. In: Proceedings of the 26th ACM Symposium on Principles of Database Systems (PODS), pp. 133–142 (2007)

22. van Ditmarsch, H., Halpern, J.Y., van der Hoek, W., Kooi, B. (eds.): Handbook of Epistemic Logic. College Publications, Kolkata (2015)

23. Fagin, R., Kolaitis, P.G., Popa, L., Tan, W.C.: Composing schema mappings: second-order dependencies to the rescue. In: Proceedings of the 23rd ACM Symposium on Principles of Database Systems (PODS) (2004)

24. Fuxman, A., Kolaitis, P.G., Miller, R., Tan, W.C.: Peer data exchange. In: Proceedings of the 24th ACM Symposium on Principles of Database Systems (PODS), pp. 160–171 (2005)

25. Ghidini, C., Serafini, L.: Distributed first order logic. Artif. Intell. **253**, 1–39 (2017). https://doi.org/10.1016/j.artint.2017.08.008

26. Halevy, A., Ives, Z., Suciu, D., Tatarinov, I.: Schema mediation in peer data management systems. In: Proceedings of the 19th IEEE International Conference on Data Engineering (ICDE), pp. 505–516 (2003)

27. Halevy, A.Y.: Theory of answering queries using views. SIGMOD Rec. **29**(4), 40–47 (2000)
28. Halevy, A.Y.: Answering queries using views: a survey. Very Large Database J. **10**(4), 270–294 (2001)
29. Hughes, G.E., Cresswell, M.J.: A Companion to Modal Logic. Methuen, London (1984)
30. Hull, R., Benedikt, M., Christophides, V., Su, J.: E-services: a look behind the curtain. In: Proceedings of the 22nd ACM Symposium on Principles of Database Systems (PODS), pp. 1–14. ACM Press and Addison Wesley (2003). https://doi.org/10.1145/773153.773154
31. Ives, Z.G.: Updates and transactions in peer-to-peer systems. In: Liu, L., Özsu, M.T. (eds.) Encyclopedia of Database Systems, 2nd edn. Springer, New York (2018). https://doi.org/10.1007/978-1-4614-8265-9_1222
32. Karvounarakis, G., Green, T.J., Ives, Z.G., Tannen, V.: Collaborative data sharing via update exchange and provenance. ACM Trans. Database Syst. **38**(3), 19:1–19:42 (2013). https://doi.org/10.1145/2500127
33. Kolaitis, P.G., Pichler, R., Sallinger, E., Savenkov, V.: Limits of schema mappings. Theory Comput. Syst. **62**(4), 899–940 (2018). https://doi.org/10.1007/s00224-017-9812-7
34. Kontchakov, R., Lutz, C., Toman, D., Wolter, F., Zakharyaschev, M.: The combined approach to query answering in *DL-Lite*. In: Proceedings of the 12th International Conference on the Principles of Knowledge Representation and Reasoning (KR), pp. 247–257 (2010)
35. Kontchakov, R., Lutz, C., Toman, D., Wolter, F., Zakharyaschev, M.: The combined approach to ontology-based data access. In: Proceedings of the 22nd International Joint Conference on Artificial Intelligence (IJCAI), pp. 2656–2661 (2011)
36. Lembo, D., Lenzerini, M., Rosati, R., Ruzzi, M., Savo, D.F.: Inconsistency-tolerant query answering in ontology-based data access. J. Web Semant. **33**, 3–29 (2015). https://doi.org/10.1016/j.websem.2015.04.002
37. Lenzerini, M.: Data integration: A theoretical perspective. In: Proceedings of the 21st ACM Symposium on Principles of Database Systems (PODS), pp. 233–246 (2002). https://doi.org/10.1145/543613.543644
38. Levesque, H.J., Lakemeyer, G.: The Logic of Knowledge Bases. The MIT Press, Cambridge (2001)
39. Madhavan, J., Halevy, A.Y.: Composing mappings among data sources. In: Proceedings of the 29th International Conference on Very Large Data Bases (VLDB), pp. 572–583 (2003)
40. Mattos, N.M.: Integrating information for on demand computing. In: Proceedings of the 29th International Conference on Very Large Data Bases (VLDB), pp. 8–14 (2003)
41. Ortiz, M., Rudolph, S., Simkus, M.: Query answering in the Horn fragments of the description logics $\mathcal{SHOIQ}$ and $\mathcal{SROIQ}$. In: Proceedings of the 22nd International Joint Conference on Artificial Intelligence (IJCAI), pp. 1039–1044. IJCAI/AAAI (2011)
42. Papazoglou, M.P., Krämer, B.J., Yang, J.: Leveraging web-services and peer-to-peer networks. In: Eder, J., Missikoff, M. (eds.) CAiSE 2003. LNCS, vol. 2681, pp. 485–501. Springer, Heidelberg (2003). https://doi.org/10.1007/3-540-45017-3_33
43. Poggi, A., Lembo, D., Calvanese, D., De Giacomo, G., Lenzerini, M., Rosati, R.: Linking data to ontologies. J. Data Semant. **10**, 133–173 (2008). https://doi.org/10.1007/978-3-540-77688-8_5

44. Roth, A., Skritek, S.: Peer data management. In: Data Exchange, Integration, and Streams, Dagstuhl Follow-Ups, vol. 5, pp. 185–215. Schloss Dagstuhl-Leibniz-Zentrum für Informatik (2013). https://doi.org/10.4230/DFU.Vol5.10452.185
45. Rumbaugh, J., Jacobson, I., Booch, G.: The Unified Modeling Language Reference Manual. Addison Wesley Publishing Co., Boston (1998)
46. Serafini, L., Ghidini, C.: Using wrapper agents to answer queries in distributed information systems. In: Yakhno, T. (ed.) ADVIS 2000. LNCS, vol. 1909, pp. 331–340. Springer, Heidelberg (2000). https://doi.org/10.1007/3-540-40888-6_32
47. Staworko, S., Chomicki, J., Marcinkowski, J.: Prioritized repairing and consistent query answering in relational databases. Ann. Math. Artif. Intell. **64**(2–3), 209–246 (2012). https://doi.org/10.1007/s10472-012-9288-8
48. Ullman, J.D.: Information integration using logical views. In: Afrati, F., Kolaitis, P. (eds.) ICDT 1997. LNCS, vol. 1186, pp. 19–40. Springer, Heidelberg (1997). https://doi.org/10.1007/3-540-62222-5_34

From Model Completeness to Verification of Data Aware Processes

Diego Calvanese[1], Silvio Ghilardi[2], Alessandro Gianola[1(✉)],
Marco Montali[1], and Andrey Rivkin[1]

[1] Faculty of Computer Science,
Free University of Bozen-Bolzano, Bolzano, Italy
{calvanese,gianola,montali,rivkin}@inf.unibz.it
[2] Dipartimento di Matematica,
Università degli Studi di Milano, Milan, Italy
silvio.ghilardi@unimi.it

Abstract. Model Completeness is a classical topic in model-theoretic algebra, and its inspiration sources are areas like algebraic geometry and field theory. Yet, recently, there have been remarkable applications in computer science: these applications range from combined decision procedures for satisfiability and interpolation, to connections between temporal logic and monadic second order logic and to model-checking. In this paper we mostly concentrate on the last one: we study verification over a general model of so-called artifact-centric systems, which are used to capture business processes by giving equal important to the control-flow and data-related aspects. In particular, we are interested in assessing (parameterized) safety properties irrespectively of the initial database instance. We view such artifact systems as array-based systems, establishing a correspondence with model checking based on Satisfiability-Modulo-Theories (SMT). Model completeness comes into the picture in this framework by supplying quantifier elimination algorithms for suitable existentially closed structures. Such algorithms, whose complexity is unexpectedly low in some cases of our interest, are exploited during search and to represent the sets of reachable states. Our first implementation, built up on top of the MCMT model-checker, makes all our foundational results fully operational and quite effective, as demonstrated by our first experiments.

1 Introduction

In this introduction, we briefly review some results coming from joint work of Franz Baader with the second author during the years 2004–2012: the novel contributions of the present paper can in fact be considered as a natural continuation of such previous cooperation. In both cases, the common background is the attempt of reinterpreting a classical model-theoretic tool (namely model-completeness) inside the realm of computational logic and of automated reasoning. In former joint work the focus was related to the combination of decision

C. Lutz et al. (Eds.): Baader Festschrift, LNCS 11560, pp. 212–239, 2019.
https://doi.org/10.1007/978-3-030-22102-7_10

procedures in first order theories, in the present paper the focus is tailored to the use of decision procedures in declarative model-checking (in particular, in model-checking oriented to the emerging area of verification of data aware processes).

Finding solutions to equations is a challenge at the heart of both mathematics and computer science. Model-theoretic algebra, originating with the ground-breaking work of Robinson [55,56], cast the problem of solving equations in a logical form, and used this setting to solve algebraic problems via model theory. The central notion is that of an *existentially closed model*, which we explain now. Call a quantifier-free formula with parameters in a model $\mathcal{M}$ *solvable* if there is an extension $\mathcal{M}'$ of $\mathcal{M}$ where the formula is satisfied. A model $\mathcal{M}$ is *existentially closed* if any solvable quantifier-free formula already has a solution in $\mathcal{M}$ itself. For example, the field of real numbers is not existentially closed, but the field of complex numbers is.

Although this definition is formally clear, it has a main drawback: it is not first-order definable in general. However, in fortunate and important cases, the class of existentially closed models of a first-order theory T are exactly the models of another first-order theory T^*. In this case, the theory T^* can be characterized abstractly as the *model companion* of T. Model companions become *model completions* (cf. Definition 2.1) in the case of universal theories with the amalgamation property; in such model completions, quantifier elimination holds, unlike in the original theory T. The model companion/model completion of a theory identifies the class of those models where *all satisfiable existential statements can be satisfied*. For example, the theory of algebraically closed fields is the model companion of the theory of fields, and dense linear orders without endpoints give the model companion of linear orders.

1.1 Model Completeness in Combined Decision Problems

A first application of model completeness in computer science, more specifically in automated reasoning, was related to the area of *Satisfiability Modulo Theories* (SMT). The SMT-LIB project[1] (started in 2003) aims at bringing together people interested in developing powerful tools combining sophisticated techniques in SAT-solving with dedicated decision procedures involving specific theories used in applications (especially in software verification).

One of the main problems in the SMT area is to design algorithms for *constraint satifiability problems modulo a given theory T*: in such problems, one is given a finite set of literals and is asked to determine whether this set is satisfiable in a model of T. Theories of interests include linear (real and integer) arithmetics and its fragments, as well as theories axiomatizing datatypes like lists, arrays, etc. Very often such theories come out as *combination* of one or more component theories (arrays of integers, reals, booleans are typical examples) and one would like to obtain constraint satisfiability algorithms for combined theories in a *modular way*. The simplest way to implement this is to have a specific module for

---

[1] http://smtlib.cs.uiowa.edu/.

each component theory and to leave such modules to exchange information concerning the clauses expressible in the shared signature. This simple methodology is quite attractive, but unfortunately not complete in general. A sufficient condition for completeness was identified in [37]: the exchange procedure is complete in case the theory axiomatizing the shared signature reduct T_0 has a model completion T_0^* and each of the component theories T_i is T_0-compatible, i.e., every model of T_i embeds into a model of $T_0^* \cup T_i$. Intuitively, the reason why this condition is sufficient is the fact that one can check satisfiability of constraints in the combined signature by restricting to models whose reduct to the shared signature is a model of T_0^*, so that quantifier elimination in T_0^* guarantees that exchanging information over the quantifier-free fragment is sufficient. This result from [37] generalizes to the non-disjoint signatures case the well-known Nelson-Oppen method [51,61], because to be stably infinite in the sense of [51] means precisely to be compatible with the pure equality theory. For the above outlined exchange procedure to yield decidability of the combined constraint satisfiability problem, we need (for termination) a further hypothesis, namely that the shared theory T_0 is locally finite (which means that the total amount of information that needs to be exchanged is finite up to T_0-equivalence). All the above hypotheses apply for instance to the case of modal algebras with operators, yieldying as a by-product the well-known fusion transfer result for decidability of the global consequence relation in modal logic [63].

The results from [37] however do not supply a sufficient condition for decidability of *combined word problems*. The case of combined word problems is in a sense more challenging: we assume that the component algorithms are only able to test (un)satifiability of a single disequation and we want to conclude that the same property can be transferred to the combined theory via suitable information exchange. In the disjoint signature case, combined word problems are always decidable in case the component equational theories have a decidable word problem [52]; however, for non-disjoint signatures, combined algorithms were known only in case the component theories satisfy a kind of term factorization property [14,36]. In [12], it was proved that T_0-compatibility, joined with a special Gaussian property, yields also here a combined decidability result; the result has again a remarkable consequence in modal logic, as it implies the fusion transfer result for decidability of the *local* consequence relation (this solves a long-standing open question and, up to now and as far as we know, the proof supplied in [12] via general combination methods is the only available proof of this result).

The above methodology was further extended to cover different combination schemata for first order theories [9–11] (again having as special case a combination schema, namely E-connections [49], introduced in the framework of modal and description logics).

Model completeness has further application in automated reasoning: it has been applied to design complete algorithms for constraint satisfiability in theory extensions [59,60] and for combination transfer for quantifier-free interpolation (both for modal logics and for software verification theories) [38,39]. Another different research line used model completeness in order to discover interesting connections between monadic second order logic and its temporal logic fragments [43,44].

1.2   Towards Model Completeness in Verification

In order to see the connection between model completeness and verification, the following simple but nevertheless important observation is crucial. In declarative approaches to model-checking, the runs of a system are identified with certain definable paths in the models of a theory T: in case transition systems are represented via quantifier-free formulae and system variables are modeled as first order variables, it is easy to see that, without loss of generality and as far as safety problems are concerned, one may *restrict to paths within existentially closed models*, thus taking profit from the properties enjoyed by the model completion T^* of T whenever it exists. In particular, during forward or backward search, one can exploit quantifier elimination in order to represent sets of reachable states via quantifier-free formulae.[2]

Our intended applications are however more complex, because we need to handle transition systems whose variables are not just individual first order variables. The systems we have in mind are generically called *array-based systems*, where the term "array-based systems" is an umbrella term generically referring to infinite-state transition systems implicitly specified using a declarative, logic-based formalism comprising *second order function variables*. The formalism captures transitions manipulating arrays via logical formulae, and its precise definition depends on the specific application of interest. The first declarative formalism for array-based systems was introduced in [40,41] to handle the verification of distributed systems, and afterwards was successfully employed also to verify a wide range of infinite-state systems [4,8]. Distributed systems are parameterized in their essence: the number N of interacting processes within a distributed system is unbounded, and the challenge is that of supplying certifications that are valid for all possible values of the parameter N. The overall state of the system is typically described by means of arrays indexed by process identifiers, and used to store the content of process variables like locations and clocks. These arrays are genuine second order function variables: they map *indexes* to *data*, in a way that changes as the system evolves.

Quantifiers are then used to represent sets of system states, however the kind of formulae that are needed for this purpose obey specific syntactic restrictions. Due to these restrictions, the proof obligations generated during model checking search (usually *backward search* is implemented in these systems) can

---

[2] It is quite curious to notice that this observation (in its essence) was already present in the paper [45], where however model completeness was not mentioned at all! Instead of quantifier elimination in the model completion T^*, the authors of [45] relied on the computation of the so called 'cover' of an existential formula (such cover turns out to be equivalent to the quantifier free equivalent formula modulo T^*).

be discharged by techniques combining *instantiation algorithms* with *quantifier elimination algorithms*. Typically, quantifiers ranging over indexes are handled by instantiation and quantifiers over data are handled via quantifier elimination (whenever quantifier elimination is considered too expensive or whenever there is the need to speed up termination, other techniques like interpolation or abstraction may be preferred to quantifier elimination).

The above discussion makes the step we are planning to make in the following evident: whenever quantifier elimination for data is not available, one may *resort to model completions to handle data quantifiers* arising during search in array-bases systems. This is not an abstract plan in fact, because there is an emerging area in verification that leads precisely to this, namely the area of verification of data aware processes (see below). We just mention another crucial fact from the implementation point of view: the cost of quantifier elimination in the model completions relevant for the application area of data aware processes is surprisingly low. In fact, eliminating a tuple of quantified variables from a primitive formula requires only polynomial time and can be achieved for instance via ground Knuth-Bendix completion, see [23] for more details.[3]

1.3   Data Aware Processes: Our Contribution

During the last two decades, a huge body of research has been dedicated to the challenging problem of reconciling data and process management within contemporary organizations [33,53,54]. This requires to move from a purely control-flow understanding of business processes to a more holistic approach that also considers how data are manipulated and evolved by the process. Striving for this integration, new models were devised, with two prominent representatives: object-centric processes [48], and business artifacts [29,46].

In parallel, a flourishing series of results has been dedicated to the formalization of such integrated models, and to the boundaries of decidability and complexity for their static analysis and verification [20]. Such results are quite fragmented, since they consider a variety of different assumptions on the model and on the static analysis tasks [20,62]. Two main trends can be identified within this line. A recent series of results focuses on very general data-aware processes that evolve a full-fledged, relational database (DB) with arbitrary first-order constraints [1,15,16,21]. Actions amount to full bulk updates that may simultaneously operate on multiple tuples, possibly injecting fresh values taken from an infinite data domain. Verification is studied by fixing the initial instance of the DB, and by considering all possible evolutions induced by the process over the initial data.

A second trend of research is instead focused on the formalization and verification of artifact-centric processes. These systems are traditionally formalized using three components [28,31]: *(i)* a read-only DB that stores fixed, background

---

[3]   Again, without mentioning any specific application, this was already observed in [45], as the specialization of the cover algorithm to signatures with unary free function symbols.

information, *(ii)* a working memory that stores the evolving state of artifacts, and *(iii)* actions that update the working memory.Different variants of this model, obtained via a careful tuning of the relative expressive power of its three components, have been studied towards decidability of verification problems parameterized over the read-only DB (see, e.g., [17,28,31,32]). These are verification problems where a property is checked for every possible configuration of the read-only DB. For instance, for the working memory, radically different models are obtained depending on whether only a single artifact instance is evolved, or whether instead the co-evolution of multiple instances of possibly different artifacts is supported. In particular, early formal models for artifact systems merely considered a fixed set of so-called *artifact variables*, altogether instantiated into a single tuple of data. This, in turn, allows one to capture the evolution of a single artifact instance [31]. We call an artifact system of this form *Simple Artifact System (SAS)*. Instead, more sophisticated types of artifact systems have been studied recently in [32,50]. Here, the working memory is not only equipped with artifact variables as in SAS, but also with so-called *artifact relations*, which supports storing arbitrarily many tuples, each accounting for a different artifact instance that can be evolved on its own. We call an artifact system of this form *Relational Artifact System (RAS)*.

The overarching goal of this work is to connect, for the first time, such formal models and their corresponding verification problems, with the models and techniques of *model checking via array-based systems* described above. This is concretized through four technical contributions.

Our *first contribution* is the definition of *a general framework of so-called RASs*, in which artifacts are formalized in the spirit of array-based systems. In this setting, SASs are a particular class of RASs, where only artifact variables are allowed. RASs employ arrays to capture a very rich working memory that simultaneously accounts for artifact variables storing single data elements, and for full-fledged artifact relations storing unboundedly many tuples. Each artifact relation is captured using a collection of arrays, so that a tuple in the relation can be retrieved by inspecting the content of the arrays with a given index. The elements stored therein may be fresh values injected into the RAS, or data elements extracted from the read-only DB, whose relations are subject to key and foreign key constraints. This constitutes a big leap from the usual applications of array-based systems, because the nature of such constraints is quite different and requires completely new techniques for handling them (for instance, for quantifier elimination, as mentioned above). To attack this complexity, by relying on array-based systems, RASs encode the read-only DB using a functional, algebraic view, where relations and constraints are captured using multiple sorts and unary functions. The resulting model captures the essential aspects of the model in [50], which in turn is tightly related (though incomparable) to the sophisticated formal model for artifact-centric systems of [32].

Our *second contribution* is the development of *algorithmic techniques* for the verification of *(parameterized) safety* properties over RASs. This amounts to determining whether there exists an instance of the read-only DB that allows the RAS to evolve from its initial configuration to an *undesired* one that falsifies a given state property. To attack this problem, we build on backward reachability search [40,41]. This is a correct, possibly non-terminating technique that *regresses* the system from the undesired configuration to those configurations that reach the undesired one. This is done by iteratively computing symbolic pre-images, until they either intersect the initial configuration of the system (witnessing unsafety), or they form a fixpoint that does not contain the initial state (witnessing safety).

Adapting backward reachability to the case of RASs, by retaining soundness and completeness, requires genuinely novel research so as to eliminate new (existentially quantified) "data" variables introduced during regression. Traditionally, this is done by quantifier instantiation or elimination. However, while quantifier instantiation can be transposed to RASs, quantifier elimination cannot, since the data elements contained in the arrays point to the content of a full-fledged DB with constraints. To reconstruct quantifier elimination in this setting, which is the main technical contribution of this work, we employ the classic model-theoretic machinery of model completions: via model completions, we prove that the runs of a RAS can be faithfully lifted to richer contexts where quantifier elimination is indeed available, despite the fact that it was not available in the original structures. This allows us to recast safety problems over RASs into equivalent safety problems in this richer setting.

Our *third contribution* is the identification of *three notable classes of RASs* for which backward reachability terminates, in turn witnessing decidability of safety. The first class restricts the working memory to variables only, i.e., focuses on SASs. The second class focuses on RASs operating under the restrictions imposed in [50]: it requires acyclicity of foreign keys and ensures a sort of locality principle where different artifact tuples are not compared. Consequently, it reconstructs the decidability result exploited in [50] if one restricts the verification logic used there to safety properties only. In addition, our second class supports full-fledged bulk updates, which greatly increase the expressive power of dynamic systems [57] and, in our setting, witness the incomparability of our results and those in [50]. The third class is genuinely novel, and while it further restricts foreign keys to form a tree-shaped structure, it does not impose any restriction on the shape of updates, and consequently supports not only bulk updates, but also comparisons between artifact tuples.

Our *fourth contribution* concerns the *implementation of backward reachability techniques* for RASs. Specifically, we have extended the well-known MCMT model checker for array-based systems [42], obtaining a fully operational counterpart to all the foundational results presented in the paper. Even though implementation and experimental evaluation are not central in this paper, we note that our model checker correctly handles the examples produced to test VERIFAS [50], as well as additional examples that go beyond the verification capabilities of VERIFAS, and report some interesting cases here. The performance of MCMT to conduct verification of these examples is very encouraging, and indeed provides the first stepping stone towards effective, SMT-based verification techniques for artifact-centric systems.

This paper is essentially a survey and is meant to summarize ongoing work (cf. [22]); results are stated without proofs or with just proof sketches (proofs are all available in the extended version [24]). The rest of the paper is structured as follows. We give necessary preliminaries in Sect. 2. We present our functional view of (read-only) DBs with constraints in Sect. 3, and we introduce the RAS formal model in Sect. 4. We study safety via backward reachability in Sect. 5, and termination of backward reachability in Sect. 6. We report on our implementation effort and related experiments in Sect. 7, and conclude the paper in Sect. 8.

2   Preliminaries

We adopt the usual first-order syntactic notions of signature, term, atom, (ground) formula, and so on. We use $\underline{u}$ to represent a tuple $\langle u_1, \ldots, u_n \rangle$. Our signatures Σ are multi-sorted and include equality for every sort, which implies that variables are sorted as well. Depending on the context, we keep the sort of a variable implicit, or we indicate explicitly in a formula that variable x has sort S by employing notation $x : S$. The notation $t(\underline{x})$, $\phi(\underline{x})$ means that the term t, the formula ϕ has free variables included in the tuple $\underline{x}$. We are concerned with constants and function symbols f, each of which has *sources* $\underline{S}$ and a *target* S', denoted as $f : \underline{S} \longrightarrow S'$; similarly relation simbols R have sources, written as $R : \underline{S}$. We assume that terms and formulae are well-typed, in the sense that the sorts of variables, constants, and function sources/targets match. A formula is said to be *universal* (resp., *existential*) if it has the form $\forall \underline{x}\,(\phi(\underline{x}))$ (resp., $\exists \underline{x}\,(\phi(\underline{x})))$, where ϕ is a quantifier-free formula. Formulae with no free variables are called *sentences*.

From the semantic side, we use the standard notions of a Σ-*structure* $\mathcal{M}$ and of *truth* of a formula in a Σ-structure under an assignment to the free variables. A Σ-*theory* T is a set of Σ-sentences; a *model* of T is a Σ-structure $\mathcal{M}$ where all sentences in T are true. We use the standard notation $T \models \phi$ to say that ϕ is true in all models of T for every assignment to the free variables of ϕ. We say that ϕ is T-*satisfiable* if there is a model $\mathcal{M}$ of T and an assignment to the free variables of ϕ that make ϕ true in $\mathcal{M}$.

A Σ-formula ϕ is a Σ-*constraint* (or just a constraint) iff it is a conjunction of literals. The constraint satisfiability problem for T asks: given an existential formula $\exists \underline{y}\, \phi(\underline{x}, \underline{y})$ (with ϕ a constraint[4]), are there a model $\mathcal{M}$ of T and an assignment α to the free variables $\underline{x}$ such that $\mathcal{M}, \alpha \models \exists \underline{y}\, \phi(\underline{x}, \underline{y})$?

A theory T has *quantifier elimination* iff for every formula $\phi(\underline{x})$ in the signature of T there is a quantifier-free formula $\phi'(\underline{x})$ such that $T \models \phi(\underline{x}) \leftrightarrow \phi'(\underline{x})$. It is well-known (and easily seen) that quantifier elimination holds in case we can eliminate quantifiers from *primitive* formulae, i.e., from formulae of the kind $\exists \underline{y}\, \phi(\underline{x}, \underline{y})$, where ϕ is a constraint. Since we are interested in effective computability, we assume that *whenever* we talk about quantifier elimination, an *effective procedure* for eliminating quantifiers is given.

---

[4] For the purposes of this definition, we may cquivalently take the formula to be quantifier-free.

Let Σ be a first-order signature. The signature obtained from Σ by adding to it a set $\underline{a}$ of new constants (i.e., 0-ary function symbols) is denoted by $\Sigma^{\underline{a}}$. Analogously, given a Σ-structure $\mathcal{A}$, the signature Σ can be expanded to a new signature $\Sigma^{|\mathcal{A}|} := \Sigma \cup \{\bar{a} \mid a \in |\mathcal{A}|\}$ by adding a set of new constants $\bar{a}$ (the *name* for a), one for each element a in $\mathcal{A}$, with the convention that two distinct elements are denoted by different "name" constants. $\mathcal{A}$ can be expanded to a $\Sigma^{|\mathcal{A}|}$-structure $\mathcal{A}' := (\mathcal{A}, a)_{a \in |\mathcal{A}|}$ by just interpreting the additional constants over the corresponding elements. From now on, when the meaning is clear from the context, we will freely use the notation $\mathcal{A}$ and $\mathcal{A}'$ interchangeably: in particular, given a Σ-structure $\mathcal{A}$ and a Σ-formula $\phi(\underline{x})$ with free variables that are all in $\underline{x}$, we will write, by abuse of notation, $\mathcal{A} \models \phi(\underline{a})$ instead of $\mathcal{A}' \models \phi(\bar{\underline{a}})$.

A *Σ-homomorphism* (or, simply, a homomorphism) between two Σ-structures $\mathcal{M}$ and $\mathcal{N}$ is any mapping $\mu : |\mathcal{M}| \longrightarrow |\mathcal{N}|$ among the support sets $|\mathcal{M}|$ of $\mathcal{M}$ and $|\mathcal{N}|$ of $\mathcal{N}$ satisfying the condition $(\mathcal{M} \models \varphi \Rightarrow \mathcal{N} \models \varphi)$ for all $\Sigma^{|\mathcal{M}|}$-atoms φ (here $\mathcal{M}$ is regarded as a $\Sigma^{|\mathcal{M}|}$-structure, by interpreting each additional constant $a \in |\mathcal{M}|$ into itself, and $\mathcal{N}$ is regarded as a $\Sigma^{|\mathcal{M}|}$-structure by interpreting each additional constant $a \in |\mathcal{M}|$ into $\mu(a)$). In case the last condition holds for all $\Sigma^{|\mathcal{M}|}$-literals, the homomorphism μ is said to be an *embedding*, and if it holds for all first order formulae, the embedding μ is said to be *elementary*.

In the following (cf. Sect. 4), we specify transitions of an artifact-centric system using first-order formulae. To obtain a more compact representation, we make use there of definable extensions as a means for introducing so-called *case-defined functions*. We fix a signature Σ and a Σ-theory T; a T-*partition* is a finite set $\kappa_1(\underline{x}), \ldots, \kappa_n(\underline{x})$ of quantifier-free formulae such that $T \models \forall \underline{x} \bigvee_{i=1}^{n} \kappa_i(\underline{x})$ and $T \models \bigwedge_{i \neq j} \forall \underline{x} \neg(\kappa_i(\underline{x}) \wedge \kappa_j(\underline{x}))$. Given such a T-partition $\kappa_1(\underline{x}), \ldots, \kappa_n(\underline{x})$ together with Σ-terms $t_1(\underline{x}), \ldots, t_n(\underline{x})$ (all of the same target sort), a *case-definable extension* is the Σ'-theory T', where $\Sigma' = \Sigma \cup \{F\}$, with F a "fresh" function symbol (i.e., $F \notin \Sigma$)[5], and $T' = T \cup \bigcup_{i=1}^{n} \{\forall \underline{x} \, (\kappa_i(\underline{x}) \rightarrow F(\underline{x}) = t_i(\underline{x}))\}$. Intuitively, F represents a case-defined function, which can be reformulated using nested if-then-else expressions and can be written as $F(\underline{x}) := \text{case of } \{\kappa_1(\underline{x}) : t_1; \cdots ; \kappa_n(\underline{x}) : t_n\}$. By abuse of notation, we identify T with any of its case-definable extensions T'. In fact, it is easy to produce from a Σ'-formula ϕ' a Σ-formula ϕ equivalent to ϕ' in all models of T': just remove (in the appropriate order) every occurrence $F(\underline{v})$ of the new symbol F in an atomic formula A, by replacing A with $\bigvee_{i=1}^{n}(\kappa_i(\underline{v}) \wedge A(t_i(\underline{v})))$. We also exploit λ-abstractions (see, e.g., formula (3) below) for a more compact (still first-order) representation of some complex expressions, and always use them in atoms like $b = \lambda y . F(y, \underline{z})$ as abbreviations of $\forall y. \, b(y) = F(y, \underline{z})$ (where, typically, F is a symbol introduced in a case-defined extension as above).

We recall a standard notion in Model Theory, namely the notion of a *model completion* of a first order theory [26] (we limit the definition to universal theories, because we shall use only this case):

---

[5] Arity and source/target sorts for F can be deduced from the context (considering that everything is well-typed).

Definition 2.1. *Let T be a universal Σ-theory and let $T^\star \supseteq T$ be a further Σ-theory; we say that $T^\star$ is a model completion of T iff: (i) every model of T can be embedded into a model of $T^\star$; (ii) for every model $\mathcal{M}$ of T, we have that $T^\star \cup \Delta_\Sigma(\mathcal{M})$ is a complete theory in the signature $\Sigma^{|\mathcal{M}|}$.*

Since T is universal, condition *(ii)* is equivalent to the fact that $T^\star$ has *quantifier elimination*; on the other hand, a standard argument (based on diagrams and compactness) shows that condition *(i)* is the same as asking that T and $T^\star$ have the same universal consequences. Thus we have an equivalent definition (to be used in the following):

Proposition 2.2. *Let T be a universal Σ-theory and let $T^\star \supseteq T$ be a further Σ-theory; $T^\star$ is a model completion of T iff: (i) every Σ-constraint satisfiable in a model of T is also satisfiable in a model of $T^\star$; (ii) $T^\star$ has quantifier elimination.*

We recall also that the model completion $T^\star$ of a theory T is unique, if it exists (see [26] for these results and for examples).

3   Read-Only Database Schemas

We now provide a formal definition of (read-only) DB-schemas by relying on an algebraic, functional characterization, and derive some key model-theoretic properties.

Definition 3.1. *A DB schema is a pair $\langle \Sigma, T \rangle$, where: (i) Σ is a DB signature, that is, a finite multi-sorted signature whose only symbols are relation symbols (of any arity), equality, unary function symbols, and constants; (ii) T is a DB theory, that is, a set of universal Σ-sentences.*

Relation symbols are used to represent plain relations, whereas unary function symbols are used to represent relations endowed with primary and foreign key constraints (as will be explained in Sect. 3.1 below). We refer to a DB schema simply through its (DB) signature Σ and (DB) theory T, and denote by Σ_{srt} the set of sorts, by Σ_{rel} the set of relations, and by Σ_{fun} the set of functions in Σ. Since Σ contains only unary function symbols and equality, each atomic Σ-formula is of the form $t_1(v_1) = t_2(v_2)$ or $R(t_1(v_1), \ldots, t_n(v_n))$, where $t_1, t_2, \ldots, t_n$ are possibly complex terms, and $v_1, v_2, \ldots, v_n$ are variables or constants.

We associate to a DB signature Σ a characteristic graph $G(\Sigma)$ capturing the dependencies induced by functions over sorts. Specifically, $G(\Sigma)$ is an edge-labeled graph whose set of nodes is Σ_{srt}, and with a labeled edge $S \xrightarrow{f} S'$ for each $f : S \longrightarrow S'$ in Σ_{fun}. We say that Σ is *acyclic* if $G(\Sigma)$ is so. The *leaves* of Σ are the nodes of $G(\Sigma)$ without outgoing edges. These terminal sorts are divided into two subsets, respectively representing *unary relations* and *value sorts*. Non-value sorts (i.e., unary relations and non-leaf sorts) are called *id sorts*, and are conceptually used to represent (identifiers of) different kinds of objects. Value sorts, instead, represent datatypes such as strings, numbers, clock values, etc.

We denote the set of id sorts in Σ by Σ_{ids}, and that of value sorts by Σ_{val}, hence $\Sigma_{srt} = \Sigma_{ids} \uplus \Sigma_{val}$.

We now consider extensional data.

Definition 3.2. *A DB instance of DB schema $\langle \Sigma, T \rangle$ is a Σ-structure $\mathcal{M}$ that is a model of T and such that every id sort of Σ is interpreted in $\mathcal{M}$ on a finite set.*

Contrast this to arbitrary *models* of T, where no finiteness assumption is made. What may appear as not customary in Definition 3.2 is the fact that value sorts can be interpreted on infinite sets. This allows us, at once, to reconstruct the classical notion of DB instance as a finite model (since only finitely many values can be pointed from id sorts using functions), at the same time supplying a potentially infinite set of fresh values to be dynamically introduced in the working memory during the evolution of the artifact system. More details on this will be given in Sect. 3.1.

We respectively denote by $S^{\mathcal{M}}$, $R^{\mathcal{M}}$, $f^{\mathcal{M}}$, and $c^{\mathcal{M}}$ the interpretation in $\mathcal{M}$ of the sort S (this is a set), of the relation symbol R (this is a set of tuples), of the function symbol f (this is a set-theoretic function), and of the constant c (this is an element of the interpretation of the corresponding sort). Obviously, $f^{\mathcal{M}}$, $R^{\mathcal{M}}$, and $c^{\mathcal{M}}$ must match the sorts in Σ. E.g., if f has source S and target U, then $f^{\mathcal{M}}$ has domain $S^{\mathcal{M}}$ and range $U^{\mathcal{M}}$.

Example 3.3. The human resource (HR) branch of a company stores the following information inside a relational database: *(i)* users registered to the company website, who are potential job applicants; *(ii)* the different, available job categories; *(iii)* employees belonging to HR, together with the job categories they are competent in. To formalize these different aspects, we make use of a DB signature Σ_{hr} consisting of: *(i)* four id sorts UserId, EmpId, CompInId, and JobCatId, used to respectively identify users, employees, job categories, and the competence relationship connecting employees to job categories; *(ii)* one value sort String, containing strings used to name users and employees, and to describe job categories; and *(iii)* five function symbols, namely: *userName* and *empName*, respectively

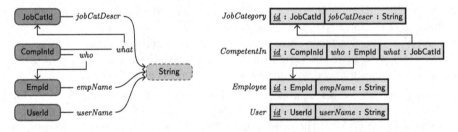

Fig. 1. On the left: characteristic graph of the human resources DB signature from Example 3.3. On the right: relational view of the DB signature; each cell denotes an attribute with its type, underlined attributes denote primary keys, and directed edges capture foreign keys.

mapping user identifiers and employee identifiers to their corresponding names; *jobCatDescr*, mapping job category identifiers to their corresponding descriptions; and *who* and *what*, mapping competence identifiers to their corresponding employees and job categories, respectively. The characteristic graph of Σ_{hr} is shown in the left part of Fig. 1.   ◁

We close the formalization of DB schemas by discussing DB theories, whose role is to encode background axioms. We illustrate a typical background axiom, required to handle the possible presence of *undefined identifiers/values* in the different sorts. This axiom is essential to capture artifact systems whose working memory is initially undefined, in the style of [32,50]. To specify an undefined value we add to every sort S of Σ a constant $\mathtt{undef}_S$ (written from now on, by abuse of notation, just as $\mathtt{undef}$, used also to indicate a tuple). Then, for each function symbol f of Σ, we add the following axiom to the DB theory:

$$\forall x \ (x = \mathtt{undef} \leftrightarrow f(x) = \mathtt{undef}) \tag{1}$$

This axiom states that the application of f to the undefined value produces an undefined value, and it is the only situation for which f is undefined.

Remark 3.4. In the artifact-centric model in the style of [32,50] that we intend to capture, the DB theory consists of Axioms (1) only. However, our technical results do not require this specific choice, and more general sufficient conditions will be discussed later. These conditions apply to natural variants of Axiom (1) (such variants might be used to model situations where we would like to have, for instance, many undefined values, see [24]).

3.1   Relational View of DB Schemas

We now clarify how the algebraic, functional characterization of DB schemas and instances can be actually reinterpreted in the classical, relational model. Definition 3.1 naturally corresponds to the definition of relational database schema equipped with single-attribute *primary keys* and *foreign keys* (plus a reformulation of constraint (1)). To technically explain the correspondence, we adopt the *named perspective*, where each relation schema is defined by a signature containing a *relation name* and a set of *typed attribute names*.

Let $\langle \Sigma, T \rangle$ be a DB schema (only for this subsection, we assume that Σ_{rel} is empty, for simplicity, because we want to concentrate on the most sophisticated part of our formal model, the part aiming at formalizing key dependencies). Each id sort $S \in \Sigma_{ids}$ corresponds to a dedicated relation R_S with the following attributes: *(i)* one identifier attribute id_S with type S; *(ii)* one dedicated attribute a_f with type S' for every function symbol $f \in \Sigma_{fun}$ of the form $f : S \longrightarrow S'$.

The fact that R_S is built starting from functions in Σ naturally induces different database dependencies in R_S. In particular, for each non-id attribute a_f of R_S, we get a *functional dependency* from id_S to a_f; altogether, such dependencies in turn witness that id_S is the *(primary) key* of R_S. In addition, for each non-id attribute a_f of R_S whose corresponding function symbol f has id sort S' as image, we get an *inclusion dependency* from a_f to the id attribute $id_{S'}$ of $R_{S'}$; this captures that a_f is a *foreign key* referencing $R_{S'}$.

Example 3.5. The diagram on the right in Fig. 1 graphically depicts the relational view corresponding to the DB signature of Example 3.3.   ◁

Given a DB instance $\mathcal{M}$ of $\langle \Sigma, T \rangle$, its corresponding *relational instance* $\mathcal{I}$ is the minimal set satisfying the following property: for every id sort $S \in \Sigma_{ids}$, let $f_1, \dots, f_n$ be all functions in Σ with domain S; then, for every identifier $o \in S^{\mathcal{M}}$, $\mathcal{I}$ contains a *labeled fact* of the form $R_S(id_S : o^{\mathcal{M}}, a_{f_1} : f_1^{\mathcal{M}}(o^{\mathcal{M}}), \dots, a_{f_n} : f_n^{\mathcal{M}}(o^{\mathcal{M}}))$. With this interpretation, the *active domain of* $\mathcal{I}$ is the set

$$\bigcup_{S \in \Sigma_{ids}} (S^{\mathcal{M}} \setminus \{\mathtt{undef}^{\mathcal{M}}\}) \cup \left\{ v \in \bigcup_{V \in \Sigma_{val}} V^{\mathcal{M}} \;\middle|\; \begin{array}{l} v \neq \mathtt{undef}^{\mathcal{M}} \text{ and there exist } f \in \Sigma_{fun} \\ \text{and } o \in dom(f^{\mathcal{M}}) \text{ s.t. } f^{\mathcal{M}}(o) = v \end{array} \right\}$$

consisting of all (proper) identifiers assigned by $\mathcal{M}$ to id sorts, as well as all values obtained in $\mathcal{M}$ via the application of some function. Since such values are necessarily *finitely many*, one may wonder why in Definition 3.2 we allow for interpreting value sorts over infinite sets. The reason is that, in our framework, an evolving artifact system may use such infinite provision to inject and manipulate new values into the working memory. From the definition of active domain above, exploiting Axioms (1) we get that the membership of a tuple $(x_0, \dots, x_n)$ to a generic $n+1$-ary relation R_S with key dependencies (corresponding to an id sort S) can be expressed in our setting by using just n unary function symbols and equality:

$$R_S(x_0, \dots, x_n) \quad \text{iff} \quad x_0 \neq \mathtt{undef} \wedge x_1 = f_1(x_0) \wedge \dots \wedge x_n = f_n(x_0) \qquad (2)$$

Hence, the representation of negated atoms is the one that directly follows from negating the formula in (2):

$$\neg R_S(x_0, \dots, x_n) \quad \text{iff} \quad x_0 = \mathtt{undef} \vee x_1 \neq f_1(x_0) \vee \dots \vee x_n \neq f_n(x_0)$$

This relational interpretation of DB schemas exactly reconstructs the requirements posed by [32,50] on the schema of the *read-only* database: *(i)* each relation schema has a single-attribute primary key; *(ii)* attributes are typed; *(iii)* attributes may be foreign keys referencing other relation schemas; *(iv)* the primary keys of different relation schemas are pairwise disjoint.

We stress that all such requirements are natively captured in our functional definition of a DB signature, and do not need to be formulated as axioms in the DB theory. The DB theory is used to express additional constraints, like the one in Axiom (1). In the following subsection, we thoroughly discuss which properties must be respected by signatures and theories to guarantee that our verification machinery is well-behaved.

One may wonder why we have not directly adopted a relational view for DB schemas. This will become clear during the technical development. We anticipate the main, intuitive reasons. First, our functional view allows us to reconstruct in a single, homogeneous framework some important results on verification of artifact systems, achieved on different models that have been unrelated so far [17, 32]. Second, our functional view makes the dependencies among different types explicit. In fact, our notion of characteristic graph, which is readily computed from a DB signature, exactly reconstructs the central notion of foreign key graph used in [32] towards the main decidability results.

3.2   Formal Properties of DB Schemas

The theory T from Definition 3.1 must satisfy a few crucial requirements for our approach to work. In this section, we define such requirements and show that they are matched in the cases we are interested in. The following proposition is motivated by the fact that in most cases the kind of axioms that we need for our DB theories T are just *one-variable universal axioms* (like Axioms (1)).

We say that T has the *finite model property* (for constraint satisfiability) iff every constraint ϕ that is satisfiable in a model of T is satisfiable in a DB instance of T.[6] The finite model property implies decidability of the constraint satisfiability problem for T if T is recursively axiomatized.

Proposition 3.6. *T has the finite model property and has a model completion in case it is axiomatized by universal one-variable formulae and Σ is acyclic.*

The proof of the above result in [24] supplies an algorithm for quantifier elimination in the model completion which is far from optimal in concrete cases. Moreover, acyclicity is not needed in general for Proposition 3.6 to hold: for instance, when $T := \emptyset$ or when T contains only Axioms (1), the proposition holds without acyclicity hypothesis. Such improvements are explained in [23], where a better quantifier elimination algorithm, based on Knuth-Bendix completion is supplied. Proposition 3.6 nevertheless motivates the following assumption:

Assumption 1. *The DB theories we consider have a decidable constraint satisfiability problem, have the finite model property, and admit a model completion.*

This assumption is matched, for instance, in the following three cases: *(i)* when T is empty; *(ii)* when T is axiomatized by Axioms (1); *(iii)* when Σ is acyclic and T is axiomatized by finitely many universal one-variable formulae (such as Axioms (1)).

Hence, the artifact-centric model in the style of [32,50] that we intend to capture *matches* Assumption 1.

---

[6] This directly implies that ϕ is satisfiable also in a DB instance that interprets value sorts into finite sets.

4 Relational Artifact Systems

We are now in the position to define our formal model of *Relational Artifact Systems* (RASs), and to study parameterized safety problems over RASs. Since RASs are array-based systems, we start by recalling the intuition behind them.

In general terms, an array-based system is described using a multi-sorted theory that contains two types of sorts, one accounting for the indexes of arrays, and the other for the elements stored therein. Since the content of an array changes over time, it is referred to by a second-order function variable, whose interpretation in a state is that of a total function mapping indexes to elements (so that applying the function to an index denotes the classical *read* operation for arrays). The definition of an array-based system with array state variable a always requires a formula $I(a)$ describing the *initial configuration* of the array a, and a formula $\tau(a, a')$ describing a *transition* that transforms the content of the array from a to a'. In such a setting, verifying whether the system can reach unsafe configurations described by a formula $K(a)$ amounts to checking whether the formula $I(a_0) \land \tau(a_0, a_1) \land \cdots \land \tau(a_{n-1}, a_n) \land K(a_n)$ is satisfiable for some n. Next, we make these ideas formally precise by grounding array-based systems in the artifact-centric setting.

Following the tradition of artifact-centric systems [17,28,31,32], a RAS consists of a read-only DB, a read-write working memory for artifacts, and a finite set of actions (also called services) that inspect the relational database and the working memory, and determine the new configuration of the working memory. In a RAS, the working memory consists of *individual* and *higher order* variables. These variables (usually called *arrays*) are supposed to model evolving relations, so-called *artifact relations* in [32,50]. The idea is to treat artifact relations in a uniform way as we did for the read-only DB: we need extra sort symbols (recall that each sort symbol corresponds to a database relation symbol) and extra unary function symbols, the latter being treated as second-order variables.

Given a DB schema Σ, an *artifact extension* of Σ is a signature Σ_{ext} obtained from Σ by adding to it some extra sort symbols[7]. These new sorts (usually indicated with letters $E, F, \ldots$) are called *artifact sorts* (or *artifact relations* by some abuse of terminology), while the old sorts from Σ are called *basic sorts*. In a RAS, artifacts and basic sorts correspond, respectively, to the index and the elements sorts mentioned in the literature on array-based systems. Below, given $\langle \Sigma, T \rangle$ and an artifact extension Σ_{ext} of Σ, when we speak of a Σ_{ext}-model of T, a DB instance of $\langle \Sigma_{ext}, T \rangle$, or a Σ_{ext}-model of T^*, we mean a Σ_{ext}-structure $\mathcal{M}$ whose reduct to Σ respectively is a model of T, a DB instance of $\langle \Sigma, T \rangle$, or a model of T^*.

An *artifact setting* over Σ_{ext} is a pair $(\underline{x}, \underline{a})$ given by a finite set $\underline{x}$ of individual variables and a finite set $\underline{a}$ of unary function variables: *the latter must have an artifact sort as source sort and a basic sort as target sort*. Variables in $\underline{x}$ are called *artifact variables*, and variables in $\underline{a}$ *artifact components*. Given a DB

---

[7] By 'signature' we always mean 'signature with equality', so as soon as new sorts are added, the corresponding equality predicates are added too.

instance $\mathcal{M}$ of Σ_{ext}, an *assignment* to an artifact setting $(\underline{x}, \underline{a})$ over Σ_{ext} is a map α assigning to every artifact variable $x_i \in \underline{x}$ of sort S_i an element $x_i^\alpha \in S_i^{\mathcal{M}}$ and to every artifact component $a_j : E_j \longrightarrow U_j$ (with $a_j \in \underline{a}$) a set-theoretic function $a_j^\alpha : E_j^{\mathcal{M}} \longrightarrow U_j^{\mathcal{M}}$. In a RAS, artifact components and artifact variables correspond, respectively, to *arrays* and *constant arrays* (i.e., arrays with all equal elements) mentioned in the literature on array-based systems.

We can view an assignment to an artifact setting $(\underline{x}, \underline{a})$ as a DB instance *extending* the DB instance $\mathcal{M}$ as follows. Let all the artifact components in $(\underline{x}, \underline{a})$ having source E be $a_{i_1} : E \longrightarrow S_1, \cdots, a_{i_n} : E \longrightarrow S_n$. Viewed as a relation in the artifact assignment $(\mathcal{M}, \alpha)$, the artifact relation E "consists" of the set of tuples $\{\langle e, a_{i_1}^\alpha(e), \ldots, a_{i_n}^\alpha(e)\rangle \mid e \in E^{\mathcal{M}}\}$. Thus each element of E is formed by an "entry" $e \in E^{\mathcal{M}}$ (uniquely identifying the tuple) and by "data" $\underline{a}_e^\alpha(e)$ taken from the read-only database $\mathcal{M}$. When the system evolves, the set $E^{\mathcal{M}}$ of entries remains fixed, whereas the components $\underline{a}_i^\alpha(e)$ may change: typically, we initially have $\underline{a}_i^\alpha(e) = \mathbf{undef}$, but these values are changed when some defined values are inserted into the relation modeled by E; the values are then repeatedly modified (and possibly also reset to $\mathbf{undef}$, if the tuple is removed and e is re-set to point to undefined values)[8].

In order to introduce verification problems in the symbolic setting of array-based systems, one first has to specify which formulae are used to represent sets of states, the system initializations, and system evolution. In such formulae, we use notations like $\phi(\underline{z}, \underline{a})$ to mean that ϕ is a formula whose free individual variables are among the $\underline{z}$ and whose free unary function variables are among the $\underline{a}$. Let $(\underline{x}, \underline{a})$ be an artifact setting over Σ_{ext}, where $\underline{x} = x_1, \ldots, x_n$ are the artifact variables and $\underline{a} = a_1, \ldots, a_m$ are the artifact components (their source and target sorts are left implicit).

- An *initial formula* is a formula $\iota(\underline{x})$ of the form[9]

$$(\textstyle\bigwedge_{i=1}^n x_i = c_i) \wedge (\bigwedge_{j=1}^m a_j = \lambda y.d_j),$$

  where c_i, d_j are constants from Σ (typically, c_i and d_j are $\mathbf{undef}$).
- A *state formula* has the form

$$\exists \underline{e}\, \phi(\underline{e}, \underline{x}, \underline{a}),$$

  where ϕ is quantifier-free and the $\underline{e}$ are individual variables of artifact sorts.
- A *transition formula* $\hat{\tau}$ has the form

$$\exists \underline{e}\, (\gamma(\underline{e}, \underline{x}, \underline{a}) \ \wedge\ \textstyle\bigwedge_i x_i' = F_i(\underline{e}, \underline{x}, \underline{a}) \ \wedge\ \bigwedge_j a_j' = \lambda y.G_j(y, \underline{e}, \underline{x}, \underline{a})) \quad (3)$$

---

[8] In accordance with MCMT conventions, we denote the application of an artifact component a to a term (i.e., constant or variable) v also as $a[v]$ (standard notation for arrays), instead of $a(v)$.

[9] Recall that $a_j = \lambda y.d_j$ abbreviates $\forall y\, a_j(y) = d_j$.

where the $\underline{e}$ are individual variables (of *both* basic and artifact sorts), γ (the 'guard') is quantifier-free, $\underline{x}'$, $\underline{a}'$ are renamed copies of $\underline{x}$, $\underline{a}$, and the F_i, G_j (the 'updates') are case-defined functions.

Transition formulae as above can express, e.g., *(i)* insertion (with/without duplicates) of a tuple in an artifact relation, *(ii)* removal of a tuple from an artifact relation, *(iii)* transfer of a tuple from an artifact relation to artifact variables (and vice-versa), and *(iv)* bulk removal/update of *all* the tuples satisfying a certain condition from an artifact relation. All the above operations can also be constrained. Our framework is more expressive than, e.g., the one in [50], as shown in [24].

Definition 4.1. *A* Relational Artifact System *(RAS) is*

$$ \mathcal{S} = \langle \Sigma, T, \Sigma_{ext}, \underline{x}, \underline{a}, \iota(\underline{x}, \underline{a}), \tau(\underline{x}, \underline{a}, \underline{x}', \underline{a}') \rangle $$

where: (i) $\langle \Sigma, T \rangle$ is a (read-only) DB schema, (ii) Σ_{ext} is an artifact extension of Σ, (iii) $(\underline{x}, \underline{a})$ is an artifact setting over Σ_{ext}, (iv) ι is an intitial formula, and (v) τ is a disjunction of transition formulae.

Example 4.2. We present here a RAS $\mathcal{S}_{hr}$ containing a multi-instance artifact accounting for the evolution of *job applications*. Each job category may receive multiple applications from registered users. Such applications are then evaluated, finally deciding which to accept or reject. The example is inspired by the job hiring process presented in [58] to show the intrinsic difficulties of capturing real-life processes with many-to-many interacting business entities using conventional process modeling notations (e.g., BPMN). An extended version of this example is presented in [24].

As for the read-only DB, $\mathcal{S}_{hr}$ works over the DB schema of Example 3.3, extended with a further value sort Score used to score job applications. Score contains 102 values in the range $[-1, 100]$, where -1 denotes the non-eligibility of the application, and a score from 0 to 100 indicates the actual one assigned after evaluating the application. For readability, we use as syntactic sugar the usual predicates $<$, $>$, and $=$ to compare variables of type Score. As for the working memory, $\mathcal{S}_{hr}$ consists of two artifacts. The first single-instance *job hiring* artifact employs a dedicated *pState* variable to capture main phases that the running process goes through: initially, hiring is disabled (*pState* = undef), and, if there is at least one registered user in the HR DB, *pState* becomes enabled. The second multi-instance artifact accounts for the evolution of *user applications*. To model applications, we take the DB signature Σ_{hr} of the read-only HR DB, and enrich it with an artifact extension containing an artifact sort applIndex used to *index* (i.e., *"internally" identify*) job applications. The management of job applications is then modeled by an artifact setting with: *(i)* artifact components with domain applIndex capturing the artifact relation storing different job applications; *(ii)* additional individual variables as temporary memory to manipulate the artifact relation. Specifically, each application consists of a job category, the identifier of the applicant user and that of an HR employee responsible for the application, the application score, and the final result (indicating whether the application is accepted or not). These information slots are encapsulated into dedicated artifact components, i.e., function variables with domain appIndex that collectively realize the application artifact relation:

$$appJobCat: \text{appIndex} \longrightarrow \text{JobCatId} \qquad appScore: \text{appIndex} \longrightarrow \text{Score}$$
$$applicant \ : \text{appIndex} \longrightarrow \text{UserId} \qquad appResp \ : \text{appIndex} \longrightarrow \text{EmpId}$$
$$appResult \ : \text{appIndex} \longrightarrow \text{String}$$

We now discuss the relevant transitions for inserting and evaluating job applications. When writing transition formulae, we make the following assumption: if an artifact variable/component is not mentioned at all, it means that it is updated identically; otherwise, the relevant update function will specify how it is updated.[10] The insertion of an application into the system can be executed when the hiring process is enabled, and consists of two consecutive steps. To indicate when a step can be applied, also ensuring that the insertion of an application is not interrupted by the insertion of another one, we manipulate a string artifact variable $aState$. The first step is executable when $aState$ is undef, and aims at loading the application data (user ID, job category ID, and employee ID) into dedicated artifact variables (uId, jId, eId, respectively) and evolves $aState$ into state received.

The second step transfers the application data into the application artifact relation (using its corresponding function variables), and resets all application-related artifact variables to undef (including $aState$, so that new applications can be inserted). For the insertion, a "free" index (i.e., an index pointing to an undefined applicant) is picked. The newly inserted application gets a default score of -1 ("not eligible"), and an undef final result:

$$\exists i{:}\text{appIndex} \ \big(pState = \text{enabled} \land aState = \text{received} \land applicant[i] = \text{undef} \land$$
$$pState' = \text{enabled} \land aState' = \text{undef} \land cId' = \text{undef} \land$$
$$appJobCat' = \lambda j. \, (\text{if } j = i \text{ then } jId \text{ else } appJobCat[j]) \land$$
$$applicant' = \lambda j. \, (\text{if } j = i \text{ then } uId \text{ else } applicant[j]) \land$$
$$appResp' = \lambda j. \, (\text{if } j = i \text{ then } eId \text{ else } appResp[j]) \land$$
$$appScore' = \lambda j. \, (\text{if } j = i \text{ then } -1 \text{ else } appScore[j]) \land$$
$$appResult' = \lambda j. \, (\text{if } j = i \text{ then } \text{undef} \text{ else } appResult[j]) \land$$
$$jId' = \text{undef} \land uId' = \text{undef} \land eId' = \text{undef} \big)$$

Notice that such a transition does not prevent the possibility of inserting exactly the same application twice, at different indexes. If this is not wanted, the transition can be suitably changed so as to guarantee that no two identical applications can coexist in the same artifact relation (see [24] for an example).

---

[10] Non-deterministic updates can be formalized using existentially quantified variables in the transition.

Each application currently considered as not eligible can be made eligible by assigning a proper score to it:

$$\exists i{:}\mathsf{applIndex}, s{:}\mathsf{Score}\,\big(pState = \mathtt{enabled} \wedge appScore[i] = \mathtt{-1} \wedge s \geq 0 \wedge$$
$$pState' = \mathtt{enabled} \wedge appScore'[i] = s\big)$$

Finally, application results are computed when the process moves to state `notified`. This is handled by the *bulk* transition:

$$pState = \mathtt{enabled} \wedge pState' = \mathtt{notified} \wedge$$
$$appResult' = \lambda j.\,(\text{if } appScore[j] > 80 \text{ then } \mathtt{winner} \text{ else } \mathtt{loser})$$

which declares applications with a score above 80 as winning, and the others as losing.   ◁

5   Parameterized Safety via Backward Reachability

A *safety* formula for $\mathcal{S}$ is a state formula $\upsilon(\underline{x})$ describing undesired states of $\mathcal{S}$. As usual in array-based systems, we say that $\mathcal{S}$ is *safe with respect to* υ if intuitively the system has no finite run leading from ι to υ. Formally, there is no DB-instance $\mathcal{M}$ of $\langle \Sigma_{ext}, T \rangle$, no $k \geq 0$, and no assignment in $\mathcal{M}$ to the variables $\underline{x}^0, \underline{a}^0, \ldots, \underline{x}^k, \underline{a}^k$ such that the formula

$$\iota(\underline{x}^0, \underline{a}^0) \wedge \tau(\underline{x}^0, \underline{a}^0, \underline{x}^1, \underline{a}^1) \wedge \cdots \wedge \tau(\underline{x}^{k-1}, \underline{a}^{k-1}, \underline{x}^k, \underline{a}^k) \wedge \upsilon(\underline{x}^k, \underline{a}^k)$$

is true in $\mathcal{M}$ (here $\underline{x}^i$, $\underline{a}^i$ are renamed copies of $\underline{x}$, $\underline{a}$). The *safety problem* for $\mathcal{S}$ is the following: *given a safety formula υ decide whether $\mathcal{S}$ is safe with respect to υ.*

Example 5.1. The following property expresses the undesired situation that, in the RAS from Example 4.2, once the evaluation is notified there is an applicant with unknown result:

$$\exists i{:}\mathsf{applIndex}\,\big(pState = \mathtt{notified} \wedge applicant[i] \neq \mathtt{undef} \wedge$$
$$appResult[i] \neq \mathtt{winner} \wedge appResult[i] \neq \mathtt{loser}\big)$$

The job hiring RAS $\mathcal{S}_{hr}$ turns out to be safe with respect to this property (cf. Sect. 7).   ◁

We shall introduce an algorithm that semi-decides safety problems for $\mathcal{S}$, and in the next section we shall examine some interesting cases where the algorithm terminates and gives a decision procedure. Before introducing the algorithm, we need some technical results specifying how far we can extend the T^*-quantifier elimination procedure and the T-satisfiability procedure for Σ-constraints to a larger class of quantified formulae in the enriched signature of our artifact settings.

An integral part of the algorithm is to compute *symbolic* preimages. For that purpose, we define for any $\phi_1(\underline{z}, \underline{a}, \underline{z}', \underline{a}')$ and $\phi_2(\underline{z}, \underline{a})$, $Pre(\phi_1, \phi_2)$ as the formula $\exists \underline{z}' \exists \underline{a}' (\phi_1(\underline{z}, \underline{a}, \underline{z}', \underline{a}') \wedge \phi_2(\underline{z}', \underline{a}'))$. The *preimage* of the set of states described by a state formula $\phi(\underline{x}, \underline{a})$ is the set of states described by $Pre(\tau, \phi)$ (notice that, when $\tau = \bigvee \hat{\tau}$, we have $Pre(\tau, \phi) = \bigvee Pre(\hat{\tau}, \phi)$).

Let us call *extended state formulae* the formulae of the kind $\exists \underline{e}\ \phi(\underline{e}, \underline{x}, \underline{a})$, where ϕ is quantifier-free and the $\underline{e}$ are individual variables of *both* artifact and basic sorts. The next two lemmas are proved via syntactic manipulations:

Algorithm 1. Schema of the backward reachability algorithm

Function BReach(v)

1   $\phi \longleftarrow v;\ B \longleftarrow \bot;$

2   **while** $\phi \wedge \neg B$ is T-satisfiable **do**

3 | **if** $\iota \wedge \phi$ is T-satisfiable **then**
    | └ **return** unsafe

4 | $B \longleftarrow \phi \vee B;$

5 | $\phi \longleftarrow Pre(\tau, \phi);$

6 | └ $\phi \longleftarrow QE(T^*, \phi);$

   └ **return** (safe, B);

Lemma 5.2. *The preimage of a state formula is logically equivalent to an extended state formula.*

Lemma 5.3. *For every extended state formula ϕ there is a state formula $QE(T^*, \phi)$ equivalent to ϕ in all Σ_{ext}-models of T^*.*

We underline that Lemmas 5.2 and 5.3 both give an explicit effective procedure for computing equivalent (extended) state formulae: such effective procedures will be an essential part of our backward reachability algorithm. Notice that Lemma 5.3 relies on quantifier elimination in T^*, in fact it is meant to eliminate existentially quantified variables *ranging over basic sorts*. Existentially quantified variables over artifact sorts, on the contrary, cannot be eliminated as they occur as arguments of artifact components.

Let us call $\exists \forall$-formulae the formulae of the kind

$$\exists \underline{e}\ \forall \underline{i}\ \phi(\underline{e}, \underline{i}, \underline{x}, \underline{a})$$

where the variables $\underline{e}, \underline{i}$ are variables whose sort is an artifact sort and ϕ is quantifier-free. The crucial point for the following lemma to hold is that the quantified variables in $\exists \forall$-formulae are all of artifact sorts (the lemma is proved by syntactic manipulations followed by suitable instantiations):

Lemma 5.4. *The satisfiability of an $\exists \forall$-formula in a Σ_{ext}-model of T is decidable. Moreover, an $\exists \forall$-formula is satisfiable in a Σ_{ext}-model of T iff it is satisfiable in a DB-instance of $\langle \Sigma_{ext}, T \rangle$ iff it is satisfiable in a Σ_{ext}-model of T^*.*

Algorithm 1 describes the *backward reachability algorithm* (or, *backward search*) for handling the safety problem for $\mathcal{S}$. It computes iterated preimages of v and applies to them the procedures from Lemmas 5.2 and 5.3, until a fixpoint is reached or until a set intersecting the initial states (i.e., satisfying ι) is found. The satisfiability tests from Lines 2 and 3 can be effectively discharged by

Lemma 5.4 (in fact, the procedure of Lemma 5.4 reduces them to T-constraint satisfiability problems).

To sum up, we obtain the following theorem (to understand the statement of the theorem, notice that by *partial correctness* we mean that, when the algorithm terminates, it gives a correct answer, and by *effectiveness* we mean that all subprocedures in the algorithm can be effectively executed):

Theorem 5.5. *Backward search (cf. Algorithm 1) is effective and partially correct for solving safety problems for RASs.*

Theorem 5.5 shows that backward search is a semi-decision procedure: if the system is unsafe, backward search always terminates and discovers it; if the system is safe, the procedure can diverge (but it is still correct). Notice that the role of quantifier elimination (Line 6 of Algorithm 1) is twofold: *(i)* It allows to discharge the fixpoint test of Line 2 (see Lemma 5.4); *(ii)* it ensures termination in significant cases, namely those where *(strongly) local formulae*, introduced in the next section, are involved.

6   Termination Results for RASs

We now present three termination results, two relating RASs to previous fundamental results, and one genuinely novel.

Termination for "Simple" Artifact Systems. An interesting class of RASs is the one where the working memory consists *only* of artifact variables (without artifact relations). We call systems of this type SASs (*Simple Artifact Systems*). For SASs, the following termination result holds.

Theorem 6.1. *Let $\langle \Sigma, T \rangle$ be a DB schema with Σ acyclic. Then, for every SAS $\mathcal{S} = \langle \Sigma, T, \underline{x}, \iota, \tau \rangle$, backward search terminates and decides safety problems for $\mathcal{S}$ in* PSPACE *in the combined size of $\underline{x}$, ι, and τ.*

It is worth noticing that the decidability part of Theorem 6.1 can be easily extended to *locally finite theories* T (thus, in particular to *arbitrary relational signatures*) whenever T has the amalgamation property and is closed under substructures. Thanks to these observations, Theorem 6.1 is reminiscent of an analogous result in [17], i.e., Theorem 5, the crucial hypotheses of which are exactly amalgamability and closure under substructures, although the setting in that paper is different (there, key dependencies are not discussed, but there is no limitation to elementarily definable classes of structures). Notice also that a distinctive feature of our framework is that it remains well-behaved even in the presence of key dependencies (a naive representation of primary key dependencies with partially functional relations would cause amalgamability to fail). Another important point is that we perform verification in a purely symbolic way, using decision procedures provided by SMT-solvers.

Termination with Local Updates. Consider an *acyclic* signature Σ not containing relation symbols, a DB theory T (satisfying our Assumption 1), and an artifact setting $(\underline{x}, \underline{a})$ over an artifact extension Σ_{ext} of Σ. We call a state formula *local* if it is a disjunction of the formulae

$$\exists e_1 \cdots \exists e_k \left(\delta(e_1, \ldots, e_k) \wedge \bigwedge_{i=1}^{k} \phi_i(e_i, \underline{x}, \underline{a})\right), \tag{4}$$

and *strongly local* if it is a disjunction of the formulae

$$\exists e_1 \cdots \exists e_k \left(\delta(e_1, \ldots, e_k) \wedge \psi(\underline{x}) \wedge \bigwedge_{i=1}^{k} \phi_i(e_i, \underline{a})\right). \tag{5}$$

In (4) and (5), δ is a conjunction of variable equalities and inequalities, ϕ_i, ψ are quantifier-free, and $e_1, \ldots, e_k$ are individual variables ranging over artifact sorts. The key limitation of local state formulae is that they cannot compare entries from different tuples of artifact relations: each ϕ_i in (4) and (5) can contain only the existentially quantified variable e_i.

A transition formula $\hat{\tau}$ is *local* (resp., *strongly local*) if whenever a formula ϕ is local (resp., strongly local), so is $Pre(\hat{\tau}, \phi)$ (modulo the axioms of T^*). Examples of (strongly) local $\hat{\tau}$ are discussed in [24].

Theorem 6.2. *If Σ is acyclic and does not contain relation symbols, backward search (cf. Algorithm 1) terminates when applied to a local safety formula in a RAS whose τ is a disjunction of local transition formulae.*

Proof (sketch). Let $\tilde{\Sigma}$ be $\Sigma_{ext} \cup \{\underline{a}, \underline{x}\}$, i.e., Σ_{ext} expanded with function symbols $\underline{a}$ and constants $\underline{x}$ ($\underline{a}$ and $\underline{x}$ are treated as symbols of $\tilde{\Sigma}$, but not as variables anymore). We call a $\tilde{\Sigma}$-structure *cyclic*[11] if it is generated by one element belonging to the interpretation of an artifact sort. Since Σ is acyclic, so is $\tilde{\Sigma}$, and then one can show that there are only finitely many cyclic $\tilde{\Sigma}$-structures $\mathcal{C}_1, \ldots, \mathcal{C}_N$ up to isomorphism. With a $\tilde{\Sigma}$-structure $\mathcal{M}$ we associate the tuple of numbers $k_1(\mathcal{M}), \ldots, k_N(\mathcal{M}) \in \mathbb{N} \cup \{\infty\}$ counting the numbers of elements generating (as singletons) the cyclic substructures isomorphic to $\mathcal{C}_1, \ldots, \mathcal{C}_N$, respectively. Then we show that, if the tuple associated with $\mathcal{M}$ is componentwise bigger than the one associated with $\mathcal{N}$, then $\mathcal{M}$ satisfies all the local formulae satisfied by $\mathcal{N}$. Finally we apply Dikson Lemma [13]. ⊣

Note that Theorem 6.2 can be used to reconstruct the decidability results of [50] concerning safety problems. Specifically, one needs to show that transitions in [50] are strongly local which, in turn, can be shown using quantifier elimination (see [24] for more details). Interestingly, Theorem 6.2 can be applied to more cases not covered in [50]. For example, one can provide transitions enforcing *updates over unboundedly many tuples* (bulk updates) that are strongly local. One can also see that the safety problem for our running example is decidable since all its transitions are strongly local. Another case considers coverability problems for broadcast protocols [30, 35], which can be encoded using local formulae over the trivial one-sorted signature containing just one basic sort, finitely many constants, and one artifact sort with one artifact component. These problems can

---

[11] This is unrelated to cyclicity of Σ defined in Sect. 3, and comes from universal algebra terminology.

be decided with a non-primitive recursive lower bound [57] (whereas the problems in [50] have an EXPSPACE upper bound). Recalling that [50] handles verification of LTL-FO, thus going beyond safety problems, this shows that the two settings are incomparable. Notice that Theorem 6.2 implies also the decidability of the safety problem for SASs, in case of acyclic Σ.

Termination for Tree-like Signatures. Σ is *tree-like* if it is acyclic, does not contain relation symbols, and all non-leaf nodes have outdegree 1. An artifact setting over Σ is tree-like if $\tilde{\Sigma} := \Sigma_{ext} \cup \{\underline{a}, \underline{x}\}$ is tree-like. In tree-like artifact settings, artifact relations have a single "data" component, and basic relations are unary or binary.

Theorem 6.3 *Backward search (cf. Algorithm 1) terminates when applied to a safety problem in a RAS with a tree-like artifact setting.*

Proof (sketch). The crux is to show, using Kruskal's Tree Theorem [47], that the finitely generated $\tilde{\Sigma}$-structures are a well-quasi-order w.r.t. the embeddability partial order.   ⊣

While tree-like RAS restrict artifact relations to be unary, their transitions are not subject to any locality restriction. This allows for expressing rich forms of updates, including general bulk updates (which allow us to capture non-primitive recursive verification problems) and transitions comparing at once different tuples in artifact relations. Notice that tree-like RASs are incomparable with the "tree" classes of [17], since the former use artifact relations, whereas the latter only individual variables. In [24] we show the power of such advanced features in a flight management process example.

7   First Experiments

We implemented a prototype of the backward reachability algorithm for RASs on top of the MCMT model checker for array-based systems. Starting from its first version [42], MCMT was successfully applied to a variety of settings: cache coherence and mutual exclusions protocols [41], timed [25] and fault-tolerant [5,6] distributed systems, and imperative programs [7,8]. Interesting case studies concerned waiting time bounds synthesis in parameterized timed networks [19] and internet protocols [18]. Further related tools include SAFARI [3], ASASP [2], and CUBICLE [27]. The latter relies on a parallel architecture with further powerful extensions. The work principle of MCMT is rather simple: the tool generates the proof obligations arising from the safety and fixpoint tests in backward search (Lines 2–3 of Algorithm 1) and passes them to the background SMT-solver (currently it is YICES [34]). In practice, the situation is more complicated because SMT-solvers are quite efficient in handling satisfiability problems in combined theories at quantifier-free level, but may encounter difficulties with quantifiers. For this reason, MCMT implements modules for *quantifier elimination* and *quantifier instantiation*. A *specific module* for the quantifier elimination problems mentioned in Line 6 of Algorithm 1 has been added to Version 2.8 of MCMT.

Table 1. Experimental results. The input system size is reflected by columns #AC, #AV, #T, indicating, resp., the number of artifact components, artifact variables, and transitions.

Exp.	#AC	#AV	#T	Prop.	Res.	Time (sec)	Exp.	#AC	#AV	#T	Prop.	Res.	Time (sec)
E1	9	18	15	E1P1	SAFE	0.06	E4	9	11	21	E4P1	SAFE	0.12
				E1P2	UNSAFE	0.36					E4P2	UNSAFE	0.13
				E1P3	UNSAFE	0.50	E5	6	17	34	E5P1	SAFE	4.11
				E1P4	UNSAFE	0.35					E5P2	UNSAFE	0.17
E2	6	13	28	E2P1	SAFE	0.72	E6	2	7	15	E6P1	SAFE	0.04
				E2P2	UNSAFE	0.88					E6P2	UNSAFE	0.08
				E2P3	UNSAFE	1.01	E7	2	28	38	E7P1	SAFE	1.00
				E2P4	UNSAFE	0.83					E7P2	UNSAFE	0.20
E3	4	14	13	E3P1	SAFE	0.05	E8	3	20	19	E8P1	SAFE	0.70
				E3P2	UNSAFE	0.06					E8P2	UNSAFE	0.15

We produced a benchmark consisting of eight realistic business process examples and ran it in MCMT (detailed explanations and results are given in [24]). The examples are partially made by hand and partially obtained from those supplied in [50]. A thorough comparison with VERIFAS [50] is matter of future work, and is non-trivial for a variety of reasons. In particular, as already mentioned in Sect. 6, the two systems tackle incomparable verification problems: on the one hand, we deal with safety problems, whereas VERIFAS handles more general LTL-FO properties; on the other hand, we tackle features not available in VERIFAS, like bulk updates and comparisons between artifact tuples. Moreover, the two verifiers implement completely different state space construction strategies: MCMT is based on backward reachability and makes use of declarative techniques that rely on decision procedures, while VERIFAS employs forward search via VASS encoding.

The benchmark set is available as part of the last distribution 2.8 of MCMT.[12] Table 1 shows the very encouraging results (the first row tackles Example 5.1). While a systematic evaluation is out of scope of this paper, MCMT effectively solves the benchmarks with a comparable performance shown in other well-studied areas, with verification times below 1s in most cases.

8 Conclusions

We have laid the foundations of SMT-based verification for artifact systems, focusing on safety problems and relying on array-based systems as underlying formal model. We have exploited the model-theoretic machinery of model completion to overcome the main technical difficulty arising from this approach, i.e., showing how to reconstruct quantifier elimination in the rich setting of artifact systems. On top of this framework, we have identified three classes of systems

---

[12] http://users.mat.unimi.it/users/ghilardi/mcmt/, subdirectory /examples/dbdriven of the distribution. The user manual contains a new section (pages 36–39) on how to encode RASs in MCMT specifications.

for which safety is decidable, which impose different combinations of restrictions on the form of actions and the shape of DB constraints. The presented techniques have been implemented on top of the well-established MCMT model checker, making our approach fully operational.

We consider the present work as the starting point for a full line of research dedicated to SMT-based techniques for the effective verification of data-aware processes, addressing richer forms of verification beyond safety (such as liveness, fairness, or full LTL-FO) and richer classes of artifact systems, (e.g., with concrete data types and arithmetics), while identifying novel decidable classes (e.g., by restricting the structure of the DB and of transition and state formulae). Concerning implementation, we plan to further develop our tool to incorporate in it the plethora of optimizations and sophisticated search strategies available in infinite-state SMT-based model checking. Finally, we plan to tackle more conventional process modeling notations, concerning in particular data-aware extensions of the de-facto standard BPMN[13].

References

1. Abdulla, P.A., Aiswarya, C., Atig, M.F., Montali, M., Rezine, O.: Recency-bounded verification of dynamic database-driven systems. In: Proceedings of the PODS, pp. 195–210 (2016)
2. Alberti, F., Armando, A., Ranise, S.: ASASP: automated symbolic analysis of security policies. In: Bjørner, N., Sofronie-Stokkermans, V. (eds.) CADE 2011. LNCS (LNAI), vol. 6803, pp. 26–33. Springer, Heidelberg (2011). https://doi.org/10.1007/978-3-642-22438-6_4
3. Alberti, F., Bruttomesso, R., Ghilardi, S., Ranise, S., Sharygina, N.: SAFARI: SMT-based abstraction for arrays with interpolants. In: Madhusudan, P., Seshia, S.A. (eds.) CAV 2012. LNCS, vol. 7358, pp. 679–685. Springer, Heidelberg (2012). https://doi.org/10.1007/978-3-642-31424-7_49
4. Alberti, F., Bruttomesso, R., Ghilardi, S., Ranise, S., Sharygina, N.: An extension of lazy abstraction with interpolation for programs with arrays. Formal Methods Syst. Des. 45(1), 63–109 (2014)
5. Alberti, F., Ghilardi, S., Pagani, E., Ranise, S., Rossi, G.P.: Brief announcement: automated support for the design and validation of fault tolerant parameterized systems - a case study. In: Lynch, N.A., Shvartsman, A.A. (eds.) DISC 2010. LNCS, vol. 6343, pp. 392–394. Springer, Heidelberg (2010). https://doi.org/10.1007/978-3-642-15763-9_36
6. Alberti, F., Ghilardi, S., Pagani, E., Ranise, S., Rossi, G.P.: Universal guards, relativization of quantifiers, and failure models in model checking modulo theories. J. Satisfiability Boolean Model. Comput. 8(1/2), 29–61 (2012)
7. Alberti, F., Ghilardi, S., Sharygina, N.: Booster: an acceleration-based verification framework for array programs. In: Cassez, F., Raskin, J.-F. (eds.) ATVA 2014. LNCS, vol. 8837, pp. 18–23. Springer, Cham (2014). https://doi.org/10.1007/978-3-319-11936-6_2
8. Alberti, F., Ghilardi, S., Sharygina, N.: A framework for the verification of parameterized infinite-state systems. Fundam. Inf. 150(1), 1–24 (2017)

[13] http://www.bpmn.org/.

9. Baader, F., Ghilardi, S.: Connecting many-sorted structures and theories through adjoint functions. In: Gramlich, B. (ed.) FroCoS 2005. LNCS (LNAI), vol. 3717, pp. 31–47. Springer, Heidelberg (2005). https://doi.org/10.1007/11559306_2

10. Baader, F., Ghilardi, S.: Connecting many-sorted theories. In: Nieuwenhuis, R. (ed.) CADE 2005. LNCS (LNAI), vol. 3632, pp. 278–294. Springer, Heidelberg (2005). https://doi.org/10.1007/11532231_21

11. Baader, F., Ghilardi, S.: Connecting many-sorted theories. J. Symbolic Logic **72**(2), 535–583 (2007)

12. Baader, F., Ghilardi, S., Tinelli, C.: A new combination procedure for the word problem that generalizes fusion decidability results in modal logics. Inf. Comput. **204**(10), 1413–1452 (2006)

13. Baader, F., Nipkow, T.: Term Rewriting and All That. Cambridge University Press, Cambridge (1998)

14. Baader, F., Tinelli, C.: Deciding the word problem in the union of equational theories. Inf. Comput. **178**(2), 346–390 (2002)

15. Bagheri Hariri, B., Calvanese, D., De Giacomo, G., Deutsch, A., Montali, M.: Verification of relational data-centric dynamic systems with external services. In: Proceedings of the PODS, pp. 163–174 (2013)

16. Belardinelli, F., Lomuscio, A., Patrizi, F.: An abstraction technique for the verification of artifact-centric systems. In: Proceedings of the KR (2012)

17. Bojańczyk, M., Segoufin, L., Toruńczyk, S.: Verification of database-driven systems via amalgamation. In: Proceedings of the PODS, pp. 63–74 (2013)

18. Bruschi, D., Di Pasquale, A., Ghilardi, S., Lanzi, A., Pagani, E.: Formal verification of ARP (address resolution protocol) through SMT-based model checking - a case study. In: Polikarpova, N., Schneider, S. (eds.) IFM 2017. LNCS, vol. 10510, pp. 391–406. Springer, Cham (2017). https://doi.org/10.1007/978-3-319-66845-1_26

19. Bruttomesso, R., Carioni, A., Ghilardi, S., Ranise, S.: Automated analysis of parametric timing-based mutual exclusion algorithms. In: Goodloe, A.E., Person, S. (eds.) NFM 2012. LNCS, vol. 7226, pp. 279–294. Springer, Heidelberg (2012). https://doi.org/10.1007/978-3-642-28891-3_28

20. Calvanese, D. ., De Giacomo, G., Montali, M.: Foundations of data aware process analysis: a database theory perspective. In: Proceedings of the PODS, pp. 1–12 (2013)

21. Calvanese, D., De Giacomo, G., Montali, M., Patrizi, F.: First-order mu-calculus over generic transition systems and applications to the situation calculus. Inf. Comput. **259**, 328–347 (2017)

22. Calvanese, D., Ghilardi, S., Gianola, A., Montali, M., Rivkin, A.: Model completeness for the verification of data-aware processes. Manuscript submitted for publication (2018)

23. Calvanese, D., Ghilardi, S., Gianola, A., Montali, M., Rivkin, A.: Quantifier elimination for database driven verification. Technical report arXiv:1806.09686, arXiv.org (2018)

24. Calvanese, D., Ghilardi, S., Gianola, A., Montali, M., Rivkin, A.: Verification of data-aware processes via array-based systems (extended version). Technical report arXiv:1806.11459, arXiv.org (2018)

25. Carioni, A., Ghilardi, S., Ranise, S.: MCMT in the land of parametrized timed automata. In: Proceedings of the VERIFY. EPiC Series in Computing, vol. 3, pp. 47–64 (2010)

26. Chang, C.-C., Keisler, J.H.: Model Theory. North-Holland Publishing Co. (1990)

27. Conchon, S., Goel, A., Krstić, S., Mebsout, A., Zaïdi, F.: Cubicle: a parallel SMT-based model checker for parameterized systems. In: Madhusudan, P., Seshia, S.A. (eds.) CAV 2012. LNCS, vol. 7358, pp. 718–724. Springer, Heidelberg (2012). https://doi.org/10.1007/978-3-642-31424-7_55

28. Damaggio, E., Deutsch, A., Vianu, V.: Artifact systems with data dependencies and arithmetic. ACM TODS **37**(3), 22 (2012)

29. Damaggio, E., Hull, R., Vaculín, R.: On the equivalence of incremental and fixpoint semantics for business artifacts with Guard-Stage-Milestone lifecycles. Inf. Syst. **38**(4), 561–584 (2013)

30. Delzanno, G., Podelski, A., Esparza, J.: Constraint-based analysis of broadcast protocols. In: Flum, J., Rodriguez-Artalejo, M. (eds.) CSL 1999. LNCS, vol. 1683, pp. 50–66. Springer, Heidelberg (1999). https://doi.org/10.1007/3-540-48168-0_5

31. Deutsch, A., Hull, R., Patrizi, F., Vianu, V.: Automatic verification of data-centric business processes. In: Proceedings of the ICDT, pp. 252–267. ACM (2009)

32. Deutsch, A., Li, Y., Vianu, V.: Verification of hierarchical artifact systems. In: Proceedings of the PODS, pp. 179–194 (2016)

33. Dumas, M.: On the convergence of data and process engineering. In: Eder, J., Bielikova, M., Tjoa, A.M. (eds.) ADBIS 2011. LNCS, vol. 6909, pp. 19–26. Springer, Heidelberg (2011). https://doi.org/10.1007/978-3-642-23737-9_2

34. Dutertre, B., De Moura, L.: The YICES SMT solver. Technical report, SRI International (2006)

35. Esparza, J., Finkel, A., Mayr, R.: On the verification of broadcast protocols. In: Proceedings of the LICS, pp. 352–359. IEEE Computer Society (1999)

36. Fiorentini, C., Ghilardi, S.: Combining word problems through rewriting in categories with products. TCS **294**(1–2), 103–149 (2003)

37. Ghilardi, S.: Model theoretic methods in combined constraint satisfiability. JAR **33**(3–4), 221–249 (2004)

38. Ghilardi, S., Gianola, A.: Interpolation, amalgamation and combination (the non-disjoint signatures case). In: Dixon, C., Finger, M. (eds.) FroCoS 2017. LNCS (LNAI), vol. 10483, pp. 316–332. Springer, Cham (2017). https://doi.org/10.1007/978-3-319-66167-4_18

39. Ghilardi, S., Gianola, A.: Modularity results for interpolation, amalgamation and superamalgamation. Ann. Pure Appl. Logic **169**(8), 731–754 (2018)

40. Ghilardi, S., Nicolini, E., Ranise, S., Zucchelli, D.: Towards SMT model checking of array-based systems. In: Armando, A., Baumgartner, P., Dowek, G. (eds.) IJCAR 2008. LNCS (LNAI), vol. 5195, pp. 67–82. Springer, Heidelberg (2008). https://doi.org/10.1007/978-3-540-71070-7_6

41. Ghilardi, S., Ranise, S.: Backward reachability of array-based systems by SMT solving: termination and invariant synthesis. Log. Methods Comput. Sci. **6**(4) (2010)

42. Ghilardi, S., Ranise, S.: MCMT: a model checker modulo theories. In: Giesl, J., Hähnle, R. (eds.) IJCAR 2010. LNCS (LNAI), vol. 6173, pp. 22–29. Springer, Heidelberg (2010). https://doi.org/10.1007/978-3-642-14203-1_3

43. Ghilardi, S., van Gool, S.J.: Monadic second order logic as the model companion of temporal logic. In: Proceedings of the LICS, pp. 417–426. ACM (2016)

44. Ghilardi, S., van Gool, S.J.: A model-theoretic characterization of monadic second order logic on infinite words. J. Symbolic Logic **82**(1), 62–76 (2017)

45. Gulwani, S., Musuvathi, M.: Cover algorithms and their combination. In: Drossopoulou, S. (ed.) ESOP 2008. LNCS, vol. 4960, pp. 193–207. Springer, Heidelberg (2008). https://doi.org/10.1007/978-3-540-78739-6_16

46. Hull, R.: Artifact-centric business process models: brief survey of research results and challenges. In: Meersman, R., Tari, Z. (eds.) OTM 2008. LNCS, vol. 5332, pp. 1152–1163. Springer, Heidelberg (2008). https://doi.org/10.1007/978-3-540-88873-4_17

47. Kruskal, J.B.: Well-quasi-ordering, the Tree Theorem, and Vazsonyi's conjecture. Trans. Amer. Math. Soc. **95**, 210–225 (1960)

48. Künzle, V., Weber, B., Reichert, M.: Object-aware business processes: fundamental requirements and their support in existing approaches. Int. J. Inf. Syst. Model. Des. **2**(2), 19–46 (2011)

49. Kutz, O., Lutz, C., Wolter, F., Zakharyaschev, M.: E-connections of abstract description systems. AIJ **156**(1), 1–73 (2004)

50. Li, Y., Deutsch, A., Vianu, V.: VERIFAS: a practical verifier for artifact systems. PVLDB **11**(3), 283–296 (2017)

51. Nelson, G., Oppen, D.C.: Simplification by cooperating decision procedures. ACM TOPLAS **1**(2), 245–257 (1979)

52. Pigozzi, D.: The join of equational theories. Colloq. Math. **30**, 15–25 (1974)

53. Reichert, M.: Process and data: two sides of the same coin? In: Meersman, R., et al. (eds.) OTM 2012. LNCS, vol. 7565, pp. 2–19. Springer, Heidelberg (2012). https://doi.org/10.1007/978-3-642-33606-5_2

54. Richardson, C.: Warning: don't assume your business processes use master data. In: Hull, R., Mendling, J., Tai, S. (eds.) BPM 2010. LNCS, vol. 6336, pp. 11–12. Springer, Heidelberg (2010). https://doi.org/10.1007/978-3-642-15618-2_3

55. Robinson, A.: On the Metamathematics of Algebra. North-Holland (1951)

56. Robinson, A.: Introduction to model theory and to the metamathematics of algebra. In: Studies in Logic and the Foundations of Mathematics. North-Holland (1963)

57. Schmitz, S., Schnoebelen, P.: The power of well-structured systems. In: D'Argenio, P.R., Melgratti, H. (eds.) CONCUR 2013. LNCS, vol. 8052, pp. 5–24. Springer, Heidelberg (2013). https://doi.org/10.1007/978-3-642-40184-8_2

58. Silver, B.: BPMN Method and Style. 2nd edn. Cody-Cassidy (2011)

59. Sofronie-Stokkermans, V.: On interpolation and symbol elimination in theory extensions. In: Olivetti, N., Tiwari, A. (eds.) IJCAR 2016. LNCS (LNAI), vol. 9706, pp. 273–289. Springer, Cham (2016). https://doi.org/10.1007/978-3-319-40229-1_19

60. Sofronie-Stokkermans, V.: On interpolation and symbol elimination in theory extensions. Log. Methods Comput. Sci. **14**(3) (2018)

61. Tinelli, C., Harandi, M.: A new correctness proof of the nelson-oppen combination procedure. In: Baader, F., Schulz, K.U. (eds.) Frontiers of Combining Systems. ALS, vol. 3, pp. 103–119. Springer, Dordrecht (1996). https://doi.org/10.1007/978-94-009-0349-4_5

62. Vianu, V.: Automatic verification of database-driven systems: a new frontier. In: Proceedings of the ICDT, pp. 1–13. ACM (2009)

63. Wolter, f.: Fusions of modal logics revisited. In: Advances in Modal Logic. CSLI Lecture Notes, vol. 1, pp. 361–379 (1996)

Situation Calculus Meets Description Logics

Jens Claßen[1](✉), Gerhard Lakemeyer[2], and Benjamin Zarrieß[3]

[1] School of Computing Science, Simon Fraser University, Burnaby, Canada
jens_classen@sfu.ca
[2] Knowledge-Based Systems Group, RWTH Aachen University, Aachen, Germany
gerhard@cs.rwth-aachen.de
[3] Institute of Theoretical Computer Science, Technische Universität Dresden,
Dresden, Germany
benjamin.zarriess@tu-dresden.de

Abstract. For more than six years, the groups of Franz Baader and
Gerhard Lakemeyer have collaborated in the area of decidable verification of GOLOG programs. GOLOG is an action programming language,
whose semantics is based on the Situation Calculus, a variant of full
first-order logic. In order to achieve decidability, the expressiveness of
the base logic had to be restricted, and using a Description Logic was
a natural choice. In this chapter, we highlight some of the main results
and insights obtained during our collaboration.

Keywords: Situation Calculus · Description Logics · Verification

Prologue

We begin our contribution to celebrate Franz' 60th birthday with some personal remarks by the second author, written as a first-person account. As these
remarks are largely historical, they will also shed light on how the technical work
described later came into being and how it is intimately connected to the work
by Franz and his group in Dresden.

Franz and I first met, I believe, in 1990, when we both gave talks at AAAI
in Boston. Indeed, in those early days, we mainly met at conferences, either
at AAAI, IJCAI or KR. But apart from that, each of us was minding his own
business, Franz working on Description Logics (DLs) and me on the Situation
Calculus and the related action programming language GOLOG. This is not to
say that I stayed away completely from DLs. While I was still in Bonn, Franz
was nice enough to share his course notes with me so that I could teach a
DL course, which I did exactly once! In 1994, I even published a paper on an
epistemic version of CLASSIC, an early variant of modern DLs, at the German
AI conference. But I soon realized that other people, in particular Franz, were
much better at this, and I left DL to them without any intention to ever return,
or so I thought.

In 1997, Franz and I became colleagues at RWTH Aachen University. Research-wise we continued our separate ways, but at least we now met regularly at (often boring) faculty meetings. It was only when Franz moved to Dresden that things took a different turn. Michael Thielscher, also at TU Dresden at the time, had the brilliant idea to gather researchers from different areas in KR and combine work on action formalisms with work on Description Logics, planning, and nonmonotonic reasoning. In the end, a DFG-funded Research Cluster on Logic-Based Knowledge Representation was established, initially started by Franz, Michael, Bernhard Nebel and myself, and later joined by Gerd Brewka. While I, together with my then Ph.D. student Jens Claßen, collaborated most closely with Bernhard's group during this time, the meetings and workshops of the entire Research Cluster not only helped us to get to know each other better personally but to also appreciate each other's research and the connections between the different areas much more.

At the time Hongkai Liu, a former Ph.D. student of Franz, worked on updating ABoxes, which is meant to reflect how a world changes. As the external examiner of Hongkai's thesis I got to know his work quite well, and I was particularly intrigued by his chapter on decidable verification of infinite sequences of updates. At the same time, Jens had started work on the verification of nonterminating GOLOG programs. When the time came to re-apply for funding from DFG, this time in the form of a Research Unit on "Hybrid Reasoning for Intelligent Systems" [9], Franz had the idea that we should join forces and explore the verification of GOLOG programs when the underlying logic is restricted to a DL fragment with the aim of arriving at decidable forms of verification. When we received funding for our Research Unit, Jens joined our project on the Aachen side and Benjamin on the Dresden side. The rest, as they say, is history. We have collaborated now for almost seven years, and it has been a lot of fun. In the following, we highlight some of the main results obtained during this time, but before we begin: Happy Birthday, Franz!

1   Introduction

The agent language GOLOG [19,33] allows one to describe an agent's behaviour in terms of a program containing both imperative and nondeterministic aspects. Its basic building blocks are the primitive actions that are defined in a theory of some action logic, typically the Situation Calculus [40,44] or its modal variant [31], but also formalisms based on Description Logics. Among GOLOG's most promising application areas is the control of autonomous, mobile robots [11,24].

As a very simple, illustrating example, consider a robot whose task it is to remove dirty dishes from a number of rooms in a building. A program for it might look like this:

loop :   **while** $(\exists x.\,OnRobot(x))$ **do**
                    $\pi x{:}Dish\;\{unload(x)\}$ **endWhile**;
             $\pi y{:}Room\;\{\;goto(y);$
                    while $(\exists x\,Dirty(x,y))$ **do**
                            $\pi x{:}Dish\;load(x,y)$ **endWhile** $\}$;
             $goto(kitchen)$

The robot is initially in the kitchen. In each iteration of the infinite outer loop, it first unloads all dishes it carries, selects a room in the building, moves there, collects all dirty dishes from there, and returns to the kitchen. Here, $Dirty(x,y)$ means "dirty dish x is in room y", and $load(x,y)$ stands for "load dish x from room y." Constructs of the form $\pi x{:}Dish$ moreover are to be read as "nondeterministically choose one object from the $Dish$ domain and do the following with it". We furthermore assume that at any time during operation, some new dish x to be removed from room y may appear, which is represented through a special, "exogenous" $newdish(x,y)$ action. Now before deploying such a program onto the real robot, it is often desirable to verify it against some temporal specification, e.g. to make sure that every dish will eventually be removed.

While a large variety of temporal verification methods have been developed in the field of Model Checking [7,12] over the last decades, the problem of verifying (typically non-terminating) GOLOG programs received surprisingly little attention among Situation Calculus researchers. Note that Model Checking is not directly applicable due to the fact that even though nowadays implicit, symbolic representations of state spaces are used, their input formalisms are very restricted in expressivity. GOLOG on the other hand relies on action descriptions in terms of (first-order) logical theories that correspond to a very large, if not infinite number of possible models. Instead of simply *checking* the property in question against a single model, *theorem proving* within the underlying logic is hence required.

De Giacomo, Ternovska and Reiter [21] were the first to address the verification of non-terminating GOLOG programs. They express programs and their properties using inductive definitions and fixpoint logics, thus heavily resorting to second-order quantification. They then do manual, meta-theoretic proofs to show that the program satisfies the desired properties. While this work was an important first step, an automated verification would be obviously much more preferable to a manual one since the latter tends to be tedious and error-prone.

Claßen and Lakemeyer [16] proposed such a method for properties expressed in a temporal logic that resembles the Computational Tree Logic CTL, but that allows for unrestricted first-order quantification. The algorithm is inspired by the classical symbolic model checking techniques for propositional CTL in the sense that it does a similar fixpoint computation to systematically explore the system's state space. The difference however is that it does not work on a single finite model, but, as explained above, uses a logical first-order action theory together with the GOLOG program (which possibly contains further first-order quantification). The method relies on regression-based reasoning, a newly proposed graph-based representation of the input program, and theorem proving for detecting convergence.

The overall verification problem for GOLOG is highly undecidable due to unrestricted first-order quantification in the underlying base logic, the kind and range of actions' effects, and GOLOG being a Turing-complete programming language. Consequently, in [16] only soundness of the method was proved, but a termination guarantee could not be given. A natural next step is to try to identify restricted, yet non-trivial fragments of GOLOG where verification becomes decidable, while a great deal of expressiveness is retained.

A natural choice for a decidable base logic with first-order expressivity is a Description Logic. Baader, Liu and ul Mehdi [4] considered actions specified in an action formalism based on the Description Logic $\mathcal{ALC}$ [5], and furthermore abstracted from the actual execution sequences of a non-terminating program by considering infinite sequences of actions defined by a Büchi automaton. They expressed properties by a variant of LTL over $\mathcal{ALC}$ axioms [2] and could show that under these restrictions, verification reduces to a decidable reasoning task within the underlying DL.

Their work was an important first step in the search for a way to overcome the above mentioned three "sources of undecidability" (i.e. undecidable base logic, range of action effects, Turing-complete program constructs), even though the restrictions employed were comparably harsh. In particular, their $\mathcal{ALC}$-based action formalism only allows for basic STRIPS-style addition and deletion of literals, and the very simple over-approximation of programs through Büchi automata loses important features such as the non-deterministic choice of argument and test conditions. Baader and Zarrieß [6] later showed that these results can indeed be lifted to a more expressive fragment of GOLOG that includes test conditions. They obtained decidability by proving that the potentially infinite transition system induced by the GOLOG program can always be represented by a finite one that admits the exact same execution traces. This was the start of a complementary line of research based on the approach of applying restrictions that allows one to compute a finite, propositional *abstraction* of the infinite state space, and then use a classical model checker to decide the query.

In this paper we want to give a brief, yet concrete impression of research conducted on both approaches, the GOLOG-specific fixpoint method as well as abstraction methods, within the aforementioned Research Unit on "Hybrid Reasoning for Intelligent Systems". The following section introduces some formal preliminaries. Sections 3 and 4 then present the GOLOG-specific fixpoint method and the abstraction technique, respectively. DL-based representations, their relation to Situation Calculus formalizations, as well as computational complexities are then discussed in Sect. 5. Section 6 gives a survey of further research we conducted, followed by a conclusion in Sect. 7.

2 Preliminaries

2.1 The Logic $\mathcal{ES}$

We use a fragment of the first-order modal Situation Calculus variant $\mathcal{ES}$ [31], and corresponding *Basic Action Theories* (BATs) [44].

Syntax: There are *terms* of sort *object* and *action*. Variables of sort object are denoted by symbols $x, y, \ldots$, and a denotes a variable of sort action. N_O is a countably infinite set of *object constant symbols* and N_A a countably infinite set of *action function symbols* with arguments of sort object. We denote the set of all ground terms (also called *standard names*) of sort object by $\mathcal{N}_O$, and those of sort action by $\mathcal{N}_A$.

Formulas are built using *fluent* predicate symbols (predicates that may vary as the result of actions) of any arity and equality, using the usual logical connectives and quantifiers. In addition we have two modalities for referring to future situations, where $\Box\phi$ says that ϕ holds after any sequence of actions, and $[t]\phi$ means that ϕ holds after executing action t.

A formula without $\Box$ and $[\cdot]$ is called *fluent formula*, one without $\Box$ *bounded*, and one without free variables a *sentence*.

Semantics: Let $\mathcal{Z} := \mathcal{N}_A^*$ be the set of all finite action sequences (including the empty sequence $\langle\rangle$) and $\mathcal{P}_F$ the set of all *primitive formulas* $F(n_1, \ldots, n_k)$, where F is a k-ary fluent and the n_i are object standard names. A *world* w maps primitive formulas and situations to truth values: $w : \mathcal{P}_F \times \mathcal{Z} \to \{0, 1\}$.

The set of all worlds is denoted by $\mathcal{W}$.

Definition 1 (Truth of Formulas). *Given a world $w \in \mathcal{W}$ and a sentence ψ, we define $w \models \psi$ as $w, \langle\rangle \models \psi$, where for any $z \in \mathcal{Z}$:*

1. *$w, z \models F(n_1, \ldots, n_k)$ iff $w[F(n_1, \ldots, n_k), z] = 1$;*
2. *$w, z \models (n_1 = n_2)$ iff n_1 and n_2 are identical;*
3. *$w, z \models \psi_1 \wedge \psi_2$ iff $w, z \models \psi_1$ and $w, z \models \psi_2$;*
4. *$w, z \models \neg\psi$ iff $w, z \not\models \psi$;*
5. *$w, z \models \forall x.\phi$ iff $w, z \models \phi_n^x$ for all $n \in \mathcal{N}_x$;*
6. *$w, z \models \Box\psi$ iff $w, z \cdot z' \models \psi$ for all $z' \in \mathcal{Z}$;*
7. *$w, z \models [t]\psi$ iff $w, z \cdot t \models \psi$.*

Above, $\mathcal{N}_x$ refers to the set of all standard names of the same sort as x. We moreover use ϕ_n^x to denote the result of simultaneously replacing all free occurrences of x in ϕ by n. Note that by rule 2 above, the unique names assumption for actions and object constants is part of our semantics. We understand $\vee, \exists, \supset, \equiv$ and $\top$ and $\bot$ as the usual abbreviations.

Definition 2 (Basic Action Theory). *A basic action theory (BAT) $\mathcal{D} = \mathcal{D}_0 \cup \mathcal{D}_{post}$ is a set of axioms consisting of:*

1. *$\mathcal{D}_0$, the initial theory, a finite set of fluent sentences describing the initial state of the world;*

2. $\mathcal{D}_{post}$ *a finite set of* successor state axioms *(SSAs), one for each fluent relevant to the application domain, incorporating Reiter's* [43] *solution to the frame problem, for encoding action effects. They have the form*[1]

$$\Box[a]F(\boldsymbol{x}) \equiv \gamma_F^+ \vee F(\boldsymbol{x}) \wedge \neg\gamma_F^-, \tag{1}$$

where the positive (negative) *effect condition* γ_F^+ (γ_F^-) *is a fluent formula with free variables a and $\boldsymbol{x}$.*

Normally, BATs also feature action precondition axioms, which we ignore here for simplicity.

Example 1. For the aforementioned dish robot we may have

$$\mathcal{D}_0 = \{\neg\exists x, yDirty(x,y),\ \neg\exists x OnRobot(x)\}.$$

Also, let $\mathcal{D}_{post}$ consist of the following SSAs (we abstract from the robot's location for simplicity):

$$\Box[a]Dirty(x,y) \quad \equiv a = newdish(x,y) \vee Dirty(x,y) \wedge a \neq load(x,y)$$
$$\Box[a]OnRobot(x) \quad \equiv \exists y.\, a = load(x,y) \vee OnRobot(x) \wedge a \neq unload(x).$$

2.2 GOLOG Programs and Verification

The primitive actions defined in the BAT can be used as basic building blocks for GOLOG programs as follows.

Definition 3 (GOLOG Program). *A program δ is built according to the following grammar:*

$$\delta ::= t \mid \psi? \mid \delta;\delta \mid \delta|\delta \mid \delta^* \mid \delta\|\delta.$$

A program can thus be an action t, a test $\psi?$ for some fluent formula ψ, or constructed from subprograms by means of sequence $\delta;\delta$, *non-deterministic choice* $\delta|\delta$, *non-deterministic iteration δ^*, and* interleaving $\delta\|\delta$. *We treat **if**, **while**, **loop** and the finitary non-deterministic choice of argument ("pick") as abbreviations:*

$$\textbf{\textit{if }} \phi \textbf{\textit{ then }} \delta_1 \textbf{\textit{ else }} \delta_2 \textbf{\textit{ endIf }} \overset{def}{=} [\phi?;\delta_1] \mid [\neg\phi?;\delta_2]$$

$$\textbf{\textit{while }} \phi \textbf{\textit{ do }} \delta \textbf{\textit{ endWhile }} \overset{def}{=} [\phi?;\delta]^*; \neg\phi?$$

$$\textbf{\textit{loop }} \delta \overset{def}{=} \textbf{\textit{while }} \top \textbf{\textit{ do }} \delta \textbf{\textit{ endWhile}}$$

$$\pi x{:}\{c_1,\dots,c_k\}.\ \delta \overset{def}{=} \delta_{c_1}^x \mid \cdots \mid \delta_{c_k}^x$$

---

[1] Free variables are understood as universally quantified from the outside; $\Box$ has lower syntactic precedence than the logical connectives, $[t]$ has higher precedence than the logical connectives. So $\Box[a]F(\boldsymbol{x}) \equiv \gamma_F$ abbreviates $\forall a, \boldsymbol{x}.\Box(([a]F(\boldsymbol{x})) \equiv \gamma_F)$.

An example for a program was presented in the introduction. Exogenous actions can be incorporated by having a loop that, in each cycle, executes one such action with non-deterministically chosen arguments

$$\delta_{exo} = \mathbf{loop} \ \pi x{:}Dish \ \pi y{:}Room \ newdish(x,y)$$

run concurrently with the actual control program δ_{ctl}, i.e. in the verification one analyzes the behaviour of $\delta_{ctl} \parallel \delta_{exo}$.

Following [16] we define the transition semantics of programs meta-theoretically. A *configuration* $\langle z, \rho \rangle$ consists of an action sequence $z \in \mathcal{Z}$ (that has already been performed) and a program ρ (that remains to be executed). Execution of a program in a world $w \in \mathcal{W}$ yields a *transition relation* $\xrightarrow{w}$ *among configurations* that is defined inductively over program expressions:

1. $\langle z, t \rangle \xrightarrow{w} \langle z \cdot t, \langle \rangle \rangle$;
2. $\langle z, \delta_1; \delta_2 \rangle \xrightarrow{w} \langle z \cdot t, \gamma; \delta_2 \rangle$, if $\langle z, \delta_1 \rangle \xrightarrow{w} \langle z \cdot t, \gamma \rangle$;
3. $\langle z, \delta_1; \delta_2 \rangle \xrightarrow{w} \langle z \cdot t, \delta' \rangle$, if $\langle z, \delta_1 \rangle \in \mathcal{F}^w$ and $\langle z, \delta_2 \rangle \xrightarrow{w} \langle z \cdot t, \delta' \rangle$;
4. $\langle z, \delta_1 | \delta_2 \rangle \xrightarrow{w} \langle z \cdot t, \delta' \rangle$, if $\langle z, \delta_1 \rangle \xrightarrow{w} \langle z \cdot t, \delta' \rangle$ or $\langle z, \delta_2 \rangle \xrightarrow{w} \langle z \cdot t, \delta' \rangle$;
5. $\langle z, \delta^* \rangle \xrightarrow{w} \langle z \cdot t, \gamma; \delta^* \rangle$, if $\langle z, \delta \rangle \xrightarrow{w} \langle z \cdot t, \gamma \rangle$;
6. $\langle z, \delta_1 \| \delta_2 \rangle \xrightarrow{w} \langle z \cdot t, \delta' \| \delta_2 \rangle$, if $\langle z, \delta_1 \rangle \xrightarrow{w} \langle z \cdot t, \delta' \rangle$;
7. $\langle z, \delta_1 \| \delta_2 \rangle \xrightarrow{w} \langle z \cdot t, \delta_1 \| \delta' \rangle$, if $\langle z, \delta_2 \rangle \xrightarrow{w} \langle z \cdot t, \delta' \rangle$.

For the set of final configurations $\mathcal{F}^w$ wrt. a world w we have:

1. $\langle z, \langle \rangle \rangle \in \mathcal{F}^w$;
2. $\langle z, \psi? \rangle \in \mathcal{F}^w$, if $w, z \models \psi$;
3. $\langle z, \delta_1; \delta_2 \rangle \in \mathcal{F}^w$, if $\langle z, \delta_1 \rangle \in \mathcal{F}^w$ and $\langle z, \delta_2 \rangle \in \mathcal{F}^w$;
4. $\langle z, \delta_1 | \delta_2 \rangle \in \mathcal{F}^w$, if $\langle z, \delta_1 \rangle \in \mathcal{F}^w$ or $\langle z, \delta_2 \rangle \in \mathcal{F}^w$;
5. $\langle z, \delta^* \rangle \in \mathcal{F}^w$;
6. $\langle z, \delta_1 \| \delta_2 \rangle \in \mathcal{F}^w$, if $\langle z, \delta_1 \rangle \in \mathcal{F}^w$ and $\langle z, \delta_2 \rangle \in \mathcal{F}^w$.

Definition 4 (Transition System of a Program). *Let δ be a program and $w \in \mathcal{W}$. Execution of δ in w yields the transition system wrt. w, δ given by $\mathsf{T}_\delta^w = (\mathsf{S}, \rightarrow)$, where the set of states $\mathsf{S} = \{ \langle z', \delta' \rangle \mid \langle \langle \rangle, \delta \rangle \xrightarrow{w}{}^* \langle z', \delta' \rangle \} \cup \{\mathfrak{e}, \mathfrak{f}\}$ consists of configurations reachable from $\langle \langle \rangle, \delta \rangle$ plus two special "sink" states for program termination and failure, and $\rightarrow$ is a transition relation such that $\mathsf{s} \rightarrow \mathsf{s}'$ iff one of the following holds:*

1. $\mathsf{s} \xrightarrow{w} \mathsf{s}'$;
2. $\mathsf{s}' = \mathfrak{e}$ *and* $(\mathsf{s} \in \mathcal{F}^w$ *or* $\mathsf{s} = \mathfrak{e})$;
3. $\mathsf{s}' = \mathfrak{f}$ *and* $($*no* s'' *with* $\mathsf{s} \xrightarrow{w} \mathsf{s}''$ *and* $\mathsf{s} \notin \mathcal{F}^w$ *or* $\mathsf{s} = \mathfrak{f})$.

Definition 5 (Temporal Properties of Programs). *The syntax for temporal formulas is the same as for propositional CTL*, but in place of propositions we allow fluent sentences ψ in Boolean combinations with the special symbols Succ and Fail (for program termination and failure, respectively):*

$$\Psi ::= \psi \mid Succ \mid Fail \mid \neg\Psi \mid \Psi \wedge \Psi \mid \boldsymbol{E\Phi} \tag{2}$$

$$\Phi ::= \Psi \mid \neg\Phi \mid \Phi \wedge \Phi \mid \boldsymbol{X}\Psi \mid \Psi \boldsymbol{U} \Psi \tag{3}$$

Formulas according to (2) are temporal state formulas, *and according to (3)* temporal path formulas. *We use the usual abbreviations* $\boldsymbol{A\Phi}$ *(Φ holds on* all *paths) for* $\neg\boldsymbol{E}\neg\Phi$, $\boldsymbol{F\Phi}$ *(eventually Φ) for* $\top \boldsymbol{U} \Phi$ *and* $\boldsymbol{G\Phi}$ *(globally Φ) for* $\neg\boldsymbol{F}\neg\Phi$.

Now let Ψ be a temporal state formula, T^w_δ the transition system wrt. w, δ, and $\mathsf{s} \in \mathsf{S}$. For an infinite path

$$\pi = \mathsf{s}_0 \to \mathsf{s}_1 \to \mathsf{s}_2 \to \cdots$$

in T^w_δ, we denote for any $j \geq 0$ the state s_j by $\pi[j]$ and the suffix $\mathsf{s}_j \to \mathsf{s}_{j+1} \to \cdots$ by $\pi[j..]$. $\mathsf{Paths}(\mathsf{s}, \mathsf{T}^w_\delta)$ denotes the *set of all paths* starting in s. Truth of Ψ in $\mathsf{T}^w_\delta, \mathsf{s}$ (written $\mathsf{T}^w_\delta, \mathsf{s} \models \Psi$) is given by:

- $\mathsf{T}^w_\delta, \mathsf{s} \models \psi$ iff $\mathsf{s} = \langle z', \delta' \rangle$ and $w, z' \models \psi$;
- $\mathsf{T}^w_\delta, \mathsf{s} \models Succ$ iff $\mathsf{s} = \mathfrak{e}$;
- $\mathsf{T}^w_\delta, \mathsf{s} \models Fail$ iff $\mathsf{s} = \mathfrak{f}$;
- $\mathsf{T}^w_\delta, \mathsf{s} \models \neg\Psi$ iff $\mathsf{T}^w_\delta, \mathsf{s} \not\models \Psi$;
- $\mathsf{T}^w_\delta, \mathsf{s} \models \Psi_1 \wedge \Psi_2$ iff $\mathsf{T}^w_\delta, \mathsf{s} \models \Psi_1$ and $\mathsf{T}^w_\delta, \mathsf{s} \models \Psi_2$;
- $\mathsf{T}^w_\delta, \mathsf{s} \models \boldsymbol{E\Phi}$ iff $\pi \in \mathsf{Paths}(\mathsf{s}, \mathsf{T}^w_\delta)$ with $\mathsf{T}^w_\delta, \pi \models \Phi$.

Let Φ be a temporal path formula, T^w_δ and s as above, and $\pi \in \mathsf{Paths}(\mathsf{s}, \mathsf{T}^w_\delta)$. Truth of Φ in T^w_δ, π (written $\mathsf{T}^w_\delta, \pi \models \Phi$) is given by:

- $\mathsf{T}^w_\delta, \pi \models \Psi$ iff $\mathsf{T}^w_\delta, \pi[0] \models \Psi$;
- $\mathsf{T}^w_\delta, \pi \models \neg\Phi$ iff $\mathsf{T}^w_\delta, \pi \not\models \Phi$;
- $\mathsf{T}^w_\delta, \pi \models \Phi_1 \wedge \Phi_2$ iff $\mathsf{T}^w_\delta, \pi \models \Phi_1$ and $\mathsf{T}^w_\delta, \pi \models \Phi_2$;
- $\mathsf{T}^w_\delta, \pi \models \boldsymbol{X\Phi}$ iff $\mathsf{T}^w_\delta, \pi[1..] \models \Phi$;
- $\mathsf{T}^w_\delta, \pi \models \Phi_1 \boldsymbol{U} \Phi_2$ iff $\exists k \geq 0 : \mathsf{T}^w_\delta, \pi[k..] \models \Phi_2$
  and $\forall j, 0 \leq j < k : \mathsf{T}^w_\delta, \pi[j..] \models \Phi_1$.

The sink states $\mathfrak{e}$, $\mathfrak{f}$ and the corresponding special symbols *Succ*, *Fail* allow us to treat terminating programs simply as special cases of non-terminating ones, where once a program terminates successfully or due to failure, the program will indefinitely loop through $\mathfrak{e}$ or $\mathfrak{f}$, respectively. Furthermore, we can analyze the termination behaviour of a program simply by verifying appropriate temporal properties, e.g. $\boldsymbol{AFSucc}$ (the program is guaranteed to terminate) or $\boldsymbol{EFFail}$ (the program may fail).

Example 2. Some temporal properties for the (non-terminating) dish robot are:

$\boldsymbol{EF}Dirty(d_1, room)$ "Is it possible that d_1 ends up dirty in *room*?"
$\boldsymbol{AG}\neg\exists x\, Dirty(d_1, x)$ "Will d_1 always remain cleaned?"
$\boldsymbol{EG}\exists x, y\, Dirty(x, y)$ "Will there forever be a dirty dish in some room?"

In the following, we will use a restricted subset of temporal formulas that resembles CTL without nesting of path quantifiers (but still with fluent sentences instead of propositions):

$$\varphi ::= \psi \mid Succ \mid Fail \mid \neg\varphi \mid \varphi \wedge \varphi \tag{4}$$

$$\Psi ::= \varphi \mid \neg\Psi \mid \Psi \wedge \Psi \mid EX\varphi \mid EG\varphi \mid E(\varphi \, U \, \varphi) \tag{5}$$

CTL formulas according to (5) are obviously a subset of temporal state formulas. Note that the properties from Example 2 are all part of this subset using $AG\varphi \equiv \neg EF\neg\varphi$.

Definition 6 (Verification Problem). *A temporal state formula Ψ is valid in a program δ for a BAT $\mathcal{D}$ iff for all worlds $w \in \mathcal{W}$ with $w \models \mathcal{D}$ it holds that $\mathsf{T}_\delta^w, \langle\langle\rangle, \delta\rangle \models \Psi$.*

3 Verification by Fixpoint Computation

The first approach [13,16,17] is inspired by classical symbolic model checking [41] in the sense that a systematic exploration of the state space is made using a fixpoint computation of preimages of state sets, however now involving first-order reasoning about actions. For this purpose, an $\mathcal{ES}$ variant [31] of Reiter's [43] regression operator is employed, which replaces fluent atoms in the scope of a $[t]$ by the right-hand side of the corresponding SSA:

Definition 7 (Regression). *Let ψ be a bounded formula. We define $\mathcal{R}[\psi] = \mathcal{R}[\langle\rangle, \psi]$, where for any $z \in \mathcal{Z}$,*

1. $\mathcal{R}[\langle\rangle, F(t)] = F(t)$ *and* $\mathcal{R}[z \cdot t, F(t)] = (\gamma_F^+ \vee F(x) \wedge \neg\gamma_F^-)_t^x{}_t^a$;
2. $\mathcal{R}[z, (t_1 = t_2)] = (t_1 = t_2)$;
3. $\mathcal{R}[z, \psi_1 \wedge \psi_2] = \mathcal{R}[z, \psi_1] \wedge \mathcal{R}[z, \psi_2]$;
4. $\mathcal{R}[z, \neg\psi] = \neg\mathcal{R}[z, \psi]$;
5. $\mathcal{R}[z, \forall x\psi] = \forall x\mathcal{R}[z, \psi]$;
6. $\mathcal{R}[z, [t]\psi] = \mathcal{R}[z \cdot t, \psi]$.

Theorem 1. *If $\mathcal{D}$ is a BAT and ψ a bounded formula, then $\mathcal{D} \models \Box(\psi \equiv \mathcal{R}[\psi])$.*

$\mathcal{R}[\psi]$ is hence equivalent to the original ψ wrt. $\mathcal{D}$, but contains no $[\cdot]$ and only talks about the initial (current) situation.

In addition to regression, another ingredient for the verification method are *characteristic graphs*, which are used to encode the reachable subprogram configurations. For any program δ, the graph $\mathcal{G}_\delta = \langle V, E, v_0 \rangle$ consists of a set of vertices V, each of which corresponds to one reachable subprogram δ', or ϵ or $\mathfrak{f}$. The initial node v_0 corresponds to the overall program δ. Edges E are labelled with tuples t/ψ, where t is an action term and ψ a fluent formula (omitted when $\top$) denoting the condition required to take that transition. We omit the formal definition; the interested reader is referred to [13]. As an example, Fig. 1 shows

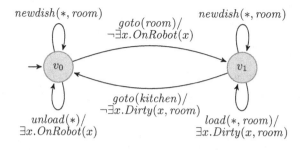

Fig. 1. Characteristic graph for the dish-cleaning robot

the graph for the control program of the dish robot as presented in the introduction, assuming that the *Room* domain only contains a single *room*. The asterisks in edge annotations such as *newdish*(∗, *room*) indicates that there is one such edge instance for every element in the *Dish* domain. (Graphs for programs where the *Room* domain is larger contain one "copy" of node v_1 for each room, with similar connections to v_0 and itself.) The algorithm uses a set of *labels* $\langle v, \psi \rangle$, one for each node $v \in V$, where ψ is a fluent formula. Intuitively, if $v = \delta'$, then $\langle v, \psi \rangle$ represents all combinations of worlds w and configurations $\langle z, \delta \rangle$ with $w, z \models \psi$. Below is the procedure for formulas of form $\textbf{\textit{EG}}\phi$, similar ones exist for $\textbf{\textit{EX}}$ and $\textbf{\textit{EU}}$ [13]:

Procedure 1. CHECKEG$[\delta, \phi]$

1: $L' := \text{LABEL}[\mathcal{G}_\delta, \bot]; \quad L := \text{LABEL}[\mathcal{G}_\delta, \phi];$
2: **while** $L \not\equiv L'$ **do**
3: $\quad \{ L' := L; \quad L := L' \text{ AND } \text{PRE}[\mathcal{G}_\delta, L'] \};$
4: **return** INITLABEL$[\mathcal{G}_\delta, L]$

That is to say first the "old" labelling L' is initialized to label every node with $\bot$ and the "current" labelling L marks every vertex with ϕ. While L and L' are not equivalent ($\psi \equiv \psi'$ for every $\langle v, \psi \rangle \in L$, $\langle v, \psi' \rangle \in L'$), L is conjoined according to

$$L_1 \text{ AND } L_2 \overset{def}{=} \{\langle v, \psi_1 \wedge \psi_2 \rangle \mid \langle v, \psi_1 \rangle \in L_1, \langle v, \psi_2 \rangle \in L_2\}$$

with its pre-image

$$\text{PRE}[\langle V, E, v_0 \rangle, L] \overset{def}{=} \{\langle v, \text{PRE}[v, L] \rangle \mid v \in V\}$$

where $\text{PRE}[v, L]$ stands for

$$\bigvee \{\mathcal{R}[\phi \wedge [t]\psi] \mid v \xrightarrow{t/\phi} v' \in E, \langle v', \psi \rangle \in L\}.$$

Note the use of regression to eliminate the action term t. Once the label set has converged, the method returns $\text{INITLABEL}[\mathcal{G}_\delta, L]$, the label formula at the initial node v_0. The algorithm is sound as follows:

Theorem 2. *Let $\mathcal{D}$ be a BAT, δ a program, and ϕ a fluent formula. If the procedure terminates, then $\psi := \text{CHECKEG}[\delta, \phi]$ is a fluent formula and $\textbf{EG}\phi$ is valid in δ for $\mathcal{D}$ iff $\mathcal{D}_0 \models \psi$.*

Example 3. Suppose we want to verify whether a run of the program for the dish robot presented in the introduction (including possible exogenous actions) is possible where there is always some dirty dish x in some room y, i.e. whether it satisfies property $\textbf{EG}\exists x, y\, Dirty(x, y)$. We hence call Procedure 1 with $\delta = \delta_{ctl} \| \delta_{exo}$ being the overall program and the axiom $\phi = \exists x, y\, Dirty(x, y)$. It starts with the following label set L:

$$L_0 = \{\langle v_0, \exists x, y\, Dirty(x, y)\rangle, \; \langle v_1, \exists x, y\, Dirty(x, y)\rangle\}.$$

For determining the pre-image for a node in the characteristic graph, each of its outgoing edges has to be considered. Recall that we have multiple instances of each $newdish(d_i, room)$ with different dishes d_i. One of the disjuncts of $\text{PRE}[v_0, L_0]$ thus is

$$\mathcal{R}[[newdish(d_1, room)]\exists x, y\, Dirty(x, y)]$$

which (using unique names of actions) reduces to

$$\exists x, y.\; x = d_1 \vee Dirty(x, y).$$

Using similar reductions for the other edges we get $\text{PRE}[v_0, L_0]$ and $\text{PRE}[v_1, L_0]$ both being equivalent to

$$\exists x, y.\; x = d_1 \vee x = d_2 \vee Dirty(x, y)$$

if there are two dishes in total. Then $L_1 = (L_0 \text{ AND } \text{PRE}[\mathcal{G}_\delta, L_0])$, which reduces to

$$\{\langle v_0, \exists x, y\, Dirty(x, y)\rangle, \langle v_1, \exists x, y\, Dirty(x, y)\rangle\}$$

Hence $L_0 \equiv L_1$, i.e. the algorithm terminates and returns $\exists x, y\, Dirty(x, y)$. Thus, there is a run where there is always some dirty dish just in case there is some dirty dish somewhere initially. Intuitively, this is correct because $G\phi$ means that ϕ persists to hold during the *entire* run, including the initial situation. Therefore, only if a dish is dirty initially it may happen that never all of them get cleaned. All we have to do now is to check whether $\mathcal{D}_0 \models \exists x, y\, Dirty(x, y)\rangle$, which is not the case according to the $\mathcal{D}_0$ from Example 1.

Naturally, the next interesting question is under what circumstances it can be guaranteed that the procedure terminates as it did here, thus rendering the verification problem decidable. As first-order logic is already undecidable, the first step is to ensure that the basic, first-order reasoning tasks of checking the equivalence of label formulas and whether the output of CHECKEG$[\delta, \phi]$ is entailed by the BAT's initial theory can be decided. We do so restricting the base logic to FO^2, the two-variable fragment of FOL. We hence require that

- fluents have at most two arguments;
- $\mathcal{D}_0$, tests ϕ? in the program δ as well as axioms ψ in temporal properties Ψ are formulas where x and y are the only variable symbols;
- the instantiations of γ_F^+ and γ_F^- by any ground action from δ are formulas where x and y are the only variable symbols.

Note that in this paper we assume all actions in a program are ground (our definition of GOLOG does not include the general nondeterministic choice of argument π, but a finitary version where each π only ranges over some finite domain). It can be shown [51] by means of a reduction of the Halting Problem for Turing Machines that otherwise, the GOLOG Verification Problem remains undecidable, even under all other restrictions we discuss here. The BAT from Example 1 and the properties from Example 2 all fulfill these requirements.

While there are other decidable fragments of first-order predicate calculus, note that not all are equally suited for our purposes. In particular, decidable quantifier prefix fragments for instance have the disadvantage that they are not closed under regression, i.e. the regression of such a formula may not be expressable by a formula in the same fragment. This is the case for FO^2 on the other hand, provided that the regression operator is slightly modified to rename variables where needed [26]. Moreover, note that the FO^2 fragment subsumes many Description Logics, so its choice paves the way for a method where representation and reasoning is handled entirely within a DL.

Now that we restricted the base logic and the class of programs we consider, the last restriction is on the range of action effects we allow. Again, it can be shown that without any such restriction, verification remains undecidable [51]. A popular subclass of action theories (originally studied within the context of progression) is that where actions only have local effects [37]:

Definition 8 (Local-Effect). *An SSA is local-effect if the conditions γ_F^+ and γ_F^- are disjunctions of formulas $\exists z[a = A(y) \wedge \phi]$, where A is an action function, y contains x, z are the remaining variables of y, and ϕ is a fluent formula with free variables y. A BAT is local-effect if all its SSAs are.*

Intuitively, an action $A(c)$ is local-effect if it only changes fluents $F(d)$ all of whose arguments d are among the action's parameters c, i.e. all objects affected have to be mentioned in the action. Note that the SSAs in Example 1 are local-effect. In [17] it was shown that under these assumptions, the verification procedure becomes complete:

Theorem 3. *The procedure* CHECKEG$[\delta, \phi]$ *terminates if the BAT is local-effect and FO^2 is used as base logic.*

There are similar theorems for the other cases **EX** and **EU**.

4 Verification by Abstraction

The second method we consider is verification by abstraction. Zarrieß and Claßen [52] show that for a GOLOG program δ with a local-effect BAT $\mathcal{D}$, the verification of a temporal formula Ψ can be reduced to classical model checking by constructing a finite, bisimilar abstraction of the original infinite transition system induced by δ and $\mathcal{D}$ wrt. Ψ. This is achieved by identifying finitely many equivalence classes for worlds, whose computation reduces to consistency checks in the underlying decidable base logic FO^2.

4.1 Regression with Sets of Effects

The first ingredient is the observation that by unique names of actions, the instantiation of a local-effect SSA on a ground action $t = A(c)$ can be significantly simplified [38], as any $\gamma_{F\,t}^{+a}$ or $\gamma_{F\,t}^{-a}$ is equivalent to

$$x = c_1 \wedge \phi_1 \vee \cdots \vee x = c_n \wedge \phi_n$$

where c_i is a vector of names contained in c, and ϕ_i is a quantifier-free sentence. We use the notation $(c_i, \phi_i) \in \gamma_{F\,t}^{+a}$ and $(c_i, \phi_i) \in \gamma_{F\,t}^{-a}$ to express that there is a disjunct of the form $x = c_i \wedge \phi_i$ in $\gamma_{F\,t}^{+a}$ or $\gamma_{F\,t}^{-a}$, respectively. Let L be the set of all positive and negative ground fluent literals:

$$\mathsf{L} = \{F(c), \neg F(c) \mid t \in \delta : (c, \phi) \in \gamma_{F\,t}^{+a} \text{ or } (c, \phi) \in \gamma_{F\,t}^{-a}\}$$

One can then define a variant of regression wrt. an effect set, given a consistent set of fluent literals and a fluent sentence.

Definition 9 (Regression with Effects). *If $F(v)$ is a fluent atom where v is a vector of variables or constants, and $E \subseteq \mathsf{L}$ a consistent set of fluent literals, then the* regression *of $F(v)$ through E, written as $\mathcal{R}[E, F(v)]$ is given by:*

$$\mathcal{R}[E, F(v)] = \left(F(v) \wedge \bigwedge_{\neg F(c) \in E} (v \neq c) \right) \vee \bigvee_{F(c) \in E} (v = c)$$

For any fluent sentence α, $\mathcal{R}[E, \alpha]$ denotes the result of replacing any occurrence of a fluent $F(v)$ by $\mathcal{R}[E, F(v)]$.

Example 4. For the dish robot, the ground actions to consider are all instances of $newdish(*, room)$, $unload(*)$, $load(*, room)$, $goto(room)$, and $goto(kitchen)$. For $t = newdish(d_1, room)$ we have:

$$\gamma_{Dirty\,t}^{+\,a} = (x = d_1 \wedge y = room)$$
$$\gamma_{Dirty\,t}^{-\,a} = \bot$$

The literals that are possible effects of the ground actions hence are:

$$\mathsf{L} = \{\ (\neg)Dirty(d_1, room),\ (\neg)Dirty(d_2, room),$$
$$(\neg)OnRobot(d_1),\ (\neg)OnRobot(d_2)\ \}$$

Let $E = \{Dirty(d_1, room)\}$. Then for example,

$$\mathcal{R}[E, \neg\exists x, y\ Dirty(x, y)] = \exists x, y\ \neg\big(Dirty(x, y) \vee (x = d_1 \wedge y = room)\big)$$

Clearly, the regression result $\mathcal{R}[E, \alpha]$ is again a fluent sentence. We note that an iterated application of the regression operator can be reduced to an application of the operator for a single set of fluent literals. For a set $E \subseteq \mathsf{L}$ we define $\neg E := \{\neg l \mid l \in E\}$ (modulo double negation). For two consistent subsets E, E' of L and a sentence α it holds that

$$\mathcal{R}[E, \mathcal{R}[E', \alpha]] \equiv \mathcal{R}[(E \setminus \neg E' \cup E'), \alpha]. \tag{6}$$

The idea is that any such set of literals then represents a class of action sequences that all bring about the same set of accumulated effects.

4.2   Finite Abstraction

To construct the abstract transition system, we identify the *context* $\mathcal{C}(\mathcal{D}, \delta)$, the set of all relevant fluent sentences

- in the initial theory $\mathcal{D}_0$;
- in all tests ψ? occurring in the program δ;
- all ϕ with $(\boldsymbol{c}, \phi) \in \gamma_{Ft}^{+a}$ or $(\boldsymbol{c}, \phi) \in \gamma_{Ft}^{-a}$ for some t in δ;
- all $F(\boldsymbol{c})$ with $(\boldsymbol{c}, \phi) \in \gamma_{Ft}^{+a}$ or $(\boldsymbol{c}, \phi) \in \gamma_{Ft}^{-a}$ for some t in δ;
- all axioms occurring in temporal properties.

Furthermore, $\mathcal{C}(\mathcal{D}, \delta)$ is assumed to be closed under negation. Intuitively, worlds satisfying the same maximal consistent set of context formulas are considered to be members of the same equivalence class, called a *type*. To incorporate actions, also the regressions of these formulas wrt. all consistent $E' \subseteq \mathsf{L}$ have to be taken into account. States belonging to the same equivalence class can be shown to simulate one another, i.e. they are indistinguishable through temporal properties. This bisimulation justifies the construction of the corresponding quotient system as an abstraction, which can be obtained as follows. Abstract states are tuples $\langle v,\ \Gamma,\ E \rangle$, where v is a node of the characteristic graph of δ, Γ is a consistent set of (regressed) context formulas representing worlds, and $E \subseteq \mathsf{L}$ is a consistent set of accumulated effects representing situations. There is a transition $\langle v,\ \Gamma,\ E \rangle \xrightarrow{t} \langle v',\ \Gamma,\ E' \rangle$ between abstract states in case

1. there is an edge $v \xrightarrow{t/\psi} v'$ in δ's characteristic graph,
2. $\Gamma \models \mathcal{R}[E, \psi]$, and
3. $E' = (E \setminus \neg E^* \cup E^*)$, where $E^* = \mathcal{E}(\Gamma, E, t)$.

Above, $\mathcal{E}(\Gamma, E, t)$ denotes the set of effects induced by action t wrt. the type given by Γ and E:

$$\mathcal{E}(\Gamma, E, t) = \{ \ F(\boldsymbol{c}) \mid (\boldsymbol{c}, \phi) \in \gamma_{Ft}^{+a}, \ \Gamma \models \mathcal{R}[E, \phi] \} \cup$$
$$\{ \neg F(\boldsymbol{c}) \mid (\boldsymbol{c}, \phi) \in \gamma_{Ft}^{-a}, \ \Gamma \models \mathcal{R}[E, F(\boldsymbol{c}) \wedge \phi] \}$$

Example 5. In our running example, the relevant fluent sentences are

$(\neg)\exists x, y\, Dirty(x, y)$, $(\neg)\exists x\, OnRobot(x)$, $(\neg)\exists x\, Dirty(x, room)$,
$(\neg)Dirty(d_1, room)$, $(\neg)Dirty(d_2, room)$,
$(\neg)OnRobot(d_1)$, $(\neg)OnRobot(d_2)$, $(\neg)\exists x\, Dirty(d_1, x)$

One possible type (in fact the only one consistent with the initial theory of the BAT from Example 1) is given by

$$\Gamma_0 = \{\neg\exists x, y\, Dirty(x, y),\ \neg\exists x\, OnRobot(x)\}$$

One abstract state is

$$\mathsf{s}_1 = \langle v_0, \Gamma_0, \{Dirty(d_1, room)\}\rangle,$$

which intuitively represents any configuration where the overall program $\delta = \delta_{ctl} \parallel \delta_{exo}$ remains to be executed, whose initial situation was as described by $\mathcal{D}_0$, and where a sequence of actions has been performed that caused $Dirty(d_1, room)$ to come about (e.g. a single $newdish(d_1, room)$, but also any sequence where other dirty dishes except d_1 have been removed already).

The characteristic graph depicted in Fig. 1 shows three kinds of outgoing edges for v_0, all of which correspond to potential transitions from s_1. $newdish(d_i, room)$ edges have no transition condition. Their effects are given by

$$\mathcal{E}(\Gamma_0, \{Dirty(d_1, room)\}, newdish(d_i, room)) = \{Dirty(d_i, room)\},$$

hence we have $\mathsf{s}_1 \to \mathsf{s}_i$ for

$$\mathsf{s}_i = \langle v_0, \Gamma_0, \{Dirty(d_1, room), Dirty(d_i, room)\}\rangle$$

(i.e. for $i = 1$ we remain in s_1). For any $unload(d_i)$ edge, condition $\exists x\, OnRobot(x)$ regresses to

$$\mathcal{R}[\{Dirty(d_1, room)\}, \exists x\, OnRobot(x)] = \exists x\, OnRobot(x)$$

(adding a dirty dish has no effect on whether the robot is holding something). Since $\Gamma_0 \not\models \exists x\, OnRobot(x)$, there is no $unload(d_i)$ transition from s_1. Finally, for the $goto(room)$ edge, the transition condition similarly regresses to

$$\mathcal{R}[\{Dirty(d_1, room)\}, \neg\exists x\, OnRobot(x)] = \neg\exists x\, OnRobot(x).$$

As $\Gamma_0 \models \neg\exists x\, OnRobot(x)$, there is a transition $\mathsf{s}_1 \to \mathsf{s}_1'$ with

$$\mathsf{s}_1' = \langle v_1, \Gamma_0, \{Dirty(d_1, room)\}\rangle.$$

Recall that in this simple encoding, *goto* actions do not have any effect, i.e.

$$\mathcal{E}(\Gamma_0, \{Dirty(d_1, room)\}, goto(room)) = \emptyset.$$

There are only finitely many nodes in the characteristic graph, relevant fluent sentences, and ground fluent literals as effects. The abstract transition system is hence finite, and can be effectively computed due to the fact that the necessary consistency and entailment checks in FO^2 are all decidable.

Finally, we can replace every relevant fluent sentence with a propositional atom (both in abstract states and temporal properties) and then call a propositional CTL model checker.

The complexity of this decision procedure is mainly determined by the complexity of consistency checks in FO^2, which we have to do for exponentially large knowledge bases. Knowledge base consistency in FO^2 is NEXPTIME-complete [25], so determining a single type can be done in N2EXPTIME. It turns out that CO-N2EXPTIME is an upper bound for the overall complexity.

Theorem 4. *The verification problem is decidable for a temporal state formula Ψ, a program δ over ground actions and a local-effect BAT $\mathcal{D}$ in* CO-N2EXPTIME.

5 Golog Programs over Description Logic Actions

Motivated by the idea of obtaining a decidable yet expressive fragment of the Situation Calculus, a first *DL-based action formalism* was introduced by Baader and his colleagues in [5]. Next, we briefly review some of the basic definitions of a simple formalism that we have used in [50,54] to analyze the complexity of the verification problem in a DL-based setting. It is a bit different from the one in [5] but adopts its main ideas.

The expressive DL $\mathcal{ALCQIO}$, which can be viewed as a fragment of the *two variable fragment of first-order logic with counting*, is the underlying logic. It is used for representing an incomplete initial situation (ABox part of the KB), general domain knowledge (TBox) and for formulating pre-conditions and effect conditions of primitive actions by means of *action descriptions*.

The signature for describing complex concepts consists of pairwise disjoint sets of *concept names* N_C (unary predicates), *role names* N_R (binary predicates) and *individual names* N_I. Several constructors can be used to form *complex concepts* from $A \in N_C$, $s \in N_R \cup \{r^- \mid r \in N_R\}$ (a role name or the inverse thereof), $a \in N_I$ and $n \in \mathbb{N}$ as shown in the first two columns of Table 1. Fragments of $\mathcal{ALCQIO}$ are obtained by restricting the available constructors for building concepts. For example, the basic DL $\mathcal{ALC}$ is obtained by disallowing at-most and at-least restrictions (the letter $\mathcal{Q}$ in the name of the DL indicates that those restrictions are allowed), inverse roles ($\mathcal{I}$) and nominals ($\mathcal{O}$).

As usual, *axioms* are grouped in boxes. The *TBox* is a finite set of *concept inclusions* and the *ABox* a finite set of *concept and role assertions* as shown in the first two columns of Table 2.

Table 1. Syntax and semantics of roles and concepts

Name	Syntax	Semantics under $\mathcal{I} = (\Delta_{\mathcal{I}}, \cdot^{\mathcal{I}})$
Role name	r	$r^{\mathcal{I}}$
Inverse role	r^-	$\{(e,d) \mid (d,e) \in r^{\mathcal{I}}\}$
Concept name	A	$A^{\mathcal{I}}$
Top concept	$\top$	$\Delta_{\mathcal{I}}$
Negation	$\neg C$	$\Delta \setminus C^{\mathcal{I}}$
Conjunction	$C \sqcap D$	$C^{\mathcal{I}} \cap D^{\mathcal{I}}$
Disjunction	$C \sqcup D$	$C^{\mathcal{I}} \cup D^{\mathcal{I}}$
Existential restriction	$\exists s.C$	$\{d \mid \exists e.(d,e) \in s^{\mathcal{I}}, e \in C^{\mathcal{I}}\}$
Value restriction	$\forall s.C$	$\{d \mid (d,e) \in s^{\mathcal{I}} \text{ implies } e \in C^{\mathcal{I}}\}$
At-most restriction	$\leq n\, s.C$	$\{d \mid \sharp\{e \mid (d,e) \in s^{\mathcal{I}} \wedge e \in C^{\mathcal{I}}\} \leq n\}$
At-least restriction	$\geq n\, s.C$	$\{d \mid \sharp\{e \mid (d,e) \in s^{\mathcal{I}} \wedge e \in C^{\mathcal{I}}\} \geq n\}$
Nominal	$\{a\}$	$\{a^{\mathcal{I}}\}$

Table 2. Syntax and semantics of axioms

	Name	*Axiom* ϱ	$\mathcal{I} \models \varrho$, iff
TBox $\mathcal{T}$	Concept inclusion	$C \sqsubseteq D$	$C^{\mathcal{I}} \subseteq D^{\mathcal{I}}$
ABox $\mathcal{A}$	Concept assertion	$a : C$	$a^{\mathcal{I}} \in C^{\mathcal{I}}$
	Role assertion	$(a,b) : s$	$(a,b) \in s^{\mathcal{I}}$

The semantics is defined in terms of an *interpretation* $\mathcal{I} = (\Delta_{\mathcal{I}}, \cdot^{\mathcal{I}})$, where $\Delta_{\mathcal{I}}$ is the non-empty domain of $\mathcal{I}$, and $\cdot^{\mathcal{I}}$ a function that maps concept names to subsets of the domain, role names to binary relations, individual names to elements, and is extended to complex concepts as shown in Table 1. Satisfaction of an axiom in an interpretation is defined as shown in Table 2. An interpretation $\mathcal{I}$ is a *model* of an ABox $\mathcal{A}$, a TBox $\mathcal{T}$ or a KB $\mathcal{K}$ iff all axioms in $\mathcal{A}$, $\mathcal{T}$ or $\mathcal{K}$, respectively, are satisfied in $\mathcal{I}$.

Obviously, a DL like $\mathcal{ALCQIO}$ is too inexpressive to formulate basic action theories as a whole like the ones in Definition 2. An alternative approach would be to take an axiomatization in form of a BAT and a program formulated in $\mathcal{ES}$ and restrict the formulas used for domain specific knowledge in the initial theory, the successor state axioms, and tests in the program to be $\mathcal{ALCQIO}$-axioms. However, for the DL-based formalism in [5] and its variants and successors, a different approach was taken which we briefly review below.

The overall idea is not to axiomatise the meaning of actions using quantification, but introduce *action descriptions* meta-theoretically. The *syntax* is similar to planning languages like STRIPS or ADL: the domain designer explicitly provides a complete list of effects for each primitive action name. The *semantics* of an action is defined in terms of a transition relation between interpretations such that the frame assumption is respected. First, we define the syntax of an *effect*.

Definition 10 (Effect Description). *Let $\mathcal{L}$ be a sub-DL of $\mathcal{ALCQIO}$ and let $A \in \mathsf{N_C}$, $r \in \mathsf{N_R}$ and $a, b \in \mathsf{N_I}$ and φ an $\mathcal{L}$-axiom (or Boolean combination of axioms). An $\mathcal{L}$-effect description ($\mathcal{L}$-effect for short) has one of the following forms*

$$\varphi \triangleright \langle A(o) \rangle^+, \varphi \triangleright \langle r(o, o') \rangle^+ \text{ (called add-effect),}$$

$$\varphi \triangleright \langle A(o) \rangle^-, \varphi \triangleright \langle r(o, o') \rangle^- \text{ (called delete-effect),}$$

where φ is called effect condition. In case the effect condition φ is a tautology like for example $\top \sqsubseteq \top$, the effect is called unconditional *and is written without the effect condition.*

For a set of effects and an interpretation, the corresponding *updated interpretation* is defined in a straightforward way.

Definition 11 (Interpretation Update). *Let $\mathcal{I} = (\Delta_{\mathcal{I}}, \cdot^{\mathcal{I}})$ be an interpretation and E a set of unconditional effects. The* update *of $\mathcal{I}$ with E is an interpretation denoted by $\mathcal{I}^{\mathsf{E}} = (\Delta_{\mathcal{I}^{\mathsf{E}}}, \cdot^{\mathcal{I}^{\mathsf{E}}})$ and is defined as follows:*

$$\Delta_{\mathcal{I}^{\mathsf{E}}} := \Delta_{\mathcal{I}};$$

$$A^{\mathcal{I}^{\mathsf{E}}} := A^{\mathcal{I}} \setminus \{a^{\mathcal{I}} \mid \langle A(a) \rangle^- \in \mathsf{E}\} \cup \{b^{\mathcal{I}} \mid \langle A(b) \rangle^+ \in \mathsf{E}\} \text{ for all } A \in \mathsf{N_C};$$

$$r^{\mathcal{I}^{\mathsf{E}}} := r^{\mathcal{I}} \setminus \{(a^{\mathcal{I}}, b^{\mathcal{I}}) \mid \langle r(a, b) \rangle^- \in \mathsf{E}\} \cup \{(a^{\mathcal{I}}, b^{\mathcal{I}}) \mid \langle r(a, b) \rangle^+ \in \mathsf{E}\} \text{ for all } r \in \mathsf{N_R};$$

$$a^{\mathcal{I}^{\mathsf{E}}} := a^{\mathcal{I}} \text{ for all } a \in \mathsf{N_I}.$$

Let E be a set of (possibly conditional) effects. The update *of $\mathcal{I}$ with E, denoted by $\mathcal{I}^{\mathsf{E}}$, is given by the update*

$$\mathcal{I}^{\mathsf{E}(\mathcal{I})} \text{ with } \mathsf{E}(\mathcal{I}) := \{\mathsf{l} \mid (\varphi \triangleright \mathsf{l}) \in \mathsf{E}, \mathcal{I} \models \varphi\}.$$

An *action theory* in this setting just consists of an initial KB and a finite set of actions, each associated with a finite set of effects.

Definition 12. *An $\mathcal{L}$-action theory Σ is a tuple $\Sigma = (\mathcal{K}, \mathsf{Act}, \mathsf{Eff})$, where $\mathcal{K}$ is an $\mathcal{L}$-KB describing the initial state, Act is a finite set of action names, and for each $\alpha \in \mathsf{Act}$, the effects of α, denoted by $\mathsf{Eff}(\alpha)$, is a finite set of $\mathcal{L}$-effects.*

We can now extend the definition of an interpretation update to sequences of actions. Let $\mathcal{I}$ be an interpretation, $\Sigma = (\mathcal{K}, \mathsf{Act}, \mathsf{Eff})$ an action theory, and $\sigma \in \mathsf{Act}^*$ a sequence of action names. The *update of $\mathcal{I}$ with σ* is an interpretation $\mathcal{I}^{\sigma}$ defined by induction on the length of σ as follows: $\mathcal{I}^{\langle \rangle} := \mathcal{I}$ for the empty sequence and $\mathcal{I}^{\sigma' \cdot \alpha} = \mathcal{I}^{\sigma'^{\mathsf{E}}}$, where $\mathsf{E} = \mathsf{Eff}(\alpha)(\mathcal{I}^{\sigma'})$.

Definition 13. *The* projection problem *is a simple instance of the verification problem that can now be defined as follows. Let $\Sigma = (\mathcal{K}, \mathsf{Act}, \mathsf{Eff})$ be an action theory, $\sigma \in \mathsf{Act}^*$ a sequence of action names, and ϱ an axiom. We say that ϱ is* true after doing σ in Σ *iff for all models $\mathcal{I}$ of $\mathcal{K}$ we have that ϱ is satisfied in $\mathcal{I}^{\sigma}$.*

For the formalism in [5] it was shown that the projection problem can be solved by a polynomial reduction to a standard KB consistency task. Thus, the complexity is the same as for standard reasoning.

As an example consider an action theory

$$\Sigma = (\mathcal{K} = \mathcal{A} \cup \mathcal{T}, \text{Act} = \{discon(d,p),\, turn\text{-}on(d)\}, \text{Eff})$$

for a domain with concept names *DishWasher*, *PowerSupply*, *On* and a role name *connected*. The following concept assertions describe an initial situation involving individuals d and p:

$$\mathcal{A} = \{d : (DishWasher \sqcap \exists connected.PowerSupply)\,,\, p : PowerSupply\}. \quad (7)$$

As a concept inclusion we can express that a dish washer that is powered on must be connected to some power supply:

$$\mathcal{T} = \{DishWasher \sqcap On \sqsubseteq \exists connected.PowerSupply\}. \quad (8)$$

We consider two actions $discon(d,p)$ (for disconnecting d and p) and $turn\text{-}on(d)$ with the following conditional effects:

$\text{Eff}(turn\text{-}on(d)) := \{(d : (\exists connected.PowerSupply)) \triangleright \langle On(d)\rangle^+\};$

$\text{Eff}(discon(d,p)) := \{\, \langle connected(d,p)\rangle^-,$

$\qquad\qquad (d : (\forall connected.(\{p\} \sqcup \neg PowerSupply))) \triangleright \langle On(d)\rangle^-\}.$

The action $turn\text{-}on(d)$ only is effective if d is connected to some power supply, and $discon(d,p)$ in addition to disconnecting d and p makes sure that d is no longer an instance of *On* in case p was the only power supply connected to d before the disconnection. Note that the effect conditions make sure that axiom (8) is never violated due to an action execution.

It is rather straightforward to provide a Situation Calculus semantics for $\mathcal{ALCQIO}$-action theories. That is, for an action theory $\Sigma = (\mathcal{K}, \text{Act}, \text{Eff})$, sequence $\sigma \in \text{Act}^*$, and axiom ϱ as projection query, one can construct a corresponding basic action theory $\mathcal{D}_\Sigma$ such that ϱ is true after doing σ in Σ iff the following entailment

$$\mathcal{D}_\Sigma \models_{\mathcal{ES}} [\sigma]\text{fol}(\varrho),$$

holds in $\mathcal{ES}$, where $\text{fol}(\cdot)$ denotes the translation from DL syntax to FOL syntax. For instance, the effect condition of the delete effect $\langle On(d)\rangle^-$ of $discon(d,p)$ is equivalent to the FOL sentence

$$\forall x.\,(connected(d,x) \rightarrow (x = p \vee \neg PowerSupply(x))).$$

Since actions only affect named individuals in effect descriptions, the resulting BAT is one with only *local effects*. We can now describe GOLOG programs over Description Logic actions as follows.

Definition 14. *The base logic $\mathcal{L}$ is a DL between $\mathcal{ALC}$ and $\mathcal{ALCQIO}$, and the initial knowledge and effects of atomic actions are described in an $\mathcal{L}$-action theory $\Sigma = (\mathcal{K}, \mathsf{Act}, \mathsf{Eff})$. A program expression δ over Σ is obtained as in Definition 3 with the following restrictions for atomic actions and tests: we require $t \in \mathsf{Act}$ and that ψ in a test ψ? is a Boolean combination of $\mathcal{L}$-axioms. We call (Σ, δ) an $\mathcal{L}$-GOLOG program.*

To describe properties of programs, CTL* temporal formulas over $\mathcal{L}$-axioms are used. Validity of a CTL* state formula over $\mathcal{L}$-axioms in an $\mathcal{L}$-GOLOG program is then defined according to Definitions 5 and 6. For the verification problem we have obtained the following tight complexity bounds:

Theorem 5. *Let (Σ, δ) be an $\mathcal{L}$-GOLOG program and Ψ a CTL* temporal state formula over $\mathcal{L}$-axioms. Checking validity is*

- *2ExpTime-complete if $\mathcal{L} \in \{\mathcal{ALCO}, \mathcal{ALCIO}, \mathcal{ALCQO}\}$, and*
- *co-N2ExpTime-complete if $\mathcal{L} = \mathcal{ALCQIO}$.*

The upper bounds are obtained by using the abstraction technique described in Sect. 4. For more details and for the proofs of the lower bounds we refer to [50]. Note that in our formalism there is no direct interaction between the TBox that is part of the initial KB $\mathcal{K}$ and the transition semantics of actions and programs. It is possible that TBox axioms are violated in some program states, which might seem not desirable because usually TBox axioms are assumed to be *global state constraints*. However, the property that a global TBox is satisfied in all program states can be expressed as a CTL* formula, and the corresponding check is an instance of the verification problem. Different approaches with a tighter integration of TBox axioms as state constraints and action semantics have been investigated, for instance, in [3,5,36].

6   Further Results

The results summarized in the previous sections laid the foundation for extensions in various directions.

Knowledge-based programs, which are suited for more realistic scenarios where the agent possesses only incomplete information about its surroundings and has to use sensing in order to acquire additional knowledge at run-time, were considered in [53,54]. As opposed to classical GOLOG, knowledge-based programs [15,45] contain explicit references to the agent's knowledge, thus enabling it to choose its course of action based on what it knows and does not know. The work introduces a new epistemic action formalism based on the basic Description Logic $\mathcal{ALC}$, obtained by combining and extending earlier proposals for DL action formalisms [5] and epistemic DLs [23]. It turned out that the corresponding verification problem is in general again undecidable in the presence of pick operators, even under severe restrictions on the knowledge base and actions. Decidability can however be obtained by syntactically limiting the domain of pick operators to contain named objects only, yielding a 2ExpSpace upper bound.

Actions with non-local effects were considered in [55], where two new classes of action theories are introduced that generalize the previously discussed local-effect ones. Instead of imposing any bound on the number of affected objects, decidability of verification is obtained by restricting certain dependencies between fluents in the successor state axioms. This allows for a much wider range of application domains, including classical examples such as the briefcase domain [42] and exploding a bomb [35]: When a briefcase is moved, all (unboundedly many, unmentioned) objects that are currently in it are being moved along, and if a bomb explodes, everything in its vicinity is destroyed.

Decision-theoretic GOLOG (DTGOLOG) extends classical GOLOG by decision-theoretic aspects in the form of stochastic actions and reward functions [8, 48]. Here, a *stochastic* action refers to an operator that can have a limited number of possible outcomes, each of which is a regular, deterministic action with an associated probability. The program together with the action theory and reward function thus essentially induce an infinite-state *Markov Decision Process* (MDP), and the objective is to verify properties expressed in first-order variants of probabilistic temporal logics such as PCTL [28] or PRCTL [1]. Using similar techniques and restrictions as discussed above, Claßen and Zarrieß [18] showed that the infinite-state MDP can effectively be abstracted to a finite one, which then can be fed into any state-of-the-art probabilistic model checker such as PRISM [30] and STORM [22].

Probabilistic beliefs constitute a more involved notion of uncertainty. Instead of just having stochastic actions affect the objective truth of world-state fluents as above, an agent is now considered to have a certain *probabilistic degree of belief*. For such a setting, Zarrieß [49] studied the complexity of the *projection problem*, a subproblem of verification where one wants to determine whether a formula (here: about the agent's probabilistic beliefs) is true after executing a given sequence of (here: stochastic) actions. He proposed a formalization where deterministic actions (the possible outcomes of stochastic ones) are once again described similar to [5], and where initial beliefs as well as queries refer to subjective probabilities applied to ABox facts and TBox statements formulated in the DL $\mathcal{ALCO}$, which can be seen as a member of the Prob-$\mathcal{ALC}$ family of probabilistic DLs [39] and is a decidable fragment of Halpern's Type 2 probabilistic first-order logic [27]. It turned out that the combination of including *both* stochastic actions and probabilistic beliefs increases the complexity from EXPTIME-complete to 2EXPTIME-complete, while the problem remains EXPTIME-complete when only deterministic actions are used.

Timed GOLOG is a variant where instead of employing a qualitative notion of time, the more realistic assumption is made that actions may have a certain (discrete-time) duration. Consequently, verification is with respect to properties expressed in a *metric* temporal logic such as TCTL* [32]. Koopmann and Zarrieß [29] studied the complexity of verification under these assumptions over a DL representation of actions for various DLs in the $\mathcal{ALC}$ family as well as a lightweight

DL. They were able to establish 2-ExpTime-completeness for almost all variants, except for the case of full $\mathcal{ALCQIO}$ ($\mathcal{ALC}$ with qualified number restrictions, inverse roles, and nominals), which yields co-N2ExpTime-completeness. They also show that these tight complexity bounds apply for the non-metric, untimed case (cf. Theorem 4), and the corresponding abstraction techniques are indeed worst-case optimal.

A *prototypical implementation* of the methods from Sects. 3 and 4 was presented in [14], with the motivation that most results so far on Golog verification remained purely theoretical. In particular the very high worst-case complexities mentioned above are widely considered intractable, and hence may appear discouraging. On the other hand, experience from other areas (such as classical planning and SAT solving) is that in practical cases, most instances are comparatively easy to solve, and only few exhaust the full theoretical complexity. A prototype was hence implemented which uses a new Golog interpreter that supports full first-order reasoning by means of an embedded theorem prover [47]. One major challenge was that regression as is used by the fixpoint method causes a severe, exponential blow-up of formulas, and therefore – once again drawing inspiration from symbolic propositional model checking – a representation based on a first-order variant [46] of ordered binary decision diagrams [10] was used. A subsequent experimental evaluation showed that the fixpoint method is often preferable since it only explores parts of the state space that are relevant for the query property, while constructing a complete, bisimilar abstraction (which can be up to double-exponential in size) is often too expensive.

7 Conclusion

We presented an overview of our work on the temporal verification of Golog programs, both from a (modal) Situation Calculus and a Description Logic perspective. The problem can be approached in two different ways, namely by means of a Golog-specific fixpoint computation method based on characteristic program graphs and regression-based reasoning, or by determining a finite abstraction and applying a classical model checker. Golog's high expressiveness renders the general verification problem highly undecidable, and so the main challenge has been to identify restrictions on the input formalism that yield decidable, yet non-trivial fragments.

Other groups of researchers have conducted work that complements ours. To name but a few, Li and Liu [34] also present a sound, but incomplete verification method based on first-order theorem proving, however addressing the (somewhat different) task of proving Hoare-style partial correctness of terminating Golog programs. De Giacomo et al. [20] study the class of theories that have an infinite overall domain of objects, but where fluent extensions in each situation are bounded, which also admits finite abstractions and thus renders temporal verification decidable. Achieving decidable projection through a Description Logic representation in the Situation Calculus is analyzed in depth by Gu and Soutchanski [26].

While through this work, we have gained a deeper understanding of the problem and possible approaches to it, many avenues for future work remain. Probably the most interesting, and also most challenging, are those that strive to advance the current state of the art further towards realistic applications in robotics (or cyber-physical systems in general), where representations of quantitative, dynamic, and probabilistic aspects are needed, and that, beyond what was presented above, require e.g. notions of continuous change, noisy sensing, and uncertain beliefs.

Acknowledgements. This work was supported by the German Research Foundation (DFG), research unit FOR 1513 on Hybrid Reasoning for Intelligent Systems, project A1.

References

1. Andova, S., Hermanns, H., Katoen, J.-P.: Discrete-time rewards model-checked. In: Larsen, K.G., Niebert, P. (eds.) FORMATS 2003. LNCS, vol. 2791, pp. 88–104. Springer, Heidelberg (2004). https://doi.org/10.1007/978-3-540-40903-8_8
2. Baader, F., Ghilardi, S., Lutz, C.: LTL over description logic axioms. In: Proceedings of the Eleventh International Conference on the Principles of Knowledge Representation and Reasoning (KR 2008), pp. 684–694. AAAI Press (2008)
3. Baader, F., Lippmann, M., Liu, H.: Using causal relationships to deal with the ramification problem in action formalisms based on description logics. In: Fermüller, C.G., Voronkov, A. (eds.) LPAR 2010. LNCS, vol. 6397, pp. 82–96. Springer, Heidelberg (2010). https://doi.org/10.1007/978-3-642-16242-8_7
4. Baader, F., Liu, H., ul Mehdi, A.: Verifying properties of infinite sequences of description logic actions. In: Proceedings of the Nineteenth European Conference on Artificial Intelligence (ECAI 2010). Frontiers in Artificial Intelligence and Applications, vol. 215, pp. 53–58. IOS Press (2010)
5. Baader, F., Lutz, C., Miličić, M., Sattler, U., Wolter, F.: Integrating description logics and action formalisms: first results. In: Proceedings of the Twentieth National Conference on Artificial Intelligence (AAAI 2005), pp. 572–577. AAAI Press (2005)
6. Baader, F., Zarrieß, B.: Verification of Golog programs over description logic actions. In: Fontaine, P., Ringeissen, C., Schmidt, R.A. (eds.) FroCoS 2013. LNCS (LNAI), vol. 8152, pp. 181–196. Springer, Heidelberg (2013). https://doi.org/10.1007/978-3-642-40885-4_12
7. Baier, C., Katoen, J.P.: Principles of Model Checking. MIT Press, Cambridge (2008)
8. Boutilier, C., Reiter, R., Soutchanski, M., Thrun, S.: Decision-theoretic, high-level agent programming in the situation calculus. In: Proceedings of the Seventeenth National Conference on Artificial Intelligence (AAAI 2000). pp. 355–362. AAAI Press (2000)
9. Brewka, G., Lakemeyer, G.: Hybrid reasoning for intelligent systems: a focus of KR research in Germany. AI Mag. **39**(4), 80–83 (2018)
10. Bryant, R.E.: Graph-based algorithms for Boolean function manipulation. IEEE Trans. Comput. **35**(8), 677–691 (1986)
11. Burgard, W., et al.: Experiences with an interactive museum tour-guide robot. Artif. Intell. **114**(1–2), 3–55 (1999)

12. Clarke, E.M., Grumberg, O., Peled, D.A.: Model Checking. MIT Press, Cambridge (1999)
13. Claßen, J.: Planning and verification in the agent language Golog. Ph.D. thesis, Department of Computer Science, RWTH Aachen University (2013). http://darwin.bth.rwth-aachen.de/opus3/volltexte/2013/4809/
14. Claßen, J.: Symbolic verification of Golog programs with first-order BDDs. In: Proceedings of the Sixteenth International Conference on the Principles of Knowledge Representation and Reasoning (KR 2018), pp. 524–529. AAAI Press (2018)
15. Claßen, J., Lakemeyer, G.: Foundations for knowledge-based programs using $\mathcal{ES}$. In: Proceedings of the Tenth International Conference on the Principles of Knowledge Representation and Reasoning (KR 2006), pp. 318–328. AAAI Press (2006)
16. Claßen, J., Lakemeyer, G.: A logic for non-terminating Golog programs. In: Proceedings of the Eleventh International Conference on the Principles of Knowledge Representation and Reasoning (KR 2008), pp. 589–599. AAAI Press (2008)
17. Claßen, J., Liebenberg, M., Lakemeyer, G., Zarrieß, B.: Exploring the boundaries of decidable verification of non-terminating Golog programs. In: Proceedings of the Twenty-Eighth AAAI Conference on Artificial Intelligence (AAAI 2014), pp. 1012–1019. AAAI Press (2014)
18. Claßen, J., Zarrieß, B.: Decidable verification of decision-theoretic GOLOG. In: Dixon, C., Finger, M. (eds.) FroCoS 2017. LNCS (LNAI), vol. 10483, pp. 227–243. Springer, Cham (2017). https://doi.org/10.1007/978-3-319-66167-4_13
19. De Giacomo, G., Lespérance, Y., Levesque, H.J.: ConGolog, a concurrent programming language based on the situation calculus. Artif. Intell. 121(1–2), 109–169 (2000)
20. De Giacomo, G., Lespérance, Y., Patrizi, F., Sardiña, S.: Verifying ConGolog programs on bounded situation calculus theories. In: Proceedings of the Thirtieth AAAI Conference on Artificial Intelligence (AAAI 2016), pp. 950–956. AAAI Press (2016)
21. De Giacomo, G., Ternovska, E., Reiter, R.: Non-terminating processes in the situation calculus. In: Working Notes of "Robots, Softbots, Immobots: Theories of Action, Planning and Control", AAAI 1997 Workshop (1997)
22. Dehnert, C., Junges, S., Katoen, J.-P., Volk, M.: A STORM is coming: a modern probabilistic model checker. In: Majumdar, R., Kunčak, V. (eds.) CAV 2017. LNCS, vol. 10427, pp. 592–600. Springer, Cham (2017). https://doi.org/10.1007/978-3-319-63390-9_31
23. Donini, F.M., Lenzerini, M., Nardi, D., Nutt, W., Schaerf, A.: An epistemic operator for description logics. Artif. Intell. 100(1–2), 225–274 (1998)
24. Ferrein, A., Niemueller, T., Schiffer, S., Lakemeyer, G.: Lessons learnt from developing the embodied AI platform CAESAR for domestic service robotics. In: Papers from the AAAI 2013 Spring Symposium on Designing Intelligent Robots: Reintegrating AI II. Technical report SS-13-04, AAAI Press (2013)
25. Grädel, E., Kolaitis, P.G., Vardi, M.Y.: On the decision problem for two-variable first-order logic. Bull. Symb. Log. 3(1), 53–69 (1997)
26. Gu, Y., Soutchanski, M.: A description logic based situation calculus. Ann. Math. Artif. Intell. 58(1–2), 3–83 (2010)
27. Halpern, J.Y.: An analysis of first-order logics of probability. Artif. Intell. 46(3), 311–350 (1990). https://doi.org/10.1016/0004-3702(90)90019-V
28. Hansson, H., Jonsson, B.: A logic for reasoning about time and reliability. Formal Asp. Comput. 6(5), 512–535 (1994)

29. Koopmann, P., Zarrieß, B.: On the complexity of verifying timed Golog programs over description logic actions. In: Proceedings of the 2018 Workshop on Hybrid Reasoning and Learning (HRL 2018) (2018)
30. Kwiatkowska, M., Norman, G., Parker, D.: PRISM 4.0: verification of probabilistic real-time systems. In: Gopalakrishnan, G., Qadeer, S. (eds.) CAV 2011. LNCS, vol. 6806, pp. 585–591. Springer, Heidelberg (2011). https://doi.org/10.1007/978-3-642-22110-1_47
31. Lakemeyer, G., Levesque, H.J.: A semantic characterization of a useful fragment of the situation calculus with knowledge. Artif. Intell. **175**(1), 142–164 (2010)
32. Laroussinie, F., Markey, N., Schnoebelen, P.: Efficient timed model checking for discrete-time systems. Theor. Comput. Sci. **353**(1–3), 249–271 (2006). https://doi.org/10.1016/j.tcs.2005.11.020
33. Levesque, H.J., Reiter, R., Lespérance, Y., Lin, F., Scherl, R.B.: GOLOG: a logic programming language for dynamic domains. J. Log. Program. **31**(1–3), 59–83 (1997)
34. Li, N., Liu, Y.: Automatic verification of partial correctness of Golog programs. In: Proceedings of the Twenty-Fourth International Joint Conference on Artificial Intelligence (IJCAI 2015), pp. 3113–3119. AAAI Press (2015)
35. Lin, F., Reiter, R.: How to progress a database. Artif. Intell. **92**(1–2), 131–167 (1997)
36. Liu, H., Lutz, C., Miličić, M., Wolter, F.: Reasoning about actions using description logics with general TBoxes. In: Fisher, M., van der Hoek, W., Konev, B., Lisitsa, A. (eds.) JELIA 2006. LNCS (LNAI), vol. 4160, pp. 266–279. Springer, Heidelberg (2006). https://doi.org/10.1007/11853886_23
37. Liu, Y., Lakemeyer, G.: On first-order definability and computability of progression for local-effect actions and beyond. In: Proceedings of the Twenty-First International Joint Conference on Artificial Intelligence (IJCAI 2009), pp. 860–866. AAAI Press (2009)
38. Liu, Y., Levesque, H.J.: Tractable reasoning with incomplete first-order knowledge in dynamic systems with context-dependent actions. In: Proceedings of the Nineteenth International Joint Conference on Artificial Intelligence (IJCAI 2005), pp. 522–527. Professional Book Center (2005)
39. Lutz, C., Schröder, L.: Probabilistic description logics for subjective uncertainty. In: Proceedings of the Twelfth International Conference on the Principles of Knowledge Representation and Reasoning (KR 2010). AAAI Press (2010)
40. McCarthy, J., Hayes, P.: Some philosophical problems from the standpoint of artificial intelligence. In: Meltzer, B., Michie, D. (eds.) Machine Intelligence, vol. 4, pp. 463–502. American Elsevier, New York (1969)
41. McMillan, K.L.: Symbolic Model Checking. Kluwer Academic Publishers, Norwell (1993)
42. Pednault, E.P.D.: Synthesizing plans that contain actions with context-dependent effects. Comput. Intell. **4**, 356–372 (1988)
43. Reiter, R.: The frame problem in the situation calculus: A simple solution (sometimes) and a completeness result for goal regression. Artificial Intelligence and Mathematical Theory of Computation: Papers in Honor of John McCarthy pp. 359–380 (1991)
44. Reiter, R.: Knowledge in Action: Logical Foundations for Specifying and Implementing Dynamic Systems. MIT Press, Cambridge (2001)
45. Reiter, R.: On knowledge-based programming with sensing in the situation calculus. ACM Trans. Comput. Log. **2**(4), 433–457 (2001)

46. Sanner, S., Boutilier, C.: Practical solution techniques for first-order MDPs. Artif. Intell. **173**(5–6), 748–788 (2009)

47. Schulz, S.: System description: E 1.8. In: McMillan, K., Middeldorp, A., Voronkov, A. (eds.) LPAR 2013. LNCS, vol. 8312, pp. 735–743. Springer, Heidelberg (2013). https://doi.org/10.1007/978-3-642-45221-5_49

48. Soutchanski, M.: An on-line decision-theoretic Golog interpreter. In: Proceedings of the Seventeenth International Joint Conference on Artificial Intelligence (IJCAI 2001), pp. 19–26. Morgan Kaufmann Publishers Inc. (2001)

49. Zarrieß, B.: Complexity of projection with stochastic actions in a probabilistic description logic. In: Proceedings of the Sixteenth International Conference on the Principles of Knowledge Representation and Reasoning (KR 2018), pp. 514–523. AAAI Press (2018)

50. Zarrieß, B.: Verification of Golog programs over description logic actions. Ph.D. thesis, Dresden University of Technology, Germany (2018). http://d-nb.info/116636531X

51. Zarrieß, B., Claßen, J.: On the decidability of verifying LTL properties of Golog programs. In: Proceedings of the AAAI 2014 Spring Symposium: Knowledge Representation and Reasoning in Robotics (KRR 2014). AAAI Press, Palo Alto (2014)

52. Zarrieß, B., Claßen, J.: Verifying CTL* properties of Golog programs over local-effect actions. In: Proceedings of the Twenty-First European Conference on Artificial Intelligence (ECAI 2014), pp. 939–944. IOS Press (2014)

53. Zarrieß, B., Claßen, J.: Decidable verification of knowledge-based programs over description logic actions with sensing. In: Proceedings of the Twenty-Eighth International Workshop on Description Logics (DL 2015). CEUR Workshop Proceedings, vol. 1350. CEUR-WS.org (2015)

54. Zarrieß, B., Claßen, J.: Verification of knowledge-based programs over description logic actions. In: Proceedings of the Twenty-Fourth International Joint Conference on Artificial Intelligence (IJCAI 2015), pp. 3278–3284. AAAI Press (2015)

55. Zarrieß, B., Claßen, J.: Decidable verification of Golog programs over non-local effect actions. In: Proceedings of the Thirtieth AAAI Conference on Artificial Intelligence (AAAI 2016), pp. 1109–1115. AAAI Press (2016)

Provenance Analysis: A Perspective for Description Logics?

Katrin M. Dannert and Erich Grädel$^{(\boxtimes)}$

RWTH Aachen University, Aachen, Germany
{dannert,graedel}@logic.rwth-aachen.de

*For Franz Baader on the occasion of his
60th birthday*

Abstract. Provenance analysis aims at understanding how the result of a computational process with a complex input, consisting of multiple items, depends on the various parts of this input. In database theory, provenance analysis based on interpretations in commutative semirings has been developed for positive database query languages, to understand which combinations of the atomic facts in a database can be used for deriving the result of a given query. In joint work with Val Tannen, we have recently proposed a new approach for the provenance analysis of logics *with negation*, such as first-order logic and fixed-point logic. It is based on new semirings of dual-indeterminate polynomials or dual-indeterminate formal power series, which are obtained by taking quotients of traditional provenance semirings by congruences that are generated by products of positive and negative provenance tokens. This provenance approach has also been applied to fragments of first-order logics such as modal and guarded logics. In this paper, we explore the question whether, and to what extent, the provenance approach might be useful in the field of description logics.

1 Introduction

This paper is intended as an account, written for the description logics community, of recent developments in semiring provenance, that make provenance analysis applicable to logical formalisms with negation. In particular, we discuss the question whether provenance analysis could be a fruitful perspective for description logics.

Provenance analysis is an algebraic approach to abstract from a computation with multiple input items, such as the evaluation of a database query, mathematical information on how the result of the computation depends on the various input data. In database theory, provenance analysis based on interpretations in commutative semirings has been successfully developed for query languages such as unions of conjunctive queries, positive relational algebra,

K. M. Dannert—Supported by the DFG RTG 2236 UnRAVeL.

nested relations, Datalog, XQuery, SQL-aggregates and several others, and it has been implemented in software systems such as Orchestra and Propolis, see e.g. [2,5,6,11,13,17]. In this approach, atomic facts are interpreted not just by true or false, but by values in an appropriate semiring, where 0 is the value of false statements, whereas any element $a \neq 0$ of the semiring stands for some shade of truth. These values are then propagated from the atomic facts to arbitrary queries in the language, which permits to answer questions such as the minimal cost of a query evaluation, the confidence one can have that the result is true, the number of different ways in which the result can be computed, or the clearance level that is required for obtaining the output, under the assumption that some facts are labelled as confidential, secret, top secret, etc. We refer to [14] for a recent account on the semiring framework for database provenance.

We argue that provenance analysis may have a strong potential for useful applications also in the context of description logics. We shall propose notions of provenance semantics for ABoxes and TBoxes where concept and role assertions take values in a commutative semiring, and concept inclusions $C \sqsubseteq D$ translate into comparisons of such provenance values. The common reasoning problems in description logics, such as subsumption, consistency, or query answering get a new twist, generalizing Boolean reasoning to algebraic reasoning in a commutative semiring. Potential applications of this approach include *cost computations* of concept assertions (by means of provenance evaluations in the tropical semiring), the study of required *clearance levels* for accessing confidential or secret data (using valuation in an access control semiring), or reasoning about *confidences* achievable in ontology-mediated query evaluations. We shall discuss these notions in more detail in Sects. 4 and 5 below.

For a long time, an essential limitation of the semiring provenance approach has been its confinement to *positive* query languages. There have been algebraically interesting attempts to cover difference of relations [1,7,8,12] but they have not resulted in systematic tracking of *negative information*, and until recently there has been no convincing provenance analysis for languages with full negation. For applications to description logics, the inability to deal with negation and absent information would certainly be a major obstacle. However, a new approach for the provenance analysis of logics with negation, such as first-order logic and fixed-point logic, has now been proposed in [9,10] based on the following ingredients:

– Negation is dealt with by transformation to negation normal form. This is a common approach in logic, but while this is often just a matter of convenience and done for simplification, its seems indispensable for provenance semantics. Indeed, beyond Boolean semantics, negation is not a compositional logical operation: the provenance value of $\neg\varphi$ is not necessarily determined by the provenance value of φ.

– On the algebraic side, new provenance semirings of polynomials and formal power series have been introduced, which take negation into account. They are obtained by taking quotients of traditional provenance semirings by congruences generated by products of positive and negative provenance tokens; they are called semirings of dual-indeterminate polynomials or dual-indeterminate power series.

- Provenance analysis of logics is closely connected to provenance analysis of games. In [10], the provenance approach to logics with negation is described from the perspective of the associated model checking games. In fact provenance analysis of games is of independent interest, and provenance values of positions in a game provide detailed information about the number and properties of the strategies of the players, far beyond the question whether or not a player has a winning strategy from a given position. However, in the interest of a reasonably compact presentation, we do not use the game perspective in this paper, but describe the approach in purely algebraic and logical terms.

In this paper we propose to study the potential of the semiring provenance approach as a perspective for description logics. Although we ourselves are certainly not experts in description logics and their applications, we believe that there are good reasons why this might be interesting and useful. Given that most description logics use negation in an essential way, the new provenance approach for dealing with negation could help to combine provenance analysis and description logics in a fruitful way. A point in favour is that description logics are, as it is put in the textbook [3], 'cousins of modal logics', and that the new approach to provenance analysis has already been applied to modal and guarded logics in [4]. On the other side, the application of provenance to description logics certainly also poses nontrivial problems. Indeed the standard scenario of provenance analysis is formula evaluation in a fixed finite structure. In most applications of description logics, however, a knowledge base is considered that, logically speaking, axiomatizes a class of structures, and the main reasoning problems are variants of satisfiability, validity, and entailment problems. Nevertheless, notions developed in [9] of provenance tracking interpretations by means of dual-indeterminate polynomials permit to deal with multiple models, and with reverse provenance analysis, constructing appropriate models from a given specification, at least in the case of a fixed universe. Further it also seems a quite promising project to generalize the tableaux-based reasoning techniques that are so popular in description logic to provenance semantics based on semirings. Thus, while differences and difficulties exist, they do not seem unsolvable. We thus hope that the description logic community will take an interest in these new developments in provenance analysis, and that a fruitful collaboration between the two fields will emerge.

This paper does not assume that the reader is already familiar with semiring provenance. However, we do assume that the reader knows basic definitions and results about description logics. Our notation and terminology is largely based on [3].

2   Commutative Semirings

Definition 1. A *commutative semiring* is an algebraic structure $(K, +, \cdot, 0, 1)$, with $0 \neq 1$, such that $(K, +, 0)$ and $(K, \cdot, 1)$ are commutative monoids, $\cdot$ distributes over $+$, and $0 \cdot a = a \cdot 0 = 0$. A semiring is *+-positive* if $a + b = 0$ implies

$a = 0$ and $b = 0$. This excludes rings. A semiring is *root-integral* if $a \cdot a = 0$ implies $a = 0$. All semirings considered in this paper are commutative, +-positive and root-integral. Further, a commutative semiring is *positive* if it is +-positive and has no divisors of 0. The standard semirings considered traditionally in provenance analysis are positive, but for the treatment of negation we need semirings (of dual-indeterminate polynomials or power series) that have divisors of 0.

Notice that a semiring K is positive if, and only if, the unique function $h : K \to \{0,1\}$ with $h^{-1}(0) = \{0\}$ is a homomorphism from K into the Boolean semiring $\mathbb{B} = (\{0,1\}, \vee, \wedge, 0, 1)$. A semiring K is (+)-idempotent if $a + a = a$, for all $a \in K$, and $(+, \cdot)$-idempotent if, in addition, $a \cdot a = a$ for all a. Further, K is *absorptive* if $a + ab = a$, for all $a, b \in K$. Obviously, every absorptive semiring is (+)-idempotent.

In provenance analysis, elements of a commutative semiring are used as truth values for logical statements. The intuition is that + describes the *alternative use* of information, as in disjunctions or existential quantifications whereas · stands for the *joint use* of information, as in conjunctions or universal quantifications. Further, 0 is the value of false statements, whereas any element $a \neq 0$ of a semiring K stands for a 'nuanced' interpretation of true.

2.1 Application Semirings

We briefly discuss some specific semirings that provide interesting information about about a logical statement.

- The *Boolean semiring* $\mathbb{B} = (\{0,1\}, \vee, \wedge, 0, 1)$ is the domain of standard logical truth values.
- The semiring $\mathbb{N} = (\mathbb{N}, +, \cdot, 0, 1)$ can be used for counting successful strategies for query evaluation. It also plays an important role for *bag semantics* in databases.
- $\mathbb{T} = (\mathbb{R}_+^\infty, \min, +, \infty, 0)$ is called the *tropical* semiring. It has many applications for cost computations, for instance for query evaluation.
- The *Viterbi* semiring $\mathbb{V} = ([0,1], \max, \cdot, 0, 1)$ is isomorphic to $\mathbb{T}$ via $x \mapsto e^{-x}$ and $y \mapsto -\ln y$. We will think of the elements of $\mathbb{V}$ as *confidence scores* and use it to describe the confidence assigned to a logical statement.
- The *access control* semiring is $\mathbb{A} = (\{P < C < S < T < 0\}, \min, \max, 0, P)$ where P is 'public', C is 'confidential', S is 'secret', T is 'top secret', and 0 is 'so secret that nobody can access it!'. The valuation of a statement in $\mathbb{A}$ describes the *minimal clearance level* that is needed to establish it.
- The *max-min* semiring on a totally ordered set $(A, \leq)$ with least element a and greatest element b is the semiring $(A, \max, \min, a, b)$. The class of max-min semirings includes, of course, the Boolean semiring and the access control semiring but also infinite ones, for instance the one on the real interval $[0,1]$ which is sometimes called the fuzzy semiring.

2.2 Provenance Semirings

Beyond such application semirings, there are important universal provenance semirings of polynomials and formal power series that are used for a general provenance analysis. They admit to compute provenance values once in a general semiring and then to specialise these via homomorphisms to specific application semirings as needed.

Let X be a set of abstract *provenance tokens*, i.e. variables that we use to label atomic data (such as concept or role assertions in description logics). The commutative semiring that is freely generated by the set X is $\mathbb{N}[X] = (\mathbb{N}[X], +, \cdot, 0, 1)$, the semiring of multivariate polynomials in indeterminates from X and with coefficients from $\mathbb{N}$.

Computing provenance values of a statement φ (from some appropriate logical formalism) in $\mathbb{N}[X]$ gives us precise information about which combinations of the atomic facts can be used to derive φ. Indeed, each monomial $c x_1^{e_1} \ldots x_k^{e_k}$ that occurs in the provenance polynomial $\pi[\![\varphi]\!] \in \mathbb{N}[X]$ indicates that we have c different evaluation strategies that make use of precisely those atomic facts that are labelled by $x_1, \ldots x_k$ and use the fact labelled by x_i precisely e_i times. Evaluation strategies can be understood either as 'proof trees' (as in [9,13]) or as winning strategies in the model checking game associated with φ (as in [10]).

There are a number of other polynomial semirings that can be obtained from $\mathbb{N}[X]$ by dropping coefficients, dropping exponents, or absorption laws, in which provenance polynomials are less informative, but possibly easier to compute. This includes the $+$-idempotent semiring $\mathbb{B}[X]$, the so-called why semiring $\mathbb{W}[X]$, the absorptive semiring $\mathbb{S}[X]$ and the free distributive lattice $\mathsf{PosBool}(X)$, see e.g. [10,13,14] for more information.

However, in none of these semirings there is an adequate treatment of negation, or tracking of missing information, because either negative atoms are not represented at all, or an atom and its negation are labelled by two different tokens without any algebraic connection between them. To address this issue, a new approach has been proposed in [9], and further developed in [10].

2.3 Dual-Indeterminate Polynomials and Formal Power Series

Here is the algebraic construction to make provenance analysis available for logics with negation. Let $X, \bar{X}$ be two disjoint sets of provenance tokens, together with a bijection $X \to \bar{X}$, that maps each 'positive' token $p \in X$ to a corresponding 'negative' token $\bar{p} \in \bar{X}$. We call p and $\bar{p}$ complementary tokens. By convention, if we annotate an atomic fact by p then $\bar{p}$ can only be used to annotate its negation, and vice versa.

Definition 2. The semiring $\mathbb{N}[X, \bar{X}]$ of *dual-indeterminate polynomials* is the quotient of the semiring of polynomials $\mathbb{N}[X \cup \bar{X}]$ by the congruence generated by the equalities $p \cdot \bar{p} = 0$ for all $p \in X$. This is the same as quotienting by the ideal generated by the polynomials $p\bar{p}$ for all $p \in X$. Two polynomials $f, g \in \mathbb{N}[X \cup \bar{X}]$ are congruent if, and only if, they become identical after deleting from each of them the monomials that contain complementary tokens. Hence, the congruence classes in $\mathbb{N}[X, \bar{X}]$ are in one-to-one correspondence with the polynomials in $\mathbb{N}[X \cup \bar{X}]$ such that none of their monomials contains complementary tokens.

Note that the semirings $\mathbb{N}[X \cup \bar{X}]$ are +-positive and root-integral, but not positive, since they obviously admit divisors of 0.

The semirings $\mathbb{N}[X, \bar{X}]$ turn out to be adequate for a general provenance analysis of full first-order logic (with negation) [9], and hence also for full relational algebra (not just its positive fragment). This extends to fragments of first order logic such as modal and guarded logics [4] and (as we propose in this paper) description logics. However for logics with fixed points or with mechanisms of unbounded iteration, polynomial semirings are not sufficient. Even for a formalism as simple as datalog (avoiding all complications arising from universal quantification and negation) one has to impose additional conditions on the semirings to guarantee the existence of least fixed points [5]. Of particular importance are ω-continuous semirings. Many application semirings are ω-continuous, but $\mathbb{N}$, and the polynomial semirings $\mathbb{N}[X]$ and $\mathbb{N}[X, \bar{X}]$ are not. The ω-continuous completion of $\mathbb{N}$ is $\mathbb{N}^{\infty} := \mathbb{N} \cup \{\infty\}$ (with $a + \infty = a \cdot \infty = \infty$), but the completion of $\mathbb{N}[X]$ is $\mathbb{N}^{\infty}[X]$ which is not a semiring of polynomials, but of formal power series (possibly infinite sums of monomials), with coefficients in $\mathbb{N}^{\infty}$ and indeterminates in X, with addition and multiplication defined in the standard way. We combine this with our approach for dealing with negation by taking quotients.

Definition 3. The semiring $\mathbb{N}^{\infty}[\![X, \bar{X}]\!]$ is the quotient of the semiring of power series $\mathbb{N}^{\infty}[\![X \cup \bar{X}]\!]$ by the congruence generated by the equalities $p \cdot \bar{p} = 0$ for all $p \in X$. The congruence classes in $\mathbb{N}^{\infty}[\![X, \bar{X}]\!]$ are in one-to-one correspondence with the power series in $\mathbb{N}^{\infty}[\![X \cup \bar{X}]\!]$ such that none of their monomials contain complementary tokens. We call these *dual-indeterminate power series*.

Every function $f : X \cup \bar{X} \to K$ into an ω-continuous semiring K with the property that $f(p) \cdot f(\bar{p}) = 0$ for all $p \in X$ extends uniquely to an ω-continuous semiring homomorphism $h : \mathbb{N}^{\infty}[\![X, \bar{X}]\!] \to K$ that coincides with f on $X \cup \bar{X}$.

3   Provenance for Model Checking Problems

Provenance analysis has been developed for query evaluation and, more generally, model checking problems in logic, in particular first-order logic and its fragments.

Let τ be a vocabulary, which in the case of description logics contains only unary predicates (concept names) and binary predicates (role names), and fix a finite universe Δ. We denote by $\mathrm{Atoms}_{\Delta}(\tau)$ the set of all atoms $R\bar{a}$ with $R \in \tau$ and $\bar{a} \in \Delta^k$. Further, let $\mathrm{NegAtoms}_{\Delta}(\tau)$ be the set of all negated atoms $\neg R\bar{a}$ where $R\bar{a} \in \mathrm{Atoms}_{\Delta}(\tau)$, and consider the set of all τ-literals on A, $\mathrm{Lit}_{\Delta}(\tau) := \mathrm{Atoms}_{\Delta}(\tau) \cup \mathrm{NegAtoms}_{\Delta}(\tau) \cup \{a \text{ op } b : a, b \in A\}$, where op stands for $=$ or $\neq$.

Definition 4. Given any commutative semiring K, a K-*interpretation* (for τ and Δ) is a function $\pi : \mathrm{Lit}_{\Delta}(\tau) \to K$ that maps equalities and inequalities to their truth values 0 or 1. A K-interpretation is *sound for negation* if $\pi[\alpha] \cdot \pi[\neg\alpha] = 0$ for every atom $\alpha \in \text{Atoms}_\Delta(\tau)$. In this paper, all K-interpretations are assumed to be sound for negation.

The equality and inequality atoms are interpreted in K as 0 or 1, i.e., their provenance is not tracked. One could give a similar treatment to other relations with a fixed meaning, e.g., assuming a linear order on A. However, we do not pursue this in this paper.

We have defined in [9] how a semiring interpretation extends to a full valuation $\pi : \text{FO}(\tau) \to K$ mapping any fully instantiated formula $\psi(\bar{a})$ to a value $\pi[\psi]$, by setting

$$\pi[\psi \vee \varphi] := \pi[\psi] + \pi[\varphi]] \qquad \pi[\psi \wedge \varphi] := \pi[\psi] \cdot \pi[\varphi]$$

$$\pi[\exists x \varphi(x)] := \sum_{a \in \Delta} \pi[\varphi(a)] \qquad \pi[\forall x \varphi(x)] := \prod_{a \in \Delta} \pi[\varphi(a)].$$

For negation, we set $\pi[\neg\varphi] := \pi[\text{nnf}(\neg\varphi)]$ where $\text{nnf}(\varphi)$ is the negation normal form of φ.

As shown in [9], for positive semirings, and also for the interpretations in semirings of dual indeterminate polynomials that we are interested in, the soundness for negation extends from atoms to arbitrary first-order formulae and implies that $\pi[\varphi] \cdot \pi[\neg\varphi] = 0$ for all $\varphi \in \text{FO}$. However, since we admit semirings with divisors of 0, soundness for negation does not necessarily imply that one of $\pi[\varphi]$ and $\pi[\neg\varphi]$ must be 0.

For modal and guarded logic similar definitions of provenance interpretations have been given and analysed in [4]. It is not difficult to adapt these definitions for description logics. Here is one for $\mathcal{ALC}$. For simplicity of notation we identify individual names with elements of the universe. Further, all concept assertions of form $a{:}C$ where C is a concept name or the negation of a concept name, and all role assertions $(a,b){:}r$ for a role name r are viewed as literals in some set $\text{Lit}_\Delta(\tau)$.

Definition 5. Let $\pi : \text{Lit}_\Delta(\tau) \to K$ be a K-interpretation for a finite universe Δ and a vocabulary τ of concept names and role names. Given a role name r and an element $a \in \Delta$, let $r(a) := \{b : \pi((a,b){:}r) \neq 0\}$. For shortness we define $\pi(rab) := \pi((a,b){:}r)$. We extend π to *concept assertions* $a{:}C$ consisting of an $\mathcal{ALC}$ concept description C, assumed to be given in negation normal form, and an element $a \in \Delta$ by

$$\pi[a{:}\bot] := 0 \qquad\qquad\qquad \pi[a{:}\top] := 1$$
$$\pi[a{:}C \sqcup D] := \pi[a{:}C] + \pi[a{:}D] \qquad \pi[a{:}C \sqcap D] := \pi[a{:}C] \cdot \pi[a{:}D]$$
$$\pi[a{:}\exists r.C] := \sum_{b \in r(a)} (\pi(rab) \cdot \pi[b{:}C]) \qquad \pi[a{:}\forall r.C] := \prod_{b \in r(a)} (\pi(rab) \cdot \pi[b{:}C]).$$

The close relationship between description logics and modal logics admits to carry over the complexity results for computing provenance values of modal formulae [4] to this setting.

Proposition 1. *Let K be an arbitrary semiring. Given a concept description C in $\mathcal{ALC}$, a K-interpretation $\pi : \text{Lit}_\Delta(\tau) \to K$, and an element $a \in \Delta$, the provenance value $\pi[\![a\!:\!C]\!]$ can be computed with $O(|C| \cdot |\pi|)$ semiring operations.*

Notice that for concept descriptions in full first-order logic rather than $\mathcal{ALC}$, the number of semiring operations needed to compute provenance values may be much higher. Indeed, the straightforward approach requires an exponential number of operations with respect to the length of a first-order concept description, and since even in the Boolean case, the model checking problem for first-order logic is PSPACE-complete, it is unlikely that polynomial bounds are possible.

Nevertheless, despite the relatively small *number* of semiring operations that are needed to compute provenance values for $\mathcal{ALC}$, the *complexity* of such computations may, depending on the *costs* of representing elements in the given semiring and the costs of addition and multiplication, still be rather high, in fact doubly exponential in the length of the concept description. See [4] for a detailed complexity analysis for the case of modal and guarded logic.

4   Provenance Semantics for ABoxes and TBoxes

We have described basic observations about the definition and computation of provenance values for concept assertions in $\mathcal{ALC}$. However the important reasoning tasks associated with description logics are not so much the evaluation of a concept assertion in a given interpretation. Description logics are used as knowledge representation languages. A *knowledge base* typically consists of a TBox $\mathcal{T}$ which is a finite set of *general concept inclusions* $C \sqsubseteq D$, describing conceptual knowledge about the domain of application, and an ABox $\mathcal{A}$, which is a finite set of *concept assertions* $a\!:\!C$ and *role assertions* $(a,b)\!:\!r$, describing specific data. Relevant questions, given an $\mathcal{ALC}$ knowledge base $(\mathcal{A}, \mathcal{T})$, concern for instance the subsumption and equivalence of two given concepts in all models of $\mathcal{T}$, the consistency of the knowledge base, or the question whether a given concept assertion $a\!:\!C$ is entailed by the knowledge base.

Can semiring provenance provide any additional insights for knowledge representation by description logics? To discuss such questions, we first discuss what provenance semantics might mean for ABoxes and TBoxes.

Provenance Semantics for an ABox. Since an ABox defines a set of statements that are asserted to be true, a natural possibility to define its provenance semantics could be to assign to every assertion in the ABox a non-zero value in the semiring, defining its precise 'shade of truth'. However, we propose a definition that is a little more general, which gives us also the possibility to declare that $a\!:\!C$ just has *some* shade of truth $\geq k$ or $> k$ without a commitment to a precise value.

Definition 6. A K-*valued ABox* is a finite set of statements of form $\pi[\![\alpha]\!]$ op k where α is a concept assertion or role assertion, k is an element of the semiring K, and op is $-, \geq$, or $>$.

In DL one sometimes restricts attention to *simple ABoxes* admitting only concept assertions $a:C$ where C is a concept name, and simple K-valued ABoxes are defined analogously. This comes with no loss of expressive power since one can replace each assertion $a:C$ by $a:A_C$, where A_C is a new concept name, and then add an equivalence $A_C \equiv C$ to the TBox.

Provenance Semantics for a TBox. For a given TBox $\mathcal{T}$, let τ be a vocabulary containing all concept names and role names appearing in $\mathcal{T}$, let Δ be a finite universe and K a commutative semiring. We want to discuss what it means that a K-interpretation $\pi : \mathrm{Lit}_\Delta(\tau) \to A$ is *consistent* with $\mathcal{T}$.

There are two main possibilities. For the stronger one we assume, without loss of generality, that $\mathcal{T}$ is given as a finite set of concept inclusions $C \sqsubseteq D$.

Definition 7. A K-interpretation $\pi : \mathrm{Lit}_\Delta(\tau) \to K$ is *strongly consistent* with $\mathcal{T}$, if for every concept inclusion $C \sqsubseteq D$ in $\mathcal{T}$ and every $a \in \Delta$, we have that $\pi[\![a:C]\!] \leq \pi[\![a:D]\!]$.

Recall that the natural order in a semiring K is defined by $x \leq y :\Longleftrightarrow \exists z(x + z = y)$. The requirement that our semirings are naturally ordered means that $\leq$ is antisymmetric (i.e. $x \leq y \wedge y \leq x$ only for $x = y$). Hence, if $\mathcal{T}$ contains both $C \sqsubseteq D$ and $D \sqsubseteq C$, and thus imposes an equivalence $C \equiv D$, strong consistency means that $\pi[\![a:C]\!] = \pi[\![a:D]\!]$ for all a.

This strong notion of consistency is rather restrictive. In many applications it may not be adequate to require that a subsumption between two concepts translates in this precise way into an ordering between their truth values. A less restrictive possibility is to view a concept inclusion $C \sqsubseteq D$ as a requirement that whenever $a:C$ has a positive 'shade of truth' then so has $a:D$. On the other side, this does not seem right in the case of concept definitions $A \equiv C$, where A is a concept name; in this case we should, of course, require that all provenance values of A and C are the same.

For the weaker notion of consistency that we have in mind we therefore rewrite a TBox as a disjoint union $\mathcal{T} = \mathcal{T}_0 \cup \mathcal{T}_1$ where $\mathcal{T}_0$ is an *acyclic* TBox, consisting of concept definitions $A \equiv C$, without cyclic dependencies among them, and $\mathcal{T}_1$ is written as a finite set of equations $C \sqcap D = \bot$. Notice that in the Boolean case, this is just an equivalent rewriting because any concept inclusion $C \sqsubseteq D$ is equivalent to $C \sqcap \neg D = \bot$.

Definition 8. A K-interpretation $\pi : \mathrm{Lit}_\Delta(\tau) \to K$ is *weakly consistent* with a TBox $\mathcal{T}_0 \cup \mathcal{T}_1$, if

(1) for every concept definition $A \equiv C$ in $\mathcal{T}_0$ and every $a \in \Delta$, we have that $\pi[\![a:A]\!] = \pi[\![a:C]\!]$, and
(2) for every equation $C \sqcap D = \bot$ in $\mathcal{T}_1$ we have that

$$\sum_{a \in \Delta} \pi[\![a:C]\!] \cdot \pi[\![a:D]\!] = 0.$$

As a sanity check for these definitions, we prove

Proposition 2. *If π is strongly consistent with a TBox $\mathcal{T} = \mathcal{T}_0 \cup \mathcal{T}_1$, then it is also weakly consistent with $\mathcal{T}$.*

Proof. We just have to show that, for every equation $C \sqcap D = \bot$ in $\mathcal{T}_1$ and every $a \in \Delta$, we have that $\pi[\![a:C]\!] \leq \pi[\![a:\neg D]\!]$ implies $\pi[\![a:C]\!] \cdot \pi[\![a:D]\!] = 0$. But $\pi[\![a:C]\!] \leq \pi[\![a:\neg D]\!]$ implies that also $\pi[\![a:C]\!] \cdot \pi[\![a:D]\!] \leq \pi[\![a:\neg D]\!] \cdot \pi[\![a:D]\!] = 0$ by distributivity and soundness for negation. Further, since the semiring is assumed to be $+$-positive, it follows that $\pi[\![a:C]\!] \cdot \pi[\![a:D]\!] = 0$. □

Definition 9. A *provenance knowledge base* consists of a K-valued ABox $\mathcal{A}$ and a TBox $\mathcal{T}$. We say that a K-interpretation $\pi : \mathrm{Lit}_\Delta(\tau) \to K$ is strongly (or weakly) consistent with $(\mathcal{A}, \mathcal{T})$ if τ contains all role names and concept names occuring in $\mathcal{A}$ and $\mathcal{T}$ and $\mathcal{T}$, if Δ contains all individual names occurring in $\mathcal{A}$ and

(1) π satisfies all assertions occurring in $\mathcal{A}$,
(2) π is strongly (or weakly) consistent with $\mathcal{T}$.

Such a K-interpretation is also called a K-model (or a weak K-model) of $(\mathcal{A}, \mathcal{T})$.

5 Reasoning Problems for Provenance Knowledge Bases

The distinction between strong and weak consistency corresponds with a distinction between strong and weak subsumption between two concept descriptions. We say that C is *strongly subsumed* by D, in a K-interpretation π, in symbols $C \sqsubseteq_\pi D$, if $\pi[\![a:C]\!] \leq \pi[\![a:D]\!]$ for all elements a of π. Similarly C is *weakly subsumed* by D in π, in symbols $C \sqsubseteq_\pi^w D$ if $\pi[\![a:C]\!] \cdot \pi[\![a:\neg D]\!] = 0$ for all a. This also implies two notions of strong and weak equivalence between two concept descriptions, denoted $C \equiv_\pi D$, and $C \equiv_\pi^w D$. Further, we write $C \sqsubseteq_\mathcal{T} D$ and $C \sqsubseteq_{(\mathcal{A}, \mathcal{T})} D$ to denote that such a subsumption holds in all models of a TBox $\mathcal{T}$ or in all models of a provenance knowledge base $(\mathcal{A}, \mathcal{T})$, and analogously for the other subsumption and equivalence properties.

In analogy to and generalisation of the standard reasoning problems in DL we propose the following problems, for a given provenance knowledge base $(\mathcal{A}, \mathcal{T})$.

Subsumption. What kind of subsumption and equivalence properties hold between concept descriptions in K-models of $\mathcal{T}$? In particular, describe the subsumption hierarchy and the weak subsumption hierarchy entailed by $\mathcal{T}$.

Consistency. Do there exist K-interpretations that are (strongly or weakly) consistent with $(\mathcal{A}, \mathcal{T})$?

Provenance values. Given a concept assertion $a : C$, what are the possible provenance values $\pi[\![a:C]\!]$ in (weak) models of $(\mathcal{A}, \mathcal{T})$? In particular is there a possible provenance value $\pi[\![a:C]\!] \neq 0$ in some such model; this generalizes the satisfiability problem.

Query answering. Given a (Boolean) query q, formulated in some appropriate query language, what are the possible provenance values $\pi[\![q]\!]$ in models of $(\mathcal{A}, \mathcal{T})$? In particular, is $\pi[\![q]\!] \neq 0$ in all such models?

Depending on the choice of the semiring, this permits to answer questions about issues such as cost, confidences, or required clearance levels for statements that we derive from the knowledge base. Here are a few examples:

(1) Consider a provenance knowledge base $(\mathcal{A}, \mathcal{T})$ with interpretations in the tropical semiring $\mathbb{T} = (\mathbb{R}_+^\infty, \min, +, \infty, 0)$. We view $\pi[\![a : A]\!]$ as the cost of using the assertion $a : A$. If $(\mathcal{A}, \mathcal{T})$ entails a strong subsumption $C \sqsubseteq D$ then this means that for all a, it is less expensive to establish the assertion $a : C$ than $a : D$. If $(\mathcal{A}, \mathcal{T})$ entail such a subsumption only in the weak sense, then this means that whenever $a : D$ can be established for free (with cost 0), then this is also the case for $a : C$.

(2) Given a TBox $\mathcal{T}$ and an $\mathbb{A}$-valued ABox $\mathcal{A}$ (i.e. with valuations in the access control semiring), the consistency of the provenance knowledge base $(\mathcal{A}, \mathcal{T})$ means that the clearance levels required by the $\mathcal{A}$ are compatible with the hierarchy of access restrictions as imposed by the TBox. For instance, $(\mathcal{A}, \mathcal{T})$ would be inconsistent if the TBox imposes a subsumption $C \sqsubseteq D$, but $\mathcal{A}$ declares $a : C$ to be top secret and $a : D$ only confidential.

(3) Given a provenance knowledge base $(\mathcal{A}, \mathcal{T})$ with interpretations in the Viterbi semiring of confidence scores, the maximal provenance value $\pi[\![q]\!]$ of a Boolean query q in models π of $(\mathcal{A}, \mathcal{T})$ describes the confidence we can have that q holds in *some* model of $(\mathcal{A}, \mathcal{T})$.

The question arises to what extent, with what algorithmic and complexity theoretic consequences, the common reasoning techniques, such as tableaux, automata based methods, query rewriting, and so on extend to the semiring provenance setting.

6   Tableaux Rules for Provenance Knowledge Bases

A standard approach in description logics for checking the consistency of a knowledge base or an ABox is based on tableaux. A tableaux algorithm uses a system of rules to extend a given ABox by more and more assertions; for instance if an ABox $\mathcal{A}$ contains the assertion $a : C \sqcap D$, but not both $a : C$ and $a : D$, then one extends $\mathcal{A}$ to $\mathcal{A}' = \mathcal{A} \cup \{a : C, a : D\}$. This process of adding new assertions is iterated until one can either read off a model from the incremented ABox, or it contains a clash of the form $a : C$ and $a : \neg C$, so that that one can conclude that the original ABox is inconsistent. See for instance [3] for a full description of a tableaux algorithm for $\mathcal{ALC}$.

The question arises whether the tableaux approach also works for provenance knowledge bases. We show that this is indeed the case if we restrict ourselves to the class of absorptive semirings for which the natural order is a linear order. For this class, we can present a tableaux algorithm which correctly determines whether the given provenance knowledge base is consistent, provided the ABox does not contain equality statements. Moreover, for the subclass of *max-min-semirings* our tablaux rules do not only check consistency but also produce more detailed descriptions of the K-models. Additionally for max-min semirings we can allow equality statements in the ABox.

We call a K-valued ABox $\mathcal{A}$ *normalized* if each assertion in $\mathcal{A}$ is in negation normal form and for each $\alpha = a : C$ or $\alpha = (a, b) : r$ there is at most one statement about the K-value of α in $\mathcal{A}$. Additionally we disallow trivial statements $\pi[\![\alpha]\!] \geq 0$. We can normalize any K-valued ABox $\mathcal{A}$ by simply deleting all assertions $\pi[\![\alpha]\!] \geq k$ and $\pi[\![\alpha]\!] > k$ for which k is not maximal and by deleting $\pi[\![\alpha]\!] \geq k$ if $\pi[\![\alpha]\!] > j \in \mathcal{A}$ for some $j \geq k$ or if $k = 0$.

Tableaux Rules for K-valued ABox Consistency. For simplicity we will not define rules for assertions of the form $\pi[\![\alpha]\!] > k$ because they are easy adaptations of the rules for assertions of the form $\pi[\![\alpha]\!] \geq k$. However we have to exclude assertions of the form $\pi[\![\alpha]\!] = k$, because for most semirings we cannot guarantee to satisfy for instance $p \cdot q = k$ by requirements on p and q that do not depend on the value of the respective other factor. Though this is a significant restriction, it is fair to assume that in many cases it suffices to require that a concept or role assertion has 'at least truth value k' instead of requiring the provenance value to be an exact $k \in K$.

So let $\mathcal{A}$ be a K-valued ABox consisting of assertions of the form $\pi[\![a : C]\!] \geq k$ or $\pi[\![(a, b) : r]\!] \geq k$, where C is not necessarily atomic and k is a value from a provenance semiring K, which we assume to be absorptive and totally ordered by its natural order. In particular this implies that addition in K is max and that multiplication in K is deflationary in both arguments with respect to the natural order, i.e. $a \cdot c = c \cdot a \leq a$ for any $a, c \in K$. The reason for this requirement is that we would like to be able to deduce form $\pi[\![a : C \sqcup D]\!] \geq k$ that one of the assertions $\pi[\![a : C]\!] \geq k$ and $\pi[\![a : D]\!] \geq k$ also has to hold, and from $\pi[\![a : C \sqcap D]\!] \geq k$ that both of them are true. In a general semiring, this is not necessarily the case and in fact we might not get any useful information about $\pi[\![a : C]\!]$ and $\pi[\![a : D]\!]$ from $\pi[\![a : C \sqcap (\sqcup)D]\!] \geq k$. With these restrictions we are able to define tableaux rules for consistency checking of K-valued ABoxes:

$\sqcap$**-rule:** if
    1. $\pi[\![a : C \sqcap D]\!] \geq k \in \mathcal{A}$, and
    2. $\{\pi[\![a : C]\!] \geq i, \pi[\![a : D]\!] \geq j\} \not\subseteq \mathcal{A}$ for all $i, j \geq k$
    then $\mathcal{A} \longrightarrow \mathcal{A} \cup \{\pi[\![a : C]\!] \geq k, \pi[\![a : D]\!] \geq k\}$
$\sqcup$**-rule:** if
    1. $\pi[\![a : C \sqcup D]\!] \geq k \in \mathcal{A}$, and
    2. $\{\pi[\![a : C]\!] \geq j, \pi[\![a : D]\!] \geq j\} \cap \mathcal{A} = \emptyset$ for all $j \geq k$
    then $\mathcal{A} \longrightarrow \mathcal{A} \cup \{\pi[\![a : X]\!] \geq k\}$ for some $X \in \{C, D\}$
$\exists$**-rule:** if
    1. $\pi[\![a : \exists r.C]\!] \geq k \in \mathcal{A}$, and
    2. there is no b and no $i, j \geq k$ such that $\{\pi[\![(a, b) : r]\!] \geq i, \pi[\![b : C]\!] \geq j\} \subseteq \mathcal{A}$
    then $\mathcal{A} \longrightarrow \mathcal{A} \cup \{\pi[\![(a, d) : r]\!] \geq k, \pi[\![d : C]\!] \geq k\}$, where d is new in $\mathcal{A}$
$\forall$**-rule:** if

1. $\{\pi[\![a\!:\!\forall r.C]\!] \geq k, \pi[\![(a,b)\!:\!r]\!] \geq \ell\} \subseteq \mathcal{A}$ for some $\ell \in K, \ell > 0$, and
2. there are no $i, j \geq k$ such that $\{\pi[\![(a,b)\!:\!r]\!] \geq i, \pi[\![b\!:\!C]\!] \geq j\} \subseteq \mathcal{A}$

then $\mathcal{A} \longrightarrow \mathcal{A} \cup \{\pi[\![(a,b)\!:\!r]\!] \geq k, \pi[\![b\!:\!C]\!] \geq k\}$

The tableaux rules are then applied in an algorithm which works as follows. It receives a normalized K-valued ABox as input and chooses one applicable rule. Then it applies that rule, creating an extended ABox which is then transformed into a normalized one. This continues until either a clash occurs, i.e. $\mathcal{A}$ contains assertions $\pi[\![a\!:\!C]\!] \geq j$ and $\pi[\![a\!:\!\neg C]\!] \geq k$ for $j, k > 0$, or no more tableaux rules are applicable. If the algorithm registers a clash, it returns 'inconsistent', and if it does not and no more rules are applicable, it returns 'consistent' and the ABox that has been constructed.

Similarily to the algorithm for a classical (non-provenance) ABox described in [3] this algorithm is non-deterministic in two ways. Firstly, it does not specify in which order the rules are applied. This is not a problem, since these choices do not affect the outcome of the algorithm, nor the ABox that is returned. The other form of non-determinism lies in choosing the concept X in the ⊔-rule. This is a relevant choice but one can determinize the algorithm by simultaneously tracking all ABoxes one could construct at once and checking that not all of them contain a clash.

The tableaux rules are based on the implications that in K if $p \cdot q \geq k$, then $p \geq k$ and $q \geq k$ and if $p + q \geq k$, then $p \geq k$ or $q \geq k$, which hold in absorptive semirings with linear natural order. Thus it is easy to check that if the algorithm observes a clash, then the original ABox was already inconsistent. However the implication for multiplication is not an equivalence. As a consequence, not every K-model of the set of atomic assertions in the final ABox $\mathcal{A}$ will be a K-model of the original ABox. Still we can construct a K-model from these atomic assertions by setting $\pi[\![\alpha]\!] = 1$ and $\pi[\![\overline{\alpha}]\!] = 0$ if $\pi[\![\alpha]\!] \geq k \in \mathcal{A}$ for some k. Here, $\overline{\alpha}$ describes the complementary statement (in negation normal form) to α, for instance $\overline{a\!:\!\neg C} = a : C$. If there is no k such that either $\pi[\![\alpha]\!] \geq k \in \mathcal{A}$ or $\pi[\![\overline{\alpha}]\!] \geq k \in \mathcal{A}$, we assign 0 to non-negated statements α and 1 to negated ones. It is important to note that in an absorptive semiring, 1 is always the maximal element with respect to the natural order. And since $1 + 1 = 1$ and $1 \cdot 1 = 1$ in these semirings, all nonatomic β which occur in assertions in $\mathcal{A}$ will also have K-value 1 and thus satisfy their respective assertions. Thus the tableaux algorithm is sound and complete and it terminates because each step simplifies the formulae which can only be done a finite number of times.

Notice that if the algorithm returns 'consistent', we also return $\mathcal{A}$. This is because while not every K-model of the atomic assertions in $\mathcal{A}$ is a K-model of the ABox, they still are necessary conditions for satisfying the ABox. Thus the new K-valued ABox gives us some information about the K-models of the original one. This is of course not new information as the new ABox has exactly the same K-models as the old one, but it gives us some requirements for the K-values of the atomic statements.

In max-min-semirings this information on the atomic statements is even more useful. In these semirings $p \cdot q \geq k$ is equivalent to $(p \geq k$ and $q \geq k)$ and

$p+q \geq k$ is equivalent to $(p \geq k$ or $q \geq k)$. Hence we do not lose any information by applying the tableaux rules and discarding the initial assertion while keeping the added ones. It follows that if our tableaux algorithm returns 'consistent', any K-model of the atomic assertions in the newly constructed ABox that sets all K-values of positive statements not in the ABox to 0 and negative statements to 1 will also be a K-model of the assertions from the original ABox. Thus we do not only get the one model where every relevant fact is set to 1, but possibly many more K-models. Additionally for max-min semirings we can define rules for assertions of the form $\pi[\![\alpha]\!] = k$ by introducing assertions using $\leq k$ for which we can in turn define rules. This is because any addition and multiplication will always take the value of one of its summands or factors. For example $p+q = k$ is equivalent to $(p = k$ and $q \leq k)$ or $(q = k$ and $p \leq k)$. We will not write down the resulting rules here, but they can be easily constructed from such equivalences.

Tableaux Rules for Provenance Knowledge Base Consistency. A more general problem than K-valued ABox consistency, is the consistency of a given provenance knowledge base $(\mathcal{A}, \mathcal{T})$. For an acyclic TBox $\mathcal{T}$ we can do this by adding a rule for $\sqsubseteq$. This rule depends on wether we require weak or strong consistency with $\mathcal{T}$ with respect to $\sqsubseteq$. For strong consistency this looks as follows. Again we require $\mathcal{A}$ to be normalized and K to be absorptive and to have a linear natural order.

strong $\sqsubseteq$-rule: if
    1. $\pi[\![a\!:\!C]\!] \geq k \in \mathcal{A}$, $C \sqsubseteq D \in \mathcal{T}$, and
    2. $\pi[\![a\!:\!D]\!] \geq j \notin \mathcal{A}$ for all $j \geq k$
    then $\mathcal{A} \longrightarrow \mathcal{A} \cup \{\pi[\![a\!:\!D]\!] \geq k\}$

If we now adjust the tableaux algorithm to check a provenance knowledge base instead of an ABox and add the strong $\sqsubseteq$-rule to the tableaux rules, we get an algorithm that checks consistency for acyclic knowledge bases.

If we consider weak consistency, we first need an equivalence rule.

$\equiv$-rule: if
    1. $\pi[\![a\!:\!C]\!] \geq k \in \mathcal{A}$, $\{C \equiv D, D \equiv C\} \cap \mathcal{T} \neq \emptyset$, and
    2. $\pi[\![a\!:\!D]\!] \geq j \notin \mathcal{A}$ for all $j \geq k$
    then $\mathcal{A} \longrightarrow \mathcal{A} \cup \{\pi[\![a\!:\!D]\!] \geq k\}$

For the weak $\sqsubseteq$-rule we encounter a small issue, which has to do with the fact that we restricted ourselves to assertions of the form $\pi[\![\alpha]\!] \geq k$ instead of also allowing $> k$. As mentioned, this restriction is not necessary and it is easy to define the corresponding rules for $>$ for all tableaux rules defined so far. So if we allow assertions $\pi[\![\alpha]\!] > k$, the $\sqsubseteq$-rule for weak consistency looks like this:

weak $\sqsubseteq$-rule: if
    1. $\{\pi[\![a\!:\!C]\!] \geq k, \pi[\![a\!:\!C]\!] > k\} \cap \mathcal{A} \neq \emptyset$, $C \sqcap D = \bot \in \mathcal{T}$, and
    2. $\{\pi[\![a\!:\!\neg D]\!] \text{ op } j, \pi[\![a\!:\!\neg D]\!] > 0 \mid \text{op} \in \{\geq, >\}\} \cap \mathcal{A} = \emptyset$ for all $j \in K$
    then $\mathcal{A} \longrightarrow \mathcal{A} \cup \{\pi[\![a\!:\!D]\!] > 0\}$

The problem with defining this rule with only $\geq$ is that we cannot express that some value is non-zero. While using $\geq$ might seem more intuitive at first glance, this is a reasonable argument for using $>$ if one wants to restrict to only one kind of comparison. It is still possible to define a weak $\sqsubseteq$-rule using only $\geq$ but this adds some additional non-determinism. This time, it does not lie in the choice of the concept, as in the $\sqcup$-rule, but in the choice of semiring value.

weak $\sqsubseteq$-rule, $\geq$-version: if
1. $\pi[\![a\!:\!C]\!] \geq k \in \mathcal{A}$, $C \sqcap D = \bot \in \mathcal{T}$, and
2. $\pi[\![a\!:\!\neg D]\!] \geq \varepsilon \notin \mathcal{A}$ for all $\varepsilon \in K$
then $\mathcal{A} \longrightarrow \mathcal{A} \cup \{\pi[\![a\!:\!D]\!] \geq \varepsilon\}$ for some $\varepsilon \in K$

With one of the weak $\sqsubseteq$-rules added to the algorithm in place of the strong $\sqsubseteq$-rule we again get a consistency checking algorithm for acyclic provenance knowledge bases. This time it checks weak consistency within the TBox. If we use the $\geq$-version of the rule however, this algorithm is not only non-deterministic but it can in general not be determinised in the same way as the algorithm containing only the $\sqcup$-rule. The reason is that unlike for the $\sqcup$-rule we might have infinitely many choices for ε in the weak $\sqsubseteq$-rule which we cannot track all at once. With the weak $\sqsubseteq$-rule allowing $>$ we do not run into this issue.

Lastly, we can consider general TBoxes, which are not necessarily acyclic. Here we encounter the same challenge as in the Boolean case that we have to guarantee termination. Consider for instance strong TBox consistency and assume we use the rules as they are defined right now. If $\mathcal{T}$ contains $C \sqsubseteq \exists r.C$ and $\pi[\![a\!:\!C]\!] \geq k \in \mathcal{A}$ then we will add $\pi[\![a\!:\!\exists r.C]\!] \geq k$ to $\mathcal{A}$. After that we will apply their $\exists$-rule and add $\pi[\![(a,d)\!:\!r]\!] \geq k, \pi[\![d\!:\!C]\!] \geq k$ for a new symbol d and then we will repeat the same process with d. This will repeat over and over again and never terminate.

In order to avoid this issue, we need to introduce an additional termination condition for the $\exists$-rule. In the Boolean case this is done by the concept of a blocked individual name (see for instance [3]). We call a an *ancestor* of b if there is a sequence of relations $r_1, \ldots, r_l$ and of individual names $c_1, \ldots, c_{l-1}$ such that $(a, c_1)\!:\!r_1 \in \mathcal{A}, (c_1, c_2)\!:\!r_2 \in \mathcal{A}, \ldots, (c_{l-1}, b)\!:\!r_l \in \mathcal{A}$. An individual name b is called *blocked* by a if a is an ancestor of b and $\{C \mid b\!:\!C \in \mathcal{A}\} \subseteq \{C \mid a\!:\!C \in \mathcal{A}\}$. To put it less technically this means that b can be reached from a via some relation assertions in $\mathcal{A}$ and a has to satisfy any concept assertion that b has to satisfy. If we think of constructing a model, this means that if we reach such a point with the Boolean tableaux rules, we can set $b = a$ and form a loop at that point. A detailed explanation on why this is possible can be found in [3].

Now we need to adapt this termination condition to the provenance setting. We call a a K-*ancestor* of b if there is a sequence of relations $r_1, \ldots, r_l$ and of individual names $c_1, \ldots, c_{l-1}$ such that $\pi[\![(a, c_1)\!:\!r_1]\!] \geq k_1 \in \mathcal{A}, \pi[\![(c_1, c_2)\!:\!r_2]\!] \geq k_2 \in \mathcal{A}, \ldots, \pi[\![(c_{l-1}, b)\!:\!r_l]\!] \geq k_l \in \mathcal{A}$ for some $k_1, \ldots, k_l > 0$. We define an individual name b to be K-*blocked* by a if a is a K-ancestor of b and for each C such that $\pi[\![b\!:\!C]\!] \geq k \in \mathcal{A}$ we have $\pi[\![a\!:\!C]\!] \geq j \in \mathcal{A}$ for some $j \geq k$. Again the intuition is that a has to satisfy all constraints on b, also taking into account the lower bound on the K-value. We say that b is K-*blocked* if b is K-blocked by some a. Again if b is blocked by a this makes it possible to form a loop. Hence we can define the new $\exists$-rule as follows.

$\exists$-rule: if
1. $\pi[\![a\!:\!\exists r.C]\!] \geq k \in \mathcal{A}$, and
2. there is no b and no $i, j \geq k$ such that $\{\pi[\![(a, b)\!:\!r]\!] \geq i, \pi[\![b\!:\!C]\!] \geq j\} \subseteq \mathcal{A}$, and
3. a is not K-blocked

then $\mathcal{A} \longrightarrow \mathcal{A} \cup \{\pi[\![(a, d)\!:\!r]\!] \geq k, \pi[\![d\!:\!C]\!] \geq k\}$, where d is new in $\mathcal{A}$

In order to ensure termination with the help of this rule, we need to check that none of the rules will again and again increase the lower bounds that occur in $\mathcal{A}$. This would avoid the blocking condition as it would further and further restrict the conditions on the individual names that are introduced. But almost all rules do not introduce a bound which is larger than the bound of the original assertion. Only the weak $\sqsubseteq$-rule in the $\geq$-version has to be adapted slightly. Intuitively the ε which can be chosen as lower bound in that rule should be as small as possible but theoretically it may be set to a value larger than k. We can simply fix this issue by requiring that $\varepsilon \leq k$ since the only information we want to reflect is that the value is larger than 0. In this new version, the algorithm is guaranteed to terminate both for strong and weak consistency because as in the Boolean case, there will be only finitely many assertions about non-blocked names if the values from the semiring, which are introduced, do not grow. Soundness and completeness can also be proved similarly to the Boolean case for which a proof can be found in [3].

7   Provenance-Tracking Interpretations

Also for classical reasoning problems in DL, for purely Boolean knowledge bases $\mathcal{K} = (\mathcal{A}, \mathcal{T})$ a provenance approach might be helpful, at least over a fixed universe. Provenance interpretations in polynomial semirings can track precisely which combinations of atomic facts are responsible for the truth and falsity of a statement, and thus may help to 'repair' an interpretation that is inconsistent with some requirement.

Definition 10. *An* $\mathbb{N}[X, \bar{X}]$-*interpretation is* provenance-tracking *if it is induced by a mapping* $\pi : \mathrm{Lit}_\Delta(\tau) \to X \cup \bar{X} \cup \{0, 1\}$ *such that* $\pi(\mathrm{Atoms}_\Delta(\tau)) \subseteq X \cup \{0, 1\}$ *and* $\pi(\mathrm{NegAtoms}_\Delta(\tau)) \subseteq \bar{X} \cup \{0, 1\}$. *Further,* π *maps equalities and inequalities to their truth values 0 or 1.*

The idea is that if π annotates a positive or negative atom with a token, then we wish to track that literal through the model-checking computation. On the other hand annotating with 0 or 1 is done when we do not track the literal, yet we need to recall whether it holds or not in the model. See [9] for more details and potential applications of provenance-tracking interpretations.

Consider now a simple ABox $\mathcal{A}$ and some fixed, but sufficiently large, universe Δ that in particular contains all individual constants appearing in $\mathcal{A}$. Any concept or role assertion in $\mathcal{A}$ is identified with an atom $\alpha \in \text{Atoms}_\Delta(\tau)$ for an appropriate vocabulary τ. Further, let X be the set of provenance tokens p_α, for $\alpha \in \text{Atoms}_\Delta(\tau)$, and let $\bar{X}$ be the corresponding set of negative tokens $\bar{p}_\alpha$. We say that a knowledge base $\mathcal{K} = (\mathcal{A}, \mathcal{T})$ is consistent over Δ if it has a model with universe Δ.

We define the provenance tracking interpretation $\pi_\mathcal{A} : \text{Lit}_\Delta(\tau) \to \mathbb{N}[X, \bar{X}]$ by

$$\pi_\mathcal{A}(\alpha) := \begin{cases} 1 & \text{if } \alpha \in \mathcal{A} \\ p_\alpha & \text{otherwise} \end{cases}$$

$$\pi_\mathcal{A}(\neg\alpha) := \begin{cases} 0 & \text{if } \alpha \in \mathcal{A} \\ \bar{p}_\alpha & \text{otherwise} \end{cases}$$

Notice that for each assertion $a : C$ the provenance value $\pi_\mathcal{A}[\![a : C]\!]$ is a polynomial in $\mathbb{N}[X, \bar{X}]$ with indeterminates p_α and $\bar{p}_\alpha$ for $\alpha \notin \mathcal{A}$. An equation system in $\mathbb{N}[X, \bar{X}]$ is a set E of equations of form $f = 0$ with $f \in \mathbb{N}[X, \bar{X}]$. A solution of E in a semiring K is a function $h : X \cup \bar{X} \to K$, making all equations in E true, such that for each token $p \in X$ we have that $h(p) = 0$ if, and only if, $h(\bar{p}) \neq 0$. In particular, such a solution is a model-defining K-interpretation [9], defining the unique structure over Δ making precisely those atoms $\alpha \in \text{Atoms}_\Delta(\tau)$ true for which $h(p_\alpha) \neq 0$.

Definition 11. We associate with every knowledge base $\mathcal{K} = (\mathcal{A}, \mathcal{T})$ and every universe Δ the equation system $E_\mathcal{K}^\Delta$ consisting of the equations

$$\pi_\mathcal{A}[\![a : C]\!] \cdot \pi_\mathcal{A}[\![a : \neg D]\!] = 0$$

for all concept inclusions $C \sqsubseteq D \in \mathcal{T}$ and all $a \in \Delta$.

Proposition 3. *A knowledge base $\mathcal{K} = (\mathcal{A}, \mathcal{T})$ is consistent over Δ if, and only if, the equation system $E_\mathcal{K}^\Delta$ has a solution (in any semiring K).*

Due to the assumption that our semirings are +-positive, we can expand the equation system $E_\mathcal{K}^\Delta$ into a single polynomial

$$f_\mathcal{K}^\Delta(X, \bar{X}) := \sum_{C \sqsubseteq D \in \mathcal{T}} \sum_{a \in \Delta} \pi_\mathcal{A}[\![a : C]\!] \cdot \pi_\mathcal{A}[\![a : \neg D]\!]$$

and we have that the solutions of the equation $f_\mathcal{K}^\Delta(X, \bar{X}) = 0$ are in correspondence with the models of the knowledge base $\mathcal{K}$ on the universe Δ. Notice that for just finding the zeros of $f_\mathcal{K}^\Delta(X, \bar{X})$, it makes no difference whether we write it as a polynomial in $\mathbb{N}[X, \bar{X}]$, or in a simpler semiring such as $\mathbb{B}[X, \bar{X}]$, $\mathbb{W}[X, \bar{X}]$, $\mathbb{S}[X, \bar{X}]$, or even the semiring of positive Boolean functions. Notice further, that the problem whether such zeros exist is NP-complete.

However, provenance polynomials allow us to do more. We can compare solutions, and we can use this approach to find solutions that describe models that are close to a given interpretation. Assume for instance that we have an interpretation $\mathcal{I}$ that is a model of a given knowledge base, but then, after adding further facts to the to ABox and/or making changes to the TBox, it happens that $\mathcal{I}$ is no longer consistent with $(\mathcal{A}, \mathcal{T})$. We may want to get back a model by a set of changes that has minimal costs in some sense. This approach is related to work in [15] on missing query answers and integrity repairs for databases.

By dualizing $f_{\mathcal{K}}^{\Delta}(X, \bar{X})$, we obtain the polynomial

$$g_{\mathcal{K}}^{\Delta}(X, \bar{X}) := \prod_{C \sqsubseteq D \in \mathcal{T}} \prod_{a \in \Delta} (\pi_{\mathcal{A}}[\![a : \neg C]\!] + \pi_{\mathcal{A}}[\![a : D]\!])$$

and we have that $g_{\mathcal{K}}^{\Delta}(X, \bar{X}) = 0$ (as a polynomial in $\mathbb{N}[X, \bar{X}]$) if, and only if, $\mathcal{K}$ is inconsistent on Δ. More interestingly, if this is not the case, then by writing out $g_{\mathcal{K}}^{\Delta}(X, \bar{X})$ as a sum of monomials $p_1^{e_1} \ldots p_k^{e_k}$, we see that, for each such monomial, every interpretation that makes all those literals true that are associated with the tokens $p_1, \ldots p_k$ is a model of $\mathcal{K}$. In general, such a monomial does not define a specific model, but a whole class of models, because those literals α for which neither p_α nor $\bar{p}_\alpha$ occur in the monomial can be interpreted in any way. Choices between different classes of models can then be made on the basis of any (partial) order between monomials in $\mathbb{N}[X, \bar{X}]$, and this can then be refined on the basis of selection criteria between different interpretations that make the same monomial true.

Coming back to the example of defining a model that is close to a given interpretation $\mathcal{I}$ (that itself is not anymore consistent with $\mathcal{K}$) we may for instance define a *cost interpretation* $\rho : \mathrm{Lit}_\Delta(\tau) \rightarrow \mathbb{T}$ into the tropical semiring $\mathbb{T} = (\mathbb{R}_+^\infty, \min, +, \infty, 0)$ that associates with the addition of a fact to $\mathcal{I}$ a cost $c \in \mathbb{R}$, and with the deletion of a fact a cost $d \in \mathbb{R}$. More precisely, for each atom $\alpha \in \mathrm{Atoms}_\Delta(\tau)$, we would put $\rho(\alpha) = 0$ and $\rho(\neg \alpha) = d$ if $\mathcal{I} \models \alpha$, and $\rho(\alpha) = c$ and $\rho(\neg \alpha) = 0$ if $\mathcal{I} \models \neg \alpha$. By setting $\hat{\rho}(p_\alpha) := \rho(\alpha)$ and $\hat{\rho}(\bar{p}_\alpha) := \rho(\neg \alpha)$, we obtain a semiring homomorphism $\hat{\rho} : \mathbb{N}[X, \bar{X}] \rightarrow \mathbb{T}$. We would then select the monomial m in $g_{\mathcal{K}}^{\Delta}(X, \bar{X})$ with minimal value $\hat{\rho}[\![m]\!]$; notice that this coincides with the provenance value $\hat{\rho}[\![g_{\mathcal{K}}^{\Delta}(X, \bar{X})]\!]$. Given the original interpretation $\mathcal{I}$ and the monomial m, we can then define a new interpretation $\mathcal{I}(m)$ with $\mathcal{I}(m) \models \alpha$ whenever p_α occurs in m, $\mathcal{I}(m) \models \neg \alpha$ whenever $\bar{p}_\alpha$ occurs in m, and $\mathcal{I}(m) \models \alpha \Longleftrightarrow \mathcal{I} \models \alpha$ for all other atoms $\alpha \in \mathrm{Atoms}_\Delta(\tau)$.

We can view $\mathcal{I}(m)$ as a model of $(\mathcal{A}, \mathcal{T})$ which, among all interpretations with universe Δ, is obtained from $\mathcal{I}$ by a set of additions and deletions of facts that leads to minimal costs for establishing the consistency with $(\mathcal{A}, \mathcal{T})$. Notice in this context, that in case $\mathcal{K} = (\mathcal{A}, \mathcal{T})$ is inconsistent, and hence $g_{\mathcal{K}}^{\Delta}(X, \bar{X})$ is the zero polynomial, then $\hat{\rho}[\![g_{\mathcal{K}}^{\Delta}(X, \bar{X})]\!] = \infty$.

Instead of such a cost based choice, by means of an interpretation in the tropical semiring, the semiring framework permits also choices by other criteria, for instance by maximizing consistency scores, using an interpretation into the Viterbi semiring $\mathbb{V}$, or by minimizing the required clearance level, by an interpretation into the access control semiring $\mathbb{A}$.

Notice that all this is algorithmically nontrivial. First of all it assumes that we have determined a universe Δ on which we evaluate the provenance polynomials. This is a separate, nontrivial, problem, but for most description logics, we can determine bounds on the size of minimal models without too much effort, so this seems not infeasible. Second, neither the problem of finding zeros, nor the computation of a provenance polynomial in standard form, as a sum of monomials, are computationally easy, in general. However, it is a fact, that at least for reasonably expressive description logics, the common reasoning problems do have a rather high complexity anyway. It is thus not at all the case that provenance analysis makes easy problems complicated. To the contrary, we hope that it actually may help to provide a more principled approach to a number of interesting questions.

8   Conclusion and Outlook

We have reported on an algebraic framework for the provenance analysis of logics with negation that we believe to be suitable and interesting also for applications in description logics. As a first step, we have seen that provenance values of concept assertions from $\mathcal{ALC}$ on a fixed interpretation can be computed with a moderate number of semiring operations. We have then discussed which variations of the traditional reasoning problems for description logics may be interesting when we evaluate concept and role assertions in a commutative semiring, and what kind of new questions might be investigated with such an approach. We have further discussed the issue of extending the familiar tableaux based algorithmic methods to provenance knowledge basis, and we have illustrated this for certain specific cases. Finally we have investigated how provenance tracking interpretations in semirings of dual-indeterminate polynomials may also help to give a new approach to traditional (purely Boolean) reasoning problems such as the consistency of a knowledge base, by means of provenance polynomials that describe multiple models, and allow us to repair inconsistencies and to make choices between different models on a principled basis. Of course, this work so far is rather preliminary, and proposes more definitions and questions than that it provides answers.

An interesting area that we have left largely untouched so far is query rewriting. This is the problem of rewriting a (say, conjunctive or first-order) query q for a given TBox $\mathcal{T}$ as a new query $q_{\mathcal{T}}$ that evaluated on any given ABox $\mathcal{A}$ should provide the same answers as the (certain) answers of the original query q on (models of) the knowledge base $(\mathcal{A}, \mathcal{T})$. First-order rewritings are only possible for rather inexpressive description logics, but for certain somewhat more expressive ones, rewritings in Datalog are possible (see [3, Chap. 7]). A provenance approach to this problem has recently been explored in [16], but it is rather different from our methods and does not make use of dual-indeterminate polynomials. It should be interesting to combine these methods with ours, taking also into account the semirings of dual-indeterminate formal power series that provide the algebraic framework for a provenance analysis of languages that include both recursion and negation.

References

1. Amsterdamer, Y., Deutch, D., Tannen, V.: On the limitations of provenance for queries with difference. In: 3rd Workshop on the Theory and Practice of Provenance, TaPP 2011 (2011). CoRR abs/1105.2255
2. Amsterdamer, Y., Deutch, D., Tannen, V.: Provenance for aggregate queries. In: Principles of Database Systems, PODS, pp. 153–164 (2011). CoRR abs/1101.1110
3. Baader, F., Horrocks, I., Lutz, C., Sattler, U.: An Introduction to Description Logic. Cambridge University Press, Cambridge (2017)
4. Dannert, K., Grädel, E.: Semiring Provenance for Guarded Logics (submitted for publication)
5. Deutch, D., Milo, T., Roy, S., Tannen, V.: Circuits for datalog provenance. In: Proceedings of 17th International Conference on Database Theory ICDT, pp. 201–212 (2014)
6. Foster, J., Green, T., Tannen, V.: Annotated XML: queries and provenance. In: Principles of Database Systems, PODS, pp. 271–280 (2008)
7. Geerts, F., Poggi, A.: On database query languages for K-relations. J. Appl. Logic 8(2), 173–185 (2010)
8. Geerts, F., Unger, T., Karvounarakis, G., Fundulaki, I., Christophides, V.: Algebraic structures for capturing the provenance of SPARQL queries. J. ACM 63(1), 7:1–7:63 (2016)
9. Grädel, E., Tannen, V.: Semiring provenance for first-order model checking. arXiv:1712.01980 [cs.LO] (2017)
10. Grädel, E., Tannen, V.: Provenance analysis for logic and games (2019, submitted for publication)
11. Green, T.: Containment of conjunctive queries on annotated relations. Theory Comput. Syst. 49(2), 429–459 (2011)
12. Green, T., Ives, Z., Tannen, V.: Reconcilable differences. In: Database Theory - ICDT 2009, pp. 212–224 (2009)
13. Green, T., Karvounarakis, G., Tannen, V.: Provenance semirings. In: Principles of Database Systems PODS, pp. 31–40 (2007)
14. Green, T., Tannen, V.: The semiring framework for database provenance. In: Proceedings of PODS, pp. 93–99 (2017)
15. Xu, J., Zhang, W., Alawini, A., Tannen, V.: Provenance analysis for missing answers and integrity repairs. IEEE Data Eng. Bull. 41(1), 39–50 (2018)
16. Ozaki, A., Penaloza, R.: Provenance in ontology-based data access. In: Proceedings of the 31st International Workshop on Description Logics (2018)
17. Tannen, V.: Provenance propagation in complex queries. In: Tannen, V., Wong, L., Libkin, L., Fan, W., Tan, W.-C., Fourman, M. (eds.) In Search of Elegance in the Theory and Practice of Computation. LNCS, vol. 8000, pp. 483–493. Springer, Heidelberg (2013). https://doi.org/10.1007/978-3-642-41660-6_26

Extending $\mathscr{EL}^{++}$ with Linear Constraints on the Probability of Axioms

Marcelo Finger[(✉)]

Department of Computer Science, University of São Paulo, São Paulo, Brazil
mfinger@ime.usp.br

Abstract. One of the main reasons to employ a description logic such as $\mathscr{EL}^{++}$ is the fact that it has efficient, polynomial-time algorithmic properties such as deciding consistency and inferring subsumption. However, simply by adding negation of concepts to it, we obtain the expressivity of description logics whose decision procedure is ExpTime-complete. Similar complexity explosion occurs if we add probability assignments on concepts. To lower the resulting complexity, we instead concentrate on assigning probabilities to Axioms/GCIs. We show that the consistency detection problem for such a probabilistic description logic is NP-complete, and present a linear algebraic deterministic algorithm to solve it, using the column generation technique. We also examine and provide algorithms for the probabilistic extension problem, which consists of inferring the minimum and maximum probabilities for a new axiom, given a consistent probabilistic knowledge base.

1 Introduction

The logic $\mathscr{EL}^{++}$ is one of the most expressive description logics in which the complexity of inferential reasoning is tractable (Baader et al. 2005a). A direct consequence of this expressivity is that, by adding extra features to this language, its complexity easily grows exponentially. By inferential complexity we mean the complexity of decision problems such as consistency detection, finding a model that satisfies a set of constraints, or Axiom subsumption. All such problems are tractable in $\mathscr{EL}^{++}$.

In this work we are interested in adding probabilistic reasoning capabilities to $\mathscr{EL}^{++}$; however, depending on how those reasoning capabilities are added to the language, the inferential complexity can explode beyond exponential time. As shown in Sect. 3.1, by extending $\mathscr{EL}^{++}$ with probabilistic constraints over concepts, inferential reasoning becomes ExpTime-hard. Such an approach was employed in many times in the literature, either by enhancing expressive description logics such as $\mathscr{ALC}$ (Heinsohn 1994; Lukasiewicz 2008; Gutiérrez-Basulto

This study was financed in part by the Coordenação de Aperfeiçoamento de Pessoal de Nível Superior – Brasil (CAPES) – Finance Code 001.
M. Finger—Partly supported by Fapesp projects 2015/21880-4 and 2014/12236-1 and CNPq grant PQ 303609/2018-4.

et al. 2011; Jung et al. 2011), or by adding probabilistic capabilities to the family of $\mathscr{EL}$-like logics (Lutz and Schröder 2010; Gutiérrez-Basulto et al. 2017).

In this work, we study a different way of extending description logics with probabilistic reasoning capabilities, namely by applying probabilities to GCI Axioms. One of our goals is to reduce the complexity of probabilistic reasoning in description logics. Another goal is to deal with the modelling situation in which a GCI Axiom is not always true, but one can assign (subjectively) a probability to its validity. Consider the following example describing one such situation.

Example 1. Consider the following medical situation, in which a patient may have symptoms which are caused buy a disease. However, some diseases cause only very nonspecific symptoms, such as high fever, skin rash and joint pain, which may also be caused by several other diseases. Dengue is one such desease with mostly nonspecific symptoms. Dengue is a mosquito-borne viral disease and more than half of the world population lives at risk of contracting it. Among its symptoms are high fever, joint pains and skin eruptions (rash). These symptoms are common but not all patients present all symptoms. Such an uncertain situation allows for probabilistic modelling.

In a certain hospital, joint pains are caused by dengue in 20% of the cases; in the remaining 80% of the cases, there is a patient whose symptoms include joint pains whose cause is *not* attributable to dengue. Also, a patient having high fever has some probability having dengue, which increases 5% if the patient also has a rash. If those probabilistic constraints are satisfiable, one can also ask the minimum and maximum probability that a given patient is a suspect of suffering from dengue.

By adding probability constraints to axioms, we hope to model such a situation. Furthermore we will show that the inferential complexity in this case remains "only" NP-complete. In fact, our approach extends some previous results which considered adding probabilistic capabilities only to ABox statements (Finger et al. 2011). By using $\mathscr{EL}^{++}$ as the underlying formalism, ABox statements can be formulated as a particular case of GCI Axioms, so the approach here has that of (Finger et al. 2011) as a particular case, but with inferential reasoning remaining in the same complexity class.

The rest of the paper proceeds as follows. Section 2 presents the formal $\mathscr{EL}^{++}$-framework and Sect. 3 introduces probabilities over axioms, and define the probabilistic satisfiability and probabilistic extension problems. Section 4 presents an algorithm for probabilistic satisfiability that combines $\mathscr{EL}^{++}$-solving with linear algebraic methods, such as column generation. Finally, Sect. 5 presents an algorithm for the probabilistic extension problem, and then we present our conclusions in Sect. 6.

2   Preliminaries

We concentrate on the description language $\mathscr{EL}^{++}$ but without concrete domains (Baader et al. 2005a). We start with a signature consisting of a triple of countable sets $N = \langle N_C, N_R, N_I \rangle$ where N_C is a set of *concept names*, N_R is a set of *role names* and N_I is a set of *individual names*. The basic *concept description* are recursively defined as follows:

- $\top$, $\bot$ and concept names in N_C are (simple) concept descriptions;
- if C, D are concept descriptions, $C \sqcap D$ is a (conjunctive) concept description;
- if C is a concept description and $r \in N_R$, $\exists r.C$ is an (existential) concept description;
- if $a \in N_I$, $\{a\}$ is a (nominal) concept description;

If C, D are concept descriptions an *axiom*, also called a *general concept inclusion* (GCI), is an expression of the form $C \sqsubseteq D$. If $r, r_1, \ldots, r_k \in N_R$ then $r_1 \circ \cdots \circ r_k \sqsubseteq r$ is a *role inclusion* (RI). A finite set of axioms is called a *TBox* and a finite set of axioms and RIs is called a *constraint box* (CBox).

A *concept assertion* is an expression of the form $C(a)$, where $a \in N_I$ and C is a concept description; a *role assertion* is an expression of the form $r(a, b)$, where $a, b \in N_I$ and $r \in N_R$. A finite set of concept and role assertions forms an assertion box (*ABox*).

Semantically, we consider an *interpretation* $\mathcal{I} = \langle \Delta^{\mathcal{I}}, \cdot^{\mathcal{I}} \rangle$. The domain $\Delta^{\mathcal{I}}$ is a non-empty set of individuals and the interpretation function $\cdot^{\mathcal{I}}$ maps each concept name $A \in N_C$ to a subset $A^{\mathcal{I}} \subseteq \Delta^{\mathcal{I}}$, each role name $r \in N_R$ to a binary relation $r^{\mathcal{I}} \subseteq \Delta^{\mathcal{I}} \times \Delta^{\mathcal{I}}$ and each individual name $a \in N_I$ to an individual $a^{\mathcal{I}} \in \Delta^{\mathcal{I}}$. The extension of $\cdot^{\mathcal{I}}$ to arbitrary concept descriptions is inductively defined as follows.

- $\top^{\mathcal{I}} = \Delta^{\mathcal{I}}$, $\bot^{\mathcal{I}} = \varnothing$;
- $(C \sqcap D)^{\mathcal{I}} = C^{\mathcal{I}} \cap D^{\mathcal{I}}$;
- $(\exists r.C)^{\mathcal{I}} = \{x \in \Delta^{\mathcal{I}} | \exists y \in C^{\mathcal{I}}, \langle x, y \rangle \in r^{\mathcal{I}}\}$;
- $(\{a\})^{\mathcal{I}} = \{a^{\mathcal{I}}\}$.

The interpretation $\mathcal{I}$ *satisfies* an axiom $C \sqsubseteq D$ if $C^{\mathcal{I}} \subseteq D^{\mathcal{I}}$ (represented as $\mathcal{I} \models C \sqsubseteq D$); the RI $r_1 \circ \cdots \circ r_k \sqsubseteq r$ is satisfied by $\mathcal{I}$ (represented as $\mathcal{I} \models r_1 \circ \cdots \circ r_k \sqsubseteq r$) if $r_1^{\mathcal{I}} \circ \cdots \circ r_k^{\mathcal{I}} \subseteq r^{\mathcal{I}}$. A model $\mathcal{I}$ satisfies the assertion $C(a)$ (represented as $\mathcal{I} \models C(a)$) if $a^{\mathcal{I}} \in C^{\mathcal{I}}$ and satisfies the assertion $r(a, b)$ (represented as $\mathcal{I} \models r(a, b)$) if $\langle a^{\mathcal{I}}, b^{\mathcal{I}} \rangle \in r^{\mathcal{I}}$. Given a CBox $\mathcal{C}$, we write $\mathcal{I} \models \mathcal{C}$ if $\mathcal{I} \models C \sqsubseteq D$ for every axiom $C \sqsubseteq D \in \mathcal{C}$ and $\mathcal{I} \models r_1 \circ \cdots \circ r_k \sqsubseteq r$ for every role inclusion in $\mathcal{C}$. Similarly, given an ABox $\mathcal{A}$, we write $\mathcal{I} \models \mathcal{A}$ if $\mathcal{I}$ satisfies all its assertions.

Given a CBox $\mathcal{C}$, we say that it *logically entails* an axiom $C \sqsubseteq D$, represented as $\mathcal{C} \models C \sqsubseteq D$, if for every interpretation $\mathcal{I} \models \mathcal{C}$ we have that $\mathcal{I} \models C \sqsubseteq D$.

Note that in $\mathcal{EL}^{++}$ there is no need for an explicit ABox, for we have that $\mathcal{I} \models C(a)$ iff $\mathcal{I} \models \{a\} \sqsubseteq C$; and $\mathcal{I} \models r(a, b)$ iff $\mathcal{I} \models \{a\} \sqsubseteq \exists r.\{b\}$.

Given a CBox, one of the important problems for $\mathcal{EL}^{++}$ is to determine its *consistency*, namely the existence of a common model which jointly validates all expressions in the CBox. There is a polynomial algorithm which decides $\mathcal{EL}^{++}$-consistency (Baader et al. 2005b).

This decision process can be used to provide a PTIME *classification* of an $\mathcal{EL}$ CBox. Given a CBox $\mathcal{C}$, the set $\mathsf{BC}_\mathcal{C}$ of *basic concepts descriptions for $\mathcal{C}$* is given by

$$\mathsf{BC}_\mathcal{C} = \{\top, \bot\} \cup \Big\{ C \in \mathsf{N_C} | C \text{ used in } \mathcal{C} \Big\} \cup \Big\{ \{a_i\} | a_i \in \mathsf{N_I} \text{ used in } \mathcal{C} \Big\}.$$

Example 2. Consider a CBox representing the situation described in Example 1; this modelling is adapted from (Finger et al. 2011).

The following TBox $\mathcal{T}_0$ describes basic knowledge on deseases:

And the following ABox presents John's symptoms.

High-fever ⊑ Symptom	Patient(john) [≡ {john} ⊑ Patient]
Joint-pain ⊑ Symptom	High-fever(s₁) [≡ {s₁} ⊑ High-fever]
Rash ⊑ Symptom	hasSymptom(john, s₁) [≡ {john} ⊑ ∃hasSymptom.{s₁}]
Dengue ⊑ Disease	Joint-pain(s₂) [≡ {s₂} ⊑ Joint-pain]
Symptom ⊑ ∃hasCause.Disease	hasSymptom(john, s₂) [≡ {john} ⊑ ∃hasSymptom.{s₂}]
Patient ⊑ ∃suspectOf.Disease	
Patient ⊑ ∃hasSymptom.Symptom	
∃hasSymptom.(∃hasCause.Dengue)⊑	
∃suspectOf.Dengue	

Note that the uncertain information on dengue and its symptoms is not represented by the CBox above.

3 Extending $\mathcal{EL}^{++}$ with Probabilistic Constraints

One of the main reasons to employ a description logic such as $\mathcal{EL}^{++}$ is the fact that it has polynomial-time algorithmic properties such as deciding and inferring subsumption. However, it is well known that simply by adding negation of concepts to $\mathcal{EL}^{++}$, we obtain the expressivity of description logic $\mathcal{ALC}$ whose decision procedure is EXPTIME-complete (Baader et al. 2017). This complexity blow up can also be expected when adding probabilistic constraints.

3.1 Why Not Assign Probability to Concepts?

When we are dealing with probabilistic constraints on description logic, one of the first ideas is to apply conditional or unconditional probability constraints to concepts. In fact, such an approach was employed in several enhancements of description logics with probabilistic reasoning capabilities, e.g. as (Heinsohn 1994; Lukasiewicz 2008; Lutz and Schröder 2010; Gutiérrez-Basulto et al. 2017).

However, one can see how such an approach would lead to problems if applied to $\mathcal{EL}^{++}$. For each concept C one can define an associated concept $\bar{C}$ subject to the following constraints:

$$P(C) + P(\bar{C}) = 1$$
$$P(C \sqcap \bar{C}) = 0$$

Without going into the (non-trivial) semantic details of concept probabilities, it is intuitively clear that those statements force $\bar{C}$ to be the negation of C.

In fact, the first statement expresses that C and $\bar{C}$ are complementary and the second statement expresses that they are disjoint; together they mean that interpretation of C and $\bar{C}$ form a partition of the domain, and thus $\bar{C}$ is the negation of C. As a consequence, the expressivity provided by probabilities over concepts adds to $\mathscr{EL}^{++}$ the expressivity of $\mathscr{ALC}$, and as a consequence the complexity of deciding axiom subsumption becomes ExpTime-hard. Detailed complexity analysis can be found in (Gutiérrez-Basulto et al. 2017).

To lower the resulting complexity, we refrain from assigning probabilities to concepts and instead concentrate on assigning probabilities to axioms.

3.2   Probability Constraints over Axioms

Assume there is a finite number of interpretations, $\mathcal{I}_1, \ldots, \mathcal{I}_m$; let P be a mapping that attributes to each $\mathcal{I}_i$ a positive value $P(\mathcal{I}_i) \geq 0$ such that $\sum_{i=1}^{m} P(\mathcal{I}_i) = 1$.

Then given an axiom $C \sqsubseteq D$, its probability is given by:

$$P(C \sqsubseteq D) = \sum_{\mathcal{I}_i \models C \sqsubseteq D} P(\mathcal{I}_i). \tag{1}$$

Note that this definition contemplates the probability of ABox elements; for example the probability $P(C(a)) = P(\{a\} \sqsubseteq C)$.

Given axioms $C_1 \sqsubseteq D_1, \ldots, C_\ell \sqsubseteq D_\ell$ and rational numbers $b_1, \ldots, b_\ell; q$, a *probabilistic constraint* consist of the linear combination:

$$b_1 \cdot P(C_1 \sqsubseteq D_1) + \cdots b_\ell \cdot P(C_\ell \sqsubseteq D_\ell) \bowtie q, \tag{2}$$

where $\bowtie \in \{\leq, \geq, =\}$. A *PBox* is a set of probabilistic constraints. A *probabilistic knowledge base* is a pair $\langle \mathcal{C}, \mathcal{P} \rangle$, where $\mathcal{C}$ is a CBox and $\mathcal{P}$ a PBox. Note that the axioms occurring in the PBox need not occur in the CBox, and in general they do not occur in it.

The intuition behind the probability of a GCI can perhaps be better understood if seen by its complement. So the probability of an axiom $C \sqsubseteq D$ is p if the probability of its failure is $1 - p$, that is, the probability of finding a model $\mathcal{I}$ in which there exists an individual a that is in concept C but not in concept D, $\mathcal{I} \models C(a)$ and $\mathcal{I} \not\models D(a)$. Under this point of view, $P(C \sqsubseteq D) = p$ if there is a probability p of finding a model in which either no individual instantiates concept C or all individual instances of concept C are also individual instances of concept D. This has as a consequence the following, somewhat unintuitive behavior: if C is a "rare" concept in the sense that most models have no instances of C, then the probability $P(C \sqsubseteq D)$ tends to be quite high for any D, for it has as lower bound the probability of a model not having any instances of C.

Note that this intuitive view also covers ABox statements, which can be expressed as axioms of the form $\{a\} \sqsubseteq C$ and $\{a\} \sqsubseteq \exists r.\{b\}$. But in these cases, all models always satisfy the nominal $\{a\}$, so e.g. $P(\{a\} \sqsubseteq C) = p$ simply means that the probability of finding a model in which individual a is an instance of concept C is p.

3.3   Probabilistic Satisfaction and Extension Problems

A probabilistic knowledge base $\langle \mathcal{C}, \mathcal{P} \rangle$ is satisfied by interpretations $\mathcal{I}_1, \ldots, \mathcal{I}_m$ if there exists a probability distribution P over the interpretations such that

- if $P(\mathcal{I}_i) > 0$ then $\mathcal{I}_i \models \mathcal{C}$;
- all probabilistic constraints in $\mathcal{P}$ hold.

This means that an interpretation can have a positive probability mass only if it satisfies CBox $\mathcal{C}$, and the composition of all those interpretations must verify the probability of constraints in $\mathcal{P}$. A knowledge base is *satisfiable* if there exists a set of interpretations and a probability distribution over them that satisfy it.

Definition 1. *The* probabilistic satisfiability problem *for the logic $\mathscr{EL}^{++}$ consists of, given a probabilistic knowledge base $\langle \mathcal{C}, \mathcal{P} \rangle$, decide if it is satisfiable.*

Definition 2. *The* probabilistic extension problem *for the logic $\mathscr{EL}^{++}$ consists of, given a satisfiable probabilistic knowledge base $\langle \mathcal{C}, \mathcal{P} \rangle$ and an axiom $C \sqsubseteq D$, find the minimum and maximum values of $P(C \sqsubseteq D)$ that are satisfiable with $\langle \mathcal{C}, \mathcal{P} \rangle$.*

Example 3. We create a probabilistic knowledge base by extending the CBox presented in Example 2 with the uncertain information described in Example 1.

Dengue symptoms are nonspecific, so in some cases the high fever is actually caused by dengue, represented by *Ax1 := High-fever $\sqsubseteq$ $\exists$hasCause.Dengue*, and in some other cases we may have a combination of high fever and rash being caused by dengue, represented by *Ax2 := High-fever $\sqcap$ Rash $\sqsubseteq$ $\exists$hasCause.Dengue*. And the fact that joint pains are caused by dengue is represented by *Ax3 := Joint-pain $\sqsubseteq$ $\exists$hasCause.Dengue*. None of the axioms *Ax1, Ax2* or *Ax3* is always the case, but there is a probability that dengue is, in fact, the cause. The following probabilistic statements represents uncertain knowledge on the relationship between dengue and its symptoms, as observed in a hospital.

$P(Ax2) - P(Ax1) = 0.05$ The probability of dengue being the cause is 5% higher when both high fever and rash are symptoms, over just having high fever;

$P(Ax3) = 0.2$ 20% of cases of joint pain are caused by dengue.

We want to know if this probabilistic database is consistent and, in case it is, we want to find upper and lower bounds for the probability that John is a suspect of having dengue, $p_{lb} \leq P(\exists suspectOf.Dengue(john)) \leq p_{ub}$.

In order to provide algorithms that tackle both the decision and the extension problems, we provide a linear algebra formulation of those problems.

3.4   A Linear Algebraic View of Probabilistic Satisfaction and Extension Problems

Initially, let us consider only restricted probabilistic constraints of the form $P(C_i \sqsubseteq D_i) = p_i$. Consider a restricted probabilistic knowledge base $\langle \mathcal{C}, \mathcal{P} \rangle$ in which the number of probabilistic constraints is $|\mathcal{P}| = k$. Let p be a vector of size k of probabilistic constraint values. Consider a finite number of interpretations, $\mathcal{I}_1, \ldots, \mathcal{I}_m$, and let us build a $k \times m$ matrix A of $\{0,1\}$ elements a_{ij} such that

$$a_{ij} = 1 \text{ iff } \mathcal{I}_j \models C_i \sqsubseteq D_i$$

Note that column A^j contains the evaluations by interpretation $\mathcal{I}_j$ of the axioms submitted to probabilistic constraints. Given a CBox $\mathcal{C}$ and sequence of n axioms $C_1 \sqsubseteq D_1, \ldots, C_n \sqsubseteq D_n$, a $\{0,1\}$-vector u of size n *represents* a $\mathcal{C}$-satisfiable interpretation $\mathcal{I}$ if $\mathcal{I} \models \mathcal{C}$, and $c_i = 1$ iff $\mathcal{I} \models C_i \sqsubseteq D_i$ for $1 \leq i \leq n$. The idea is to assign positive probability mass $pi_j > 0$ only if A^j represents a $\mathcal{C}$-satisfiable interpretation.

Let π be a vector of size m representing a probability distribution. Consider the following set of constraints associated to $\langle \mathcal{C}, \mathcal{P} \rangle$, expressing the fact that π is a probability distribution that respects the constraints given by matrix A:

$$A \cdot \pi = p$$

$$\sum_{j=1}^{m} \pi_j = 1 \tag{3}$$

$$\pi \geq 0$$

The fact that constraints (3) actually represent satisfiability is given by the following.

Lemma 1. *A probabilistic knowledge base $\langle \mathcal{C}, \mathcal{P} \rangle$ with restricted probabilistic constraints is satisfiable iff there is a vector π that satisfies its associated constraints* (3).

When the probabilistic knowledge base is satisfiable, the number m of interpretations associated to the columns of matrix A may be exponentially large with respect to the number k of constraints in $\mathcal{P}$. However, Carathéodory's Theorem (Eckhoff 1993) guarantees that if there is a solution to (3) then there is also a small solution, namely one with at most $k + 1$ positive values.

Lemma 2. *If constraints* (3) *have a solution then there exists a solution π with at most $k + 1$ values such that $\pi_j > 0$.*

Now instead of considering only a restricted form of probability constraints, let us consider constraints of the form (2) as defined in Sect. 3, namely

$$b_{i1} \cdot P(C_1 \sqsubseteq D_1) + \cdots + b_{i\ell} \cdot P(C_\ell \sqsubseteq D_\ell) \bowtie q_i,$$

where $b_{ij}, q_i \in \mathbb{Q}$, $\bowtie \in \{\leq, \geq, =\}$ and $i = 1, \ldots k$.

We assume there are at most ℓ axioms mentioned in $\mathcal{P}$, such that $b_{i,j} = 0$ if $P(C_j \sqsubseteq D_j)$ does not occur at constraint i. Consider a matrix $B_{k \times \ell}$ and a vector x of size ℓ. We now have the following set of associated constraints to the probabilistic knowledge base $\langle \mathcal{C}, \mathcal{P} \rangle$, extending (3):

$$B \cdot x = q$$
$$A \cdot \pi = x \qquad (4)$$
$$\sum_{j=1}^{m} \pi_j = 1$$
$$x, \pi \geq 0$$

As before, A's columnns are $\{0,1\}$-representations of the validity of the axioms occurring in $\mathcal{P}$ under the interpretation $\mathcal{I}_j$. Constraints (4) are *solvable* if there are vectors x and π that verify all conditions. Analogously, the solvability of constraints (4) characterize the satisfiability of probabilistic knowledge bases with unrestricted constraints.

Lemma 3. *A probabilistic knowledge base $\langle \mathcal{C}, \mathcal{P} \rangle$ is satisfiable if and only if its associated set of constraints* (4) *are solvable.*

Example 4. Consider four interpretations for the knowledge base described in Example 3. Interpretation $\mathcal{I}_1$ satisfies CBox $\mathcal{C}$ of Example 2 and also axioms *Ax1, Ax2, Ax3*. Interpretation $\mathcal{I}_2$ satisfies $\mathcal{C}$ and axioms *Ax2, Ax3* but not *Ax1*. Interpretation $\mathcal{I}_3$ satisfies $\mathcal{C}$ and only axiom *Ax3*. Interpretation $\mathcal{I}_4$ satisfies only $\mathcal{C}$ but none of the axioms. We then consider a probability distribution π, such that $\pi(\mathcal{I}_1) = 5\%$, $\pi(\mathcal{I}_2) = 5\%$, $\pi(\mathcal{I}_3) = 10\%$, $\pi(\mathcal{I}_4) = 80\%$. The following shows that all probabilistic restrictions are satisfied.

$$
\begin{array}{c}
Ax1 \\
Ax2 \\
Ax3 \\
1
\end{array}
\begin{bmatrix}
1 & 0 & 0 & 0 \\
1 & 1 & 0 & 0 \\
1 & 1 & 1 & 0 \\
1 & 1 & 1 & 1
\end{bmatrix}
\cdot
\begin{bmatrix}
0.05 \\
0.05 \\
0.10 \\
0.80
\end{bmatrix}
=
\begin{bmatrix}
0.05 \\
0.10 \\
0.20 \\
1.00
\end{bmatrix}
$$

So $P(Ax2) - P(Ax1) = 0.05$ and $P(Ax3) = 0.2$.

When constraints (4) are *solvable*, vector x has size $\ell = O(k)$, but vector π can be exponentially large in k. By a simple linear algebraic trick, constraints of the form (4) can he presented in the following form:

$$C \cdot \pi^x = d$$
$$\pi^x \geq 0 \qquad (5)$$

In fact, it suffices to make:

$$
C = \begin{bmatrix} 0 & B \\ A & -I_\ell \\ 1 & 0 \end{bmatrix} ; \qquad
d = \begin{bmatrix} q \\ 0 \\ 1 \end{bmatrix} ; \qquad
\pi^x = \begin{bmatrix} \pi \\ x \end{bmatrix}
$$

where I_ℓ is the identity matrix, and $\mathbf{1}$ is a row of $|\pi|$ 1's. When we say that the column C^j represents a $\mathcal{C}$-satisfiable interpretation, we actually mean that the part of C^j that corresponds to some column A^j that represents a $\mathcal{C}$-satisfiable interpretation, its k-initial positions are 0 and its last element is 1. Note that C has $k+\ell+1$ rows and $|\pi|+\ell$ columns. Again, Carathéodory's Theorem guarantees small solutions.

Lemma 4. *If constraints* (4) *have a solution then there exists a solution* π^x *with at most* $k + \ell + 1$ *values such that* $\pi^x_j > 0$.

We now show that probabilistic satisfiability is NP-hard.

Lemma 5. *The satisfiability problem for probabilistic knowledge bases is NP-hard.*

Proof. We reduce SAT to probabilistic satisfiability over $\mathcal{EL}^{++}$; unlike PSAT[1], it does not suffice to set all probabilities to 1, as $\mathcal{EL}^{++}$ is decidable in polynomial time. Instead, we show how to represent 3-SAT clauses (i.e. disjunction of three literals) as a set of probabilistic axioms, basically probabilistic ABox statements. For that, consider a set of propositional variables $x_1, \ldots, x_n$ upon which the set Γ of clauses of the SAT problem are built. On the probabilistic knowledge base side, consider a single individual a and $2n$ basic concepts $X_1, \ldots, X_n$ and $\overline{X}_1, \ldots, \overline{X}_n$, subject to the following $2n$ restrictions:

$$P(a \sqsubseteq X_i) + P(a \sqsubseteq \overline{X}_i) = 1 \tag{6}$$
$$P(a \sqsubseteq X_i \sqcap \overline{X}_i) = 0$$

The idea is to represent the propositional atomic information x_i by the axiom $a \sqsubseteq X_i$, its negation by $a \sqsubseteq \overline{X}_i$, and the fact that a clause $y_i \vee \ldots \vee y_m$ holds is represented by the probabilistic statement

$$P(a \sqsubseteq \overline{Y}_i \sqcap \ldots \sqcap \overline{Y}_m) = 0. \tag{7}$$

Given Γ, we build a probabilistic knowledge base $\langle \varnothing, \mathcal{P} \rangle$ by the representation (7) of the clauses in Γ plus $2n$ assertions of the form (6). We claim that Γ is satisfiable iff $\langle \varnothing, \mathcal{P} \rangle$ is. In fact, suppose Γ is satisfiable by valuation v, make a $\mathcal{EL}^{++}$ model $\mathcal{I}$ such that $\mathcal{I} \models a \sqsubseteq X_i$ iff $v(x_i) = 1$ and assign probability 1 to $\mathcal{I}$; clearly $\langle \varnothing, \mathcal{P} \rangle$ is satisfiable. Now suppose $\langle \varnothing, \mathcal{P} \rangle$ is satisfiable, so there exists an $\mathcal{EL}^{++}$ model $\mathcal{I}$ which is assigned probability strictly bigger than 0. Construct a valuation v such that $v(x_i) = 1$ iff $\mathcal{I} \models a \sqsubseteq X_i$. Clearly $v(\Gamma) = 1$, otherwise there is a clause $y_i \vee \ldots \vee y_m$ in Γ such that $v(y_i \vee \ldots \vee y_m) = 0$ and thus $\mathcal{I} \models a \sqsubseteq \overline{Y}_i$ for $i = 1, \ldots, m$; then $P(a \sqsubseteq \overline{Y}_i \sqcap \ldots \sqcap \overline{Y}_m) \geq P(\mathcal{I}) > 0$, contradicting (7).

Theorem 1. *The satisfiability problem for probabilistic knowledge bases is NP-complete.*

Proof. Lemma 4 provides a small witness for every problem, such that by guessing that witness we can show in polynomial time that the constraints are solvable; so the problem is in NP. Lemma 5 provides NP-hardness.

---

[1] PSAT, or Probabilistic SATisfiability, consists of determining the satisfiability of a set of probabilistic assertions on classical propositional formulas (Finger and Bona 2011; Finger and De Bona 2015; Bona et al. 2014).

4   Column Generation Algorithm for Probabilistic Knowledge Base Satisfiability

An algorithm for deciding probabilistic knowledge base satisfiability has to provide a means to find a solution for restrictions (4) if one exists; otherwise determine no solution is possible. Furthermore, we will assume that the constraints are presented in format (3).

We now provide a method similar to PSAT-solving to decide the satisfiability of probabilistic knowledge base $\langle \mathcal{C}, \mathcal{P} \rangle$. We construct a vector c of costs whose size is the same as size of π^x such that $c_j \in \{0, 1\}$, $c_j = 1$ if column C^j satisfies the following condition: either the first k positions are not 0, or the next ℓ cells representing A^j correspond to an interpretation that *does not* satisfy the CBox $\mathcal{C}$, or the last position of C^j is not 1; if C^j is one of the last ℓ columns, or its first k elements are 0 and the next ℓ elements are a representation of an interpretation A^j that is $\mathcal{C}$-satisfiable and its last element is 1, then $c_j = 0$. Then we generate the following optimization problem associated to (3).

$$
\begin{aligned}
\min \quad & c' \cdot \pi^x \\
\text{subject to } & C \cdot \pi^x = d \\
& \pi^x \geq 0
\end{aligned}
\tag{8}
$$

Lemma 6. *Given a probabilistic knowledge base $\langle \mathcal{C}, \mathcal{P} \rangle$ and its associated linear algebraic restrictions (4), $\langle \mathcal{C}, \mathcal{P} \rangle$ is satisfiable if, and only if, minimization problem (8) has a minimum such that $c'\pi = 0$.*

Condition $c'\pi = 0$ means that only the columns of A^j corresponding to $\mathcal{C}$-satisfiable interpretations can be attributed probability $\pi_j > 0$, which immediately leads to solution of (8). Minimization problem (8) can be solved by an adaptation of the simplex method with column generation such that the columns of C corresponding to columns of A are generated on the fly. The simplex method is a stepwise method which at each step considers a basis consisting of $k + \ell + 1$ columns of matrix C and computes its associated cost (Bertsimas and Tsitsiklis 1997). The processing proceeds by finding a column of C outside the basis, creating a new basis by substituting one of the basis columns by this new column such that the associated cost never increases. To guarantee the cost never increases, the new column C^j to be inserted in the basis has to obey a restriction called reduced cost given by $\tilde{c}_j = c_j - c_{B_a} B_a^{-1} C^j \leq 0$, where c_j is the cost of column C^j, B_a is the basis and c_{B_a} is the cost associated to the basis. Note that in our case, we are only inserting columns that represent $\mathcal{C}$-satisfiable interpretations, so that we only insert columns of matrix C and their associated cost $c_j = 0$. Therefore, every new column C^j to be inserted in the basis has to obey the inequality

$$
c_{B_a} B_a^{-1} C^j \geq 0.
\tag{9}
$$

Note that the first k positions in C^j are 0 and the last one is always 1.

A column C^j representing a $\mathcal{C}$-satisfying interpretation may or may not satisfy condition (9). We call an interpretation that does satisfy (9) as *cost reducing interpretation*. Our strategy for column generation is given by finding cost reducing interpretations for a given basis.

Lemma 7. *There exists an algorithm that decides the existence of cost reducing interpretations whose complexity is in NP.*

Proof. Since we are dealing with a CBox in $\mathcal{EL}^{++}$, the existence of satisfying interpretations is polynomial-time and thus in NP, we can guess one such equilibrium and in polynomial time both verify it is a $\mathcal{C}$-satisfying interpretation and that is satisfies (9).   □

We can actually build a deterministic algorithm for Lemma 7 by reducing it to a SAT problem. In fact, computing $\mathcal{EL}^{++}$ satisfiability can be encoded in a 3-SAT formula φ; the condition (9) can also be encoded by a 3-SAT formula ψ in linear time, e.g. by Warners algorithm (Warners 1998), such that the SAT problem consisting of deciding $\varphi \cup \psi$ is satisfiable if, and only if, there exists a cost reducing interpretation. Furthermore its valuation provides the desired column C^j, after prefixing it with k 0's and appending a 1 at its end. This SAT-based algorithm we call the $\mathcal{EL}^{++}$-*Column Generation Method*. In practice, column generation tries *first* to output one of the last ℓ columns in C; if the insertion of one such column causes $det(B_a) = 0$ or $\pi^x \ngeq 0$, or if all the last ℓ $\mathcal{C}$-columns are in the basis, the proper$\mathcal{EL}^{++}$-Column Generation Method is invoked.

Algorithm 4.1. PKBSAT-CG: a probabilistic knowledge base solver via Column Generation

---

Input: A probabilistic knowledge base $\langle \mathcal{C}, \mathcal{P} \rangle$ and its associated set of restrictions in format (3).
Output: No, if $\langle \mathcal{C}, \mathcal{P} \rangle$ is unsatisfiable. Or a solution $\langle B_a, \pi^x \rangle$ that minimizes (8).

1: $B_a^{(0)} := I_{k+\ell+1}$;
2: $s := 0$, $\pi^{x(s)} = (B_a^{(0)})^{-1} \cdot d$ and $c^{(s)} = [1 \cdots 1]'$;
3: **while** $c^{(s)\prime} \cdot \pi^{x(s)} \neq 0$ **do**
4:      $y^{(s)} = GenerateColumn(B_a^{(s)}, \mathcal{C}, c^{(s)})$;
5:      **if** Column generation failed **then**
6:          **return** No;   {probabilistic knowledge base is unsatisfiable}
7:      **else**
8:          $B_a^{(s+1)} = merge(B_a^{(s)}, y^{(s)})$;
9:          $s\!+\!+$, recompute $\pi^{x(s)} := (B_a^{(s-1)})^{-1} \cdot d$; $c^{(s)}$ the costs of $B_a^{(s)}$ columns;
10:     **end if**
11: **end while**
12: **return** $\langle B_a^{(s)}, \pi^{x(s)} \rangle$;   {probabilistic knowledge base is satisfiable}

Algorithm 4.1 presents the top level probabilistic knowledge base decision procedure. Lines 1–2 present the initialization of the algorithm. We assume the vector p is in descending order. At the initial step we make $B^{(0)} = U_{K+1}$, this forces $\pi_{K+1}^{(0)} = p_{K+1} \geq 0$, $\pi_j^{(0)} = p_j - p_{j+1} \geq 0, 1 \leq j \leq K$; and $c^{(0)} = [c_1 \cdots c_{K+1}]'$, where $c_j = 0$ if column j in $B^{(0)}$ is an interpretation; otherwise $c_j = 1$. Thus the initial state $s = 0$ is a feasible solution.

Algorithm 4.1 main loop covers lines 3–11 which contains the column generation strategy at beginning of the loop (line 4). If column generation fails the process ends with failure in line 6; the correctness of unsatisfiability by failure is guaranteed by Lemma 6. Otherwise a column is removed and the generated column is inserted in a process we called *merge* at line 8. The loop ends successfully when the objective function (total cost) $c^{(s)'} \cdot \pi^{x(s)}$ reaches zero and the algorithm outputs a probability distribution π^x and the set of interpretations columns in B_a, at line 12.

Column generation first tries to insert a cost decreasing column from the last ℓ columns in C; if no such column is found, $\mathcal{EL}^{++}$-Column Generation Method described above is invoked.

The procedure *merge* is part of the simplex method which guarantees that given a column y and a feasible solution $\langle B_a, \pi^x \rangle$ there always exists a column j in B_a such that if $B_a[j := y]$ is obtained from B_a by replacing column j with y, then there is $\tilde{\pi}^x \geq 0$ such that $\langle B_a[j := y], \tilde{\pi}^x \rangle$ is a feasible solution. We have thus proved the following result.

Example 5. Suppose at some point for the execution of Algorithm 4.1, we have the following situation.

$$B_a = \begin{bmatrix} 0 & 1 & 0 & 1 & -1 \\ 1 & 1 & 0 & 0 & 0 \\ 0 & 1 & 0 & 0 & -1 \\ 1 & 0 & 0 & -1 & 0 \\ 1 & 1 & 1 & 0 & 0 \end{bmatrix} \quad \pi = \begin{bmatrix} 0.05 \\ 0.15 \\ 0.80 \\ 0.05 \\ 0.15 \end{bmatrix} \quad d = \begin{bmatrix} 0.05 \\ 0.20 \\ 0 \\ 0 \\ 1.00 \end{bmatrix} \begin{matrix} P(Ax2) - P(Ax1) \\ P(Ax3) \\ P(Ax1) \\ P(Ax2) \\ 1 \end{matrix} \quad c = \begin{bmatrix} 0 \\ 1 \\ 0 \\ 0 \\ 0 \end{bmatrix}$$

B_a's last two columns are C's last two columns. Columns 1 and 3 represent C-satisfiable models. Column 2 has a non-zero initial position, and thus $c_2 = 1$ is the only non-zero element of cost vector c; total cost is 0.15. Thus, $c_{B_a} B_a^{-1} C^j \geq 0$ leads to $Ax3 + Ax1 - Ax2 \geq 0$. A model that satisfies all 3 axioms is C-satisfiable and verifies the inequality; the corresponding column returned by column generation is $[0\ 1\ 1\ 1\ 1]'$. The *merge* procedure will insert it in the second column of the basis, with cost 0, which leads to a new value of $\pi = [0.05\ 0.15\ 0.80\ 0.20\ 0.15]'$; the total cost is 0, the minimum of (8) os achieved, the first 3 positions of π are a probability distribution over B_a's first 3 columns, the last ones are $P(Ax2)$ and $P(Ax1)$ in such a probabilistic model.

Theorem 2. *Algorithm 4.1 decides probabilistic knowledge base satisfiability using column generation.*

5   Algorithm for the Probabilistic Extension Problem

We now analyse the problem of probabilistic knowledge base extension. Given a satisfiable knowledge base, our aim is to find the maximum and minimum probabilistic constraints for some axiom $C \sqsubseteq D$ maintaining satisfiability. Given a precision $\varepsilon = 2^{-k}$, the algorithm works by making a binary search through the binary representation of the possible constraints to $C \sqsubseteq D$, solving a probabilistic knowledge base satisfiability problem in each step.

Algorithm 5.1 presents a procedure to solve the maximum extension problem. We invoke PKBSAT – CG($\langle C, P \rangle$) several times in the process. Obtaining the minimum extension is easily adaptable from Algorithm 5.1.

Algorithm 5.1. PKBEx-BS: a solver for probabilistic knowledge base extension via Binary Search

---

Input: A satisfiable probabilistic knowledge base $\langle C, P \rangle$, an axiom $C \sqsubseteq D$, and a precision $\varepsilon > 0$.
Output: Maximum $P(C \sqsubseteq D)$ value with precision ε.
1: $k := \lceil |\log \varepsilon| \rceil$;
2: $j := 1$, $v_{min} := 0$, $v_{max} := 1$;
3: **if** $PKBSAT\text{-}CG(C, P \cup \{P(C \sqsubseteq D) = 1\}) = $ Yes **then**
4:    $v_{min} := 1$;
5: **else**
6:    **while** $j \leq k$ **do**
7:       $v_{max} = v_{min} + \frac{1}{2^j}$;
8:       **if** $PKBSAT\text{-}CG(C, P \cup \{P(C \sqsubseteq D) \geq v_{max}\}) = $ Yes **then**
9:          $v_{min} := v_{max}$;
10:       **end if**
11:       $j{+}{+}$;
12:    **end while**
13: **end if**
14: **return** v_{min};

Suppose the goal is to find the maximum possible value for constraining $C \sqsubseteq D$. Iteration 1 solves PKBSAT for $P(C \sqsubseteq D) = 1$; if it is satisfiable, $\overline{P}(C \sqsubseteq D) = 1$, else $\overline{P}(C \sqsubseteq D) = 0$ with precision $2^0 = 1$, and it can be refined by solving PKBSAT for $P(C \sqsubseteq D) = 0.5$; if it is satisfiable, $\overline{P}(C \sqsubseteq D) = 0.5$, else $\overline{P}(C \sqsubseteq D) = 0$, both cases with precision $2^{-1} = 0.5$. One more iteration gives precision $2^{-2} = 0.25$, and it consists of solving PKBSAT for $P(C \sqsubseteq D) = 0.75$ in case the former iteration was satisfiable, otherwise $P(C \sqsubseteq D) = 0.25$. The proceeds until the desired precision is reached, which takes $|\log 2^{-k}| + 1 = k + 1$ iterations.

Theorem 3. *Given a precision $\varepsilon > 0$, probabilistic knowledge base extension can be obtained with $O(|\log \varepsilon|)$ iterations of probabilistic knowledge base satisfiability.*

Example 6. If we continue he previous examples, by applying Algorithm 5.1, we obtain that

$$0.20 \leq P(\exists suspectOf.Dengue(john)) \leq 0.95.$$

that is, the probability of John having Dengue lies between twenty percent and ninety five percent. Such a high spread means that knowing lower and upper bounds for probability is not really informative.

6 Conclusions and Further Work

In this paper we have extended the logic $\mathcal{ELS}^{++}$ with probabilistic reasoning capabilities over GCI axioms, without causing an exponentially-hard complexity blow up in reasoning tasks. We have provided deterministic algorithms based on logic and linear algebra for the problems of probabilistic satisfiability and probabilistic extension, and we have demonstrated that the decision problems are NP-complete.

In the future, we plan to explore more informative probabilistic measures, such as probabilities under minimum entropy distributions and the dealing of conditional probabilities, instead of only focusing on probabilities of $\sqsubseteq$-axioms, as was done here. We also plan to study fragments of the logics presented here in the search for tractable fragments of probabilistic description logics.

References

Baader, F., Brandt, S., Lutz, C.: Pushing the EL envelope. In: Proceedings of IJCAI 2005, San Francisco, CA, USA, pp. 364–369. Morgan Kaufmann Publishers Inc. (2005a)

Baader, F., Brandt, S., Lutz, C.: Pushing the EL envelope. Technical report LTCS-Report LTCS-05-01 (2005b)

Baader, F., Horrocks, I., Lutz, C., Sattler, U.: An Introduction to Description Logic. Cambridge University Press, Cambridge (2017)

Bertsimas, D., Tsitsiklis, J.N.: Introduction to Linear Optimization. Athena Scientific, Belmont (1997)

Bona, G.D., Cozman, F.G., Finger, M.: Towards classifying propositional probabilistic logics. J. Appl. Logic **12**(3), 349–368 (2014)

Eckhoff, J.: Helly, Radon, and Carathéodory type theorems. In: Handbook of Convex Geometry, pp. 389–448. Elsevier (1993)

Finger, M., Bona, G.D.: Probabilistic satisfiability: logic-based algorithms and phase transition. In: IJCAI 2011, pp. 528–533 (2011)

Finger, M., De Bona, G.: Probabilistic satisfiability: algorithms with the presence and absence of a phase transition. Ann. Math. Artif. Intell. **75**(3), 351–379 (2015)

Finger, M., Wassermann, R., Cozman, F.G.: Satisfiability in EL with sets of probabilistic ABoxes. In: Rosati et al. (2011)

Gutiérrez-Basulto, V., Jung, J.C., Lutz, C., Schröder, L.: A closer look at the probabilistic description logic Prob-EL. In: AAAI 2011 (2011)

Gutiérrez-Basulto, V., Jung, J.C., Lutz, C., Schröder, L.: Probabilistic description logics for subjective uncertainty. JAIR **58**, 1–66 (2017)

Heinsohn, J.: Probabilistic description logics. In: Proceedings of UAI 1994, pp. 311–318 (1994)

Jung, J.C., Gutiérrez-Basulto, V., Lutz, C., Schröder, L.: The complexity of probabilistic EL. In: Rosati et al. (2011)

Lukasiewicz, T.: Expressive probabilistic description logics. Artif. Intell. **172**(6), 852–883 (2008)

Lutz, C., Schröder, L.: Probabilistic description logics for subjective uncertainty. In: KR 2010. AAAI Press (2010)

Rosati, R., Rudolph, S., Zakharyaschev, M. (eds.): Proceedings of DL 2011. CEUR Workshop Proceedings, vol. 745. CEUR-WS.org (2011)

Warners, J.P.: A linear-time transformation of linear inequalities into conjunctive normal form. Inf. Process. Lett. **68**(2), 63–69 (1998)

Effective Query Answering
with Ontologies and DBoxes

Enrico Franconi[(✉)] and Volha Kerhet

KRDB Research Centre for Knowledge and Data,
Free University of Bozen-Bolzano, Bolzano, Italy
{franconi,kerhet}@inf.unibz.it
http://krdb.eu

Abstract. The goal of this chapter is to survey the formalisation of a precise and uniform integration between first-order ontologies, first-order queries, and classical relational databases (DBoxes) We include here non-standard variants of first-order logic, such as the one with active domain semantics and standard name assumption, used typically in database theory. We present a general framework for the rewriting of a domain independent first-order query in presence of an arbitrary domain independent first-order logic ontology over a signature extending a database signature with additional predicates. The framework supports deciding the existence of a logically equivalent and – given the ontology – safe-range first-order reformulation (called exact reformulation) of a domain independent first-order query in terms of the database signature, and if such a reformulation exists, it provides an effective approach to construct the reformulation based on interpolation using standard theorem proving techniques (i.e., tableau). Since the reformulation is a safe-range formula, it is effectively executable as an SQL query. We finally present an application of the framework with the very expressive $\mathcal{ALCHOI}$ and $\mathcal{SHOQ}$ description logics ontologies, by providing effective means to compute safe-range first-order exact reformulations of queries.

1 Introduction

We address the problem of query reformulation with expressive ontologies over databases. An ontology provides a conceptual view of the database and it is composed by constraints on a vocabulary *extending* the basic vocabulary of the data. Querying a database using the terms in such a richer ontology allows for more flexibility than using only the basic vocabulary of the relational database directly.

In this chapter we study and develop a query rewriting framework applicable to knowledge representation systems where data is stored in a classical finite relational database, in a way that in the literature has been called the *locally-closed world* assumption [12], *exact views* [13,25,26], or *DBox* [16,31]. A DBox is a set of ground atoms which semantically behaves like a database, i.e., the interpretation of the database predicates in the DBox is exactly equal to the database

© Springer Nature Switzerland AG 2019
C. Lutz et al. (Eds.): Baader Festschrift, LNCS 11560, pp. 301–328, 2019.
https://doi.org/10.1007/978-3-030-22102-7_14

relations in any model. The DBox predicates are *closed*, i.e., their extensions are the same in every interpretation, whereas the other predicates in the ontology are *open*, i.e., their extensions may vary among different interpretations. We do not consider here the *open* interpretation for the database predicates (also called *ABox* or *sound views*). In an ABox the interpretation of database predicates contains the database relations and possibly more data coming from the non-database predicates. This notion is less faithful in the representation of a database semantics since it would allow for spurious interpretations of database predicates with additional unwanted tuples not present in the original database.

In our general framework an ontology is a set of first-order formulas, and queries are (possibly open) first-order formulas. Within this setting, the framework provides precise semantic conditions to decide the existence of a safe-range first-order equivalent reformulation of a query in terms of the database signature. It also provides an constructive approach to build the reformulation with sufficient conditions. We are interested in safe-range reformulations of queries because their range-restricted syntax is needed to reduce the original query answering problem to a relational algebra evaluation (e.g., via SQL) over the original database [1]. Our framework points out several conditions on the ontologies and the queries to guarantee the existence of a safe-range reformulation. We show that these conditions are feasible in practice and we also provide an implementable method to ensure their validation. Standard theorem proving techniques can be used to compute the reformulation.

In order to be complete, our framework is applicable to ontologies and queries expressed in any fragment of first-order logic enjoying finitely controllable determinacy [26], a stronger property than the finite model property of the logic. If the employed logic does not enjoy finitely controllable determinacy our approach would become sound but incomplete, but still effectively implementable using standard theorem proving techniques. We have explored non-trivial applications where the framework is complete; in this chapter, the application with $\mathcal{ALCHOI}$ and $\mathcal{SHOQ}$ ontologies and concept queries is discussed. We show how (i) to check whether the answers to a given query with an ontology are *solely* determined by the extension of the DBox (database) predicates and, if so, (ii) to find an equivalent rewriting of the query in terms of the DBox predicates to allow the use of standard database technology (SQL) for answering the query. This means we benefit from the low computational complexity in the size of the data for answering queries on relational databases. In addition, it is possible to reuse standard techniques of description logics reasoning to find rewritings, such as in the paper by [31].

The query reformulation problem has received strong interest in classical relational database research as well as modern knowledge representation studies. Differently from the mainstream research on query reformulation [21], which is mostly based on perfect or maximally contained rewritings with sound views under relatively inexpressive constraints (see, e.g., the DL-Lite approach in [2]), we focus here on exact rewritings with exact views, since it characterises precisely the query answering problem with ontologies and databases, in the case when the exact semantics of the database must be preserved. As an example, consider a ground negative query over a given standard relational database; by adding an ontology on top of it, its answer is not supposed to change—since the query uses only the signature of the database and additional constraints are not supposed to change the meaning of the query—whereas if the database were treated as an ABox (sound views) the answer may change in presence of an ontology. This may be important from the application perspective: a DBox preserves the behaviour of the legacy application queries over a relational database. Moreover, by focussing on exact reformulations of *definable* queries (as opposed to considering the certain answer semantics to arbitrary queries, such as in DL-Lite), we guarantee that answers to queries can be subsequently composed in an arbitrary way: this may be important to legacy database applications. A comprehensive summary comparing the ABox- and DBox-based approaches to data representation in description logics appears in [8].

The approach to query reformulation with first-order theories based on exact rewritings was first thoroughly analysed in [26] by Nash, Segoufin and Vianu. They addressed the question whether a query can be answered using a set of (exact) views by means of an exact rewriting over a database represented as a DBox. The authors defined and investigated the notions of *determinacy* of a query by a set of views and its connection to exact rewriting. Nash, Segoufin and Vianu also studied several combinations of query and view languages trying to understand the expressivity of the language required to express the exact rewriting, and, thus, they obtained results on the *completeness of rewriting languages*. They investigated languages ranging from full first-order logic to conjunctive queries. In a more practical database settings, Toman and Weddell have long advocated the use of exact reformulations for automatic generation of plans that implement user queries under system constraints–a process they called *query compilation* [32]. The exact rewriting framework has also been applied to devise the formal foundations of the problems of *view update* and of characterising *unique solutions* in data exchange. In the former problem, a target view of some source database is updatable if the source predicates have an exact reformulation given the view over the target predicates [14]. In the latter problem, unique solutions exist if the target predicates have an exact reformulation given the data exchange mappings over the source predicates [27]. Another application of DBoxes has been in the context of constraints representation in ontologies [29].

The chapter is organised as follows: Sect. 2 provides the necessary formal background and definitions; Sect. 3 introduces the notion of a query determined by a database; Sect. 4 introduces a characterisation of the query reformulation problem; in Sects. 5 and 6 the conditions allowing for an effective reformulation are analysed, and a sound and complete algorithm to compute the reformulation is introduced. Finally, we present the case of $\mathcal{ALCHOI}$ and $\mathcal{SHOQ}$ ontologies in Sect. 7 and conclude in Sect. 8. This chapter extends the work first presented in [17,18].

2 Preliminaries

Let $\mathcal{FOL}(\mathbb{C}, \mathbb{P})$ be a classical function-free first-order language with equality over a signature $\Sigma = (\mathbb{C}, \mathbb{P})$, where $\mathbb{C}$ is a set of *constants* and $\mathbb{P}$ is a set of *predicates* with associated arities. The arity of a predicate P we denote by $\text{AR}(P)$. In the rest we will refer to an arbitrary fragment of $\mathcal{FOL}(\mathbb{C}, \mathbb{P})$, which will be called $\mathcal{L}$. We denote by $\sigma(\phi)$ the signature of the formula ϕ, that is all the predicates and constants occurring in ϕ. We denote with $\mathbb{P}_{\{\phi_1, \ldots, \phi_n\}}$ the set of all predicates occurring in the formulas $\phi_1, \ldots, \phi_n$, with $\mathbb{C}_{\{\phi_1, \ldots, \phi_n\}}$ the set of all constants occurring in the formulas $\phi_1, \ldots, \phi_n$; for the sake of brevity, instead of $\mathbb{P}_{\{\phi\}}$ (resp. $\mathbb{C}_{\{\phi\}}$) we write $\mathbb{P}_\phi$ (resp. $\mathbb{C}_\phi$). We denote with $\sigma(\phi_1, \ldots, \phi_n)$ the signature of the formulas $\phi_1, \ldots, \phi_n$, namely the union of $\mathbb{P}_{\{\phi_1, \ldots, \phi_n\}}$ and $\mathbb{C}_{\{\phi_1, \ldots, \phi_n\}}$. We denote the set of all variables appearing in ϕ as $\text{VAR}(\phi)$, and the set of the free variables appearing in ϕ as $\text{FREE}(\phi)$; we may use for ϕ the notation $\phi(\bar{x})$, where $\bar{x} = \text{FREE}(\phi)$ is the (possibly empty) set of free variables of the formula. The notation $\phi(\bar{x}, \bar{y})$ means $\text{FREE}(\phi) = \bar{x} \cup \bar{y}$. A formula in $\mathcal{FOL}(\mathbb{C}, \mathbb{P})$ is in *prenex normal form*, if it is written as a string of quantifiers followed by a quantifier-free part. Every formula is equivalent to a formula in prenex normal form and can be converted into it in polynomial time [23].

Let $\mathbb{X}$ be a countable set of variables we use. We define a *substitution* Θ to be a total function $\mathbb{X} \mapsto \mathbb{S}$ assigning an element of the set $\mathbb{S}$ to each variable in $\mathbb{X}$. We can see substitution as a countable set of assignments of elements from $\mathbb{S}$ to elements from $\mathbb{X}$. That is, if $\mathbb{X} = \{x_1, x_2, \ldots\}$, then $\Theta := \{x_1 \rightarrow s_1, x_2 \rightarrow s_2, \ldots\}$, where $s_1, s_2, \ldots$ are elements from $\mathbb{S}$ assigned to corresponding variables from $\mathbb{X}$ by Θ.

As usual, an *interpretation* $\mathcal{I} = \langle \Delta^{\mathcal{I}}, \cdot^{\mathcal{I}} \rangle$ includes a non-empty set – the domain $\Delta^{\mathcal{I}}$ – and an interpretation function $\cdot^{\mathcal{I}}$ defined over constants and predicates of the signature. We say that interpretations $\mathcal{I} = \langle \Delta^{\mathcal{I}}, \cdot^{\mathcal{I}} \rangle$ and $\mathcal{J} = \langle \Delta^{\mathcal{J}}, \cdot^{\mathcal{J}} \rangle$ are *equal*, written $\mathcal{I} = \mathcal{J}$, if $\Delta^{\mathcal{I}} = \Delta^{\mathcal{J}}$ and $\cdot^{\mathcal{I}} = \cdot^{\mathcal{J}}$. We use standard definitions of validity, satisfiability and entailment of a formula. An *extension* of $\phi(\bar{x})$ in interpretation $\mathcal{I} = \langle \Delta^{\mathcal{I}}, \cdot^{\mathcal{I}} \rangle$, denoted $(\phi(\bar{x}))^{\mathcal{I}}$, is the set of substitutions from the variable symbols to elements of $\Delta^{\mathcal{I}}$ which satisfy ϕ in $\mathcal{I}$. That is,

$$(\phi(\bar{x}))^{\mathcal{I}} = \{\Theta : \mathbb{X} \mapsto \Delta^{\mathcal{I}} \mid \mathcal{I}, \Theta \models \phi(\bar{x})\}.$$

If ϕ is closed, then the extension depends on whether ϕ holds in $\mathcal{I} = \langle \Delta^{\mathcal{I}}, \cdot^{\mathcal{I}} \rangle$ or not. Thus, for a closed formula ϕ, $(\phi)^{\mathcal{I}} = \{\Theta \mid \Theta : \mathbb{X} \mapsto \Delta^{\mathcal{I}}\}$ – the set of all possible substitutions assigning elements from the domain $\Delta^{\mathcal{I}}$ to variables $\mathbb{X}$ – if $\mathcal{I} \models \phi$, and $(\phi)^{\mathcal{I}} = \emptyset$, if $\mathcal{I} \not\models \phi$.

Given an interpretation $\mathcal{I} = \langle \Delta^{\mathcal{I}}, \cdot^{\mathcal{I}} \rangle$, we denote by $\mathcal{I}|_{\mathbb{S}}$ the interpretation restricted to a smaller signature $\mathbb{S} \subseteq \mathbb{P} \cup \mathbb{C}$, i.e., the interpretation with the same domain $\Delta^{\mathcal{I}}$ and the same interpretation function $\cdot^{\mathcal{I}}$ defined only for the constants and predicates from the set $\mathbb{S}$. The *semantic active domain of the signature* $\sigma' \subseteq \mathbb{P} \cup \mathbb{C}$ *in an interpretation* $\mathcal{I}$, denoted $adom(\sigma', \mathcal{I})$, is the set of all elements of the domain $\Delta^{\mathcal{I}}$ occurring in interpretations of predicates and constants from σ' in $\mathcal{I}$:

$$adom(\sigma', \mathcal{I}) := \bigcup_{P \in \sigma'} \bigcup_{(a_1,...,a_n) \in P^{\mathcal{I}}} \{a_1, \ldots, a_n\} \cup \bigcup_{c \in \sigma'} \{c^{\mathcal{I}}\}.$$

If $\sigma' = \sigma(\phi)$, where ϕ is a formula, we call $adom(\sigma(\phi), \mathcal{I})$ a *semantic active domain of the formula* ϕ in an interpretation $\mathcal{I}$.

Let $X \subseteq \mathbb{X}$ be a set of variables and $\mathbb{S}$ a set. Let us consider the restriction of a substitution to a set of variables from $\mathbb{X}$. That is, we consider a function $\Theta|_X$ assigning an element in $\mathbb{S}$ to each variable in X. We abuse the notation and call such restriction simply *substitution*. Thus, hereafter substitution is a function from a set of variables $X \subseteq \mathbb{X}$ to a set S: $\Theta : X \mapsto S$, including the empty substitution ϵ when $X = \emptyset$. *Domain* and *image* (*range*) of a substitution Θ are written as $dom(\Theta)$ and $rng(\Theta)$ respectively.

Given a subset of the set of constants $\mathbb{C}' \subseteq \mathbb{C}$, we write that a formula $\phi(\bar{x})$ is true in an interpretation $\mathcal{I}$ with its free variables substituted according to a substitution $\Theta : \bar{x} \mapsto \mathbb{C}'$ as $\mathcal{I} \models \phi(\bar{x}/\Theta)$. Given an interpretation $\mathcal{I} = \langle \Delta^{\mathcal{I}}, \cdot^{\mathcal{I}} \rangle$ and a subset of its domain $\Delta \subseteq \Delta^{\mathcal{I}}$, we write that a formula $\phi(\bar{x})$ is true in $\mathcal{I}$ with its free variables interpreted according to a substitution $\Theta : \bar{x} \mapsto \Delta$ as $\mathcal{I}, \Theta \models \phi(\bar{x})$.

A (possibly empty) finite set $\mathcal{KB}$ of closed formulas will be called an *ontology*. As usual, an interpretation in which a closed formula is true is called a *model* for the formula; the set of all models of a formula ϕ (respectively $\mathcal{KB}$) is denoted by $M(\phi)$ (respectively $M(\mathcal{KB})$).

2.1   DBoxes

A *DBox* $\mathcal{DB}$ is a *finite* set of ground atoms of the form $P(c_1, \ldots, c_n)$, where $P \in \mathbb{P}$, n-ary predicate, and $c_i \in \mathbb{C}$ ($1 \leq i \leq n$). DBox can be seen as a variant of database representation. The set of all predicates appearing in a DBox $\mathcal{DB}$ is denoted by $\mathbb{P}_{\mathcal{DB}}$, and the set of all constants appearing in $\mathcal{DB}$ is called the *active domain of* $\mathcal{DB}$, and is denoted by $\mathbb{C}_{\mathcal{DB}}$.

An interpretation $\mathcal{I}$ *embeds* a DBox $\mathcal{DB}$, if $a^{\mathcal{I}} = a$ for every DBox constant $a \in \mathbb{C}_{\mathcal{DB}}$ (the *standard name assumption (SNA)*, customary in databases, see [1]) and that, for any $o_1, \ldots, o_n \in \Delta^{\mathcal{I}}$, $(o_1, \ldots, o_n) \in P^{\mathcal{I}}$ if and only if $P(o_1, \ldots, o_n) \in \mathcal{DB}$. We denote the set of all interpretations embedding a DBox $\mathcal{DB}$ as $E(\mathcal{DB})$. A DBox $\mathcal{DB}$ is *legal for an ontology* $\mathcal{KB}$ if there exists a model of $\mathcal{KB}$ embedding $\mathcal{DB}$.

In other words, in every interpretation embedding $\mathcal{DB}$ the interpretation of any DBox predicate is always the same and it is given exactly by its content in the DBox; this is, in general, not the case for the interpretation of the non-DBox predicates. We say that all the DBox predicates are *closed*, while all the other predicates are *open* and may be interpreted differently in different interpretations. We do not consider here the *open world* assumption (the *ABox*) for embedding a DBox in an interpretation. In an open world, an interpretation $\mathcal{I}$ soundly embeds a DBox if it holds that $(c_1, \ldots, c_n) \in P^{\mathcal{I}}$ if (but *not* only if) $P(c_1, \ldots, c_n) \in \mathcal{DB}$.

In order to allow for an arbitrary DBox to be embedded, we generalise the *standard name assumption* to all the constants in $\mathbb{C}$; this implies that the domain of any interpretation necessarily includes the set of all the constants $\mathbb{C}$, which we assume to be finite. The finiteness of $\mathbb{C}$ corresponds to the finite ability of a database system to represent distinct constant symbols; $\mathbb{C}$ is meant to be unknown in advance, since different database systems may have different limits. We will see that the framework introduced here will not depend on the choice of $\mathbb{C}$.

If $\sigma' \subseteq \mathbb{P}_{DB} \cup \mathbb{C}$, then for any interpretations $\mathcal{I}$ and $\mathcal{J}$ embedding $\mathcal{DB}$ we have: $adom(\sigma', \mathcal{I}) = adom(\sigma', \mathcal{J})$; so, for such a case we introduce the notation $adom(\sigma', \mathcal{DB}) := adom(\sigma', \mathcal{I})$, where $\mathcal{I}$ is any interpretation embedding the DBox $\mathcal{DB}$, and call it a *semantic active domain of the signature σ' in a DBox $\mathcal{DB}$*. Intuitively, $adom(\sigma', \mathcal{DB})$ includes the constants from σ' and from $\mathcal{DB}$ appearing in the relations corresponding to the predicates from σ'. If $\sigma' = \sigma(\phi)$, where ϕ is a formula expressed in terms of only DBox predicates from $\mathbb{P}_{DB}$ (and possibly some constants), we call $adom(\sigma(\phi), \mathcal{DB})$ a *semantic active domain of the formula ϕ in a DBox $\mathcal{DB}$*.

2.2 Queries

A *query* is a (possibly closed) formula. Given a query $\mathcal{Q}(\bar{x})$. We define the *certain answer* to $\mathcal{Q}(\bar{x})$ over a DBox $\mathcal{DB}$ and under an ontology $\mathcal{KB}$ as follows:

Definition 1 (Certain answer). *The (certain) answer to a query $\mathcal{Q}(\bar{x})$ over a DBox $\mathcal{DB}$ under an ontology $\mathcal{KB}$ is the set of substitutions with constants:*

$$\{\Theta \mid dom(\Theta) = \bar{x}, \ rng(\Theta) \subseteq \mathbb{C}, \ \forall \mathcal{I} \in M(\mathcal{KB}) \cap E(\mathcal{DB}) : \mathcal{I} \models \mathcal{Q}(\bar{x}/\Theta)\}.$$

Query answering is defined as an *entailment* problem, and as such it is going to have the same (high) complexity as entailment.

Note that if a query $\mathcal{Q}$ is closed (i.e., a *boolean* query), then the certain answer is $\{\epsilon\}$ if $\mathcal{Q}$ is true in all the models of the ontology embedding the DBox, and $\emptyset$ otherwise. In the following, we assume that the closed formula $\mathcal{Q}(\bar{x}/\Theta)$ is neither a valid fromula nor an inconsistent formula under the ontology $\mathcal{KB}$ – with Θ a substitution $\Theta : \bar{x} \mapsto \mathbb{C}$ assigning to variables *distinct* constants not appearing in $\mathcal{Q}$, nor in $\mathcal{KB}$, nor in $\mathbb{C}_{DB}$; this assumption is needed in order to avoid trivial reformulations.

One can see that if an ontology is inconsistent or a DBox is illegal for an ontology, then the certain answer to any query over the DBox under the ontology is a set of all possible substitutions. Also, if an ontology is a tautology, we actually have a simple case of query answering over a database (DBox) without an ontology. Thus, we can discard these cases and assume to have only consistent non-tautological ontologies and legal DBoxes.

We now show that we can weaken the standard name assumption for the constants by just assuming *unique names*, without changing the certain answers. As we said before, an interpretation $\mathcal{I}$ satisfies the standard name assumption if $c^{\mathcal{I}} = c$ for any $c \in \mathbb{C}$. Alternatively, an interpretation $\mathcal{I}$ satisfies the unique name assumption (UNA) if $a^{\mathcal{I}} \neq b^{\mathcal{I}}$ for any different $a, b \in \mathbb{C}$. We denote the set of all interpretations satisfying the standard name assumption as $I(SNA)$. We denote the set of all interpretations satisfying the unique name assumption as $I(UNA)$.

The following proposition allows us to freely interchange the standard name and the unique name assumptions with interpretations embedding DBoxes. This is of practical advantage, since we can encode the unique name assumption in classical first-order logic reasoners, and many description logics reasoners do support natively the unique name assumption as an extension to OWL.

Proposition 1 (SNA vs UNA). *For any query $\mathcal{Q}(\bar{x})$, ontology $\mathcal{KB}$ and DBox $\mathcal{DB}$,*

$$\{\Theta \mid dom(\Theta) = \bar{x}, rng(\Theta) \subseteq \mathbb{C}, \forall \mathcal{I} \in I(SNA) \cap M(\mathcal{KB}) \cap E(\mathcal{DB}) : \mathcal{I} \models \mathcal{Q}(\bar{x}/\Theta)\} =$$

$$\{\Theta \mid dom(\Theta) = \bar{x}, rng(\Theta) \subseteq \mathbb{C}, \forall \mathcal{I} \in I(UNA) \cap M(\mathcal{KB}) \cap E(\mathcal{DB}) : \mathcal{I} \models \mathcal{Q}(\bar{x}/\Theta)\}.$$

Since a query can be an arbitrary first-order formula, its answer may depend on the domain, which we do not know in advance. For example, the query $Q(x) = \neg Student(x)$ over the database (DBox) $\{Student(a), Student(b)\}$, with domain $\{a, b, c\}$ has the answer $\{x \rightarrow c\}$, while with domain $\{a, b, c, d\}$ has the answer $\{x \rightarrow c, \ x \rightarrow d\}$. Therefore, the notion of *domain independent* queries has been introduced in relational databases. Here we adapt the classical definitions [1,3] to our framework: we need a more general version of domain independence, namely domain independence w.r.t an ontology, i.e., restricted to the models of an ontology.

Definition 2 (Domain independence). *A formula $\mathcal{Q}(\bar{x})$ is domain independent with respect to an ontology $\mathcal{KB}$ if and only if for every two models $\mathcal{I}$ and $\mathcal{J}$ of $\mathcal{KB}$ (i.e., $\mathcal{I} = \langle \Delta^{\mathcal{I}}, \cdot^{\mathcal{I}} \rangle$ and $\mathcal{J} = \langle \Delta^{\mathcal{J}}, \cdot^{\mathcal{J}} \rangle$) which have the same interpretations for all the predicates and constants, and for every substitution $\Theta : \bar{x} \mapsto \Delta^{\mathcal{I}} \cup \Delta^{\mathcal{J}}$ we have:*

$$rng(\Theta) \subseteq \Delta^{\mathcal{I}} \ and \ \mathcal{I}, \Theta \models \mathcal{Q}(\bar{x}) \quad iff$$
$$rng(\Theta) \subseteq \Delta^{\mathcal{J}} \ and \ \mathcal{J}, \Theta \models \mathcal{Q}(\bar{x}).$$

The above definition reduces to the classical definition of domain independence whenever the ontology is empty. A weaker version of domain independence – which is relevant for open formulas – is the following.

Definition 3 (Ground domain independence). *A formula $\mathcal{Q}(\bar{x})$ is ground domain independent if and only if $\mathcal{Q}(\bar{x}/\Theta)$ is domain independent for every substitution $\Theta : \bar{x} \mapsto \mathbb{C}$.*

For example, the formula $\neg P(x)$ is ground domain independent, but it is not domain independent.

The problem of checking whether a FOL formula is domain independent is undecidable [1]. That is why we consider a well known domain independent syntactic fragment of FOL introduced by Codd, namely the *safe-range* fragment. We recall the formal definition [1] of a safe-range formula. First, a formula should be transformed to a *safe-range normal form*, denoted by SRNF. A formula ϕ in $\mathcal{FOL}(\mathbb{C}, \mathbb{P})$ can be transformed to SRNF(ϕ) by the following steps [1]:

- Variable substitution: no distinct pair of quantifiers may employ same variable;
- Remove universal quantifiers;
- Remove implications;
- Push negation;
- Flatten 'and's and 'or's.

Definition 4 (Range restriction of a formula). *Range restriction of a formula ϕ in a safe-range normal form, denoted $rr(\phi)$, is a subset of* FREE(ϕ) *or $\bot$ recursively defined as follows:*

- $\phi = R(t_1, \ldots, t_n)$, *where each t_i is either a variable or a constant: $rr(\phi)$ is a set of variables in $t_1, \ldots, t_n$;*
- $\phi = (x = c)$ *or* $\phi = (c = x)$, *where c is a constant: $rr(\phi) = \{x\}$;*
- $\phi = (x = y)$: $rr(\phi) = \emptyset$;
- $\phi = \phi_1 \wedge \phi_2$: $rr(\phi) = rr(\phi_1) \cup rr(\phi_2)$;
- $\phi = \phi_1 \vee \phi_2$: $rr(\phi) = rr(\phi_1) \cap rr(\phi_2)$;
- $\phi = \phi_1 \wedge (x = y)$: $rr(\phi) = rr(\phi_1)$ *if* $\{x, y\} \cap rr(\phi_1) = \emptyset$; $rr(\phi) = rr(\phi_1) \cup \{x, y\}$ *otherwise;*
- $\phi = \neg\phi_1$: $rr(\phi) = \emptyset \cap rr(\phi_1)$;
- $\phi = \exists x \phi_1$: $rr(\phi) = rr(\phi_1) \setminus \{x\}$ *if $x \in rr(\phi_1)$; $rr(\phi) = \bot$ otherwise,*

where $\bot \cup Z = \bot \cap Z = \bot \setminus Z = Z \setminus \bot = \bot$ for any range restriction of a formula Z.

Definition 5 (Safe-range formula). *A formula ϕ in $\mathcal{FOL}(\mathbb{C}, \mathbb{P})$ is safe-range if and only if $rr(\text{SRNF}(\phi)) = $ FREE(ϕ).*

Definition 6 (Ground safe-range formula). *A formula $\mathcal{Q}(\bar{x})$ is ground safe-range if and only if $\mathcal{Q}(\bar{x}/\Theta)$ is safe-range for every substitution $\Theta : \bar{x} \mapsto \mathbb{C}$.*

It was proved in [1] that a safe-range fragment is *equally expressive* to a domain independent fragment; indeed a well-known Codd's theorem states that any safe-range formula is domain independent, and any domain independent formula can be easily transformed into a logically equivalent safe-range formula.

Intuitively, a formula is safe-range if and only if its variables are bounded by positive predicates or equalities. For example, the formula $\neg A(x) \wedge B(x)$ is safe-range, while queries $\neg A(x)$ and $\forall x.\, A(x)$ are not. An ontology $\mathcal{KB}$ is safe-range (domain independent), if every formula in $\mathcal{KB}$ is safe-range (domain independent). The safe-range fragment of first-order logic with the standard name assumption is equally expressive to the relational algebra, which is the core of SQL [1].

3 Determinacy

The certain answer to a query includes all the substitutions which make the query true in *all* the models of the ontology embedding the DBox: so, if a substitution would make the query true only in some model, then it would be discarded from the certain answer. In other words, it may be the case that the answer to the query is not necessarily the same among all the models of the ontology embedding the DBox. In this case, the query is not fully determined by the given source data; indeed, given the DBox, there is some answer which is possible, but not certain. Due to the indeterminacy of the query with respect to the data, the complexity to compute the certain answer in general increases up to the complexity of entailment in the fragment of first-order logic used to represent the ontology. We focus on the case when a query has the same answer over all the models of the ontology embedding the DBox, namely, when the information requested by the query is fully available from the source data without ambiguity. In this way, the indeterminacy disappears, and the complexity of the process may decrease (see Sect. 4). The *determinacy* of a query w.r.t. a DBox (source database) [13, 25, 26] has been called *implicit definability* of a formula (the query) from a set of predicates (the DBox predicates) by Beth [7].

Definition 7 (Finite Determinacy or Implicit Definability). *A query* $\mathcal{Q}(\bar{x})$ *is (finitely) determined by (or implicitly definable from) the DBox predicates* $\mathbb{P}_{DB}$ *under* $\mathcal{KB}$ *if and only if for any two models* $\mathcal{I}$ *and* $\mathcal{J}$ *of the ontology* $\mathcal{KB}$ — *both with a finite interpretation to the DBox predicates* $\mathbb{P}_{DB}$ — *whenever* $\mathcal{I}|_{\mathbb{P}_{DB} \cup \mathbb{C}} = \mathcal{J}|_{\mathbb{P}_{DB} \cup \mathbb{C}}$ *then for every substitution* $\Theta : \bar{x} \mapsto \Delta^{\mathcal{I}}$ *we have:* $\mathcal{I}, \Theta \models \mathcal{Q}(\bar{x})$ *if and only if* $\mathcal{J}, \Theta \models \mathcal{Q}(\bar{x})$.

Intuitively, the answer to an implicitly definable query does not depend on the interpretation of non-DBox predicates. Once the DBox and a domain are fixed, it is never the case that a substitution would make the query true in some model of the ontology and false in others, since the truth value of an implicitly definable query depends only on the interpretation of the DBox predicates and constants and on the domain (which are fixed). In practice, by focusing on finite determinacy of queries we guarantee that the user can always interpret the answers as being not only certain, but also *exact* – namely that whatever is not in the answer can never be part of the answer in any possible world.

In the following we focus on ontologies and queries in those fragments of $\mathcal{FOL}(\mathbb{C}, \mathbb{P})$ for which determinacy under models with a finite interpretation of DBox predicates (finite determinacy) and determinacy under models with an unrestricted interpretation of DBox predicates (unrestricted determinacy) coincide. We say that these fragments have *finitely controllable determinacy*. Sometimes it may be the case that we consider ontology in one fragment ($\mathcal{F}_1$) and query in another one ($\mathcal{F}_2$). Then we say that fragment $\mathcal{F}_1$ has *finitely controllable determinacy of queries from fragment* $\mathcal{F}_2$ if for every query expressed in $\mathcal{F}_2$ and for every ontology expressed in $\mathcal{F}_1$ finite determinacy of the query under the ontology coincides with unrestricted determinacy.

We require that whenever a query is finitely determined then it is also determined in unrestricted models (the reverse is trivially true). Indeed, the results we obtained would fail if finite determinacy and unrestricted determinacy do not coincide: it can be shown [20] that Theorem 3 below fails if we consider only models with a finite interpretation of DBox predicates.

Example 1 (Example from database theory). Let $\mathbb{P} = \{P, R, A\}$, $\mathbb{P}_{DB} = \{P, R\}$,

$$\mathcal{KB} = \{\forall x, y, z.\, R(x, y) \land R(x, z) \to y = z,$$
$$\forall x, y.\, R(x, y) \to \exists z.\, R(z, x),$$
$$(\forall x, y.\, R(x, y) \to \exists z.\, R(y, z)) \to (\forall x.\, A(x) \leftrightarrow P(x))\}.$$

$\mathcal{KB}$ is domain independent. The formula $\forall x, y.\, R(x, y) \to \exists z.\, R(y, z)$ is entailed from the first two formulas *only* over finite interpretations of R. The query $\mathcal{Q} = A(x)$ is domain independent and finitely determined by P (it is equivalent to $P(x)$ under the models with a finite interpretation of R), but it is not determined by any DBox predicate under models with an unrestricted interpretation of R. The fragment in which $\mathcal{KB}$ and $\mathcal{Q}$ are expressed does not enjoy finitely controllable determinacy.

The next theorem immediately follows from the example above.

Theorem 1. *Domain independent fragment does not have finitely controllable determinacy.*

Let $\mathcal{Q}$ be any formula in $\mathcal{FOL}(\mathbb{C}, \mathbb{P})$ and $\widetilde{\mathcal{Q}}$ be the formula obtained from it by uniformly replacing every occurrence of each non-DBox predicate P with a new predicate $\widetilde{P}$. We extend this renaming operator $\widetilde{\cdot}$ to any set of formulas in a natural way. One can check whether a query is implicitly definable by using the following theorem.

Theorem 2 (Testing determinacy, [7]). *A query $\mathcal{Q}(\bar{x})$ is implicitly definable from the DBox predicates $\mathbb{P}_{DB}$ under the ontology $\mathcal{KB}$ if and only if $\mathcal{KB} \cup \widetilde{\mathcal{KB}} \models \forall \bar{x}.\, \mathcal{Q}(\bar{x}) \leftrightarrow \widetilde{\mathcal{Q}}(\bar{x})$.*

This theorem means, that the problem of checking whether a query is implicitly definable reduces to the problem of checking entailment in first-order logic.

The *exact reformulation* of a query [26] (also called *explicit definition* by [7]) is a formula logically equivalent to the query which makes use *only* of DBox predicates and constants.

Definition 8 (Exact reformulation or explicit definability). *A query $\mathcal{Q}(\bar{x})$ is explicitly definable from the DBox predicates $\mathbb{P}_{DB}$ under the ontology $\mathcal{KB}$ if and only if there is some formula $\widehat{\mathcal{Q}}(\bar{x})$ in $\mathcal{FOL}(\mathbb{C}, \mathbb{P})$, such that $\mathcal{KB} \models \forall \bar{x}.\, \mathcal{Q}(\bar{x}) \leftrightarrow \widehat{\mathcal{Q}}(\bar{x})$ and $\sigma(\widehat{\mathcal{Q}}) \subseteq \mathbb{P}_{DB}$. We call this formula $\widehat{\mathcal{Q}}(\bar{x})$ an exact reformulation of $\mathcal{Q}(\bar{x})$ under $\mathcal{KB}$ over $\mathbb{P}_{DB}$.*

Determinacy of a query is completely characterised by the existence of an exact reformulation of the query: it is well known that a first-order query is determined by DBox predicates *if and only if* there exists a first-order exact reformulation.

Theorem 3 (Projective Beth definability, [7]). *A query $\mathcal{Q}(\bar{x})$ is implicitly definable from the DBox predicates $\mathbb{P}_{DB}$ under an ontology $\mathcal{KB}$, if and only if it is explicitly definable as a formula $\widehat{\mathcal{Q}}(\bar{x})$ in $\mathcal{FOL}(\mathbb{C}, \mathbb{P})$ over $\mathbb{P}_{DB}$ under $\mathcal{KB}$.*

3.1 Finite Controllability of Determinacy for GNFO

We consider a *guarded negation first-order logic* (GNFO) [4,5] – a fragment of FOL in which all occurrences of negation are *guarded* by an atomic predicate. Formally it consists of all formulas generated by the following recursive definition:

$$\phi ::= R(t_1, \ldots, t_n) \mid t_1 = t_2 \mid \phi_1 \wedge \phi_2 \mid \phi_1 \vee \phi_2 \mid \exists x. \phi \mid \alpha \wedge \neg \phi \qquad (1)$$

where each t_i is either a variable or a constant, α in $\alpha \wedge \neg \phi$ is an atomic formula (possibly an equality statement) containing all free variables of ϕ. This fragment is "good" in a sense that it is decidable and has *finite model property*, that we use to prove Theorem 4.

Definition 9 (Answer-guarded formula). *A first-order logic formula is answer-guarded if it has a form*

$$Atom(\bar{x}) \wedge \varphi(\bar{x}),$$

where $\varphi(\bar{x})$ is some first-order logic formula and Atom is a predicate which arity is equal to the number of free variables of the formula.

The following theorem holds.

Theorem 4. *GNFO has finitely controllable determinacy of*

- *answer-guarded GNFO queries;*
- *boolean GNFO queries;*
- *GNFO queries with one free variable.*

This result is interesting as it is and also important for us because we consider GNFO subfragments of DLs for application of our query reformulation framework. Queries in these subfragments are either boolean or with one free variable (concept queries) (Sect. 7).

4   Exact Safe-Range Query Reformulation

In this section we analyse the conditions under which the original query answering problem corresponding to an entailment problem can be reduced systematically to a model checking problem of a safe-range formula over the database (e.g., using a database system with SQL).

Given a DBox signature $\mathbb{P}_{DB}$, an ontology $\mathcal{KB}$, and a query $\mathcal{Q}(\bar{x})$ expressed in some fragment of $\mathcal{FOL}(\mathbb{C},\mathbb{P})$ and determined by the DBox predicates, our goal is to find a safe-range reformulation $\widehat{\mathcal{Q}}(\bar{x})$ of $\mathcal{Q}(\bar{x})$ in $\mathcal{FOL}(\mathbb{C},\mathbb{P})$, that when evaluated as a relational algebra expression over a legal DBox, gives the same answer as the certain answer to $\mathcal{Q}(\bar{x})$ over the DBox under $\mathcal{KB}$. This can be reformulated as the following problem:

Problem 1 (Exact safe-range query reformulation). Find an exact reformulation $\widehat{\mathcal{Q}}(\bar{x})$ of $\mathcal{Q}(\bar{x})$ under $\mathcal{KB}$ as a safe-range query in $\mathcal{FOL}(\mathbb{C},\mathbb{P})$ over $\mathbb{P}_{DB}$.

Since an exact reformulation is equivalent under the ontology to the original query, the certain answer to the original query and to the reformulated query are identical. More precisely, the following proposition holds.

Proposition 2. *Given a DBox $\mathcal{DB}$, let $\mathcal{Q}(\bar{x})$ be implicitly definable from $\mathbb{P}_{DB}$ under $\mathcal{KB}$ and let $\widehat{\mathcal{Q}}(\bar{x})$ be an exact reformulation of $\mathcal{Q}(\bar{x})$ under $\mathcal{KB}$ over $\mathbb{P}_{DB}$, then:*

$$\{\Theta \mid dom(\Theta) = \bar{x},\ rng(\Theta) \subseteq \mathbb{C},\ \forall \mathcal{I} \in M(\mathcal{KB}) \cap E(\mathcal{DB}) : \mathcal{I} \models \mathcal{Q}(\bar{x}/\Theta)\} =$$
$$\{\Theta \mid dom(\Theta) = \bar{x},\ rng(\Theta) \subseteq \mathbb{C},\ \forall \mathcal{I} \in M(\mathcal{KB}) \cap E(\mathcal{DB}) : \mathcal{I} \models \widehat{\mathcal{Q}}(\bar{x}/\Theta)\}.$$

From the above equation it is clear that in order to answer to an exactly reformulated query, one may still need to consider all the models of the ontology embedding the DBox, i.e., we still have an entailment problem to solve. The following theorem states the condition to reduce the original query answering problem – based on entailment – to the problem of checking the validity of the exact reformulation over a *single* model: the condition is that the reformulation should be domain independent.

Theorem 5 (Adequacy of exact safe-range query reformulation). *Let $\mathcal{DB}$ be a DBox which is legal for $\mathcal{KB}$, and let $\mathcal{Q}(\bar{x})$ be a query. If $\widehat{\mathcal{Q}}(\bar{x})$ is an exact domain independent (or safe-range) reformulation of $\mathcal{Q}(\bar{x})$ under $\mathcal{KB}$ over $\mathbb{P}_{DB}$, then:*

$$\{\Theta \mid dom(\Theta) = \bar{x},\ rng(\Theta) \subseteq \mathbb{C},\ \forall \mathcal{I} \in M(\mathcal{KB}) \cap E(\mathcal{DB}) : \mathcal{I} \models \mathcal{Q}(\bar{x}/\Theta)\} =$$
$$\{\Theta \mid dom(\Theta) = \bar{x},\ rng(\Theta) \subseteq adom(\sigma(\widehat{\mathcal{Q}}), \mathcal{DB}),\ \forall \mathcal{I} = \langle \mathbb{C}, \cdot^{\mathcal{I}} \rangle \in E(\mathcal{DB}) :$$
$$\mathcal{I}|_{\mathbb{P}_{DB} \cup \mathbb{C}} \models \widehat{\mathcal{Q}}(\bar{x}/\Theta)\}.$$

Since, given a DBox $\mathcal{DB}$, for all interpretations $\mathcal{I} = \langle \mathbb{C}, \cdot^{\mathcal{I}} \rangle$ embedding $\mathcal{DB}$ there is only one interpretation $\mathcal{I}|_{\mathbb{P}_{DB} \cup \mathbb{C}}$ with the signature restricted to the DBox predicates, this theorem reduces the entailment problem to a model checking problem.

A safe-range reformulation is *necessary* to transform a first-order query to a relational algebra query which can then be evaluated by using SQL techniques. The theorem above shows in addition that being safe-range is also a *sufficient* property for an exact reformulation to be correctly evaluated as an SQL query.

Let us now see an example in which we cannot reduce the problem of answering an exact reformulation to model checking over a DBox, if the exact reformulation is not safe-range.

Example 2. Let $\mathbb{P} = \{P, A\}$, $\mathbb{P}_{\mathcal{DB}} = \{P\}$, $\mathbb{C} = \{a\}$, $\mathcal{DB} = \{P(a, a)\}$, $\mathcal{KB} = \{\forall y. P(a, y) \vee A(y)\}$, $\mathcal{Q}(\bar{x}) = \widehat{\mathcal{Q}}(\bar{x}) = \forall y. P(x, y)$ (i.e., $\bar{x} = \{x\}$).

- $\mathbb{C}$ includes the active domain $\mathbb{C}_{\mathcal{DB}}$ (it is actually equal).
- $\mathcal{DB}$ is legal for $\mathcal{KB}$ because there is $\mathcal{I} = \langle \{a\}, \cdot^{\mathcal{I}} \rangle$ such that $P^{\mathcal{I}} = \{(a, a)\}$, $A^{\mathcal{I}} = \emptyset$ and obviously, $\mathcal{I} \in M(\mathcal{KB})$.
- $\{\Theta \mid dom(\Theta) = \bar{x}, \ rng(\Theta) \subseteq \mathbb{C}, \ \forall \mathcal{I} \in M(\mathcal{KB}) \cap E(\mathcal{DB}) : \mathcal{I} \models \mathcal{Q}(\bar{x}/\Theta)\} = \emptyset$
  because one can take $\mathcal{I} = \langle \{a, b\}, \cdot^{\mathcal{I}} \rangle$ such that $P^{\mathcal{I}} = \{(a, a)\}$,
  $A^{\mathcal{I}} = \{b\}$; then $\mathcal{I} \in M(\mathcal{KB}) \cap E(\mathcal{DB})$, but for the only possible substitution $\{x \to a\}$ we have: $\mathcal{I} \not\models \forall y \, P(a, y)$.
- However,
  $\{\Theta \mid dom(\Theta) = \bar{x}, \ rng(\Theta) \subseteq adom(\sigma(\widehat{\mathcal{Q}}), \mathcal{DB}), \ \forall \mathcal{I} = \langle \mathbb{C}, \cdot^{\mathcal{I}} \rangle \in E(\mathcal{DB}) : \mathcal{I}|_{\mathbb{P}_{\mathcal{DB}} \cup \mathbb{C}} \models \widehat{\mathcal{Q}}(\bar{x}/\Theta)\} = \{x \to a\}$

As we have seen, answers to a query for which a reformulation exists contain only constants from the active domain of the DBox and the query (Theorem 5); therefore, ground statements in the ontology involving non-DBox predicates and non-active domain constants (for example, as ABox statements) will not play any role in the final evaluation of the reformulated query over the DBox.

5 Conditions for an Exact Safe-Range Reformulation

We have just seen the importance of getting an exact safe-range query reformulation. In this section we are going to study the conditions under which an exact safe-range query reformulation exists.

First of all, we will focus on the semantic notion of safe-range, namely domain independence. While implicit definability is – as we already know – a sufficient condition for the existence of an exact reformulation, it does not guarantee alone the existence of a domain independent reformulation.

Example 3. Let $\mathbb{P} = \{A, B\}$, $\mathbb{P}_{\mathcal{DB}} = \{B\}$, $\mathcal{KB} = \{\forall x. B(x) \leftrightarrow A(x)\}$, $\mathcal{Q}(x) = \neg A(x)$. Then $\mathcal{Q}(x)$ is implicitly definable from $\mathbb{P}_{\mathcal{DB}}$ under $\mathcal{KB}$, and every exact reformulation of $\mathcal{Q}(x)$ over $\mathbb{P}_{\mathcal{DB}}$ under $\mathcal{KB}$ is logically equivalent to $\neg A(x)$ and not domain independent.

By looking at the example, it seems that the reason for the non domain independent reformulation lies in the fact that the ontology, which is domain independent, cannot guarantee existence of an exact domain independent reformulation of the non domain independent query. However, let us consider the following examples.

Example 4. Let $\mathbb{P}_{DB} = \{A, C\}$, $\mathcal{KB} = \{\neg A(a), \forall x. A(x) \leftrightarrow B(x)\}$ and let a query $\mathcal{Q}(x) = \exists y \neg B(y) \wedge C(x)$. It is easy to see that $\mathcal{KB}$ is domain independent and $\mathcal{Q}(x)$ is not. $\mathcal{Q}(x)$ is implicitly definable from $\mathbb{P}_{DB}$ under $\mathcal{KB}$, and $\widehat{\mathcal{Q}}(x) = \neg A(a) \wedge C(x)$ is an exact domain independent reformulation of $\mathcal{Q}(x)$.

Example 5. Let $\mathbb{P}_{DB} = \{B\}$, $\mathcal{KB} = \{\neg A(a), \forall x. A(x) \leftrightarrow B(x)\}$ and let a query $\mathcal{Q} = \neg A(a)$. $\mathcal{KB}$ and $\mathcal{Q}$ are domain independent. $\mathcal{Q}$ is implicitly definable from $\mathbb{P}_{DB}$ under $\mathcal{KB}$, and $\widehat{\mathcal{Q}} = \exists y \neg B(y)$ is an exact reformulation of $\mathcal{Q}$, which is not domain independent.

That is, the following proposition holds.

Proposition 3. *Domain independence of an ontology and an original query does not guarantee domain independence of an exact reformulation of the query under the ontology over any set of DBox predicates.*

It is obvious that in spite of the fact that the query $\mathcal{Q}(x)$ form the Example 4 is not domain independent, it is domain independent with respect to the ontology $\mathcal{KB}$. In other words, in this case the ontology guarantees the existence of an exact domain independent reformulation. With queries that are domain independent with respect to an ontology, the following theorem holds, giving the *semantic* requirements for the existence of an exact domain independent reformulation.

Theorem 6 (Semantic characterisation). *Given a set of DBox predicates $\mathbb{P}_{DB}$, a domain independent ontology $\mathcal{KB}$, and a query $\mathcal{Q}(\bar{x})$. A domain independent exact reformulation $\widehat{\mathcal{Q}}(\bar{x})$ of $\mathcal{Q}(\bar{x})$ over $\mathbb{P}_{DB}$ under $\mathcal{KB}$ exists if and only if $\mathcal{Q}(\bar{x})$ is implicitly definable from $\mathbb{P}_{DB}$ under $\mathcal{KB}$ and it is domain independent with respect to $\mathcal{KB}$.*

The above theorem shows us the semantic conditions to have an exact domain independent reformulation of a query, but it does not give us a method to compute such reformulation and its equivalent safe-range form. The following theorem gives us sufficient conditions for the existence of an exact safe-range reformulation in any decidable fragment of $\mathcal{FOL}(\mathbb{C}, \mathbb{P})$ where finite and unrestricted determinacy coincide (i.e., a fragment with finitely controllability of determinacy), and gives us a constructive way to compute it, if it exists.

Theorem 7 (Constructive). *Given a DBox $\mathcal{DB}$. If:*

1. *$\mathcal{KB}$ is a safe-range ontology (that is, $\mathcal{KB}$ is domain independent),*
2. *$\mathcal{Q}(\bar{x})$ is a safe-range query (that is, $\mathcal{Q}(\bar{x})$ is domain independent),*
3. *$\mathcal{KB} \cup \widetilde{\mathcal{KB}} \models \forall \bar{x}. \mathcal{Q}(\bar{x}) \leftrightarrow \widetilde{\mathcal{Q}}(\bar{x})$ (that is, $\mathcal{Q}(\bar{x})$ is implicitly definable from $\mathbb{P}_{DB}$ under $\mathcal{KB}$),*

then there exists an exact reformulation $\widehat{\mathcal{Q}}(\bar{x})$ of $\mathcal{Q}(\bar{x})$ as a safe-range query in $\mathcal{FOL}(\mathbb{C}, \mathbb{P})$ over $\mathbb{P}_{\mathcal{DB}}$ under $\mathcal{KB}$ that can be obtained constructively.

In order to constructively compute the exact safe-range query reformulation we use the tableau based method to find the Craig's interpolant [15] to compute $\widehat{\mathcal{Q}}(\bar{x})$ from a validity proof of the implication $(\mathcal{KB} \wedge \mathcal{Q}(\bar{x})) \to (\widetilde{\mathcal{KB}} \to \widetilde{\mathcal{Q}}(\bar{x}))$. See Sect. 6 for full details.

Let us now consider a fully worked out example, adapted from the paper by [26].

Example 6. Given: $\mathbb{P} = \{R, V_1, V_2, V_3\}$, $\mathbb{P}_{\mathcal{DB}} = \{V_1, V_2, V_3\}$,

$$\mathcal{KB} = \{\forall x, y. V_1(x, y) \leftrightarrow \exists z, v. R(z, x) \wedge R(z, v) \wedge R(v, y),$$
$$\forall x, y. V_2(x, y) \leftrightarrow \exists z. R(x, z) \wedge R(z, y),$$
$$\forall x, y. V_3(x, y) \leftrightarrow \exists z, v. R(x, z) \wedge R(z, v) \wedge R(v, y)\},$$
$$\mathcal{Q}(x, y) = \exists z, v, u. R(z, x) \wedge R(z, v) \wedge R(v, u) \wedge R(u, y).$$

The conditions of the theorem are satisfied: $\mathcal{Q}(x, y)$ is implicitly definable from $\mathbb{P}_{\mathcal{DB}}$ under $\mathcal{KB}$ (since $\sigma(\mathcal{Q}) \subseteq \mathbb{P}_{\mathcal{DB}}$); $\mathcal{Q}(x, y)$ is safe-range; $\mathcal{KB}$ is safe-range. Therefore, with the tableau method one finds the Craig's interpolant to compute $\widehat{\mathcal{Q}}(x, y)$ from a validity proof of the implication $(\mathcal{KB} \wedge \mathcal{Q}(\bar{x})) \to (\widetilde{\mathcal{KB}} \to \widetilde{\mathcal{Q}}(\bar{x}))$ and obtain $\widehat{\mathcal{Q}}(x, y) = \exists z. V_1(x, z) \wedge \forall v. (V_2(v, z) \to V_3(v, y))$ – an exact ground safe-range reformulation.

Since the answer to $\widehat{\mathcal{Q}}(x, y)$ is in the semantic active domain of the signature $\sigma(\widehat{\mathcal{Q}}) \subseteq \mathbb{P}_{\mathcal{DB}} \cup \mathbb{C}_{\widehat{\mathcal{Q}}}$ in the DBox $\mathcal{DB}$ (it follows from Theorem 5), all fee variables in $\widehat{\mathcal{Q}}(x, y)$ can be "guarded" by some DBox predicates or constants. Note that $\mathcal{Q}(x, y) = \exists z, v, u. R(z, x) \wedge R(z, v) \wedge R(v, u) \wedge R(u, y) \equiv^{\mathcal{KB}} \exists z, v. R(z, x) \wedge R(z, v) \wedge V_2(v, y) \equiv^{\mathcal{KB}} \mathcal{Q}(x, y) \wedge V_2(v, y)$ (where '$\equiv^{\mathcal{KB}}$' means "logically equivalent with respect to $\mathcal{KB}$"). Then $\mathcal{KB} \models \mathcal{Q}(x, y) \leftrightarrow \widehat{\mathcal{Q}}(x, y) \wedge \exists v. V_2(v, y)$. Therefore, $\widehat{\mathcal{Q}}(x, y) \wedge \exists v. V_2(v, y) = (\exists z. V_1(x, z) \wedge \forall v. (V_2(v, z) \to V_3(v, y))) \wedge \exists v. V_2(v, y)$ is an exact safe-range reformulation of $\mathcal{Q}(x, y)$ from $\mathbb{P}_{\mathcal{DB}}$ under $\mathcal{KB}$.

6   Constructing the Safe-Range Reformulation

In this section we introduce a method to compute a safe-range reformulation of an implicitly definable query when conditions in Theorem 7 are satisfied. The method is based on the notion of interpolant introduced by [11].

Definition 10 (Interpolant). *The sentence χ in $\mathcal{FOL}(\mathbb{C}, \mathbb{P})$ is an interpolant for the sentence $\phi \to \psi$ in $\mathcal{FOL}(\mathbb{C}, \mathbb{P})$, if all predicate and constant symbols of χ are in the set of predicate and constant symbols of both ϕ and ψ, and both $\phi \to \chi$ and $\chi \to \psi$ are valid sentences in $\mathcal{FOL}(\mathbb{C}, \mathbb{P})$.*

Theorem 8 (Craig's interpolation). *If $\phi \to \psi$ is a valid sentence in $\mathcal{FOL}(\mathbb{C}, \mathbb{P})$, and neither ϕ nor ψ are valid, then there exists an interpolant.*

Note that the Beth definability (Theorem 3) and Craig's interpolation theorem do not hold for all fragments of $\mathcal{FOL}(\mathbb{C}, \mathbb{P})$: an interpolant may not always be expressed in the fragment itself, but obviously it is in $\mathcal{FOL}(\mathbb{C}, \mathbb{P})$ (because of Theorem 8).

An interpolant is used to find an exact reformulation of a given implicitly definable query as follows.

Theorem 9 (Interpolant as definition). *Let $\mathcal{Q}(\bar{x})$ be a query with $n \geq 0$ free variables implicitly definable from the DBox predicates $\mathbb{P}_{DB}$ under the ontology $\mathcal{KB}$. Then, the closed formula with $c_1, \ldots, c_n$ distinct constant symbols in $\mathbb{C}$ not appearing in $\mathcal{KB}$ or $\mathcal{Q}(\bar{x})$:*

$$((\bigwedge \mathcal{KB}) \wedge \mathcal{Q}(\bar{x}/c_1, \ldots, c_n)) \to ((\bigwedge \widetilde{\mathcal{KB}}) \to \widetilde{\mathcal{Q}}(\bar{x}/c_1, \ldots, c_n)) \tag{2}$$

is valid, and its interpolant $\widehat{\mathcal{Q}}(c_1, \ldots, c_n/\bar{x})$ is an exact reformulation of $\mathcal{Q}(\bar{x})$ under $\mathcal{KB}$ over $\mathbb{P}_{DB}$.

Therefore, to find an exact reformulation of an implicitly definable query in terms of DBox predicates it is enough to find an interpolant of the implication (2) and then to substitute all the constants $c_1, \ldots, c_n$ back with the free variables $\bar{x}$ of the original query. An interpolant can be constructed from a validity proof of (2) by using automated theorem proving techniques such as tableau or resolution. In order to guarantee the safe-range property of the reformulation, we use a tableau method as in the book by [15].

6.1 Tableau-Based Method to Compute an Interpolant

In this section we recall in our context the tableau based method to compute an interpolant. This method was described and its correctness was proved in [15].

Assume $\phi \to \psi$ is valid, therefore $\phi \wedge \neg\psi$ is unsatisfiable. Then there is a closed tableau corresponding to $\phi \wedge \neg\psi$. In order to compute an interpolant from this tableau one needs to modify it to a *biased tableau*.

Definition 11 (Biased tableau). *A biased tableau for formulas $\phi \wedge \neg\psi$ is a tree $T = (V, E)$ where:*

- *V is a set of nodes, each node is labelled by a set of biased formulas. A biased formula is an expression in the form of $L(\varphi)$ or $R(\varphi)$ where φ is a formula. For each node n, $S(n)$ denotes the set of biased formulas labelling n.*
- *The root of the tree is labelled by $\{L(\phi), R(\neg\psi)\}$*
- *E is a set of edges. Given 2 nodes n_1 and n_2, $(n_1, n_2) \in E$ iff there is a biased completion rule from n_1 to n_2. We say there is a biased completion rule from n_1 to n_2 if*
  - *$Y(\mu)$ is the result of applying a rule to $X(\varphi)$, where X and Y refer to L or R (for some rules, there are two possibilities of choosing $Y(\mu)$), and*
  - *$S(n_2) = (S(n_1) \setminus \{X(\varphi)\}) \cup \{Y(\mu)\}$.*

Let C be the set of all constants in the input formulas of the tableau. C^{par} extends C with an infinite set of new constants. A constant is new if it does not occur anywhere in the tableau. With these notations, we have the following rules:

– Propositional rules

Negation rules			α–rule	β–rule
$\dfrac{X(\neg\neg\varphi)}{X(\varphi)}$	$\dfrac{X(\neg\top)}{X(\bot)}$	$\dfrac{X(\neg\bot)}{X(\top)}$	$\dfrac{X(\varphi_1 \wedge \varphi_2)}{\begin{array}{c} X(\varphi_1) \\ X(\varphi_2) \end{array}}$	$\dfrac{X(\neg(\neg\varphi_1 \wedge \neg\varphi_2))}{X(\varphi_1) \mid X(\varphi_2)}$

– First order rules

γ–rule	σ–rule
$\dfrac{X(\forall x.\varphi)}{X(\varphi(t))}$ for any $t \in C^{par}$	$\dfrac{X(\exists x.\varphi)}{X(\varphi(c))}$ for a new constant c

– Equality rules

reflexivity rule	replacement rule
$\dfrac{X(\varphi)}{X(t=t)}$ $t \in C^{par}$ occurs in φ	$\dfrac{\begin{array}{c} X(t=u) \\ Y(\varphi(t)) \end{array}}{Y(\varphi(u))}$

A node in the tableau is *closed* if it contains $X(\varphi)$ and $Y(\neg\varphi)$. If a node is closed, no rule is applied. In the other words, it becomes a leaf of the tree. A branch is closed if it contains a closed node and a tableau is closed if all of its branches are closed. Obviously, if the standard tableau for first-order logic is closed then so is the biased tableau and vice versa.

Given a closed biased tableau, the interpolant is computed by applying *interpolant rules*. An interpolant rule is written as $S \xrightarrow{int} I$, where I is a formula and $S = \{L(\phi_1), L(\phi_2), ..., L(\phi_n), R(\psi_1), R(\psi_2), ..., R(\psi_m)\}$.

– Rules for closed branches

r1. $S \cup \{L(\varphi), L(\neg\varphi)\} \xrightarrow{int} \bot$ r2. $S \cup \{R(\varphi), R(\neg\varphi)\} \xrightarrow{int} \top$

r3. $S \cup \{L(\bot)\} \xrightarrow{int} \bot$ r4. $S \cup \{R(\bot)\} \xrightarrow{int} \top$

r5. $S \cup \{L(\varphi), R(\neg\varphi)\} \xrightarrow{int} \varphi$ r6. $S \cup \{R(\varphi), L(\neg\varphi)\} \xrightarrow{int} \neg\varphi$

– Rules for propositional cases

p1. $\dfrac{S \cup \{X(\varphi)\} \xrightarrow{int} I}{S \cup \{X(\neg\neg\varphi)\} \xrightarrow{int} I}$
p2. $\dfrac{S \cup \{X(\top)\} \xrightarrow{int} I}{S \cup \{X(\neg\bot)\} \xrightarrow{int} I}$

p3. $\dfrac{S \cup \{X(\bot)\} \xrightarrow{int} I}{S \cup \{X(\neg\top)\} \xrightarrow{int} I}$
p4. $\dfrac{S \cup \{X(\varphi_1), X(\varphi_2)\} \xrightarrow{int} I}{S \cup \{X(\varphi_1 \wedge \varphi_2)\} \xrightarrow{int} I}$

p5. $\dfrac{S \cup \{L(\varphi_1)\} \xrightarrow{int} I_1 \ \ S \cup \{L(\varphi_2)\} \xrightarrow{int} I_2}{S \cup \{L(\neg(\neg\varphi_1 \wedge \neg\varphi_2))\} \xrightarrow{int} I_1 \vee I_2}$

p6. $\dfrac{S \cup \{R(\varphi_1)\} \xrightarrow{int} I_1 \ \ S \cup \{R(\varphi_2)\} \xrightarrow{int} I_2}{S \cup \{R(\neg(\neg\varphi_1 \wedge \neg\varphi_2))\} \xrightarrow{int} I_1 \wedge I_2}$

– Rules for first order cases:

f1. $\dfrac{S \cup \{X(\varphi(p))\} \xrightarrow{int} I}{S \cup \{X(\exists x.\varphi(x))\} \xrightarrow{int} I}$ where p is a parameter that does not occur in S or φ

f2. $\dfrac{S \cup \{L(\varphi(c))\} \xrightarrow{int} I}{S \cup \{L(\forall x.\varphi(x))\} \xrightarrow{int} I}$ if c occurs in $\{\phi_1, ..., \phi_n\}$

f3. $\dfrac{S \cup \{R(\varphi(c))\} \xrightarrow{int} I}{S \cup \{R(\forall x.\varphi(x))\} \xrightarrow{int} I}$ if c occurs in $\{\psi_1, ..., \psi_m\}$

f4. $\dfrac{S \cup \{L(\varphi(c))\} \xrightarrow{int} I}{S \cup \{L(\forall x.\varphi(x))\} \xrightarrow{int} \forall x.I[c/x]}$ if c does not occur in $\{\phi_1, ..., \phi_n\}$

f5. $\dfrac{S \cup \{R(\varphi(c))\} \xrightarrow{int} I}{S \cup \{R(\forall x.\varphi(x))\} \xrightarrow{int} \exists x.I[c/x]}$ if c does not occur in $\{\psi_1, ..., \psi_m\}$

– Rules for equality cases

e1. $\dfrac{S \cup \{X(\varphi(p)), X(t = t)\} \xrightarrow{int} I}{S \cup \{X(\varphi(p))\} \xrightarrow{int} I}$
e2. $\dfrac{S \cup \{X(\varphi(u)), X(t = u)\} \xrightarrow{int} I}{S \cup \{X(\varphi(t)), X(t = u)\} \xrightarrow{int} I}$

e3. $\dfrac{S \cup \{L(\varphi(u)), R(t = u)\} \xrightarrow{int} I}{S \cup \{L(\varphi(t)), R(t = u)\} \xrightarrow{int} t = u \to I}$ if u occurs in $\varphi(t), \psi_1, ..., \psi_m$

e4. $\dfrac{S \cup \{R(\varphi(u)), L(t = u)\} \xrightarrow{int} I}{S \cup \{R(\varphi(t)), L(t = u)\} \xrightarrow{int} t = u \wedge I}$ if u occurs in $\varphi(t), \psi_1, ..., \psi_m$

e5. $\dfrac{S \cup \{L(\varphi(u)), R(t = u)\} \xrightarrow{int} I}{S \cup \{L(\varphi(t)), R(t = u)\} \xrightarrow{int} I[u/t]}$ if u does not occur in $\varphi(t), \psi_1, ..., \psi_m$

e6. $\dfrac{S \cup \{R(\varphi(u)), L(t = u)\} \xrightarrow{int} I}{S \cup \{R(\varphi(t)), L(t = u)\} \xrightarrow{int} I[u/t]}$ if u does not occur in $\varphi(t), \psi_1, ..., \psi_m$

In summary, in order to compute an interpolant of ϕ and ψ, one first need to generate a biased tableaux proof of unsatisfiability of $\phi \wedge \neg\psi$ using biased completion rules and then apply interpolant rules from bottom leaves up to the root. Let us consider an example to demonstrate how the method works.

Example 7. Let $\mathbb{P} = \{S, G, U\}$, $\mathbb{P}_{DB} = \{S, U\}$,

$$\mathcal{KB} = \{\forall x(S(x) \rightarrow (G(x) \vee U(x)))$$
$$\forall x(G(x) \rightarrow S(x))$$
$$\forall x(U(x) \rightarrow S(x))$$
$$\forall x(G(x) \rightarrow \neg U(x))\}$$
$$\mathcal{Q}(x) = G(x)$$

Obviously, $\mathcal{Q}$ is implicitly definable from S and U, since the ontology states that G and U partition S. Now we will follow the tableau method to find its exact reformulation. For compactness, we use the notation S^I instead of $S \xrightarrow{int} I$.

$$S_0 = \{L(\forall x(S(x) \rightarrow (G(x) \vee U(x)))),$$
$$L(\forall x(G(x) \rightarrow S(x))),$$
$$L(\forall x(U(x) \rightarrow S(x))),$$
$$L(\forall x(G(x) \rightarrow \neg U(x))),$$
$$L(G(c)),$$
$$R(\forall x(S(x) \rightarrow (G_1(x) \vee U(x)))),$$
$$R(\forall x(G_1(x) \rightarrow S(x))),$$
$$R(\forall x(U(x) \rightarrow S(x))),$$
$$R(\forall x(G_1(x) \rightarrow \neg U(x))),$$
$$R(\neg G_1(c))\}$$

By applying the rule for $\forall$ and removing the implication, we have:

$$S_1 = \{L(\neg S(c) \vee G(c) \vee U(c)),$$
$$L(\neg G(c) \vee S(c))),$$
$$L(\neg U(c) \vee S(c)),$$
$$L(\neg G(c) \vee \neg U(c)),$$
$$L(G(c)),$$
$$R(\neg S(c) \vee G_1(c) \vee U(c)),$$
$$R(\neg G_1(c) \vee S(c)),$$
$$R(\neg U(c) \vee S(c)),$$
$$R(\neg G_1(c) \vee \neg U(c)),$$
$$R(\neg G_1(c))\};$$

and the interpolant of S_1 can be computed as follows:

$$\dfrac{\dfrac{\dfrac{\dfrac{S_4 \cup \{R(\neg S(c))\}^{S(c)} \quad S_4 \cup \{R(U(c))\}^{\neg U(c)}}{S_4 = S_3 \cup \{R(\neg S(c) \vee U(c))\}^{(S(c) \wedge \neg U(c))}} \vee \quad S_3 \cup \{R(G_1(c))\}^{\top}}{S_3 = S_2 \cup \{L(\neg U(c))\}^{(S(c) \wedge \neg U(c))}} \; B.7 \quad S_2 \cup \{L(\neg G(c))\}^{\perp}}{S_2 = S_1 \cup \{L(S(c))\}^{(S(c) \wedge \neg U(c))}} \; B.5 \quad S_1 \cup \{L(\neg G(c))\}^{\perp}}{S_1^{(S(c) \wedge \neg U(c))}} \; B.3$$

Therefore, $S(c) \wedge \neg U(c)$ is the interpolant and $\widehat{\mathcal{Q}}(x) = S(x) \wedge \neg U(x)$ is an exact reformulation of $\mathcal{Q}(x)$.

6.2   A Safe-Range Reformulation

Now we want to show that the reformulation computed by the above tableau based method under the condition of Theorem 7 generates a ground safe-range query.

Theorem 10 (Ground safe-range reformulation). *Let $\mathcal{KB}$ be an ontology, and let $\mathcal{Q}(\bar{x})$ be a query which is implicitly definable from $\mathbb{P}_{DB}$ under $\mathcal{KB}$. If $\mathcal{KB}$ and $\mathcal{Q}(\bar{x})$ are safe-range then a reformulation $\widehat{\mathcal{Q}}(\bar{x})$ obtained using the tableau method described in Sect. 6.1 is ground safe-range.*

In other words, the conditions of Theorem 10 guarantee that all quantified variables in the reformulation are range-restricted. We need to consider now the still unsafe free variables. The theorem below will help us deal with non-range-restricted free variables. Let us first define the *active domain predicate of a signature* σ' as the formula:

$$Adom_{\sigma'}(x) := \bigvee_{P \in \mathbb{P} \cap \sigma'} (\exists x_1, \ldots, x_{\mathrm{AR}(P)-1}. \, P(x, x_1, \ldots, x_{\mathrm{AR}(P)-1}) \vee \ldots \vee$$

$$\vee P(x_1, \ldots, x_{\mathrm{AR}(P)-1}, x)) \vee \bigvee_{c \in \mathbb{C} \cap \sigma'} (x = c).$$

If $\sigma' = \sigma(\phi)$, where ϕ is a formula, then instead of $Adom_{\sigma(\phi)}$ we simply write $Adom_\phi$ and call it *active domain predicate of the formula* ϕ.

Theorem 11 (Range of the query). *Let $\mathcal{KB}$ be a domain independent ontology, and let $\mathcal{Q}(x_1, \ldots, x_n)$ be a query which is domain independent with respect to $\mathcal{KB}$. Then*

$$\mathcal{KB} \models \forall x_1, \ldots, x_n. \, \mathcal{Q}(x_1, \ldots, x_n) \rightarrow Adom_\mathcal{Q}(x_1) \wedge \ldots \wedge Adom_\mathcal{Q}(x_n).$$

Given a safe-range ontology, a safe-range and implicitly definable query is obviously domain independent with respect to the ontology (by definition). By Theorem 10 there exists a ground safe-range exact reformulation obtained using the tableau method. This reformulation is also domain independent with respect to the ontology (by definition). And then Theorem 11 says that the answer to the reformulation can only include semantic active domain elements of the reformulation. Therefore, the active domain predicate of the reformulation can be used as a "guard" for free variables which are not bounded by any positive predicate. In this way we obtain a new safe-range reformulation from the ground safe-range one.

Based on Theorems 10 and 11, we propose a complete procedure to construct a safe-range reformulation in Algorithm 1.

Algorithm 1. Safe-range reformulation

Input: a safe-range $\mathcal{KB}$, a safe-range and implicitly definable query $\mathcal{Q}(\bar{x})$.
Output: an exact safe-range reformulation $\widehat{\mathcal{Q}}(\bar{x})$.

1: Compute the interpolant $\widehat{\mathcal{Q}}(\bar{x})$ as in Theorem 9
2: For each free variable x which is not bounded by any positive predicate in $\widehat{\mathcal{Q}}(\bar{x})$ do
   $\widehat{\mathcal{Q}}(\bar{x}) := \widehat{\mathcal{Q}}(\bar{x}) \wedge Adom_{\widehat{\mathcal{Q}}}(x)$
3: Return $\widehat{\mathcal{Q}}(\bar{x})$

Syntax	Semantics
A	$A^{\mathcal{I}} \subseteq \Delta^{\mathcal{I}}$
$\{o\}$	$\{o^{I}\} \subseteq \Delta^{I}$
P	$P^{\mathcal{I}} \subseteq \Delta^{\mathcal{I}} \times \Delta^{\mathcal{I}}$
P^{-}	$\{(y,x)\|(x,y) \in P^{\mathcal{I}}\}$
$\neg C$	$\Delta^{\mathcal{I}} \backslash C^{\mathcal{I}}$
$C \sqcap D$	$C^{\mathcal{I}} \cap D^{\mathcal{I}}$
$C \sqcup D$	$C^{\mathcal{I}} \cup D^{\mathcal{I}}$
$\exists R$	$\{x\|\{y\|(x,y) \in R^{\mathcal{I}}\} \neq \emptyset\}$
$\exists R.C$	$\{x\|\{y\|(x,y) \in R^{\mathcal{I}}\} \cap C^{\mathcal{I}} \neq \emptyset\}$
$\forall R.C$	$\{x\|$ if $(x,y) \in R^{\mathcal{I}}$ then $y \in C^{\mathcal{I}}\}$

Fig. 1. Syntax and semantics of $\mathcal{ALCHOI}$ concepts and roles

7 The Guarded Negation Fragment of $\mathcal{ALCHOI}$ and $\mathcal{SHOQ}$

$\mathcal{ALCHOI}$ is an extension of the description logic $\mathcal{ALC}$ with role hierarchies, individuals and inverse roles: it corresponds to the $\mathcal{SHOI}$ description logic without transitive roles. $\mathcal{ALCHOI}$ without inverse roles and with qualified cardinality

Syntax	Semantics
A	$A^{\mathcal{I}} \subseteq \Delta^{\mathcal{I}}$
P	$P^{\mathcal{I}} \subseteq \Delta^{\mathcal{I}} \times \Delta^{\mathcal{I}}$
$C \sqcap D$	$C^{\mathcal{I}} \cap D^{\mathcal{I}}$
$C \sqcup D$	$C^{\mathcal{I}} \cup D^{\mathcal{I}}$
$\neg C$	$\Delta^{\mathcal{I}} \backslash C^{\mathcal{I}}$
$\{o\}$	$\{o\}^{\mathcal{I}} \subseteq \Delta^{\mathcal{I}}$
$\geq nP$	$\{x \mid \#(\{y \mid (x,y) \in P^{\mathcal{I}}\}) \geq n\}$
$\leq nP$	$\{x \mid \#(\{y \mid (x,y) \in P^{\mathcal{I}}\}) \leq n\}$
$\geq nP.C$	$\{x \mid \#(\{y \mid (x,y) \in P^{\mathcal{I}}\} \cap C^{\mathcal{I}}) \geq n\}$
$\leq nP.C$	$\{x \mid \#(\{y \mid (x,y) \in P^{\mathcal{I}}\} \cap C^{\mathcal{I}}) \leq n\}$

Fig. 2. Syntax and semantics of $\mathcal{SHOQ}$ concepts and roles

restrictions and transitive roles forms the description logic $\mathcal{SHOQ}$. For more details see, e.g., [22]. The syntax and semantics of $\mathcal{ALCHOI}$ and $\mathcal{SHOQ}$ concept expressions and roles is summarised in the Figs. 1 and 2 respectively, where A is an atomic concept, C and D are concepts, o is an individual name, P is an atomic role, and R is either P or P^-. A TBox in $\mathcal{ALCHOI}$ is a set of concept inclusion axioms $C \sqsubseteq D$ and role inclusion axioms $R \sqsubseteq S$ (where C, D are concepts and R, S are roles) with the usual description logics semantics. A TBox in $\mathcal{SHOQ}$ is defined in the same way without a possibility to express inverse roles and with additional possibility to express transitivity axioms $Trans(P)$.

In this section, we present an application of Theorem 7, by introducing the $\mathcal{ALCHOI}_{GN}$ description logic, the *guarded negation* syntactic fragment of $\mathcal{ALCHOI}$ (Fig. 3), and $\mathcal{SHOQ}_{GN+}$, the *extended guarded negation* syntactic fragment of $\mathcal{SHOQ}$ (Fig. 4). $\mathcal{ALCHOI}_{GN}$ and $\mathcal{SHOQ}_{GN+}$ restrict $\mathcal{ALCHOI}$ and $\mathcal{SHOQ}$ respectively by just prescribing that negated concepts should be *guarded* by some generalised atom – an atomic concept, a nominal, an unqualified existential restriction (for $\mathcal{ALCHOI}$) or an unqualified *atleast* number restriction (for $\mathcal{SHOQ}$), i.e., absolute negation is forbidden. $\mathcal{ALCHOI}_{GN}$ is actually at the intersection of the GNFO fragment and $\mathcal{ALCHOI}$ (by definition).

Each of these fragments has the very important property of *coinciding* (being equally expressive to) with the domain independent and safe-range fragments of the corresponding description logic, therefore providing an excellent candidate language for ontologies and queries satisfying the conditions of Theorem 7.

$$R ::= P \mid P^-$$
$$B ::= A \mid \{o\} \mid \exists R$$
$$C ::= B \mid \exists R.C \mid \exists R.\neg C \mid B \sqcap \neg C \mid C \sqcap D \mid C \sqcup D$$

Fig. 3. Syntax of $\mathcal{ALCHOI}_{GN}$ concepts and roles

$$B ::= A \mid \{o\} \mid \ \geq nP$$
$$C ::= B \mid \ \geq nP.C \mid \ \geq nP.\neg C \mid B \sqcap \neg C \mid C \sqcap D \mid C \sqcup D$$

Fig. 4. Syntax of $\mathcal{SHOQ}_{GN+}$ concepts

Theorem 12 (Expressive power equivalence). *The domain independent (safe-range) fragment of $\mathcal{ALCHOI}$ and $\mathcal{ALCHOI}_{GN}$ are equally expressive.*

Theorem 13 (Expressive power equivalence). *The domain independent (safe-range) fragment of $\mathcal{SHOQ}$ and $\mathcal{SHOQ}_{GN+}$ are equally expressive.*

In other words the first theorem says that any domain independent (or safe-range) TBox axiom and any domain independent (or safe-range) concept query in $\mathcal{ALCHOI}$ is logically equivalent, respectively, to a TBox axiom and a concept query in $\mathcal{ALCHOI}_{GN}$, and vice-versa. And the second theorem says that any domain independent (or safe-range) TBox axiom and any domain independent (or safe-range) concept query in $\mathcal{SHOQ}$ is logically equivalent, respectively, to a TBox axiom and a concept query in $\mathcal{SHOQ}_{GN+}$, and vice-versa.

7.1 Applying the Constructive Theorem

We want to reformulate concept queries over an ontology with a DBox so that the reformulated query can be evaluated as an SQL query over the database represented by the DBox. We consider applications of the Constructive Theorem 7 in the fragments $\mathcal{ALCHOI}_{GN}$ and $\mathcal{SHOQ}_{GN+}$. In this context, the database is a DBox, the ontology is an $\mathcal{ALCHOI}_{GN}$ ($\mathcal{SHOQ}_{GN+}$) TBox, and the query is an $\mathcal{ALCHOI}_{GN}$ ($\mathcal{SHOQ}_{GN+}$) concept query. A concept query is either an $\mathcal{ALCHOI}_{GN}$ ($\mathcal{SHOQ}_{GN+}$) concept expression denoting an open formula with one free variable, or an $\mathcal{ALCHOI}_{GN}$ ($\mathcal{SHOQ}_{GN+}$) ABox concept assertion denoting a boolean query.

As expected, a DBox includes ground atomic statements of the form $A(a)$ and $P(a, b)$ (where A is an atomic concept and P is an atomic role). It is easy to prove the following propositions:

Proposition 4. $\mathcal{ALCHOI}_{GN}$ *TBoxes, concept queries are safe-range (domain independent).*

Proposition 5. $\mathcal{SHOQ}_{GN+}$ *TBoxes, concept queries are safe-range (domain independent).*

We also proved the following theorems.

Theorem 14. $\mathcal{ALCHOI}_{GN}$ *TBoxes have finitely controllable determinacy of concept queries.*

Theorem 15. $\mathcal{SHOQ}_{GN+}$ *TBoxes have finitely controllable determinacy of concept queries.*

Therefore, we satisfy the conditions of Theorem 7, with a language which is like the very expressive $\mathcal{ALCHOI}$ description logic, but with *guarded* negation. And we also satisfy the conditions of Theorem 7, with a language which is like the very expressive $\mathcal{SHOQ}$ description logic, but with *extended guarded* negation ("extended" here means that cardinality restrictions and transitivity axioms are allowed in $\mathcal{SHOQ}_{GN+}$ in spite of the fact that they are not expressible in GNFO).

We argue that non-guarded negation should not appear in a cleanly designed ontology, and, if present, should be fixed. Indeed, the use of *absolute* negative information – such as, e.g., in "a non-`male` is a `female`" ($\neg$ `male` $\sqsubseteq$ `female`) – should be discouraged by a clean design methodology, since the subsumer would include *all sorts* of objects in the universe (but the ones of the subsumee type) without any obvious control. Only *guarded* negative information in the subsumee should be allowed – such as in the axiom "a non-`male` `person` is a `female`" (`person` $\sqcap \neg$ `male` $\sqsubseteq$ `female`).

This observation suggests a fix for non-guarded negations: for every non-guarded negation users will be asked to replace it by a guarded one, where the guard may be an arbitrary atomic concept, or nominal, or unqualified existential restriction (in the case of $\mathcal{ALCHOI}$) or unqualified *atleast* number restriction (in the case of $\mathcal{SHOQ}$). Therefore, the user is asked to make explicit the *type* of that concept, in a way to make it domain independent (i.e. belonging to $\mathcal{ALCHOI}_{GN}$ or $\mathcal{SHOQ}_{GN+}$). Note that the type could be also a fresh new atomic concept. We believe that the fix we are proposing for $\mathcal{ALCHOI}$ and $\mathcal{SHOQ}$ is a reasonable one, and would make all $\mathcal{ALCHOI}$ and $\mathcal{SHOQ}$ ontologies eligible to be used with our framework.

7.2 A Complete Procedure

$\mathcal{ALCHOI}_{GN}$ and $\mathcal{SHOQ}_{GN+}$ are decidable logics (as a fragments of $\mathcal{ALCHOI}$ and $\mathcal{SHOQ}$ respectively) and they are feasible applications of our general framework. Given an $\mathcal{ALCHOI}_{GN}$ ($\mathcal{SHOQ}_{GN+}$) ontology $\mathcal{KB}$ and a concept query $\mathcal{Q}$ in $\mathcal{ALCHOI}_{GN}$ ($\mathcal{SHOQ}_{GN+}$), we can apply the procedure below to generate a safe-range reformulation over the DBox concepts and roles (based on the constructive theorem, all the conditions of which are satisfied), if it exists.

Note that the procedure for checking determinacy and computing the reformulation could be run in offline mode at compile time. Indeed, it could be run for each atomic concept in the ontology, and store persistently the outcome for each of them if the reformulation has been successful. This pre-computation may be an expensive operation, since – as we have seen – it is based on entailment, but the complexity involves only the size of the ontology and not of the data.

In order to get an idea about the size of the reformulations of concept queries, for the $\mathcal{ALCFI}$ description logic there is a tableau-based algorithm computing explicit definitions of at most double exponential size [9,10]; this algorithm is optimal because it is also shown that the smallest explicit definition of an implicitly defined concept *may* be double exponentially long in the size of the input TBox.

Input: An $\mathcal{ALCHOI}_{GN}$ ($\mathcal{SHOQ}_{GN+}$) TBox $\mathcal{KB}$, a concept query $\mathcal{Q}$ in $\mathcal{ALCHOI}_{GN}$ ($\mathcal{SHOQ}_{GN+}$), and a DBox signature (DBox atomic concepts and roles).

Output: A safe-range reformulation $\widehat{\mathcal{Q}}$ expressed over the DBox signature.

1: Check the implicit definability of the query $\mathcal{Q}$ by testing if $\mathcal{KB} \cup \widetilde{\mathcal{KB}} \models \mathcal{Q} \equiv \widetilde{\mathcal{Q}}$ using a standard OWL2 reasoner ($\mathcal{ALCHOI}_{GN}$ and $\mathcal{SHOQ}_{GN+}$ are sublanguages of OWL2). Continue if this holds.

2: Compute a safe-range reformulation $\widehat{\mathcal{Q}}$ from the tableau proof generated in step 1 (see Section 6). This can be implemented as a simple extension of a standard description logic reasoner even in the presence of the most important optimisation techniques such as semantic branching, absorption, and backjumping as explained by [31] and [9].

Clearly, similarly to DL-Lite reformulations, more research is needed in order to optimise the reformulation step in order to make it practical. However, note that the framework presented here has a clear advantage from the point of view of *conceptual modelling* since implicit definitions (that is, queries) under general TBoxes can be double exponentially more succinct than acyclic concept definitions (that is, explicit queries over the DBox).

The case of query answering with unrestricted description logics and DBoxes, when the rewriting may not be possible at all, has been studied thoroughly by [24, 28] regarding both data and combined complexity.

8 Conclusions and Future Work

We have introduced a framework to compute the exact reformulation of first-order queries to a database (DBox) under constraints. We have found the exact conditions which guarantee that a safe-range reformulation exists, and we show that it can be evaluated as an SQL query over the DBox to give the same answer as the original query under the constraints. Non-trivial case studies have been presented in the field of description logics, with the $\mathcal{ALCHOI}$ and $\mathcal{SHOQ}$ languages.

This framework is useful in data exchange-like scenarios, where the target database (made by determined relations) should be materialised as a proper database, over which arbitrary queries should be performed. This is not achieved in a context with non-exact reformulations preserving the certain answers. In our scenario with description logics ontologies reformulations of concept queries are pre-computed offline once. We have shown that our framework works in theory also in the case of arbitrary safe-range first-order queries.

Next, we would like to study optimisations of reformulations. From the practical perspective, since there might be many rewritten queries from one original query, the problem of selecting an optimised query in terms of query evaluation is very important. In fact, one has to take into account which criteria should be used to optimise, such as: the size of the reformulations, the numbers of used predicates, the priority of predicates, the number of relational operators, and clever usage of duplicates. In this context one may also want to control the process of formula proving to make it produce an optimal reformulation. For instance, using the tableau method, one may prefer one order of expansion rules application to another and, hence, build another interpolant.

Concurrently, we are exploring the problem of *fixing* real ontologies in order to enforce definability when it is known it should be the case [19]. This happens when it is intuitively obvious that the answer of a query can be found from the available data (that is, the query is definable from the database), but the mediating ontology does not entail the definability. We introduce the novel problem of *definability abduction* and we solve it completely in the data exchange scenario.

There is also another interesting open problem about checking that a given DBox is legal with respect to a given ontology. Remember that a DBox $\mathcal{DB}$ is legal for an ontology $\mathcal{KB}$ if there exists a model of $\mathcal{KB}$ embedding $\mathcal{DB}$. This check involves heavy computations for which an optimised algorithm is still unknown: as a matter of fact, the only known method today is to reduce the problem to a satisfiability problem where the DBox is embedded in a TBox using nominals [16]. More research is needed in order to optimise the reasoning with nominals in this special case.

In the case of description logics, we would like to work on extending the theoretical framework with conjunctive queries: we need finitely controllable determinacy with conjunctive queries, which for some description logics seems to follow from the works by [6,30].

References

1. Abiteboul, S., Hull, R., Vianu, V.: Foundations of Databases. Addison-Wesley, Boston (1995)
2. Artale, A., Calvanese, D., Kontchakov, R., Zakharyaschev, M.: The DL-Lite family and relations. J. Artif. Intell. Res. (JAIR) **36**, 1–69 (2009)
3. Avron, A.: Constructibility and decidability versus domain independence and absoluteness. Theor. Comput. Sci. **394**, 144–158 (2008). https://doi.org/10.1016/j.tcs.2007.12.008. http://dl.acm.org/citation.cfm?id=1351194.1351447
4. Bárány, V., ten Cate, B., Otto, M.: Queries with guarded negation. PVLDB **5**(11), 1328–1339 (2012)
5. Bárány, V., ten Cate, B., Segoufin, L.: Guarded negation. In: Aceto, L., Henzinger, M., Sgall, J. (eds.) ICALP 2011. LNCS, vol. 6756, pp. 356–367. Springer, Heidelberg (2011). https://doi.org/10.1007/978-3-642-22012-8_28
6. Bárány, V., Gottlob, G., Otto, M.: Querying the guarded fragment. In: Proceedings of the 25th Annual IEEE Symposium on Logic in Computer Science (LICS 2010), pp. 1–10 (2010)
7. Beth, E.: On Padoa's method in the theory of definition. Indagationes Math. **15**, 330–339 (1953)
8. Calvanese, D., Franconi, E.: First-order ontology mediated database querying via query reformulation. In: Flesca, S., Greco, S., Masciari, E., Saccà, D. (eds.) A Comprehensive Guide Through the Italian Database Research Over the Last 25 Years. SBD, vol. 31, pp. 169–185. Springer, Cham (2018). https://doi.org/10.1007/978-3-319-61893-7_10

9. ten Cate, B., Franconi, E., Seylan, İ.: Beth definability in expressive description logics. In: Proceedings of the 22nd International Joint Conference on Artificial Intelligence (IJCAI 2011), pp. 1099–1106 (2011)
10. ten Cate, B., Franconi, E., Seylan, I.: Beth definability in expressive description logics. J. Artif. Intell. Res. (JAIR) **48**, 347–414 (2013). https://doi.org/10.1613/jair.4057
11. Craig, W.: Three uses of the Herbrand-Gentzen theorem in relating model theory and proof theory. J. Symb. Log. **22**(3), 269–285 (1957)
12. Etzioni, O., Golden, K., Weld, D.S.: Sound and efficient closed-world reasoning for planning. Artif. Intell. **89**, 113–148 (1997). https://doi.org/10.1016/S0004-3702(96)00026-4. http://dl.acm.org/citation.cfm?id=249678.249685
13. Fan, W., Geerts, F., Zheng, L.: View determinacy for preserving selected information in data transformations. Inf. Syst. **37**, 1–12 (2012). https://doi.org/10.1016/j.is.2011.09.001
14. Feinerer, I., Franconi, E., Guagliardo, P.: Lossless selection views under conditional domain constraints. IEEE Trans. Knowl. Data Eng. **27**(2), 504–517 (2015). https://doi.org/10.1109/TKDE.2014.2334327
15. Fitting, M.: First-Order Logic and Automated Theorem Proving, 2nd edn. Springer, New York (1996). https://doi.org/10.1007/978-1-4612-2360-3
16. Franconi, E., Ibanez-Garcia, Y.A., Seylan, İ.: Query answering with DBoxes is hard. Electron. Notes Theor. Comput. Sci. **278**, 71–84 (2011)
17. Franconi, E., Kerhet, V., Ngo, N.: Exact query reformulation with first-order ontologies and databases. In: del Cerro, L.F., Herzig, A., Mengin, J. (eds.) JELIA 2012. LNCS (LNAI), vol. 7519, pp. 202–214. Springer, Heidelberg (2012). https://doi.org/10.1007/978-3-642-33353-8_16
18. Franconi, E., Kerhet, V., Ngo, N.: Exact query reformulation over databases with first-order and description logics ontologies. J. Artif. Intell. Res. (JAIR) **48**, 885–922 (2013). https://doi.org/10.1613/jair.4058
19. Franconi, E., Ngo, N., Sherkhonov, E.: The definability abduction problem for data exchange. In: Krötzsch, M., Straccia, U. (eds.) RR 2012. LNCS, vol. 7497, pp. 217–220. Springer, Heidelberg (2012). https://doi.org/10.1007/978-3-642-33203-6_18
20. Gurevich, Y.: Toward logic tailored for computational complexity. In: Börger, E., Oberschelp, W., Richter, M.M., Schinzel, B., Thomas, W. (eds.) Computation and Proof Theory. LNM, vol. 1104, pp. 175–216. Springer, Heidelberg (1984). https://doi.org/10.1007/BFb0099486
21. Halevy, A.Y.: Answering queries using views: a survey. VLDB J. **10**, 270–294 (2001). https://doi.org/10.1007/s007780100054
22. Horrocks, I., Sattler, U.: Ontology reasoning in the SHOQ(D) description logic. In: Proceedings of the 17th International Joint Conference on Artificial Intelligence (IJCAI 2001), pp. 199–204 (2001)
23. Kleene, S.C.: Mathematical Logic. Dover, New York (2002)
24. Lutz, C., Seylan, I., Wolter, F.: Ontology-mediated queries with closed predicates. In: Proceedings of the Twenty-Fourth International Joint Conference on Artificial Intelligence, IJCAI 2015, Buenos Aires, Argentina, 25–31 July 2015, pp. 3120–3126 (2015). http://ijcai.org/Abstract/15/440
25. Marx, M.: Queries determined by views: pack your views. In: Proceedings of the 26th ACM symposium on Principles of Database Systems, PODS 2007, pp. 23–30 (2007). https://doi.org/10.1145/1265530.1265534
26. Nash, A., Segoufin, L., Vianu, V.: Views and queries: determinacy and rewriting. ACM Trans. Database Syst. **35**, 211–2141 (2010). https://doi.org/10.1145/1806907.1806913

27. Ngo, N., Franconi, E.: Unique solutions in data exchange under STS mappings. In: Proceedings of the 10th Alberto Mendelzon International Workshop on Foundations of Data Management (AMW-2016) (2016). http://ceur-ws.org/Vol-1644/paper5.pdf

28. Ngo, N., Ortiz, M., Simkus, M.: Closed predicates in description logics: results on combined complexity. In: Principles of Knowledge Representation and Reasoning: Proceedings of the Fifteenth International Conference, KR 2016, Cape Town, South Africa, 25–29 April 2016, pp. 237–246 (2016). http://www.aaai.org/ocs/index.php/KR/KR16/paper/view/12906

29. Patel-Schneider, P.F., Franconi, E.: Ontology constraints in incomplete and complete data. In: Cudré-Mauroux, P., et al. (eds.) ISWC 2012. LNCS, vol. 7649, pp. 444–459. Springer, Heidelberg (2012). https://doi.org/10.1007/978-3-642-35176-1_28

30. Rosati, R.: On the finite controllability of conjunctive query answering in databases under open-world assumption. J. Comput. Syst. Sci. **77**(3), 572–594 (2011)

31. Seylan, İ., Franconi, E., de Bruijn, J.: Effective query rewriting with ontologies over DBoxes. In: Proceedings of the 21st International Joint Conference on Artificial Intelligence (IJCAI 2009), pp. 923–925 (2009)

32. Toman, D., Weddell, G.E.: Fundamentals of physical design and query compilation. Synth. Lect. Data Manag. **3**, 1–124 (2011). https://doi.org/10.2200/S00363ED1V01Y201105DTM018

Checking the Data Complexity of Ontology-Mediated Queries: A Case Study with Non-uniform CSPs and Polyanna

Olga Gerasimova[1]([✉]), Stanislav Kikot[2], and Michael Zakharyaschev[3]

[1] National Research University Higher School of Economics, Moscow, Russia
olga.g3993@gmail.com
[2] University of Oxford, Oxford, UK
[3] Birkbeck, University of London, London, UK

Abstract. It has recently been shown that first-order- and datalog-rewritability of ontology-mediated queries (OMQs) with expressive ontologies can be checked in NExpTime using a reduction to CSPs. In this paper, we present a case study for OMQs with Boolean conjunctive queries and a fixed ontology consisting of a single covering axiom $A \sqsubseteq F \sqcup T$, possibly supplemented with a disjointness axiom for T and F. The ultimate aim is to classify such OMQs according to their data complexity: AC^0, L, NL, P or coNP. We report on our experience with trying to distinguish between OMQs in P and coNP using the reduction to CSPs and the Polyanna software for finding polymorphisms.

1 Introduction

Description logics (DLs) [4] have been tailored—by carefully picking and restricting various constructs that are relevant to intended applications—to make sure that reasoning with all ontologies in a given DL can uniformly be done in a given complexity class. For example, concept subsumption can be checked in ExpTime for all $\mathcal{ALC}$-ontologies, in P for all $\mathcal{EL}$-ontologies, and in NL for all DL-Lite-ontologies.

In ontology-mediated query (OMQ) answering, a typical reasoning problem is to check whether a Boolean query q holds in every model of an ontology $\mathcal{T}$ and a data instance $\mathcal{D}$. In the context of ontology-based data access (OBDA) and management [28,35], this problem is solved by reducing answering the OMQ $Q = (\mathcal{T}, q)$ over $\mathcal{D}$ to standard database query evaluation over $\mathcal{D}$. If the target database query language is first-order logic, then such a reduction is possible for Q just in case answering it can be done in AC^0 for data complexity. If a reduction to first-order queries with (deterministic) transitive closure is acceptable, then answering Q should be done in NL (respectively, L) for data complexity. In terms of the data complexity measure, OMQ answering with conjunctive queries (CQs) can uniformly be done in coNP for all $\mathcal{ALC}$-ontologies, in P for all $\mathcal{EL}$-ontologies, and in AC^0 for all DL-Lite-ontologies.

© Springer Nature Switzerland AG 2019
C. Lutz et al. (Eds.): Baader Festschrift, LNCS 11560, pp. 329–351, 2019.
https://doi.org/10.1007/978-3-030-22102-7_15

In OBDA practice, an ontology $\mathcal{T}$ is designed by a domain expert to capture the natural vocabulary of the intended end-users, who formulate their queries in terms of that vocabulary and execute them using an OBDA system such as Mastro [10] or Ontop [11,30]. Thus, from the user's point of view, the ontology $\mathcal{T}$ is fixed. Moreover, the class of queries the user is interested in could also be limited. For example, the NPD FactPages ontology[1], used for testing OBDA in industry [20,22], contains covering axioms of the form $A \sqsubseteq B_1 \sqcup \cdots \sqcup B_n$, which are not allowed in *DL-Lite* as there exist coNP-hard OMQs with such axioms. However, all of the practically important OMQs with the NPD FactPages ontology we know of can be answered in AC^0. Also answering CQs mediated by covering axioms can model some attacks in the setting of sensitive information disclosure [7].

These observations have lead to the following non-uniform problems: (i) What is the worst-case data complexity of answering OMQs with a fixed ontology and arbitrary CQs? (ii) What is the data complexity of answering a given single OMQ?

A systematic investigation of these problems was launched in [8,24,25]. In particular, [8] discovered a remarkable connection between OMQs and constraint satisfaction problems (CSPs) and used it to show that deciding FO-rewritability and datalog-rewritability of OMQs with $\mathcal{SHIU}$ ontologies is NExpTime-complete.

In this article, we are concerned with a very special case of the non-uniform problem (ii) above: classify the OMQs with the fixed ontology $Cov_A = \{A \sqsubseteq F \sqcup T\}$ and arbitrary CQs according to their data complexity. We also consider three variants of Cov_A, namely, $Cov_\top = \{\top \sqsubseteq F \sqcup T\}$, $Cov_\top^\perp = \{\top \sqsubseteq F \sqcup T, \ F \sqcap T \sqsubseteq \perp\}$ and $Cov_A^\perp = \{A \sqsubseteq F \sqcup T, \ F \sqcap T \sqsubseteq \perp\}$ with top $\top$ and bottom $\perp$ concepts. It turns out that a single covering axiom, possibly supplemented with a disjointness axiom, gives rise to a surprisingly non-trivial and diverse class of OMQs. To illustrate, we show and discuss a few simple examples.

Appetisers

Suppose we are interested in querying digraphs of social network users, in which only some of the users have specified their gender. Let F mean 'female', T 'male' and R the 'follows' relation.

Example 1. Our first OMQ $\boldsymbol{Q} = (Cov_\top, \boldsymbol{q})$ with $\boldsymbol{q} = \exists y, z \, (T(y) \wedge R(y, z) \wedge F(z))$ is supposed to check whether one can claim with certainty that, in given a data instance, there is always a man who follows a woman. We draw the CQ $\boldsymbol{q}$ as the labelled digraph

---

[1] http://sws.ifi.uio.no/project/npd-v2/.

Now, consider the data instance $\mathcal{D} = \{T(u_0), R(u_0, u_1), R(u_1, u_2), F(u_2)\}$, which can be depicted as the labelled digraph

In every model of $\mathcal{C}ov_T$ extending $\mathcal{D}$, we must have $T(u_1)$ or $F(u_1)$. In the former case, q is satisfied by the assignment $y \mapsto u_1, z \mapsto u_2$, in the latter one by $y \mapsto u_0, z \mapsto u_1$. It follows that the certain answer to Q over $\mathcal{D}$ is yes.

More generally, it is readily seen that the certain answer to Q over any given $\mathcal{D}$ is yes iff $\mathcal{D}$ contains an R-path from a T-vertex to an F-vertex. As known from the basic computational complexity theory [3], the reachability problem in digraphs is NL-complete (NL stands for Nondeterministic Logarithmic-space), and so answering Q is NL-complete for data complexity.

Example 2. Consider next the OMQ $Q = (\mathcal{C}ov_T, q)$ with q given by the digraph

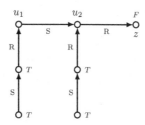

which checks if one can claim with certainty that there are a man and a woman who follow each other. In this case, answering Q is L-complete (L stands for deterministic Logarithmic-space) for data complexity, that is, as complex as reachability in undirected graphs. Indeed, to show that Q can be answered in L, with each data instance $\mathcal{D}$ we associate the undirected graph G with the same vertices as $\mathcal{D}$ connecting u and v by an edge iff $R(u, v)$ and $R(v, u)$ are both in $\mathcal{D}$. Then the answer to Q over $\mathcal{D}$ is yes iff G contains a path from a T-vertex to an F-vertex. To prove L-hardness, with any undirected graph G and a pair s, t of its vertices we associate a data instance $\mathcal{D}$ obtained by replacing each edge (u, v) in G by $R(u, v)$ and $R(v, u)$ and adding atoms $T(s)$ and $F(t)$. It is readily seen that the certain answer to Q over $\mathcal{D}$ is yes iff t is reachable from s in G.

Example 3. Now suppose $Q = (\mathcal{C}ov_T, q)$ and q looks as in the picture below

where S is another binary relation between the users. Consider the following data instance $\mathcal{D}$:

The certain answer to Q over D is yes. Indeed, in any model of Cov_T based on D, we have either $T(u_i)$ or $F(u_i)$, $i = 1, 2$. If $F(u_1)$ holds, then q maps onto the left vertical section of the model. Similarly, if $F(u_2)$ holds, then q maps onto the right vertical section. Otherwise, we have $T(u_1)$ and $T(u_2)$, in which case q maps onto the horizontal section.

In general, for any data instance D, the certain answer to Q over D is yes iff the following monadic datalog query with goal G, encoding our argument above, returns the answer yes over D:

$$P(x) \leftarrow T(x)$$
$$P(z) \leftarrow P(x) \wedge S(x, y) \wedge P(y) \wedge R(y, z)$$
$$G \leftarrow P(x) \wedge S(x, y) \wedge P(y) \wedge R(y, z) \wedge F(z)$$

It follows that answering Q can be done in P (Polynomial time). It is not hard to show that this OMQ is P-hard. The proof is by reduction of the monotone circuit evaluation problem, which is known to be P-complete [27]. Without any loss of generality we assume that AND-nodes have two inputs and that the circuit consists of alternating layers of OR- and AND-nodes with an AND node at the top; an example of such a circuit is shown in the picture below. Now, given such a circuit C and an input α for it, we define a data instance D_C^α as the set of the following atoms:

- $R(g, h)$, if a gate g is an input of a gate h;
- $S(g, h)$, if g and h are distinct inputs of some AND-gate;
- $S(g, g)$, if g is an input gate or a non-output AND-gate;
- $T(g)$, if g is an input gate with 1 under α;
- $F(g)$, for the only output gate g;
- $A(g)$, for those g that are neither inputs nor the output.

To illustrate, the picture below shows a monotone circuit C, an input α for it, and the data instance D_C^α, where the solid arrows represent R and the dashed ones S:

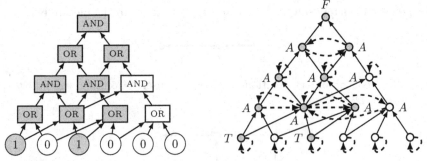

The reader can check that $C(\alpha) = 1$ iff the answer to Q over D_C^α is yes. Indeed, the datalog program above computes the value of the circuit by placing P on those nodes which are evaluated into 1.

Example 4. Curiously enough, the OMQ $Q = (Cov_T, q')$ with q' obtained from q in the previous example by changing S to R

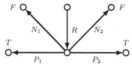

is NL-complete for data complexity, showing which is an instructive exercise. (Hint: prove that a data instance validates Q iff it has a path that starts with T, which is followed by T, and ends with F).

Example 5. It has been known since Schaerf's paper [32] that answering the OMQ $Q = (Cov_T, q)$ with q below is CONP-complete for data complexity.

Schaerf showed CONP-hardness by encoding the satisfiability problem for $2+2$-CNFs that consist of clauses of the form $P_1 \vee P_2 \vee \neg N_1 \vee \neg N_2$. The reader may find entertaining the task of showing that the OMQ (Cov_A, q) with q below is also CONP-complete.

One possible solution will be given in Sect. 6 below.

The remainder of this article is organised as follows. In the next section, we provide definitions of the basic notions we require later on. Then, in Sect. 3, we give a brief survey of related work. In Sect. 4, we explain by means of a simple example how detecting tractability of an path-OMQ can be reduced to checking tractability of a CSP. Then, in Sect. 5, we discuss how the program Polyanna [13], which was designed to check tractability of CSPs, can be used in the context of our case study for detecting whether answering a given OMQ with a 4-variable path CQ can be done in P or is CONP-hard. In Sect. 6, we sketch direct proofs of CONP-hardness using a reduction of 3SAT. In Sect. 7, we show how Polyanna can be used for constructing monadic datalog rewritings of tractable OMQs. Finally, in the appendix, we summarise what we know about the data complexity of answering the OMQs in the framework of our case study.

2 Preliminaries

In this paper, a *Boolean conjunctive query* (*CQ*) is any first-order (FO) sentence of the form $q = \exists x\, \varphi(x)$, where φ is a conjunction of unary or binary atoms whose variables are all among x. We often regard CQs as *sets* of their atoms, depict them as labelled digraphs, and assume that all of our CQs are *connected* as graphs. By a *solitary occurrence* of F in a CQ q we mean any occurrence of $F(x)$ in q, for some variable x, such that $T(x) \notin q$; likewise, a *solitary occurrence* of T in q is any occurrence $T(x) \in q$ such that $F(x) \notin q$. We say that q is a *path CQ* if all the variables $x_0, \ldots, x_n$ in q are ordered so that

- the binary atoms in q form a chain $R_1(x_0, x_1), \ldots, R_n(x_{n-1}, x_n)$;

– the unary atoms in q are of the form $T(x_i)$ and $F(x_j)$, for some i and j with $0 \le i, j \le n$.

By *answering* an *ontology-mediated query* (OMQ) $Q = (T, q)$, where T is one of the following ontologies (or TBoxes)

$$Cov_A = \{A \sqsubseteq F \sqcup T\}, \qquad Cov_A^{\perp} = \{A \sqsubseteq F \sqcup T,\ F \sqcap T \sqsubseteq \perp\},$$
$$Cov_T = \{\top \sqsubseteq F \sqcup T\}, \qquad Cov_{\mp}^{\pm} = \{\top \sqsubseteq F \sqcup T,\ F \sqcap T \sqsubseteq \perp\},$$

we understand the problem of checking, given a data instance (or ABox) $\mathcal{A}$, whether q holds in every model of $T \cup \mathcal{A}$, in which case we write $T, \mathcal{A} \models q$. For every Q, this problem is clearly in CONP for data complexity. It is in the complexity class AC^0 if there is an FO-sentence q', called an *FO-rewriting* of Q, such that $T, \mathcal{A} \models q$ iff $\mathcal{A} \models q'$, for any ABox $\mathcal{A}$.

A *datalog program*, Π, is a finite set of *rules* $\forall \boldsymbol{x} (\gamma_0 \leftarrow \gamma_1 \wedge \cdots \wedge \gamma_m)$, where each γ_i is an atom $P(\boldsymbol{y})$ with $\boldsymbol{y} \subseteq \boldsymbol{x}$. (As usual, we omit $\forall \boldsymbol{x}$.) The atom γ_0 is the *head* of the rule, and $\gamma_1, \ldots, \gamma_m$ its *body*. All the variables in the head must occur in the body. The predicates in the head of rules are *IDB predicates*, the rest *EDB predicates* [1].

A *datalog query* is a pair (Π, G), where Π is a datalog program and G an 0-ary atom, the *goal*. The *answer* to (Π, G) over an ABox $\mathcal{A}$ is 'yes' if G holds in the FO-structure obtained by closing $\mathcal{A}$ under Π, in which case we write $\Pi, \mathcal{A} \models G$. A datalog query (Π, G) is a *datalog rewriting* of an OMQ $Q = (T, q)$ in case $T, \mathcal{A} \models q$ iff $\Pi, \mathcal{A} \models G$, for any ABox $\mathcal{A}$. The *answering problem* for (Π, G)—i.e., checking, given an ABox $\mathcal{A}$, whether $\Pi, \mathcal{A} \models G$—is clearly in P. Answering a datalog query with a *linear* program, whose rules have at most one IDB predicate in the body, can be done in NL. A datalog query is *monadic* if all of its IDB predicates are of arity at most 1.

3 Related Work

We begin by putting our case study problem into the context of more general investigations of (*i*) boundedness (i.e., equivalence to an FO-query) and linearisability of datalog programs and (*ii*) the data complexity of answering OMQs with expressive ontologies.

The decision problem whether a given datalog program is bounded (equivalent to an FO-query) has been a hot research topic in database theory since the late 1980s. Thus, it was shown that boundedness is undecidable already for linear datalog programs with binary IDB predicates [34] and single rule programs (aka *sirups*) [26]. On the other hand, deciding boundedness is 2ExpTime-complete for *monadic* datalog programs [6,12] and PSPACE-complete for linear monadic programs [12]; for linear sirups, it is even NP-complete [34].

The last two results are relevant to deciding FO-rewritability of OMQs (Cov_A, q), where q has a single solitary F (see Sect. 2) and is called a 1-*CQ*.

Indeed, suppose that $F(x)$ and $T(y_1), \ldots, T(y_n)$ are all the solitary occurrences of F and T in $\boldsymbol{q}$. Let $\Pi_{\boldsymbol{q}}$ be a monadic datalog program with three rules

$$G \leftarrow F(x), \boldsymbol{q}', P(y_1), \ldots, P(y_n), \tag{1}$$

$$P(x) \leftarrow T(x), \tag{2}$$

$$P(x) \leftarrow A(x), \boldsymbol{q}', P(y_1), \ldots, P(y_n), \tag{3}$$

where $\boldsymbol{q}' = \boldsymbol{q} \setminus \{F(x), T(y_1), \ldots, T(y_n)\}$ and P is a fresh predicate symbol that never occurs in our ABoxes. Then, for any ABox $\mathcal{A}$, we have $\mathcal{C}ov_A, \mathcal{A} \models \boldsymbol{q}$ iff $\Pi_{\boldsymbol{q}}, \mathcal{A} \models G$. Thus, FO-rewritability of $(\mathcal{C}ov_A, \boldsymbol{q})$ is clearly related to boundedness of the sirup (3).

The problem of linearising datalog programs, that is, transforming them into equivalent linear datalog programs, which are known to be in NL for data complexity, has also attracted much attention [2,29,31,36] after the Ullman and van Gelder pioneering paper [33]. Here, Example 4 is very instructive: it is easy to construct $\Pi_{\boldsymbol{q}}$ automatically (either directly or by using the *markability* technique from [19] for disjunctive datalog), but clearly some additional artificial intelligence is required to notice the Hint and use it to produce a linear program. In Sect. 7, we show a datalog program for a similar query, which is produced automatically from an arc consistency procedure; again, this program is not linear but linearisable.

By establishing a remarkable connection to CSPs, it was shown in [8] that deciding FO- and datalog-rewritability of OMQs with a $\mathcal{SHIU}$ ontology is NExpTime-complete. This result is obviously applicable to our case study, and we shall discuss it in detail in the next section.

An $AC^0/NL/P$ trichotomy for the data complexity of answering OMQs with an $\mathcal{EL}$ ontology and atomic query, which can be checked in ExpTime, was established in [23]. This result is applicable to OMQs $(\mathcal{C}ov_A, \boldsymbol{q})$, in which $\boldsymbol{q}$ is an F-*tree* having a single solitary $F(x)$ such that the binary atoms in $\boldsymbol{q}$ form a ditree with root x. Indeed, denote by $\mathcal{T}_Q$ the $\mathcal{EL}$ TBox with concept inclusions $F \sqcap C_{\boldsymbol{q}} \sqsubseteq G'$, $T \sqsubseteq P$ and $A \sqcap C_{\boldsymbol{q}} \sqsubseteq P$, where $C_{\boldsymbol{q}}$ is an $\mathcal{EL}$-concept representing $\boldsymbol{q} \setminus \{F(x)\}$ with P for T (so for $\boldsymbol{q}$ of the form

$C_{\boldsymbol{q}} = \exists R_1.(F \sqcap P \sqcap \exists R_2.\exists R_3.P))$. Then, for any ABox $\mathcal{A}$ that does not contain G', we have $\Pi_Q, \mathcal{A} \models G$ iff $\mathcal{T}_Q, \mathcal{A} \models \exists x \, G'(x)$.

Yet, despite all of these efforts and results (implying, in view of the recent positive solution to the Feder-Vardi conjecture [9,37], that there is a P/CONP dichotomy for OMQs with $\mathcal{SHIU}$ ontologies, which is decidable in NExpTime), we are still lacking simple and transparent, in particular syntactic, conditions guaranteeing this or that data complexity or type of rewritability. Some results in this direction were obtained in [17,19]. That a transparent classification of monadic sirups according to their data complexity has not been found so far and the close connection to CSPs indicate that this problem is extremely hard in general.

In the next section, we illustrate how OMQs of the form $(Cov^\perp_\top, q)$ with a path CQ q can be reduced to CSPs.

4   Converting Path OMQs to CSPs

In the context of our case study, we are interested in *non-uniform CSPs*. Let $\mathcal{B}$ be a fixed relational structure which in this setting is called a *template*. Each template $\mathcal{B}$ gives rise to the decision problem $\mathrm{CSP}(\mathcal{B})$ which is to decide, given an ABox $\mathcal{A}$, whether there is a homomorphism from $\mathcal{A}$ to $\mathcal{B}$, in which case we write $\mathcal{A} \to \mathcal{B}$. We show, following [8], how given an OMQ $Q = (Cov^\perp_\top, q)$ with a path CQ q, one can construct a template $\mathcal{B}_q$ such that, for any data instance $\mathcal{A}$, we have $\mathcal{A} \to \mathcal{B}_q$ iff $Cov^\perp_\top, \mathcal{A} \not\models q$. We illustrate the construction of $\mathcal{B}_q$ using the CQ q below:

$$\overset{T}{\underset{R}{\circ\!\!\longrightarrow}} \overset{T}{\underset{R}{\circ\!\!\longrightarrow}} \overset{F}{\underset{R}{\circ\!\!\longrightarrow}} \overset{F}{\circ}$$

The construction generalises to arbitrary path CQs in the obvious way and can be further extended to OMQs with tree-shaped CQs.

First, we assign labels A, B and C to the first three vertices of q in the following way:

$$\overset{T}{\underset{C}{\circ}} \underset{R}{\longrightarrow} \overset{T}{\underset{B}{\circ}} \underset{R}{\longrightarrow} \overset{F}{\underset{A}{\circ}} \underset{R}{\longrightarrow} \overset{F}{\circ}$$

Then we construct the following disjunctive datalog program Π such that, for any $\mathcal{A}$, we have $Cov^\perp_\top, \mathcal{A} \models q$ iff $\mathcal{A}, \Pi \models \exists x\, C(x)$:

$$A(x) \leftarrow F(x), R(x,y), F(y) \tag{4}$$

$$B(x) \leftarrow T(x), R(x,y), F(y), A(y) \tag{5}$$

$$C(x) \leftarrow T(x), R(x,y), T(y), B(y) \tag{6}$$

$$T(x) \lor F(x) \leftarrow \tag{7}$$

$$\perp \leftarrow T(x), F(x) \tag{8}$$

(Informally, the labels A, B and C are used in Π to detect the query pattern in a data instance step-by-step.)

We now construct the CSP template $\mathcal{B}_q$ using unary *types* for Π, which are sets t of unary predicates in Π such that t contains either T or F, but not both of them, and $C \notin t$. Thus in this example the types are 8 specific subsets of $\{A, B, C, T, F\}$ that are used as vertex labels in Fig. 1 (talking about types, we often omit curly brackets and commas, so, for example, FAB stands for $\{F, A, B\}$). In general, for a path CQ with n variables there are 2^{n-1} types. Two types t_1 and t_2 are called R-*compatible* if the data instance

$$\{R(t_1, t_2)\} \cup \{X(t_1) \mid X \in t_1\} \cup \{Y(t_2) \mid Y \in t_2\}$$

is a model of Π. The domain of the template $\mathcal{B}_q$ consists of the types for Π, unary relations are interpreted in the natural way by $T^{\mathcal{B}_q} = \{t \mid T \in t\}$,

$F^{\mathcal{B}_q} = \{t \mid F \in t\}$, and the binary relation is specified by $R^{\mathcal{B}_q} = \{(t_1, t_2) \mid t_1$ and t_2 are R-compatible$\}$.

For the q above we obtain the template $\mathcal{B}_q$ shown on the left-hand side of Fig. 1, where the vertices are labelled by the corresponding types. For example, there are no edges between F and FB since the data instances $\{F(x), R(x, y), F(y), B(y)\}$ and $\{F(x), B(x), R(x, y), F(y)\}$ do not satisfy the rule $A(x) \leftarrow F(x), R(x, y), F(y)$, but there is an edge from FA to F since the data instance $\{F(x), A(x), R(x, y), F(y)\}$ is a model of Π. One can see that $\mathcal{A}, \Pi \not\models \exists x\, C(x)$ iff $\mathcal{A} \rightarrow \mathcal{B}_q$, and so $\mathcal{B}_q$ is as required.

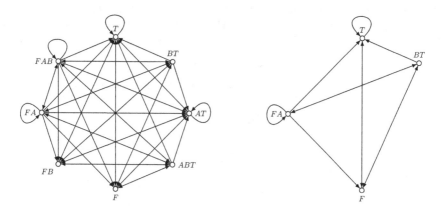

Fig. 1. CSP template $\mathcal{B}_q$ (left) and its core (right).

5 Enter Polyanna

To check whether $\mathrm{CSP}(\mathcal{B}_q)$ is in P or CONP-hard, one can use the program Polyanna [13]. Polyanna proceeds in two stages. First, it finds a core of the template $\mathcal{B}_q$. We remind the reader that a relational structure $\mathcal{D}'$ is a *core* of a relational structure $\mathcal{D}$ if $\mathcal{D}'$ is a minimal substructure of $\mathcal{D}$ that is homomorphically equivalent to $\mathcal{D}$ (in the sense that $\mathcal{D}' \rightarrow \mathcal{D}$ and $\mathcal{D} \rightarrow \mathcal{D}'$). The process of constructing such a core by Polyanna is called 'squashing'. For example, it squashes the 8-vertex template $\mathcal{B}_q$ in Fig. 1 into the core template with 4 vertices shown on the right-hand side of the figure.

Then Polyanna decides tractability or CONP-hardness of $\mathrm{CSP}(\mathcal{B}_q)$ by checking whether the core template has polymorphisms of certain types by constructing and solving the corresponding 'indicator problems' [18]. While doing this, Polyanna uses different decomposition techniques to reduce computation in the case when the indicator problem has symmetries. The indicator problem for polymorphisms of arity k and cores with d vertices, for a signature Γ, has $k \cdot d^k$ variables and $\Sigma_{R \in \Gamma} |R|^k$ constraints. Given a core template of size d, Polyanna considers polymorphisms of arity up to $\max(3, d)$. In practice, for our use case, this implies that it can handle cores of size up to 4, but runs out of memory for some cores of size ≥ 5.

It has recently been shown that tractability of CSP can be determined by considering special polymorphisms of arity 4 that satisfy the identity $f(y, x, y, z) = f(x, y, z, x)$; they are called *Siggers polymorphisms* in [5]. However, we could not find any publicly available implementations for these polymorphisms, and so Polyanna can be considered as top-edge technology even if it is more than 15 years old. We conjecture that the range of its applicability can be significantly extended by enhancing it with the ability to search for Siggers polymorphisms.

We used Polyanna to classify the data complexity of OMQs $\boldsymbol{Q} = (\mathcal{C}ov_{\mp}^{\pm}, \boldsymbol{q})$ with path CQs $\boldsymbol{q}$ of length up to 4 (that is, with at most 4 variables).[2]

She correctly determined that all of them but four OMQs with two T-nodes and two F-nodes are in P. For example, the OMQ with the CQ

and the core CSP template $\mathcal{B}$ shown below was classified as tractable.

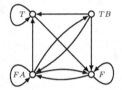

In Sect. 7, we illustrate how Polyanna's output can be used to construct a monadic datalog-rewritings of OMQs.

Polyanna also managed to determine that the CQs $\boldsymbol{q}_1$ and $\boldsymbol{q}_2$ below give coNP-hard OMQs:

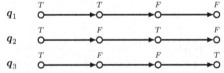

In fact, the CSP templates for $\boldsymbol{q}_1$ and $\boldsymbol{q}_2$ have a 4-vertex core, while $\mathrm{CSP}(\mathcal{B}_{\boldsymbol{q}_3})$ has the following core with five vertices:

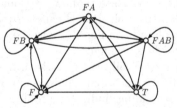

which turned out to be too hard for Polyanna.

---

[2] Path CQs of length 5 produce templates of size 16, and it takes over an hour to detect its core.

In the next section, we sketch direct proofs of CONP-hardness of answering the three OMQs $\boldsymbol{Q}_i = (\mathcal{C}ov\frac{1}{\top}, \boldsymbol{q}_i)$, for $i = 1, 2, 3$, by reduction of 3SAT. The conaisseurs might find it instructive to compare the gadgets used in those proofs with the general machinery of pp-interpretations coming from universal algebra.

6   Three CONP-hard OMQs

Consider first the OMQ $\boldsymbol{Q}_1 = (\mathcal{C}ov\frac{1}{\top}, \boldsymbol{q}_1)$. For every propositional variable p in a given 3CNF ψ, we construct a 'gadget' shown in the picture below, where the number of vertices above each of the circles matches the number of clauses in ψ; we refer to these vertices as p-*contacts* and, respectively, $\neg p$-*contacts*:

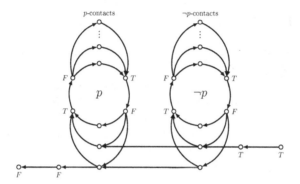

Observe that, for any model $\mathcal{I}$ of $\mathcal{C}ov\frac{1}{\top}$ based on the constructed gadget for p, if $\mathcal{I} \not\models \boldsymbol{q}_1$ then either (i) the p-contacts are all in $F^{\mathcal{I}}$ and the $\neg p$-contacts are all in $T^{\mathcal{I}}$, or (ii) the p-contacts are all in $T^{\mathcal{I}}$ and the $\neg p$-contacts are all in $F^{\mathcal{I}}$.

Now, for every clause $c = (l_1 \vee l_2 \vee l_3)$ in ψ with literals l_i, we add to the constructed gadgets the atoms $T(c)$, $R(c, a^c_{\neg l_1})$, $R(a^c_{\neg l_1}, a^c_{l_2})$, $R(a^c_{l_2}, a^c_{l_3})$, where c is a new individual, $a^c_{\neg l_1}$ a fresh $\neg l_1$-contact, $a^c_{l_2}$ a fresh l_2-contact, and $a^c_{l_3}$ a fresh l_3-contact. For example, for the clause $c = (p \vee q \vee r)$, we obtain the fragment below:

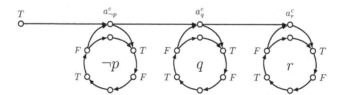

The resulting ABox is denoted by $\mathcal{A}_\psi$. The reader can check that ψ is satisfiable iff $\mathcal{C}ov\frac{1}{\top}, \mathcal{A}_\psi \not\models \boldsymbol{q}_1$. It follows that answering $\boldsymbol{Q}_1$ is CONP-hard.

Next, consider the OMQ $\boldsymbol{Q}_2 = (\mathcal{C}ov\frac{1}{\top}, \boldsymbol{q}_2)$. Similarly to the previous case, for every variable p occurring in ψ, we take the following p-*gadget*, where n is the number of clauses in ψ:

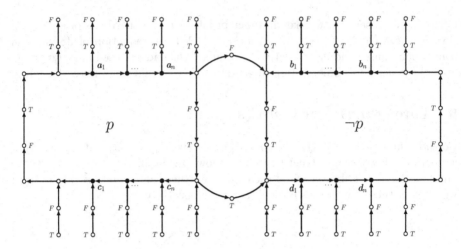

The key property of the p-gadget is that, for any model $\mathcal{I}$ of $Cov^{\perp}_{\top}$ based on this gadget, if $\mathcal{I} \not\models q_2$ then either the vertices without labels on left-hand side of the gadget are all in $T^{\mathcal{I}}$ and the vertices without labels on the right-hand side are all in $F^{\mathcal{I}}$, or the other way round. We refer to the a_i and b_i as $p^{\uparrow}$- and $\neg p^{\uparrow}$-contacts, and to the c_i and d_i as $p^{\downarrow}$- and $\neg p^{\downarrow}$-contacts, respectively.

Now, for every clause $c = (l_1 \vee l_2 \vee l_3)$ in ψ, we add to the constructed gadgets for the variables in ψ the atoms $R(u^c_{\neg l_1}, v^c_{l_2})$, $R(v^c_{l_2}, c)$, $T(c)$, $R(c, w^c_{l_3})$, where c is a new individual, $u^c_{\neg l_1}$ a fresh $\neg l_1^{\uparrow}$-contact, $v^c_{l_2}$ a fresh $l_2^{\downarrow}$-contact, and $w^c_{l_3}$ a fresh $l_3^{\downarrow}$-contact. For example, for the clause $c = (p \vee q \vee \neg r)$, we obtain the fragment below:

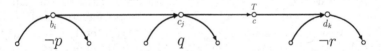

The resulting ABox $\mathcal{A}_\psi$ is such that ψ is satisfiable iff $Cov^{\perp}_{\top}, \mathcal{A}_\psi \not\models q_2$.

Finally, for the OMQ $\boldsymbol{Q}_3 = (Cov^{\perp}_{\top}, q_3)$, we use the following p-gadget:

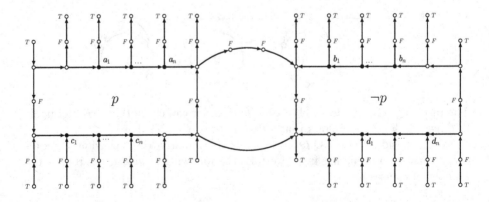

and a similar encoding of clauses. The only difference occurs at the individual c in the clause gadget where we place an F-atom instead of a T-atom.

7 Monadic Datalog Rewritings Based on Arc Consistency

For tractable OMQs, Polyanna outputs semilattice functions, which guarantee that the classical arc-consistency check gives correct answers to these OMQs; see [21] for a survey of different arc-consistency routines. The classical arc-consistency procedure can be interpreted as a monadic datalog program, which gives a monadic datalog-rewriting of the OMQ in question. It may be instructive to have a look at the monadic datalog program for the CQ

$$
\begin{array}{ccc}
\overset{T}{\underset{x}{\circ}} \xrightarrow{\quad R \quad} & \overset{F}{\underset{y}{\circ}} \xrightarrow{\quad R \quad} & \overset{F}{\underset{z}{\circ}}
\end{array}
$$

which is dual to the one in Example 4. Below is a 3-vertex core of the template for this query:

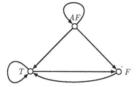

The idea of arc-consistency is to introduce an IDB predicate P for each subset of vertices in the template, with the intuitive meaning that P should hold of a constant c in a given data instance iff it follows that c must be mapped by a homomorphism within the set corresponding to P. Using semilattice function, one can prove that the CSP does not have a solution iff we can deduce the predicate that corresponds to the empty set.

To make our monadic datalog program more readable, we associate predicate names P, Q, T', F', F_0 etc. to sets of types according to the table below. The table also shows the R-image of any set X (the set of all vertices that are R-accessible from X) and its R-pre-image (the set of all vertices from which X is R-accessible) encoded as predicate names:

predicate	subset of vertices	R-image	R-pre-image
P	$\{T, F\}$	P	1
T'	$\{T\}$	P	1
F'	$\{AF, F\}$	1	Q
Q	$\{T, AF\}$	1	1
F_0	$\{F\}$	T	Q
F_1	$\{AF\}$	1	F_1
1	$\{AF, A, T\}$	1	1

Then we have the following program for arc-consistency. The first two rules say that T and F should be preserved. The next rules say 'if X is the R-pre-image of Y, then we should have the rule $X(x) \leftarrow R(x, y), Y(y)$' (second group), 'if

Z is the R-image of Y, then we should have the rule $X(z) \leftarrow Y(y), R(y, z)$ (third group). We also have rules for Boolean reasoning: if $X \subseteq Y$, then we add $Y(z) \leftarrow X(z)$; and if $Z = X \cap Y$, we have the rule $Z(z) \leftarrow X(z), Y(z)$. Finally, we have rules with the goal predicate in the head for disjoint X and Y. Here goes the program:

$$T'(x) \leftarrow T(x) \tag{9}$$
$$F'(x) \leftarrow F(x) \tag{10}$$

$$P(y) \leftarrow T'(x), R(x, y) \tag{11}$$
$$Q(x) \leftarrow R(x, y), F_0(y) \tag{12}$$
$$F_1(x) \leftarrow R(x, y), F_1(y) \tag{13}$$
$$P(y) \leftarrow P(x), R(x, y) \tag{14}$$
$$Q(x) \leftarrow R(x, y), F'(y) \tag{15}$$

$$F'(x) \leftarrow F_1(x) \tag{16}$$
$$F'(x) \leftarrow F_0(x) \tag{17}$$
$$Q(x) \leftarrow F_1(x) \tag{18}$$
$$P(x) \leftarrow F_0(x) \tag{19}$$
$$Q(x) \leftarrow T'(x) \tag{20}$$
$$P(x) \leftarrow T'(x) \tag{21}$$

$$T'(x) \leftarrow Q(x), P(x) \tag{22}$$
$$F_0(x) \leftarrow P(x), F'(x) \tag{23}$$
$$F_1(x) \leftarrow Q(x), F'(x) \tag{24}$$
$$G \leftarrow F_1(x), P(x) \tag{25}$$
$$G \leftarrow T'(x), F'(x) \tag{26}$$
$$G \leftarrow Q(x), F_0(x) \tag{27}$$

To illustrate, consider the following data instance

Then rule (9) produces $T'(x_0)$, rule (11) produces $P(x_1)$, and rule (14) produces $P(x_2)$ and $P(x_3)$. On the other hand, rule (10) produces $F'(x_3)$ and $F'(x_2)$, and so rule (23) produces $F_0(x_2)$ and $F_0(x_3)$. Now rule (12) produces $Q(x_2)$, which gives us G by rule (27).

However, the program returns 'no' on the following data instance (it still produces $P(x_3)$ and $Q(x_2)$, but now instead of G it produces $T'(x_2)$):

8   Appendix

We conclude this article with a brief appendix that summarises what is known about the data complexity of answering the OMQs in the framework of our case study. For more details and proofs the reader is referred to [14–16].

We classify the OMQs according to the number of occurrences of solitary F in their CQs (the case of solitary T is symmetric).

8.1   0-CQs

By a 0-CQ we mean any CQ that does not contain a solitary F. A *twin* in a CQ q is any pair $F(x), T(x) \in q$. Here is an encouraging syntactic criterion of FO-rewritability for OMQs of the form $(Cov_A^{\perp}, q)$:

Theorem 1. *(i) If q is a 0-CQ, then answering both $(Cov_A^{\perp}, q)$ and (Cov_A, q) is in AC^0, with q being an FO-rewriting of these OMQs.*

(ii) If q is not a 0-CQ and does not contain twins, then answering both $(Cov_{\bar{T}}^{\perp}, q)$ and (Cov_T, q) is L-hard.

Corollary 1. *An OMQ $(Cov_A^{\perp}, q)$ is in AC^0 iff q is a 0-CQ, which can be decided in linear time.*

Theorem 1 (*i*) generalises to OMQs with ontologies $Cov_n = \{A \sqsubseteq B_1 \sqcup \cdots \sqcup B_n\}$, for $n \geq 2$:

Theorem 2. *Suppose q is any CQ that does not contain an occurrence of B_i, for some i ($1 \leq i \leq n$). Then answering the OMQ $Q = (Cov_n, q)$ is in AC^0.*

Thus, only those CQs can 'feel' Cov_n as far as FO-rewritability is concerned that contain all the B_n (which makes them quite complex in practice).

If twins can occur in CQs (that is, F and T are not necessarily disjoint), the picture becomes more complex. On the one hand, we have the following criterion for OMQs (Cov_A, q) with a *path CQ* q whose variables $x_0, \ldots, x_n$ in q are ordered so that the binary atoms in q form a chain $R_1(x_0, x_1), \ldots, R_n(x_{n-1}, x_n)$.

Theorem 3. *An OMQ (Cov_A, q) with a path CQ q is in AC^0 iff q is a 0-CQ. If q contains both solitary F and T, then (Cov_A, q) is NL-hard.*

On the other hand, this AC^0/NL criterion collapses for path CQs with loops:

Proposition 1. *The OMQ (Cov_A, q), where q is shown below, is in AC^0.*

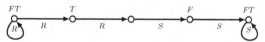

Note that the CQ q above is *minimal* (not equivalent to any of its proper sub-CQs). Note also that, if a minimal 1-CQ q contains both a solitary F and a solitary T, then FO-rewritability of (Cov_A, q) implies that q contains at least one twin (FT) and at least one y with $T(y) \notin q$ and $F(y) \notin q$ (which can be shown using Theorem 7 (*i*)).

8.2 1-CQs

1-CQs have exactly one solitary F. We now complement the sufficient conditions of L- and NL-hardness from Sect. 8.1 with sufficient conditions of OMQ answering in L- and NL.

A *1-CQ* $q'(x, y)$ is *symmetric* if the CQs $q'(x, y)$ and $q'(y, x)$ are equivalent in the sense that $q'(a, b)$ holds in $\mathcal{A}$ iff $q'(b, a)$ holds in $\mathcal{A}$, for any ABox $\mathcal{A}$ and $a, b \in \mathrm{ind}(\mathcal{A})$.

Theorem 4. *Let* $Q = (Cov_A, q)$ *be any OMQ such that*

$$q = \exists x, y \, (F(x) \land q_1'(x) \land q'(x, y) \land q_2'(y) \land T(y)),$$

for some connected CQs $q'(x, y)$, $q_1'(x)$ *and* $q_2'(y)$ *that do not contain solitary* T *and* F, *and* $q'(x, y)$ *is symmetric. Then answering* Q *can be done in* L.

If we do not require $q'(x, y)$ to be symmetric, the complexity upper bound increases to NL:

Theorem 5. *Let* $Q = (Cov_A, q)$ *be any OMQ such that*

$$q = \exists x, y \, (F(x) \land T(y) \land q'(x, y)),$$

for some connected CQ $q'(x, y)$ *without solitary occurrences of* F *and* T. *Then answering* Q *can be done in* NL.

By an *F-tree CQ* we mean a CQ q having a single solitary $F(x)$ such that the binary atoms in q form a ditree with root x.

Theorem 6. (*i*) *Answering any OMQ* (Cov_A, q) *with a* 1-CQ q *can be done in* P.

(*ii*) *Answering any OMQ* (Cov_A, q) *with an F-tree* q *is either in* AC^0 *or* NL-*complete or* P-*complete. The trichotomy can be decided in* ExpTime.

Theorem 6 (*ii*) was proved by a reduction to the $\mathrm{AC}^0/\mathrm{NL}/\mathrm{P}$-trichotomy of [23]. It is to be noted, however, that applying the algorithm from [23] in our case is tricky because the input ontology must first be converted to a normal form. As a result, we do not obtain transparent syntactic criteria on the shape of q that would guarantee that the OMQ (Cov_A, q) belongs to the desired complexity class.

We now give a semantic sufficient condition for an OMQ with a 1-CQ to lie in NL. This condition uses ideas and constructions from [12,23]. Let $Q = (Cov_A, q)$ be an OMQ with a 1-CQ q having a solitary $F(x)$. Define by induction a class $\mathfrak{K}_Q$ of ABoxes that will be called *cactuses for* Q. We start by setting $\mathfrak{K}_Q := \{q\}$, regarding q as an ABox, and then recursively apply to $\mathfrak{K}_Q$ the following two rules:

(bud) if $T(y) \in \mathcal{A} \in \mathfrak{K}_Q$ with solitary $T(y)$, then we add to $\mathfrak{K}_Q$ the ABox obtained by replacing $T(y)$ in $\mathcal{A}$ with $(q \setminus F(x)) \cup \{A(x)\}$, in which x is renamed to y and all of the other variables are given *fresh* names;

(**prune**) if $Cov_A, \mathcal{A}' \models \mathbf{Q}$, where $\mathcal{A}' = \mathcal{A} \setminus \{T(y)\}$ and $T(y)$ is solitary, we add to $\mathfrak{K}_{\mathbf{Q}}$ the ABox obtained by removing $T(y)$ from $\mathcal{A} \in \mathfrak{K}_{\mathbf{Q}}$.

It is readily seen that, for any ABox $\mathcal{A}'$, we have $Cov_A, \mathcal{A}' \models \mathbf{Q}$ iff there exist $\mathcal{A} \in \mathfrak{K}_{\mathbf{Q}}$ and a homomorphism $h\colon \mathcal{A} \to \mathcal{A}'$. Denote by $\mathfrak{K}_{\mathbf{Q}}^{\dagger}$ the set of minimal cactuses in $\mathfrak{K}_{\mathbf{Q}}$ (that have no proper sub-cactuses in $\mathfrak{K}_{\mathbf{Q}}$).

For a cactus $\mathcal{C} \in \mathfrak{K}_{\mathbf{Q}}$, we refer to the copies of (maximal subsets of) $\mathbf{q}$ that comprise $\mathcal{C}$ as *segments*. The *skeleton* $\mathcal{C}^s$ of $\mathcal{C}$ is the ditree whose nodes are the segments $\mathfrak{s}$ of $\mathcal{C}$ and edges $(\mathfrak{s}, \mathfrak{s}')$ mean that $\mathfrak{s}'$ was attached to $\mathfrak{s}$ by budding. The atoms $T(y) \in \mathfrak{s}$ are called the *buds* of $\mathfrak{s}$. The *rank* $r(\mathfrak{s})$ of $\mathfrak{s}$ is defined by induction: if $\mathfrak{s}$ is a leaf, then $r(\mathfrak{s}) = 0$; for non-leaf $\mathfrak{s}$, we compute the maximal rank m of its children and then set

$$r(\mathfrak{s}) = \begin{cases} m + 1, & \text{if } \mathfrak{s} \text{ has } \geq 2 \text{ children of rank } m; \\ m, & \text{otherwise.} \end{cases}$$

The *width* of $\mathcal{C}$ and $\mathcal{C}^s$ is the rank of the root in $\mathcal{C}^s$. We say that $\mathfrak{K}_{\mathbf{Q}}^{\dagger}$ is of *width* k if it contains a cactus of width k but no cactus of greater width. The *depth* of $\mathcal{C}$ and $\mathcal{C}^s$ is the number of edges in the longest branch in $\mathcal{C}^s$.

We illustrate the definition by an example. Denote by $\mathbf{q}_{TnT}$, for $n \geq 0$, the 1-CQ shown below, where all the binary predicates are R and the n variables without labels do not occur in F- or T-atoms:

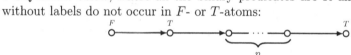

Example 6. Let $\mathbf{Q} = (Cov_\top, \mathbf{q}_{T1T})$. In the picture below, we show a cactus $\mathcal{C}$ obtained by applying (**bud**) twice to $\mathbf{q}_{T1T}$ (with $A = \top$ omitted):

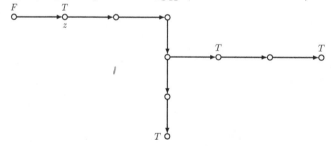

One can check that $Cov_\top, \mathcal{C} \setminus \{T(z)\} \models \mathbf{q}_{T1T}$, and so an application of (**prune**) will remove $T(z)$ from $\mathcal{C}$. Using this observation, one can show that $\mathfrak{K}_{\mathbf{Q}}^{\dagger}$ is of width 1. On the other hand, if $\mathbf{Q} = (Cov_A, \mathbf{q}_{T1T})$ then $\mathfrak{K}_{\mathbf{Q}}^{\dagger}$ is of unbounded width as follows from Theorem 9 below.

Theorem 7. *Let $\mathbf{Q} = (Cov_A, \mathbf{q})$ be an OMQ with a 1-CQ $\mathbf{q}$. Then*

(i) $\mathbf{Q}$ is in AC^0 iff for every $\mathcal{C} \in \mathfrak{K}_{\mathbf{Q}}^{\dagger}$, there is a homomorphism $h\colon \mathbf{q} \to \mathcal{C}$;

(ii) $\mathbf{Q}$ is rewritable in linear datalog, and so is in NL, if $\mathfrak{K}_{\mathbf{Q}}^{\dagger}$ is of bounded width.

It is worth noting that, for $Q = (Cov_T, q)$ with q from Proposition 1, $\mathfrak{K}_Q^\dagger$ consists of q and the cactus of depth 1, in which the only solitary T is removed by **(prune)**. Clearly, there is a homomorphism from q into this cactus, and so Q is FO-rewritable. However, for the 1-CQ q in the picture below (where all edges are bidirectional), (Cov_T, q) is not FO-rewritable, but there is a homomorphism from q to both cactuses of depth 1. We do not know whether, in general, there is an upper bound N_q such that the existence of homomorphisms $h\colon q \to C$, for all $C \in \mathfrak{K}_Q$ of depth N_q, would ensure FO-rewritability of (Cov_A, q). For 1-CQs q with a single solitary T, one can take $N_q = |q| + 1$. Neither do we know the exact complexity of deciding FO-rewritability of OMQs with 1-CQs. As mentioned in Sect. 3, this problem is reducible to the boundedness problem for monadic datalog programs, which is known to be in 2ExpTime.

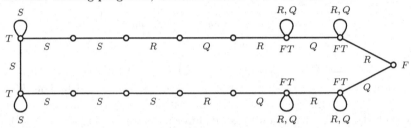

Theorem 7 (ii) allows us to obtain a sufficient condition for linear-datalog rewritability of OMQs (Cov_A, q) with an F-path CQ q, that is, a path CQ with a single solitary F at its root. We represent such a q as shown in the picture below, which indicates *all* the solitary occurrences of F and T:

$$q = \quad \overset{F}{\underset{x}{\circ}} \!\!\to\! \circ \cdots \circ \to \overset{T}{\underset{y_1}{\circ}} \to \circ \cdots \circ \to \overset{T}{\underset{y_i}{\circ}} \to \circ \cdots \circ \to \overset{T}{\underset{y_m}{\circ}} \to \circ \cdots \circ \to \underset{y_{m+1}}{\circ}$$

We require the following sub-CQs of q:

- q_i is the suffix of q that starts at y_i, but without $T(y_i)$, for $1 \le i \le m$;
- q_i^* is the prefix of q that ends at y_i, but without $F(x)$ and $T(y_i)$, for $1 \le i \le m$;
- q_{m+1}^* is q without $F(x)$,

and write $f_i\colon q_i \twoheadrightarrow q$ if f_i is a homomorphism from q_i into q with $f_i(y_i) = x$.

Theorem 8. *If for each $1 \le i \le m$ there exist $f_i\colon q_i \twoheadrightarrow q$, then (Cov_A, q) is rewritable into a linear datalog program, and so is NL-complete.*

For F-path CQs q without twins, we extend Theorem 8 to a NL/P dichotomy (provided that NL $\neq$ P). Given such a CQ q, we denote by N_q the set of the numbers indicating the length of the path from x to each of the y_i, $i = 1, \ldots, m+1$.

Theorem 9. *Let $Q = (Cov_A, q)$ be an OMQ where q is an F-path CQ without twins having a single binary relation. The following are equivalent unless* NL $=$ P:

(i) Q *is* NL-*complete;*
(ii) $\{0\} \cup N_q$ *is an arithmetic progression;*
(iii) there exist $f_i \colon q_i \twoheadrightarrow q$ *for every* $i = 1, \ldots, m.$

If these conditions do not hold, then Q *is* P-*complete.*

Note that the proof of P-hardness in Theorem 9 does not go through for $A = \top$. Thus, for $(Cov_\top, q_{T1T})$, we are in the framework of Example 6 and, by Theorem 7 *(ii)*, this OMQ is in NL. In fact, we have the following NL/P dichotomy for the OMQs of the form $Q = (Cov_\top, q_{TnT})$:

– either n is equal to 1, and answering Q is in NL,
– or $n \geq 2$, and answering Q is P-hard.

Proposition 2. *Answering the OMQ* $(Cov_\top, q_{T1T})$ *is* NL-*complete.*

Theorem 10. *The OMQs* $(Cov_\top, q_{TnT})$ *(and* (Cov_A, q_{TnT})*), for* $n \geq 2$, *are* P-*complete.*

On the other hand we have:

Proposition 3. *Answering the OMQ* (Cov_A, q_{T1T}) *is* P-*complete.*

We now apply Theorem 7 *(ii)* to the class of *TF-path CQs* of the form

$$q_{TF} = \begin{array}{ccccccc} T & & F & & T & & T \\ \circ{\longrightarrow}\circ \cdots \circ{\longrightarrow}\circ{\longrightarrow}\circ \cdots \circ{\longrightarrow}\circ{\longrightarrow}\circ \cdots \circ{\longrightarrow}\circ \\ y_0 & & x & & y_1 & & y_m & & y_{m+1} \end{array}$$

where the $T(y_i)$ and $F(x)$ are all the solitary occurrences of T and F in q_{TF}. We represent this CQ as

$$q_{TF} = \{T(y_0)\} \cup q_0 \cup q,$$

where q_0 is the sub-CQ of q_{TF} between y_0 and x with $T(y_0)$ removed and q is the same as in Theorem 8 (and q_{m+1}^* is q without $F(x)$).

Theorem 11. *If* q *satisfies the condition of Theorem 8 and there is a homomorphism* $h \colon q_{m+1}^* \to q_0$ *such that* $h(x) = y_0$, *then answering* (Cov_A, q_{TF}) *is* NL-*complete.*

For example, the OMQ (Cov_A, q) with q shown below is NL-complete:

$$\begin{array}{cccc} T & FT & F & T \\ \circ{\longrightarrow}\circ{\longrightarrow}\circ{\longrightarrow}\circ \end{array}$$

On the other hand, we have the following:

Proposition 4. *Answering the OMQs* (Cov_A, q) *and* $(Cov_\top, q)$ *is* P-*complete for* q *of the forms*

8.3 2-CQs

A *2-CQ* has at least two solitary F and at least two solitary T. We have the following generalisations of the coNP-hardness results from Sect. 6 for the OMQs $Q_j = (Cov_T^{\neq}, q_j)$, for $j \in \{1, 2, 3\}$:

Consider the *2-2-CQs*, which are path 2-CQs where all the F are located after all the T, and every occurrence of T or F is solitary. We represent any given 2-2-CQ q as shown below

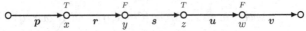

where p, r, u and v do not contain F and T, while s may contain solitary occurrences of both T and F (in other words, the T shown in the picture are the first two occurrences of T in q and the F are the last two occurrences of F in q). Denote by q_r the suffix of q that starts from x but without $T(x)$; similarly, q_u is the suffix of q starting from z but without $F(z)$. Denote by q_r^- the prefix of q that ends at y but without $T(y)$; similarly, q_u^- is the prefix of q ending at w but without $F(w)$. Using the construction from [15], one can show the following:

Theorem 12. *Any OMQ (Cov_A, q) with a 2-2-CQ q is* coNP-*complete provided the following conditions are satisfied: (i) there is no homomorphism* $h_1 \colon q_u \to q_r$ *with* $h_1(z) = x$, *and (ii) there is no homomorphism* $h_2 \colon q_r^- \to q_u^-$ *with* $h_2(y) = w$.

We do not know yet whether this theorem holds for Cov_T in place of Cov_A.

In Theorems 13 and 14, we assume that p and v do not contain F and T, while r and u may only contain solitary occurrences of T ($F \notin r, u$), and s only solitary occurrences of F ($T \notin s$).

Theorem 13. *Any OMQ (Cov_A, q) with q of the form*

$$\circ \xrightarrow{\quad p \quad} \overset{T}{\bullet} \xrightarrow{\quad x \quad} \overset{}{\circ} \xrightarrow{\quad r \quad} \overset{F}{\bullet} \xrightarrow{\quad y \quad} \overset{}{\circ} \xrightarrow{\quad s \quad} \overset{F}{\bullet} \xrightarrow{\quad z \quad} \overset{}{\circ} \xrightarrow{\quad u \quad} \overset{T}{\bullet} \xrightarrow{\quad w \quad} \overset{}{\circ} \xrightarrow{\quad v \quad} \circ$$

is coNP-*complete provided the following conditions are satisfied: (i) there is no homomorphism* $h_1 \colon r_t \to u$ *with* $h_1(y) = w$, *and (ii) there is no homomorphism* $h_2 \colon u_t \to r$ *with* $h_2(z) = x$, *where r is the sub-CQ of q between x and y without $T(x)$, $F(y)$, and similarly for u, r_t is r with $T(x)$ and u_t is u with $T(w)$.*

In Theorem 14, we use $r^{ext} = r(x, y) \wedge T(y) \wedge s_1(y, y_1) \wedge F(y_1)$, where s_1 is the part of s such that $s(y, z) = s_1(y, y_1) \wedge F(y_1) \wedge s_2(y_1, z)$ and $s_1(y, y_1)$ does not contain any occurrences of F. In other words, the variable y_1 corresponds to the first appearance of F in s, where s is the sub-CQ of q between y and z without $F(y)$, $T(z)$.

Theorem 14. *Any OMQ (Cov_A, q) with q of the form*

$$\circ \xrightarrow{\quad p \quad} \overset{T}{\bullet} \xrightarrow{\quad x \quad} \overset{}{\circ} \xrightarrow{\quad r \quad} \overset{F}{\bullet} \xrightarrow{\quad y \quad} \overset{}{\circ} \xrightarrow{\quad s \quad} \overset{T}{\bullet} \xrightarrow{\quad z \quad} \overset{}{\circ} \xrightarrow{\quad u \quad} \overset{F}{\bullet} \xrightarrow{\quad w \quad} \overset{}{\circ} \xrightarrow{\quad v \quad} \circ$$

is coNP-*complete provided the following conditions hold: (i) there is no homomorphism* $g_1 \colon r_t \to u$ *with* $g_1(y) = w$, *and (ii) there is no homomorphism* $g_2 \colon u \to r^{ext}$ *with* $g_2(z) = x$ *and* $g_2(w) = y_1$.

Acknowledgements. The work of O. Gerasimova and M. Zakharyaschev was carried out at the National Research University Higher School of Economics and supported by the Russian Science Foundation under grant 17-11-01294. We are grateful to Peter Jeavons and Standa Živný for helpful discussions of Polyanna and arc consistency.

References

1. Abiteboul, S., Hull, R., Vianu, V.: Foundations of Databases. Addison-Wesley, Reading (1995)
2. Afrati, F.N., Gergatsoulis, M., Toni, F.: Linearisability on datalog programs. Theor. Comput. Sci. **308**(1–3), 199–226 (2003). https://doi.org/10.1016/S0304-3975(02)00730-2
3. Arora, S., Barak, B.: Computational Complexity: A Modern Approach, 1st edn. Cambridge University Press, New York (2009)
4. Baader, F., Horrocks, I., Lutz, C., Sattler, U.: An Introduction to Description Logic. Cambridge University Press, Cambridge (2017). http://www.cambridge.org /de/academic/subjects/computer-science/knowledge-management-databases-and-data-mining/introduction-description-logic?format=PB#17zVGeWD2TZUeu6s.97
5. Barto, L., Krokhin, A., Willard, R.: Polymorphisms, and how to use them. In: Dagstuhl Follow-Ups. vol. 7. Schloss Dagstuhl-Leibniz-Zentrum fuer Informatik (2017)
6. Benedikt, M., ten Cate, B., Colcombet, T., Vanden Boom, M.: The complexity of boundedness for guarded logics. In: 30th Annual ACM/IEEE Symposium on Logic in Computer Science, LICS 2015, Kyoto, Japan, 6–10 July 2015, pp. 293–304. IEEE Computer Society (2015). https://doi.org/10.1109/LICS.2015.36
7. Benedikt, M., Grau, B.C., Kostylev, E.V.: Logical foundations of information disclosure in ontology-based data integration. Artif. Intell. **262**, 52–95 (2018). https://doi.org/10.1016/j.artint.2018.06.002
8. Bienvenu, M., ten Cate, B., Lutz, C., Wolter, F.: Ontology-based data access: a study through disjunctive datalog, CSP, and MMSNP. ACM Trans. Database Syst. **39**(4), 33:1–33:44 (2014)
9. Bulatov, A.A.: A dichotomy theorem for nonuniform CSPs. In: Umans, C. (ed.) 58th IEEE Annual Symposium on Foundations of Computer Science, FOCS 2017, Berkeley, CA, USA, 15–17 October 2017, pp. 319–330. IEEE Computer Society (2017). https://doi.org/10.1109/FOCS.2017.37
10. Calvanese, D., et al.: The MASTRO system for ontology-based data access. Semant. Web **2**(1), 43–53 (2011)
11. Calvanese, D., et al.: Ontop: answering SPARQL queries over relational databases. Semant. Web **8**(3), 471–487 (2017)
12. Cosmadakis, S.S., Gaifman, H., Kanellakis, P.C., Vardi, M.Y.: Decidable optimization problems for database logic programs (preliminary report). In: STOC, pp. 477–490 (1988)
13. Gault, R., Jeavons, P.: Implementing a test for tractability. Constraints **9**(2), 139–160 (2004)
14. Gerasimova, O., Kikot, S., Podolskii, V., Zakharyaschev, M.: More on the data complexity of answering ontology-mediated queries with a covering axiom. In: Różewski, P., Lange, C. (eds.) KESW 2017. CCIS, vol. 786, pp. 143–158. Springer, Cham (2017). https://doi.org/10.1007/978-3-319-69548-8_11

15. Gerasimova, O., Kikot, S., Podolskii, V.V., Zakharyaschev, M.: On the data complexity of ontology-mediated queries with a covering axiom. In: Artale, A., Glimm, B., Kontchakov, R. (eds.) Proceedings of the 30th International Workshop on Description Logics. CEUR Workshop Proceedings, Montpellier, France, 18–21 July 2017, vol. 1879. CEUR-WS.org (2017). http://ceur-ws.org/Vol-1879/paper39.pdf

16. Gerasimova, O., Kikot, S., Zakharyaschev, M.: Towards a data complexity classification of ontology-mediated queries with covering. In: Ortiz, M., Schneider, T. (eds.) Proceedings of the 31st International Workshop on Description Logics co-located with 16th International Conference on Principles of Knowledge Representation and Reasoning (KR 2018), Tempe, Arizona, US. CEUR Workshop Proceedings, 27–29 October 2018, vol. 2211. CEUR-WS.org (2018). http://ceur-ws.org/Vol-2211/paper-36.pdf

17. Hernich, A., Lutz, C., Ozaki, A., Wolter, F.: Schema.org as a description logic. In: Calvanese, D., Konev, B. (eds.) Proceedings of the 28th International Workshop on Description Logics. CEUR Workshop Proceedings, Athens, Greece, 7–10 June 2015, vol. 1350. CEUR-WS.org (2015). http://ceur-ws.org/Vol-1350/paper-24.pdf

18. Jeavons, P., Cohen, D., Gyssens, M.: A test for tractability. In: Freuder, E.C. (ed.) CP 1996. LNCS, vol. 1118, pp. 267–281. Springer, Heidelberg (1996). https://doi.org/10.1007/3-540-61551-2_80

19. Kaminski, M., Nenov, Y., Grau, B.C.: Datalog rewritability of disjunctive datalog programs and non-Horn ontologies. Artif. Intell. **236**, 90–118 (2016). https://doi.org/10.1016/j.artint.2016.03.006

20. Kharlamov, E., et al.: Ontology based data access in Statoil. J. Web Semant. **44**, 3–36 (2017). https://doi.org/10.1016/j.websem.2017.05.005

21. Kozik, M.: Weak consistency notions for all the CSPs of bounded width. In: Grohe, M., Koskinen, E., Shankar, N. (eds.) Proceedings of the 31st Annual ACM/IEEE Symposium on Logic in Computer Science, LICS 2016, New York, NY, USA, 5–8 July 2016, pp. 633–641. ACM (2016). https://doi.org/10.1145/2933575.2934510

22. Lanti, D., Rezk, M., Xiao, G., Calvanese, D.: The NPD benchmark: reality check for OBDA systems. In: Alonso, G., et al. (eds.) Proceedings of the 18th International Conference on Extending Database Technology, EDBT 2015, Brussels, Belgium, 23–27 March 2015, pp. 617–628. OpenProceedings.org (2015). https://doi.org/10.5441/002/edbt.2015.62

23. Lutz, C., Sabellek, L.: Ontology-mediated querying with the description logic EL: trichotomy and linear datalog rewritability. In: Sierra, C. (ed.) Proceedings of the Twenty-Sixth International Joint Conference on Artificial Intelligence, IJCAI 2017, Melbourne, Australia, 19–25 August 2017, pp. 1181–1187. ijcai.org (2017). https://doi.org/10.24963/ijcai.2017/164

24. Lutz, C., Wolter, F.: Non-uniform data complexity of query answering in description logics. In: Brewka, G., Eiter, T., McIlraith, S.A. (eds.) Principles of Knowledge Representation and Reasoning: Proceedings of the Thirteenth International Conference, KR 2012, Rome, Italy, 10–14 June 2012. AAAI Press (2012). http://www.aaai.org/ocs/index.php/KR/KR12/paper/view/4533

25. Lutz, C., Wolter, F.: The data complexity of description logic ontologies. Log. Methods Comput. Sci. **13**(4) (2017). https://doi.org/10.23638/LMCS-13(4:7)2017

26. Marcinkowski, J.: DATALOG SIRUPs uniform boundedness is undecidable. In: Proceedings, 11th Annual IEEE Symposium on Logic in Computer Science, New Brunswick, New Jersey, USA, 27–30 July 1996, pp. 13–24. IEEE Computer Society (1996). https://doi.org/10.1109/LICS.1996.561299

27. Papadimitriou, C.: Computational Complexity. Addison-Wesley, Boston (1994)

28. Poggi, A., Lembo, D., Calvanese, D., De Giacomo, G., Lenzerini, M., Rosati, R.: Linking data to ontologies. In: Spaccapietra, S. (ed.) Journal on Data Semantics X. LNCS, vol. 4900, pp. 133–173. Springer, Heidelberg (2008). https://doi.org/10.1007/978-3-540-77688-8_5

29. Ramakrishnan, R., Sagiv, Y., Ullman, J.D., Vardi, M.Y.: Proof-tree transformation theorems and their applications. In: Proceedings of the Eighth ACM SIGACT-SIGMOD-SIGART Symposium on Principles of Database Systems, pp. 172–181. ACM (1989)

30. Rodríguez-Muro, M., Kontchakov, R., Zakharyaschev, M.: Ontology-based data access: *Ontop* of databases. In: Alani, H., et al. (eds.) ISWC 2013. LNCS, vol. 8218, pp. 558–573. Springer, Heidelberg (2013). https://doi.org/10.1007/978-3-642-41335-3_35

31. Saraiya, Y.P.: Linearizing nonlinear recursions in polynomial time. In: Silberschatz, A. (ed.) Proceedings of the Eighth ACM SIGACT-SIGMOD-SIGART Symposium on Principles of Database Systems, Philadelphia, Pennsylvania, USA, 29–31 March 1989, pp. 182–189. ACM Press (1989). https://doi.org/10.1145/73721.73740

32. Schaerf, A.: On the complexity of the instance checking problem in concept languages with existential quantification. J. Intell. Inf. Syst. **2**, 265–278 (1993)

33. Ullman, J.D., Gelder, A.V.: Parallel complexity of logical query programs. Algorithmica **3**, 5–42 (1988). https://doi.org/10.1007/BF01762108

34. Vardi, M.Y.: Decidability and undecidability results for boundedness of linear recursive queries. In: Edmondson-Yurkanan, C., Yannakakis, M. (eds.) Proceedings of the Seventh ACM SIGACT-SIGMOD-SIGART Symposium on Principles of Database Systems, Austin, Texas, USA, 21–23 March 1988, pp. 341–351. ACM (1988). http://doi.acm.org/10.1145/308386.308470

35. Xiao, G., et al.: Ontology-based data access: a survey. In: Lang, J. (ed.) Proceedings of the Twenty-Seventh International Joint Conference on Artificial Intelligence, IJCAI 2018, Stockholm, Sweden, 13–19 July 2018, pp. 5511–5519. ijcai.org (2018). https://doi.org/10.24963/ijcai.2018/777

36. Zhang, W., Yu, C.T., Troy, D.: Necessary and sufficient conditions to linearize double recursive programs in logic databases. ACM Trans. Database Syst. **15**(3), 459–482 (1990). https://doi.org/10.1145/88636.89237

37. Zhuk, D.: A proof of CSP dichotomy conjecture. In: Umans, C. (ed.) 58th IEEE Annual Symposium on Foundations of Computer Science, FOCS 2017, Berkeley, CA, USA, 15–17 October 2017, pp. 331–342. IEEE Computer Society (2017). https://doi.org/10.1109/FOCS.2017.38

Perceptual Context in Cognitive Hierarchies

Bernhard Hengst, Maurice Pagnucco, David Rajaratnam, Claude Sammut, and Michael Thielscher[✉]

School of Computer Science and Engineering,
UNSW, Sydney, Australia
{bernhardh,morri,daver,claude,mit}@cse.unsw.edu.au

Abstract. Cognition does not only depend on bottom-up sensor feature abstraction, but also relies on contextual information being passed top-down. Context is higher level information that helps to predict belief states at lower levels. The main contribution of this paper is to provide a formalisation of perceptual context and its integration into a new process model for cognitive hierarchies. Several simple instantiations of a cognitive hierarchy are used to illustrate the role of context. Notably, we demonstrate the use context in a novel approach to visually track the pose of rigid objects with just a 2D camera.

1 Introduction

There is strong evidence that intelligence necessarily involves hierarchical structures [1–4,6,9–11,15,17,18,20,22,24,25,27]. We recently addressed the formalisation of cognitive hierarchies that allow for the integration of disparate representations, including symbolic and sub-symbolic representations, in a general framework for cognitive robotics [8]. Sensory information processing is upward-feeding, progressively abstracting more complex state features, while behaviours are downward-feeding progressively becoming more concrete, ultimately controlling robot actuators.

However, neuroscience suggests that the brain is subject to top-down cognitive influences for attention, expectation and perception [12]. Higher level signals carry important information to facilitate scene interpretation. For example, the recognition of the Dalmatian, and the disambiguation of the symbol ⋏ in Fig. 1 intuitively show that higher level context is necessary to correctly interpret these images[1]. Furthermore, the human brain is able to make sense of dynamic 3D scenes from light falling on our 2D retina in varying lighting conditions. Replicating this ability is still a challenge in artificial intelligence and computer vision, particularly when objects move relative to each other, can occlude each other, and are without texture. Prior, more abstract contextual knowledge is important to help segment images into objects or to confirm the presence of an object from faint or partial edges in an image.

---

[1] Both of these examples appear in [16] but are also well-known in the cognitive psychology literature.

© Springer Nature Switzerland AG 2019
C. Lutz et al. (Eds.): Baader Festschrift, LNCS 11560, pp. 352–366, 2019.
https://doi.org/10.1007/978-3-030-22102-7_16

TAE CAT

Fig. 1. The image on the left would probably be indiscernible without prior knowledge of Dalmations. The ambiguous symbol $\wedge$ on the right can be interpreted as either an "H" or an "A" depending on the word context.

In this paper we extend the existing cognitive hierarchy formalisation [8] by introducing the notion of perceptual context, which modifies the beliefs of a child node given the beliefs of its parent nodes. It is worth emphasising that defining the role of context as a top-down predictive influence on a node's belief state and the corresponding process model that defines how the cognitive hierarchy evolves over time is non-trivial. Our formalisation captures the dual influences of context and behaviour as a predictive update of a node's belief state. Consequently, *the main contribution of this paper is the inclusion and formalisation of contextual influences as a predictive update within a cognitive hierarchy.*

As a meta-framework, the cognitive hierarchy requires instantiation. We provide two simple instantiation examples to help illustrate the formalisation of context. The first is a running example using a small belief network. The second example involves visual servoing to track a moving object. This second example quantifies the benefit of context and demonstrates the role of context in a complete cognitive hierarchy including behaviour generation.

As a third, realistic and challenging example that highlights the importance of context we consider the tracking of the 6° of freedom pose of multiple, possibly occluded, marker-less objects with a 2D camera. We provide a novel instantiation of a cognitive hierarchy for a real robot using the context of a spatial cognitive node modelled using a 3D physics simulator. Note, this formalisation is provided in outline only due to space restrictions.

2 The Architectural Framework

For the sake of brevity the following presentation both summarises and extends the formalisation of cognitive hierarchies as introduced in [8]. We shall, however, highlight how our contribution differs from their work. The essence of this framework is to adopt a meta-theoretic approach, formalising the interaction between abstract cognitive nodes, while making no commitments about the representation and reasoning mechanism within individual nodes.

2.1 Motivating Example

As an explanatory aid to formalising the use of context in a hierarchy we will use the disambiguation of the symbol $\not\vdash\!\!\backslash$ in Fig. 1 as a simple running example. This system can be modelled as a two layer causal tree updated according to Pearl's Bayesian belief propagation rules [26]. The lower-level layer disambiguates individual letters while the higher-level layer disambiguates complete words (Fig. 2). We assume that there are only two words that are expected to be seen, with equal probability: "THE" and "CAT".

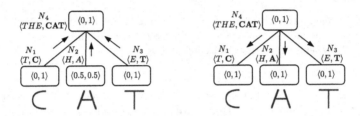

Fig. 2. Disambiguating the symbol $\not\vdash\!\!\backslash$ requires context from the word recognition layer.

There are three independent letter sensors with the middle sensor being unable to disambiguate the observed symbol $\not\vdash\!\!\backslash$ represented by the conditional probabilities $p(\not\vdash\!\!\backslash|H) = 0.5$ and $p(\not\vdash\!\!\backslash|A) = 0.5$. These sensors feed into the lower-level nodes (or *processors* in Pearl's terminology), which we label as N_1, N_2, N_3. The results of the lower level nodes are combined at N_4 to disambiguate the observed word.

Each node maintains two state variables; the *diagnostic* and *causal* supports (displayed as the pairs of values in Fig. 2). Intuitively, the diagnostic support represents the knowledge gathered through sensing while the causal support represents the contextual bias. A node's overall belief is calculated by the combination of these two state variables.

While sensing data propagates up the causal tree, the example highlights how node N_2 is only able to resolve the symbol $\not\vdash\!\!\backslash$ in the presence of contextual feedback from N_4.

2.2 Nodes

A cognitive hierarchy consists of a set of nodes. Nodes are tasked to achieve a goal or maximise future value. They have two primary functions: world-modelling and behaviour-generation. World-modelling involves maintaining a *belief state*, while behaviour-generation is achieved through *policies*, where a policy maps states to sets of actions. A node's belief state is modified by sensing or by the combination of actions and higher-level context. We refer to this latter as *prediction update* to highlight how it sets an expectation about what the node is expecting to observe in the future.

Definition 1. *A* cognitive language *is a tuple* $\mathcal{L} = (\mathcal{S}, \mathcal{A}, \mathcal{T}, \mathcal{O}, \mathcal{C})$, *where* $\mathcal{S}$ *is a set of belief states,* $\mathcal{A}$ *is a set of actions,* $\mathcal{T}$ *is a set of task parameters,* $\mathcal{O}$ *is a set of observations, and* $\mathcal{C}$ *is a set of contextual elements. A* cognitive node *is a tuple* $N = (\mathcal{L}, \Pi, \lambda, \tau, \gamma, s^0, \pi^0)$ *s.t:*

- $\mathcal{L}$ *is the cognitive language for* N, *with initial belief state* $s^0 \in \mathcal{S}$.
- Π *a set of policies such that for all* $\pi \in \Pi$, $\pi : \mathcal{S} \rightarrow 2^{\mathcal{A}}$, *with initial policy* $\pi^0 \in \Pi$.
- *A policy selection function* $\lambda : 2^{\mathcal{T}} \rightarrow \Pi$, *s.t.* $\lambda(\{\}) = \pi^0$.
- *An observation update operator* $\tau : 2^{\mathcal{O}} \times \mathcal{S} \rightarrow \mathcal{S}$.
- *A prediction update operator* $\gamma : 2^{\mathcal{C}} \times 2^{\mathcal{A}} \times \mathcal{S} \rightarrow \mathcal{S}$.

Definition 1 differs from [8] in two ways: the introduction of a set of context elements in the cognitive language, and the modification of the *prediction* update operator, previously called the *action* update operator, to include context elements when updating the belief state.

This definition can now be applied to the motivating example to instantiate the nodes in the Bayesian causal tree. We highlight only the salient features for this instantiation.

Example. *Let* $E = \{\langle x, y \rangle \mid 0 \leq x, y \leq 1.0\}$ *be the set of probability pairs, representing the recognition between two distinct features. For node* N_2, *say (cf. Fig. 2), these features are the letters "H" and "A" and for* N_4 *these are the words "THE" and "CAT". The set of belief states for* N_2 *is* $\mathcal{S}_2 = \{\langle \langle d \rangle, c \rangle \mid d, c \in E\}$, *where* d *is the* diagnostic *support and* c *is the* causal *support. Note, the vector-in-vector format allows for structural uniformity across nodes. Assuming equal probability over letters, the initial belief state is* $\langle \langle \langle 0.5, 0.5 \rangle \rangle, \langle 0.5, 0.5 \rangle \rangle$. *For* N_4 *the set of belief states is* $\mathcal{S}_4 = \langle \langle d_1, d_2, d_3 \rangle, c \rangle \mid d_1, d_2, d_3, c \in E\}$, *where* d_i *is the contribution of node* N_i *to the diagnostic support of* N_4.

For N_2 *the context is the causal supports from above,* $\mathcal{C}_2 = E$, *while the observations capture the influence of the "H"-"A" sensor,* $\mathcal{O}_2 = \{\langle d \rangle \mid d \in E\}$. *In contrast the observations for* N_4 *need to capture the influence of the different child diagnostic supports, so* $\mathcal{O}_4 = \{\langle d_1, d_2, d_3 \rangle \mid d_1, d_2, d_3 \in E\}$.

The observation update operators need to replace the diagnostic supports of the current belief with the observation, which is more complicated for N_4 *due to its multiple children,* $\tau_2(\{\mathbf{d_1, d_2, d_3}\}, \langle d, c \rangle) = \langle \Sigma_{i=1}^3 \mathbf{d_i}, c \rangle$. *Ignoring the influence of actions, the prediction update operator simply replaces the causal support of the current belief with the context from above, so* $\gamma_2(\{c'\}, \emptyset, \langle \langle d \rangle, c \rangle) = \langle \langle d \rangle, c' \rangle$.

2.3 Cognitive Hierarchy

Nodes are interlinked in a hierarchy, where sensing data is passed up the *abstraction hierarchy*, while actions and context are sent down the hierarchy (Fig. 3).

Definition 2. *A* cognitive hierarchy *is a tuple* $H = (\mathcal{N}, N_0, F)$ *s.t:*

- $\mathcal{N}$ *is a set of cognitive nodes and* $N_0 \in \mathcal{N}$ *is a distinguished node corresponding to the external world.*

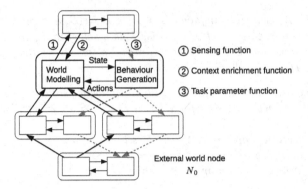

Fig. 3. A cognitive hierarchy, highlighting internal interactions as well as the sensing, action, and context graphs.

- F *is a set of function triples* $\langle \phi_{i,j}, \psi_{j,i}, \varrho_{j,i} \rangle \in F$ *that connect nodes* $N_i, N_j \in \mathcal{N}$ *where:*
  - $\phi_{i,j} : \mathcal{S}_i \to 2^{\mathcal{O}_j}$ *is a sensing function, and*
  - $\psi_{j,i} : 2^{\mathcal{A}_j} \to 2^{\mathcal{T}_i}$ *is a task parameter function.*
  - $\varrho_{j,i} : \mathcal{S}_j \to 2^{\mathcal{C}_i}$ *is a context enrichment function.*
- *Sensing graph: each* $\phi_{i,j}$ *represents an edge from node* N_i *to* N_j *and forms a directed acyclic graph (DAG) with* N_0 *as the unique source node of the graph.*
- *Prediction graph: the set of task parameter functions (equivalently, the context enrichment functions) forms a converse to the sensing graph such that* N_0 *is the unique sink node of the graph.*

Definition 2 differs from the original with the introduction of the *context enrichment* functions and the naming of the *prediction graph* (originally the *action graph*). The connection between nodes consists of triples of sensing, task parameter and context functions. The sensing function extracts observations from a lower-level node in order to update a higher level node, while the context enrichment function performs the converse. The task parameter function translates a higher-level node's actions into a set of task parameters, which is then used to select the active policy for a node.

Finally, the external world is modelled as a distinguished node, N_0. Sensing functions allow other nodes to observe properties of the external world, and task parameter functions allow actuator values to be modified, but N_0 doesn't "sense" properties of other nodes, nor does it generate task parameters for those nodes. Similarly, context enrichment functions connected to N_0 would simply return the empty set, unless one wanted to model unusual properties akin to the quantum effects of observations on the external world. Beyond this, the internal behaviour of N_0 is considered to be opaque.

The running example can now be encoded formally as a cognitive hierarchy, again with the following showing only the salient features of the encoding.

Example. *We construct a hierarchy* $H = (\mathcal{N}, N_0, F)$, *consisting of five nodes* $\mathcal{N} = \{N_0, N_1, \ldots, N_4\}$. *The function triples in* F *will include* $\phi_{0,2}$ *for the visual sensing of the middle letter, and* $\phi_{2,4}$ *and* $\varrho_{4,2}$ *for the sensing and context between* N_2 *and* N_4.

The function $\phi_{0,2}$ *returns the probability of the input being the characters* *"H" and "A". Here* $\phi_{0,2}(\land) = \{\langle 0.5, 0.5 \rangle\}$.

Defining $\phi_{2,4}$ *and* $\varrho_{4,2}$ *requires a conditional probability matrix* $M = \begin{bmatrix} 1 & 0 \\ 0 & 1 \end{bmatrix}$ *to capture how the letters "H" and "A" contribute to the recognition of "THE" and "CAT".*

For sensing from N_2 *we use zeroed vectors to prevent influence from the diagnostic support components from* N_1 *and* N_2. *Hence* $\phi_{2,4}(\langle\langle d\rangle, c\rangle) = \{\langle\langle 0, 0\rangle, \eta \cdot M \cdot d^T, \langle 0, 0\rangle\rangle\}$, *where* d^T *is the transpose of vector* d, *and* η *is a normalisation constant.*

For context we capture how N_4's *causal support and its diagnostic support components from* N_1 *and* N_2 *influences the causal support of* N_2. *Note that this also prevents any feedback from* N_2's *own diagnostic support to its causal support. So,* $\varrho_{4,2}(\langle\langle d_1, d_2, d_3\rangle, c\rangle) = \{\eta \cdot (d_1 \cdot d_3 \cdot c) \cdot M\}$.

2.4 Active Cognitive Hierarchy

The above definitions capture the static aspects of a system but require additional details to model its operational behaviour. Note, the following definitions are unmodified from the original formalism and are presented here because they are necessary to the developments of later sections.

Definition 3. *An active cognitive node is a tuple* $Q = (N, s, \pi, a)$ *where: (1)* N *is a cognitive node with* $\mathcal{S}$, Π, *and* $\mathcal{A}$ *being its set of belief states, set of policies, and set of actions respectively, (2)* $s \in \mathcal{S}$ *is the current belief state,* $\pi \in \Pi$ *is the current policy, and* $a \in 2^{\mathcal{A}}$ *is the current set of actions.*

Essentially an active cognitive node couples a (static) cognitive node with some dynamic information; in particular the current belief state, policy and set of actions.

Definition 4. *An active cognitive hierarchy is a tuple* $\mathcal{X} = (H, \mathcal{Q})$ *where* H *is a cognitive hierarchy with set of cognitive nodes* $\mathcal{N}$ *such that for each* $N \in \mathcal{N}$ *there is a corresponding active cognitive node* $Q = (N, s, \pi, a) \in \mathcal{Q}$ *and vice-versa.*

The active cognitive hierarchy captures the dynamic state of the system at a particular instance in time. Finally, an *initial active cognitive hierarchy* is an active hierarchy where each node is initialised with the initial belief state and policy of the corresponding cognitive node, as well as an empty set of actions.

2.5   Cognitive Process Model

The *process model* defines how an active cognitive hierarchy evolves over time and consists of two steps. Firstly, sensing observations are passed up the hierarchy, progressively updating the belief state of each node. Next, task parameters and context are passed down the hierarchy updating the active policy, the actions, and the belief state of the nodes.

We do not present all definitions here, in particular we omit the definition of the *sensing update* operator as this remains unchanged in our extension. Instead we define a *prediction update* operator, replacing the original *action update*, with the new operator incorporating both context and task parameters in its update. First, we characterise the updating of the beliefs and actions for a single active cognitive node.

Definition 5. *Consider an active cognitive hierarchy* $\mathcal{X} = (H, \mathcal{Q})$ *where* $H = (\mathcal{N}, N_0, F)$. *The prediction update of* $\mathcal{X}$ *with respect to an active cognitive node* $Q_i = (N_i, s_i, \pi_i, a_i) \in \mathcal{Q}$, *written* **PredUpdate'**$(\mathcal{X}, Q_i)$, *is an active cognitive hierarchy* $\mathcal{X}' = (H, \mathcal{Q}')$ *where* $\mathcal{Q}' = \mathcal{Q} \setminus \{Q_i\} \cup \{Q_i'\}$ *and* $Q_i' = (Q_i, \gamma_i(C, a_i', s_i), \pi_i', a_i')$ *s.t:*

– *if there is no node* N_x *where* $\langle \phi_{i,x}, \psi_{x,i}, \varrho_{x,i} \rangle \in F$ *then:* $\pi_i' = \pi_i$, $a_i' = \pi_i(s_i)$ *and* $C = \emptyset$,
– *else:*

$$\pi_i' = \lambda_i(T) \text{ and } a_i' = \pi_i'(s_i)$$
$$T = \bigcup \{\psi_{x,i}(a_x) \mid \langle \phi_{i,x}, \psi_{x,i}, \varrho_{x,i} \rangle \in F \text{ where } Q_x = (N_x, s_x, \pi_x, a_x) \in \mathcal{Q}\}$$
$$C = \bigcup \{\varrho_{x,i}(s_x) \mid \langle \phi_{i,x}, \psi_{x,i}, \varrho_{x,i} \rangle \in F \text{ where } Q_x = (N_x, s_x, \pi_x, a_x) \in \mathcal{Q}\}$$

The intuition for Definition 5 is straightforward. Given a cognitive hierarchy and a node to be updated, the update process returns an identical hierarchy except for the updated node. This node is updated by first selecting a new active policy based on the task parameters of all the connected higher-level nodes. The new active policy is applied to the existing belief state to generate a new set of actions. Both these actions and the context from the connected higher-level nodes are then used to update the node's belief state.

Using the single node update, updating the entire hierarchy simply involves successively updating all its nodes.

Definition 6. *Consider an active cognitive hierarchy* $\mathcal{X} = (H, \mathcal{Q})$ *where* $H = (\mathcal{N}, N_0, F)$, *and let* Ψ *be the prediction graph induced by the task parameter functions in* F. *The action process update of* $\mathcal{X}$, *written* **PredUpdate**$(\mathcal{X})$, *is an active cognitive model:*

$$\mathcal{X}' = \textbf{PredUpdate}'(\dots \textbf{PredUpdate}'(\mathcal{X}, Q_n), \dots Q_0)$$

where the sequence $[Q_n, \dots, Q_0]$ *consists of all active cognitive nodes of the set* $\mathcal{Q}$ *such that the sequence satisfies the partial ordering induced by the prediction graph* Ψ.

Importantly, the update ordering in Definition 6 satisfies the partial ordering induced by the prediction graph, thus guaranteeing that the prediction update is well-defined.

Lemma 1. *For any active cognitive hierarchy $\mathcal{X}$ the prediction process update of $\mathcal{X}$ is well-defined.*

Proof. Follows from the DAG structure.

The final part of the process model, which we omit here, is the combined operator, **Update**, that first performs a sensing update followed by a prediction update. This operation follows exactly the original, and similarly the theorem that the process model is well-defined also follows.

We can now apply the update process (sensing then prediction) to show how it operates on the running example.

Example. *When N_2 senses the symbol $\land$, $\phi_{0,2}$ returns that "A" and "H" are equally likely, so τ_2 updates the diagnostic support of N_2 to $\langle\langle 0.5, 0.5 \rangle\rangle$. On the other hand N_1 and N_2 unambiguously sense "C" and "T" respectively, so N_4's observation update operator, τ_4, will update its diagnostic support components to $\langle\langle 0,1 \rangle, \langle 0.5, 0.5 \rangle, \langle 0,1 \rangle\rangle$. The nodes overall belief, $\langle 0,1 \rangle$, is the normalised product of the diagnostic support components and the causal support, indicating here the unambiguous recognition of "CAT".*

Next, during prediction update, context from N_4 is passed back down to N_2, through $\phi_{4,2}$ and γ_2, updating the causal support of N_2 to $\langle 0,1 \rangle$. Hence, N_2 is left with the belief state $\langle\langle\langle 0.5, 0.5 \rangle\rangle, \langle 0,1 \rangle\rangle$, which when combined, indicates that the symbol $\land$ should be interpreted as an "A".

We next appeal to another simple example to illustrate the use of context to improve world modelling and in turn behaviour generation in a cognitive hierarchy.

3 A Simple Visual Servoing Example

Consider a mobile camera tasked to track an object sliding down a frictionless inclined plane. The controller is constructed as a three-node cognitive hierarchy. Figure 4 depicts the cognitive hierarchy and the scene.

The performance of the controller will be determined by how well the camera keeps the object in the centre of its field-of-view, specifically the average error in the tracking distance over a time period of 3 s.

The details of the instantiation of the cognitive hierarchy controller follow. The cognitive hierarchy is $H = (\mathcal{N}, N_0, F)$ with $\mathcal{N} = \{N_0, N_1, N_2\}$. N_0 is the unique opaque node representing the environment. The cognitive language for N_1 is a tuple $\mathcal{L}_1 = (\mathcal{S}_1, \mathcal{A}_1, \mathcal{T}_1, \mathcal{O}_1, \mathcal{C}_1)$, and for N_2 it is $\mathcal{L}_2 = (\mathcal{S}_2, \mathcal{A}_2, \mathcal{T}_2, \mathcal{O}_2, \mathcal{C}_2)$. The cognitive nodes are $N_1 = (\mathcal{L}_1, \Pi_1, \lambda_1, \tau_1, \gamma_1, s_1^0, \pi_1^0)$ and $N_2 = (\mathcal{L}_2, \Pi_2, \lambda_2, \tau_2, \gamma_2, s_2^0, \pi_2^0)$. For brevity we only describe the material functions.

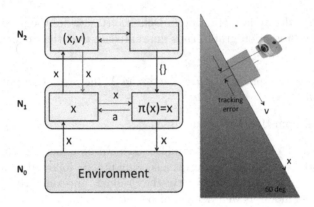

Fig. 4. A three-node cognitive hierarchy controller tasked to visually follow an object. Context flow is shown in red. (Color figure online)

The belief state of N_1 is the position of the object: $\mathcal{S}_1 = \{x \mid x \in \mathbb{R}\}$. The belief state of N_2 is both the position and velocity of the object: $\mathcal{S}_2 = \{\langle x, v \rangle \mid x, v \in \mathbb{R}\}$. The object starts at rest on the inclined plane at the origin: $s_1^0 = 0.0$ and $s_2^0 = \langle 0.0, 0.0 \rangle$.

N_1 receives object position observations from the environment: $\mathcal{O}_1 = \{x \mid x \in \mathbb{R}\}$. These measurements are simulated from the physical properties of the scene and include a noise component to represent errors in the sensor measurements: $\phi_{0,1}(\cdot) = \{0.5kt^2 + \nu\}$, with constant acceleration $k = 8.49\,\mathrm{m/s^2}$, t the elapsed time and ν zero mean Gaussian random noise with a standard deviation of 0.1. The acceleration assumes an inclined plane of $60°$ in a $9.8\,\mathrm{m/s^2}$ gravitational field. The N_1 observation update operator implements a Kalman filter with a fixed gain of 0.25: $\tau_1(\langle\{x\}, y\rangle) = (1.0 - 0.25)y + 0.25x$.

N_2 receives observations $\mathcal{O}_2 = \{x \mid x \in \mathbb{R}\}$ from N_1: $\phi_{1,2}(x) = \{x\}$. In turn it updates its position estimate accepting the value from N_1: $\tau_2(\langle\{x\}, \langle y, v \rangle\rangle) = \langle x, v \rangle$. The prediction update operator uses a physics model to estimate the new position and velocity of the object after time-step $\delta t = 0.05\,\mathrm{s}$: $\gamma_2(\langle\{\}, \{\}, \langle x, v \rangle\rangle) = \langle x + v\delta t + 0.5k\delta t^2, v + k\delta t \rangle$ with known acceleration $k = 8.49$.

Both N_1 and N_2 have one policy function each. The N_2 policy selects the N_1 policy. The effect of the N_1 policy: $\pi_1(x) = \{x\}$, is to move the camera to the estimated position of the object via the task parameter function connecting the environment: $\psi_{1,0}(\{x\}) = \{x\}$.

We consider two versions of the N_1 prediction update operator. Without context the next state is the commanded policy action: $\gamma_1(\langle\{x\}, \{y\}, z\rangle) = y$. With context the context enrichment function passes the N_2 estimate of the position of the object to N_1: $\varrho_{2,1}(\langle x, v \rangle) = \{x\}$, where $\mathcal{C}_1 = \{x \mid x \in \mathbb{R}\}$. The update operator becomes: $\gamma_1(\langle\{x\}, \{y\}, z\rangle) = x$.

When we simulate the dynamics and the repeated update of the cognitive hierarchy at $1/\delta t$ Hertz for 3 s, we find that without context the average tracking error is 2.004 ± 0.009. Using context the average tracking error reduces to 0.125 ± 0.015—a 94% error reduction.[2]

4   Using Context to Track Objects Visually

Object tracking has applications in augmented reality, visual servoing, and human-machine interfaces. We consider the problem of on-line monocular model-based tracking of multiple objects without markers or texture, using the 2D RGB camera built into the hand of a Baxter robot. The use of natural object features makes this a challenging problem.

Current practice for tackling this problem is to use 3D knowledge in the form of a CAD model, from which to generate a set of edge points (control points) for the object [21]. The idea is to track the corresponding 2D camera image points of the visible 3D control points as the object moves relatively to the camera. The new pose of the object relative to the camera is found by minimising the perspective re-projection error between the control points and their corresponding 2D image.

However, when multiple objects are tracked, independent CAD models fail to handle object occlusion. In place of the CAD models we use the machinery provided by a 3D physics simulator. The object-scene and virtual cameras from a simulator are ideal to model the higher level context for vision. We now describe how this approach is instantiated as a cognitive hierarchy with contextual feedback. It is important to note that the use of the physics simulator is not to replace the real-world, but is used as mental imagery to efficiently represent the spatial belief state of the robot.

4.1   Cognitive Hierarchy for Visual Tracking

We focus on world-modelling in a two-node cognitive hierarchy (Fig. 5). The external world node that includes the Baxter robot, streams the camera pose and RGB images as sensory input to the arm node. The arm node belief state is $s = \{p^a\} \cup \{\langle p_a^i, c^i \rangle | \text{object } i\}$, where p^a is the arm pose, and for all recognised objects i in the field of view of the arm camera, p_a^i is the object pose relative to the arm camera, and c^i is the set of object edge lines and their depth. The objects in this case include scattered cubes on a table. Information from the arm node is sent to the spatial node that employs a Gazebo physics simulator [19] as mental imagery to model the objects.

A novel feature of the spatial node is that it simulates the robot's arm camera as an object aware depth camera. No such camera exists in ,reality, but the Gazebo spatial belief state of the robot is able to not only provide a depth image, but one that segments the depth image by object. This object aware depth image provides the context to the arm node to generate the required control points.

---

[2] It is of course intuitive in this simple example that as N_2 has the benefit of the knowledge of the transition dynamics of the object it can better estimate its position and provide this context to direct the camera.

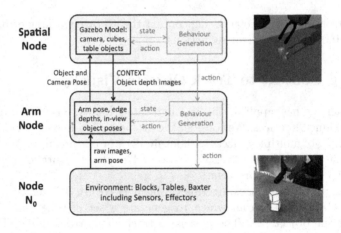

Fig. 5. Cognitive hierarchy comprising an arm node and a spatial node. Context from the spatial node is in the form of an object segmented depth image from a simulated special camera that shadows the real camera.

4.2 Update Functions and Process Update

We now describe the update functions and a single cycle of the process update for this cognitive hierarchy.

The real monocular RGB arm camera is simulated in Gazebo with an object aware depth camera with identical characteristics (i.e. the same intrinsic camera matrix). The simulated camera then produces depth and an object segmentation images from the simulated objects that corresponds to the actual camera image. This vital contextual information is then used for correcting the pose of the visible objects.

The process update starts with the sensing function $\phi_{N_0,Arm}$ that takes the raw camera image and observes all edges in the image, represented as a set of line segments, l.

$$\phi_{N_0,Arm}(\{rawImage\}) = \{l\}$$

The observation update operator τ_{Arm} takes the expected edge lines c^i for each object i and transforms the lines to best match the image edge lines l [21]. The update function uses the OpenCV function $solvePnP$ to find a corrected pose p_a^i for each object i relative to the arm-camera a[3].

$$\tau_{Arm}(\{l, c^i | \text{object } i\}) = \{p_a^i | \text{object } i\}$$

The sensing function from the arm to spatial node takes the corrected pose p_a^i for each object i, relative to the camera frame a, and transforms it into the Gazebo reference frame via the Baxter's reference frame given the camera pose p^a.

$$\phi_{Arm,Spatial}(\{p^a, \langle p_a^i, c^i \rangle | \text{object } i\}) = \{g_a^i | \text{object } i\}$$

---

[3] The pose of a rigid object in 3D space has 6° of freedom, three describing its translated position, and three the rotation or orientation, relative to a reference pose.

The spatial node observation update $\tau_{Spatial}$, updates the pose of all viewed objects g_a^i in the Gazebo physics simulator. Note $\{g_a^i|\text{object } i\} \subset \text{gazebo state}$.

$$\tau_{Spatial}(\{g_a^i|\text{object } i\}) = \text{gazebo.move}(i, g_a^i) \quad \forall i$$

The update cycle now proceeds down the hierarchy with prediction updates. The prediction update for the spatial node $\gamma_{Spatial}$ consists of predicting the interaction of objects in the simulator under gravity. Noise introduced during the observation update may result in objects separating due to detected collisions or settling under gravity.

$$\gamma_{Spatial}(\text{gazebo state}) = \text{gazebo.simulate}(\text{gazebo state}))$$

We now turn to the context enrichment function $\varrho_{Spatial,Arm}$ that extracts predicted camera image edge lines and depth data for each object in view of the simulator.

$$\varrho_{Spatial,Arm}(\text{gazebo state}) = \{c^i|\text{object } i\}$$

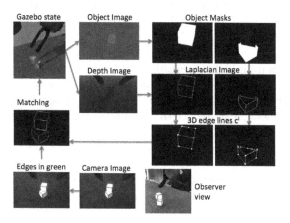

Fig. 6. The process update showing stages of the context enrichment function and the matching of contextual information to the real camera to correct the arm and spatial node belief state.

The stages of the context enrichment function $\varrho_{Spatial,Arm}$ are shown in Fig. 6. The simulated depth camera extracts an object image that identifies the object seen at every pixel location. It also extracts a depth image that gives the depth from the camera of every pixel. The object image is used to mask out each object in turn. Applying a Laplacian function to the part of the depth image masked out by the object yields all visible edges of the object. A Hough line transform identifies line end points in the Laplacian image and finds the depth of their endpoints from the depth image, producing c^i.

Figure 7 shows the cognitive hierarchy tracking several different cube configurations. This is only possible given the context from the spatial belief state. Keeping track of the pose of objects allows behaviours to be generated that for example pick up a cube with appropriately oriented grippers.

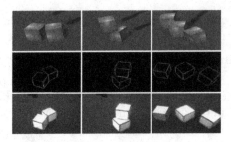

Fig. 7. Tracking several cube configurations. Top row: Gazebo GUI showing spatial node state. 2nd row: matching real image edges in green to simulated image edges in red. Bottom row: camera image overlaid with edges in green. (Color figure online)

5 Related Support and Conclusion

There is considerable evidence supporting the existence and usefulness of top-down contextual information. Reliability [5] and speed [7] of scene analysis provide early evidence.

These observations are further supported by neuroscience, suggesting that feedback pathways from higher more abstract processing areas of the brain down to areas closer to the sensors are greater than those transmitting information upwards [13]. The authors summarise the process—"what is actually happening flows up, and what you expect to happen flows down". It has been argued that the traditional idea that the processing of visual information consists of a sequence of feedforward operations needs to be supplemented by top-down contextual influences [12].

In the field of robotics, recent work in online interactive perception shows the benefit of predicted measurements from one level being passed to the next-lower level as state predictions [23].

This paper has included and formalised the essential element of context in the meta framework of cognitive hierarchies. The process model of an active cognitive hierarchy has been revised to include context updates satisfying the partial order induced by the prediction graph. We have illustrated the role of context with two simple examples and a novel way to track the pose of texture-less objects with a single 2D camera. Our motivating example highlighted the use of context in a cognitive hierarchy inspired by belief propagation in causal trees. In fact, as a general result it can be proved [14] that any Bayesian causal tree can be encoded as a cognitive hierarchy, testifying to the representation versatility of our framework.

Acknowledgments. The alphabetically last author wants to thank Franz Baader for many years of successful collegial collaboration, in particular within the joint research cluster on Logic-Based Knowledge Representation that was funded by the German Research Foundation (DFG) over many years. This centre would never have come to fruition without Franz's relentless pursuit of excellence not only in the projects that he was directly involved in but also with the cluster as a whole. One of the most pleasant experiences that we shared during that time was our journey to IJCAI'09 in Pasadena, which I am sure Franz remembers.

This material is based upon work supported by the Asian Office of Aerospace Research and Development (AOARD) under Award No: FA2386-15-1-0005. This research was also supported under Australian Research Council's (ARC) *Discovery Projects* funding scheme (project number DP 150103035).

Disclaimer. Any opinions, findings, and conclusions or recommendations expressed in this publication are those of the authors and do not necessarily reflect the views of the AOARD.

References

1. Albus, J.S., Meystel, A.M.: Engineering of Mind: An Introduction to the Science of Intelligent Systems. Wiley, London (2001). http://eu.wiley.com/Wiley CDA/WileyTitle/productCd-0471438545.html
2. Ashby, W.R.: Design for a Brain. Chapman and Hall, London (1952). https://archive.org/details/designforbrainor00ashb
3. Bakker, B., Schmidhuber, J.: Hierarchical reinforcement learning based on subgoal discovery and subpolicy specialization. In: Proceedings of the 8-th Conference on Intelligent Autonomous Systems, IAS-8, pp. 438–445 (2004)
4. Beer, S.: Decision and Control. Wiley, London (1966)
5. Biederman, I., Kubovy, M., Pomerantz, J.: On the semantics of a glance at a scene. In: Perceptual Organization, pp. 213–263. Lawrence Erlbaum, NJ (1981)
6. Brooks, R.A.: A robust layered control system for a mobile robot. IEEE J. Robot. Autom. **2**(1), 14–23 (1986). https://doi.org/10.1109/JRA.1986.1087032
7. Cavanagh, P.: What's up in top-down processing? In: Gorea, A. (ed.) Representations of Vision: Trends and Tacit Assumptions in Vision Research, pp. 295–304. Cambridge University Press, Cambridge (1991)
8. Clark, K., et al.: A framework for integrating symbolic and sub-symbolic representations. In: 25th Joint Conference on Artificial Intelligence (IJCAI -16) (2016)
9. Dayan, P., Hinton, G.E.: Feudal reinforcement learning. In: Advances in Neural Information Processing Systems 5 (NIPS) (1992)
10. Dietterich, T.G.: Hierarchical reinforcement learning with the MAXQ value function decomposition. J. Artif. Intell. Res. (JAIR) **13**, 227–303 (2000)
11. Drescher, G.L.: Made-up Minds: A Constructionist Approach to Artificial Intelligence. MIT Press, Cambridge (1991)
12. Gilbert, C.D., Li, W.: Top-down influences on visual processing. Nat. Rev. Neurosci. **14**(5) (2013). https://doi.org/10.1038/nrn3476
13. Hawkins, J., Blakeslee, S.: On Intelligence. Times Books, Henry Holt and Company, New York (2004)
14. Hengst, B., Pagnucco, M., Rajaratnam, D., Sammut, C., Thielscher, M.: Perceptual context in cognitive hierarchies. CoRR abs/1801.02270 (2018). http://arxiv.org/abs/1801.02270
15. Hubel, D.H., Wiesel, T.N.: Brain mechanisms of vision. In: A Scientific American Book: The Brain, pp. 84–96 (1979)
16. Johnson, J.: Designing with the Mind in Mind: Simple Guide to Understanding User Interface Design Rules. Morgan Kaufmann Publishers Inc., San Francisco (2010)

17. Jong, N.K.: Structured exploration for reinforcement learning. Ph.D. thesis, University of Texas at Austin (2010)
18. Kaelbling, L.P.: Hierarchical learning in stochastic domains: preliminary results. In: Machine Learning Proceedings of the Tenth International Conference. pp. 167–173. Morgan Kaufmann, San Mateo (1993)
19. Koenig, N., Howard, A.: Design and use paradigms for Gazebo, an open-source multi-robot simulator. In: International Conference on Intelligent Robots and Systems (IROS), pp. 2149–2154. IEEE/RSJ (2004)
20. Konidaris, G., Kuindersma, S., Grupen, R., Barto, A.: Robot learning from demonstration by constructing skill trees. Int. J. Robot. Res. (2011). https://doi.org/10.1177/0278364911428653
21. Lepetit, V., Fua, P.: Monocular model-based 3D tracking of rigid objects. Found. Trends. Comput. Graph. Vis. 1(1), 1–89 (2005). https://doi.org/10.1561/0600000001
22. Marthi, B., Russell, S., Andre, D.: A compact, hierarchical q-function decomposition. In: Proceedings of the Proceedings of the Twenty-Second Conference Annual Conference on Uncertainty in Artificial Intelligence (UAI-06), pp. 332–340. AUAI Press, Arlington (2006)
23. Martin, R.M., Brock, O.: Online interactive perception of articulated objects with multi-level recursive estimation based on task-specific priors. In: IROS, pp. 2494–2501. IEEE (2014)
24. Minsky, M.: The Society of Mind. Simon & Schuster Inc., New York (1986)
25. Nilsson, N.J.: Teleo-reactive programs and the triple-tower architecture. Electron. Trans. Artif. Intell. 5, 99–110 (2001)
26. Pearl, J.: Probabilistic Reasoning in Intelligent Systems: Networks of Plausible Inference. Morgan Kaufmann, San Francesco (1988). Revised second printing edn
27. Turchin, V.F.: The Phenomenon of Science. Columbia University Press, New York (1977)

Do Humans Reason with $\mathcal{E}$-Matchers?

Steffen Hölldobler[1,2]([⊠])

[1] Technische Universität Dresden, Dresden, Germany
sh@iccl.tu-dresden.de
[2] North Caucasus Federal University, Stavropol, Russian Federation

Abstract. The *Weak Completion Semantics* is a novel, integrated and computational cognitive theory. Recently, it has been applied to ethical decision making. To this end, it was extended by equational theories as needed by the fluent calculus. To compute least models equational matching problems have to be solved. Do humans consider equational matching in reasoning episodes?

Keywords: Logic programming · Weak Completion Semantics ·
Fluent calculus · Ethical decision making

1 The Weak Completion Semantics

The *Weak Completion Semantics* is a novel cognitive theory. It is based on ideas initially presented by Stenning and van Lambalgen [37,38], but is mathematically sound [19]. Under the *Weak Completion Semantics* scenarios are modeled by:

1 reasoning towards a (logic) program,
2 weakly completing the program,
3 computing its least model under Łukasiewicz logic,
4 reasoning with respect to the least model,
5 if necessary, applying skeptical abduction.

Steps 1–5 have been applied to different human reasoning tasks like suppression [7] and the abstract as well as the social version of the selection task [15,42]. In human syllogistic reasoning, the *Weak Completion Semantics* [31] has outperformed the twelve cognitive theories compared in [24] (see e.g. [11] for an overview).

Recently, the *Weak Completion Semantics* has been applied to ethical decision making [17]. In particular, so-called trolley problems [14] were modelled. This line of research was inspired by Pereira and Saptawijaya. In their book [33], they have implemented various ethical problems as logic programs and have queried them for moral permissibility. However, their approach does not provide a general method to model ethical dilemmas and is not integrated into a cognitive theory about human reasoning, nor it intends to do so.

In order to model ethical decision problems within the *Weak Completion Semantics* we had to select a method for reasoning about actions and causality.

© Springer Nature Switzerland AG 2019
C. Lutz et al. (Eds.): Baader Festschrift, LNCS 11560, pp. 367–384, 2019.
https://doi.org/10.1007/978-3-030-22102-7_17

We opted for the fluent calculus because we liked the ease with which counterfactuals could be modelled in it although the situation calculus [35] or the event calculus [25] might have been used as well. The original idea underlying the fluent calculus was published together with Schneeberger in [20]. The name *fluent calculus* was coined later by Thielscher in [40].

The fluent calculus is based on equational reasoning. Thus, we had to extend the *Weak Completion Semantics* by equational theories in much the same way as Jaffar, Lassez and Maher extended the theory of definite logic programs [22]. Luckily, as shown together with Dietz Saldanha, Schwarz and Stefanus in [13], the main features of the *Weak Completion Semantics* prevail in the presence of an equational theory: each weakly completed program still has a least model under Łukasiewicz logic [27] which is the least fixed point of an appropriately specified semantic operator.

The application of this operator requires to solve an $\mathcal{E}$-matching problem [4]. The focus of this paper is on this particular problem, how it arises in ethical decision making, how an existing matching algorithm can be adapted, and whether it is reasonable to assume that humans actually make use of such an $\mathcal{E}$-matching algorithm.

2   Basics

We assume the reader to be familiar with logic programs and their semantics [1,26], equational reasoning [34] and unification theory [4] and will mention only some basics in this section.

Let 1 be a constant symbol, $\circ$ a binary function symbol written infix, and X, Y, Z variables. Then, the (universally closed) equations

$$X \circ 1 \approx X, \tag{1}$$

$$X \circ Y \approx Y \circ X, \tag{2}$$

$$(X \circ Y) \circ Z \approx X \circ (Y \circ Z) \tag{3}$$

state that 1 is a unit with respect to $\circ$ and that $\circ$ is commutative as well as associative. In other words, Eqs. (1)–(3) specify an *AC1-theory*. A set of equations $\mathcal{E}$ together with the axioms of equality define a finest congruence relation $=_{\mathcal{E}}$ on the set of ground terms. Let $[t]$ denote the congruence class containing the ground term t. Let $[p(t_1, \ldots, t_n)]$ be an abbreviation for $p([t_1], \ldots, [t_n])$, where p is an n-ary relation symbol and t_i, $1 \leq i \leq n$, are ground terms. $[p(s_1, \ldots, s_m)] = [q(t_1, \ldots, t_n)]$ iff $p = q$, $n = m$, and for all $1 \leq i \leq n$ we find $[s_i] = [t_i]$.

A term which is not a variable and does not contain the symbols 1 and $\circ$ is called a *fluent*. The set of *fluent terms* is the smallest set such that 1 and each fluent is a fluent term, and if s and t are fluent terms, then so is $s \circ t$. There is a one-to-one correspondence between fluent terms and multisets of fluents. A fluent term $t_1 \circ \ldots \circ t_n \circ 1$ corresponds to the multiset $\{t_1, \ldots, t_n\}$. In fact, all fluent terms which are in the congruence class represented by $t_1 \circ \ldots \circ t_n \circ 1$ and defined by the AC1-theory correspond to $\{t_1, \ldots, t_n\}$. One should also note that 1 corresponds to the empty multiset.

A *(normal logic) program* $\mathcal{P}$ is a finite or countably infinite set of (universally closed) clauses of the form $A \leftarrow Body$, where the *head* A is an atom and *Body* is either a non-empty, finite conjunction of literals, or $\top$, or $\bot$. We assume that programs do not contain equations. Clauses of the form $A \leftarrow \top$ and $A \leftarrow \bot$ are called *(positive) facts* and *(negative) assumptions*, respectively. All other clauses are called *rules*.

Let $\mathcal{P}$ be a ground program. Let A be a ground atom. A is *defined* in $\mathcal{P}$ iff $\mathcal{P}$ contains a clause of the form $A \leftarrow Body$; otherwise A is said to be *undefined*. The set of all atoms that are defined in $\mathcal{P}$ is denoted by $def\,\mathcal{P}$.[1] $\neg A$ is *assumed* in $\mathcal{P}$ iff $\mathcal{P}$ contains an assumption $A \leftarrow \bot$ and $\mathcal{P}$ does neither contain a fact nor a rule with head A. Consider the following transformation for a given ground program $\mathcal{P}$:

1. For all $A \in def\,\mathcal{P}$, replace all clauses of the form $A \leftarrow Body_1$, $A \leftarrow Body_2$, ... occurring in $\mathcal{P}$ by $A \leftarrow Body_1 \vee Body_2 \vee \ldots$.
2. Replace all occurrences of $\leftarrow$ by $\leftrightarrow$.

The resulting set of equivalences is called the *weak completion* of $\mathcal{P}$. The weak completion of a program differs from the completion of a program as defined by Clark in [8] in that undefined atoms are not mapped to false.

Let $\mathcal{P}$ be a program, $g\mathcal{P}$ the set of ground instances of clauses occurring in $\mathcal{P}$, and $\mathcal{E}$ an equational theory. As shown in [13], the weak completion of $\mathcal{P}$ has a least Herbrand $\mathcal{E}$-model under Łukasiewicz three-valued logic which can be computed as the least fixed point of the following semantic operator. Let I be an interpretation represented by $\langle I^\top, I^\bot \rangle$, where $I^\top$ is the set of all ground atoms mapped to true and $I^\bot$ is the set of all ground atoms mapped to false. We define $\Phi_{\mathcal{P}}(I) = \langle J^\top, J^\bot \rangle$, where

$$J^\top = \{[A] \mid \text{there exists } A \leftarrow Body \in g\mathcal{P} \text{ and } I(Body) = \top\},$$
$$J^\bot = \{[A] \mid \text{there exists } A \leftarrow Body \in g\mathcal{P}$$
$$\text{and for all } A' \leftarrow Body \in g\mathcal{P} \text{ with } [A] = [A'] \text{ we find } I(Body) = \bot\}.$$

Furthermore, let $\Phi_{\mathcal{P}} \uparrow 0 = \langle \emptyset, \emptyset \rangle$ and $\Phi_{\mathcal{P}} \uparrow (i+1) = \Phi_{\mathcal{P}}(\Phi_{\mathcal{P}} \uparrow i)$ for all $i \geq 0$.

3 Ethical Decision Problems

3.1 The Bystander Case

A trolley whose conductor has fainted is headed towards two people walking on the main track.[2] The banks of the track are so steep that these two people will not be able to get off the track in time. Hank is standing next to a switch which can turn the trolley onto a side track, thereby preventing it from killing the two people. However, there is a man standing on the side track. Hank can change the switch, killing him. Or he can refrain from doing so, letting the two die. Is it morally permissible for Hank to change the switch?

---

[1] We omit parentheses if a relation symbol is unitary and is applied only to constant symbols or variables.

[2] Note that in the original trolley problem, five people are on the main track. For the sake of simplicity, we assume that only two people are on the main track.

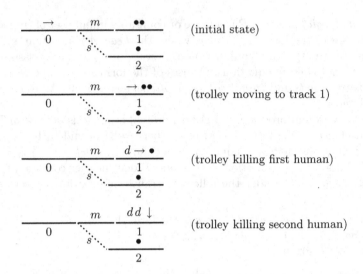

Fig. 1. The bystander case (initial state) and its ramifications if Hank decides to do nothing, where ↓ denotes that no further action is applicable.

The case is illustrated in Fig. 1 (initial state). The tracks are divided into segments 0, 1, and 2, the arrow represents that the trolley t is moving forward and that the track is clear (c), the switch is in position m (main) but can be changed into position s (side), and a bullet above a track segment represents a human (h) on this track. Let t, c, and h be indexed to denote the track to which they apply. In addition, we need a fluent d denoting a dead human.

We choose to represent a state by a pair of multisets consisting of the casualties in its second element and all other fluents in its first element. Thus, the initial state in Fig. 1 is

$$(\{t_0, c_0, m, h_1, h_1, h_2\}, \mathring{\emptyset})$$

which is represented by the pair of fluent terms

$$(t_0 \circ c_0 \circ m \circ h_1 \circ h_1 \circ h_2, 1) \tag{4}$$

in the fluent calculus.

There are two kinds of actions: the ones which can be performed by Hank (the direct actions *donothing* and *change*), and the actions which are performed by the trolley (the indirect actions *downhill* and *kill*). We will represent the actions by the trolley explicitly with the help of a five-place relation symbol *action* specifying the preconditions in the first two positions, the name in the third position, and the immediate effects of an action in the forth and fifth position. As a state is represented by two multisets, the preconditions and the immediate effects have also two parts:

$$action(t_0 \circ c_0 \circ m, 1, downhill, t_1 \circ c_0 \circ m, 1) \leftarrow \top, \tag{5}$$

$$action(t_0 \circ c_0 \circ s, 1, downhill, t_2 \circ c_0 \circ s, 1) \leftarrow \top, \tag{6}$$

$$action(t_1 \circ h_1, 1, kill, t_1, d) \leftarrow \top, \tag{7}$$

$$action(t_2 \circ h_2, 1, kill, t_2, d) \leftarrow \top. \tag{8}$$

If the trolley is on track 0, this track is clear, and the switch is in position m, then it will run downhill onto track 1 whereas track 0 remains clear and the switch will remain in position m; if, however, the switch is in position s, the trolley will run downhill onto track 2. If the trolley is on either track 1 or 2 and there is a human on this track, it will kill the human leading to a casualty.

In the original version of the fluent calculus, causality is expressed by the ternary predicate *plan* or *causes* stating that the execution of a plan transfers an initial state into a goal state. Its base case is of the form $causes(X, [\,], X)$, i.e., an empty plan does not change any state X. Generating models bottom up using a semantic operator one has to consider all ground instances of this atom. This set is usually too large to consider as a base case for modeling human reasoning episodes. The solution presented herein overcomes this problem in that we only have a small number of base cases depending on the number of options an agent like Hank may consider.

In fact, we are not going to solve planning problems like whether there exists a plan such that its execution transforms the initial state (4) into a goal state meeting certain constraints. Rather we want to compare the outcomes, i.e., the indirect effects, of the actions Hank can possibly perform. In other words, we want to compare the ramifications of either doing nothing or throwing the switch in the bystander scenario.

To this end, we will use a ternary relation symbol *ramify* whose first argument is the name of an action and whose second and third argument are the state obtained when executing the action. The possible actions of Hank are the base cases in the definition of *ramify*:

$$ramify(donothing, t_0 \circ c_0 \circ m \circ h_1 \circ h_1 \circ h_2, 1) \leftarrow \top, \tag{9}$$

$$ramify(change, t_0 \circ c_0 \circ s \circ h_1 \circ h_1 \circ h_2, 1) \leftarrow \top. \tag{10}$$

Further actions can be applied to the second and third argument of *ramify* given the actions specified in (5)–(8):

$$ramify(A, E_1 \circ Z_1, E_2 \circ Z_2) \leftarrow action(P_1, P_2, A', E_1, E_2) \; \wedge \tag{11}$$
$$ramify(A, P_1 \circ Z_1, P_2 \circ Z_2) \; \wedge$$
$$\neg ab_{ramify} \; A'.$$

It checks whether an action A' is applicable in a given state $(P_1 \circ Z_1, P_2 \circ Z_2)$. This is the case if the preconditions (P_1, P_2) are contained in the given state. If this holds, then the action is executed leading to the successor state $(E_1 \circ Z_1, E_2 \circ Z_2)$, where (E_1, E_2) are the direct effects of the action A'. In other words, if an action is applied, then its preconditions are consumed and its direct effects are produced. Such an action application is considered to be a ramification [41] with respect to the initial, direct action performed by Hank. Hence, the first argument A of *ramify* is not changed.

The execution of an action is also conditioned by $\neg ab_{ramify}\, A'$, where ab_{ramify} is an abnormality predicate. Such abnormalities were introduced by Stenning and van Lambalgen in [37,38] to represent conditionals as licenses for inference. In this example, there is nothing abnormal known with respect to the actions *downhill* and *kill* and, consequently, the assumptions

$$ab_{ramify}\, downhill \leftarrow \bot, \tag{12}$$

$$ab_{ramify}\, kill \leftarrow \bot \tag{13}$$

are added. But we can imagine situations where the trolley will only cross the switch if the switch is not broken. If the switch is broken, the trolley may derail. If such an abnormality becomes known, then the assumption (12) may be over-ridden.

Let

$$\mathcal{P}_0 = \{(5), (6), (7), (8), (11), (12), (13)\}$$

and consider the AC1-equational theory (1)–(3). Hank has the choice to do nothing or to change the switch. The indirect effects of Hank's decision are computed as ramifications in the fluent calculus [41].

If Hank does nothing, then let

$$\mathcal{P}_1 = \mathcal{P}_0 \cup \{(9)\}.$$

The least model of the weak completion of $\mathcal{P}_1$ – which is equal to the least fixed point of $\Phi_{\mathcal{P}_1}$ – is computed by iterating $\Phi_{\mathcal{P}_1}$ starting with the empty interpretation $\langle \emptyset, \emptyset \rangle$. The following equivalence classes will be mapped to true in the subsequent steps of this iteration:

$$[ramify(donothing, t_0 \circ c_0 \circ m \circ h_1 \circ h_1 \circ h_2, 1)], \tag{14}$$

$$[ramify(donothing, t_1 \circ c_0 \circ m \circ h_1 \circ h_1 \circ h_2, 1)], \tag{15}$$

$$[ramify(donothing, t_1 \circ c_0 \circ m \circ h_1 \circ h_2, d)], \tag{16}$$

$$[ramify(donothing, t_1 \circ c_0 \circ m \circ h_2, d \circ d)]. \tag{17}$$

They correspond precisely to the four states shown in Fig. 1. No further action is applicable to the final state $(t_1 \circ c_0 \circ m \circ h_2, d \circ d)$. The two people on the main track will be killed.

But one problem remains: The least fixed point of the $\Phi_{\mathcal{P}_1}$ operator contains (14)–(17), and we would like to identify the instance of the *ramify* predicate to which no further action is applicable. Only this instance will be compared to the corresponding instance if Hank is changing the switch. The other instances are only intermediate states. To this end we specify

$$aa(A, P_1 \circ Z_1, P_2 \circ Z_2) \leftarrow action(P_1, P_2, A', E_1, E_2)\, \wedge \tag{18}$$
$$ramify(A, P_1 \circ Z_1, P_2 \circ Z_2)\, \wedge$$
$$\neg ab_{ramify}\, A'.$$

Table 1. The computation of the least model of $wc\mathcal{P}_1'$, i.e., the program obtained if Hank is doing nothing. In each step, only the atoms are listed which are newly added.

$\Phi_{\mathcal{P}_1'}$	$I^\top$	$I^\perp$
1	$[ramify(donothing, t_0 \circ c_0 \circ m \circ h_1 \circ h_1 \circ h_2, 1)]$	$[ab_{ramify} \, downhill]$
	$[action(t_0 \circ c_0 \circ m, 1, downhill, t_1 \circ c_0 \circ m, 1)]$	$[ab_{ramify} \, kill]$
	$[action(t_0 \circ c_0 \circ s, 1, downhill, t_2 \circ c_0 \circ s, 1)]$	
	$[action(t_1 \circ h_1, 1, kill, t_1, d)]$	
	$[action(t_2 \circ h_2, 1, kill, t_2, d)]$	
2	$[ramify(donothing, t_1 \circ c_0 \circ m \circ h_1 \circ h_1 \circ h_2, 1)]$	
	$[aa(donothing, t_0 \circ c_0 \circ m \circ h_1 \circ h_1 \circ h_2, 1)]$	
3	$[ramify(donothing, t_1 \circ c_0 \circ m \circ h_1 \circ h_2, d)]$	
	$[aa(donothing, t_1 \circ c_0 \circ m \circ h_1 \circ h_1 \circ h_2, 1)]$	
4	$[ramify(donothing, t_1 \circ c_0 \circ m \circ h_2, d \circ d)]$	
	$[aa(donothing, t_1 \circ c_0 \circ m \circ h_1 \circ h_2, d)]$	

Informally, $aa(A, X_1, X_2)$ is true if there is an action A' which is applicable in the state (X_1, X_2). Comparing (18) and (11) we find that the bodies of these rules are identical. Thus, whenever a truth value is assigned to the head of (11), the same truth value will be assigned to the corresponding head of (18). Formally, let

$$\mathcal{P}_1' = \mathcal{P}_1 \cup \{(18)\}.$$

The computation of the least fixed point of $\Phi_{\mathcal{P}_1'}$ is shown in Table 1.

Let us return to Hank's choices. If Hank is changing the switch, then let

$$\mathcal{P}_2' = \mathcal{P}_0 \cup \{(10), (18)\}.$$

The least fixed point of $\Phi_{\mathcal{P}_2'}$ contains

$$[ramify(change, t_2 \circ c_0 \circ s \circ h_1 \circ h_1, d)] \tag{19}$$

and there is no further action applicable to the state $(t_2 \circ c_0 \circ s \circ h_1 \circ h_1, d)$. The two people on the main track will be saved but the person on the side track will be killed. This case is illustrated in Fig. 2.

The two cases (17) and (19) can be compared by means of a *prefer* rule:

$$prefer(A_1, A_2) \leftarrow ramify(A_1, Z_1, D_1) \, \wedge \tag{20}$$
$$\neg ctxt\,aa(A_1, Z_1, D_1) \, \wedge$$
$$ramify(A_2, Z_2, D_1 \circ d \circ D_2) \, \wedge$$
$$\neg ctxt\,aa(A_2, Z_2, D_1 \circ d \circ D_2) \, \wedge$$
$$\neg ab_{prefer} \, A_1,$$
$$ab_{prefer} \, change \leftarrow \perp, \tag{21}$$
$$ab_{prefer} \, donothing \leftarrow \perp. \tag{22}$$

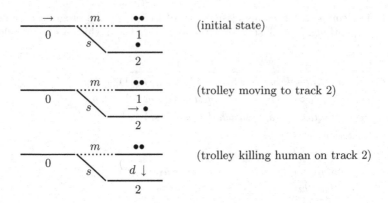

Fig. 2. The bystander case (initial state) and its ramifications if Hank decides to change the switch. One should observe that now the switch points to the side track.

prefer compares only states to which no further action is applicable. In the bystander case these are the states

$$(t_1 \circ c_0 \circ m \circ h_2, d \circ d)$$

and

$$(t_2 \circ c_0 \circ s \circ h_1 \circ h_1, d)$$

They can be identified in the least fixed points of $\Phi_{\mathcal{P}_1'}$ and $\Phi_{\mathcal{P}_2'}$ because there is no corresponding tuple of the aa relation. Thus,

$$aa(donothing, t_1 \circ c_0 \circ m \circ h_2, d \circ d)$$

and

$$aa(change, t_2 \circ c_0 \circ s \circ h_1 \circ h_1, d)$$

are mapped to unknown by the least fixed points of $\Phi_{\mathcal{P}_1'}$ and $\Phi_{\mathcal{P}_2'}$. The *ctxt* operator [12] will map these unknowabilities to false (see Table 2) and the negations thereof will be mapped to true. Comparing D_1 and $D_1 \circ d \circ D_2$, action A_2 leads to at least one more dead person than action A_1. Hence, A_1 is preferred over A_2 if nothing abnormal is known about A_1.

Under an *utilitarian* point of view [6], the *change* action is preferable to the *donothing* action as it will kill fewer humans. On the other hand, a purely utilitarian view is morally questionable in case of human casualties. Hank may ask himself: *Would I still save the humans on the main track if there were no human on the side track and I changed the switch?* This is a counterfactual because its antecedent is false in the given scenario. But we can easily deal with it by starting a new computation with the additional fact

$$ramify(change, t_0 \circ c_0 \circ s \circ h_1 \circ h_1 \circ c_2, 1) \leftarrow \top. \qquad (23)$$

Table 2. The truth table of the *ctxt* operator, where F is a formula.

F	$ctxt\,F$
$\top$	$\top$
$\bot$	$\bot$
U	$\bot$

Comparing (23) and (10), h_2 has been replaced by c_2. There is no human on track 2 anymore and, hence, this track is clear. This is a minimal change necessary to satisfy the precondition of the counterfactual. In this case, the least model of the extended program will contain

$$[ramify(change, t_0 \circ c_0 \circ s \circ h_1 \circ h_1 \circ c_2, 1)]$$

and no further action is applicable in state $(t_0 \circ c_0 \circ s \circ h_1 \circ h_1 \circ c_2, 1)$. This case is illustrated in Fig. 3. Using

$$perm\_double\ change \leftarrow prefer(change, donothing)\ \wedge \tag{24}$$
$$ramify(change, t_2 \circ c_0 \circ s \circ h_1 \circ h_1 \circ c_2, 1)\ \wedge$$
$$\neg ctxt\,aa(change, t_2 \circ c_0 \circ s \circ h_1 \circ h_1 \circ c_2, 1)\ \wedge$$
$$\neg ab_{perm\_double}\ change,$$
$$ab_{perm\_double}\ change \leftarrow \bot \tag{25}$$

allows Hank to conclude that changing the switch is permissible according to the *Doctrine of Double Effect* [2]. This principles states that sometimes it is permissible to cause a harm as a side effect (or "double effect") of bringing about a good result even though it would not be permissible to cause such a harm as a means to bringing about the same good end [28].

3.2 The Footbridge Case

The case is similar to the bystander case except that instead of the switch a footbridge lies accross the main track. Ian is standing on the footbridge next to a heavy human, whom he can throw on the track in the path of the trolley to stop it. Is it morally permissible for Ian to throw the human down?

This case is illustrated in Fig. 4. The track is again segmented. We use b_1 to denote that there is a heavy human on the footbridge crossing segment 1 of the track. Ian has two possibilities: *donothing* and *throw*. They are represented as the base cases in the definition of *ramify*:

$$ramify(donothing, t_0 \circ c_0 \circ c_1 \circ b_1 \circ h_2 \circ h_2, 1) \leftarrow \top, \tag{26}$$
$$ramify(throw, t_0 \circ c_0 \circ h_2 \circ h_2, d) \leftarrow \top. \tag{27}$$

One should observe that in the case of *donothing* track 1 is clear (c_1), whereas this does not hold if Ian has decided to throw down the heavy human. In the latter case, a dead body is blocking track 1.

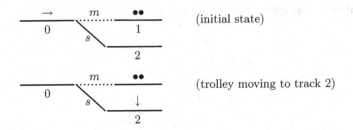

(initial state)

(trolley moving to track 2)

Fig. 3. The bystander case (initial state) and its ramifications if Hank is considering the counterfactual.

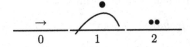

Fig. 4. The footbridge case.

As in the bystander case, one is tempted to reason that the *throw* action is preferable to the *donothing* action as it will kill fewer humans. But throwing down a heavy human involves an intentional direct kill, and intentional kills are not allowed under the *Doctrine of Double Effect*. This can be modeled with the help of the abnormality predicate ab_{prefer} by the clauses:

$$ab_{prefer}\ throw \leftarrow \bot, \tag{28}$$

$$ab_{prefer}\ throw \leftarrow intent\_direct\_kill\ throw, \tag{29}$$

$$intent\_direct\_kill\ throw \leftarrow ramify(donothing, t_2 \circ c_0 \circ c_1 \circ b_1, d \circ d)\ \wedge$$
$$\neg ctxt\,aa(donothing, t_2 \circ c_0 \circ c_1 \circ b_1, d \circ d). \tag{30}$$

Hence, throwing down the heavy human is not preferred and, thus, not permissible. The example demonstrates again the way abnormalities are used in the *Weak Completion Semantics*. If nothing is known, then a negative assumption about the abnormality is made by (28). This assumption can be overridden once additional knowledge becomes available. In this case we learn that an intentional direct kill overrides the negative assumption, which is expressed in (29). Moreover, from the specification of the *throw* action we can derive that the killing of the heavy human is a direct effect of this action. We may further derive that the intention to kill the heavy human was to save the humans on the main track by asking the counterfactual: *Would Ian still save the humans on the main track if he does not throw down the heavy human?* This counterfactual can be answered by considering the *donothing* action. As doing nothing will lead to two dead bodies on the main track, throwing down the heavy human was an intentional direct kill (see (30)).

3.3 The Loop Case

The case is similar to the bystander case. Ned is standing next to a switch which he can throw which will temporarily turn the trolley onto a loop side track. There is a heavy human on the side track. If the trolley hits the heavy human then this will slow down the trolley, giving the two people on the main track sufficient time to escape. But it will kill the heavy human. Is it morally permissible for Ned to change the switch?

This case is illustrated in Fig. 5. Ned can reason that if he does nothing, then the humans on the main track will be killed. Likewise, if he changes the switch, then the humans on the main track will be saved whereas the human on the side track will be killed. But the counterfactual *if there were no human on the side track and he changes the switch, then he would still save the humans on the main track* will be false. Hence, according to the *Doctrine of Double Effect* changing the switch is impermissible. However, the *Doctrine of Triple Effect* [23] allows to distinguish between an action *in order* that an effect occurs and an action *because* that effect will occur. Whereas the former is classified as impermissible, the latter is permissible. In the loop case, changing the switch will save the humans on the main track *because* the killing of the heavy man on the side track will slow down the trolley. It is an indirect intentional kill. The *change* action is permissible under the *Doctrine of Triple Effect*. In the footbridge case, throwing down and killing the heavy man was in order to slow down the trolley. It is a direct intentional kill. The *throw* action is impermissible under the *Doctrine of Triple Effect*.

This example can also be modeled under the *Weak Completion Semantics*. Because killing a human is not a direct effect of the *change* action we may add:

$$ab_{prefer}\ change \leftarrow intent\_direct\_kill\ change, \tag{31}$$

$$intent\_direct\_kill\ change \leftarrow \bot. \tag{32}$$

Consequently, the *change* action will be preferred over the *donothing* action. A properly revised definition for permissibility will allow Ned to conclude that changing the switch is permissible under the *Doctrine of Triple Effect*:

$$perm\_triple\ change \leftarrow prefer(change, donothing)\ \wedge \tag{33}$$
$$\neg\ intent\_direct\_kill\ change$$
$$\neg\ ab_{perm\_triple}\ change,$$

$$ab_{perm\_triple}\ change \leftarrow \bot. \tag{34}$$

3.4 Summary

Dominic Deckert discusses additional trolley problems in [9]:

- The *loop-push* case is a variant of the loop case: *besides changing the switch, a heavy human has to be pushed on the looping side track in order to save the humans on the main track.* Thus, a direct intentional kill is needed to stop the trolley and, consequently, neither the *Doctrine of Double Effect* nor the *Doctrine of Triple Effect* permit the *change* action.

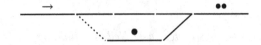

Fig. 5. The loop case.

- Another variant of the loop-case is the *man-in-front* case: *a heavy object is blocking the sidetrack behind the heavy human such that if the trolley hits the heavy object, it will stop.* Hence, the killing of the heavy human is no longer intended in order to save the humans on the main track and the *change* action is permissible under the *Doctrines of Double and Triple Effect.*
- The *collapse-bridge* case is a variant of the footbridge case: *instead of throwing the heavy human from the bridge, the bridge is collapsed in its entirety. This places the heavy human and the debris of he bridge on the track, effectively stopping the trolley.* Hence, the killing of the heavy human is not intentional and the collapse of the bridge becomes permissible under the *Doctrines of Double and Triple Effect.*

In all cases counterfactuals are necessary to determine whether the death of a human was intented in order to save the humans on the main track. Within the *Weak Completion Semantics* all these problems can be modelled as well. Table 3 summarizes the results.

4   Fluent Matching Problems

Let us discuss the computation of the least fixed point of the semantic operator $\Phi_{\mathcal{P}_1'}$ shown in Table 1 in more detail.

$$\Phi_{\mathcal{P}_1'}(\langle \emptyset, \emptyset \rangle) = \Phi_{\mathcal{P}_1'} \uparrow 1 = \langle I_1^\top, I_1^\perp \rangle,$$

where
$$I_1^\top = \{\ [ramify(donothing, t_0 \circ c_0 \circ m \circ h_1 \circ h_1 \circ h_2, 1)],$$
$$[action(t_0 \circ c_0 \circ m, 1, downhill, t_1 \circ c_0 \circ m, 1)],$$
$$[action(t_0 \circ c_0 \circ s, 1, downhill, t_2 \circ c_0 \circ s, 1)],$$
$$[action(t_1 \circ h_1, 1, kill, t_1, d)],$$
$$[action(t_2 \circ h_2, 1, kill, t_2, d)]\ \},$$
$$I_1^\perp = \{\ [ab_{ramify}\ downhill],$$
$$[ab_{ramify}\ kill]\ \}.$$

Table 3. The six cases and the permissible actions according to the different views.

	Bystander	Loop	Footbridge	Loop-Push	Man-in-Front	Collapse
Double effect	*change*	-	-	-	*change*	*collapse*
Triple effect	*change*	*change*	-	-	*change*	*collapse*
Utilitarianism	*change*	*change*	*throw*	*change throw*	*change*	*collapse*

Considering the body of (11) we find that both possible ground instances of $ab_{ramify}\, A'$, viz. $ab_{ramify}\, downhill$ and $ab_{ramify}\, kill$, are false under $\Phi_{\mathcal{P}'_1} \uparrow 1$ and their negations are true under $\Phi_{\mathcal{P}'_1} \uparrow 1$. The only ground instance of

$$ramify(A, P_1 \circ Z_1, P_2 \circ Z_2) \tag{35}$$

being true under $\Phi_{\mathcal{P}'_1} \uparrow 1$ is

$$ramify(donothing, t_0 \circ c_0 \circ m \circ h_1 \circ h_1 \circ h_2, 1). \tag{36}$$

Hence, we are searching for a ground instance of

$$action(P_1, P_2, A', E_1, E_2)$$

being true under $\Phi_{\mathcal{P}'_1} \uparrow 1$ such that

- the ground instance of P_1 is contained in $t_0 \circ c_0 \circ m \circ h_1 \circ h_1 \circ h_2$ and
- the ground instance of P_2 is contained in 1.

There are four candidates in $\Phi_{\mathcal{P}'_1} \uparrow 1$. The only possible ground instance of an action meeting the conditions is

$$action(t_0 \circ c_0 \circ m, 1, downhill, t_1 \circ c_0 \circ m, 1). \tag{37}$$

Comparing the second arguments of (35) and (36) with the first argument of (37) we find that

$$P_1 = t_0 \circ c_0 \circ m \quad \text{and} \quad Z_1 = h_1 \circ h_1 \circ h_2.$$

Likewise, comparing the third arguments of (35) and (36) with the second argument of (37) we find that

$$P_2 = 1 \quad \text{and} \quad Z_2 = 1.$$

Combining Z_1 with the fourth argument of (37) and, likewise, combining Z_2 with the fifth argument of (37) we learn that

$$ramify(donothing, t_1 \circ c_0 \circ m \circ h_1 \circ h_1 \circ h_2, 1)$$

must be true under $\Phi_{\mathcal{P}'_1} \uparrow 2$.

Likewise, we can compute that

$$[ramify(donothing, t_1 \circ c_0 \circ m \circ h_1 \circ h_2, d)]$$

must be true under $\Phi_{\mathcal{P}'_1} \uparrow 3$ and

$$[ramify(donothing, t_1 \circ c_0 \circ m \circ h_2, d \circ d)]$$

must be true under $\Phi_{\mathcal{P}'_1} \uparrow 4$.

Hence, in order to compute the semantic operator we have to solve AC1-matching problems of the form

$$t_0 \circ c_0 \circ m \circ Z_1 =_{AC1} t_0 \circ c_0 \circ m \circ h_1 \circ h_1 \circ h_2 \quad \text{and} \quad 1 \circ Z_2 =_{AC1} 1$$

or of the form

$$t_0 \circ c_0 \circ s \circ Z_1 =_{AC1} t_0 \circ c_0 \circ m \circ h_1 \circ h_1 \circ h_2 \quad \text{and} \quad 1 \circ Z_2 =_{AC1} 1.$$

Whereas the latter has no solution, the former does. In general, we need to solve so-called *fluent matching problems* of the form

$$s \circ Z =_{AC1} t \tag{38}$$

where s and t are ground fluent terms and Z is a variable.

Such problems have been considered in [21,39], where s was a fluent term. It was shown that fluent matching is decidable, finitary, and there always exists a minimal and complete set of matchers. The fluent matching algorithm presented in [21,39] can be easily adapted to the fact that s is ground:

(i) If $s =_{AC1} 1$ then return $\{Z \mapsto t\}$.
(ii) Don't care non-deterministically select a fluent u occurring in s and remove u from s.
(iii) If u occurs in t, then delete u from t and goto (i), else stop with failure.

Hence, with s being a ground fluent term, fluent matching becomes unitary.

Using the correspondence between fluent terms and multisets, let $\mathcal{S}$ and $\mathcal{T}$ be the multisets corresponding to the fluent terms s and t. Then, the fluent matching problem (38) has a solution iff

$$\mathcal{S} \,\dot{\subseteq}\, \mathcal{T}.$$

If (38) has a solution, then Z is mapped onto the fluent term corresponding to

$$\mathcal{T} \,\dot{\setminus}\, \mathcal{S}.$$

5   On the Adequateness of the Approach

Do humans reason with $AC1$-matchers in the limited form described in the previous section? Obviously, the multisets should not be large as there is compelling evidence that humans cannot deal with many different objects at the same time [29]. In the trolley problems discussed in this paper the maximal number of fluents was six. Even if we increase the number of humans on the main track to five as in the original version of the bystander case [14], the size of the multisets becomes only nine. Moreover, the actions did not increase the number of fluents in that the number of immediate effects was always equal to the number of preconditions.

In Germany, small children in Kindergarden are asked to solve puzzles of the following form: given several fruits like, for example, four apples and three peas, they are asked how many pieces are left after they would give some, say, two apples and one pea, away. The puzzles are presented in pictures. In most cases, the children are crossing out the pieces given away and, afterwards, are counting the remaining ones. In other words, they seem to solve exactly the AC1-matching problems discussed in the previous section. But to the best of my knowledge, there are almost no experimental data on how humans deal with multisets (see e.g. [16,32]). Hence, we hypothesize that humans can solve such matching problems although we must be careful as the ethical decision problems considered herein are more abstract than the puzzles solved by the children and it is well-known that humans solve less abstract problems differently than abstract ones (see e.g. [15,30,42]). Thus, the hypothesis must be experimentally tested.

6 Future Work

In the specification of the *prefer* relation in Subsect. 3.1, the *ctxt* operator was used. Having been introduced in [12] it was shown that the *ctxt* operator destroys the monotonicity property of the semantic operator. Consequently, the semantic operator may not have fixed points anymore. If, however, the program is acyclic, then the semantic operator has been shown to be a contraction [12,18] and Banach's contraction mapping theorem [5] can be applied to obtain the fixed point. The programs considered in this paper are acyclic. The trolley is either moving forward on finite tracks or a human gets killed. The *prefer* relation as well as the permissibility relations are acyclic. But a small gap remains in the chain of argumentation: We need to show that the semantic operator is a contraction in the presence of equational theories.

We would like to apply ethical decision making to much larger classes of problems, as presented in [3] for example. Although the problems discussed in [3] are similar to the ones presented in this paper, the data gathered in the moral machine experiment contains the cultural background of the actors and permissibility of actions depends also on the cultural background. In addition, the counterfactuals needed to determine the permissibility of actions were encoded by the author. We need a theory to automatically generate and test counterfactuals. This includes a specification of minimal change.

The *Weak Completion Semantics* has been implemented as a connectionist network [10]. However, the connectionist solution to the AC1-matching problems discussed in this paper needs to be added to the implementation. We envision a solution where a finite multiset of fluents is encoded by a finite array of units, which are activated using a dynamic binding mechanism like temporal synchrony [36]. In an AC1-matching problem, two arrays of units are compared. If a unit in the first array is activated in the same phase as a unit in the second array then both units shall be deactivated.

Last but not least, we need experiments to test our hypothesis that humans reason with AC1-matchers.

Acknowledgements. This work would not have been possible without the inspiration and tremendous help of Dominic Deckert, Emmanuelle-Anna Dietz Saldanha, Sibylle Schwarz, and Lim Yohanes Stefanus. Many thanks to Marco Ragni, Luís Moniz Pereira, and the anonymous referees for valuable comments.

References

1. Apt, K.R.: From Logic to Logic Programming. Prentice Hall, London (1997)
2. Aquinas, T.: Summa theologica II-II, q. 64, art. 7, "of Killing". In: Baumgarth, W.P., Regan, R. (eds.) On Law, Morality, and Politics, pp. 226–227. Hackett Publishing Co., Indianapolis (1988)
3. Awad, E., et al.: The moral machine experiment. Nature **563**, 59–64 (2018)
4. Baader, F., Siekmann, J.: Unification theory. In: Gabbay, D.M., Hogger, C.J., Robinson, J.A. (eds.) Handbook of Logic in Artificial Intelligence and Logic Programming, vol. 2, pp. 41–125. Oxford University Press (1994)
5. Banach, S.: Sur les opérations dans les ensembles abstraits et leur application aux équations intégrales. Fund. Math. **3**, 133–181 (1922)
6. Bentham, J.: An Introduction to the Principles of Morals and Legislation. Dover Publications Inc. (2009)
7. Byrne, R.: Suppressing valid inferences with conditionals. Cognition **31**, 61–83 (1989)
8. Clark, K.: Negation as failure. In: Gallaire, H., Minker, J. (eds.) Logic and Databases, pp. 293–322. Plenum, New York (1978)
9. Deckert, D.: A formalization of the trolley problem with the fluent calculus. TU Dresden, Informatik (2018)
10. Dietz Saldanha, E.A., Hölldobler, S., Kencana Ramli, C.D.P., Palacios Medinacelli, L.: A core method for the weak completion semantics with skeptical abduction. J. Artif. Intell. Res. **63**, 51–86 (2018)
11. Dietz Saldanha, E.A., Hölldobler, S., Lourêdo Rocha, I.: The weak completion semantics. In: Schon, C., Furbach, U. (eds.) Proceedings of the Workshop on Bridging the Gap Between Human and Automated Reasoning - Is Logic and Automated Reasoning a Foundation for Human Reasoning?, vol. 1994, pp. 18–30. CEUR-WS.org (2017). http://ceur-ws.org/Vol-1994/
12. Dietz Saldanha, E.-A., Hölldobler, S., Pereira, L.M.: Contextual reasoning: usually birds can abductively fly. In: Balduccini, M., Janhunen, T. (eds.) LPNMR 2017. LNCS (LNAI), vol. 10377, pp. 64–77. Springer, Cham (2017). https://doi.org/10.1007/978-3-319-61660-5_8
13. Dietz Saldanha, E.A., Hölldobler, S., Schwarz, S., Stefanus, L.Y.: The weak completion semantics and equality. In: Barthe, G., Sutcliffe, G., Veanes, M. (eds.) Proceedings of the 22nd International Conference on Logic for Programming, Artificial Intelligence, and Reasoning, LPAR-22, vol. 57, pp. 326–242. EPiC series in Computing (2018)
14. Foot, P.: The Problem of Abortion and the Doctrine of Double Effect, vol. 5. Oxford Review (1967)
15. Griggs, R., Cox, J.: The elusive thematic materials effect in the Wason selection task. Br. J. Psychol. **73**, 407–420 (1982)
16. Halford, G.S.: Children's Understanding. Psychology Press, New York (1993)

17. Hölldobler, S.: Ethical decision making under the weak completion semantics. In: Schon, C. (ed.) Proceedings of the Workshop on Bridging the Gap between Human and Automated Reasoning, vol. 2261, pp. 1–5. CEUR-WS.org (2018). http://ceur-ws.org/Vol-2261/
18. Hölldobler, S., Kencana Ramli, C.D.P.: Contraction properties of a semantic operator for human reasoning. In: Li, L., Yen, K.K. (eds.) Proceedings of the Fifth International Conference on Information, pp. 228–231. International Information Institute (2009)
19. Hölldobler, S., Kencana Ramli, C.D.P.: Logic programs under three-valued Łukasiewicz semantics. In: Hill, P.M., Warren, D.S. (eds.) ICLP 2009. LNCS, vol. 5649, pp. 464–478. Springer, Heidelberg (2009). https://doi.org/10.1007/978-3-642-02846-5_37
20. Hölldobler, S., Schneeberger, J.: A new deductive approach to planning. New Gener. Comput. **8**, 225–244 (1990)
21. Hölldobler, S., Schneeberger, J., Thielscher, M.: AC1-unification/matching in linear logic programming. In: Baader, F., Siekmann, J., Snyder, W. (eds.) Proceedings of the Sixth International Workshop on Unification. BUCS Tech Report 93–004, Boston University, Computer Science Department (1993)
22. Jaffar, J., Lassez, J.L., Maher, M.J.: A theory of complete logic programs with equality. In: Proceedings of the International Conference on Fifth Generation Computer Systems, pp. 175–184. ICOT (1984)
23. Kamm, F.M.: Intricate Ethics: Rights, Responsibilities, and Permissible Harm. Oxford University Press, Oxford (2006)
24. Khemlani, S., Johnson-Laird, P.N.: Theories of the syllogism: a meta-analysis. Psychol. Bull. **138**(3), 427–457 (2012)
25. Kowalski, R., Sergot, M.: A logic-based calculus of events. New Gener. Comput. **4**, 67–95 (1986)
26. Lloyd, J.W.: Foundations of Logic Programming. Springer, Heidelberg (1984). https://doi.org/10.1007/978-3-642-83189-8
27. Łukasiewicz, J.: O logice trójwartościowej. Ruch Filozoficzny **5**, 169–171 (1920). English translation: On three-valued logic. In: Borkowski, L. (ed.) Jan Łukasiewicz Selected Works, pp. 87–88. North Holland (1990)
28. McIntyre, A.: Doctrine of double effect. In: Zalta, E.N. (ed.) The Stanford Encyclopedia of Philosophy, Spring 2019 edn. (2019). https://plato.stanford.edu/archives/spr2019/entries/double-effect/
29. Miller, G.A.: The magical number seven, plus or minus two: some limits on our capacity for processing information. Psychol. Rev. **63**(2), 81–97 (1956)
30. Nickerson, R.S.: Conditional Reasoning. Oxford University Press, Oxford (2015)
31. Oliviera da Costa, A., Dietz Saldanha, E.A., Hölldobler, S., Ragni, M.: A computational logic approach to human syllogistic reasoning. In: Gunzelmann, G., Howes, A., Tenbrink, T., Davelaar, E.J. (eds.) Proceedings of the 39th Annual Conference of the Cognitive Science Society, pp. 883–888. Cognitive Science Society, Austin (2017)
32. Osherson, D.N.: Logical Abilities in Children, vol. 1. Routledge, London (1974)
33. Pereira, L.M., Saptawijaya, A.: Programming Machine Ethics. Springer, Heidelberg (2016). https://doi.org/10.1007/978-3-319-29354-7
34. Plaisted, D.A.: Equational reasoning and term rewriting system. In: Gabbay, D.M., Hogger, C.J., Robinson, J.A. (eds.) Handbook of Logic in Artificial Intelligence and Logic Programming, vol. 1, chap. 5. Oxford University Press, Oxford (1993)

35. Reiter, R.: The frame problem in the situation calculus: a simple solution (sometimes) and a completeness result for goal regression. In: Lifschitz, V. (ed.) Artificial Intelligence and Mathematical Theory of Computation—Papers in Honor of John McCarthy, pp. 359–380. Academic Press (1991)
36. Shastri, L., Ajjanagadde, V.: From associations to systematic reasoning: a connectionist representation of rules, variables and dynamic bindings using temporal synchrony. Behav. Brain Sci. **16**(3), 417–494 (1993)
37. Stenning, K., van Lambalgen, M.: Semantic interpretation as computation in nonmonotonic logic: the real meaning of the suppression task. Cogn. Sci. **29**, 919–960 (2005)
38. Stenning, K., van Lambalgen, M.: Human Reasoning and Cognitive Science. MIT Press, Cambridge (2008)
39. Thielscher, M.: AC1-Unifikation in der linearen logischen Programmierung. Master's thesis, Intellektik, Informatik, TH Darmstadt (1992)
40. Thielscher, M.: Introduction to the fluent calculus. Electron. Trans. Artif. Intell. **2**(3–4), 179–192 (1998)
41. Thielscher, M.: Controlling semi-automatic systems with FLUX. In: Palamidessi, C. (ed.) ICLP 2003. LNCS, vol. 2916, pp. 515–516. Springer, Heidelberg (2003). https://doi.org/10.1007/978-3-540-24599-5_49
42. Wason, P.C.: Reasoning about a rule. Q. J. Exp. Psychol. **20**, 273–281 (1968)

Pseudo-contractions as Gentle Repairs

Vinícius Bitencourt Matos[1], Ricardo Guimarães[1], Yuri David Santos[2],
and Renata Wassermann[1(✉)]

[1] Universidade de São Paulo, São Paulo, Brazil
{vbm,ricardof,renata}@ime.usp.br
[2] Rijksuniversiteit Groningen, Groningen, The Netherlands
y.david.santos@rug.nls

Abstract. Updating a knowledge base to remove an unwanted consequence is a challenging task. Some of the original sentences must be either
deleted or weakened in such a way that the sentence to be removed is
no longer entailed by the resulting set. On the other hand, it is desirable
that the existing knowledge be preserved as much as possible, minimising the loss of information. Several approaches to this problem can be
found in the literature. In particular, when the knowledge is represented
by an ontology, two different families of frameworks have been developed in the literature in the past decades with numerous ideas in common but with little interaction between the communities: applications of
AGM-like Belief Change and justification-based Ontology Repair. In this
paper, we investigate the relationship between pseudo-contraction operations and gentle repairs. Both aim to avoid the complete deletion of sentences when replacing them with weaker versions is enough to prevent
the entailment of the unwanted formula. We show the correspondence
between concepts on both sides and investigate under which conditions
they are equivalent. Furthermore, we propose a unified notation for the
two approaches, which might contribute to the integration of the two
areas.

Keywords: Belief change · Pseudo-contraction · Justification ·
Ontology repair · Knowledge representation

1 Introduction

In computer science, ontologies are shareable representations of a domain's
knowledge. Ontology development and maintenance tasks involve ontology engineers, domain specialists and other professionals. In this collaborative process,
one complicating aspect is that even a small modification may impact considerably what the ontology entails. To facilitate the execution of those tasks, techniques to aid in repairing and evolving ontologies were created. Two fields, in
particular, provide such methods: Ontology Repair and Belief Change.

While there are many distinct definitions of ontology in the literature, as
well as many ways to represent them, we will focus on ontologies represented

© Springer Nature Switzerland AG 2019
C. Lutz et al. (Eds.): Baader Festschrift, LNCS 11560, pp. 385–403, 2019.
https://doi.org/10.1007/978-3-030-22102-7_18

in Description Logics (or DLs, for short), decidable fragments of First-Order Logic. For an overview, please refer to the introductory texts [3,4]. In this paper, ontologies are finite sets of formulas (or axioms) in some Description Logic. This logical background allows users to obtain extended information from the explicit representation using reasoners. We employ the usual notation for DL formulas [3].

The area of *Ontology Repair* groups together a set of formal definitions and tools devised to help ontology maintainers in the task of debugging and getting rid of unwanted inferences. Different approaches have been proposed, depending on whether one is interested in repairing only the ABox [7,20], i.e., the part of the ontology dealing with instances, while leaving the terminological part (TBox) fixed, or considering the ontology as a whole [15,16,18].

Belief Change is a research area that aims at solving problems related to changing knowledge bases/logical theories, especially in the face of new, possibly conflicting, information. The work of Alchourrón, Gärdenfors and Makinson [1] is widely recognized as the initial hallmark of this area of research, and gave rise to what is known as the *AGM paradigm*. Initially developed having propositional logic in mind, in the last decade the AGM theory has been adapted to several other formalisms, including Description Logics [9,23,29]. Therefore, we can model problems in ontology maintenance using this framework (with a few modifications).

In this work, we show that Belief Change and Ontology Repair are closely related. Both have different notations for similar, and sometimes the same, concepts. Moreover, they share similar techniques to solve the same problem. We also show how both fields proposed similar solutions to obtain repairs that are fine-grained, that is, repairs that avoid removing whole formulas. Most of the results are quite straightforward, nonetheless, we found that both communities could benefit from them.[1]

2   Background

In this section, we briefly introduce the notation and the concepts used in the areas of Belief Change and Ontology Repair. We will denote a language by $\mathfrak{L}$, and we will use Cn to refer explicitly to a consequence operator. In this way, $\text{Cn}(X)$ denotes the logical consequences of Cn over X, where X is a set of formulas in $\mathfrak{L}$. As we will be dealing with a family of logics, the DLs, we will assume that in each case Cn is associated to the smallest DL that can represent the set of formulas given as argument. Moreover we assume it to be:

- **monotonic:** if $X \subseteq Y$, then $\text{Cn}(X) \subseteq \text{Cn}(Y)$;
- **compact:** for any X if $\alpha \in \text{Cn}(X)$, there is a finite $X' \subseteq X$ such that $\alpha \in \text{Cn}(X')$;
- **idempotent:** $\text{Cn}(X) = \text{Cn}(\text{Cn}(X))$.

---

[1] Actually, the idea to put together the main definitions used in Ontology Repair and Belief Revision for DLs appeared in many discussions with Franz during the last decade, so we deemed the paper as an appropriate Birthday present.

2.1   Ontology Repair

Ontology Repair consists in transforming an ontology so that it does not imply a certain formula. In what follows, we define the main concepts based on the presentation given by Baader et al. [5]. Consider that $\mathcal{O} = \langle \mathcal{O}_s, \mathcal{O}_r \rangle$ is an ontology consisting of a static and a refutable part ($\mathcal{O}_s$ and $\mathcal{O}_r$, respectively), which are assumed to be disjoint.[2] The static part contains those axioms which we want to preserve when we repair the ontology, while the refutable part contains those which we are willing to give up if needed. We assume that the separation into a static and refutable part is given as part of the input, be it a decision of an ontology engineer or obtained via some (semi-)automatic process.

Definition 1 (Repair). *Let* $\mathcal{O} = \langle \mathcal{O}_s, \mathcal{O}_r \rangle$ *be an ontology and let* α *be a sentence entailed by* $\mathcal{O}$ *but not by* $\mathcal{O}_s$. *An ontology* $\mathcal{O}'$ *is a* repair *of* $\mathcal{O}$ *with respect to* α *if* $\mathrm{Cn}(\mathcal{O}_s \cup \mathcal{O}') \subseteq \mathrm{Cn}(\mathcal{O}) \setminus \{\alpha\}$.

Classically, a repair consists of a subset of the refutable part of the ontology:

Definition 2 (Classical repair). *A repair* $\mathcal{O}'$ *of the ontology* $\mathcal{O}$ *with respect to the sentence* α *is a* classical repair *if it is contained in* $\mathcal{O}_r$.

And usually, we try to preserve as much knowledge as possible, looking for an optimal repair:

Definition 3 (Optimal repair). *A repair* $\mathcal{O}'$ *of the ontology* $\mathcal{O}$ *with respect to the sentence* α *is an* optimal repair *if no other repair* $\mathcal{O}''$ *(of* $\mathcal{O}$ *w.r.t.* α*) is such that* $\mathrm{Cn}(\mathcal{O}_s \cup \mathcal{O}') \subset \mathrm{Cn}(\mathcal{O}_s \cup \mathcal{O}'')$.

An *optimal classical repair* is a classical repair which is optimal in the sense that there is no classical repair which contains it.

In order to find classical repairs, a construction based on the ideas of justifications and hitting sets can be used. Justifications are minimal subsets of an ontology that imply the unwanted sentence:

Definition 4 (Justification [18]). *Let* $\mathcal{O} = \langle \mathcal{O}_s, \mathcal{O}_r \rangle$ *be an ontology and* α *a sentence entailed by* $\mathcal{O}$ *but not by* $\mathcal{O}_s$. *A* justification *for* α *in* $\mathcal{O}$ *is an inclusion-minimal subset* J *of* $\mathcal{O}_r$ *such that* $\alpha \in \mathrm{Cn}(\mathcal{O}_s \cup J)$. *We will denote the set of all justifications for* α *in* $\mathcal{O}$ *as* $\mathrm{Just}(\mathcal{O}, \alpha)$.

Schlobach [26] has proposed an algorithm to debug incoherent ontologies inspired by Reiter's hitting set tree [22]. Other authors, such as Kalyanpur et al. [18,19] and Horridge [15], extended and generalised this algorithm to find all justifications for any given entailment.

Definition 5 (Hitting set [22]). *Given a set* $\mathcal{J}$ *of justifications for a sentence in an ontology, a* hitting set *of* $\mathcal{J}$ *is a set* H *of sentences contained in* $\bigcup \mathcal{J}$ *such that* $H \cap J \neq \varnothing$ *for every* $J \in \mathcal{J}$.

---

[2] The notation $\langle \mathcal{O}_s, \mathcal{O}_r \rangle$ is meant to represent the set $\mathcal{O}_s \cup \mathcal{O}_r$ in such a way that it is possible to tell whether a sentence is in the static part or in the refutable part.

Regarding the actual repair, done by removing at least one formula from each justification, a simple description is presented in Algorithm 1 from Baader et al. [5]. We assume the existence of a function $\texttt{Justifications}(\mathcal{O}, \alpha)$ that computes $\text{Just}(\mathcal{O}, \alpha)$ and a function $\texttt{MinimalHittingSet}(\mathcal{J})$ that computes an inclusion-minimal hitting set of $\mathcal{J}$.

Algorithm 1. Classical repair algorithm

Input: An ontology $\mathcal{O} = \langle \mathcal{O}_s, \mathcal{O}_r \rangle$ and a formula α
Output: A classical repair $\mathcal{O}'$ of $\mathcal{O}$ w.r.t. α

1 **Function** *ClassicalRepair(O, α)*
2     $\mathcal{J} \leftarrow \texttt{Justifications } (\mathcal{O}, \alpha)$
3     $H \leftarrow \texttt{MinimalHittingSet}(\mathcal{J})$
4     $\mathcal{O}' \leftarrow \mathcal{O}_r$
5   **for** $\beta \in H$ **do**
6       $\lfloor \;\; \mathcal{O}' \leftarrow \mathcal{O}' \setminus \{\beta\}$
7   **return** $\mathcal{O}'$

A special case of Ontology Repair is *ABox Repair*, where the TBox is fixed, i.e., the TBox is contained in $\mathcal{O}_s$:

Definition 6 (ABox Repair [20]). *Let $\mathcal{O}$ be an ontology, with TBox $\mathcal{T}$ and ABox $\mathcal{A}$. An ABox Repair of $\mathcal{O}$ is an inclusion-maximal subset $\mathcal{A}'$ of $\mathcal{A}$ such that the ontology $\mathcal{O}'$ consisting of $\mathcal{T} \cup \mathcal{A}'$ is consistent.*

It is easy to see that when $\mathcal{T} = \mathcal{O}_s$ and $\mathcal{A} = \mathcal{O}_r$, an ABox repair is an optimal repair according to Definition 3.

2.2   Belief Change

In the classical AGM paradigm [1], an agent's knowledge is represented by a *theory* or *belief set* — a set closed under a consequence operator. In this paper, we consider an alternative representation, *belief bases* [10], dropping the closure requirement. Three change operations on the agent's knowledge B with respect to a sentence α are considered:

- Expansion $(B + \alpha)$: the sentence α is incorporated to B possibly leading to an inconsistency.
- Revision $(B * \alpha)$: the sentence α is incorporated to B in a way that the resulting set is consistent.
- Contraction $(B - \alpha)$: the sentence α is removed from B and must not be entailed by the contracted set.

In [1], a construction for a contraction operation known as *partial meet contraction* was proposed, based on the idea of selecting maximal subsets of the agent's knowledge.

Definition 7 (Remainder and remainder set [1]**).** *Let $B \subseteq \mathfrak{L}$ and $\alpha \in \mathfrak{L}$. The* remainder set *of B with respect to α, denoted by $B \perp \alpha$, is the set of all $X \subseteq B$ such that $\alpha \notin \mathrm{Cn}(X)$ and there is no Y such that $X \subset Y \subseteq B$ and $\alpha \notin \mathrm{Cn}(Y)$. Each such X is an α-remainder of B.*

In order to compute the contraction, at least one of the α-remainders is selected, according to some preference criteria, encoded as a function:

Definition 8 (Selection function [1]**).** *Let $B \subseteq \mathfrak{L}$. A function γ is a* selection function *for B if, for every $\alpha \in \mathfrak{L}$, it is the case that $\varnothing \neq \gamma(B \perp \alpha) \subseteq B \perp \alpha$ if $B \perp \alpha$ is nonempty, or $\gamma(B \perp \alpha) = \{B\}$ otherwise.*

The selected α-remainders are then joined to form the resulting contracted set:

Definition 9 (Partial meet contraction [1]**).** *Let $B \subseteq \mathfrak{L}$, and let γ be a selection function for B. The* partial meet contraction *of B by a sentence α, denoted by $B -_\gamma \alpha$, is defined as $\bigcap \gamma(B \perp \alpha)$.*

Hansson has characterised the operation of partial meet base contraction by means of the following *rationality postulates*, where α and α' are sentences and B is a belief base [12]:

- **inclusion:** $B - \alpha \subseteq B$;
- **relevance:** if $\alpha' \in B \backslash (B - \alpha)$, then there is some B' such that $B - \alpha \subseteq B' \subseteq B$ and $\alpha \in \mathrm{Cn}(B' \cup \{\alpha'\}) \setminus \mathrm{Cn}(B')$;
- **success:** $\alpha \notin B - \alpha$ unless $\alpha \in \mathrm{Cn}(\varnothing)$;
- **uniformity:** if α and α' are such that, for every $B' \subseteq B$, it is the case that $\alpha \in \mathrm{Cn}(B')$ if and only if $\alpha' \in \mathrm{Cn}(B')$, then $B - \alpha = B - \alpha'$.

Theorem 1 [12]**.** *An operation $\dot{-}$ is a partial meet contraction for a belief base B if and only if $\dot{-}$ satisfies inclusion, relevance, success and uniformity.*

Another construction for contraction was proposed in [13], which relies on minimal sets implying the undesirable sentence:

Definition 10 (Kernel and kernel set [13]**).** *Let $B \subseteq \mathfrak{L}$ and $\alpha \in \mathfrak{L}$. The* kernel set *of B with respect to α, denoted by $B \perp\!\!\!\perp \alpha$, is such that a set X is in $B \perp\!\!\!\perp \alpha$ if and only if $X \subseteq B$, $\alpha \in \mathrm{Cn}(X)$, and there is no $Y \subset X$ such that $\alpha \in \mathrm{Cn}(Y)$. Each such X is an α-kernel.*

In order to contract α from B, at least one element of each α-kernel must be removed:

Definition 11 (Incision function [13]**).** *Let $B \subseteq \mathfrak{L}$. A function σ is an* incision function *for B if, for every $\alpha \in \mathfrak{L}$, it is the case that $\sigma(B \perp\!\!\!\perp \alpha) \subseteq \bigcup(B \perp\!\!\!\perp \alpha)$ and $\sigma(B \perp\!\!\!\perp \alpha) \cap X \neq \varnothing$ for every $X \in B \perp\!\!\!\perp \alpha$.*

Definition 12 (Kernel contraction [13]**).** *Let* $B \subseteq \mathfrak{L}$, *and let* σ *be an incision function for* B. *The* kernel contraction *of* B *by a sentence* α, *denoted by* $B -_\sigma \alpha$, *is defined as* $B \setminus \sigma(B \perp\!\!\!\perp \alpha)$.

Kernel contraction can also be characterised by a set of rationality postulates:

Theorem 2 [13]. *An operation* $\dot{-}$ *is a kernel contraction for a belief base* B *if and only if* $\dot{-}$ *satisfies inclusion, success, uniformity and:*
*(**core-retainment**) if* $\alpha' \in B \setminus (B - \alpha)$, *then there is some* B' *such that* $B' \subseteq B$ *and* $\alpha \in \mathrm{Cn}(B' \cup \{\alpha'\}) \setminus \mathrm{Cn}(B')$.

From this theorem, as core-retainment is slightly more general than relevance, we see that all partial meet contractions can be constructed as a kernel contraction, but the opposite does not hold, i.e., kernel contractions are more general than partial meet contractions. This will be further explored in Sect. 4.

An algorithm for computing kernel contraction based on the idea of hitting sets was proposed in [28]. Note that the same idea underlies the computation of repairs, presented above. Although the theory of Belief Change is mostly used for propositional logic, the constructions and characterisation results were extended in [14] to any compact and monotonic logic.

3 Correspondence Between Belief Change and Repairs in Description Logics

In this section, we will analyse the close relationship between the concepts and constructions presented for Ontology Repair and Belief Change. Also, we propose a unified notation for Belief Change and Ontology Repair operations.

Definition 13 (Maximal Non-Implying Subsets). *Let* B *be a knowledge base,* α *a sentence, and* Φ *a set of static sentences (i.e. which should be preserved in any operation). The set of* maximal α-non-implying subsets *of* B *with respect to* Φ, *denoted by* $\mathrm{MaxNon}(B, \alpha, \Phi)$, *is such that* $X \in \mathrm{MaxNon}(B, \alpha, \Phi)$ *if and only if* $X \subseteq B$, $\alpha \notin \mathrm{Cn}(\Phi \cup X)$, *and there is no* Y *such that* $X \subset Y \subseteq B$ *and* $\alpha \notin \mathrm{Cn}(\Phi \cup Y)$.

For brevity, we shall omit the last argument whenever it is empty: $\mathrm{MaxNon}(B, \alpha) = \mathrm{MaxNon}(B, \alpha, \varnothing)$.

Remark 1. If $\Phi \subseteq B$, then the maximal α-non-implying subsets of B with respect to Φ contain all of the elements of Φ, i.e., $X \supseteq \Phi$ for every $X \in \mathrm{MaxNon}(B, \alpha, \Phi)$.

Proof. If there is some $X \in \mathrm{MaxNon}(B, \alpha, \Phi)$ such that $X \not\supseteq \Phi$, then the set $Y = X \cup \Phi$ is such that $X \subset Y \subseteq B$, and since $\Phi \cup Y = \Phi \cup X$, we have that $\alpha \notin \mathrm{Cn}(\Phi \cup Y) = \mathrm{Cn}(\Phi \cup X)$, violating the definition of MaxNon. $\qquad \square$

Definition 13 corresponds to Definition 7 if $\Phi = \varnothing$, i.e., $\mathrm{MaxNon}(B, \alpha) = B \perp \alpha$.

Definition 14 (Minimal Implying Subsets). *Let B be a knowledge base, α a sentence, and Φ a set of static sentences. The set of* minimal α-implying *subsets of B with respect to Φ, denoted by* $\mathrm{MinImp}(B, \alpha, \Phi)$, *is such that $X \in \mathrm{MinImp}(B, \alpha, \Phi)$ if and only if $X \subseteq B$, $\alpha \in \mathrm{Cn}(\Phi \cup X)$, and there is no $Y \subset X$ such that $\alpha \in \mathrm{Cn}(\Phi \cup Y)$.*

As in the previous definition, the last argument will be omitted if empty: $\mathrm{MinImp}(B, \alpha) = \mathrm{MinImp}(B, \alpha, \varnothing)$.

Remark 2. The minimal α-implying subsets of B with respect to Φ do not contain elements of Φ, i.e., $X \cap \Phi = \varnothing$ for every $X \in \mathrm{MinImp}(B, \alpha, \Phi)$.

Proof. If there is some $X \in \mathrm{MinImp}(B, \alpha, \Phi)$ such that $X \cap \Phi \neq \varnothing$, then the set $Y = X \setminus \Phi$ is such that $Y \subset X$, and since $\Phi \cup Y = \Phi \cup X$, we have that $\alpha \in \mathrm{Cn}(\Phi \cup Y) = \mathrm{Cn}(\Phi \cup X)$, which contradicts the definition of MinImp. $\square$

If $\Phi = \varnothing$, Definition 14 corresponds to Definition 10, i.e., $\mathrm{MinImp}(B, \alpha) = B \perp\!\!\!\perp \alpha$. Definition 14 is also closely related to Definition 4: $\mathrm{MinImp}(B, \alpha, \Phi) = \mathrm{Just}(\langle \Phi, B \setminus \Phi \rangle, \alpha)$, or conversely, $\mathrm{Just}(\langle \mathcal{O}_s, \mathcal{O}_r \rangle, \alpha) = \mathrm{MinImp}(\mathcal{O}_s \cup \mathcal{O}_r, \alpha, \mathcal{O}_s)$. Our definitions of MaxNon and MinImp also correspond, respectively, to the sets of MaNAs (maximal non-axiom sets) and MinAs (minimal axiom sets) in the literature [6].

Since we usually represent sentences by lowercase Greek letters, we propose to replace γ and σ by g and f, respectively, to represent selection and incision functions (Definitions 8 and 11).

The usual notations for partial meet contraction and kernel contraction overlap, making it impossible to distinguish between the two without the context (i.e. $B -_\delta \alpha$ could be either construction depending on what δ is). We propose a clearer notation for these constructions: $\mathrm{PMC}_g(B, \alpha)$ for partial meet contraction and $\mathrm{KC}_f(B, \alpha)$ for kernel contraction. Similarly, a generic contraction operation will be represented by $\mathrm{C}(B, \alpha)$.

Let $B \subseteq \mathfrak{L}$ and $\alpha \in \mathfrak{L}$. The following two properties follow straightly from Definitions 4 and 10.

Proposition 1 (Kernel $\sim$ Justification). *If $\alpha \in \mathrm{Cn}(B)$, then a set X is an α-kernel of B with respect to α if and only if X is a justification for α in $\langle \varnothing, B \rangle$.*

In our notation, the set of all such sets X is denoted by $\mathrm{MinImp}(B, \alpha)$, which unifies the concepts of the following proposition:

Proposition 2 (Kernel set $\sim$ Set of all justifications). *If $\alpha \in \mathrm{Cn}(B)$, then $B \perp\!\!\!\perp \alpha = \mathrm{Just}(\langle \varnothing, B \rangle, \alpha)$.*

A classical repair (Definition 2) can be seen as a contraction operation that satisfies two of Hansson's postulates for base contraction.

Proposition 3 (Classical Repair $\implies$ Postulates for base contraction).
Let Rep *be an operation that yields a classical repair. Define the operation* C_{Rep}
as

$$C_{Rep}(B, \alpha) = \begin{cases} Rep(\langle \varnothing, B \rangle, \alpha), & \text{if } B \models \alpha; \\ B, & \text{otherwise.} \end{cases}$$

Then, C_{Rep} *satisfies success and inclusion.*

In Sect. 4 we show a special case of classical repair that satisfies the other two postulates needed for characterising base contraction, namely uniformity and relevance, and also discuss a result similar to Proposition 3 for optimal classical repairs.

The following proposition, which is an immediate consequence of the Upper Bound Property [2], will be useful to show the connection between partial meet base contraction and classical repairs.

Proposition 4 (Existence of α-remainder preserving $\mathcal{O}_s$). *Let* $\mathcal{O} = \langle \mathcal{O}_s, \mathcal{O}_r \rangle$ *be an ontology and* α *be a sentence entailed by* $\mathcal{O}$ *but not by* $\mathcal{O}_s$. *Then, there is at least one* α-remainder X *of* $\mathcal{O}_s \cup \mathcal{O}_r$ *such that* $\mathcal{O}_s \subseteq X$.

Now we can show that partial meet base contractions that include the static part of the ontology yield classical repairs.

Proposition 5 (Partial meet base contraction $\implies$ Classical repair).
Under the conditions of Proposition 4, if the selection function g *is such that* $\mathcal{O}_s \subseteq X$ *for every* $X \in g(\mathcal{O} \perp \alpha)$, *then the operation* Rep_g *defined as*

$$Rep_g(\mathcal{O}, \alpha) = PMC_g(\mathcal{O}, \alpha) \setminus \mathcal{O}_s$$

yields a classical repair.

Proof. Let $\mathcal{O}' = Rep_g(\mathcal{O}, \alpha)$. Since g only selects α-remainders including $\mathcal{O}_s$, we have that $\mathcal{O}_s \subseteq PMC_g(\mathcal{O}, \alpha)$, which implies that $\mathcal{O}_s \cup \mathcal{O}' = PMC_g(\mathcal{O}, \alpha)$. Hence, from the inclusion postulate, we have that $\mathcal{O}_s \cup \mathcal{O}' \subseteq \mathcal{O}$, and monotonicity of Cn gives $Cn(\mathcal{O}_s \cup \mathcal{O}') \subseteq Cn(\mathcal{O})$. This is sufficient to show that the result of Rep_g is a repair. From the inclusion postulate, we have that $PMC_g(\mathcal{O}, \alpha) \subseteq \mathcal{O}$, which proves that $\mathcal{O}' = PMC_g(\mathcal{O}, \alpha) \setminus \mathcal{O}_s \subseteq \mathcal{O}_r$. Therefore, Rep_g yields a classical repair. $\square$

To conclude this section, we summarise the relationships between essential concepts in both Belief Change and Ontology Repair using the diagram in Fig. 1. In the diagram, we represent each concept with an ellipse and the areas by solid rectangles. Moreover, in the Belief Change area, we separate the concepts that constitute the partial meet approach from those that are part of the kernel approach as discussed in Sect. 2. We can note how most of the concepts have direct connections; the only exceptions are MIPS (set of Minimal Incoherence-Preserving Sub-TBoxes) and MUPS (set of Minimal Unsatisfiability-Preserving

Sub-TBoxes) [27], which are variants of the usual justifications, and as such cannot be mapped directly to our definition of MinImp. More specifically, if $\mathcal{T}$ is a TBox and A a concept name, then the set of MUPSes of A in $\mathcal{T}$ is equivalent to MinImp($\mathcal{T}, A \sqsubseteq \bot$); and the set of all MIPSes in $\mathcal{T}$ is given by MIPS($\mathcal{T}$) = $\min_{\subseteq} \left\{ \bigcup_{A \in \mathcal{N}_C(\mathcal{T})} \text{MinImp}(\mathcal{T}, A \sqsubseteq \bot) \right\}$, where $\mathcal{N}_C(\mathcal{T})$ is the set of concept names in $\mathcal{T}$.

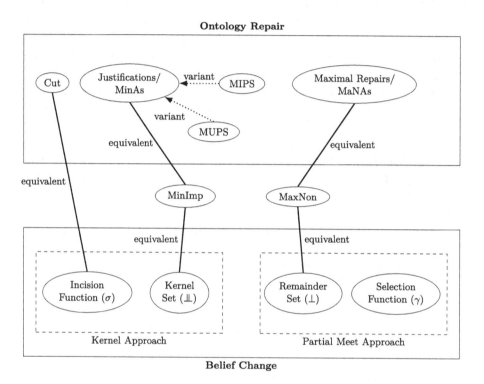

Fig. 1. Relationship between concepts in Belief Change and Ontology Repair

With these results we can see that repairs in general already satisfy some of the postulates for base contraction. In the next section we also show that whenever a classical repair is optimal in the sense of Definition 3, it also satisfies other two postulates: uniformity and relevance.

4   Minimal Change in Classical Repair and Contractions

Most approaches in Ontology Repair and Debugging rely on minimal implying sets for both definition and implementation. As examples, we have the work of Schlobach and Cornet [27], Kalyanpur [18] and Qi et al. [21]. The desire to remove formulas only when strictly necessary is also common in these approaches [17, 18,21,26], and in terms of Belief Change, in particular of kernel revision, this desideratum can be expressed using the notion of minimal incision functions (Definition 15). In this section, we consider that the whole ontology is refutable as this will allow us to cover approaches in Ontology Repair which do not consider a static part and use the classical theory of Belief (Base) Change which also assumes that all axioms are retractable.

Definition 15 (Minimal Incision Function [8]). *Let f be an incision function for the ontology $\mathcal{O}$. We say that f is minimal if no other incision function f' over $\mathrm{MinImp}(\mathcal{O}, \alpha)$ is such that $f'(\mathrm{MinImp}(\mathcal{O}, \alpha)) \subset f(\mathrm{MinImp}(\mathcal{O}, \alpha))$.*

This requirement of preserving as many axioms as possible can also be expressed using the partial meet construction, more specifically with maxichoice contraction functions (Definition 16). This coincidence between the two methods occurs in this case because kernel base contraction functions are equivalent to maxichoice base contraction functions, in DLs, whenever the incision function is minimal.

Definition 16 (Maxichoice Contraction Function [1]). *We say that a partial meet contraction function PMC_g over $\mathrm{MaxNon}(\mathcal{O}, \alpha)$ is a maxichoice contraction function if it is based on a selection function g such that $|g(\mathrm{MaxNon}(\mathcal{O}, \alpha))| = 1$.*

Note that the Maxichoice Contraction Functions give as result an optimal classical repair as in Definition 3 (elements of $\mathrm{MaxNon}(\mathcal{O}, \alpha)$), whenever a repair exists, because DLs are monotonic.

Lemma 1. *An incision function f is minimal if and only if $\mathcal{O} \setminus f(\mathrm{MinImp}(\mathcal{O}, \alpha))$ is an element of $\mathrm{MaxNon}(\mathcal{O}, \alpha)$.*

Lemma 1 is directly adapted from Falappa et al. [8], where it was stated for propositional logic. The proof, however, can be transported without issues to Description Logics.

Given a kernel contraction KC_f, it can be constructed as a partial meet construction if and only if it satisfies the relevance postulate. This postulate constrains which formulas a contraction is allowed to remove, hence it expresses a "minimal change" requirement on the outcome.

We can straightforwardly extend this mapping to DL belief bases. According to Hansson and Wassermann [14], the only property that distinguishes partial meet from kernel base contraction is relevance, as kernel operations only guarantee its more general counterpart called core-retainment. In other words, the class of kernel base contraction functions contain that of partial meet contraction functions.

Fallappa et al. [8] also showed that whenever a kernel contraction satisfies the relevance postulate, it is equivalent to some partial meet function. The importance of this postulate is to rule out contraction functions that would remove axioms that are not responsible for the undesired entailment, expressing a "minimal change" requirement on the outcome. This point is clarified in Example 1:

Example 1. Consider the following DL ontology, with only the refutable part:

$$\mathcal{O}_{\text{rel}} = \{\mathsf{c} : \mathsf{A}, \ \mathsf{c} : \mathsf{B}, \ \mathsf{A} \sqcup \mathsf{B} \sqsubseteq \mathsf{D}\}.$$

Suppose that we want to remove the entailment $\mathsf{c} : \mathsf{D}$. We can do this by computing $\text{MinImp}(\mathcal{O}_{\text{rel}}, \mathsf{c} : \mathsf{D})$ and removing elements using an incision function.

$$\text{MinImp}(\mathcal{O}_{\text{rel}}, \ \mathsf{c} : \mathsf{D}) = \{\{\mathsf{c} : \mathsf{A}, \ \mathsf{A} \sqcup \mathsf{B} \sqsubseteq \mathsf{D}\}, \{\mathsf{c} : \mathsf{B}, \ \mathsf{A} \sqcup \mathsf{B} \sqsubseteq \mathsf{D}\}\}.$$

Now, we consider an incision function f such that $f(\mathcal{O}_{\text{rel}}, \ \mathsf{c} : \mathsf{D}) = \{\mathsf{c} : \mathsf{A}, \ \mathsf{A} \sqcup \mathsf{B} \sqsubseteq \mathsf{D}\}$. The kernel contraction based on this function does not satisfy relevance, as there is no $\mathcal{O}'$ with $\{\mathsf{c} : \mathsf{B}\} \subseteq \mathcal{O}' \subseteq \mathcal{O}_{\text{rel}}$ such that $\mathcal{O}' \nvDash \mathsf{c} : \mathsf{D}$ but $\mathcal{O}' \cup \{\mathsf{c} : \mathsf{A}\} \vDash \mathsf{c} : \mathsf{D}$.

This essentially means that removing $\mathsf{c} : \mathsf{A}$ represents a redundancy, since it always depends on other members of $f(\mathcal{O}_{\text{rel}}, \mathsf{c} : \mathsf{D})$ to produce the undesired entailment. In other words, it only occurs in minimal implying sets where $\mathsf{A} \sqcup \mathsf{B} \sqsubseteq \mathsf{D}$ already occurs.

If we consider an incision function f' with $f'(\mathcal{O}_{\text{rel}}, \mathsf{c} : \mathsf{D}) = \{\mathsf{A} \sqcup \mathsf{B} \sqsubseteq \mathsf{D}\}$, then we can take $\mathcal{O}' = \{\mathsf{c} : \mathsf{A}, \mathsf{c} : \mathsf{B}\}$ and the conditions of the relevance postulate will be satisfied (in this particular case, f' is also minimal, which is sufficient but not necessary to obtain relevance).

What we show here is a way to generate relevance-complying kernel base contractions. Before proceeding to the actual construction, we need an auxiliary result regarding the union of incision functions:

Proposition 6. *If $\{f_i\}_{i \in I \subset \mathbb{N}}$ is a set of incision functions over $\mathcal{O}$, then $f(\mathcal{O}, \alpha) := \bigcup_{i \in I} f_i(\mathcal{O}, \alpha)$ is an incision function.*

Proof. Let $\{f_i\}_{i \in I \subset \mathbb{N}}$ and f be as in the proposition statement. Then, for all α:

1. Since $f_i(\mathcal{O}, \alpha) \subseteq \bigcup \text{MinImp}(\mathcal{O}, \alpha)$ for all i, we have that $f(\mathcal{O}, \alpha) \subseteq \bigcup \text{MinImp}(\mathcal{O}, \alpha)$;
2. If $\varnothing \neq X \in \text{MinImp}(\mathcal{O}, \alpha)$, then for all f_i: $X \cap f_i(\mathcal{O}, \alpha) \neq \varnothing$. And this is sufficient to ensure that f satisfies the same condition.

Therefore, f is also an incision function. $\square$

Now, we can use Proposition 6 to define a particular class of incision functions:

Definition 17 (Regular Incision Function). *We say that an incision function f is* regular *if it can be equivalently expressed as the union of minimal incision functions, in symbols: $f(\mathcal{O}, \alpha) = \bigcup_{i \in I \subseteq \mathbb{N}} f_i(\mathcal{O}, \alpha)$, where each f_i is a minimal incision function. Incision functions that are not regular will be called* irregular. *We say that a kernel base contraction function is (ir)regular if it is based on a(n) (ir)regular incision function.*

Finally, we show that regularity is equivalent to relevance for kernel base contraction functions. To simplify the proof we use the intermediate result stated in Lemma 2.

Lemma 2. *If a contraction* $C(\mathcal{O}, \alpha)$ *satisfies relevance, then for each* $\beta \in \mathcal{O} \setminus C(\mathcal{O}, \alpha)$ *there is at least one set* $\mathcal{O}' \in \mathrm{MaxNon}(\mathcal{O}, \alpha)$ *such that* $\beta \notin \mathcal{O}'$ *and* $C(\mathcal{O}, \alpha) \subseteq \mathcal{O}'$.

Proof. Let $\mathcal{O}$ be a DL ontology, α a DL formula and C a contraction function satisfying relevance. The relevance postulate states that for each $\beta \in \mathcal{O} \setminus C(\mathcal{O}, \alpha)$ there is a set $\mathcal{O}'$ with $C(\mathcal{O}, \alpha) \subseteq \mathcal{O}' \subseteq \mathcal{O}$ such that $\mathcal{O} \nvDash \alpha$, but $\mathcal{O} \cup \{\beta\} \vDash \alpha$. Note that for every $\mathcal{O}' \in \mathrm{MaxNon}(\mathcal{O}, \alpha)$ we already have that $C(\mathcal{O}, \alpha) \subseteq \mathcal{O}'$. Therefore, we just need to show that for each $\beta \in \mathcal{O} \setminus C(\mathcal{O}, \alpha)$ there is at least one set $\mathcal{O}' \in \mathrm{MaxNon}(\mathcal{O}, \alpha)$ such that $\beta \notin \mathcal{O}'$.

Let us assume that for some $\beta \in \mathcal{O} \setminus C(\mathcal{O}, \alpha)$ there is no $\mathcal{O}' \in \mathrm{MaxNon}(\mathcal{O}, \alpha)$ such that $\beta \notin \mathcal{O}'$, i.e., $\beta \in \bigcap \mathrm{MaxNon}(\mathcal{O}, \alpha)$. Due to monotonicity, this means that for any $\mathcal{O}'' \subseteq \mathcal{O}$ such that $\mathcal{O}'' \nvDash \alpha$ we have that $\mathcal{O}'' \cup \{\beta\} \nvDash \alpha$, as $\mathcal{O}'' \cup \{\beta\} \subseteq \mathcal{O}'$ for some $\mathcal{O}' \in \mathrm{MaxNon}(\mathcal{O}, \alpha)$. However, this contradicts the relevance assumption. Hence, there must be at least one set $\mathcal{O}' \in \mathrm{MaxNon}(\mathcal{O}, \alpha)$ such that $\beta \notin \mathcal{O}'$. $\square$

Corollary 1. *The class of kernel contraction functions that satisfy relevance and that of regular kernel base contraction functions are the same.*

Proof. Let KC_f be a regular kernel base contraction function. As f is regular we can write $f(\mathcal{O}, \alpha) = \bigcup\limits_{i \in I \subseteq \mathbb{N}} f_i(\mathcal{O}, \alpha)$, where each f_i is a minimal incision function. For each $\beta \in \mathcal{O} \setminus \mathrm{KC}_f(\mathcal{O}, \alpha)$ we have that there is at least one f_{i_β} with $i_\beta \in I$ such that $\beta \in f_{i_\beta}(\mathcal{O}, \alpha)$. Finally, note that for each β we have at least one set $\mathcal{O} \setminus f_{i_\beta}(\mathcal{O}, \alpha)$ such that $\mathrm{KC}_f(\mathcal{O}, \alpha) \subseteq \mathcal{O} \setminus f_{i_\beta}(\mathcal{O}, \alpha) \subset \mathcal{O}$. Moreover, $\mathcal{O} \setminus f_{i_\beta}(\mathcal{O}, \alpha) \in \mathrm{MaxNon}(\mathcal{O}, \alpha)$, hence it does not imply α, but it does if we add β. Therefore, the regular contraction function satisfies relevance.

To prove the other direction, consider a kernel base contraction function KC_f that satisfies relevance. From Lemma 2, we know that for each $\beta \in f(\mathcal{O}, \alpha)$ there is a $\mathcal{O}_\beta \in \mathrm{MaxNon}(\mathcal{O}, \alpha)$ such that $\beta \notin \mathcal{O}_\beta$. Also from Lemma 2, we know that for each $\mathcal{O}_\beta$ we have $\mathrm{KC}_f(\mathcal{O}, \alpha) \subseteq \mathcal{O}_\beta$, and as such $\mathrm{KC}_f(\mathcal{O}, \alpha) \subseteq \bigcap\limits_{\beta \in f(\mathcal{O}, \alpha)} \mathcal{O}_\beta$.

Then, using Lemma 1 we can write $\mathrm{KC}_f(\mathcal{O}, \alpha) \subseteq \bigcap\limits_{\beta \in f(\mathcal{O}, \alpha)} (\mathcal{O} \setminus f_\beta(\mathcal{O}, \alpha)) = \mathcal{O} \setminus \bigcup\limits_{\beta \in f_\beta(\mathcal{O}, \alpha)} f_\beta(\mathcal{O}, \alpha)$, where each f_β is a minimal incision function defined as $f_\beta = (\mathcal{O} \setminus \mathcal{O}_\beta)$.

Now we show that $\mathrm{KC}_f(\mathcal{O}, \alpha) \supseteq \mathcal{O} \setminus \bigcup\limits_{\beta \in f_\beta(\mathcal{O}, \alpha)} f_\beta(\mathcal{O}, \alpha)$, where each f_β is defined as in the previous paragraph. Suppose for a contradiction that there is a $\delta \in \mathcal{O} \setminus \bigcup\limits_{\beta \in f_\beta(\mathcal{O}, \alpha)} f_\beta(\mathcal{O}, \alpha)$, but $\delta \notin \mathrm{KC}_f(\mathcal{O}, \alpha)$. We have that $\delta \in f(\mathcal{O}, \alpha)$

and hence, due to Lemma 2 there must be an $\mathcal{O}_\delta \in \mathrm{MaxNon}(\mathcal{O}, \alpha)$ such that $\delta \notin \mathcal{O}_\delta$. But as we proved earlier, $\delta \notin \mathrm{KC}_f(\mathcal{O}, \alpha) \subseteq \mathcal{O}_\delta$. However, we assumed $\delta \in \mathcal{O} \setminus \bigcup\limits_{\beta \in f_\beta(\mathcal{O}, \alpha)} f_\beta(\mathcal{O}, \alpha) = \bigcap\limits_{\beta \in f(\mathcal{O}, \alpha)} (\mathcal{O} \setminus f_\beta(\mathcal{O}, \alpha)) \subseteq \mathcal{O}_\delta$ what implies $\delta \in \mathcal{O}_\delta$. As the contradiction shows, such δ cannot exist, and hence $\mathrm{KC}_f(\mathcal{O}, \alpha) \supseteq \mathcal{O} \setminus \bigcup\limits_{\beta \in f_\beta(\mathcal{O}, \alpha)} f_\beta(\mathcal{O}, \alpha)$.

Finally, $\mathrm{KC}_f(\mathcal{O}, \alpha) = \mathcal{O} \setminus \bigcup\limits_{\beta \in f_\beta(\mathcal{O}, \alpha)} f(\mathcal{O}, \alpha)$, concluding the proof. □

In this section, we highlighted a mapping between the algorithms that obtain classical repairs by finding minimal hitting sets for justifications and the kernel base contraction functions based on minimal incisions (e.g. with Lemma 1). Furthermore, we also showed how to build kernel base contraction functions (and thus, justification-based repairs) that also satisfy relevance.

5   Generalisations of Classical Operations

Both in classical repairs and in contraction operations, sentences will either be kept or be removed altogether, as shown in the following example:

Example 2. [24,25] Consider the following DL knowledge base, which states that Cleopatra has a son and a daughter:

- Concepts: Person, Man and Woman.
- Role: hasChild.
- Individuals (assumed different): cleopatra, c_1 and c_2.
- TBox axioms: Man $\sqsubseteq$ Person, Woman $\sqsubseteq$ Person.
- ABox axioms: c_1 : Man, c_2 : Woman, cleopatra : Woman, (cleopatra, c_1) : hasChild, (cleopatra, c_2) : hasChild.

If we want to contract by cleopatra : ∃hasChild.Man (i.e. if we do not want "Cleopatra has a son" to be entailed by our ontology), then both a classical repair and a base contraction would necessarily remove either our belief that c_1 is a man or our belief that c_1 is Cleopatra's child (or both). In the first case, the resulting ontology has no information left about the classes that the individual c_1 belongs to, which means that the fact that c_1 is a person is no longer known. In contrast, if we apply a pseudo-contraction or a gentle repair, it is possible to *replace* the sentence c_1 : Man with the weaker sentence c_1 : Person, which is enough to prevent the entailment of cleopatra : ∃hasChild.Man.

In Belief Change, a pseudo-contraction is a generalisation of contraction that satisfies the following postulate instead of inclusion:

- **logical inclusion:** $\mathrm{Cn}(B - \alpha) \subseteq \mathrm{Cn}(B)$

Logical inclusion lifts the requirement that a sentence must be either kept or removed when contracting a belief base and leaves room for adding new sentences as long as they already followed from the original base.

Partial meet and kernel pseudo-contractions will be denoted, respectively, by $\text{PMPC}_g(B, \alpha)$ and $\text{KPC}_f(B, \alpha)$, and a generic pseudo-contraction will be represented by $\text{PC}(B, \alpha)$.

Definition 18 (Pseudo-contraction [11]). *A pseudo-contraction is an operation PC that satisfies success and logical inclusion.*

Recently, a very similar idea was introduced by Baader et al. [5] in Ontology Repair, that of weakening axioms instead of deleting them completely.

Definition 19 (Weakening [5]). *A sentence α_1 is weaker than a sentence α_2 if $\text{Cn}(\{\alpha_1\}) \subset \text{Cn}(\{\alpha_2\})$.*

In a gentle repair, one can either remove an axiom or substitute it with a weaker version, retaining part of the information.

Definition 20 (Gentle Repair).[3] *Let $\mathcal{O} = \langle \mathcal{O}_s, \mathcal{O}_r \rangle$ be an ontology and let α be a sentence entailed by $\mathcal{O}$ but not by $\mathcal{O}_s$. An ontology $\mathcal{O}'$ is a gentle repair of $\mathcal{O}$ with respect to α if $\text{Cn}(\mathcal{O}_s \cup \mathcal{O}') \subseteq \text{Cn}(\mathcal{O}) \setminus \{\alpha\}$ and, for every $\varphi \in \mathcal{O}'$, either $\varphi \in \mathcal{O}_r$ or φ is weaker than ψ for some $\psi \in \mathcal{O}_r \setminus \mathcal{O}'$.*

Algorithm 2 is very similar to Algorithm 1, but for every sentence in the hitting set H, a weaker sentence is used to replace it.

Algorithm 2. Gentle repair algorithm

Input: An ontology $\mathcal{O} = \langle \mathcal{O}_s, \mathcal{O}_r \rangle$ and a formula α
Output: A gentle repair $\mathcal{O}'$ of $\mathcal{O}$ w.r.t. α

1 **Function** *GentleRepair($\mathcal{O}, \alpha$)*
2     $\mathcal{O}' \leftarrow \mathcal{O}_r$
3   **while** $\alpha \in \text{Cn}(\mathcal{O}_s \cup \mathcal{O}')$ **do**
4        $\mathcal{J} \leftarrow$ Justifications($\langle \mathcal{O}_s, \mathcal{O}' \rangle, \alpha$)
5        $H \leftarrow$ MinimalHittingSet($\mathcal{J}$)
6   **for** $\beta \in H$ **do**
7           $\beta' \leftarrow$ GetWeakerSentence($\mathcal{O}_s, \mathcal{J}, \alpha, \beta$)
8           $\mathcal{O}' \leftarrow (\mathcal{O}' \setminus \{\beta\}) \cup \{\beta'\}$
9   **return** $\mathcal{O}'$

A modified version of Algorithm 2 was proposed by Baader et al. [5] where instead of weakening each element of the minimal hitting set, only a single formula in each justification needs to be changed:

---

[3] In [5], the concept of gentle repair has not been formally defined, only explained in intuitive terms. This is the definition which will be used here.

Algorithm 3. Modified gentle repair algorithm

Input: An ontology $\mathcal{O} = \langle \mathcal{O}_s, \mathcal{O}_r \rangle$ and a formula α
Output: A gentle repair $\mathcal{O}'$ of $\mathcal{O}$ w.r.t. α

1 **Function** *ModifiedGentleRepair($\mathcal{O}$, α)*
2      $\mathcal{O}' \leftarrow \mathcal{O}_r$
3   **while** $\alpha \in \mathrm{Cn}(\mathcal{O}_s \cup \mathcal{O}')$ **do**
4          $J \leftarrow$ OneJustification($\langle \mathcal{O}_s, \mathcal{O}' \rangle$, α)
5          $\beta' \leftarrow$ GetWeakerSentence($\mathcal{O}_s$, $\{J\}$, α, β)
6          $\mathcal{O}' \leftarrow (\mathcal{O}' \setminus \{\beta\}) \cup \{\beta'\}$
7   **return** $\mathcal{O}'$

Baader et al. [5] remark that as the unmodified version requires the computation of minimal hitting sets, which is expensive, the modified version has an important advantage, even though both are prone to consume exponential time on $|\mathcal{O}_r|$.

Algorithms 2 and 3 require a function GetWeakerSentence which, given $\mathcal{O}_s$, $\mathcal{J}$, α and β, returns a sentence β' weaker than β such that $\alpha \notin \mathrm{Cn}(\mathcal{O}_s \cup (J \setminus \{\beta\}) \cup \{\beta'\})$ for every $J \in \mathcal{J}$ such that $\beta \in J$. Such a β' always exists: a tautology satisfies the requirements. However, replacing a sentence with a tautology is logically equivalent to removing it, which means that a classical repair is obtained if this function only returns tautologies. Algorithm 3 needs a function OneJustification($\mathcal{O}$, α) that computes an element J of Just($\mathcal{O}, \alpha$).

Theorem 3 [5]. *Algorithms 2 and 3 always stop after a finite number of iterations (of lines 3–8 and 3–6, respectively), which is at most exponential in $|\mathcal{O}_r|$, regardless of the DL of $\mathcal{O} = \langle \mathcal{O}_s, \mathcal{O}_r \rangle$.*

We can now proceed to analyse the relation between gentle repairs and pseudo-contractions.

Proposition 7 (Gentle Repair $\implies$ Pseudo-contraction). *Let* GRep *be an operation that yields a gentle repair. Define the operation* $\mathrm{PC}_{\mathrm{GRep}}$ *as*

$$\mathrm{PC}_{\mathrm{GRep}}(B, \alpha) = \begin{cases} \mathrm{GRep}(\langle \varnothing, B \rangle, \alpha), & \text{if } B \models \alpha; \\ B, & \text{otherwise.} \end{cases}$$

Then, $\mathrm{PC}_{\mathrm{GRep}}$ *is a pseudo-contraction operation.*

The result above is similar to Proposition 3 and follows from Definition 20, which guarantees that $\mathrm{PC}_{\mathrm{GRep}}$ satisfies success and logical inclusion.

For the other direction (pseudo-contractions as gentle repairs), we will introduce the notion of two-place pseudo-contractions. Regular pseudo-contractions allow the result to contain some weakened versions of formulas that were originally in the belief base. This can be achieved by applying a partial meet operation on a "weak closure" of the belief base (the original set plus *some* of its classical consequences) [25]. However, as this weak closure does not depend on the sentence that is being contracted, we are not able to add only weakenings of formulas that would be removed. Two-place pseudo-contractions employ a consequence operator that depends on both the set of beliefs and the input sentence. Before defining it, we need the following concept:

Definition 21 (Extension of a selection function [24]). *Let g be a selection function for B, and let $B \subseteq B^*$. We say that g' is an extension of g to B^* if g' is such that for every $X \in g(\mathrm{MaxNon}(B, \alpha))$ there is a $Y \in g'(\mathrm{MaxNon}(B^*, \alpha))$ such that $X \subseteq Y$.*

Now we can define the two-place pseudo-contraction:

Definition 22 (Two-place pseudo-contraction [24,25]). *Let $\alpha \in \mathfrak{L}$, Cn' be a consequence relation, g be a selection function for $B \subseteq \mathfrak{L}$, $\mathrm{Cn}^*(B, \alpha) = \mathrm{Cn'}(B \setminus \bigcap g(\mathrm{MaxNon}(B, \alpha))) \cup B$ and g' be an extension of g to $\mathrm{Cn}^*(B, \alpha)$. The two-place pseudo-contraction of B by α, denoted by $\mathrm{TPPC}(B, \alpha)$, is the set $\bigcap g'(\mathrm{MaxNon}(\mathrm{Cn}^*(B, \alpha), \alpha))$.*

Notice that $B \subseteq \mathrm{Cn}^*(B, \alpha)$ for all α, and so g' can be an extension of g. The construction above was proposed by Ribeiro and Wassermann [24] (and generalised by Santos et al. [25]) as a way to weaken sentences in belief base pseudo-contractions, instead of removing them.

Consider one more property for selection functions:

Definition 23. *We say that a selection function g for B satisfies A-inclusion, with $A \subseteq B$ if, for all α, if $\alpha \notin \mathrm{Cn}(A)$ and $X \in g(\mathrm{MaxNon}(B, \alpha))$, then $A \subseteq X$.*

In the following, take $B = \mathcal{O}_s \cup \mathcal{O}_r$ (we might abuse notation and say that $B = \mathcal{O}$).

Lemma 3. *Consider a two-place contraction as in Definition 22. Then, for all $\varphi \in B \setminus \bigcap g(\mathrm{MaxNon}(B, \alpha))$, there is an $X \in g'(\mathrm{MaxNon}(\mathrm{Cn}^*(B, \alpha), \alpha))$ such that $\varphi \notin X$.*

Proof. Assume $\varphi \in B \setminus \bigcap g(\mathrm{MaxNon}(B, \alpha))$. We need to show that there is some $X \in g'(\mathrm{MaxNon}(\mathrm{Cn}^*(B, \alpha), \alpha))$ such that $\varphi \notin X$. Since $\varphi \notin \bigcap g(\mathrm{MaxNon}(B, \alpha))$, there is a $Y \in g(\mathrm{MaxNon}(B, \alpha))$ such that $\varphi \notin Y$. So $Y \subseteq B$, $\alpha \notin \mathrm{Cn}(Y)$ and for any $Y' \subseteq B$ such that $Y \subset Y'$, $\alpha \in \mathrm{Cn}(Y')$. So, since g' is an extension of g to $\mathrm{Cn}^*(B, \alpha)$, there is an $X \in g'(\mathrm{MaxNon}(\mathrm{Cn}^*(B, \alpha), \alpha))$ such that $Y \subseteq X$. And since $\varphi \notin Y$, $\varphi \in B$ and for any $Y' \subseteq B$ such that $Y \subset Y'$ we have $\alpha \in \mathrm{Cn}(Y')$, we conclude that $\varphi \notin X$. $\square$

Now we can show under which conditions a TPPC yields a gentle repair.

Definition 24. *We say a consequence operator Con is strictly weakening if $\varphi \in \mathrm{Con}(B)$ iff $\varphi \in B$ or $\mathrm{Con}(\{\varphi\}) \subset \mathrm{Con}(\{\psi\})$, for some $\psi \in B$.*

Proposition 8 (Two-place Pseudo-Contraction $\implies$ Gentle Repair).
Let $\mathrm{TPPC}(B, \alpha)$ *and* Cn^* *be as in Definition 22,* Cn^* *based on a consequence relation* Cn' *that satisfies subclassicality (that is,* $\mathrm{Cn'}(X) \subseteq \mathrm{Cn}(X)$*, where* Cn *is defined as in Sect. 2), g satisfy $\mathcal{O}_s$-inclusion, and* Cn' *be monotonic and strictly weakening. If* $\alpha \notin \mathrm{Cn}(\mathcal{O}_s)$*, then* $\mathcal{O}' = \mathrm{TPPC}(B, \alpha) \setminus \mathcal{O}_s$ *is a gentle repair of* $\mathcal{O}$ *w.r.t.* α.

Proof. First we show that $\mathrm{Cn}(\mathcal{O}_s \cup \mathcal{O}') \subseteq \mathrm{Cn}(\mathcal{O})$. By subclassicality of Cn', it follows that $\mathrm{TPPC}(B, \alpha) \subseteq \mathrm{Cn}(B)$ (and we abuse notation with $\mathcal{O} = \mathcal{O}_s \cup \mathcal{O}_r = B$), so we have $\mathrm{TPPC}(B, \alpha) \subseteq \mathrm{Cn}(\mathcal{O})$, and by monotonicity and idempotence of Cn we get $\mathrm{Cn}(\mathrm{TPPC}(B, \alpha)) \subseteq \mathrm{Cn}(\mathcal{O})$. Now by $\mathcal{O}_s$-inclusion of g, we will have that $\mathcal{O}_s \subseteq \mathrm{TPPC}(B, \alpha)$, and therefore $\mathcal{O}_s \cup \mathcal{O}' = \mathcal{O}_s \cup (\mathrm{TPPC}(B, \alpha) \setminus \mathcal{O}_s) = \mathrm{TPPC}(B, \alpha)$, so $\mathrm{Cn}(\mathcal{O}_s \cup \mathcal{O}') \subseteq \mathrm{Cn}(\mathcal{O})$.

Now we show that $\alpha \notin \mathrm{Cn}(\mathcal{O}_s \cup \mathcal{O}')$. We already found that $\mathcal{O}_s \cup \mathcal{O}' = \mathrm{TPPC}(B, \alpha)$. By success of TPPC, it follows that $\alpha \notin \mathrm{Cn}(\mathrm{TPPC}(B, \alpha))$.

Now we have to show that for all $\varphi \in \mathcal{O}'$, either $\varphi \in \mathcal{O}_r$ or $\mathrm{Cn}(\{\varphi\}) \subset \mathrm{Cn}(\{\psi\})$, for some $\psi \in \mathcal{O}_r \setminus \mathcal{O}'$. Take some $\varphi \in \mathcal{O}'$. If $\varphi \in \mathcal{O}_r$ we are done, so let us assume that $\varphi \notin \mathcal{O}_r$. So $\varphi \in \mathrm{TPPC}(B, \alpha)$, but $\mathrm{TPPC}(B, \alpha) \subseteq \mathrm{Cn}^*(B, \alpha) = \mathrm{Cn}'$ $(B \setminus \bigcap g(\mathrm{MaxNon}(B, \alpha))) \cup B$. But $\varphi \notin B$, so $\varphi \in \mathrm{Cn}'(B \setminus \bigcap g(\mathrm{MaxNon}(B, \alpha)))$. Since Cn' is strictly weakening, either $\varphi \in B \setminus \bigcap g(\mathrm{MaxNon}(B, \alpha))$ or there is a $\psi \in B \setminus \bigcap g(\mathrm{MaxNon}(B, \alpha))$ such that $\mathrm{Cn}(\{\varphi\}) \subset \mathrm{Cn}(\{\psi\})$. Since g has $\mathcal{O}_s$-inclusion, $\mathcal{O}_s \subseteq \bigcap g(\mathrm{MaxNon}(B, \alpha))$, so $B \setminus \bigcap g(\mathrm{MaxNon}(B, \alpha)) \subseteq \mathcal{O}_r$, and therefore $\psi \in \mathcal{O}_r$. Now it is left to show that $\psi \notin \mathcal{O}'$, i.e., $\psi \notin \mathrm{TPPC}(B, \alpha)$ or $\psi \in \mathcal{O}_s$. Since $\psi \in \mathcal{O}_r$ and $\mathcal{O}_s$ and $\mathcal{O}_r$ are assumed to be disjoint, $\psi \notin \mathcal{O}_s$. So we have to show that $\psi \notin \mathrm{TPPC}(B, \alpha)$. But $\psi \in B \setminus \bigcap g(\mathrm{MaxNon}(B, \alpha))$, which together with Lemma 3 gives us $\psi \notin \mathrm{TPPC}(B, \alpha)$. □

6   Conclusions and Future Work

In this work, we illustrate the relationship between Ontology Repair and Belief Change in general, as well as between particular constructions. Not only do we exhibit a conceptual mapping between definitions in both fields, but we also show under which conditions both approaches are equivalent.

In order to support collaboration between Belief Change and Ontology Repair specialists, we present a new notation. The main objectives are the elimination of traditional but cumbersome symbols (such as $\bot\!\!\!\bot$ for MinImp and $\bot$ for MaxNon) and avoid hiding certain parameters, problems that hinder the comprehension of texts in Belief Change.

Moreover, we proved that maxichoice base contraction (Definition 16) is equivalent to Ontology Repair approaches that produce optimal classical repairs. Additionally, we also gave a construction using the kernel based approach that yields relevance-compliant contraction functions.

We also studied two solutions to repair ontologies while avoiding removing whole sentences: gentle repairs from the Ontology Repair area and pseudo-contractions from the Belief Change area. Our results show that their similarities go beyond their motivation: pseudo-contraction and gentle repairs are two sides of the same coin.

As future work, we include the identification of mapping between Ontology Evolution approaches and the base revision operation. Furthermore, devising different forms of weakening is essential for both pseudo-contraction and gentle repair.

Acknowledgements. We would like to thank Franz for the many discussions that led to this paper and the two anonymous referees who contributed substantially to improve it.

The first author was supported by CNPq through grant 131803/2018-2. The second author was supported by FAPESP through grant 2017/04410-0.

References

1. Alchourrón, C., Gärdenfors, P., Makinson, D.: On the logic of theory change: partial meet contraction and revision functions. J. Symb. Log. **50**(2), 510–530 (1985)
2. Alchourrón, C.E., Makinson, D.: Hierarchies of regulations and their logic. In: Hilpinen, R. (ed.) New Studies in Deontic Logic. SYLI, vol. 152, pp. 125–148. Springer, Dordrecht (1981). https://doi.org/10.1007/978-94-009-8484-4_5
3. Baader, F., Calvanese, D., McGuinness, D.L., Nardi, D., Patel-Schneider, P.F. (eds.): The Description Logic Handbook: Theory, Implementation, and Applications. Cambridge University Press, Cambridge (2003)
4. Baader, F., Horrocks, I., Lutz, C., Sattler, U.: An Introduction to Description Logic. Cambridge University Press, Cambridge (2017)
5. Baader, F., Kriegel, F., Nuradiansyah, A., Peñaloza, R.: Making repairs in description logics more gentle. In: Principles of Knowledge Representation and Reasoning: Proceedings of the Sixteenth International Conference, KR 2018, Tempe, Arizona, pp. 319–328 (2018)
6. Baader, F., Peñaloza, R.: Axiom pinpointing in general tableaux. In: Olivetti, N. (ed.) TABLEAUX 2007. LNCS (LNAI), vol. 4548, pp. 11–27. Springer, Heidelberg (2007). https://doi.org/10.1007/978-3-540-73099-6_4
7. Du, J., Qi, G.: Tractable computation of representative abox repairs in description logic ontologies. In: Zhang, S., Wirsing, M., Zhang, Z. (eds.) KSEM 2015. LNCS (LNAI), vol. 9403, pp. 28–39. Springer, Cham (2015). https://doi.org/10.1007/978-3-319-25159-2_3
8. Falappa, M.A., Fermé, E.L., Kern-Isberner, G.: On the logic of theory change: Relations between incision and selection functions. In: Brewka, G., Coradeschi, S., Perini, A., Traverso, P. (eds.) Proceedings of the 17th European Conference on Artificial Intelligence (ECAI 2006) Including Prestigious Applications of Intelligent Systems (PAIS 2006), Riva del Garda, Italy, 29 August–1 September 2006. Frontiers in Artificial Intelligence and Applications, vol. 141, pp. 402–406. IOS Press (2006)
9. Flouris, G., Plexousakis, D., Antoniou, G.: On applying the AGM theory to DLs and OWL. In: Gil, Y., Motta, E., Benjamins, V.R., Musen, M.A. (eds.) ISWC 2005. LNCS, vol. 3729, pp. 216–231. Springer, Heidelberg (2005). https://doi.org/10.1007/11574620_18
10. Hansson, S.O.: In defense of base contraction. Synthese **91**, 239–245 (1992)
11. Hansson, S.O.: Changes of disjunctively closed bases. J. Log. Lang. Inf. **2**(4), 255–284 (1993). https://doi.org/10.1007/BF01181682

12. Hansson, S.O.: Reversing the levi identity. J. Philos. Log. **22**(6), 637–669 (1993). https://doi.org/10.1007/BF01054039
13. Hansson, S.O.: Kernel contraction. J. Symb. Log. (1994)
14. Hansson, S.O., Wassermann, R.: Local change. Studia Logica **70**(1), 49–76 (2002). https://doi.org/10.1023/A:1014654208944
15. Horridge, M.: Justification based explanation in ontologies. Ph.D. thesis, University of Manchester (2011)
16. Ji, Q., Haase, P., Qi, G., Hitzler, P., Stadtmüller, S.: RaDON—repair and diagnosis in ontology networks. In: Aroyo, L., et al. (eds.) ESWC 2009. LNCS, vol. 5554, pp. 863–867. Springer, Heidelberg (2009). https://doi.org/10.1007/978-3-642-02121-3_71
17. Ji, Q., Qi, G., Haase, P.: A relevance-directed algorithm for finding justifications of DL entailments. In: Gómez-Pérez, A., Yu, Y., Ding, Y. (eds.) ASWC 2009. LNCS, vol. 5926, pp. 306–320. Springer, Heidelberg (2009). https://doi.org/10.1007/978-3-642-10871-6_21
18. Kalyanpur, A.: Debugging and repair of OWL ontologies. Ph.D. thesis, University of Maryland at College Park, College Park, MD, USA (2006)
19. Kalyanpur, A., Parsia, B., Horridge, M., Sirin, E.: Finding all justifications of OWL DL entailments. In: Aberer, K., et al. (eds.) ASWC/ISWC -2007. LNCS, vol. 4825, pp. 267–280. Springer, Heidelberg (2007). https://doi.org/10.1007/978-3-540-76298-0_20
20. Lembo, D., Lenzerini, M., Rosati, R., Ruzzi, M., Savo, D.F.: Inconsistency-tolerant semantics for description logics. In: Hitzler, P., Lukasiewicz, T. (eds.) RR 2010. LNCS, vol. 6333, pp. 103–117. Springer, Heidelberg (2010). https://doi.org/10.1007/978-3-642-15918-3_9
21. Qi, G., Haase, P., Huang, Z., Ji, Q., Pan, J.Z., Völker, J.: A kernel revision operator for terminologies - algorithms and evaluation. In: Sheth, A., et al. (eds.) ISWC 2008. LNCS, vol. 5318, pp. 419–434. Springer, Heidelberg (2008). https://doi.org/10.1007/978-3-540-88564-1_27
22. Reiter, R.: A theory of diagnosis from first principles. Artif. Intell. **32**(1), 57–95 (1987). https://doi.org/10.1016/0004-3702(87)90062-2
23. Ribeiro, M.M., Wassermann, R.: Base revision for ontology debugging. J. Log. Comput. **19**(5), 721–743 (2009). https://doi.org/10.1093/logcom/exn048
24. Ribeiro, M.M., Wassermann, R.: Degrees of recovery and inclusion in belief base dynamics. In: 12th International Workshop on Non-Monotonic Reasoning (NMR 2008) (2008)
25. Santos, Y.D., Matos, V.B., Ribeiro, M.M., Wassermann, R.: Partial meet pseudo-contractions. Int. J. Approx. Reason. **103**, 11–27 (2018). https://doi.org/10.1016/j.ijar.2018.08.006
26. Schlobach, S.: Debugging and semantic clarification by pinpointing. In: Gómez-Pérez, A., Euzenat, J. (eds.) ESWC 2005. LNCS, vol. 3532, pp. 226–240. Springer, Heidelberg (2005). https://doi.org/10.1007/11431053_16
27. Schlobach, S., Cornet, R.: Non-standard reasoning services for the debugging of description logic terminologies. In: Proceedings of the 18th International Joint Conference on Artificial Intelligence (IJCAI 2003), Acapulco, Mexico, 9–15 August 2003, pp. 355–362. Morgan Kaufmann (2003)
28. Wassermann, R.: An algorithm for belief revision. In: Principles of Knowledge Representation and Reasoning Proceedings of the Seventh International Conference, KR 2000, Breckenridge, Colorado, USA, 11–15 April 2000, pp. 345–352 (2000)
29. Wassermann, R.: On AGM for non-classical logics. J. Philos. Log. **40**(2), 271–294 (2011). https://doi.org/10.1007/s10992-011-9178-2

FunDL
A Family of Feature-Based Description Logics, with Applications in Querying Structured Data Sources

Stephanie McIntyre, David Toman$^{(\boxtimes)}$, and Grant Weddell

Cheriton School of Computer Science, University of Waterloo, Waterloo, Canada
{srmcinty,david,gweddell}@uwaterloo.ca

Abstract. Feature-based description logics replace the notion of *roles*, interpreted as binary relations, with *features*, interpreted as unary functions. Another notable feature of these logics is their use of path functional dependencies that allow for complex identification constraints to be formulated. The use of features and path functional dependencies makes the logics particularly well suited for capturing and integrating data sources conforming to an underlying object-relational schema that include a variety of common integrity constraints. We first survey expressive variants of feature logics, including the boundaries of decidability. We then survey a restricted tractable family of feature logics suited to query answering, and study the limits of tractability of reasoning.

1 Introduction

We survey the work we have done on developing FunDL, a family of description logics that can be used to address a number of problems in querying structured data sources, with a particular focus on data sources that have an underlying object-relational schema. All member dialects of this family have two properties in common. First, each is *feature based*: the usual notion of *roles* in description logic that are interpreted as binary relations is replaced with the notion of features that are interpreted as unary functions. We have found features to be a better fit with object-relational schema, e.g., for capturing the ubiquitous notion of *attributes*. And second, each dialect includes a concept constructor for capturing a variety of equality generating dependencies: so-called *path functional dependencies* (PFDs) that generalize the notions of primary keys, uniqueness constraints and functional dependencies that are again ubiquitous in object-relational schema. PFDs also ensure member dialects do not forgo the ability to capture roles or indeed n-ary relations in general. This can be accomplished by the simple expedient of reification via features, and then by employing PFDs to ensure a set semantics for reified relations. Indeed, the dialect $\mathcal{DLFD}$, introduced in the first part of our survey, can capture very expressive role-based dialects of description logics, including dialects with so-called qualified number restrictions, inverse roles, role hierarchies, and so on [29].

© Springer Nature Switzerland AG 2019
C. Lutz et al. (Eds.): Baader Festschrift, LNCS 11560, pp. 404–430, 2019.
https://doi.org/10.1007/978-3-030-22102-7_19

Our survey consists of three general parts, with the first two parts focusing on the problem of logical implication for FunDL dialects with EXPTIME and PTIME complexity, respectively, and in which the dialects assume features are interpreted as *total* functions. In the third part of our survey, we begin with a review of more recent work on how such dialects may be adapted to support features that are instead *partial* functions. We then consider how *role hierarchies* can be captured as concept hierarchies in which the concepts are introduced as reifications of roles. Part three concludes with a review of other reasoning problems, in particular, on knowledge base consistency for FunDL dialects, and on query answering for dialects surveyed in part two.

We begin in the next section with a general introduction to FunDL: what features are, what the various concept constructors are, basic notational conventions, *the grammar protocol we follow to define the various dialects*, and so on. Our survey concludes with a brief overview of related work.

2   Background and Definitions

Here, we define a nameless all inclusive member dialect of the FunDL family for the purpose of introducing a space of concept constructors that we then use for defining all remaining dialects in our survey. We also say how a theory is defined by a so-called *terminology* (or TBox) consisting of a finite set of sentences expressing *inclusion dependencies*, and introduce the problem of logical implication of an inclusion dependency by a TBox. Indeed, we focus exclusively on the problem of logical implication throughout the first two parts of our survey.

Definition 1 (Feature-Based DLs). Let F and PC be sets of feature names and primitive concept names, respectively. A *path expression* is defined by the grammar $\mathsf{Pf} ::= f.\mathsf{Pf} \mid id$, for $f \in \mathsf{F}$. We define derived *concept descriptions* by the grammar on the left-hand-side of Fig. 1.

An *inclusion dependency* $\mathcal{C}$ is an expression of the form $C_1 \sqsubseteq C_2$. A *terminology* (TBox) $\mathcal{T}$ consists of a finite set of inclusion dependencies. A *posed question* $\mathcal{Q}$ is a single inclusion dependency.

The *semantics* of expressions is defined with respect to a structure $\mathcal{I} = (\triangle, \cdot^{\mathcal{I}})$, where $\triangle$ is a domain of objects or entities and $(\cdot)^{\mathcal{I}}$ an interpretation function that fixes the interpretations of primitive concept names A to be subsets of $\triangle$ and feature names f to be total functions $(f)^{\mathcal{I}} : \triangle \rightarrow \triangle$. The interpretation is extended to path expressions, $(id)^{\mathcal{I}} = \lambda x.x$, $(f.\mathsf{Pf})^{\mathcal{I}} = (\mathsf{Pf})^{\mathcal{I}} \circ (f)^{\mathcal{I}}$ and derived concept descriptions C as defined in the centre column of Fig. 1.

An interpretation $\mathcal{I}$ *satisfies an inclusion dependency* $C_1 \sqsubseteq C_2$ if $(C_1)^{\mathcal{I}} \subseteq (C_2)^{\mathcal{I}}$ and is a *model of* $\mathcal{T}$ ($\mathcal{I} \models \mathcal{T}$) if it satisfies all inclusion dependencies in $\mathcal{T}$. The *logical implication problem* asks if $\mathcal{T} \models \mathcal{Q}$ holds, that is, if $\mathcal{Q}$ is satisfied in all models of $\mathcal{T}$. □

We shall see that the logical implication problem for this logic is undecidable for a variety of reasons. For example, the value restriction, top and same-as concept constructors are all that are needed to encode the uniform word problem [24].

Syntax	Semantics: Defn of $(\cdot)^{\mathcal{I}}$	
$C ::= A$	$(A)^{\mathcal{I}} \subseteq \triangle$	(primitive concept; $A \in \mathsf{PC}$)
$\mid C_1 \sqcap C_2$	$(C_1)^{\mathcal{I}} \cap (C_2)^{\mathcal{I}}$	(conjunction)
$\mid \forall \mathsf{Pf}.C$	$\{x \mid (\mathsf{Pf})^{\mathcal{I}}(x) \in (C)^{\mathcal{I}}\}$	(value restriction)
$\mid C_1 \sqcup C_2$	$(C_1)^{\mathcal{I}} \cup (C_2)^{\mathcal{I}}$	(disjunction)
$\mid \neg C$	$\triangle \setminus (C)^{\mathcal{I}}$	(negation)

$$\mid C : \mathsf{Pf}_1, ..., \mathsf{Pf}_k \to \mathsf{Pf} \quad \{x \mid \forall y \in (C)^{\mathcal{I}}. \textstyle\bigwedge_{i=1}^{k}(\mathsf{Pf}_i)^{\mathcal{I}}(x) = (\mathsf{Pf}_i)^{\mathcal{I}}(y) \qquad \text{(PFD)}$$
$$\to (\mathsf{Pf})^{\mathcal{I}}(x) = (\mathsf{Pf})^{\mathcal{I}}(y)\}$$

$\mid \exists f^{-1}$	$\{x \mid \exists y \in \triangle : (f)^{\mathcal{I}}(y) = x\}$	(feature inverse)
$\mid \exists f^{-1}.C$	$\{x \mid \exists y \in (C)^{\mathcal{I}} : (f)^{\mathcal{I}}(y) = x\}$	(qualified feature inverse)
$\mid \top$	$\triangle$	(top)
$\mid \bot$	$\emptyset$	(bottom)
$\mid (\mathsf{Pf}_1 = \mathsf{Pf}_2)$	$\{x \mid (\mathsf{Pf}_1)^{\mathcal{I}}(x) = (\mathsf{Pf}_2)^{\mathcal{I}}(x)\}$	(same-as)

Fig. 1. Concept constructors in feature-based description logics.

Thus, each dialect of the FunDL family in our survey will correspond to some fragment of this logic. Grammars defining a dialect use the non-terminals C and D to characterize concept constructors permitted on left-hand-sides and right-hand-sides of inclusion dependencies occurring in a TBox, respectively, and the non-terminal E to characterize concept constructors permitted in posed questions. We also assume, when an explicit definition of non-terminal D (resp. E) is missing, that D concept descriptions align with C concept descriptions (resp. E concept descriptions align with D concept descriptions).

To see how FunDL dialects are useful in capturing structured data sources, consider a visualization of a hypothetical object-relational university schema in Fig. 2. Here, nodes are classes, labelled directed edges are attributes, thick edges denote inheritance, and underlined attributes denote primary keys. Introducing a primitive concept and a feature for each class and attribute then enables attribute typing, inheritance, primary keys and a variety of other data dependencies to be captured as inclusion dependencies in a university TBox:

1. (*disjoint classes*) PERSON $\sqsubseteq \neg$DEPT,
2. (*attribute typing*) PERSON $\sqsubseteq \forall name$.STRING,
3. (*unary primary key*) PERSON $\sqsubseteq$ PERSON : $name \to id$,
4. (*disjoint attribute values*) PERSON $\sqsubseteq$ DEPT : $name \to id$,
5. (*inheritance*) PROF $\sqsubseteq$ PERSON,
6. (*views*) $\forall reports$.CHAIR $\sqsubseteq$ PROF,
7. (*mandatory participation*) $\exists head^{-1} \sqsubseteq$ CHAIR,
8. (*binary primary key*) CLASS $\sqsubseteq$ CLASS : $dept, num \to id$, and
9. (*cover*) PERSON $\sqsubseteq$ (STUDENT $\sqcup$ PROF).

Allowing path expressions to occur in PFD concepts turns out to be quite useful in capturing additional varieties of equality generating dependencies, as in the

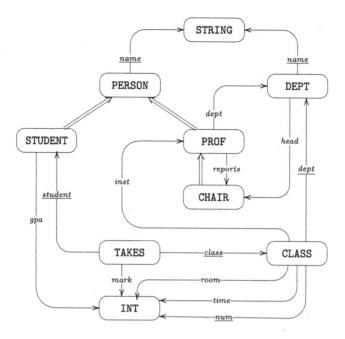

Fig. 2. An object-relational schema.

following:

$$\text{TAKES} \sqsubseteq \text{TAKES} : student, \, class.room, \, class.time \rightarrow class.$$

This inclusion dependency expresses a constraint induced by the interaction of time and space, that *no student can take two different classes in the same room at the same time* or, to paraphrase, that *no pair of classes with at least one student taking them can be in the same room at the same time.* The second reading illustrates how so-called *identification constraints* in DL-Lite dialects can also be captured [11].

In the third part of our survey, we review work on how features may be interpreted as partial functions. This leads to the addition of the concept constructor $\exists f$ for capturing domain elements for which feature f is defined. Consequently, it becomes possible to say, e.g., that a DEPT does not have a *gpa* by adding the inclusion dependency

$$\text{DEPT} \sqsubseteq \neg \exists gpa$$

to the university TBox.

Note that any logical implication problem for the university TBox defined thus far can be solved by appeal to one of the expressive FunDL dialects, and, notwithstanding *cover* constraints, can be solved by one of the tractable dialects in PTIME. An ability to do this has many applications in information systems technology. For example, early work on FunDL has shown how to reduce the problem of determining when a SQL query can be reformulated without mentioning the DISTINCT keyword to a logical consequence problem [20]. More recent applications allow one to resolve fundamental issues in reasoning about identity in conceptual modelling and SQL programming [6], and in ontology-based data access [7,26].

2.1   Ackerman Decision Problems

Our complexity reductions are tied to the classical *Ackermann case of the decision problem* [1].

Definition 2 (Monadic Ackerman Formulae). Let P_i be monadic predicate symbols and x, y_i, z_i variables. A *monadic first-order formula in the Ackermann class* is a formula of the form $\exists z_1 \ldots \exists z_k \forall x \exists y_1 \ldots \exists y_l . \varphi$ where φ is a quantifier-free formula over the symbols P_i.   □

Every formula with the Ackermann prefix can be converted to *Skolem normal form* by replacing variables z_i by Skolem constants and y_i by unary Skolem functions not appearing in the original formula. This, together with standard Boolean equivalences, yields a finite set of universally-quantified clauses containing at most one variable (x).

Proposition 3 ([16]). The Ackermann decision problem is complete for EXP-TIME.

The lower bound holds even for the Horn fragment of the decision problem called Datalog$_{nS}$ [15]. A Datalog$_{nS}$ program is a finite set of *definite Horn* Datalog$_{nS}$ clauses. A *recognition problem* for a Datalog$_{nS}$ program Π and a ground atom Q is to determine if Q is true in all models of Π (i.e., if $\Pi \cup \{\neg Q\}$ is unsatisfiable).

3   Expressive FunDL Dialects

In this first part of our survey, we consider the logical implication problem for an expressive Boolean complete dialect with value restrictions on features. We begin by presenting a lower bound for a fragment of this dialect and then follow with upper bounds. We subsequently consider extensions to the dialect that admit additional concept constructors, namely PFDs and inverse features.

3.1   Logical Implication in $\mathcal{DLF}$

The dialect $\mathcal{DLF}_0$ of FunDL is defined by the following grammar (and recall our protocol whereby right-hand-sides of inclusion dependencies and posed questions are also defined by non-terminal C):

$$C ::= A \mid C_1 \sqcap C_2 \mid \forall f.C$$

Observe that $\mathcal{DLF}_0$ is a Horn fragment that only allows primitive concepts, conjunctions and value restrictions. We show that every Datalog$_{nS}$ recognition problem can be simulated by a $\mathcal{DLF}_0$ implication problem [29]. For this reduction, each monadic predicate symbol is assumed to also qualify as a primitive concept name in $\mathcal{DLF}_0$. Given an instance of a Datalog$_{nS}$ recognition problem in the form of a Datalog$_{nS}$ program Π and a ground goal atom $G = P(\overline{\mathsf{Pf}}(0))$, we construct an implication problem for $\mathcal{DLF}_0$ as follows: in Π,

$$
\begin{aligned}
\mathcal{T}_\Pi \ &= \{\forall \mathsf{Pf}'_1 . Q'_1 \sqcap \ldots \sqcap \forall \mathsf{Pf}'_k . Q'_k \sqsubseteq \forall \mathsf{Pf}' . P' : \\
&\quad P'(\overline{\mathsf{Pf}}'(x)) \leftarrow Q'_1(\overline{\mathsf{Pf}}'_1(x)), \ldots, Q'_k(\overline{\mathsf{Pf}}'_k(x)) \in \Pi\}, \\
\mathcal{Q}_{\Pi,G} &= \forall \mathsf{Pf}_1 . Q_1 \sqcap \cdots \sqcap \forall \mathsf{Pf}_k . Q_k \sqsubseteq \forall \mathsf{Pf} . P,
\end{aligned}
$$

where the $\overline{\mathsf{Pf}}(x)$ terms in Datalog$_{nS}$ naturally correspond to path functions Pf in $\mathcal{DLF}_0$, and where the posed question $\mathcal{Q}_{\Pi,G}$ is formed from ground facts $Q_i(\overline{\mathsf{Pf}}_i(0)) \in \Pi$, and the ground goal atom $G = P(\overline{\mathsf{Pf}}(0))$.

Theorem 4 ([30]). Let Π be a Datalog$_{nS}$ program and G a ground atom. Then

$$
\Pi \models G \iff \mathcal{T}_\Pi \models \mathcal{Q}_{\Pi,G}.
$$

For the reduction to work, one needs two features. (Unlike the case with $\mathcal{ALC}$ style logics, the problem becomes PSPACE-complete with one feature.) This result was later used to show EXPTIME-hardness for $\mathcal{FL}_0$ [3].

We now show a matching upper bound for the Boolean complete dialect with value restrictions, as defined by the following:

$$
C ::= A \mid C_1 \sqcap C_2 \mid C_1 \sqcup C_2 \mid \forall f . C \mid \neg C
$$

We first show how the *semantics* of $\mathcal{DLF}$ constructors can be captured by Ackermann formulae: let C, C_1, and C_2 range over concept descriptions and f over attribute names. We introduce a unary predicate subscripted by a description that simulates that description in our reduction:

$$
\begin{aligned}
&\forall x. (P_C(x) \lor P_{\neg C}(x)), \forall x. \neg(P_C(x) \land P_{\neg C}(x)) \\
&\forall x. P_{C_1 \sqcap C_2}(x) \leftrightarrow (P_{C_1}(x) \land P_{C_2}(x)) \\
&\forall x. P_{C_1 \sqcup C_2}(x) \leftrightarrow (P_{C_1}(x) \lor P_{C_2}(x)) \\
&\forall x. P_{\forall f . C}(x) \leftrightarrow P_C(f(x))
\end{aligned} \tag{$*$}
$$

To complete the translation of a $\mathcal{DLF}$ implication problem $\mathcal{T} \models \mathcal{Q}$, for $\mathcal{Q}$ of the form $C \sqsubseteq D$, what remains is the translation of the inclusion dependencies in $\mathcal{T} \cup \{\mathcal{Q}\}$:

- $\Phi_{\mathcal{DLF}} = \bigwedge_{\varphi \in \mathrm{Semantics}(\mathcal{T}, \mathcal{Q})} \varphi$,
- $\Phi_\mathcal{T} = \bigwedge_{C' \sqsubseteq D' \in \mathcal{T}} \forall x. P_{C'}(x) \rightarrow P_{D'}(x)$, and
- $\Phi_\mathcal{C} = P_C(0) \land P_{\neg D}(0)$ (a Skolemized negation of the posed question $\mathcal{Q}$),

where $\mathrm{Semantics}(\mathcal{T}, \mathcal{Q})$ is the set of all formulae $(*)$ whose subscripts range over concepts and subconcepts that appear in $\mathcal{T} \cup \{\mathcal{Q}\}$.

Theorem 5 ([30]). Let $\mathcal{T}$ and $\mathcal{Q} = C \sqsubseteq D$ be a terminology and inclusion dependency in $\mathcal{DLF}$, respectively. Then $\mathcal{T} \models \mathcal{C}$ iff $\Phi_{\mathcal{DLF}} \land \Phi_\mathcal{T} \land \Phi_\mathcal{Q}$ is not satisfiable.

Theorems 4 and 5 establish a tight EXPTIME complexity bound for the $\mathcal{DLF}$ logical implication problem.

3.2   Adding Path Functional Dependencies to $\mathcal{DLF}$

Allowing unrestricted use of the PFD concept constructor leads to undecidable implication problems, as in the case of a description logic defined by the following grammar:

$$C ::= A \mid C_1 \sqcap C_2 \mid C_1 \sqcup C_2 \mid \forall f.C \mid \neg C \mid C : \mathrm{Pf}_1, ..., \mathrm{Pf}_k \to \mathrm{Pf}$$

This remains true even for very simple varieties of PFD concept constructors.

The undecidability results are based on a reduction of the unrestricted tiling problem [4,5] to the logical implication problem. The crux of the reduction is the use of the PFD constructor under negation or, equivalently, on the left-hand-side of inclusion dependencies. For example, the dependency

$$A \sqsubseteq \neg(B : f, g \to id)$$

states that, for some A object, there must be a distinct B object that agrees with this A object on features f and g, i.e., there must be a square in a model of the above inclusion dependency. Such squares can then be connected into a grid using additional PFDs and the Boolean structure of the logic in a way that enables tiling to be simulated.

This idea can be sharpened to the following three borderline cases, where *simple*, *unary* and *key* refer, respectively, to conditions in which path expressions correspond to individual features or to *id*, in which left-hand-sides of PFDs consist of a single path expression, and in which the right-hand-side is *id* [35]:

1. PFDs are simple and key, and therefore resemble

$$C : f_1, \ldots, f_k \to id$$

   (i.e., the standard notion of *relational keys*);
2. PFDs are simple and non-key, and therefore resemble

$$C : f_1, \ldots, f_k \to f$$

   (i.e., the standard notion of relational *functional dependencies*); and
3. PFDs are simple and unary, and therefore resemble either of the following:

$$C : f \to g \quad \text{or} \quad C : f \to id \,.$$

Observe that the three cases are exhaustive: the only possibility not covered happens when all PFDs have the form $C : \mathrm{Pf} \to id$, i.e., are unary and key. However, it is a straightforward exercise in this case to map logical implication problems to alternative formulations in decidable DL dialects with inverses and functional restrictions. Notably, the reductions make no use of attribute value restrictions in the first two of these cases; they rely solely on PFDs and the standard Boolean constructors.

On Regaining Decidability. It turns out that undecidability is indeed a consequence of allowing PFDs to occur within the scope of negation (and, as a consequence, all FunDL dialects disallow this possibility). Among the first expressive and decidable dialects is $\mathcal{DLFD}$, the description logic defined by the following grammar rules:

$$C ::= A \mid C_1 \sqcap C_2 \mid C_1 \sqcup C_2 \mid \forall f.C \mid \neg C \mid \top$$
$$D ::= C \mid D_1 \sqcap D_2 \mid D_1 \sqcup D_2 \mid \forall f.D \mid C : \mathsf{Pf}_1, ..., \mathsf{Pf}_k \to \mathsf{Pf}$$

Observe that PFDs must now occur on right hand sides of inclusion dependencies at either the top level or *within the scope of monotone concept constructors*. (Allowing PFDs on left hand sides is equivalent to allowing PFDs in the scope of negation: $D_1 \sqsubseteq \neg(D_2 : f \to g)$ is equivalent to $D_1 \sqcap (D_2 : f \to g) \sqsubseteq \bot$.)

To establish the complexity lower bound, we first study the problem for a subset of $\mathcal{DLFD}$ in which all inclusion dependencies are of the form

$$\top \sqsubseteq \top : \mathsf{Pf}_1, ..., \mathsf{Pf}_k \to \mathsf{Pf}.$$

An implication problem in this subset is called a *PFD membership problem*. It will simplify matters to assume that each monadic predicate symbol P in Datalog$_{nS}$ maps to a distinct feature p in $\mathcal{DLFD}$, and that each such p differs from the attributes corresponding to unary function symbols in Datalog$_{nS}$.

We proceed similarly to the $\mathcal{DLF}$ case: Let Π be an arbitrary Datalog$_{nS}$ program and $G = P(\overline{\mathsf{Pf}}(0))$ a ground atom. We construct an implication problem for $\mathcal{DLFD}$ as follows:

$$\mathcal{T}_\Pi = \{\top \sqsubseteq \top : \mathsf{Pf}'_1 \cdot p'_1, ..., \mathsf{Pf}'_k \cdot p'_k \to \mathsf{Pf}' \cdot p' :$$
$$P'(\overline{\mathsf{Pf}'}(x)) \leftarrow P'_1(\overline{\mathsf{Pf}'_1}(x)), ..., P'_k(\overline{\mathsf{Pf}'_k}(x)) \in \Pi\},$$
$$\mathcal{C}_{\Pi,G} = \top \sqsubseteq \top : \mathsf{Pf}_1 \cdot p_1, ..., \mathsf{Pf}_k \cdot p_k \to \mathsf{Pf} \cdot p,$$

where $P_1(\overline{\mathsf{Pf}}_1(0)), ..., P_k(\overline{\mathsf{Pf}}_k(0))$ are the ground facts in Π.

Theorem 6 ([30]). Let Π be an arbitrary Datalog$_{nS}$ program and $G = P(\overline{\mathsf{Pf}}(0))$ a ground atom. Then $\Pi \models G \iff \mathcal{T}_\Pi \models \mathcal{C}_{\Pi,G}$.

The reduction establishes another source of EXPTIME-hardness for our $\mathcal{DLFD}$ fragment that originates from the PFDs only.

To establish the upper bound, we reduce logical implication in $\mathcal{DLFD}$ to logical implication in $\mathcal{DLF}$. The reduction is based on the following observations:

1. If the posed question does *not* contain the PFD concept constructor then the implication problem reduces to the implication problem in $\mathcal{DLF}$ since, due to the tree model property of the logic, the PFD inclusion dependencies in the TBox are satisfied vacuously;
2. Otherwise the posed question contains a PFD, e.g., has the form

$$A \sqsubseteq B : \mathsf{Pf}_1, ..., \mathsf{Pf}_k \to \mathsf{Pf}.$$

To falsify the posed question in this case, we need to construct a model consisting of two trees respectively rooted by A and B that obey the TBox inclusion dependencies, that agree on paths $\mathsf{Pf}_1, \ldots, \mathsf{Pf}_k$ originating from the respective roots, and that disagree on Pf. Since the two trees are identical up to node labels and the agreements always equate corresponding nodes in the two trees, the model can be simulated in $\mathcal{DLF}$ by *doubling* the primitive concepts (one for simulating concept membership in each of the two trees) and by introducing an auxiliary primitive concept to simulate path agreements. This *two trees* idea can then be generalized to account for posed questions having (possibly multiple) PFDs nested in other monotone concept constructors.

The above assumes that PFDs are not nested in other constructors in a TBox; this can be achieved by a simple conservative extension of the given TBox and appropriate reformulation of the posed question [35].

Theorem 7 ([30])**.** The implication problem for $\mathcal{DLFD}$ can be reduced to an implication problem for $\mathcal{DLF}$ with only a linear increase in size.

Theorems 6 and 7 establish a tight EXPTIME complexity bound for the $\mathcal{DLFD}$ implication problem.

3.3   Adding Inverse Features

Allowing right-hand-sides of inclusion dependencies to now employ inverse features together with PFDs, as in $\mathcal{DLFDI}$, a FunDL dialect defined by the following grammar:

$$C ::= A \mid C_1 \sqcap C_2 \mid C_1 \sqcup C_2 \mid \forall f.C \mid \neg C \mid \top$$
$$D ::= C \mid D_1 \sqcap D_2 \mid D_1 \sqcup D_2 \mid \forall f.D \mid \exists f^{-1}.C \mid C : \mathsf{Pf}_1, ..., \mathsf{Pf}_k \rightarrow \mathsf{Pf}$$

leads immediately to undecidability, similarly to [14]. Again, the reduction is from the unrestricted tiling problem in which an *initial square* is generated by the constraints

$$A \sqsubseteq \exists f^{-1}.B \sqcap \exists f^{-1}.C, \quad B \sqcap C \sqsubseteq \bot, \text{ and } B \sqsubseteq C : f \rightarrow g,$$

and further inclusion dependencies then extend it to a properly tiled grid.

Theorem 8 ([31]) Logical implication for $\mathcal{DLFDI}$ is undecidable.

On Regaining Decidability with Inverses. We review two approaches to restricting either the PFD constructor or the way inverses are allowed to be qualified to regain decidability of the logical implication problem.

Prefix-restricted PFDs. The first approach syntactically restricts the PFD constructor as follows:

Definition 9 [Prefix Restricted Terminologies]. Let D : $\mathsf{Pf}.\mathsf{Pf}_1,\ldots,$ $\mathsf{Pf}.\mathsf{Pf}_k \to \mathsf{Pf}'$ be an arbitrary PFD where Pf is the maximal common prefix of the path expressions $\{\mathsf{Pf}.\mathsf{Pf}_1,\ldots,\mathsf{Pf}.\mathsf{Pf}_k\}$. The PFD is *prefix-restricted* if either Pf' is a prefix of Pf or Pf is a prefix of Pf'. ☐

This condition applies to the argument PFDs occurring in a terminology and strengthens the results in [14]. Note that, because of *accidental common prefixes*, it is not sufficient to simply require that unary PFDs resemble keys since, for example, a k-ary PFD $A_1 \sqsubseteq A_2 : f.a_1,\ldots,f.a_k \to h$ has a logical consequence $A_1 \sqsubseteq A_2 : f \to h$, thus yielding the ability to construct tiling similar to the one outlined above.

Theorem 10 ([31]). Let $\mathcal{T}$ be $\mathcal{DLFDI}$ terminology with prefix-restricted PFDs. Then the implication problem $\mathcal{T} \models \mathcal{Q}$ is decidable and EXPTIME-complete.

Coherent Terminologies. The second of our conditions for recovering decidability is to impose a coherency condition on terminologies themselves. The main advantage of this approach is that we thereby regain the ability for unrestricted use of PFDs in terminologies. The disadvantage is roughly that there is a *single use* restriction on using feature inversions in terminologies.

Definition 11 (Coherent Terminologies). A terminology $\mathcal{T}$ is *coherent* if

$$\mathcal{T} \models (\exists f^{-1}.D) \sqcap (\exists f^{-1}.E) \sqsubseteq \exists f^{-1}(D \sqcap E)$$

for all descriptions D, E that appear as subconcepts of concepts that appear in $\mathcal{T}$, or their negations. ☐

Note that we can *syntactically guarantee* that $\mathcal{T}$ is coherent by adding inclusion dependencies of the form $(\exists f^{-1}D) \sqcap (\exists f^{-1}E) \sqsubseteq \exists f^{-1}(D \sqcap E)$ to $\mathcal{T}$ for all concept descriptions D, E appearing in $\mathcal{T}$. This restriction allows us to construct interpretations of non-PFD descriptions in which objects do not have more than one f predecessor (for all $f \in \mathsf{F}$) and thus satisfy all PFDs vacuously.

By restricting logical implication problems for $\mathcal{DLFDI}$ to cases in which terminologies are coherent, it becomes possible to apply reductions to satisfiability problems for Ackerman formulae.

Theorem 12 ([31]). Let $\mathcal{T}$ be a coherent $\mathcal{DLFDI}$ terminology. Then the implication problem $\mathcal{T} \models \mathcal{C}$ is decidable and EXPTIME-complete.

Note that *unqualified inverse features of the form* $\exists f^{-1}$ immediately imply coherency. Moreover, one can qualify an f predecessor by concept C by asserting

$$A \sqsubseteq \exists f^{-1}, \quad \forall f.A \sqsubseteq C.$$

Thus, the restriction to *unqualified* inverses does not rule out cases in which qualified inverses might be useful, and avoids the problem of allowing multiple

f predecessors (that could then interact with the PFD constructs). Hence, for the remainder of the survey, we assume unqualified inverse features in FunDL dialects.

3.4   Equational Constraints

As pointed out in our introductory comments, allowing equational (same-as) concepts in TBoxes leads immediately to undecidability via a reduction from the uniform word problem [24]. Conversely, allowing equational concepts in *posed questions* extends the capabilities of the logics, in particular allowing for capturing factual assertions (called an ABox, see Sect. 5.3). To this end we introduce the FunDL dialect $\mathcal{DLFDE}$ defined as follows:

$$
\begin{aligned}
C &::= A \mid C_1 \sqcap C_2 \mid C_1 \sqcup C_2 \mid \forall f.C \mid \neg C \mid \top \\
D &::= C \mid D_1 \sqcap D_2 \mid D_1 \sqcup D_2 \mid \forall f.D \mid C : \mathsf{Pf}_1, ..., \mathsf{Pf}_k \to \mathsf{Pf} \\
E &::= C \mid E_1 \sqcap E_2 \mid \bot \mid \neg E \mid \forall f.E \mid (\mathsf{Pf}_1 = \mathsf{Pf}_2)
\end{aligned}
$$

Undecidability. It is easy to see that the following two restricted cases have decidable decision problems:

- allowing arbitrary PFDs in terminologies, and
- allowing equational concepts in the posed question.

Unfortunately, the combination of the two cases leads again to undecidability. One can use the equational concept to create a seed square for a tiling problem (although a triangle is actually sufficient in this case, as in $A \sqsubseteq (f.g = g) \sqcap \forall f.B$ [35]) that can then be extended into an infinite grid using PFDs in a TBox (e.g., $A \sqsubseteq (B : g \to f.h) \sqcap (B : g \to k.g)$ for the triable seed case), and ultimately to an instance of a tiling problem. Hence:

Theorem 13 ([35]). *Let $\mathcal{T}$ be a $\mathcal{DLFD}$ terminology and E an equational concept. Then the problem $\mathcal{T} \models E \sqsubseteq \bot$ is undecidable.*

Decidability and a Boundary Condition. To regain decidability, we restrict the PFD constructor to adhere to a *boundary* condition, in particular, to have either of the following two forms:

- $C : \mathsf{Pf}_1, \ldots, \mathsf{Pf}.\mathsf{Pf}_i, \ldots, \mathsf{Pf}_k \to \mathsf{Pf}$; and
- $C : \mathsf{Pf}_1, \ldots, \mathsf{Pf}.\mathsf{Pf}_i, \ldots, \mathsf{Pf}_k \to \mathsf{Pf}.f$, for some primitive feature f.

We call the resulting fragment $\mathcal{DLFDE}^-$. The condition distinguishes, e.g., the PFDs $f \to id$ and $f \to g$ from the PFD $f \to g.f$. Intuitively, a simple saturation procedure that *fires* PFDs on a hypothetical database is now guaranteed to terminate as a consequence.

Notice that the boundary condition still admits PFDs that express arbitrary keys or functional dependencies in the sense of the relational model, including those occurring in all our examples. Thus, restricting PFDs in this manner does not sacrifice any ability to capture database schema for legacy data sources.

Theorem 14 ([21]). *Let $\mathcal{T}$ and $\mathcal{T}'$ be respective $\mathcal{DLF}$ and $\mathcal{DLFD}$ terminologies in which the latter contains only PFD inclusion dependencies, and let E be an equational concept. Then there is a concept E' such that*

$$\mathcal{T} \cup \mathcal{T}' \models E \sqsubseteq \bot \text{ iff } \mathcal{T} \models (E \sqcap E') \sqsubseteq \bot.$$

Moreover, E' can be constructed from $\mathcal{T}'$ and E effectively and in time polynomial in $|\mathcal{T}'|$.

The *boundary* condition on PFDs is essential for the above theorem to hold. If unrestricted PFDs are combined with either equations or an ABox, there is no limit on the *length* of paths participating in path agreements when measured from an initial object $o \in E \sqcap E'$ in the associated satisfiability problem. Moreover, any minimal relaxation of this condition, i.e., allowing only non-key PFDs of the form $C : f \rightarrow g.h$, already leads to undecidability [32, 35]:

Theorem 15 ([21]). *$\mathcal{DLFDE}^-$ logical implication and the problem of ABox consistency defined in Sect. 5.3 are decidable and complete for EXPTIME.*

The construction essentially generates a pattern (part of a model) that satisfies E (which already contains the effects of all PFDs due to the boundary condition) and then tests if this pattern can be extended to a full model using the decision procedure for $\mathcal{DLF}$. Note also that posed questions containing PFDs can be rewritten to equivalent posed questions replacing the PFDs with their semantic definitions via path agreements and disagreements.

Inverses. Finally, we conjecture that adding unqualified inverse constructor to $\mathcal{DLFDE}$ under the restrictions outlined in Sect. 4.4 preserves all the results.

4 Tractable FunDL Dialects

In this second part of our survey, we consider the logical implication problem for FunDL dialects for which the logical implication problem can be solved in PTIME. We begin by reviewing $\mathcal{CFD}$, chronologically, the first member of the FunDL family and, so far as we are aware, the first DL dialect to introduce a type constructor, PFDs, for capturing equality generating dependencies [8, 20].

Ensuring tractability requires that we somehow evade Theorems 4 and 6. This is generally achieved by requiring a TBox to satisfy the following additional conditions:

1. Interaction between value restrictions and conjunctions on the left-hand-sides of inclusion dependencies must somehow be controlled,
2. Inclusion dependencies must be Horn (which effectively disallows the use of disjunction)[1], and

---

[1] Allowing the use of conjunction at the top level on the right-hand-side is a simple syntactic sugar.

3. PFDs must satisfy an additional syntactic *boundary* condition in addition to being disallowed on the left-hand-side of inclusion dependencies.

We shall see that violating any of these conditions leads to intractability of logical implication.

4.1 Horn Inclusion Dependencies

The first way of limiting the interactions between value restrictions and conjunctions on the left-hand-sides of inclusion dependencies is by simply disallowing value restrictions entirely, and by no longer permitting posed questions to mention either negations or disjunctions. This approach underlies the FunDL dialect called $\mathcal{CFD}$ given by the following grammar:

$$C ::= A \mid C_1 \sqcap C_2$$
$$D ::= C \mid D_1 \sqcap D_2 \mid \forall f.D \mid C : \mathsf{Pf}_1, ..., \mathsf{Pf}_k \to \mathsf{Pf}$$
$$E ::= C \mid \bot \mid E_1 \sqcap E_2 \mid \forall f.E \mid (\mathsf{Pf}_1 = \mathsf{Pf}_2)$$

The main idea behind decidability and complexity of the logical implication problem is similar to the idea in Theorem 15. However, we no longer need to use the $\mathcal{DLF}$ decision procedure to verify that the partial model can be completed to a full model since, in $\mathcal{CFD}$, one can always employ complete F-trees whose nodes belong to all primitive concepts (without having to check for their existence [20,36]). Hence, the complexity reduces to the construction of the initial part of the model. This, with the help of the restrictions on E concepts, can be done in PTIME.

Theorem 16 ([36])**.** The logical implication problem for $\mathcal{CFD}$ is complete for PTIME.

The hardness follows from the fact that the PFDs alone can simulate HornSAT.

Extensions Versus Tractability. Unfortunately, extending this fragment while maintaining tractability is essentially infeasible. The following table summarizes the effects of allowing additional concept constructors in the TBox on the right-hand-side of inclusion dependencies, reading down, and in the posed question, reading across [36]:

$\mathcal{T}/\mathcal{Q}$	$\mathcal{CFD}$ or $\mathcal{CFD}_{\neq,(\neg)}$	$\mathcal{CFD}_{\neq,\sqcup}$ or $\mathcal{CFD}_{\neg}$
$\mathcal{CFD}$	P-c / in P	P-c / coNP-c
$\mathcal{CFD}^{\sqcup}$	coNP-c / coNP-c	coNP-c / coNP-c
$\mathcal{CFD}^{\bot}$	PSPACE-c / in P	PSPACE-c / coNP-c
$\mathcal{CFD}^{\sqcup,\bot}$	EXPTIME-c / coNP-c	EXPTIME-c / coNP-c

The complexities listed in the table are with respect to the size of the TBox and the size of the posed question. Note in particular that concept *disjointness,* in which $\perp$ is allowed on right-hand-sides of inclusion dependencies, leads to PSPACE-completeness. This is due to the need for checking whether a partial model can be completed, which in turn requires testing for reachability in an implicit but exponentially-sized graph.

4.2 Value Restrictions Instead of Conjunctions

An alternative that allows us to evade the ramifications of Theorem 4 is disallowing *conjunctions* on the left-hand-sides of inclusion dependencies, yielding the dialect $\mathcal{CFD}_{nc}$ [37] given by the following:

$$
\begin{aligned}
C &::= A \mid \forall f.C \\
D &::= C \mid \neg C \mid D_1 \sqcap D_2 \mid \forall f.D \mid C : \mathsf{Pf}_1, ..., \mathsf{Pf}_k \to \mathsf{Pf} \\
E &::= C \mid \perp \mid E_1 \sqcap E_2 \mid \forall f.E \mid (\mathsf{Pf}_1 = \mathsf{Pf}_2)
\end{aligned}
$$

The main idea behind tractability of $\mathcal{CFD}_{nc}$ relies on the fact that left-hand-sides of inclusion dependencies can only observe object membership in a *single atomic concept* (as opposed to a conjunction of concepts). Hence, while models of this logic require exponentially many objects labelled by conjunctions of primitive concepts in general, they can be abstracted in a polynomial way. The construction of the actual model is then similar to the standard NFA to DFA construction followed by unfolding of the resulting DFA.

Theorem 17 ([37]). *The logical implication problem for $\mathcal{CFD}_{nc}$ is complete for PTIME.*

As with the dialect $\mathcal{CFD}$, hardness follows from reducing HornSAT to reasoning with PFDs.

4.3 Value Restrictions and Limited Conjunctions

The above has shown that allowing an arbitrary use of concept conjunction on the left-hand-sides of inclusion dependences in a $\mathcal{CFD}_{nc}$ TBox immediately leads to hardness for EXPTIME (a consequence of Theorem 4). The complexity can be traced to the need for exponentially many objects labelled by different sets of primitive concepts to be generated. The following definition provides a way of controlling this need for all such objects:

Definition 18 (Restricted Conjunction). Let $k > 0$ be a constant. We say that TBox $\mathcal{T}$ is a $\mathcal{CFD}_{kc}$ TBox if, whenever $\mathcal{T} \models (A_1 \sqcap \cdots \sqcap A_n) \sqsubseteq B$ for some set of primitive concepts $\{A_1, \ldots, A_n\} \cup \{B\}$, with $n > k$, then $\mathcal{T} \models (A_{i_1} \sqcap \cdots \sqcap A_{i_k}) \sqsubseteq B$ for some k-sized subset $\{A_{i_1}, \ldots, A_{i_k}\}$ of the primitive concepts $\{A_1, \ldots, A_n\}$. $\square$

A *saturation-style* procedure based on this definition can be implemented to generate *all implied inclusion dependencies* with at most k primitive concepts (value restrictions) on left-hand-sides of inclusion dependencies [26]. The decision procedure essentially follows the procedure for $\mathcal{CFD}_{nc}$ but is exponential in k due to the need to consider sets of concepts up to size k (essentially by determining all implied inclusion dependencies that are not a trivial weakening of other inclusion dependencies) and leads to the following:

Theorem 19 ([26]). The logical implication problem for $\mathcal{CFD}_{kc}$ is complete for PTIME for a fixed value of k; the decision procedure is exponential in k.

In addition, the procedure enables an incremental means of determining the minimum k for which a given TBox is a $\mathcal{CFD}_{kc}$ TBox, that is, allows for testing if a given parameter k suffices:

Theorem 20 (**Testing for k** [26]). A TBox $\mathcal{T}$ is *not* a $\mathcal{CFD}_{kc}$ TBox if and only if there is an additional single-step inference that infers a non-trivial inclusion dependency (i.e., one that is not a weakening of an already discovered dependency) with $k+1$ conjuncts on the left hand side.

An algorithm based on iterative deepening allows one to determine the value of k for a given TBox in a *pay as you go* way. Hence the decision procedure also runs within the optimal time bound, exponential in k and polynomial in $|\mathcal{T}| + |\mathcal{Q}|$, even when k is *not* part of the input.

4.4   Adding Inverse Features

Recall from Sect. 3.3 that we consider only the (unqualified) inverse feature constructor, $\exists f^{-1}$, to be added to the D grammar rules of $\mathcal{CFD}_{nc}$ and $\mathcal{CFD}_{kc}$, yielding the respective logics $\mathcal{CFDI}_{nc}$ and $\mathcal{CFDI}_{kc}$. However, additional restrictions are still required to guarantee tractability of logical consequence [38]. We introduce the restrictions by examples:

1. *Inverses and Value Restrictions.* Interactions between these two concept constructors can be illustrated by the following inference:

$$\{A \sqsubseteq \exists f^{-1}, \forall f.A' \sqsubseteq \forall f.B\} \models A \sqcap A' \sqsubseteq B.$$

   This cannot be allowed since unrestricted use of this construction yields hardness for EXPTIME (see Theorem 4). $\mathcal{CFDI}_{nc}$ syntactically restricts TBoxes to avoid the above situation by requiring additional inclusion dependencies of the form $A \sqsubseteq A'$, $A' \sqsubseteq A$, or $A \sqcap A' \sqsubseteq \bot$ to be present in a TBox whenever the above pattern appears. Note that $\mathcal{CFDI}_{kc}$ does not require this restriction since the *testing for k* procedure we have outlined will detect the above situation (thus determining the *price*).

2. *Inverses and PFDs.* The second interaction that hinders tractability is between inverses and PFDs. In particular, a logical consequence problem of the form

$$\{A \sqsubseteq \exists f^{-1}, \forall f.A \sqsubseteq A, \ldots\} \models (\forall h_1.A) \sqcap (\forall h_2.A) \sqcap (h_1.f = h_2.f) \sqsubseteq h_1 = h_2$$

   will force two infinite f anti-chains starting from two A objects created by the left-hand-side of the posed question. We have shown how to use these anti-chains and additional PFDs in the TBox to reduce *linearly bounded DTM acceptance* [19] to logical implication in this case, yielding PSPACE-hardness, and how to repair this by further limiting the syntax of PFDs in a way that disables this kind of interaction with inverse features [38]. In particular, PFDs in a TBox must now have one of the following two forms:

- $C : \mathsf{Pf}_1, \ldots, \mathsf{Pf} . \mathsf{Pf}_i, \ldots, \mathsf{Pf}_k \to \mathsf{Pf}$; and
- $C : \mathsf{Pf}_1, \ldots, \mathsf{Pf} .g, \ldots, \mathsf{Pf}_k \to \mathsf{Pf} .f$, for some primitive features f and g.

Inverses obeying these two restrictions can then be added to both the FunDL dialects $\mathcal{CFDI}_{nc}$ and $\mathcal{CFDI}_{kc}$ while maintaining tractability:

Theorem 21 ([38]). *The logical implication problems for $\mathcal{CFDI}_{nc}$ and $\mathcal{CFDI}_{kc}$ are complete for PTIME, in the latter case for a fixed value of k.*

5   Partial Features, Roles, ABoxes and Query Answering

The third part of our survey considers how partial features and role hierarchies can be accommodated in FunDL dialects, and how to check for knowledge base consistency and to evaluate queries over FunDL knowledge bases consisting of a so-called ABox in addition to a TBox.

5.1   Partial Features

We first consider the impact of changing the semantics of features in the FunDL family to *partial features* [25,26,40,41]. The changes can be summarized as follows:

1. Features $f \in \mathsf{F}$ are now interpreted as *partial* functions on $\triangle$ (i.e., the result can be *undefined* for some elements of $\triangle$);
2. A path function Pf now denotes a partial function resulting from the composition of partial functions;
3. The syntax of C in feature-based DLs is extended with an additional concept constructor, $\exists f$, called an *existential restriction* that can now appear on both sides of inclusion dependencies;
4. The $\exists f$ concept constructor is interpreted as $\{x \mid \exists y \in \triangle . (f)^{\mathcal{I}}(x) = y\}$.
5. We adopt a *strict* interpretation of set membership and equality. This means that set membership holds only when the value exists; and equality holds only when both sides are defined and denote the same object.

In the light of these changes, we need to consider their impact on concept constructors that involve features or feature paths:

Value Restrictions. Our definition of value restriction $\forall f.C$ (see Definition 1) assumes features are total. For partial features, there is now a choice:

1. keeping the original semantics, i.e., objects in the interpretation of $\forall f.C$ *must have a feature f defined* and leading to a C object, or

2. altering the semantics to match $\mathcal{ALC}$-style semantics, i.e., the f value of objects in the interpretation of such a value restriction must be a C object, *if such a value exists*; we denote this variant $\forall f.C$.

While not equivalent, it is easy to see that many inclusion dependencies can be expressed using either variant of the value restriction, for example

$$A \sqsubseteq \widetilde{\forall} f.B \quad \text{can be expressed as} \quad A \sqcap \forall f.\top \sqsubseteq \forall f.B.$$

Note that when the original semantics is used, the existential restriction $\exists f$ is simply a synonym for $\forall f.\top$. Also, since features are still *functional*, the so-called *qualified existential restrictions* of the form $\exists f.C$, with semantics given by $(\exists f.C)^{\mathcal{I}} = \{x \mid \exists y \in \Delta.(f)^{\mathcal{I}}(x) = y \land y \in (C)^{\mathcal{I}}\}$, can be simulated by expansion to $\exists f \sqcap \forall f.C$. Indeed, hereon we write $\exists \mathsf{Pf}$ as shorthand for $\exists f_1 \sqcap \forall f_1.(\exists f_2 \sqcap \forall f_2.(\dots(\exists f_k)\dots))$.

PFDs. Our PFDs agree with the definition of identity constraints in [11], where $\mathsf{Pf}_0 = id$, which also require path values to exist. To further clarify the impact of this observation, note that a *PFD inclusion dependency* of the form $C_1 \sqsubseteq C_2 : \mathsf{Pf}_1, \ldots, \mathsf{Pf}_k \to \mathsf{Pf}_0$ is violated when (a) all path functions $\mathsf{Pf}_0, \ldots, \mathsf{Pf}_k$ are defined for a C_1 object e_1 and a C_2 object e_2, and (b) $(\mathsf{Pf}_i)^{\mathcal{I}}(e_1) = (\mathsf{Pf}_i)^{\mathcal{I}}(e_2)$ holds only for $1 \le i \le k$. Formally, and more explicitly, this leads to the following interpretation of PFDs in the presence of partial features:

$$((C : \mathsf{Pf}_1, \ldots, \mathsf{Pf}_k \to \mathsf{Pf}_0))^{\mathcal{I}} =$$
$$\{x \mid \forall y.y \in (C)^{\mathcal{I}} \land x \in ((\exists \mathsf{Pf}_0))^{\mathcal{I}} \land y \in ((\exists \mathsf{Pf}_0))^{\mathcal{I}} \land$$
$$\textstyle\bigwedge_{i=1}^{k}(x \in ((\exists \mathsf{Pf}_i))^{\mathcal{I}} \land y \in ((\exists \mathsf{Pf}_i))^{\mathcal{I}} \land (\mathsf{Pf}_i)^{\mathcal{I}}(x) = (\mathsf{Pf}_i)^{\mathcal{I}}(y))$$
$$\to (\mathsf{Pf}_0)^{\mathcal{I}}(x) = (\mathsf{Pf}_0)^{\mathcal{I}}(y) \}.$$

Equational Concepts. Similarly to PFDs, we assume the strict interpretation of equalities, i.e., an object belongs to $(\mathsf{Pf}_1 = \mathsf{Pf}_2)$ if and only if both Pf_1 and Pf_2 are defined for the object and agree.

Partiality in Expressive FunDL. In expressive FunDL dialects, partiality can be simulated by introducing an auxiliary primitive concept G that will stand for the *domain of existing objects*. Depending on our choice of semantics for value restrictions we get a mapping of a TBox under the partial semantics to a TBox under the total semantics. We first define a way to *modify concept descriptions* to capture the desired semantics of partiality:

1. $\mathsf{PtoT}(C) = C[\forall f.C \mapsto \neg G \sqcup \forall f.(C \sqcap G)$ for $f \in F]$, for the original semantics,
2. $\mathsf{PtoT}(C) = C[\exists f \mapsto \neg G \sqcup \forall f.G$, for $f \in F]$ for the $\mathcal{ALC}$-style semantics.

Now we can define a partial to total TBox mapping

$$\mathcal{T}_{\text{total}} = \{G \sqsubseteq \mathsf{PtoT}(C) \mid \top \sqsubseteq C \in \mathcal{T}_{\text{partial}}\} \cup \{\forall f.G \sqsubseteq G \mid f \in F\},$$

and show:

Theorem 22 ([41]). Let $\mathcal{T}_{\text{partial}}$ be a *partial-DLFI* TBox in which all inclusion dependencies are of the form $\top \sqsubseteq C$ with C in negation normal form. Then

$$\mathcal{T}_{\text{partial}} \models \top \sqsubseteq C \iff \mathcal{T}_{\text{total}} \models G \sqsubseteq \mathsf{PtoT}(C),$$

for G a fresh primitive concept.

To extend this construction to the full *partial-DLFDI* logic, it is sufficient to *encode* the path function existence preconditions of PFDs in terms of the auxiliary concept G as follows: if $A \sqsubseteq B : \mathsf{Pf}_1, \ldots, \mathsf{Pf}_k \to \mathsf{Pf}_0 \in \mathcal{T}_{\text{partial}}$ then

$$A \sqcap (\prod_{i=0}^{k} \forall \mathsf{Pf}_i . G) \sqsubseteq B \sqcap (\prod_{i=0}^{k} \forall \mathsf{Pf}_i . G) : \mathsf{Pf}_1, \ldots, \mathsf{Pf}_k \to \mathsf{Pf}_0 \qquad (1)$$

is added to $\mathcal{T}_{\text{total}}$. Here, we are assuming w.l.o.g. that A and B are primitive concept names (*DLFD* allows one to give such names to complex concepts).

Theorem 23 ([41]). Let $\mathcal{T}_{\text{partial}}$ be a *partial-DLFDI* TBox in which all inclusion dependencies are of the form $\top \sqsubseteq C$ or $A \sqsubseteq B : \mathsf{Pf}_1, \ldots, \mathsf{Pf}_k \to \mathsf{Pf}_0$. Then

$$\mathcal{T}_{\text{partial}} \models \top \sqsubseteq C \iff \mathcal{T}_{\text{total}} \models G \sqsubseteq \mathsf{PtoT}(C), \text{ and}$$
$$\mathcal{T}_{\text{partial}} \models A \sqsubseteq B : \mathsf{Pf}_1, \ldots, \mathsf{Pf}_k \to \mathsf{Pf} \iff \mathcal{T}_{\text{total}} \models (1),$$

for G a fresh primitive concept.

This result can also be extended to the logic $\mathcal{DLFDE}^-$ by appropriately transforming the posed question with respect to the strict interpretation of equational constraints.

Partiality in Tractable FunDL. A similar construction can be used to accommodate partial features in tractable FunDL dialects. However, there is a need to accommodate the various restrictions in these logics that guarantee tractability. Hence, we assume that we will be given a *partial-CFDI$_{kc}$* TBox $\mathcal{T}_{\text{partial}}$ in a normal form, and that the semantics of value restrictions is the same in both the partial and the total logic. We then derive a $\mathcal{CFDI}_{(k+1)c}$ TBox $\mathcal{T}_{\text{total}}$ by applying the following rules:

1. $\quad A \sqsubseteq \bot \quad \mapsto \quad A \sqcap G \sqsubseteq \bot$
2. $\quad A \sqsubseteq B \quad \mapsto \quad A \sqcap G \sqsubseteq B$
3. $\quad A \sqcap B \sqsubseteq C \quad \mapsto \quad A \sqcap B \sqcap G \sqsubseteq C$
4. $\quad A \sqsubseteq \forall f.B \quad \mapsto \quad A \sqcap G \sqsubseteq \forall f.B \sqcap \forall f.G$
5. $\quad \forall f.A \sqsubseteq B \quad \mapsto \quad \forall f.A \sqcap \forall f.G \sqsubseteq B$
6. $\quad A \sqsubseteq \exists f \quad \mapsto \quad A \sqcap G \sqsubseteq \forall f.G$
7. $\quad \exists f \sqsubseteq A \quad \mapsto \quad \forall f.G \sqsubseteq A$

and by adding the inclusion dependency $\forall f.G \sqsubseteq G$ to $\mathcal{T}_{\text{total}}$ for each feature.

Conversely, value restrictions in more traditional role-based description logics, such as $\mathcal{ALC}$, also cover the *vacuous cases*, containing objects for which f is undefined (in addition to the above). This definition unfortunately leads to computational difficulties: the *disjunctive* nature of such a value restriction, when used on left-hand-sides of inclusion dependencies, destroys the canonical model property of the logic. This leads to intractability of query answering as shown by Calvanese *et al.* [12]. To regain tractability, it becomes necessary to restrict the use of value restrictions on the left-hand-side of inclusion dependencies. In a normal form, the C grammar for left-hand-side concepts must replace $\forall f.A$ with $\forall f.A \sqcap \exists f$. This leads to alternative rules when simulating the partial-feature logic in the total-feature counterpart, i.e.,

$$
\begin{array}{lll}
4'. & A \sqsubseteq \forall f.B & \mapsto \quad A \sqcap G \sqsubseteq \forall f.B \\
5'. & (\forall f.A \sqcap \exists f) \sqsubseteq B & \mapsto \quad (\forall f.A \sqcap \forall f.G) \sqsubseteq B
\end{array}
$$

The technique for treating posed questions [40] extends to *partial-$\mathcal{CFDI}_{kc}$* and yields the following:

Theorem 24 ([40]). Let $\mathcal{T}_{\text{partial}}$ be a *partial-$\mathcal{CFDI}_{kc}$* TBox, $\mathcal{Q}_{\text{partial}}$ a posed question, and $\mathcal{T}_{\text{total}}$ be defined as above. Then $\mathcal{T}_{\text{total}}$ is a $\mathcal{CFDI}_{(k+1)c}$ TBox and

$$
\mathcal{T}_{\text{partial}} \models \mathcal{Q}_{\text{partial}} \iff \mathcal{T}_{\text{total}} \models \mathcal{Q}_{\text{total}},
$$

where $\mathcal{Q}_{\text{total}}$ is effectively constructed from $\mathcal{Q}_{\text{partial}}$ by adding appropriate conjunctions with G concepts.

Since $|\mathcal{Q}_{\text{partial}}|$ is linear in $|\mathcal{Q}_{\text{total}}|$, this provides a tractable decision procedure for logical implication in *partial-$\mathcal{CFDI}_{kc}$*. An analogous result involving *partial-$\mathcal{CFDI}_{kc}$* knowledge base reasoning was studied in [26].

5.2   Simulating Roles and Role Constructors

It is well known that unrestricted use of *role functionality* with *role hierarchies*, e.g., DL-Lite$_{\text{core}}^{\mathcal{HF}}$, leads to intractability [2,10]. Conversely, the ability to reify roles would seem to enable capturing a limited variety of *role hierarchies*.[2]

Consider roles R and S and the corresponding primitive concepts C_R and C_S, respectively, and assume that the domains and ranges of the reified roles are captured by the features *dom* and *ran* common to both the reified roles. Subsumption and disjointness of these roles can then be captured as follows:

$$
\begin{array}{lll}
R \sqsubseteq S & \mapsto & C_R \sqsubseteq C_S, C_R \sqsubseteq C_S : dom, ran \to id \quad \text{and} \\
R \sqcap S \sqsubseteq \bot & \mapsto & C_R \sqsubseteq \neg C_S, C_R \sqsubseteq C_S : dom, ran \to id,
\end{array}
$$

assuming that the reified role R (and analogously S) also satisfies the key constraint $C_R \sqsubseteq C_R : dom, ran \to id$. Such a reduction does *not* lend itself to capturing role hierarchies between roles and *inverses* of roles (due to fixing the names of the features *dom* and *ran*).

---

[2] Unlike DL-Lite$_{\text{core}}^{(\mathcal{HF})}$, that restricts the applicability of functional constraints in the presence of role hierarchies, we review what forms of role hierarchies can be captured while retaining the ability to specify arbitrary keys and functional dependencies.

Moreover, for tractable fragments of FunDL, a condition introduced earlier, governing the interactions between inverse features and value restrictions, introduces additional interactions that interfere with (simulating) role hierarchies, in particular in cases when *mandatory participation* constraints are present. Consider again roles R_1 and R_2 and the corresponding primitive concepts C_{R_1} and C_{R_2}, respectively, and associated constraints that declare typing for the roles,

$$C_{R_1} \sqsubseteq \forall dom.A_1, C_{R_1} \sqsubseteq \forall ran.B_1, C_{R_1} \sqsubseteq C_{R_1} : dom, ran \rightarrow id$$
$$C_{R_2} \sqsubseteq \forall dom.A_2, C_{R_2} \sqsubseteq \forall ran.B_2, C_{R_2} \sqsubseteq C_{R_2} : dom, ran \rightarrow id$$

originating, e.g., from an ER diagram postulating that entity sets A_i and B_i participate in a relationship R_i (for $i = 1, 2$). Now consider a situation where the participation of A_i in R_i is *mandatory* (expressed, e.g., as $A_i \sqsubseteq \exists R_i$ in DL-Lite). This leads to the following constraints:

$$A_1 \sqsubseteq \exists dom^{-1}, \forall dom.A_1 \sqsubseteq C_{R_1} \text{ and } A_2 \sqsubseteq \exists dom^{-1}, \forall dom.A_2 \sqsubseteq C_{R_2}.$$

The earlier condition governing the use of inverse roles then requires that one of

$$A_1 \sqsubseteq A_2, A_2 \sqsubseteq A_1, \text{ or } A_1 \sqsubseteq \neg A_2$$

are present in the TBox. The first (and second) conditions imply that $C_{R_1} \sqsubseteq C_{R_2}$ ($C_{R_2} \sqsubseteq C_{R_1}$, respectively). The third condition states that the domains of (the reified versions of) R_1 and R_2 are disjoint, hence the roles themselves must also be disjoint. Hence, in the presence of $C_{R_1} \sqsubseteq C_{R_2} : dom, ran \rightarrow id$, the concepts C_{R_1} and C_{R_2} must also be disjoint.

All this shows that some form of role hierarchies can be accommodated in FunDL dialects. However:

1. only primitive roles can be captured (i.e., capturing inverse roles will not be possible), and
2. when tractability is required, only *role forests* can be captured, that is, for each pair of roles participating in the same role hierarchy, one must be a super-role of the other or their domain and range features must be distinct.

The first restriction originates in the way (binary) roles are reified—by assigning canonically-named features. This prevents modelling constraints such as $R \sqsubseteq R^-$ (which would seem to require simple equational constraints for feature renaming). The second condition is essential to maintaining tractability of reasoning [38]. Note, however, that no such restriction is needed for roles that do *not* participate in the same role hierarchy; this is achieved by appropriate choice of names for the features *dom* and *ran*.

Last, our approach to role hierarchies can easily be extended to handling hierarchies of higher-arity non-homogeneous relationships (again, via reification and appropriate naming of features) that originate, e.g., from relating the aggregation constructs via inheritance in the EER model [27, 28]. The reification based approach differs from approaches to modelling higher arity relationships directly in the underlying description logic, such as $\mathcal{DLR}$ [13,14] in which only homogeneous relationships can be related in hierarchies. This is due to the positional nature of referring to components of such relationship in lieu of using arguably more flexible *keywords* (realized by features in FunDL).

5.3   ABoxes, Knowledge Bases, and Consistency

First we consider the issue of *knowledge bases*, combinations of terminological knowledge (TBoxes) with factual assertions about particular objects (ABoxes).

Definition 25 (ABoxes and Knowledge Bases). A knowledge base $\mathcal{K}$ is defined by a *TBox* $\mathcal{T}$ and an *ABox* $\mathcal{A}$ consisting of a finite set of facts in form of *concept assertions* $A(a)$, *basic function assertions* $f(a) = b$ and *path function assertions* $\mathsf{Pf}_1(a) = \mathsf{Pf}_2(b)$. $\mathcal{A}$ is called a *primitive* ABox if it consists only of concept and basic function assertions. Semantics is extended to interpret individuals a to be elements of $\triangle$. An interpretation $\mathcal{I}$ satisfies a concept assertion $A(a)$ if $(a)^{\mathcal{I}} \in (A)^{\mathcal{I}}$, a basic function assertion $f(a) = b$ if $(f)^{\mathcal{I}}((a)^{\mathcal{I}}) = (b)^{\mathcal{I}}$ and a path function assertion $\mathsf{Pf}_1(a) = \mathsf{Pf}_2(b)$ if $(\mathsf{Pf}_1)^{\mathcal{I}}((a)^{\mathcal{I}}) = (\mathsf{Pf}_2)^{\mathcal{I}}((b)^{\mathcal{I}})$. $\mathcal{I}$ satisfies a knowledge base $\mathcal{K}$ if it satisfies each inclusion dependency and assertion in $\mathcal{K}$, and also satisfies UNA if, for any individuals a and b occurring in $\mathcal{K}$, $(a)^{\mathcal{I}} \neq (b)^{\mathcal{I}}$. □

A standard reasoning problem for knowledge bases is the consistency problem, the question whether a knowledge base has a model. We relate this problem to the logical implication problems for FunDL dialects that admit equational constructs in the posed questions. It turns out that either capacity alone is sufficient: each is able to effectively simulate the other [21].

ABoxes vs. Equalities in Posed Questions. Intuitively, path equations can *enforce* that an arbitrary finite graph (with feature-labeled edges and concept description-labeled nodes) is a part of any model that satisfies the equations. Such a graph can equivalently be enforced by an ABox. Hence we have:

Theorem 26 ([21]). Let $\mathcal{T}$ be a $\mathcal{DLFD}$ terminology and $\mathcal{A}$ an ABox. Then there is a concept E such that $\mathcal{T} \cup \mathcal{A}$ is not consistent if and only if $\mathcal{T} \models E \sqsubseteq \bot$.

Conversely, it is also possible to show that ABox reasoning can be used for reasoning about equational constraints in the posed questions. However, as the equational concepts are closed under Boolean constructors, a single equational problem may need to map to several ABox consistency problems.

Theorem 27 ([21]). Let $\mathcal{T}$ be a $\mathcal{DLFD}$ terminology and E an equational concept. Then there is a finite set of ABoxes $\{\mathcal{A}_i : 0 < i \leq k\}$ such that

$$\mathcal{T} \models E \sqsubseteq \bot \text{ iff } \mathcal{T} \cup \mathcal{A}_i \text{ is not consistent for all } 0 < i \leq k.$$

Theorems 26 and 27 hold even when the terminology $\mathcal{T}$ is a $\mathcal{DLF}$ TBox (i.e., does not contain any occurrences of the PFD concept constructor) or to the tractable FunDL dialects $\mathcal{CFD}$ and $\mathcal{CFDI}_{kc}$. Here, posed question E concepts must be limited to retain a PTIME upper bound in the size of the posed question (Sect. 4 has the details).

5.4 Query Answering

Conjunctive queries (CQ) are, as usual, formed from atomic queries (or *atoms*) of the form $C(x)$ and $x.\mathsf{Pf}_1 = y.\mathsf{Pf}_2$, where x and y are variables, using conjunction and existential quantification. To simplify notation, we conflate conjunctive queries with the set of its constituent atoms and a set of *answer variables*. Given a knowledge base (KB) consisting of a TBox and ABox expressed in terms of a tractable FunDL dialect, our goal is to compute the so called *certain answers*:

Definition 28 (Certain Answer). Let $\mathcal{K}$ be a KB over a tractable FunDL dialect and $Q = \{\bar{x} \mid \varphi\}$ a CQ. A *certain answer* to Q over $\mathcal{K}$ is a substitution of constant symbols $\bar{a}$, $[\bar{x} \mapsto \bar{a}]$, such that $\mathcal{K} \models Q[\bar{x} \mapsto \bar{a}]$.   □

Computing certain answers in this case requires a combination of *perfect rewriting* [10] and of the *combined approach* [22,36]. The latter is necessary because tractable FunDL dialects are complete for PTIME and first-order rewriting alone followed by evaluating the rewritten query over the ABox will not suffice. The former is necessary to avoid the need for exponentially many anonymous objects in an ABox completion (unlike $\mathcal{EL}$ logics in which there is a need for only polynomially many such objects).

This approach was introduced for $\mathcal{CFDI}_{nc}$ in [17,18] and the two steps are realized by two procedures:

1. Completion$_\mathcal{T}(\mathcal{A})$: this procedure applies consequences of the TBox $\mathcal{T}$ to the ABox $\mathcal{A}$. In particular, concept membership is fully determined for all all ABox individuals. For example, if $\{A(a), f(a) = b, f(b) = c, \ldots\} \subseteq \mathcal{A}$ and $\mathcal{T} \models A \sqsubseteq \forall f.A$, we require $\{A(b), A(c), \ldots\} \subseteq$ Completion$_\mathcal{T}(\mathcal{A})$. (Indeed, propagating concepts along paths that exists in an ABox is the reason why perfect rewriting alone will not suffice in tractable FunDL dialects.)

2. Fold$_\mathcal{T}(Q)$: this procedure rewrites an input CQ to an union of CQs that account for the constraints in $\mathcal{T}$ that postulate existence of anonymous objects in all models of the knowledge base. A (slight simplification of a) typical rule applied during such a rewriting looks as follows:
   If $\{y.f = x, A(y)\} \subseteq \psi$ and y does not appear elsewhere in ψ nor is an answer variable, then Fold$(Q) :=$ Fold$(Q) \cup \{\{\bar{y} \mid \psi_i\}\}$ for all $\psi_i = \psi - \{y.f = x, A(y)\} \cup \{B_i(x)\}$, where B_i are all maximal primitive concepts w.r.t. $\sqsubseteq$ satisfying the logical implication conditions $\mathcal{T} \models B_i \sqsubseteq \exists f^{-1}$ and $\mathcal{T} \models \forall f.B_i \sqsubseteq A\}$.
   The rule states that whenever the variable y is connected to the rest of the query via a single feature f, it may be mapped to an anonymous individual. This is accommodated by the query ψ_i that no longer uses the variable y,

but implies ψ since the existence of the necessary individual is implied by the TBox $\mathcal{T}$ and the $B_i(x)$ atom in ψ_i.

Note that query rewriting requires a *completed* ABox. Thus, the rewriting produces fewer disjuncts since only maximal concepts need to be retained.

Theorem 29 ([40]). Let $\mathcal{K} = (\mathcal{T}, \mathcal{A})$ be a $\mathcal{CFDI}_{nc}$ knowledge base and Q a conjunctive query. Then

$$\mathcal{K} \models Q[\bar{x} \mapsto \bar{a}] \iff (\emptyset, \mathsf{Completion}_{\mathcal{T}}(\mathcal{A})) \models \mathsf{Fold}_{\mathcal{T}}(Q)[\bar{x} \mapsto \bar{a}].$$

Note that $(\emptyset, \mathsf{Completion}_{\mathcal{T}}(\mathcal{A})) \models \mathsf{Fold}_{\mathcal{T}}(Q)[\bar{x} \mapsto \bar{a}]$ reduces to evaluating the query $\mathsf{Fold}_{\mathcal{T}}(Q)$ over a finite relational structure $\mathsf{Completion}_{\mathcal{T}}(\mathcal{A})$. Tractability (in $|\mathcal{K}|$) then follows from $|\mathsf{Completion}_{\mathcal{T}}(\mathcal{A})|$ being polynomial in $|\mathcal{A}|$ and the fact that reasoning in $\mathcal{K}$ is in PTIME. This approach was later extended to other tractable dialects of FunDL including logics with partial features up to and including *partial-$\mathcal{CFDI}_{kc}$* [26].

6   Related Work

Recall that $\mathcal{CFDI}_{nc}$ is a tractable FunDL dialect in which left-hand-sides of inclusion dependencies exclude the use of negation as well as conjunction. The possibility of the Krom extension of this dialect, that readmits negation, has also been explored [39]. Tractability is still possible, but requires TBoxes to be free of non-key PFDs, requires ABoxes to be primitive, and requires the adoption of UNA. (Relaxing any of these conditions leads to intractability.)

We have also considered how concepts in FunDL dialects can replace constants in an ABox as a way of referring to entities or objects. Indeed, the judicious adoption of features instead of roles in these dialects makes it easy for an ABox to be a window on factual data in backend object-relational data sources. Coupled with the notion of *referring expression types*, this overall development pays off nicely in ontology-based data access and in relating conceptual and object-relational database design in information systems [6,7].

A short review of ways in which PFDs themselves have been generalized completes our survey.

Path Order Dependencies. PFDs can be viewed as a variety of tuple generating dependencies in which equality is the only predicate occurring on the right-hand-side. The possibility that any comparison operator can be used instead has also been investigated. In particular, so-called *guarded order dependencies* can be added to the expressive FunDL dialect $\mathcal{DLF}$ without impacting the complexity of logical implication [29]. For our introductory university TBox, a correlation between *gpa* and *mark* can be expressed by such a dependency:

$$\mathrm{TAKES} \sqsubseteq \mathrm{TAKES} : class^=, mark^< \rightarrow student.gpa^{\leq}$$

The dependency asserts that *the grade point average of a student is never greater than that of another student when there is some class they have both taken in which the latter student obtained a better grade.*

Regular Path Functional Dependencies. Left and right-hand-sides of PFDs can be viewed as instances of finite regular languages. The possibility of allowing these languages to be defined by regular expressions admitting the Kleene closure operator has also been investigated. In particular, *regular path functional dependencies* were introduced in [30], and more general *regular path order dependencies* in [33], and, in both cases, were shown to not impact the complexity of logical implication when added to $\mathcal{DLF}$. This remains the case when value restrictions are also generalized by allowing component path expressions to be given by regular expressions. For example, to ensure that *every professor eventually reports to a dean*, one can now add the inclusion dependencies

$$\text{DEAN} \sqsubseteq \text{CHAIR} \quad \text{and} \quad \text{PROF} \sqsubseteq \neg\forall reports * . \neg \text{DEAN}$$

to the university TBox.

Temporal Path Functional Dependencies. Finally, adding both a temporal variety of PFDs and a *global model operator* ($\square$) to $\mathcal{DLF}$ is also possible without impact on the complexity of logical implication [34]. This enables adding the inclusion dependency

$$\text{PERSON} \sqsubseteq (\square_{\text{forever}}\text{PERSON}) \sqcap (\text{PERSON} : id \rightarrow_{\text{forever}} name)$$

to the university TBox to ensure that *a person is always a person* and that *the name of a person never changes*. Adding the inclusion dependency

$$\text{DEPT} \sqsubseteq (\text{DEPT} : id \rightarrow_{\text{term}} head) \sqcap (\text{DEPT} : head \rightarrow_{\text{term}} id)$$

would ensure that *a professor is the unique head of a department for a fixed term*. However, it *not* possible to add any form of eventuality together with temporal PFDs to $\mathcal{DLF}$ (e.g., by also adding regular PFDs) and at the same time retain EXPTIME complexity of logical implication for $\mathcal{DLF}$ itself [34].

References

1. Ackermann, W.: Uber die Erfullbarkeit gewisser Zahlausdrucke. Math. Ann. **100**, 638–649 (1928)
2. Artale, A., Calvanese, D., Kontchakov, R., Zakharyaschev, M.: The DL-lite family and relations. J. Artif. Intell. Res. **36**, 1–69 (2009). https://doi.org/10.1613/jair. 282
3. Baader, F., Brandt, S., Lutz, C.: Pushing the EL envelope. In: Kaelbling, L.P., Saffiotti, A. (eds.) Proceedings of the Nineteenth International Joint Conference on Artificial Intelligence, IJCAI 2005, Edinburgh, Scotland, UK, 30 July–5 August 2005, pp. 364–369. Professional Book Center (2005). http://ijcai.org/Proceedings/ 05/Papers/0372.pdf
4. Berger, R.: The undecidability of the dominoe problem. Mem. Amer. Math. Soc. **66**, 1–72 (1966)
5. van Emde Boas, P.: The convenience of tilings. In: Complexity, Logic, and Recursion Theory. pp. 331–363. Marcel Dekker Inc. (1997)

6. Borgida, A., Toman, D., Weddell, G.: On referring expressions in information systems derived from conceptual modelling. In: Comyn-Wattiau, I., Tanaka, K., Song, I.-Y., Yamamoto, S., Saeki, M. (eds.) ER 2016. LNCS, vol. 9974, pp. 183–197. Springer, Cham (2016). https://doi.org/10.1007/978-3-319-46397-1_14

7. Borgida, A., Toman, D., Weddell, G.E.: On referring expressions in query answering over first order knowledge bases. In: Baral, C., Delgrande, J.P., Wolter, F. (eds.) Proceedings of the Fifteenth International Conference, Principles of Knowledge Representation and Reasoning KR 2016, Cape Town, South Africa, 25–29 April 2016, pp. 319–328. AAAI Press (2016). http://www.aaai.org/ocs/index.php/KR/KR16/paper/view/12860

8. Borgida, A., Weddell, G.: Adding uniqueness constraints to description logics. In: Bry, F., Ramakrishnan, R., Ramamohanarao, K. (eds.) DOOD 1997. LNCS, vol. 1341, pp. 85–102. Springer, Heidelberg (1997). https://doi.org/10.1007/3-540-63792-3_10

9. Brewka, G., Lang, J. (eds.): Principles of knowledge representation and reasoning. In: Proceedings of the Eleventh International Conference, KR 2008, Sydney, Australia, 16–19 September 2008. AAAI Press (2008)

10. Calvanese, D., De Giacomo, G., Lembo, D., Lenzerini, M., Rosati, R.: Tractable reasoning and efficient query answering in description logics: the DL-Lite family. J. Autom. Reasoning **39**(3), 385–429 (2007). https://doi.org/10.1007/s10817-007-9078-x

11. Calvanese, D., De Giacomo, G., Lembo, D., Lenzerini, M., Rosati, R.: Path-based identification constraints in description logics. In: Brewka and Lang [9], pp. 231–241. http://www.aaai.org/Library/KR/2008/kr08-023.php

12. Calvanese, D., De Giacomo, G., Lembo, D., Lenzerini, M., Rosati, R.: Data complexity of query answering in description logics. Artif. Intell. **195**, 335–360 (2013). https://doi.org/10.1016/j.artint.2012.10.003

13. Calvanese, D., De Giacomo, G., Lenzerini, M.: On the decidability of query containment under constraints. In: Mendelzon, A.O., Paredaens, J. (eds.) Proceedings of the Seventeenth ACM SIGACT-SIGMOD-SIGART Symposium on Principles of Database Systems, 1–3 June 1998, Seattle, Washington, USA. pp. 149–158. ACM Press (1998). https://doi.org/10.1145/275487.275504

14. Calvanese, D., De Giacomo, G., Lenzerini, M.: Identification constraints and functional dependencies in description logics. In: Nebel, B. (ed.) Proceedings of the Seventeenth International Joint Conference on Artificial Intelligence, IJCAI 2001, Seattle, Washington, USA, 4–10 August 2001. pp. 155–160. Morgan Kaufmann (2001). http://ijcai.org/proceedings/2001-1

15. Chomicki, J., Imielinski, T.: Finite representation of infinite query answers. ACM Trans. Database Syst. **18**(2), 181–223 (1993). https://doi.org/10.1145/151634.151635

16. Fürer, M.: Alternation and the ackermann case of the decision problem. L'Enseignement Math. **27**, 137–162 (1981)

17. Jacques, J.S., Toman, D., Weddell, G.E.: Object-relational queries over cfdi$_{nc}$ knowledge bases: OBDA for the SQL-Literate. In: Kambhampati, S. (ed.) Proceedings of the Twenty-Fifth International Joint Conference on Artificial Intelligence, IJCAI 2016, New York, NY, USA, 9–15 July 2016, pp. 1258–1264. IJCAI/AAAI Press (2016). http://www.ijcai.org/Abstract/16/182

18. Jacques, J.S., Toman, D., Weddell, G.E.: Object-relational queries over cfdi_nc knowledge bases: OBDA for the SQL-Literate (extended abstract). In: Lenzerini, M., Peñaloza, R. (eds.) Proceedings of the 29th International Workshop on Description Logics, Cape Town, South Africa, 22–25 April 2016. CEUR Workshop Proceedings, vol. 1577. CEUR-WS.org (2016). http://ceur-ws.org/Vol-1577/paper_10.pdf

19. Karp, R.M.: Reducibility among combinatorial problems. In: Miller, R.E., Thatcher, J.W., Bohlinger, J.D. (eds.) Complexity of Computer Computations. The IBM Research Symposia Series, pp. 85–103. Springer, Boston (1972). https://doi.org/10.1007/978-1-4684-2001-2_9

20. Khizder, V.L., Toman, D., Weddell, G.: Reasoning about duplicate elimination with description logic. In: Lloyd, J., et al. (eds.) CL 2000. LNCS (LNAI), vol. 1861, pp. 1017–1032. Springer, Heidelberg (2000). https://doi.org/10.1007/3-540-44957-4_68

21. Khizder, V.L., Toman, D., Weddell, G.E.: Adding aboxes to a description logic with uniqueness constraints via path agreements. In: Calvanese, D., et al. (eds.) Proceedings of the 2007 International Workshop on Description Logics (DL2007), Brixen-Bressanone, near Bozen-Bolzano, Italy, 8–10 June 2007. CEUR Workshop Proceedings, vol. 250. CEUR-WS.org (2007). http://ceur-ws.org/Vol-250/paper_69.pdf

22. Kontchakov, R., Lutz, C., Toman, D., Wolter, F., Zakharyaschev, M.: The combined approach to query answering in DL-Lite. In: Lin, F., Sattler, U., Truszczynski, M. (eds.) Proceedings of the Twelfth International Conference Principles of Knowledge Representation and Reasoning, KR 2010, Toronto, Ontario, Canada, 9–13 May 2010. AAAI Press (2010). http://aaai.org/ocs/index.php/KR/KR2010/paper/view/1282

23. Liu, L., Özsu, M.T. (eds.): Encyclopedia of Database Systems. Springer, US (2009). https://doi.org/10.1007/978-0-387-39940-9

24. Machtey, M., Young, P.: An Introduction to the General Theory of Algorithms. North-Holland, Amsterdam (1978)

25. McIntyre, S., Borgida, A., Toman, D., Weddell, G.E.: On limited conjunctions in polynomial feature logics, with applications in OBDA. In: Thielscher, M., Toni, F., Wolter, F. (eds.) Proceedings of the Sixteenth International Conference Principles of Knowledge Representation and Reasoning, KR 2018, Tempe, Arizona, 30 October–2 November 2018, pp. 655–656. AAAI Press (2018). https://aaai.org/ocs/index.php/KR/KR18/paper/view/18016

26. McIntyre, S., Borgida, A., Toman, D., Weddell, G.E.: On limited conjunctions and partial features in parameter tractable feature logics. In: Proceedings of the Thirty-Third AAAI Conference on Artificial Intelligence, 27 January–1 February 2019, Honolulu, Hawaii, U.S.A. (2019, in press)

27. Song, I., Chen, P.P.: Entity relationship model. In: Liu and Özsu [23], pp. 1003–1009. https://doi.org/10.1007/978-0-387-39940-9_148

28. Thalheim, B.: Extended entity-relationship model. In: Liu and Özsu [23], pp. 1083–1091. https://doi.org/10.1007/978-0-387-39940-9_157

29. Toman, D., Weddell, G.E.: On attributes, roles, and dependencies in description logics and the ackermann case of the decision problem. In: Goble, C.A., McGuinness, D.L., Möller, R., Patel-Schneider, P.F. (eds.) Proceedings of the Working Notes of the 2001 International Description Logics Workshop (DL-2001), Stanford, CA, USA, 1–3 August 2001. CEUR Workshop Proceedings, vol. 49. CEUR-WS.org (2001). http://ceur-ws.org/Vol-49/TomanWeddell-76start.ps

30. Toman, D., Weddell, G.E.: On reasoning about structural equality in XML: a description logic approach. Theor. Comput. Sci. **336**(1), 181–203 (2005). https://doi.org/10.1016/j.tcs.2004.10.036
31. Toman, D., Weddell, G.E.: On the interaction between inverse features and path-functional dependencies in description logics. In: Kaelbling, L.P., Saffiotti, A. (eds.) Proceedings of the Nineteenth International Joint Conference on Artificial Intelligence, IJCAI 2005, Edinburgh, Scotland, UK, 30 July–5 August 2005, pp. 603–608. Professional Book Center (2005). http://ijcai.org/Proceedings/05/Papers/1421.pdf
32. Toman, D., Weddell, G.: On keys and functional dependencies as first-class citizens in description logics. In: Furbach, U., Shankar, N. (eds.) IJCAR 2006. LNCS (LNAI), vol. 4130, pp. 647–661. Springer, Heidelberg (2006). https://doi.org/10.1007/11814771_52
33. Toman, D., Weddell, G.: On order dependencies for the semantic web. In: Parent, C., Schewe, K.-D., Storey, V.C., Thalheim, B. (eds.) ER 2007. LNCS, vol. 4801, pp. 293–306. Springer, Heidelberg (2007). https://doi.org/10.1007/978-3-540-75563-0_21
34. Toman, D., Weddell, G.E.: Identifying objects over time with description logics. In: Brewka and Lang [9], pp. 724–732. http://www.aaai.org/Library/KR/2008/kr08-071.php
35. Toman, D., Weddell, G.E.: On keys and functional dependencies as first-class citizens in description logics. J. Autom. Reasoning **40**(2–3), 117–132 (2008). https://doi.org/10.1007/s10817-007-9092-z
36. Toman, D., Weddell, G.E.: Applications and extensions of PTIME description logics with functional constraints. In: Boutilier, C. (ed.) Proceedings of the 21st International Joint Conference on Artificial Intelligence, IJCAI 2009, Pasadena, California, USA, 11–17 July 2009, pp. 948–954 (2009). http://ijcai.org/Proceedings/09/Papers/161.pdf
37. Toman, D., Weddell, G.: Conjunctive query answering in $\mathcal{CFD}_{nc}$: a PTIME description logic with functional constraints and disjointness. In: Cranefield, S., Nayak, A. (eds.) AI 2013. LNCS (LNAI), vol. 8272, pp. 350–361. Springer, Cham (2013). https://doi.org/10.1007/978-3-319-03680-9_36
38. Toman, D., Weddell, G.: On adding inverse features to the description logic $\mathcal{CFD}_{nc}^{\forall}$. In: Pham, D.-N., Park, S.-B. (eds.) PRICAI 2014. LNCS (LNAI), vol. 8862, pp. 587–599. Springer, Cham (2014). https://doi.org/10.1007/978-3-319-13560-1_47
39. Toman, D., Weddell, G.: On the krom extension of $\mathcal{CFDI}_{nc}^{\forall-}$. In: Pfahringer, B., Renz, J. (eds.) AI 2015. LNCS (LNAI), vol. 9457, pp. 559–571. Springer, Cham (2015). https://doi.org/10.1007/978-3-319-26350-2_50
40. Toman, D., Weddell, G.: On partial features in the $\mathcal{DLF}$ Family of Description Logics. In: Booth, R., Zhang, M.-L. (eds.) PRICAI 2016. LNCS (LNAI), vol. 9810, pp. 529–542. Springer, Cham (2016). https://doi.org/10.1007/978-3-319-42911-3_44
41. Toman, D., Weddell, G.E.: On partial features in the DLF dialects of description logic with inverse features. In: Artale, A., Glimm, B., Kontchakov, R. (eds.) Proceedings of the 30th International Workshop on Description Logics, Montpellier, France, 18–21 July 2017. CEUR Workshop Proceedings, vol. 1879. CEUR-WS.org (2017). http://ceur-ws.org/Vol-1879/paper44.pdf

Some Thoughts on Forward Induction in Multi-Agent-Path Finding Under Destination Uncertainty

Bernhard Nebel[✉]

Institut für Informatik, Albert-Ludwigs-Universität Freiburg,
Freiburg im Breisgau, Germany
nebel@uni-freiburg.de
http://www.informatik.uni-freiburg.de/~nebel

Abstract. While the notion of implicit coordination helps to design frameworks in which agents can cooperatively act with only minimal communication, it so far lacks exploiting observations made while executing a plan. In this note, we have a look at what can be done in order to overcome this shortcoming, at least in a specialized setting.

1 Introduction

In implicitly coordinated multi-agent path finding under destination uncertainty (MAPF/DU) [5] (and more generally in epistemic planning [2,3]), we have so far concentrated on generating plans in a way such that each agent tries to generate situations from which the other agents can provably find a plan that guarantees success. This means in particular that we do not make use of observations made during the execution of a plan in order to learn something about the destinations (or gain other information), something similar to what is called *forward induction* [1] in game theory or *plan recognition* [6,8] in the area of automated planning.

Not making use of observations implies that agents cannot use their actions in order to signal their intention. For these reasons, plans might be longer than necessary or an instance might not be solvable, although by making inferences about the intentions of the other players, the instance could be solvable. In this paper, we will analyze, in which situations one can make use of observations and how this can be integrated into the planning process.

In order to so, we will introduce the basic notation and terminology in the next section. In Sect. 3, we will then analyze how to modify our notion of a solution concept for MAPF/DU.

2 Background

The multi-agent path finding problem in its simplest form could be stated as follows. The environment is modelled as an undirected, simple graph $G = (V, E)$.

© Springer Nature Switzerland AG 2019
C. Lutz et al. (Eds.): Baader Festschrift, LNCS 11560, pp. 431–440, 2019.
https://doi.org/10.1007/978-3-030-22102-7_20

A *configuration* of *agents* A on the graph G is an injective function $\alpha \colon A \to V$. For $i \in A$ and $v \in V$, by $\alpha[i/v]$ we refer to the function that has the same values for all $j \neq i$ as α, but for i it has the value v: $\alpha[i/v](i) = v$.

Given a *movement action* of agent i from v to v' and a configuration α, a *successor configuration* $\alpha' = \alpha[i/v']$ is generated, provided $\alpha(i) = v$, $(\alpha(i), \alpha'(i)) \in E$, and there exists no j with $\alpha(j) = v'$. The *MAPF problem* is then to generate for a given *MAPF instance* $\mathcal{M} = \langle A, G, \alpha_0, \alpha_* \rangle$ with a given set of agents A, a given graph G, the *initial configuration* α_0, and the *goal configuration* α_*, a sequence of movements from α_0 to α_*. We always assume that such movement plans are *cycle-free*, i.e., that during the execution of such a plan no configuration is reached twice. We call a plan *successful* for a MAPF instance if it transforms α_0 into α_*. Since in the following we only consider successful movement plans, we just call them *plans*. If there exists such a plan for a given instance, we call the instance *solvable*.

For this basic version of the MAPF problem, most interesting questions concerning computational properties have been answered already in a paper by Kornhauser et al. in 1984 [4]. It is known that solvability can be decided in cubic time and the plan length and the time to find a plan is also bounded cubicly. Finally, solving the bounded planning problem (corresponding to the optimization problem) is NP-complete [7].

2.1   Generalized MAPF

Plans are usually generated in a centralized manner and the agents then follow the plan. In our generalized setting, we assume all agents plan by themselves and the goals of the agents are not common knowledge any longer. Instead only the agent itself knows its own destination. Common knowledge are the possible destinations for each agent, formalized by a *destination function* $\beta \colon A \to 2^V$, with the constraint that for all $i \in A$ either the real destination is among the possible ones, i.e., $\alpha_*(i) \in \beta(i)$, or $\beta(i) = \emptyset$, because agent i already arrived (and is not allowed to move anymore). We require further that all combinations of possible destinations are consistent, i.e., $\beta(i) \cap \beta(j) = \emptyset$ for all $i \neq j \in A$.

In the original MAPF problem, the state space for the planning process is simply the space of all configurations α of the agents in the graph. For the MAPF problem with destination uncertainty we also have to take into account the possible belief states of all the agents. For this reason, we have to make the possible destination function part of the state space as well, i.e., an *objective state* is now the tuple $s = (\alpha, \beta)$, which captures the *common knowledge* of all agents. Since the precise destinations are not common knowledge any longer, it is necessary to have some form of signal so that an agent can tell the other agents that it does not want to move any more—meaning it has reached its final destination. Only with such a *success announcement* the agents will in the end know that everyone has reached its destination.

An instance of the problem is now given by the tuple $\mathcal{M}_{DU} = \langle A, G, s_0, \alpha_* \rangle$, with the set of agents A, the graph $G = (V, E)$, the initial objective state $s_0 = (\alpha_0, \beta_0)$, and the goal configuration α_*. Movement actions change the configuration α, while success announcements change the destination function β. If an agent i makes a success announcement while being in location v, we change the destination function to $\beta[i/\emptyset]$, signaling that the agent has reached its destination and is not allowed to move anymore. The goal state is reached if for all agents i, $\alpha(i) = \alpha_*(i)$ (the destination has been reached) and $\beta(i) = \emptyset$ (success has been announced).

2.2   Branching Plans

When an agent i is starting to generate a plan, the agent knows, of course, its true destination $\alpha_*(i)$. The subjective view of the world is captured by the tuple $(\alpha, \beta, i, \alpha_*(i))$, which we call *subjective state of agent i*. Given a subjective state $(\alpha, \beta, i, \alpha_*(i))$, we call (α, β) the *corresponding objective state*. Using its subjective state, agent i can plan to make movements that eventually will lead to a goal state. Most probably, it will be necessary to plan for other agents to move out of the way or to move to their destination. So, the planning agent has to put itself into the shoes of another agent j: i must make a *perspective shift* taking j's view. Since i does not know the true destination of j, i must take all possibilities into account and plan for all of them. In other words, i must plan for j using all possible subjective states of j: $s_v^j = (\alpha, \beta, j, v)$ for $v \in \beta(j)$. When planning for each possible destination of j, the planning agent i must pretend not to know the true destination of itself because it plans with the knowledge of agent j, which is uncertain about i's destinations.

All in all, a plan in the context of MAPF with destination uncertainty is no longer a linear sequence, but a *branching plan*. Furthermore, it is not enough to reach the true goal state, but the plan has to be successful for all possible destinations of all the agents (except for the starting agent i, who knows its own destination).

Such a branching plan corresponds roughly to what has been termed *policy* in the more general context of *implicitly coordinated epistemic planing* [2,3]. In order to illustrate the concept of a branching plan, let us consider the example in Fig. 1. Here *square agent S* knows that its destination is v_3 (the solid square) and the *circle agent C* knows that its destination is v_4 (the solid circle). However both are unsure about the destinations of the other agent. So S knows that v_1 and v_4 are possible destinations for C. C in turns knows that v_2 and v_3 are possible destinations for S.

Fig. 1. Small example with square agent (S) and circle agent (C)

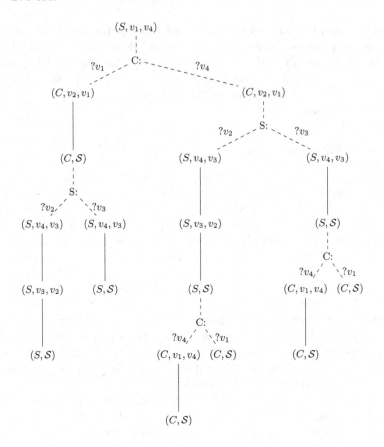

Fig. 2. Branching plan

Let us now assume that *square agent* S moves first to v_4. Now S puts itself into the shoes of *circle agent* C and reasons about what C would do, if v_1 is C's destination, and how C would continue if v_4 is C's destination. In the former case, C moves to v_1 and announces that it has reached its destination. In the other case, it will also move to v_1, offering S the possibility to move to its destination, whether it is v_2 or v_3. After that, C could move to its destination. All in all, a branching plan could look as depicted in Fig. 2. In this plan, each perspective shift to another agent is followed by branching according to the possible destinations of the agent. In general, we do not always require such branching because the agent might decide to move independently of its own destinations. One of the main results is then that all successful branching plans need to branch only on so-called *stepping stone* situations [5, Theorem 5]. These are configurations in which one agent has unblocked ways to all its possible destinations, and for those destinations, there are successful subplans after that agent has reached it. For example, in Fig. 2, after S's initial movement from v_1 to v_4, there is neither a stepping stone situation for C not for S. However, after

C had then moved from v_2 to v_1, C had created a stepping stone situation for S. S can move now uninterrupted to v_2, announce success and there is a successful plan afterwards for C. In case v_3 is the real destination, S can move there and again there is a successful plan afterwards for C. From the fact that a plan needs only to branch on stepping stones, it follows that these branching plans need to have only polynomial depth.

2.3 Joint Execution Model

After all agents have planned, we have a *family of plans* $(\pi_i)_{i \in A}$. *Joint execution* of this family of plans is then performed in an asynchronous, interleaved fashion. From all the agents i that have as their first action one of their own moves, one agent is chosen and its movement is executed. This is very similar to what happens in real-time board games, such as *Magic Maze*. The player who acts fastest carries out the action. For all the other agents the following happens: Either the movement was anticipated and then the movement is removed from the plan or the agent has to replan from the new situation. The interesting question is, whether such an asynchronous, distributed execution is guaranteed to eventually lead to the desired goal configuration and how many steps it takes to reach the common goal.

3 Exploiting Observations While Executing

As has been shown, under some reasonable conditions it is possible to guarantee success, provided that there is at least one agent which is able to come up with a plan initially [5]. However, there are also situations which look easily solvable, but it turns out that our notion of *implicit coordination* does not capture this. One such example is shown in Fig. 3. The square agent wants to go v_4 and knows that the circle agent wants to go either to v_2 or to v_5. Similarly, the circle agent wants to go to v_2 and knows that the square agent wants to go either to v_1 or to v_4.

If the square agent tries to solve the instance, it will try to create a *stepping stone* situation [5, Sect. 3.2] for the other agent. The only possible way to do that appears to be to move to v_6. Now the circle agent can move to both possible destinations. Unfortunately, after moving to v_2, the circle agent cannot any

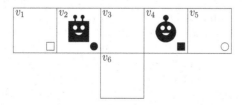

Fig. 3. MAPF/DU instance that is only solvable using inferences about observations

longer guarantee that the square agent can reach both of its possible destinations. Specifically, the square agent is blocked from reaching v_1. So, the move of the square agent to v_6 does not create a stepping stone for the circle agent. As is easy to see, all other movements of the square agents do not create a stepping stone either. Since the situation is completely symmetric, also the circle agent is unable to create a stepping stone.

If we try to explain movements by assuming that the other agents act rationally, we may assume that they always try to come up with shortest possible plans. Therefore, if the circle agents observes the square agent moving to v_6, the circle agent may rightly conclude that v_1 cannot be the actual destination of the square agent. Because if it were, then the square agent would have moved there directly announcing success, which would have led to an overall shorter plan. So, for the remaining part of the plan both agents can assume that v_1 is not the actual destination of the square agent. This implies that the circle agent can safely move to either v_2 or v_5 and afterwards the square agent can move to the only possible destination, namely v_4, which then solves the previously unsolvable instance.

Below we will discuss how to generalize this kind of reasoning.

3.1   Safe Abduction

Abduction is the inference to the best explanation, given an observation and a background theory. This is apparently what we are using when drawing the above conclusion that v_1 is not the destination of the square agent. In general, abduction is an "unsafe" inference in that the best explanation is not necessarily the correct one. For instance, often the best explanation for the malfunctioning of a device is based on a single-failure assumption, which might nevertheless not be the right explanation.

In our context, incorrect explanations could easily lead to situations, where destinations are no longer accessible, turning a solvable instance into an unsolvable one. In order to avoid that we will only accept explanations that are safe in the sense that they do not exclude a destination that is still be possible. Using the criterion of aiming for shortest plans, it may, however, still be possible to infer meaningful information.

3.2   Observations and Explanations

So what should count as an *observation* that needs an explanation? As in the example above, a meaningful observation is a sequence of movements by one agent i starting at a node v ending at node v' without interruptions by other agents. In the example, this would be the movements of the square agent from v_2 over v_3 to v_6. In this example, one might also could consider the movement from v_2 to v_3 as one observation.

In order to explain an observation, we take all possible destinations of the moving agent i into account and generate shortest plans for each of these destinations $v_{i,1}^*, \ldots, v_{i,k}^*$, starting with movements of agent i at node v not using the prefix from v to v'. Call these plans $\pi_{i,j}$. In creating these plans one has to take into account that the other agents do not know the destinations of that agent. Similarly, create shortest plans that include the prefix from v to v' and call these plans $\pi'_{i,j}$. Note that all these sub-plans may also use safe abduction!

Assuming that $|\pi|$ denotes the execution cost of plan π, we now compare the plans for all destinations $v_{i,j}$. If $|\pi_{i,j}| < |\pi'_{i,,j}|$,[1] we conclude that agent i cannot rationally try to reach destination $v^*_{i,j}$ moving from v to v'. In fact, all agents observing this behavior can conclude this and agent i (and everybody else) is aware that everybody else knows that. In other words, after agent i moved from v to v', it is common knowledge that agent i is trying to reach one of the destinations $v_{i,j}$ such that $|\pi'_{i,j}| \le |\pi_{i,j}|$. Note that we included all destinations $v^*_{i,j}$ with $|\pi'_{i,j}| = |\pi_{i,,j}|$, since there is no reason do dismiss $v^*_{i,j}$ after having reached v'.

One important prerequisite for this kind of inference to be correct is, however, that agents indeed always generate a shortest plan. And this does not only concern the overall plan, where we measure the length as the longest trace through the branching plan. Instead, this should be true also for each sub-plan at each point, where a perspective shift happens. This is something we currently do not require from our plans when giving success guarantees. Furthermore, we explicitly do not require to branch at each point where a perspective shift happens. However, it is, of course, possible to make that a requirement.

The most interesting feature of using this kind of inference is that it also can make instances solvable that were unsolvable before. For the example in Fig. 3, we showed that no stepping stone exists, so that it cannot be solved by an implicitly coordinated branching plan. However, using the notion of safe abduction, C can come up with the plan of moving to v_6. Now this is definitely not a prefix of an optimal plan for solving this instance when v_5 is C's destination. On the other hand, there exists no plan at all to solve the instance when v_2 is the destination, i.e., plan length is infinite. In other words, everybody can safely assume that C does not have v_5 as a destination. Using this assumption, the instance can then be easily solved.

However, it turns out that sometimes the agent might not have the right option to act in order to signal that a possible destination can be excluded. Let us reconsider the example from above but place C initially into cell v_6. Now the problem is that the only way to signal that v_5 is not C's destination is to do nothing. However, doing nothing cannot be observed in our asynchronous execution model. The way out here could be to introduce an observable *wait* action, which induces also execution costs. Then no successful shortest sub-plan for a particular destination could contain this action. On the other hand, if there does not exist a plan for a possible destination where the agent moves first, a wait action does not matter, because plan length is infinite in any case. In our modified example, where C is initially in cell v_6, there exists a plan for C's destination v_5 with C moving first, hence a wait action could not be part of a shortest plan. For C's destination v_2 on the other hand, there does not

---

[1] If no plan can be found, then we assume infinitely large execution costs.

exist any plan with C moving first. So, a wait action is appropriate here. So, a wait action can signal that only those destinations remain for which there exists no plan initially; in our example, this would mean that v_2 must be the actual destination.

Actually, an alternative to a wait action might be to make one move and then return to the original location. This would also signal that only those destinations are possible for which there is no initial plan where the agent moves first.

In any case, regardless of whether we use a *wait* action or a back and forth movement, we seem to violate the requirement set out earlier, namely that plans should by cycle-free, which in the case of branching plans translates into the requirement that no objective state should be visited twice on a possible execution trace. However, the movement is made in order to change the common knowledge (see below) and in so far, no cycle is created.

3.3   Forward Induction

During the MAPF/DU planning process, common knowledge over all possible destinations is maintained using the possible destination function β, i.e., $\beta(i)$ is the commonly known set of all possible destinations for agent i. This set is reduced to the empty set whenever agent i makes a success announcement. If we use safe abduction as described above, we can reduce the set of possible destinations $\beta(i)$ to all those that are still possible according to the definition in the previous subsection. Interestingly, since this is common knowledge, this reduction can be propagated to the entire sub-plan following i's movement (or inaction).

In order to illustrate that this can even proceed over more than one stage, let us consider a more complex example, where we add a third agent T, the triangle agent (Fig. 4).

Here, T could start by moving to v_4, signalling that v_{10} is not its destination. The only way for C would be to move to v_{10} (in order to help T later on), in order to allow T to move to its possible destination. Not that at this point we are not entitled to make the inference that v_7 is not the destination of C, because C could not have moved there announcing success with the guarantee

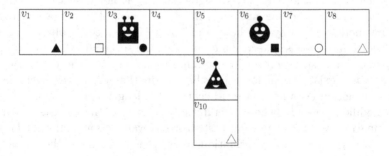

Fig. 4. More complex example

that the remaining problem could be solved. If T's goal were v_8, it could move there and the rest would be easy for C and S. However, instead it moves to v_9 signalling that v_8 is not its destination. Now it is common knowledge that v_1 is T's destination. However C and S do not have knowledge about their respective destinations. S would unblock the way for T by moving to v_6, and T could move to v_1, announcing success. C cannot make any meaningful move, so S has to move, either to v_2 or v_3, in order to allow for C either to move to its destination v_7, or to signal that this not C's destination. Now C could execute a wait action in order to signal that v_7 is not C's destination. So, S could happily move to v_6 and announce success, after which C could easily finish.

3.4   Computational Complexity

From a computational complexity point, most probably nothing changes. The construction in the proof of Theorem 11 of the original paper [5] still works. In particular, forward induction is of no help and not a hindrance in deciding the constructed MAPF/DU instance. For proving PSPACE membership, one has to prove a generalized stepping stone theorem. A generalized stepping stone is now a state such that either the agent can reach all possible destinations, announce success and the solvability of the simplified problem can be guaranteed (as usual), or the movements of the agent end in a state such that we can safely abduct and can at least eliminate one possible destination. With that, polynomial depth of the branching plan follows and then one could easily guess and check all traces iteratively. One has to guess also the depth of plans (which subsequently have to be verified) in order to allow for the verification of safe abductions.

4   Outlook

Although the complexity probably does not change, algorithmically things become more involved. In fact, one might want to consider only special cases of safe abduction inferences in order to reduce computational overhead. For example, one might only consider situations when success announcements are possible and ignored, as in the above examples. Otherwise the computational burden might be too high. In particular, it remains unclear whether we could reduce the worst case execution costs, which are what we are interested in when proving success guarantees.

An interesting question then comes up related to the omniscience problem. If an agent does something that another agent can take as a signal, then the other agent actually has to recognize that as a signal, otherwise the plan of the acting agent might not work out. In other words, all agents have to use the same level of reasoning.

All in all, the ideas spelled out here might hopefully serve as a starting point for defining a notion of implicit coordination that also takes into account the observation about actions of the other agents. Hopefully, this might also lead to generalizing these ideas to more general settings such as epistemic planning with monotonic uncertainty reduction or even general epistemic planning.

References

1. Battigalli, P., Siniscalchi, M.M.: Strong belief and forward induction reasoning. J. Econ. Theory **106**(2), 356–391 (2002). https://doi.org/10.1006/jeth.2001.2942
2. Bolander, T., Engesser, T., Mattmüller, R., Nebel, B.: Better eager than lazy? How agent types impact the successfulness of implicit coordination. In: Principles of Knowledge Representation and Reasoning: Proceedings of the Sixteenth International Conference (KR 2018), pp. 445–453 (2018). https://aaai.org/ocs/index.php/KR/KR18/paper/view/18070
3. Engesser, T., Bolander, T., Mattmüller, R., Nebel, B.: Cooperative epistemic multi-agent planning for implicit coordination. In: Proceedings of the Ninth Workshop on Methods for Modalities (M4MICLA 2017), pp. 75–90 (2017). https://doi.org/10.4204/EPTCS.243.6
4. Kornhauser, D., Miller, G.L., Spirakis, P.G.: Coordinating pebble motion on graphs, the diameter of permutation groups, and applications. In: 25th Annual Symposium on Foundations of Computer Science (FOCS 1984), pp. 241–250 (1984). https://doi.org/10.1109/SFCS.1984.715921
5. Nebel, B., Bolander, T., Engesser, T., Mattmüller, R.: Implicitly coordinated multi-agent path finding under destination uncertainty: success guarantees and computational complexity. J. Artif. Intell. Res. **64**, 497–527 (2019). https://doi.org/10.1613/jair.1.11376
6. Ramírez, M., Geffner, H.: Plan recognition as planning. In: Proceedings of the 21st International Joint Conference on Artificial Intelligence (IJCAI 2009), pp. 1778–1783 (2009). http://ijcai.org/Proceedings/09/Papers/296.pdf
7. Ratner, D., Warmuth, M.K.: Finding a shortest solution for the N × N extension of the 15-puzzle is intractable. In: Proceedings of the 5th National Conference on Artificial Intelligence (AAAI 1986), pp. 168–172 (1986). http://www.aaai.org/Library/AAAI/1986/aaai86-027.php
8. Schmidt, C.F., Sridharan, N.S., Goodson, J.L.: The plan recognition problem: an intersection of psychology and artificial intelligence. Artif. Intell. **11**(1–2), 45–83 (1978). https://doi.org/10.1016/0004-3702(78)90012-7

Temporally Attributed Description Logics

Ana Ozaki[1]([✉])[iD], Markus Krötzsch[2]([✉])[iD], and Sebastian Rudolph[2][iD]

[1] KRDB Research Centre, Free University of Bozen-Bolzano, Bolzano, Italy
ana.ozaki@unibz.it
[2] TU Dresden, Dresden, Germany
markus.kroetzsch@tu-dresden.de

Abstract. Knowledge graphs are based on graph models enriched with (sets of) attribute-value pairs, called annotations, attached to vertices and edges. Many application scenarios of knowledge graphs crucially rely on the frequent use of annotations related to *time*. Building upon attributed logics, we design description logics enriched with temporal annotations whose values are interpreted over discrete time. Investigating the complexity of reasoning in this new formalism, it turns out that reasoning in our temporally attributed description logic $\mathcal{ALCH}_{@}^{T}$ is highly undecidable; thus we establish restrictions where it becomes decidable, and even tractable.

1 Introduction

Graph-based data formats play an essential role in modern information management, since they offer schematic flexibility, ease information re-use, and simplify data integration. Ontological knowledge representation has been shown to offer many benefits to such data-intensive applications, e.g., by supporting integration, querying, error detection, or repair. However, practical *knowledge graphs*, such as Wikidata [38] or YAGO2 [21], are based on *enriched* graphs where edges are augmented with additional annotations.

Example 1. Figure 1 shows an excerpt of the information that Wikidata provides about Franz Baader. Binary relations, such as memberOf(FranzBaader, AcademiaEuropaea), are the main modelling primitive for encoding knowledge. They correspond to labelled directed edges in the graph. However, many of these edges are annotated with additional information, specifying validity times, references (collapsed in the figure), auxiliary details, and other pieces of information that pertain to this binary relationship.

A similar approach to knowledge modelling is followed in the popular *Property Graph* data model [34], and supported by modern graph stores such as Amazon Azure, BlazeGraph, and Neo4j. Other data models allowing attribute-value pairs to be associated with relations are UML, entity-relation and object-role modelling (see, e.g., [2,10,37] for works drawing the connection between these data models and DLs). Predicate logic does not have a corresponding notion

© Springer Nature Switzerland AG 2019
C. Lutz et al. (Eds.): Baader Festschrift, LNCS 11560, pp. 441–474, 2019.
https://doi.org/10.1007/978-3-030-22102-7_21

Franz Baader (Q92729)

German computer scientist

member of	Academia Europaea		edit
	affiliation	AE section Informatics	
	start time	2011	
	▸ 2 references		

employer	TU Dresden		edit
	start time	2002	
	position held	full professor	
	▸ 1 reference		
	RWTH Aachen University		edit
	start time	1993	
	end time	2002	
	position held	associate professor	
	▸ 1 reference		

educated at	University of Erlangen-Nuremberg		edit
	start time	1985	
	end time	1989	
	academic degree	doctorate	
	▸ 1 reference		

Fig. 1. Excerpt of the Wikidata page of Franz Baader; https://wikidata.org/wiki/Q92729

of enriched relationships, and established ontology languages that are based on traditional logic are therefore not readily applicable to enriched graphs [22]. To provide better modelling support, *attributed logics* have been proposed as a way of integrating annotations with logical reasoning [32]. This approach has been applied to description logics (DLs) [7] to obtain *attributed DLs* [12,23,24].

Annotations in practical knowledge graphs have many purposes, such as recording provenance, specifying context, or encoding *n*-ary relations. One of their most important uses, however, is to encode *temporal validity* of statements. In Wikidata, e.g., *start/end time* and *point in time* are among the most frequent annotations, used in 6.7 million statements overall.[1] YAGO2 introduced the

---

[1] As of March 2019, the only more common annotations are *reference* (provenance) and *determination method* (context); see https://tools.wmflabs.org/sqid/#/browse?type=properties&sortpropertyqualifiers=fa-sort-desc.

SPOTL data format that enriches *subject-property-object* triples (known from RDF) with information on *time* and *location* [21].

Reasoning with time clearly requires an adequate semantics, and many approaches were proposed. Validity time points and intervals are a classical topic in data management [17,18], and similar models of time have also been studied in ontologies [4,26]. However, researchers in ontologies have most commonly focussed on abstract models of time as used in temporal logics [8,31,39]. Temporal reasoning in $\mathcal{ALC}$ with concrete domains was proposed by Lutz et al. [29]. It is known that satisfiability of $\mathcal{ALC}$ with a concrete domain consisting of a dense domain and containing the predicates $=$ and $<$ is EXPTIME-complete [28]. In the same setting but for *discrete time*, the complexity of the satisfiability problem is open, a criterion which only guarantees decidability has been proposed by Carapelle and Turhan [15]. None of these approaches has been considered for attributed logics yet, and indeed support for temporal reasoning for knowledge graphs, such as Wikidata and YAGO2, is still missing today. In this paper, we address this shortcoming by endowing attributed description logics with a temporal semantics for annotations. Indeed, annotations are already well-suited for representing time-related data.

Example 2. We introduce temporally attributed DLs that use special temporal annotation attributes, which can refer to individual time points or to intervals of time. For example, information about Franz Baader's current employment can be expressed by an annotated DL fact as follows:

$$\mathsf{employer}(\mathsf{FranzBaader}, \mathsf{TUD})@[\mathsf{since}\colon 2002, \mathsf{position}\colon \mathsf{fullProfessor}] \qquad (1)$$

Here, the special temporal attribute since is used alongside the regular attribute position. Likewise, we can express intervals, as in the following axiom[2]

$$\mathsf{educatedAt}(\mathsf{FranzBaader}, \mathsf{FAU})@[\mathsf{during}\colon [1985, 1989], \mathsf{degree}\colon \mathsf{doctorate}] \qquad (2)$$

Some facts might also be associated with a specific time rather than with a duration. For example, we could encode some of the knowledge in Wikidata with the fact:

$$\mathsf{bornIn}(\mathsf{FranzBaader}, \mathsf{Spalt})@[\mathsf{time}\colon 1959] \qquad (3)$$

Not all people are as thoroughly documented on Wikidata, but attributed DLs also provide ways of leaving some information unspecified, as in the following fact about one of Baader's former doctoral students:

$$\big(\exists \mathsf{bornIn}@[\,\mathsf{between}\colon [1950, 2000]\,].\top\,\big)(\mathsf{Carsten}), \qquad (4)$$

which merely states that Carsten Lutz was born *somewhere* within the second half of the 20th century.

---

[2] FAU is the official abbreviation for the *Friedrich-Alexander University* in Erlangen/Nuremberg.

$$\exists employer@\lfloor position: fullProfessor\rfloor.\top \sqsubseteq Professor \tag{5}$$

$$X: \lfloor position: fullProfessor\rfloor \quad (\exists employer@X.\top \sqsubseteq Professor) \tag{6}$$

$$X: \lfloor position: fullProfessor\rfloor \quad (\exists employer@X.\top \sqsubseteq Professor@[time: X.time]) \tag{7}$$

$$\exists employer@[position: fullProfessor, time: x].\top \sqsubseteq Professor@[time: x] \tag{8}$$

Fig. 2. Examples for axioms in attributed description logics

To deal with such temporally annotated data in a semantically adequate way and to specify temporal background knowledge, we propose the temporally attributed description logic $\mathcal{ALCH}_@^\top$ that enables reasoning and querying with such information. In addition to the basic support for representing information with attributes, our logic includes a special semantics for temporal attributes, and the support for (safe) variables in DL axioms. Beyond defining syntax and semantics of $\mathcal{ALCH}_@^\top$, our contributions are the following:

- We show that the full formalism is highly undecidable using an encoding of a recurring tiling problem.
- We present three ways (of increasing reasoning complexity) for regaining decidability: disallowing variables altogether (EXPTIME), disallowing the use of variables only for temporal attributes (2EXPTIME), or disallowing the use of temporal attributes referencing time points in the future (3EXPTIME).
- Finally we single out a lightweight case based on the description logic $\mathcal{EL}$ which features PTIME reasoning.

2   Temporally Attributed DLs

We first present the syntax and underlying intuition of temporally attributed description logics. In DL, a true fact corresponds to the membership of an element in a class, or of a pair of elements in a binary relation. Attributed DLs further allow each true fact to carry a finite set of annotations [23], given as attribute-value pairs. As suggested in Example 2, the same relationship may be true with several different annotation sets, e.g., to capture that Baader has been educated at FAU Erlangen-Nuremberg during two intervals: once for his PhD and once for his *Diplom* (not shown in Fig. 1).

Example 3. To guide the reader in following the formal definitions, we first illustrate the main features of attributed DL by means of some example axioms, shown in Fig. 2. We already use time as an example annotation, but do not yet rely on any specific semantic interpretation for this attribute.

$$\exists \text{educatedAt} @ \lfloor \text{before}: x \rfloor .\top \sqsubseteq \neg \exists \text{bornIn} @ \lfloor \text{time}: x \rfloor .\top \tag{9}$$

$$\exists \text{educatedAt} @ \lfloor \text{time}: x \rfloor .\top \sqsubseteq \neg \exists \text{bornIn} @ \lfloor \text{after}: x \rfloor .\top \tag{10}$$

$$\exists \text{educatedAt} @ \lfloor \text{time}: x \rfloor .\top \sqsubseteq \exists \text{educatedAt} @ \lfloor \text{during} : [x, x] \rfloor .\top \tag{11}$$

$$\exists \text{bornIn} @ \lfloor \text{between} : [x, y] \rfloor .\top \sqsubseteq \neg \exists \text{educatedAt} @ \lfloor \text{before}: x \rfloor .\top \tag{12}$$

Fig. 3. Examples for axioms in temporally attributed description logics

The (non-temporal) attributed DL axiom (5) states that people employed as full professors are professors. The *open specifier* $\lfloor \text{position}: \text{fullProfessor} \rfloor$ requires that the given attribute is among the annotations, but allows other annotations to be there as well (denoted by the half-open brackets). Axiom (6) is equivalent to (5), but assigns the annotation set to a *set variable* X.

If the employer relation specifies a validity time, the same time would apply to Professor. This is accomplished by axiom (7), which uses the expression time: X.time to declare that *all* (zero or more) time values of X should be copied. The closed brackets in the conclusion specify that no further attribute-value pairs may occur in the annotation of the conclusion.

A subtly different meaning is captured by (8), which uses an *object variable* x as a placeholder for a single attribute value. In contrast to (8), axiom (7) (i) requires that at least one time annotation is present (rather than allowing zero or more), (ii) requires that the annotation set in the premise has exactly two attribute-value pairs (rather than being open for more), and (iii) infers distinct Professor assertions for each time x (rather than one assertion that copies all time points). Item (iii) deserves some reflection. As argued above, it is meaningful that the same fact holds true with different annotation sets, and this does not imply that it is also true with the union of these annotations. However, in the case of time, our intuition is that something is true at several times individually exactly if it is true at all of these times together. Our formal semantics will ensure that this equivalence holds.

We define our description logic $\mathcal{ALCH}_@^{\mathbb{T}}$ as a multi-sorted version of the attributed DL $\mathcal{ALCH}_@$, thereby introducing datatypes for time points and intervals. Elements of the different types are represented by members of mutually disjoint sets of *(abstract) individual names* $\mathsf{N_I}$, *time points* $\mathsf{N_T}$, and *time intervals* $\mathsf{N_T^2}$. We represent time points by natural numbers, and assume that elements of $\mathsf{N_T}$ ($\mathsf{N_T^2}$) are (pairs of) numbers in *binary* encoding. We write $[k, \ell]$ for a pair of numbers k, ℓ in $\mathsf{N_T^2}$. Moreover, we require that there are the following seven special individual names, called *temporal attributes*: time, before, after, until, since, during, between $\in \mathsf{N_I}$.

The intuitive meaning of temporal attributes is as one might expect: time describes individual times at which a statement is true, while the others describe (half-open) intervals. The meaning of before, after, and between is existential in that they require the statement to hold only at some time in the interval, while until, since, and during are universal and require something to be true throughout an interval.

Example 4. The examples in Fig. 3 illustrate the special semantics of temporal attributes. Axiom (9) states that nobody can be educated before being born. Axiom (10) is equivalent. In particular, our semantics ensures that temporal attributes like time, before, and after will be inferred even when not stated explicitly. For example, (11) is a tautology. Longer intervals of during can be inferred for any span of consecutive time points (our time model is discrete). Finally, we also allow using object variables in time intervals, as illustrated in (12), which is actually *equivalent* to (9) as well.

With these examples in mind, we continue to define the syntax of our temporal DLs formally. Axioms of $\mathcal{ALCH}_@^{\mathbb{T}}$ are further based on sets of *concept names* N_C, and *role names* N_R. Attributes are represented by individual names, and we associate a *value type* valtype(a) with each individual $a \in N_I$ for this purpose: during and between have value type N_T^2, all other temporal attributes have value type N_T, and all other individuals have value type N_I. An *attribute-value pair* is an expression $a : v$ where $a \in N_I$ and $v \in$ valtype(a). Now concept and role assertions of $\mathcal{ALCH}_@^{\mathbb{T}}$ have the following form, respectively:

$$C(a)@[a_1 : v_1, \ldots, a_n : v_n] \tag{13}$$

$$r(a, b)@[a_1 : v_1, \ldots, a_n : v_n] \tag{14}$$

where $C \in N_C$, $r \in N_R$, $a, b \in N_I$, and $a_i : v_i$ are attribute-value pairs. Note that (4) in Example 2 is not a concept assertion in the sense of (13), since it uses a complex concept expression. As usual in DLs, our language will allow us to encode such complex assertions by giving them a new name in a terminological axiom.

Role and concept inclusion axioms of $\mathcal{ALCH}_@^{\mathbb{T}}$ introduce additional expressive power to refer to partially specified and variable annotation sets. Attribute values may now also contain *object variables* taken from pairwise disjoint sets Var(N_I), Var(N_T), and Var(N_T^2). Moreover, whole annotation sets might be represented by *set variables* from a set N_V.

Definition 1. *An (annotation set) specifier can be a set variable $X \in N_V$, a closed specifier of the form $[a_1 : v_1, \ldots, a_n : v_n]$, or an open specifier of the form $\lfloor a_1 : v_1, \ldots, a_n : v_n \rfloor$, where $n \geq 0$, $a_i \in N_I$ and each v_i is an expression that is compatible with the value type of its attribute in the sense that it has one of the following forms:*

- *$v_i \in$ valtype(a_i) $\cup$ Var(valtype(a_i)), or*
- *$v_i = [v, w]$ with valtype(a_i) $= N_T^2$ and v, w in $N_T \cup$ Var(N_T), or*
- *$v_i = X.b$ with $X \in N_V$, $b \in N_I$, and valtype(a_i) $=$ valtype(b).*

The set of all specifiers is denoted $\mathbf{S}$. A specifier is ground if it does not contain variables.

Intuitively, closed specifiers define specific annotation sets whereas open specifiers provide lower bounds [23]. Object variables are used to copy values from one attribute to another, as long as the attributes have the same value type (in the same annotation set or in a new one); the expression $X.b$ is used to copy all of the zero or more b-values of annotation set X. We also allow specifiers to be empty. That is, we allow $\bigsqcup$ (meaning "any annotation set") and $\bigsqcap$ (meaning "the empty annotation set"). To simplify notation, we may omit $@\bigsqcup$ and $@\bigsqcap$ in role or concept expressions (and $@\bigsqcap$ in assertions).

Definition 2. $\mathcal{ALCH}_@^{\mathsf{T}}$ role expressions *have the form* $r@S$ *with* $r \in \mathsf{N_R}$ *and* $S \in \mathbf{S}$. $\mathcal{ALCH}_@^{\mathsf{T}}$ concept expressions C, D *are defined recursively:*

$$C, D ::= \top \mid A@S \mid \neg C \mid (C \sqcap D) \mid \exists R.C \qquad (15)$$

with $A \in \mathsf{N_C}$, $S \in \mathbf{S}$ *and* R *an* $\mathcal{ALCH}_@^{\mathsf{T}}$ *role expression.*

We use abbreviations $(C \sqcup D)$, $\bot$, and $\forall R.C$ for $\neg(\neg C \sqcap \neg D)$, $\neg\top$, and $\neg(\exists R.\neg C)$, respectively. $\mathcal{ALCH}_@^{\mathsf{T}}$ axioms are essentially just (role/concept) inclusions between $\mathcal{ALCH}_@^{\mathsf{T}}$ role and concept expressions, which may, however, share variables.

Example 5. Object variables can be used to create new intervals of time using the temporal information present on the annotations. In the following example, we illustrate a concept inclusion that allows for inferring the (minimum) period in which a person typically is a PhD student:

$$\exists \mathsf{obtainedMSc}@\lfloor \mathsf{between} : [x, x'] \rfloor.\top \sqcap \exists \mathsf{obtainedPhD}@\lfloor \mathsf{between} : [y, y'] \rfloor.\top$$
$$\sqsubseteq \mathsf{PhDStudent}@\lfloor \mathsf{during} : [x', y] \rfloor \qquad (16)$$

It is sometimes useful to represent annotations by variables while also specifying some further constraints on their possible values. This can be accommodated by adding such constraints as (optional) prefixes to axioms.

Definition 3. *An* $\mathcal{ALCH}_@^{\mathsf{T}}$ concept inclusion *is an expression of the form*

$$X_1 : S_1, \ldots, X_n : S_n \quad (C \sqsubseteq D), \qquad (17)$$

where C, D *are* $\mathcal{ALCH}_@^{\mathsf{T}}$ concept expressions, $S_1, \ldots, S_n \in \mathbf{S}$ *are closed or open specifiers, and* $X_1, \ldots, X_n \in \mathsf{N_V}$ *are set variables occurring in* C, D *or in* $S_1, \ldots, S_n$. *We require that all variables are* safe *in the following sense:*

(1) every set variable in the axiom also occurs in the left concept C,[3] *and*
(2) every object variable in the axiom also occurs either in the left concept C *or in a specifier* S_i *in a prefix* $X_i : S_i$.

---

[3] This is a simplification from previous works [24] where set variables were allowed to occur in the specifier prefix only under some circumstances. It is not hard to see that our simplification does not relinquish relevant expressivity if we permit some normalisation.

$\mathcal{ALCH}_@^T$ role inclusions *are defined analogously, but with role expressions instead of the concept expressions. An* $\mathcal{ALCH}_@^T$ *ontology is a set of* $\mathcal{ALCH}_@^T$ *assertions, and role and concept inclusions.*

Note that any $\mathcal{ALCH}$ axiom is also an $\mathcal{ALCH}_@^T$ axiom in the sense that the absence of explicit annotations can be considered to mean "$@\bigsqcup$."

3   Semantics of Temporally Attributed DLs

We first recall the general semantics of attributed DLs without temporal attributes. The semantics of $\mathcal{ALCH}_@^T$ can then be obtained as a multi-sorted extension that imposes additional restrictions on the interpretation of time.

An *interpretation* $\mathcal{I} = (\Delta^{\mathcal{I}}, \cdot^{\mathcal{I}})$ of attributed logic consists of a non-empty domain $\Delta^{\mathcal{I}}$ and a function $\cdot^{\mathcal{I}}$. Individual names $a \in \mathsf{N}_\mathsf{I}$ are interpreted as elements $a^{\mathcal{I}} \in \Delta^{\mathcal{I}}$. To interpret annotation sets, we use the set $\Phi^{\mathcal{I}}$ of all finite binary relations over $\Delta^{\mathcal{I}}$. Each concept name $C \in \mathsf{N}_\mathsf{C}$ is interpreted as a set $C^{\mathcal{I}} \subseteq \Delta^{\mathcal{I}} \times \Phi^{\mathcal{I}}$ of elements with annotations, and each role name $r \in \mathsf{N}_\mathsf{R}$ is interpreted as a set $r^{\mathcal{I}} \subseteq \Delta^{\mathcal{I}} \times \Delta^{\mathcal{I}} \times \Phi^{\mathcal{I}}$ of pairs of elements with annotations. Each element (pair of elements) may appear with multiple annotations [23].

Note that attributes are represented by domain elements in this semantics. This has no actual impact on reasoning in the context of this paper, and could be changed to use a separate *sort* for attributes or to consider them as a kind of predicate that is part of a fixed schema. While this detail is immaterial to our proofs, it is worth noting that attributes are also treated as special kinds of domain objects in important practical knowledge graphs. Both RDF-based models and Wikidata use (technically different) notions of *property* that are part of the domain and can therefore be described by facts. This ability is frequently used in practice to store annotations, to declare constraints, or to establish mappings to external vocabularies. We believe that in particular constraint information and mappings in datasets should be accessible to ontological reasoning. In contrast, the Property Graph data model represents attributes as *property keys* (plain strings) that cannot be used as objects (vertices) in the graph [33]. However, in this model, attribute values (*property values*) cannot refer to objects in the graph either. We do not consider it desirable to impose those restrictions, since our more general model can capture more real-world graphs, and is useful for expressing many natural statements (e.g., values like *full professor* in Fig. 1 refer to domain objects, which Property Graph would not allow).

3.1   Time-Sorted Interpretations

To deal with time, we define interpretation that include temporal sorts in addition to the usual abstract domain.

Definition 4. *A* time-sorted interpretation $\mathcal{I} = (\Delta^{\mathcal{I}}, \cdot^{\mathcal{I}})$ *is an interpretation with a domain* $\Delta^{\mathcal{I}}$ *that is a disjoint union of* $\Delta_I^{\mathcal{I}} \cup \Delta_T^{\mathcal{I}} \cup \Delta_{2T}^{\mathcal{I}}$, *where* $\Delta_I^{\mathcal{I}}$ *is the* abstract domain, $\Delta_T^{\mathcal{I}}$ is a finite or infinite interval,[4] called temporal domain, and $\Delta_{2T}^{\mathcal{I}} = \Delta_T^{\mathcal{I}} \times \Delta_T^{\mathcal{I}}$.

We interpret individual names $a \in \mathsf{N}_\mathsf{I}$ as elements $a^{\mathcal{I}} \in \Delta_I^{\mathcal{I}}$; time points $t \in \mathsf{N}_\mathsf{T}$ as $t^{\mathcal{I}} \in \Delta_T^{\mathcal{I}}$; and intervals $[t, t'] \in \mathsf{N}_\mathsf{T}^2$ as $[t, t']^{\mathcal{I}} = (t^{\mathcal{I}}, t'^{\mathcal{I}}) \in \Delta_{2T}^{\mathcal{I}}$. A pair $(\delta, \epsilon) \in \Delta_I^{\mathcal{I}} \times \Delta^{\mathcal{I}}$ is well-typed, if one of the following holds:

(a) $\delta = a^{\mathcal{I}}$ for an attribute a of value type N_T and $\epsilon \in \Delta_T^{\mathcal{I}}$; or
(b) $\delta = a^{\mathcal{I}}$ for an attribute a of value type N_T^2 and $\epsilon \in \Delta_{2T}^{\mathcal{I}}$; or
(c) $\delta = a^{\mathcal{I}}$ for an attribute a of value type N_I and $\epsilon \in \Delta_I^{\mathcal{I}}$.

Let $\Phi^{\mathcal{I}}$ be the set of all finite sets of well-typed pairs. The function $\cdot^{\mathcal{I}}$ maps concept names $C \in \mathsf{N}_\mathsf{C}$ to $C^{\mathcal{I}} \subseteq \Delta^{\mathcal{I}} \times \Phi^{\mathcal{I}}$ and role names $r \in \mathsf{N}_\mathsf{R}$ to $r^{\mathcal{I}} \subseteq \Delta^{\mathcal{I}} \times \Delta^{\mathcal{I}} \times \Phi^{\mathcal{I}}$.

Note that $\Delta_T^{\mathcal{I}}$ can be finite if N_T and N_T^2 are (which is always admissible, since any ontology mentions only finitely many time points). $\mathcal{I}$ satisfies a concept assertion $C(a)@[a_1 : v_1, \ldots, a_n : v_n]$ if $(a^{\mathcal{I}}, \{(a_1^{\mathcal{I}}, v_1^{\mathcal{I}}), \ldots, (a_n^{\mathcal{I}}, v_n^{\mathcal{I}})\}) \in C^{\mathcal{I}}$, and likewise for role assertions. For interpreting expressions with (object or set) variables, we need a notion of variable assignment.

Definition 5 (semantics of terms). *A variable assignment for a time-sorted interpretation $\mathcal{I}$ is a function $\mathcal{Z}$ that maps set variables $X \in \mathsf{N}_\mathsf{V}$ to finite binary relations $\mathcal{Z}(X) \in \Phi^{\mathcal{I}}$, and object variables $x \in \mathsf{Var}(\mathsf{N}_\mathsf{I}) \cup \mathsf{Var}(\mathsf{N}_\mathsf{T}) \cup \mathsf{Var}(\mathsf{N}_\mathsf{T}^2)$ to elements $\mathcal{Z}(x) \in \Delta_I^{\mathcal{I}} \cup \Delta_T^{\mathcal{I}} \cup \Delta_{2T}^{\mathcal{I}}$ (respecting their types). For (set or object) variables x, let $x^{\mathcal{I}, \mathcal{Z}} := \mathcal{Z}(x)$, and for abstract individuals, time points, or time intervals a, let $a^{\mathcal{I}, \mathcal{Z}} := a^{\mathcal{I}}$.*

Intuitively, each specifiers defines a set of annotation sets. For closed specifiers, there is just one such set (corresponding exactly to the specified attribute-value pairs), whereas for open specifiers, we obtain many sets (namely all supersets of the set that was specified). The following definition is making this formal, and also defines the semantics for all types of expressions that may occur in the value position of attributes within specifiers.

Definition 6 (semantics of specifiers). *A specifier $S \in \mathbf{S}$ is interpreted as a set $S^{\mathcal{I}, \mathcal{Z}} \subseteq \Phi^{\mathcal{I}}$ of matching annotation sets. We set $X^{\mathcal{I}, \mathcal{Z}} := \{\mathcal{Z}(X)\}$ for variables $X \in \mathsf{N}_\mathsf{V}$. The semantics of closed specifiers is defined as follows:*

(i) $[a : v]^{\mathcal{I}, \mathcal{Z}} := \{\{(a^{\mathcal{I}}, v^{\mathcal{I}, \mathcal{Z}})\}\}$, with $v \in \mathsf{valtype}(a) \cup \mathsf{Var}(\mathsf{valtype}(a))$;
(ii) $[a : [v, w]]^{\mathcal{I}, \mathcal{Z}} := \{\{(a^{\mathcal{I}}, (v^{\mathcal{I}, \mathcal{Z}}, w^{\mathcal{I}, \mathcal{Z}}))\}\}$, with $\mathsf{valtype}(a) = \mathsf{N}_\mathsf{T}^2$, and $v, w \in \mathsf{N}_\mathsf{T} \cup \mathsf{Var}(\mathsf{N}_\mathsf{T})$;
(iii) $[a : X.b]^{\mathcal{I}, \mathcal{Z}} := \{\{(a^{\mathcal{I}}, \delta) \mid (b^{\mathcal{I}}, \delta) \in \mathcal{Z}(X)\}\}$;
(iv) $[a_1 : v_1, \ldots, a_n : v_n]^{\mathcal{I}, \mathcal{Z}} := \{\bigcup_{i=1}^{n} F_i\}$ with $\{F_i\} = [a_i : v_i]^{\mathcal{I}, \mathcal{Z}}$ for all $i \in \{1, \ldots, n\}$.

---

[4] As usual for the natural numbers, a finite interval $[k, \ell]$ is $\{n \in \mathbb{N} \mid k \le n \le \ell\}$ and an infinite interval $[k, \infty)$ is $\{n \in \mathbb{N} \mid k \le n\}$.

$S^{\mathcal{I},\mathcal{Z}}$ *therefore is a singleton set for variables and closed specifiers. For open specifiers, however, we define* $\lfloor a_1\!:\!v_1,\ldots,a_n\!:\!v_n\rfloor^{\mathcal{I},\mathcal{Z}}$ *to be the set*

$$\{F \in \Phi^{\mathcal{I}} \mid F \supseteq G \text{ for } \{G\} = \lceil a_1\!:\!v_1,\ldots,a_n\!:\!v_n\rceil^{\mathcal{I},\mathcal{Z}}\}.$$

With the above definitions in place, we can now define the semantics of concepts and roles in the expected way, simply adding the appropriate condition for the additional annotation sets.

Definition 7 (semantics of concepts and roles). *For* $A \in \mathsf{N_C}$, $r \in \mathsf{N_R}$, *and* $S \in \mathbf{S}$, *let:*

$$(A@S)^{\mathcal{I},\mathcal{Z}} := \{\delta \mid (\delta, F) \in A^{\mathcal{I}} \text{ for some } F \in S^{\mathcal{I},\mathcal{Z}}\}, \tag{18}$$

$$(r@S)^{\mathcal{I},\mathcal{Z}} := \{(\delta, \epsilon) \mid (\delta, \epsilon, F) \in r^{\mathcal{I}} \text{ for some } F \in S^{\mathcal{I},\mathcal{Z}}\}. \tag{19}$$

The semantics of further concept expressions is defined as usual: $\top^{\mathcal{I},\mathcal{Z}} = \Delta^{\mathcal{I}}$, $\neg C^{\mathcal{I},\mathcal{Z}} = \Delta^{\mathcal{I}} \setminus C^{\mathcal{I},\mathcal{Z}}$, $(C \sqcap D)^{\mathcal{I},\mathcal{Z}} = C^{\mathcal{I},\mathcal{Z}} \cap D^{\mathcal{I},\mathcal{Z}}$, *and* $(\exists R.C)^{\mathcal{I},\mathcal{Z}} = \{\delta \mid$ *there is* $(\delta, \epsilon) \in R^{\mathcal{I},\mathcal{Z}}$ *with* $\epsilon \in C^{\mathcal{I},\mathcal{Z}}\}$.

$\mathcal{I}$ *satisfies* a concept inclusion of the form (17) if, for all variable assignments $\mathcal{Z}$ that satisfy $\mathcal{Z}(X_i) \in S_i^{\mathcal{I},\mathcal{Z}}$ for all $1 \leq i \leq n$, we have $C^{\mathcal{I},\mathcal{Z}} \subseteq D^{\mathcal{I},\mathcal{Z}}$. Satisfaction of role inclusions is defined analogously. $\mathcal{I}$ satisfies an ontology if it satisfies all of its axioms. As usual, $\models$ denotes both satisfaction and the induced logical entailment relation.

3.2   Semantics of Time

Time-sorted interpretations can be used to interpret $\mathcal{ALCH}_{@}^{\mathsf{T}}$ ontologies, but they do not take the intended semantics of time into account yet. For example, we might find that $A(c)@[\text{after}\!:\!1993]$ holds whereas $A(c)@[\text{time}\!:\!t]$ does not hold for any time $t \in \mathsf{N_T}$ with $t^{\mathcal{I}} > 1993$. To ensure consistency, we would like to view an interpretation with temporal domain $\Delta_T^{\mathcal{I}}$ as a sequence $(\mathcal{I}_i)_{i \in \Delta_T^{\mathcal{I}}}$ of regular (unsorted) interpretations that define the state of the world at each point in time. Such a sequence represents a *local* view of time as a sequence of events, whereas the time-sorted interpretation represents a *global* view that can explicitly refer to time points. Axioms of $\mathcal{ALCH}_{@}^{\mathsf{T}}$ refer to this global view, but it should be based on an actual sequence of events. To simplify the relationship between local and global views, we assume that the underlying abstract domain $\Delta_I^{\mathcal{I}}$ and interpretation of constants remains the same over time.

Definition 8. *Consider a temporal domain* $\Delta_T^{\mathcal{I}}$ *and an abstract domain* $\Delta_I^{\mathcal{I}}$, *and let* $(\mathcal{I}_i)_{i \in \Delta_T^{\mathcal{I}}}$ *be a sequence of (unsorted) interpretations with domain* $\Delta_I^{\mathcal{I}}$, *such that, for all* $a \in \mathsf{N_I}$, *we have* $a^{\mathcal{I}_i} = a^{\mathcal{I}_j}$ *for all* $i, j \in \Delta_T^{\mathcal{I}}$.

We define a global interpretation for $(\mathcal{I}_i)_{i \in \Delta_T^{\mathcal{I}}}$ *as a multi-sorted interpretation* $\mathcal{I} = (\Delta^{\mathcal{I}}, \cdot^{\mathcal{I}})$ *as follows. Let* $a^{\mathcal{I}} = a^{\mathcal{I}_i}$ *for all* $a \in \mathsf{N_I}$. *For any finite set*

$F \in \Phi^{\mathcal{I}}$, let $F_I := F \cap (\Delta_I^{\mathcal{I}} \times \Delta_I^{\mathcal{I}})$ denote its abstract part without any temporal attributes. For any $A \in \mathsf{N_C}$, $\delta \in \Delta^{\mathcal{I}}$, and $F \in \Phi^{\mathcal{I}}$ with $F \setminus F_I \neq \emptyset$, we have $(\delta, F) \in A^{\mathcal{I}}$ if and only if[5] $(\delta, F_I) \in A^{\mathcal{I}_i}$ for some $i \in \Delta_T^{\mathcal{I}}$, and the following conditions hold for all $(a^{\mathcal{I}}, x) \in F$:

- if $a = $ time, then $(\delta, F_I) \in A^{\mathcal{I}_x}$,
- if $a = $ before, then $(\delta, F_I) \in A^{\mathcal{I}_j}$ for some $j < x$,
- if $a = $ after, then $(\delta, F_I) \in A^{\mathcal{I}_j}$ for some $j > x$,
- if $a = $ until, then $(\delta, F_I) \in A^{\mathcal{I}_j}$ for all $j \leq x$,
- if $a = $ since, then $(\delta, F_I) \in A^{\mathcal{I}_j}$ for all $j \geq x$,
- if $a = $ between, then $(\delta, F_I) \in A^{\mathcal{I}_j}$ for some $j \in [x]$,
- if $a = $ during, then $(\delta, F_I) \in A^{\mathcal{I}_j}$ for all $j \in [x]$,

where $[x]$ for an element $x \in \Delta_{2T}^{\mathcal{I}}$ denotes the finite interval represented by the pair of numbers x, and $j \in \Delta_T^{\mathcal{I}}$. For roles $r \in \mathsf{N_R}$, we define $(\delta, \epsilon, F) \in r^{\mathcal{I}}$ analogously.

In words: in a global interpretation all tuples are consistent with the given sequence of local interpretations. One can see a global interpretation as a snapshot of a local interpretation, with timestamps encoding the information of the temporal sequence. If a global interpretation does not contain temporal attributes the characterization of Definition 8 holds vacuously for any temporal sequence, meaning that without temporal attributes the semantics is essentially the same as for $\mathcal{ALCH}_@$.

Definition 9. *An* interpretation *of* $\mathcal{ALCH}_@^{\mathsf{T}}$ *is a time-sorted interpretation* $\mathcal{I}$ *that is a global interpretation of an interpretation sequence* $(\mathcal{I}_i)_{i \in \Delta_T^{\mathcal{I}}}$ *as in Definition 8.*

A model *of an* $\mathcal{ALCH}_@^{\mathsf{T}}$ *ontology* $\mathcal{O}$ *is an* $\mathcal{ALCH}_@^{\mathsf{T}}$ *interpretation that satisfies* $\mathcal{O}$, *and* $\mathcal{O}$ *entails an axiom* α, *written* $\mathcal{O} \models \alpha$, *if* α *is satisfied by all models of* $\mathcal{O}$.

By virtue of the syntax and semantics of $\mathcal{ALCH}_@^{\mathsf{T}}$ we can express background knowledge that helps to maintain integrity of the annotated knowledge and allows us to derive new information from it.

Example 6. Recall the imprecise assertion (4). Even without investigating further into the life of Carsten Lutz, we do know that he has published papers as early as 1997 [30], hence we can assume that he was educated before that:

$$(\exists \mathsf{educatedAt@}\lfloor \mathsf{before}: 1997 \rfloor.\top)(\mathsf{Carsten}) \quad (20)$$

where we again simplify presentation by allowing a complex concept expression in an assertion. Now together with axiom (9) (or, equivalently, (10) or (12)), we can infer

$$(\exists \mathsf{bornIn@}\lfloor \mathsf{between}: [1950, 1996] \rfloor.\top)(\mathsf{Carsten}) \quad (21)$$

which, though hardly more precise, serves to illustrate entailments in $\mathcal{ALCH}_@^{\mathsf{T}}$.[6]

---

[5] 'for some $i \in \Delta_T^{\mathcal{I}}$' is useful for attributes which universally quantify time points (e.g., until).

[6] Readers who long for greater precision may consult the literature [27].

Some temporal attributes are closely related. Clearly, time can be captured by using during or between with singleton intervals. Conversely, during can be expressed by specifying all time points in the respective interval explicitly using time, but this incurs an exponential blow-up over the binary encoding of time intervals. Similarly, between could be expressed as a disjunction of statements with specific times. Since time can be infinite, since and after cannot be captured using finite intervals. It may seem as if until and before correspond to during and between using intervals starting at 0. However, it is not certain that 0 is the first element in the temporal domain of an interpretation, and the next example shows that this cannot be assumed in general.

Example 7. The ontology with the two axioms $C(a)@[\text{until}:10]$ and $C@[\text{before}:5] \sqsubseteq \bot$ is satisfiable in $\mathcal{ALCH}_@^{\mathbb{T}}$, but it does not have models that have times before 5. Replacing until:10 with during:$[0,10]$ would therefore lead to an inconsistent ontology.

4   Reasoning in $\mathcal{ALCH}_@^{\mathbb{T}}$

In our investigations, we focus on the decidability and complexity of the satisfiability problem as the central reasoning task. As usual, entailment of assertions is reducible to satisfiability. Also, our definition of assertions could be easily extended to complex concept expressions, since such assertions can be encoded using concept inclusions. Thus, all of our decidability and complexity results hold for the problem of answering instance queries, defined as the class of the assertions allowing complex concept expressions, such as that of Example 2 (Eq. 4).

In this section, we study the expressivity and decidability in $\mathcal{ALCH}_@^{\mathbb{T}}$. Our first result, Theorem 1, shows that reasoning is on the first level of the analytical hierarchy and therefore highly undecidable.

Theorem 1. *Satisfiability of $\mathcal{ALCH}_@^{\mathbb{T}}$ ontologies is Σ_1^1-hard, and thus not recursively enumerable. Moreover, the problem is Σ_1^1-hard even with at most one set variable per inclusion and with only the temporal attributes* time *and* after.

Proof. We reduce from the following tiling problem, known to be Σ_1^1-hard [20]: given a finite set of tile types T with horizontal and vertical compatibility relations H and V, respectively, and $t_0 \in T$, decide whether one can tile $\mathbb{N} \times \mathbb{N}$ with t_0 appearing infinitely often in the first row. We define an $\mathcal{ALCH}_@^{\mathbb{T}}$ ontology $\mathcal{O}_{T,t_0}$ that expresses this property. In our encoding, we use the following symbols:

- a concept name A, to mark individuals representing a grid position with a time point;
- a concept name P to keep time points associated with previous columns in the grid;
- concept names A_t, for each $t \in T$, to mark individuals with tile types;
- an individual name a, to be connected with the first row of the grid;
- an auxiliary concept name I, to mark the individual a, and a concept name B, used to create the vertical axis;

– role names r, s, to connect horizontally and vertically the elements of the grid, respectively.

We define $\mathcal{O}_{T,t_0}$ as the set of the following $\mathcal{ALCH}_{@}^{\mathrm{T}}$ assertion and concept inclusions. We start encoding the first row of the grid with an assertion $I(a)$ and the concept inclusions:

$$I \sqsubseteq \exists r.A@\lfloor \text{time}: 0 \rfloor \text{ and } \exists r.A@X \sqsubseteq \exists r.A@\lfloor \text{after}: X.\text{time} \rfloor.$$

Every element in A must be marked in at most one time point (in fact, exactly one):

$$A@X \sqsubseteq \neg A@\lfloor \text{after}: X.\text{time} \rfloor \tag{22}$$

Every element representing a grid position can be associated with exactly one tile type at the same time point:

$$A@X \sqsubseteq \bigsqcup_{t \in T} A_t@\lfloor \text{time}: X.\text{time} \rfloor,$$
$$\exists r.A_t@X \sqsubseteq \neg \exists r.A_{t'}@\lfloor \text{time}: X.\text{time} \rfloor, \text{ for } t \neq t' \in T.$$

We also have:

$$A_t@X \sqsubseteq A@\lfloor \text{time}: X.\text{time} \rfloor, \text{ for each } t \in T$$

to ensure that elements are in A_t and A at the same time point (exactly one, see Eq. 22). The condition that t_0 appears infinitely often in the first row is expressed with:

$$I \sqsubseteq \exists r.(A_{t_0}@\lfloor \text{time}: 0 \rfloor \sqcup A_{t_0}@\lfloor \text{after}: 0 \rfloor),$$
$$I \sqcap \exists r.A_{t_0}@X \sqsubseteq \exists r.A_{t_0}@\lfloor \text{after}: X.\text{time} \rfloor.$$

To vertically connect subsequent rows of the grid, we have:

$$I \sqsubseteq B \text{ and } B \sqsubseteq \exists s.B.$$

We add, for each $t \in T$, the following inclusion to ensure compatibility between vertically adjacent tile types:

$$\exists r.A_t@X \sqsubseteq \forall s.\exists r.(\bigsqcup_{(t,t') \in V} A_{t'}@\lfloor \text{time}: X.\text{time} \rfloor)$$

We also have:

$$\exists s.\exists r.A@X \sqsubseteq \exists r.A@\lfloor \text{time}: X.\text{time} \rfloor$$

to ensure that the set of time points in each row is the same. We now encode compatibility between horizontally adjacent tile types. We first state that, given a node associated with a time point p, for every sibling node d, if d is associated with a time point after p then we mark d with P and p:

$$\exists r.A@X \sqsubseteq \forall r.(\neg A@\lfloor \text{after}: X.\text{time}\rfloor \sqcup P@\lfloor \text{time}: X.\text{time}\rfloor).$$

For each node, P keeps the time points associated with previous columns in the grid (finitely many). We also have:

$$\exists r.P@X \sqsubseteq \exists r.A@\lfloor \text{time}: X.\text{time}\rfloor \text{ and } P@X \sqsubseteq A@\lfloor \text{after}: X.\text{time}\rfloor$$

to ensure that P keeps only those previous time points. Finally, for each $t \in T$, we add to $\mathcal{O}_{T,t_0}$ the inclusion:

$$\exists r.A_t@X \sqsubseteq \forall r.(\neg A@\lfloor \text{after}: X.\text{time}\rfloor \sqcup$$
$$P@\lfloor \text{after}: X.\text{time}\rfloor \sqcup \bigsqcup_{(t,t')\in H} A_{t'}).$$

Intuitively, as P keeps the time points associated with previous columns in the grid, only the node representing the horizontally adjacent grid position of a node associated with a time point p will not be marked with P after p.   □

Theorem 2 shows that even if after is only allowed in assertions reasoning is undecidable, though, in the arithmetical hierarchy [35]. For this statement, recall that Σ_1^0 is the class of recursively enumerable problems.

Theorem 2. *Satisfiability of $\mathcal{ALCH}_@^\mathbb{T}$ ontologies with the temporal attributes* time, after *and* before *but* after *only in assertions is Σ_1^0-complete. The problem is Σ_1^0-hard even with at most one set variable per inclusion.*

The detailed proof of this result can be found in the appendix.

5   Decidable Temporally Attributed DLs

To recover decidability, we need to restrict $\mathcal{ALCH}_@^\mathbb{T}$ in some way. In this section, we do so by restricting the use of variables or of temporal attributes, leading to a range of different reasoning complexities.

A straightforward approach for recovering decidability is to restrict to *ground* $\mathcal{ALCH}_@^\mathbb{T}$, where we disallow set and object variables altogether. It is clear from the known complexity of $\mathcal{ALCH}$ that reasoning is still ExpTime-hard. We establish a matching membership result by providing a satisfiability-preserving polynomial time translation to $\mathcal{ALCH}$ extended with role conjunctions and disjunctions (denoted $\mathcal{ALCH}b$), where satisfiability is known to be in ExpTime [36].

Theorem 3. *Satisfiability of ground $\mathcal{ALCH}_@^\mathbb{T}$ ontologies is ExpTime-complete.*

Proof. Consider a ground $\mathcal{ALCH}_{@}^{\mathsf{T}}$ ontology $\mathcal{O}$, and let $k_0 < \ldots < k_n$ be the ascending sequence of all numbers mentioned (in binary encoding) in time points or in time intervals in $\mathcal{O}$. We define $\mathbb{N}_{\mathcal{O}} := \{k_i \mid 0 \leq i \leq n\} \cup \{k_i + 1 \mid 0 \leq i < n\}$, and let $k_{\min} := \min(\mathbb{N}_{\mathcal{O}})$ and $k_{\max} = \max(\mathbb{N}_{\mathcal{O}})$, where we assume $k_{\min} = k_{\max} = 0$ if $\mathbb{N}_{\mathcal{O}} = \emptyset$. For a finite interval $v \subseteq \mathbb{N}$, let $\mathbb{N}_{\mathcal{O}}^v$ be the set of all finite, non-empty intervals $u \subseteq v$ with end points in $\mathbb{N}_{\mathcal{O}}$. The number of intervals in $\mathbb{N}_{\mathcal{O}}^v$ then is polynomial in the size of $\mathcal{O}$.

We translate $\mathcal{O}$ into an $\mathcal{ALCH}b$ ontology $\mathcal{O}^{\dagger}$ as follows. First, $\mathcal{O}^{\dagger}$ contains every axiom from $\mathcal{O}$, with each annotated concept name $A@S$ and each annotated role name $r@S$ replaced by a fresh concept name A_S and a fresh role name r_S, respectively.

Second, given a ground specifier S, we denote by $S(a\!:\!b)$ the result of removing all temporal attributes from S and adding the pair $a\!:\!b$. Moreover, let S_{T} be the set of temporal attribute-value pairs in S. Then, for each A_S and r_S with $S_{\mathsf{T}} \neq \emptyset$, $\mathcal{O}^{\dagger}$ contains the equivalences (as usual, $\equiv$ refers to bidirectional $\sqsubseteq$ here):

$$A_S \equiv \bigsqcap_{(a:b) \in S_{\mathsf{T}}} (A_{S(a:b)})^{\sharp} \quad \text{and} \quad r_S \equiv \bigsqcap_{(a:b) \in S_{\mathsf{T}}} (r_{S(a:b)})^{\sharp} \tag{23}$$

where the concept/role expressions $(H_{S(a:b)})^{\sharp}$ for $H \in \{A, r\}$ are defined as follows:

- $(H_{S(during:v)})^{\sharp} = \bigsqcap_{u \in \mathbb{N}_{\mathcal{O}}^v} H_{S(during:u)}$
- $(H_{S(between:v)})^{\sharp} = \bigsqcup_{k \in (v \cap \mathbb{N}_{\mathcal{O}})} H_{S(during:[k,k])}$
- $(H_{S(time:k)})^{\sharp} = (H_{S(during:[k,k])})^{\sharp}$
- $(H_{S(since:k)})^{\sharp} = (H_{S(during:[k,k_{\max}])})^{\sharp} \sqcap H_{S(since:k_{\max})}$
- $(H_{S(until:k)})^{\sharp} = (H_{S(during:[k_{\min},k])})^{\sharp} \sqcap H_{S(until:k_{\min})}$
- $(H_{S(after:k)})^{\sharp} = (H_{S(between:[k+1,k_{\max}])})^{\sharp} \sqcup H_{S(after:k_{\max})}$
- $(H_{S(before:k)})^{\sharp} = (H_{S(between:[k_{\min},k-1])})^{\sharp} \sqcup H_{S(before:k_{\min})}$

where $k \neq k_{\min}$ and $k \neq k_{\max}$. If $k \in \{k_{\min}, k_{\max}\}$ then we set $(H_{S(a:k)})^{\sharp} = H_{S(a:k)}$. Only polynomially many inclusions in the size of $\mathcal{O}$ are introduced by (23) in $\mathcal{O}^{\dagger}$.

Finally, given attribute-value pairs $a\!:\!b$ and $c\!:\!d$ for temporal attributes a and b, we say that $a\!:\!b$ *implies* $c\!:\!d$ if $A(e)@[a\!:\!b] \models A(e)@[c\!:\!d]$ for some arbitrary $A \in \mathsf{N_C}$ and $e \in \mathsf{N_I}$. Based on a given $\mathsf{N_I}$, this implication relationship is computable in polynomial time. We then extend $\mathcal{O}^{\dagger}$ with all inclusions $A_S \sqsubseteq A_T$ and $r_S \sqsubseteq r_T$, where A_S, A_T and r_S, r_T are concept and role names occurring in $\mathcal{O}^{\dagger}$, including those introduced in (23), such that for each temporal attribute-value pair $c\!:\!d$ in T there is a temporal attribute-value pair $a\!:\!b$ in S such that $a\!:\!b$ implies $c\!:\!d$ and:

- T is an open specifier and the set of non-temporal attribute-value pairs in S is a superset of the set of non-temporal attribute-value pairs in T; or
- S, T are closed specifiers and the set of non-temporal attribute-value pairs in S is equal to the set of non-temporal attribute-value pairs in T.

This finishes the construction of $\mathcal{O}^\dagger$. As shown in the appendix, $\mathcal{O}$ is satisfiable iff $\mathcal{O}^\dagger$ is satisfiable. □

While ground $\mathcal{ALCH}_{@}^{\mathbb{T}}$ can already be used for some interesting conclusions, it is still rather limited. However, satisfiability of (non-ground) $\mathcal{ALCH}_{@}$ ontologies is also decidable [23], and indeed we can regain decidability in $\mathcal{ALCH}_{@}^{\mathbb{T}}$ by restricting the use of variables to non-temporal attributes. Using a similar reasoning as in the case of $\mathcal{ALCH}_{@}$, we obtain a 2ExpTime upper bound by constructing an equisatisfiable (exponentially larger) ground $\mathcal{ALCH}_{@}^{\mathbb{T}}$ ontology. The details of this proof are given in the appendix.

Theorem 4. *Satisfiability in $\mathcal{ALCH}_{@}^{\mathbb{T}}$ is 2ExpTime-complete for ontologies without expressions of the form $X.a$; $a\!:\!x$ with x in $\mathsf{Var}(\mathsf{N_T})$; and $a\!:\![t,t']$ with one of t,t' in $\mathsf{Var}(\mathsf{N_T})$, where a is a temporal attribute.*

Another way for regaining decidability is by limiting the temporal attributes that make reference to time points in the future:

Theorem 5. *Satisfiability of $\mathcal{ALCH}_{@}^{\mathbb{T}}$ ontologies with only the temporal attributes* during, time, before *and* until *is in 3ExpTime.*

The proof of this result is found in the appendix. It is based on translating the $\mathcal{ALCH}_{@}^{\mathbb{T}}$ ontology into a ground $\mathcal{ALCH}_{@}^{\mathbb{T}}$ ontology, which, however, is double-exponential in size if we assume that time points in the temporal domain have been encoded in binary. The claimed 3ExpTime upper bound then follows from Theorem 3.

Our result in our next Theorem 6 below is that this upper bound is tight. The proof is by reduction from the word problem for double-exponentially space-bounded alternating Turing machines (ATMs) [16] to the entailment problem for $\mathcal{ALCH}_{@}^{\mathbb{T}}$ ontologies. The main challenge in this reduction is that we need a mechanism that allows us to transfer the information of a double-exponentially space bounded tape, so that each configuration following a given configuration is actually a successor configuration (i.e., tape cells are changed according to the transition relation). We encode our tape using time: we can have exponentially many time points in an interval with end points encoded in binary. So considering each time point as a bit position, we construct a counter with *exponentially many bits*, encoding the position of double-exponentially many tape cells.

Theorem 6. *Satisfiability of $\mathcal{ALCH}_{@}^{\mathbb{T}}$ ontologies with only* time *and* before *is 3ExpTime-hard.*

Our main theorem of this section completes and summarises our results regarding decidability and complexity for different combinations of temporal attributes:

Theorem 7. *In $\mathcal{ALCH}_{@}^{\mathbb{T}}$, any combination of temporal attributes containing* {time, after} *is undecidable. Moreover, the combination* {time, before} *is 3ExpTime-complete, and the combination* {time, during, since, until} *and every subset of it are 2ExpTime-complete.*

The cases of undecidability and 3ExpTime-completeness follow from (the proofs of) Theorems 1, 5, and 6. Hardness for 2ExpTime is inherited from $\mathcal{ALCH}_@$ [23], so our proof in the appendix mainly needs to establish the membership for this case.

Certain combinations referring to time points in the future, e.g., time and since, are harmless while others are highly undecidable, e.g., time and after (by Theorem 1). Essentially, what causes undecidability in $\mathcal{ALCH}_@^{\mathsf{T}}$ is a combination with the ability to refer to arbitrarily many intervals of time points in the future.

6 Lightweight Temporal Attributed DLs

The complexities of the previous section are still rather high, whereas modern description logics research has often aimed at identifying tractable DLs [9]. In this section, we therefore seek to obtain a tractable temporally attributed DL that is based on the popular $\mathcal{EL}$-family of DLs [6]. We investigate $\mathcal{ELH}_@^{\mathsf{T}}$, the fragment of $\mathcal{ALCH}_@^{\mathsf{T}}$ which uses only $\exists$, $\sqcap$, $\top$ and $\bot$ in concept expressions. It is clear that variables lead to intractable reasoning complexities, but it turns out that ground $\mathcal{ELH}_@^{\mathsf{T}}$ still remains intractable:

Theorem 8. *Satisfiability of ground $\mathcal{ELH}_@^{\mathsf{T}}$ ontologies is* ExpTime-*complete.*

Proof. The upper bound follows from Theorem 3. For the lower bound, we show how one can encode disjunctions (i.e., inclusions of the form $\top \sqsubseteq B \sqcup C$), which allow us to reduce satisfiability of ground $\mathcal{ALCH}_@^{\mathsf{T}}$ to satisfiability of ground $\mathcal{ELH}_@^{\mathsf{T}}$ ontologies. In fact, several combinations of the temporal attributes time, between, before and after suffice to encode $\top \sqsubseteq B \sqcup C$. For example, see the inclusions using the temporal attributes time and between: $\top \sqsubseteq A@\lfloor \text{between} : [1,2] \rfloor$, $A@\lfloor \text{time} : 1 \rfloor \sqsubseteq B$, $A@\lfloor \text{time} : 2 \rfloor \sqsubseteq C$.

□

It is known that the entailment problem for $\mathcal{EL}$ ontologies with concept and role names annotated with time intervals over finite models is in PTime [26]. Indeed, our temporal attribute during can be seen as a syntactic variant of the time intervals in the mentioned work and, if we restrict to the temporal attributes time, during, since and until, the complexity of the satisfiability problem for ground $\mathcal{ELH}_@^{\mathsf{T}}$ ontologies is in PTime. Our proof here (for ground $\mathcal{ELH}_@^{\mathsf{T}}$ over $\mathbb{N}$ or over a finite interval in $\mathbb{N}$) is based on a polynomial translation to $\mathcal{ELH}$ extended with role conjunction, where satisfiability is PTime-complete [36].

Theorem 9. *Satisfiability of ground $\mathcal{ELH}_@^{\mathsf{T}}$ ontologies without the temporal attributes* between, before *and* after *is* PTime-*complete.*

Proof. Hardness follows from the PTime-hardness of $\mathcal{EL}$ [6]. For membership, note that the translation in Theorem 3 for the temporal attributes during, since and until does not introduce disjunctions or negations. So the result of translating a ground $\mathcal{ELH}_@^{\mathsf{T}}$ ontology belongs to $\mathcal{ELH}$ extended with role conjunction.   □

7   Related Work

In this section, we discuss the main differences and similarities between our logic and other related formalisms. Potentially related works include classical first-order and second-order logic, temporal extensions of description logics, and temporal extensions of other logics. When setting out to compare our approach to other logics, it is important to understand that there is no immediate formal basis for doing so. Our approach differs both in syntax (structure of formulae) and in semantics (model theory) from existing logics, so that an immediate comparison is not possible. There are three distinct perspectives one might take for discussing comparisons:

(1) Translate models of temporally attributed logics to models of another logic, and investigate which classes of models can be characterised by theories of either type.
(2) Look for polynomial reductions of common inference tasks, i.e., for syntactic translations between formulae that preserve the answer to some decision problem.
(3) Compare intuitive modelling capabilities on an informal level, looking at intended usage and application scenarios.

Approach (1) can lead to the closest relationships between two distinct logical formalisms. Unfortunately, it is not obvious how to relate our temporalised model theory to classical logical formalisms. It is clear that one could capture the semantic conditions of temporally attributed DLs in second-order logic, which would lead to models that explicitly define (axiomatically) the temporal domain and that associate temporal validity with every tuple. This is close in spirit to the way in which *weak second-order logic* was related to (non-temporal) attributed logics by Marx et al. [32], although their work did in fact show a mere reduction of satisfiability in the sense of (2). Our undecidability results of Theorem 1 imply that, for any faithful translation of temporally attributed models into classical relational structures, $\mathcal{ALCH}_@^T$ can capture classes of models that are not expressible in first-order logic.

Besides the translation to models of classical logic, it might also be promising to seek direct translations to model theories of temporal logics, especially to metric temporal logics (MTL) [5,19]. So far, the combination of MTL with DLs has only been investigated considering discrete time domains. Recent works on DatalogMTL consider dense (real or rational) time domains [13,14], into which our integer time could be embedded. Note that the containment of integers in rationals and reals does not mean that there is any corresponding relationship between the expressivity of the logics (indeed, decision procedures for DatalogMTL are also based on restricting attention to a suitably defined set of discrete, non-dense time points). However, choosing a discrete domain does not mean that the complexity of the satisfiability problem is lower, neither it means that the technical results are simpler (as we have already pointed out in the introduction, the complexity of satisfiability of $\mathcal{ALC}$ with a concrete discrete domain using the predicates = and < is open). A detailed semantic comparison requires a thorough investigation of the semantic assumptions in either logic, which has to be left to future research.

Approach (2), the syntactic reduction of inference tasks, is the heart of our complexity results. Our upper complexity bounds are obtained by either grounding the ontology and then translating it to an ontology in a classical DL; or directly translating it into a classical DL. Most DLs, including $\mathcal{ALC}$ and $\mathcal{EL}$, are syntactic variants of fragments of first-order logic [7], and thus our decidable fragments can be translated into first-order logic. The difference in the complexity results for $\mathcal{ALC}$ is due to the ability of expressing certain statements in a more succinct way. For $\mathcal{EL}$, we have shown that some temporal attributes increase expressivity, allowing disjunctions (and negations) to be encoded in the logic. A similar interplay between temporal logic and $\mathcal{EL}$ has also been observed in other studies on temporal DLs [3]. Nevertheless the resulting logic is still expressible in $\mathcal{ALC}$ and, thus also in first-order logic.

Approach (3), the comparison of intuitive semantics and modelling applications, brings many further logics into the scope of investigation (not surprisingly, the motivation of modelling time has inspired many technically diverse formalisms). Some of the statements used in our examples can also be naturally expressed in temporal DLs. For instance, axiom (10) in Fig. 3 is expressible in $\mathcal{ALC}$ extended with Linear Temporal Logic [31,39] with:

$$\exists \mathsf{educatedAt}.\top \sqsubseteq \neg\Diamond\exists \mathsf{bornIn}.\top.$$

Other authors have also considered extending $\mathcal{ALC}$ with Metric Temporal Logic (MTL) [5,19], where axiom (4) of Example 2 can be expressed with:

$$\Diamond_{[1950,2000]}\exists \mathsf{bornIn}.\top\;(\mathsf{Carsten}).$$

However, axiom (16) from Example 5 cannot be naturally expressed by temporal DLs. The complexity results can also be very different, for instance, the complexity of propositional MTL is already undecidable over the reals and ExpSpace-complete over the naturals [1], whereas in Theorem 3 of this paper we show that we can enhance $\mathcal{ALC}$ with many types of time related annotations with time points encoded in *binary* while keeping the same ExpTime complexity of $\mathcal{ALC}$. Regarding temporal $\mathcal{EL}$, it is known that, if temporal operators are allowed in concept expressions then satisfiability is not easier than satisfiability for temporal $\mathcal{ALC}$ [3]; and it decreases to PSpace if temporal operators can only be applied over the axioms [11]. Our lightweight fragment based on $\mathcal{EL}$ features PTime complexity but allows only ground specifiers using particular types of temporal attributes. Syntactic restrictions on the specifiers, similar to those used for attributed $\mathcal{EL}$ [23,24], could also be applied to have a more interesting PTime fragment of temporally attributed $\mathcal{EL}$.

8   Conclusion

We investigated decidability and complexities of attributed description logics enriched with special attributes whose values are interpreted over a temporal dimension. We discussed several ways of restricting the general, undecidable setting in order to regain decidability. Our complexity results range from PTIME to 3EXPTIME.

As future work, we plan to study forms of generalising our logic to capture the semantics of other standard types of annotations in knowledge graphs, such as provenance [12] and spatial information. Another direction is to study our logic over other temporal domains such as the real numbers (see [13,14] for a combination of Datalog with MTL over the reals). It would also be interesting to investigate query answering.

Acknowledgements. This work is partly supported by the German Research Foundation (DFG) in CRC 248 (Perspicuous Systems), CRC 912 (HAEC), and Emmy Noether grant KR 4381/1-1; and by the European Research Council (ERC) Consolidator Grant 771779 (DeciGUT).

A   Proofs for Section 4

Theorem 2. *Satisfiability of $\mathcal{ALCH}_{@}^{\mathbb{T}}$ ontologies with the temporal attributes* time, after *and* before *but* after *only in assertions is Σ_1^0-complete. The problem is Σ_1^0-hard even with at most one set variable per inclusion.*

Proof. We first show hardness. We reduce the word problem for deterministic Turing machines (DTM) to satisfiability of $\mathcal{ALCH}_{@}^{\mathbb{T}}$ ontologies with the temporal attribute after occurring only in assertions. A DTM is a tuple $(Q, \Sigma, \Theta, q_0, q_f)$, where:

- Q is a finite set of states,
- Σ is a finite alphabet containing the *blank symbol* $\sqcup$,
- $\{q_0, q_f\} \subseteq Q$ are the *initial* and the *final* states, resp., and
- $\Theta : Q \times \Sigma \to Q \times \Sigma \times \{l, r\}$ is the *transition function*.

A *configuration* of $\mathcal{M}$ is a word wqw' with $w, w' \in \Sigma^*$ and q in Q. The meaning is that the (one-sided infinite) tape contains the word ww' with only blanks behind it, the machine is in state q and the head is on the left-most symbol of w'. The notion of a *successive configuration* is defined in the usual way, in terms of the transition relation Θ. A *computation* of $\mathcal{M}$ on a word w is a sequence of successive configurations $\alpha_0, \alpha_1, \ldots$, where $\alpha_0 = q_0w$ is the *initial configuration* for the input w. Let $\mathcal{M}$ be a DTM and $w = \sigma_1\sigma_2\cdots\sigma_n$ an input word. Assume w.l.o.g. that $\mathcal{M}$ never attempts to move to the left when its head is in the left-most tape position and that q_0 occurs only in the domain of Θ (but not in the range).

We construct an $\mathcal{ALCH}_{@}^{\mathbb{T}}$ ontology $\mathcal{O}_{\mathcal{M},w}$ with after occurring only in assertions that is satisfiable iff $\mathcal{M}$ accepts w. Models of $\mathcal{O}_{\mathcal{M},w}$ have a similar structure as in the proof of Theorem 1. We create a vertical chain with:

$$I(a), \quad I \sqsubseteq B \quad \text{and} \quad B \sqsubseteq \exists s.B$$

and ensure that horizontally the set of time points is the same:

$$\exists r.A@X \sqsubseteq \forall s.\exists r.A@\lfloor \text{time}: X.\text{time} \rfloor, \tag{24}$$

$$\exists s.\exists r.A@X \sqsubseteq \exists r.A@\lfloor \text{time}: X.\text{time} \rfloor. \tag{25}$$

Every element representing a tape cell is marked with A in at most one time point (in fact, it will be exactly one):

$$A@X \sqsubseteq \neg A@\lfloor \text{before}: X.\text{time} \rfloor$$

The main difference is that horizontally we do not have infinitely many sibling nodes. That is, over the naturals, adding the inclusion $\exists r.A@X \sqsubseteq \exists r.A@\lfloor \text{before}: X.\text{time} \rfloor$ would make $\mathcal{O}_{\mathcal{M},w}$ unsatisfiable and here we cannot use after in inclusions. Instead, for each $q \neq q_f$ in Q, we add to $\mathcal{O}_{\mathcal{M},w}$ the inclusions:

$$S_q \sqcap A@X \sqsubseteq S_q@\lfloor \text{time}: X.\text{time} \rfloor, \tag{26}$$

$$\exists r.S_q@X \sqsubseteq \exists r.A@\lfloor \text{before}: X.\text{time} \rfloor \tag{27}$$

where S_q is a concept name representing a state. Intuitively, each vertically aligned set of elements (w.r.t. time) represents a configuration and a sequence of configurations going backwards in time represents a computation of $\mathcal{M}$ with input w. The goal is to ensure that $\mathcal{O}_{\mathcal{M},w}$ is satisfiable iff we reach the final state, that is, w is accepted by $\mathcal{M}$.

We now add to $\mathcal{O}_{\mathcal{M},w}$ assertions to trigger the inclusions in Eqs. 24, 25, 26 and 27:

$$r(a,b), \quad S_{q_0}(b), \quad A(b)@[\text{after}: 0].$$

We also use in our encoding concepts C_σ for each symbol $\sigma \in \Sigma$. To encode the input word $w = \sigma_1 \sigma_2 \cdots \sigma_n$, we add:

$$C_\sigma \sqcap A@X \sqsubseteq C_\sigma@\lfloor \text{time}: X.\text{time} \rfloor \text{ for each } \sigma \in \Sigma,$$
$$C_{\sigma_1}(b), \quad \exists r.S_{q_0}@X \sqsubseteq \forall s^i.\exists r.C_{\sigma_{i+1}}@\lfloor \text{time}: X.\text{time} \rfloor$$

for $1 \leq i < n$. It is straightforward to add inclusions encoding that (i) the rest of the tape in the initial configuration is filled with the blank symbol, (ii) each node representing a tape cell in a configuration is associated with only one C_σ with $\sigma \in \Sigma$ and (iii) at most one S_q with $q \in Q$ (exactly the node representing the head position). Also, for each element, the time point associated with A is the same for the concepts of the form C_σ and S_q (if true in the node).

To access the 'next' configuration, we use an auxiliary concept F that keeps time points in the future. Recall that since a computation here goes backwards in time, these time points are associated with previous configurations:

$$\exists r.A@X \sqsubseteq \forall r.(\neg A@\lfloor \text{before}: X.\text{time} \rfloor \sqcup F@\lfloor \text{time}: X.\text{time} \rfloor).$$

We now ensure that tape contents are transferred to the 'next' configuration, except for the tape cell at the head position:

$$\exists r.(C_\sigma @X \sqcap S_{\overline{q}}) \sqsubseteq \forall r.(F@\lfloor \text{before}: X.\text{time}\rfloor \sqcup \neg A@\lfloor \text{before}: X.\text{time}\rfloor \sqcup C_\sigma)$$

for each $\sigma \in \Sigma$, where $S_{\overline{q}}$ is a shorthand for $\neg \bigsqcup_{q\in Q} S_q$. Finally we encode the transition function. We explain for $\Theta(q, \sigma) = (q', \tau, D)$ with $D = r$ (the case with $D = l$ can be handled analogously). We encode that the 'next' state is q':

$$\exists r.(S_q @X \sqcap C_\sigma) \sqsubseteq \forall s.\forall r.(F@\lfloor \text{before}: X.\text{time}\rfloor \sqcup \neg A@\lfloor \text{before}: X.\text{time}\rfloor \sqcup S_{q'})$$
$$(28)$$

and change to τ the tape cell at the (previous) head position:

$$\exists r.(S_q @X \sqcap C_\sigma) \sqsubseteq \forall r.(F@\lfloor \text{before}: X.\text{time}\rfloor \sqcup \neg A@\lfloor \text{before}: X.\text{time}\rfloor \sqcup C_\tau).$$

Equation 28 also increments the head position.

This finishes our reduction.

For the upper bound, we point out that if an $\mathcal{ALCH}_@^{\mathbb{T}}$ ontology $\mathcal{O}$ with after only in assertions is satisfiable then there is a satisfiable ontology $\mathcal{O}'$ that is the result of replacing each occurrence of after : k in $\mathcal{O}$ by some time : l with $k < l \in \mathbb{N}$. By Theorem 5, one can decide satisfiability of $\mathcal{O}'$ (that is, satisfiability of ontologies with only the temporal attributes time and before). As the replacements of after : k by time : l in assertions can be enumerated, it follows that satisfiability of $\mathcal{ALCH}_@^{\mathbb{T}}$ ontologies is in Σ_1^0.   □

B   Proofs for Section 5

Theorem 3. *Satisfiability of ground $\mathcal{ALCH}_@^{\mathbb{T}}$ ontologies is* ExpTime-*complete.*

Proof. The construction of an ontology $\mathcal{O}^\dagger$ was already given in the main text. It remains to show that $\mathcal{O}$ is satisfiable iff $\mathcal{O}^\dagger$ is satisfiable. Given a model $\mathcal{I}$ of $\mathcal{O}$, we directly obtain an $\mathcal{ALCH}b$ interpretation $\mathcal{J}$ over $\Delta^\mathcal{I}$ by undoing the renaming and applying $\mathcal{I}$, i.e., by mapping $A_S \in \mathsf{N_C}$ to $A@S^\mathcal{I}$, $r_S \in \mathsf{N_R}$ to $r@S^\mathcal{I}$, and $a \in \mathsf{N_I}$ to $a^\mathcal{I}$. By the semantics of $\mathcal{ALCH}_@^{\mathbb{T}}$, $\mathcal{J} \models \mathcal{O}^\dagger$. Conversely, given an $\mathcal{ALCH}b$ model $\mathcal{J}$ of $\mathcal{O}^\dagger$, we construct an interpretation $\mathcal{I} = (\Delta^\mathcal{I}, \cdot^\mathcal{I})$ of $\mathcal{ALCH}_@^{\mathbb{T}}$ with $\Delta_T^\mathcal{I} = [\max(0, k_{\min} - 2), k_{\max} + 2]$ and $\Delta_I^\mathcal{I} = \Delta^\mathcal{J} \cup \{\star\} \cup \mathbb{T}$, where $\mathbb{T}$ is the set of temporal attributes and $\star$ is a fresh individual name. We define $a^\mathcal{I} := a^\mathcal{J}$ for all $a \in \mathsf{N_I} \cup \mathsf{N_T} \cup \mathsf{N_T^2}$.

For a ground closed specifier S with $a_1 : b_1, \ldots, a_n : b_n$ as non-temporal attributes, we define:

$$F_S := \{(a_1^\mathcal{I}, b_1^\mathcal{I}), \ldots, (a_n^\mathcal{I}, b_n^\mathcal{I})\}.$$

Similarly, for a ground open specifier S with $a_1 : b_1, \ldots, a_n : b_n$ as non-temporal attribute-value pairs, we define:

$$F_S := \{(a_1^\mathcal{I}, b_1^\mathcal{I}), \ldots, (a_n^\mathcal{I}, b_n^\mathcal{I}), (\star, \star)\}.$$

To simplify the presentation, we write $a : b \in S$ if $a : b$ occurs in S. Furthermore, let $A^{\mathcal{I}_i}$ be the set of all tuples (δ, F_S) such that one of the following holds:

- $\delta \in A_S^{\mathcal{J}}$, during: $v \in S$ and $i \in v$;
- $\delta \in A_S^{\mathcal{J}}$, after: $k_{\max} \in S$ and $i = k_{\max} + 1$;
- $\delta \in A_S^{\mathcal{J}}$, since: $k_{\max} \in S$ and $k_{\max} + 1 \leq i \leq k_{\max} + 2$;
- $\delta \in A_S^{\mathcal{J}}$, before: $k_{\min} \in S$, $i = k_{\min} - 1$ and $k_{\min} > 0$;
- $\delta \in A_S^{\mathcal{J}}$, until: $k_{\min} \in S$, $\max(k_{\min} - 2, 0) \leq i \leq k_{\min} - 1$ and $k_{\min} > 0$.

We define $r^{\mathcal{I}_i}$ analogously. Given the definitions of $A^{\mathcal{I}_i}$ and $r^{\mathcal{I}_i}$, for all $i \in \mathbb{N}$, $A \in \mathsf{N_C}$ and $r \in \mathsf{N_R}$, we define $\cdot^{\mathcal{I}}$ as in Definition 8.

Claim. For all A_S, r_S occurring in $\mathcal{O}^\dagger$: (1) $A_S^{\mathcal{J}} = A@S^{\mathcal{I}}$ and (2) $r_S^{\mathcal{J}} = r@S^{\mathcal{I}}$.

Proof of the Claim. If no temporal attribute occurs in S then by definition of $\mathcal{I}$ (in particular, F_S), we clearly have that $\delta \in A_S^{\mathcal{J}}$ iff $\delta \in A@S^{\mathcal{I}}$. Also, by semantics of $\mathcal{ALCH}_@^{\mathbb{T}}$, for a ground specifier S with a non-empty set $S_{\mathbb{T}}$ of temporal attributes the following holds for any $\mathcal{I}$ and concept $A@S$:

$$A@S^{\mathcal{I}} = \bigcap_{a:b \in S_{\mathbb{T}}} A@S(a:b)^{\mathcal{I}}$$

So we can consider $A@S$ with S containing only one temporal attribute. We argue for during and between (one can give a similar argument for the other temporal attributes):

- if the temporal attribute-value pair during: v is in S then, by definition of $\mathcal{I}$ (and F_S), $\delta \in A_S^{\mathcal{J}}$ iff $\delta \in A@S^{\mathcal{I}}$;
- if the temporal attribute-value pair between: v is in S then, by Eq. 23, $\delta \in A_S^{\mathcal{J}}$ iff $\delta \in \bigcup_{k \in v \cap \mathbb{N}_{\mathcal{O}}} A_{S(\text{during}: [k,k])}^{\mathcal{J}}$. By definition of $\mathcal{I}$, $\delta \in A_{S(\text{during}: [k,k])}^{\mathcal{J}}$ iff $\delta \in A@S(\text{during}: [k,k])^{\mathcal{I}}$, for $k \in v \cap \mathbb{N}_{\mathcal{O}}$. Then,

$$\delta \in A_S^{\mathcal{J}} \text{ iff } \delta \in \bigcup_{k \in v \cap \mathbb{N}_{\mathcal{O}}} A@S(\text{during}: [k,k])^{\mathcal{I}};$$

so $\delta \in A@S^{\mathcal{I}}$.

In the definition of $\mathbb{N}_{\mathcal{O}}$, we add $k_i + 1$ for each k_i occurring in $\mathcal{O}$, to ensure that axioms such as $\top \sqsubseteq A@\lfloor \text{between}: [k,l] \rfloor \sqcap \neg A@\lfloor \text{time}: k \rfloor \sqcap \neg A@\lfloor \text{time}: l \rfloor$ with $l - k \geq 2$ remain satisfiable. Also, in the definition of $\mathcal{I}$ we use the interval $\Delta_T^{\mathcal{I}} = [\max(0, k_{\min} - 2), k_{\max} + 2]$, and so, we give a margin of two 'additional' points in each side of the interval $[k_{\min}, k_{\max}]$ used in the translation. This is to ensure that axioms such as $\top \sqsubseteq A@\lfloor \text{before}: k_{\min} \rfloor \sqcap \neg A@\lfloor \text{until}: k_{\min} \rfloor$ with $k_{\min} \geq 2$ remain satisfiable. Point (2) can be proven with an easy adaptation of Point (1).

The Claim directly implies that $\mathcal{I} \models \mathcal{O}$. Note that $\star$ ensures that axioms such as $\top \sqsubseteq A@\lfloor a:b \rfloor \sqcap \neg A@\lfloor a:b \rfloor$ remain satisfiable.   □

Theorem 4. *Satisfiability in $\mathcal{ALCH}_@^{\mathbb{T}}$ is 2ExpTime-complete for ontologies without expressions of the form $X.a$; $a:x$ with x in $\mathsf{Var}(\mathsf{N_T})$; and $a: [t, t']$ with one of t, t' in $\mathsf{Var}(\mathsf{N_T})$, where a is a temporal attribute.*

Proof. The 2ExpTime lower bound follows from the fact that satisfiability of $\mathcal{ALCH}_@$ (so without temporal attributes) is already 2ExpTime-hard [23]. Our proof strategy for the upper bound consists on defining an ontology with grounded versions of inclusion axioms. Let $\mathcal{O}$ be an $\mathcal{ALCH}_@^{\mathsf{T}}$ ontology and let $\mathsf{N} := \mathsf{N}_\mathsf{I}^\mathcal{O} \cup \mathsf{N}_\mathsf{T}^\mathcal{O} \cup \mathsf{N}_\mathsf{T}^{2\mathcal{O}}$ be the union of the sets of individual names, time points, and intervals, occurring in $\mathcal{O}$, respectively. Let $\mathcal{I}$ be an interpretation of $\mathcal{ALCH}_@^{\mathsf{T}}$ over the domain $\Delta^\mathcal{I} = \mathsf{N} \cup \{x\}$, where x is a fresh individual name, satisfying $a^\mathcal{I} = a$ for all $a \in \mathsf{N}$. Let $\mathcal{Z} : \mathsf{N}_\mathsf{V} \to \Phi_\mathcal{O}^\mathcal{I}$ be a variable assignment, where $\Phi_\mathcal{O}^\mathcal{I} := \mathcal{P}_{\mathsf{fin}}\left(\Delta^\mathcal{I} \times \Delta^\mathcal{I}\right)$. Consider a concept inclusion α of the form $X_1 : S_1, \ldots, X_n : S_n$ ($C \sqsubseteq D$). We say that $\mathcal{Z}$ is *compatible with* α if $\mathcal{Z}(X_i) \in S_i^{\mathcal{I}, \mathcal{Z}}$ for all $1 \leq i \leq n$. In this case, the $\mathcal{Z}$-*instance* $\alpha_\mathcal{Z}$ of α is the concept inclusion $C' \sqsubseteq D'$ obtained by

- replacing each X_i by $[a : b \mid (a, b) \in \mathcal{Z}(X_i)]$;
- replacing every $a : X_i.b$ occurring in some specifier (with a, b *non-temporal* attributes) by all $a : c$ such that $(b, c) \in \mathcal{Z}(X_i)$; and
- replacing each object variable x by $\mathcal{Z}(x)$.

Then, the grounding $\mathcal{O}_g$ of $\mathcal{O}$ contains all $\mathcal{Z}$-instances $\alpha_\mathcal{Z}$ for all concept inclusions α in $\mathcal{O}$ and all compatible variable assignments $\mathcal{Z}$; and analogous axioms for role inclusions.

There may be (at most) exponentially many different instances for each terminological axiom in $\mathcal{O}$, thus $\mathcal{O}_g$ is of exponential size. We show that $\mathcal{O}$ is satisfiable iff $\mathcal{O}_g$ is satisfiable. By construction, we have $\mathcal{O} \models \mathcal{O}_g$, i.e., any model of $\mathcal{O}$ is also a model of $\mathcal{O}_g$. Conversely, let $\mathcal{I} = (\Delta^\mathcal{I}, \cdot^\mathcal{I})$ be a model of $\mathcal{O}_g$. W.l.o.g., assume that there is $x \in \Delta^\mathcal{I}$ such that $x \neq a^\mathcal{I}$ for all $a \in \mathsf{N}_\mathsf{I}^\mathcal{O} \setminus \{x\}$. For an annotation set $F \in \mathcal{P}_{\mathsf{fin}}\left(\Delta^\mathcal{I} \times \Delta^\mathcal{I}\right)$, we define $\mathrm{rep}_x(F)$ to be the annotation set obtained from F by replacing any individual $\delta \notin \mathcal{I}(\mathsf{N}_\mathsf{I}^\mathcal{O})$ in F by x.

Let $\sim$ be the equivalence relation induced by $\mathrm{rep}_x(F) = \mathrm{rep}_x(G)$ and define an interpretation $\mathcal{J}$ of $\mathcal{ALCH}_@^{\mathsf{T}}$ over the domain $\Delta^\mathcal{J} := \Delta^\mathcal{I}$, where $A^\mathcal{J} := \{(\delta, F)\mid (\delta, G) \in A^\mathcal{I}$ and $F \sim G\}$ for all $A \in \mathsf{N}_\mathsf{C}$, $r^\mathcal{J} := \{(\delta, \epsilon, F) \mid (\delta, \epsilon, G) \in r^\mathcal{I}$ and $F \sim G\}$ for all $r \in \mathsf{N}_\mathsf{R}$, and $a^\mathcal{J} := a^\mathcal{I}$ for all $a \in \mathsf{N}_\mathsf{I} \cup \mathsf{N}_\mathsf{T} \cup \mathsf{N}_\mathsf{T}^2$. It remains to show that $\mathcal{J}$ is indeed a model of $\mathcal{O}$. Suppose for a contradiction that there is a concept inclusion α in $\mathcal{O}$ that is not satisfied by $\mathcal{J}$ (the case for role inclusions is analogous). Then we have some compatible variable assignment $\mathcal{Z}$ that leaves α unsatisfied. Let $\mathcal{Z}_x$ be the variable assignment $X \mapsto \mathrm{rep}_x(\mathcal{Z}(X))$ for all $X \in \mathsf{N}_\mathsf{V}$. Clearly, as expressions of the form $a : X_i.b$, $a : x$, and $a : [t, t']$ with at least one of t, t' an object variable, are not allowed for a, b being temporal attributes, $\mathcal{Z}_x$ is also compatible with α. But now we have $C^{\mathcal{J}, \mathcal{Z}} = C^{\mathcal{I}, \mathcal{Z}_x}$ for all $\mathcal{ALCH}_@^{\mathsf{T}}$ concepts C, yielding the contradiction $\mathcal{I} \not\models \alpha_{\mathcal{Z}_x}$. Thus, $\mathcal{O}$ is satisfiable iff $\mathcal{O}_g$ is satisfiable. The result then follows from Theorem 3. $\qquad \square$

Theorem 5. *Satisfiability of* $\mathcal{ALCH}_@^{\mathsf{T}}$ *ontologies with only the temporal attributes* during, time, before *and* until *is in* 3ExpTime.

Proof. The difference w.r.t. the proof of Theorem 4 is that here expressions of the form $a : X_i.b$, $a : x$, and $a : [t, t']$ with at least one of t, t' an object variable, may occur in front of the temporal attributes during, before, time and until and the other temporal attributes are not allowed (not even in assertions). Let v be the internal $[0, k]$, where k is the largest number occurring in $\mathcal{O}$ (or 0 if no number occurs). To define our ground translation, we consider variable assignments $\mathcal{Z}$: $\mathsf{N_V} \to \Phi_{\mathcal{O},v}^{\mathcal{I}}$, where $\Phi_{\mathcal{O},v}^{\mathcal{I}} := \mathcal{P}_{\mathsf{fin}}\left(\Delta^{\mathcal{I}} \times \Delta^{\mathcal{I}}\right)$ and $\Delta^{\mathcal{I}}$ is the set of all individual names in $\mathcal{O}$ plus a fresh individual name x, all time points in v and all intervals contained in v. This gives us a ground ontology $\mathcal{O}_g$ with size double-exponential in the size of $\mathcal{O}$. Clearly, $\mathcal{O}$ is satisfiable iff $\mathcal{O}_g$ is satisfiable. $\qquad\square$

Theorem 6. *Satisfiability of $\mathcal{ALCH}_@^{\mathbb{T}}$ ontologies with only* time *and* before *is* 3ExpTime-*hard.*

Proof. We reduce the word problem for double-exponentially space-bounded alternating Turing machines (ATMs) to the entailment problem for $\mathcal{ALCH}_@^{\mathbb{T}}$ ontologies. We consider w.l.o.g. ATMs with only finite computations on any input. As usual, an ATM is a tuple $\mathcal{M} = (Q, \Sigma, \Theta, q_0)$, where:

- $Q = Q_\exists \uplus Q_\forall$ is a finite set of states, partitioned into *existential states* $Q_\exists$ and *universal states* $Q_\forall$,
- Σ is a finite alphabet containing the *blank symbol* $\llcorner$,
- $q_0 \in Q$ is the *initial state*, and
- $\Theta \subseteq Q \times \Sigma \times Q \times \Sigma \times \{l, r\}$ is the *transition relation*.

We use the same notions of configuration, computation and initial configuration given in the proof of Theorem 2. We recall the acceptance condition of an ATM. A configuration $\alpha = wqw'$ is *accepting* iff

- α is a universal configuration and all its successor configurations are accepting, or
- α is an existential configuration and at least one of its successor configurations is accepting.

Note that, by the definition above, universal configurations without any successors are accepting. We assume w.l.o.g. that all configurations wqw of a computation of $\mathcal{M}$ satisfy $|ww'| \leq 2^{2^n}$. $\mathcal{M}$ *accepts* a word in $(\Sigma \setminus \{\llcorner\})^*$ (in space double-exponential in the size of the input) iff the initial configuration is accepting.

There exists a double-exponentially space bounded ATM $\mathcal{M} = (Q, \Sigma, q_0, \Theta)$ whose word problem is 3ExpTime-hard [16]. Let $\mathcal{M}$ be such a double-exponentially space bounded ATM and $w = \sigma_1\sigma_2\cdots\sigma_n$ an input word. W.l.o.g., we assume that $\mathcal{M}$ never attempts to move to the left (right) when the head is on the left-most (right-most) tape cell.

We construct an $\mathcal{ALCH}_@^{\mathbb{T}}$ ontology $\mathcal{O}_{\mathcal{M},w}$ that entails $A(a)$ iff $\mathcal{M}$ accepts w. We represent configurations using individuals in $\mathcal{O}_{\mathcal{M},w}$, which are connected to the corresponding successor configurations by roles encoding the transition. W.l.o.g., we assume that these individuals form a tree, which we call the *configuration tree*. Furthermore, each node of this tree, i.e., each configuration, is connected to 2^{2^n} individuals representing the tape cells. The main ingredients of our construction are:

- an individual a denoting the root of the configuration tree;
- an attribute bit, with values in $\{0,1\}$, used to encode double-exponentially many tape positions;
- an attribute flip which has value 1 at a (unique) time point where bit has value 0 and bit has value 1 in all subsequent time points;
- a concept A marking accepting configurations;
- a concept H marking the head position;
- a concept T marking tape cells;
- a concept I marking the initial configuration;
- concepts S_q for each state $q \in Q$;
- concepts C_σ for each symbol $\sigma \in \Sigma$;
- roles r_θ for all transitions $\theta \in \Theta$;
- a role tape connecting configurations to tape cells; and
- attributes $a_0, \ldots, a_n$ to encode the binary representation of time values.

To encode the binary representation of time values we first state that for time: $2^n - 1$ we have all a_i set to 1:

$$T \sqsubseteq T@\lfloor \text{time}: 2^n - 1, a_n:1, \ldots, a_0:1 \rfloor.$$

We now use the following intuition: if the a_i attributes represent a pattern $s{\cdot}1000$, where s is a binary sequence and $\cdot$ means concatenation, then $s{\cdot}0111$ should occur *before* that pattern in the time line. To ensure this, we add concept inclusions of the form, for all $0 \le i \le n$:

$$X:S\ (T@X \sqsubseteq T@\lfloor \text{before}: X.\text{time}, P^X_{a_i} \rfloor)$$

where S is $\lfloor a_i:1, a_{i-1}:0, \ldots, a_0:0 \rfloor$ and $P^X_{a_i}$ abbreviates

$$a_n: X.a_n, \ldots, a_{i+1}: X.a_{i+1}, a_i:0, a_{i-1}:1, \ldots, a_0:1.$$

By further adding a concept inclusion encoding that a_i can only be one of $1,0$ at the same time point we have that, in any model, the a_i attributes encode

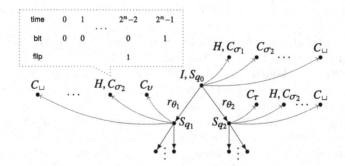

Fig. 4. A model of $\mathcal{O}_{\mathcal{M},w}$ encoding the computation tree of an ATM; blue edges (potentially grey) represent the tape role (we omit for brevity T in nodes representing tape cells) (Color figure online)

the binary representation of the corresponding time value, for time points in $[0, 2^n - 1]$. This means that, for time points in $[0, 2^n - 1]$, we can simulate the temporal attribute after by using variables and specifiers of the form $X : \lfloor a_i : 1 \rfloor$ and $\lfloor a_n : X.a_n, \ldots, a_{i-1} : X.a_{i-1} a_i : 0 \rfloor$, for all $0 \leq i \leq n$.

Remark 1. To simplify the presentation, in the following, we use the temporal attributes after and during (the latter is used to encode the initial configuration). Given the construction above it is straightforward to replace the inclusions using after and during with inclusions using the attributes a_i.

We encode the meaning of the attribute flip (i.e., it has value 1 at the time point from which bits should be flipped to increment a tape position) with the following concept inclusions:

$$T@\lfloor \mathsf{bit}\!:\!0 \rfloor \sqsubseteq T@\lfloor \mathsf{flip}\!:\!1 \rfloor \tag{29}$$

$$X : \lfloor \mathsf{flip}\!:\!1 \rfloor (T@X \sqsubseteq \neg T@\lfloor \mathsf{bit}\!:\!0, \mathsf{after}\!:\!X.\mathsf{time} \rfloor) \tag{30}$$

$$X : \lfloor \mathsf{flip}\!:\!1 \rfloor (T@X \sqsubseteq T@\lfloor \mathsf{bit}\!:\!0, \mathsf{time}\!:\!X.\mathsf{time} \rfloor) \tag{31}$$

Intuitively, in Eq. 29 we say that if there is a time point where we have bit with value 0 then there is a time point where we should flip some bit to increment the tape position, i.e., where flip is 1. In Eq. 30 we state that there is no bit with value 0 after a time point marked with flip set to 1. Finally, in Eq. 31, we state that bit has value 0 where flip has value 1. Thus, Eqs. 30 and 31 ensure that there is at most one time point where flip has value 1.

Let $\boldsymbol{\Omega}$ be a sequence with the following variables X_i^j, with $1 \leq i \leq n$ and $1 \leq j \leq 5$, and their respective specifiers:

- $X_i^1 : \lfloor \mathsf{flip}\!:\!1 \rfloor$, we look at our auxiliary attribute that indicates from which time point we should flip our bits to obtain the next tape position (this will be a time point with bit value 0);
- we also define $X_i^2 : \lfloor \mathsf{before}\!:\!X_i^1.\mathsf{time}, \mathsf{bit}\!:\!0 \rfloor$ and $X_i^3 : \lfloor \mathsf{before}\!:\!X_i^1.\mathsf{time}, \mathsf{bit}\!:\!1 \rfloor$, to filter time points with bit values 0 and 1, respectively, before the time point with flip : 1, related to X_i^1;
- we use $X_i^4 : \lfloor \mathsf{bit}\!:\!0 \rfloor$ and $X_i^5 : \lfloor \mathsf{bit}\!:\!1 \rfloor$ to filter time points bit values 0 and 1, respectively.

Basically, the first three variables are related to specifiers that filter the information needed to increment the tape position encoded with the bit attribute. The last two variables X_i^j are related to specifiers that filter the information needed to copy the tape position. We now define specifiers S_i^j, S_i, for $1 \leq i \leq n$ and $1 \leq j \leq 5$. Intuitively, the next four specifiers are used to increment the tape position, using the information given by the X_i^j variables. The last two specifiers copy the tape position, again using the information given by the X_i^j variables:

– the negation of a concept expression associated with $S_i = \lfloor\text{after} : X_i^1.\text{time}, \text{bit} : 1\rfloor$ ensures that we have $\text{bit} : 0$ in all time points after the time point marked with $\text{flip} : 1$ in the previous position;
– we use $S_i^1 = \lfloor\text{time} : X_i^1.\text{time}, \text{bit} : 1\rfloor$ to flip to 1 the bit marked with $\text{flip} : 1$ in the previous position;
– in addition, we define $S_i^2 = \lfloor\text{time} : X_i^2.\text{time}, \text{bit} : 0\rfloor$ and $S_i^3 = \lfloor\text{time} : X_i^3.\text{time}, \text{bit} : 1\rfloor$ to transfer to the next tape position bit values which should not be flipped (i.e., those that are before the time point with $\text{flip} : 1$);
– finally, we define $S_i^4 = \lfloor\text{time} : X_i^4.\text{time}, \text{bit} : 0\rfloor$ and $S_i^5 = \lfloor\text{time} : X_i^5.\text{time}, \text{bit} : 1\rfloor$, to receive a copy of the bit values.

To simplify the presentation, we define the abbreviations $P_i, P_i^+, P_i^=$ for the following concepts, respectively, to be used in concept inclusions with Ω:

– $\bigsqcap_{1 \leq j \leq 5} T@X_i^j$, we filter the bits encoding a tape position and the information of which bits should be flipped in order to increment it;
– $\bigsqcap_{1 \leq j \leq 3} T@S_i^j \sqcap \neg T@S_i$, we increment the tape position,
– $\bigsqcap_{4 \leq j \leq 5} T@S_i^j$, we copy the tape position.

We may also write $P, P^+, P^=$ if $i = 1$.

Encoding the Initial Configuration. We add assertions to $\mathcal{O}_{\mathcal{M},w}$ that encode the initial configuration of $\mathcal{M}$. We mark the root of the configuration tree with the initial state by adding $S_{q_0}(a)$ and initialise the tape cells with the input word by adding $I(a)$ and the concept inclusions:

$$\Omega\left(I \sqsubseteq \exists\text{tape}.(H \sqcap C_{\sigma_1} \sqcap T@\lfloor\text{during} : [0, 2^n - 1], \text{bit}: 0\rfloor)\right)$$

$$\Omega\left(I \sqcap \exists\text{tape}.P_i \sqsubseteq \exists\text{tape}.(C_{\sigma_{i+1}} \sqcap P_i^+)\right) \text{ for } 1 \leq i < n$$

$$\Omega\left(I \sqcap \exists\text{tape}.P_n \sqsubseteq \exists\text{tape}.(C_\sqcup \sqcap P_n^+)\right)$$

The intuition is as follows. In the first inclusion, we place the head, represented by the concept H, in the first position of the tape and fill the tape cell with the first symbol of the input word, represented by the concept C_{σ_1}. We then add the remaining symbols of the input word in their corresponding tape positions. In the last inclusion we add a blank symbol after the input word. We now add the following concept inclusion fill the remaining tape cells with blank in the initial configuration marked with the concept I:

$$\Omega\left(I \sqcap \exists\text{tape}.(C_\sqcup \sqcap P) \sqsubseteq \exists\text{tape}.(C_\sqcup \sqcap P^+)\right)$$

Synchronising Configurations. For each transition $\theta \in \Theta$, we make sure that tape contents are transferred to successor configurations, except for the tape cell at the head position:

$$\Omega\left(\exists\text{tape}.(P \sqcap \neg H \sqcap C_\sigma) \sqsubseteq \forall r_\theta.\exists\text{tape}.(P^= \sqcap C_\sigma)\right)$$

We now encode our transitions $\theta = (q, \sigma, q', \tau, D) \in \Theta$ with concept inclusions of the form (we explain for $D = r$, the case $D = l$ is analogous):

$$\Omega\Big(S_q \sqcap \exists\text{tape}.\big(H \sqcap P \sqcap C_\sigma\big) \sqcap \exists\text{tape}.\big(P^+ \sqcap C_v\big) \sqsubseteq$$

$$\exists r_\theta.(S_{q'} \sqcap \exists\text{tape}.\big(H \sqcap P^+ \sqcap C_v\big) \sqcap \exists\text{tape}.(P^= \sqcap C_\tau))\Big)$$

Essentially, if the head is at position P then, to move it to the right, we increment the head position using P^+ in the successor configuration. We use the specifiers in Ω to modify the tape cell with C_σ in the head position to C_τ in the successor configuration.

Acceptance Condition. Finally, we add concept inclusions that propagate acceptance from the leaf nodes of the configuration tree backwards to the root of the tree. For existential configurations, we add $S_q \sqcap \exists r_\theta.A \sqsubseteq A$ for each $q \in Q_\exists$, whereas to handle universal configurations, we add, for each $q \in Q_\forall$, the concept inclusion

$$S_q \sqcap \exists\text{tape}.\big(C_\sigma \sqcap H\big) \sqcap \bigsqcap_{\substack{\theta \in \Theta \\ \theta=(q,\sigma,q',\tau,D)}} \exists r_\theta.A \sqsubseteq A$$

where the conjunction may be empty if there are no suitable $\theta \in \Theta$.

With an inductive argument along the recursive definition of acceptance, we show that $\mathcal{O}_{\mathcal{M},w} \models A(a)$ iff $\mathcal{M}$ accepts w.

Given a natural number $i < 2^{2^n}$, we write $i_\mathbf{b}[j]$ for the value of the j-th bit of the binary representation of i using 2^n bits, where $i_\mathbf{b}[0]$ is the value the most significant bit. In the following, we write B_i as a shorthand for the concept:

$$\bigsqcap_{0 \leq y < 2^n} T@\lfloor \text{bit} : i_\mathbf{b}[y], \text{time} : y \rfloor.$$

Following the terminology provided in [25], given an interpretation $\mathcal{I}$ of $\mathcal{ALCH}_{@}^{\mathsf{T}}$, we say that an element $\delta \in \Delta_{\mathcal{I}}^{\mathcal{I}}$ *represents* a configuration $\tau_1 \ldots \tau_{i-1} q \tau_i \ldots \tau_m$ if $(\delta, F) \in S_q^{\mathcal{I}}$, for some $F \in \Phi^{\mathcal{I}}$, $\delta \in (\exists\text{tape}.(B_i \sqcap H))^{\mathcal{I}}$ and $\delta \in (\exists\text{tape}.(B_j \sqcap C_{\tau_j}))^{\mathcal{I}}$, for all $0 \leq j < 2^{2^n}$. We are now ready to show Claims 1 and 2.

Claim 1. If $\delta \in \Delta_I^{\mathcal{I}}$ represents a configuration α and some transition $\theta \in \Theta$ is applicable to α then δ has an r_θ-successor that represents the result of applying θ to α.

Proof of the Claim 1. Let $\delta \in \Delta_I^{\mathcal{I}}$ be an element representing a configuration α and assume $\theta \in \Theta$ is applicable to α. To synchronise configurations, we added to $\mathcal{O}_{\mathcal{M},w}$ concept inclusions that (1) ensure that tape contents other than the content at the head position are copied to all r_θ-successors of δ; and (2) create an r_θ-successor that represents the correct state, position of the head and corresponding symbols at the previous and current position of the head. Then our concept inclusions ensure that δ has an r_θ-successor that represents the result of applying θ to α.

Claim 2. w is accepted by $\mathcal{M}$ iff $\mathcal{O}_{\mathcal{M},w} \models A(a)$.

Proof of the Claim 2. Consider an arbitrary interpretation $\mathcal{I}$ of $\mathcal{ALCH}_{@}^{\mathsf{T}}$ that satisfies $\mathcal{O}_{\mathcal{M},w}$. First we show that if any element $\delta \in \Delta_I^{\mathcal{I}}$ represents an accepting configuration then $(\delta, F) \in A^{\mathcal{I}}$, for some $F \in \Phi^{\mathcal{I}}$. We make a case distinction.

- If α is a universal configuration, then all successor configurations of α must be accepting. By Claim 1, for any θ-successor configuration α' of α there is a corresponding r_θ-successor δ' of δ. By induction hypothesis for α', (δ', F') is in $A^{\mathcal{I}}$, for some $F' \in \Phi^{\mathcal{I}}$. Since this holds for all θ-successor configurations of α, our concept inclusion encoding acceptance of universal configurations implies that $(\delta, F) \in A^{\mathcal{I}}$, for some $F \in \Phi^{\mathcal{I}}$, as required. This argument covers the base case where α has no successors.
- If α is an existential configuration, then there is some accepting θ-successor configuration α' of α. By Claim 1, there is an r_θ-successor δ' of δ that represents α' and, by induction hypothesis, $(\delta', F') \in A^{\mathcal{I}}$, for some $F' \in \Phi^{\mathcal{I}}$. Then, our concept inclusion encoding acceptance of existential configurations applies and so, we conclude that $(\delta, F) \in A^{\mathcal{I}}$, for some $F \in \Phi^{\mathcal{I}}$.

Since elements in $I^{\mathcal{I}}$ represent the initial configuration of $\mathcal{M}$, this shows that $I^{\mathcal{I}} \subseteq A^{\mathcal{I}}$ when the initial configuration is accepting. As $I(a)$ is an assertion in $\mathcal{O}_{\mathcal{M},w}$, we have that $(a^{\mathcal{I}}, G) \in A^{\mathcal{I}}$, for some $G \in \Phi^{\mathcal{I}}$.

We now show that if the initial configuration is not accepting, then there is some interpretation $\mathcal{I}$ of $\mathcal{ALCH}_{@}^{\mathsf{T}}$ such that $I^{\mathcal{I}} \not\subseteq A^{\mathcal{I}}$, in particular, $(a^{\mathcal{I}}, G) \notin A^{\mathcal{I}}$, for all $G \in \Phi^{\mathcal{I}}$. To show this we construct a canonical interpretation $\mathcal{J}$ of $\mathcal{O}_{\mathcal{M},w}$ as follows. Let $\mathsf{Con}_{\mathcal{M}} := \{wqw' \mid |ww'| \leq 2^{2^n}, q \in Q, \{w, w'\} \subseteq \Sigma^*\}$ be the set of all possible $\mathcal{M}$ configurations with size bounded by 2^{2^n}. Also, we define a set $\mathsf{Tp}_{\mathcal{M}} := \{\alpha \cdot c_\sigma^i \mid \alpha \in \mathsf{Con}_{\mathcal{M}}, 0 \leq i < 2^{2^n}, \sigma \in \Sigma\}$, containing individuals that represent tape cells, related to each possible configuration of a computation of $\mathcal{M}$. The domain $\Delta^{\mathcal{J}}$ is a disjoint union of $\Delta_I^{\mathcal{J}} \cup \Delta_T^{\mathcal{J}} \cup \Delta_{2T}^{\mathcal{J}}$, where:

- $\Delta_I^{\mathcal{J}} = \mathsf{Con}_{\mathcal{M}} \cup \mathsf{Tp}_{\mathcal{M}} \cup \mathbb{T}$, where $\mathbb{T} \subseteq \mathsf{N}_{\mathsf{I}}$ is either time or before;
- $\Delta_T^{\mathcal{J}} = \{0^{\mathcal{J}}, \dots, (2^n - 1)^{\mathcal{J}}\}$; and $\Delta_{2T}^{\mathcal{J}} = \Delta_T^{\mathcal{J}} \times \Delta_T^{\mathcal{J}}$.

The extension of the concepts C_σ, H and B_j in the interpretation is defined as expected so that every element $\alpha \cdot c_\sigma^i \in \mathsf{Tp}_{\mathcal{M}}$ is in C_σ and B_i and no other C_τ or B_j, with $\tau \neq \sigma$ or $i \neq j$. Also, $\alpha \cdot c_\sigma^i$ is in H iff α is of the form wqw' and $|w| = i - 1$. We connect α to $\alpha \cdot c_\sigma^i$ using the role tape iff α has σ at position i. Moreover, α is in S_q iff α is of the form wqw'. We then have that every configuration $\alpha \in \mathsf{Con}_{\mathcal{M}}$ represents itself and no other configuration. $I^{\mathcal{J}}$ is the singleton set containing the initial configuration $a^{\mathcal{J}}$. Given two configurations α and α' and a transition $\theta \in \Theta$, we connect α to α' using the role r_θ iff there is a transition θ from α to α'. Finally, $A^{\mathcal{J}}$ is defined to be the set of tuples (α, F), for some $F \in \Phi^{\mathcal{J}}$, where α is an accepting configuration.

Now, if the initial configuration $a^{\mathcal{J}}$ is not accepting then, by construction, $(a, G) \notin A^{\mathcal{J}}$, for all $G \in \Phi^{\mathcal{J}}$. By checking the concept inclusions in $\mathcal{O}_{\mathcal{M},w}$, we can see that $\mathcal{J}$ satisfies $\mathcal{O}_{\mathcal{M},w}$. Then, $\mathcal{J}$ is a counterexample for $\mathcal{O}_{\mathcal{M},w} \models A(a)$, and so $\mathcal{O}_{\mathcal{M},w} \not\models A(a)$. $\qquad\square$

Theorem 7. *In $\mathcal{ALCH}_{@}^{\mathbb{T}}$, any combination of temporal attributes containing* $\{\text{time}, \text{after}\}$ *is undecidable. Moreover, the combination* $\{\text{time}, \text{before}\}$ *is* 3EXPTIME-*complete, and the combination* $\{\text{time}, \text{during}, \text{since}, \text{until}\}$ *and every subset of it are* 2EXPTIME-*complete.*

Proof. The proof of Theorem 1 uses only the temporal attributes time and after. Thus, any combination containing these attributes is Σ_1^1-hard. By Theorems 5 and 6 the combination $\{\text{time}, \text{before}\}$ is 3EXPTIME-complete. It remains to show that the combination $\{\text{time}, \text{during}, \text{since}, \text{until}\}$ is in 2EXPTIME (since 2EXPTIME-hardness is already known for $\mathcal{ALCH}_{@}$ [23]).

Our proof strategy consists in showing that, given an $\mathcal{ALCH}_{@}^{\mathbb{T}}$ interpretation and an $\mathcal{ALCH}_{@}^{\mathbb{T}}$ ontology that contains only the temporal attributes in $\{\text{time}, \text{during}, \text{since}, \text{until}\}$, one can always transform this interpretation so that only time points explicitly mentioned in the ontology are relevant to determine if the interpretation is a model of the ontology. Then one can check satisfiability by grounding the ontology using only those time points explicitly mentioned. We start by providing some notation.

Given an $\mathcal{ALCH}_{@}^{\mathbb{T}}$ ontology $\mathcal{O}$, we define a set $\mathbb{N}_{\mathcal{O}}$ as in Theorem 3, except that we do not need $k_i + 1$ here. To this end, let $k_0 < \ldots < k_n$ be the ascending sequence of all numbers mentioned in time points or in time intervals (as endpoints) in $\mathcal{O}$. We define $\mathbb{N}_{\mathcal{O}}$ as $\{k_i \mid 0 \leq i \leq n\}$, and let $k_{\min} := \min(\mathbb{N}_{\mathcal{O}})$ and $k_{\max} = \max(\mathbb{N}_{\mathcal{O}})$, where we assume $k_{\min} = k_{\max} = 0$ if $\mathbb{N}_{\mathcal{O}} = \emptyset$.

Let $\mathcal{I} = (\Delta^{\mathcal{I}}, \cdot^{\mathcal{I}})$ be an $\mathcal{ALCH}_{@}^{\mathbb{T}}$ interpretation. By Definition 8, $\mathcal{I}$ is a global interpretation of a sequence $(\mathcal{I}_i)_{i \in \Delta_T^{\mathcal{I}}}$ of $\mathcal{ALCH}_{@}$ interpretations with domain $\Delta_I^{\mathcal{I}}$. We now define a sequence $(\mathcal{J}_i)_{i \in \Delta_T^{\mathcal{J}}}$ of $\mathcal{ALCH}_{@}$ interpretations as follows. Let $\Delta_I^{\mathcal{J}} = \Delta_I^{\mathcal{I}}$ and let $\Delta_T^{\mathcal{J}} = \{k_{\min}^{\mathcal{J}}, \ldots, k_{\max}^{\mathcal{J}}\}$. For all $A \in \mathsf{N}_\mathsf{C}$, all $F \in \Phi^{\mathcal{I}}$ with $F \setminus F_I \neq \emptyset$ and $k \in [k_{\min}, k_{\max}]$:

$$(\delta, F_I) \in A^{\mathcal{J}_k} \text{ iff } (\delta, F_I) \in A^{\mathcal{I}_k}$$

and either:

(1) $k \in \mathbb{N}_{\mathcal{O}}$; or
(2) there is $k_i < k$ such that $k_i \in \mathbb{N}_{\mathcal{O}}$ and $(\delta, F_I) \in A^{\mathcal{I}_j}$ for all $k_i \leq j \leq k_{\max}$; or
(3) there is $k_i > k$ such that $k_i \in \mathbb{N}_{\mathcal{O}}$ and $(\delta, F_I) \in A^{\mathcal{I}_j}$ for all $k_{\min} \leq j \leq k_i$.

We analogously apply the definition above for all role names $r \in \mathsf{N}_\mathsf{R}$. We define $\mathcal{I}_{\mathcal{O}}$ as a global interpretation of the sequence $(\mathcal{J}_i)_{i \in \Delta_T^{\mathcal{J}}}$ and set $(\delta, F) \in A^{\mathcal{I}_{\mathcal{O}}}$ iff $(\delta, F) \in A^{\mathcal{I}}$ for all $A \in \mathsf{N}_\mathsf{C}$ with $F = F_I$, and similarly for all role names $r \in \mathsf{N}_\mathsf{R}$. Let $\mathcal{O}_g$ be the result of grounding $\mathcal{O}$ in the same way as in the proof of Theorem 4 using time points in $\mathbb{N}_{\mathcal{O}}$ (here $\mathcal{O}$ may have expressions of the form $X.a$, $a:x$, or $a:[t,t']$, with $a \in \{\text{time}, \text{during}, \text{since}, \text{until}\}$, and $t, t' \in \mathsf{N_T} \cup \mathsf{Var}(\mathsf{N_T})$).

Claim. For all $A@S, r@S$ occurring in $\mathcal{O}_g$: $A@S^{\mathcal{I}_{\mathcal{O}}} = A@S^{\mathcal{I}}$ and $r@S^{\mathcal{I}_{\mathcal{O}}} = r@S^{\mathcal{I}}$.

Proof of the Claim. This claim follows by definition of $(\mathcal{J}_i)_{i \in \Delta_T^{\mathcal{J}}}$ and the fact that only the temporal attributes {time, during, since, until} are allowed. Correctness for the temporal attributes time and during follows from item (1), whereas correctness for the temporal attributes since and until follows from items (2) and (3), respectively.

By definition of $\mathcal{O}_g$, we know that $\mathcal{O} \models \mathcal{O}_g$. So if $\mathcal{O}$ is satisfiable then $\mathcal{O}_g$ is satisfiable. Conversely, by the Claim, one can show with an inductive argument that $C^{\mathcal{I}_{\mathcal{O}}} = C^{\mathcal{I}}$ for all $\mathcal{ALCH}_@^{\mathsf{T}}$ concepts C occurring in $\mathcal{O}_g$. So, if an $\mathcal{ALCH}_@^{\mathsf{T}}$ interpretation $\mathcal{I}$ satisfies $\mathcal{O}_g$ then $\mathcal{I}_{\mathcal{O}}$ satisfies $\mathcal{O}$. Since $\mathcal{O}_g$ is at most exponentially larger than $\mathcal{O}$, it follows that satisfiability in this fragment is in 2ExpTime. $\square$

References

1. Alur, R., Henzinger, T.A.: Real-time logics: complexity and expressiveness. Inf. Comput. **104**(1), 35–77 (1993)
2. Artale, A., Franconi, E., Peñaloza, R., Sportelli, F.: A decidable very expressive description logic for databases. In: d'Amato, C., et al. (eds.) ISWC 2017. LNCS, vol. 10587, pp. 37–52. Springer, Cham (2017). https://doi.org/10.1007/978-3-319-68288-4_3
3. Artale, A., Kontchakov, R., Lutz, C., Wolter, F., Zakharyaschev, M.: Temporalising tractable description logics. In 14th International Symposium on Temporal Representation and Reasoning (TIME 2007), 28–30 June 2007, Alicante, Spain, pp. 11–22 (2007)
4. Artale, A., Kontchakov, R., Wolter, F., Zakharyaschev, M.: Temporal description logic for ontology-based data access. In: Proceedings of the 23rd International Joint Conference on Artificial Intelligence, IJCAI 2013, pp. 711–717 (2013)
5. Baader, F., Borgwardt, S., Koopmann, P., Ozaki, A., Thost, V.: Metric temporal description logics with interval-rigid names. In: Dixon, C., Finger, M. (eds.) FroCoS 2017. LNCS (LNAI), vol. 10483, pp. 60–76. Springer, Cham (2017). https://doi.org/10.1007/978-3-319-66167-4_4
6. Baader, F., Brandt, S., Lutz, C.: Pushing the $\mathcal{EL}$ envelope. In: Kaelbling, L.P., Saffiotti, A. (eds.) Proceedings of the 19th International Joint Conference on Artificial Intelligence (IJCAI 2005), pp. 364–369. Professional Book Center (2005)
7. Baader, F., Calvanese, D., McGuinness, D., Nardi, D., Patel-Schneider, P. (eds.): The Description Logic Handbook Theory, Implementation, and Applications, 2nd edn. Cambridge University Press (2007)
8. Baader, F., Ghilardi, S., Lutz, C.: LTL over description logic axioms. ACM Trans. Comput. Log. **13**(3) (2012)
9. Baader, F., Lutz, C., Turhan, A.-Y.: Small is again beautiful in description logics. KI **24**(1), 25–33 (2010)
10. Berardi, D., Calvanese, D., De Giacomo, G.: Reasoning on UML class diagrams. Artif. Intell. **168**(1–2), 70–118 (2005)
11. Borgwardt, S., Thost, V.: Temporal query answering in the description logic EL. In: Proceedings of the Twenty-Fourth International Joint Conference on Artificial Intelligence, IJCAI 2015, Buenos Aires, Argentina, 25–31 July 2015, pp. 2819–2825 (2015)

12. Bourgaux, C., Ozaki, A.: Querying attributed DL-Lite ontologies using provenance semirings. In: Proceedings of the 33rd AAAI Conference on Artificial Intelligence (2019)
13. Brandt, S., Kalayci, E., Kontchakov, R., Ryzhikov, V., Xiao, G., Zakharyaschev, M.: Ontology-based data access with a horn fragment of metric temporal logic. In: Proceedings of the Thirty-First AAAI Conference on Artificial Intelligence, 4–9 February 2017, San Francisco, California, USA, pp. 1070–1076 (2017)
14. Brandt, S., Kalayci, E., Ryzhikov, V., Xiao, G., Zakharyaschev, M.: Querying log data with metric temporal logic. J. Artif. Intell. Res. **62**, 829–877 (2018)
15. Carapelle, C., Turhan, A.-Y.: Description logics reasoning w.r.t. general TBoxes is decidable for concrete domains with the EHD-property. In: ECAI 2016–22nd European Conference on Artificial Intelligence, pp. 1440–1448 (2016)
16. Chandra, A.K., Kozen, D.C., Stockmeyer, L.J.: Alternation. J. ACM **28**(1), 114–133 (1981)
17. Chomicki, J.: Temporal query languages: a survey. In: Gabbay, D.M., Ohlbach, H.J. (eds.) ICTL 1994. LNCS, vol. 827, pp. 506–534. Springer, Heidelberg (1994). https://doi.org/10.1007/BFb0014006
18. Fisher, M.D., Gabbay, D.M., Vila, L. (eds.): Handbook of Temporal Reasoning in Artificial Intelligence. Elsevier (2005)
19. Gutiérrez-Basulto, V., Jung, J.C., Ozaki, A.: On metric temporal description logics. In Proceedings of the 22nd European Conference on Artificial Intelligence (ECAI 2016), pp. 837–845. IOS Press (2016)
20. Harel, D.: Effective transformations on infinite trees, with applications to high undecidability, dominoes, and fairness. J. ACM **33**(1), 224–248 (1986)
21. Hoffart, J., Suchanek, F.M., Berberich, K., Weikum, G.: YAGO2: a spatially and temporally enhanced knowledge base from wikipedia. J. Artif. Intell. **194**, 28–61 (2013)
22. Krötzsch, M.: Ontologies for knowledge graphs? In: Artale, A., Glimm, B., Kontchakov, R. (eds.) Proceedings of the 30th International Workshop on Description Logics (DL 2017). CEUR Workshop Proceedings, vol. 1879, CEUR-WS.org (2017)
23. Krötzsch, M., Marx, M., Ozaki, A., Thost, V.: Attributed description logics: ontologies for knowledge graphs. In: The Semantic Web - ISWC - 16th International Semantic Web Conference, pp. 418–435 (2017)
24. Krötzsch, M., Marx, M., Ozaki, A., Thost, V.: Attributed description logics: reasoning on knowledge graphs. In: Proceedings of the Twenty-Seventh International Joint Conference on Artificial Intelligence, IJCAI, pp. 5309–5313 (2018)
25. Krötzsch, M., Rudolph, S., Hitzler, P.: Complexities of horn description logics. ACM Trans. Comput. Log. **14**(1), 2:1–2:36 (2013)
26. Leo, J., Sattler, U., Parsia, B.: Temporalising EL concepts with time intervals. In: Proceedings of the 27th International Workshop on Description Logics (DL 2014). CEUR Workshop Proceedings, vol. 1193, pp. 620–632. CEUR-WS.org (2014)
27. Lutz, C.: The complexity of description logics with concrete domains. Ph.D. thesis, RWTH Aachen University, Germany (2002)
28. Lutz, C.: Combining interval-based temporal reasoning with general TBoxes. Artif. Intell. **152**(2), 235–274 (2004)
29. Lutz, C., Haarslev, V., Möller, R.: A concept language with role-forming predicate restrictions. Technical report, University of Hamburg (1997)
30. Lutz, C., Möller, R.: Defined topological relations in description logics. In: Brachman, R.J., et al. (eds.) Proceedings of the 1997 International Workshop on Description Logics (DL 1997). URA-CNRS, vol. 410 (1997)

31. Lutz, C., Wolter, F., Zakharyaschev, M.: Temporal description logics: a survey. In: 15th International Symposium on Temporal Representation and Reasoning, TIME, pp. 3–14 (2008)
32. Marx, M., Krötzsch, M., Thost, V.: Logic on MARS: ontologies for generalised property graphs. In: Sierra, C. (ed.) Proceedings of the 26th International Joint Conferences on Artificial Intelligence (IJCAI 2017). IJCAI, pp. 1188–1194 (2017)
33. openCypher community. Cypher Query Language Reference, Version 9. http://www.opencypher.org/resources (2019)
34. Rodriguez, M.A., Neubauer, P.: Constructions from dots and lines. Bull. Am. Soc. Inf. Sci. Technol. **36**(6), 35–41 (2010)
35. Rogers Jr., H.: Theory of Recursive Functions and Effective Computability, Paperback edn. MIT Press (1987)
36. Rudolph, S., Krötzsch, M., Hitzler, P.: Cheap boolean role constructors for description logics. In: Hölldobler, S., Lutz, C., Wansing, H. (eds.) JELIA 2008. LNCS (LNAI), vol. 5293, pp. 362–374. Springer, Heidelberg (2008). https://doi.org/10.1007/978-3-540-87803-2_30
37. Sportelli, F., Franconi, E.: Formalisation of ORM derivation rules and their mapping into OWL. In: Debruyne, C., et al. (eds.) On the Move to Meaningful Internet Systems: OTM 2016 Conferences, OTM 2016. Lecture Notes in Computer Science, vol. 10033, pp. 827–843. Springer, Cham (2016). https://doi.org/10.1007/978-3-319-48472-3_52
38. Vrandečić, D., Krötzsch, M.: Wikidata: a free collaborative knowledgebase. Commun. ACM **57**(10), 78–85 (2014)
39. Wolter, F., Zakharyaschev, M.: Temporalizing description logics. Front. Combin. Syst. **2**, 379–402 (1999)

Explaining Axiom Pinpointing

Rafael Peñaloza$^{(\boxtimes)}$ [ID]

University of Milano-Bicocca, Milan, Italy
rafael.penaloza@unimib.it

Abstract. Axiom pinpointing refers to the task of highlighting (or pinpointing) the axioms in an ontology that are responsible for a given consequence to follow. This is a fundamental task for understanding and debugging very large ontologies. Although the name axiom pinpointing was only coined in 2003, the problem itself has a much older history, even if considering only description logic ontologies. In this work, we try to explain axiom pinpointing: what it is; how it works; how it is solved; and what it is useful for. To answer this questions, we take a historic look at the field, focusing mainly on description logics, and the specific contributions stemming from one researcher, who started it all in more than one sense.

1 Introduction

One important aspect behind any (artificially) intelligent application is the availability of knowledge about the domain, which can be accessed and used effectively. In logic-based knowledge representation, this domain knowledge is expressed through a collection of logical constraints (or *axioms*) that limit how the different terms under consideration are interpreted, and related to each other. To abstract from the specific logical language used to express these constraints, we call any such representation an *ontology*.

Description logics (DLs) are a family of knowledge representation formalisms that have been successfully applied to represent the knowledge of several application domains. The family contains several different logical languages that are distinguished by their expressivity and the computational complexity of reasoning over them. They range from the extremely inexpressive DL-Lite [24] at the basis of ontology-based data access [72], to the expressive $\mathcal{SROIQ}(\mathcal{D})$ [36] underlying the standard ontology language for the semantic web OWL2 [50]. In between these two, many other DLs exist. Another prominent example is the light-weight DL $\mathcal{EL}$ [1], which allows for polynomial time reasoning [11], and is used for many ontologies within the bio-medical domain. In particular, we can mention SNOMED CT, an ontology providing a unified nomenclature for clinical terms in medicine composed of approximately half a million axioms [59].

It is hardly surprising that ontology engineering—the task of building an ontology—is a costly and error-prone task. As the size of an ontology increases, it becomes harder to have a global perspective of its constraints and their relationships. For this reason, it becomes more likely to be surprised by some of

C. Lutz et al. (Eds.): Baader Festschrift, LNCS 11560, pp. 475–496, 2019.
https://doi.org/10.1007/978-3-030-22102-7_22

the consequences that can be derived from them. A classical example arose with SNOMED, from which at some point it was possible to derive that every amputation of a finger was in fact an amputation of a hand. Finding the six axioms—out of approximately 500,000—that cause this error without the help of an automated method would have been almost impossible.

Here is where axiom pinpointing comes into play. Essentially, axiom pinpointing refers to the task of highlighting (or pinpointing) the specific axioms that are responsible for a consequence to follow from an ontology. Considering that the underlying ontology language is monotonic, this task corresponds to finding classes of minimal subontologies still entailing the consequence. The term itself was coined by Schlobach and Cornet in 2003, in a work that triggered several follow-up approaches dealing with logics of varying expressivity. However, the underlying method was proposed almost a decade earlier by Baader and Hollunder in the context of default reasoning.[1]

By 2006, the field of axiom pinpointing was starting to mature to a point that begged for general solutions, beyond the specific attempts known at the time. With the advice of Franz, I embarked then on a trip to produce these general solutions. That trip was supposed to end in 2009 with the defense of my dissertation, and search for new research topics. However, the repercussions of the work continue today.

In this paper, we attempt to present a historical perspective on axiom pinpointing, explaining the existing methods, its applications, and newer extensions that have been developed throughout the years. We note that although axiom pinpointing has been studied—with different names—for other knowledge representation formalisms, and by several people, our focus here is constrained to DLs and specifically to the contributions that Franz made to the field either directly, or indirectly through the people working in his group. This work is intended as a basic, non-technical introduction to the field, but relevant references are provided for the reader interested in a deeper understanding.

2 Description Logics in a Nutshell

We briefly introduce the notions on description logics (DLs) [2,6] that will be useful for understanding the rest of this work. Although there exist more complex and expressive DLs, here we will focus on the basic $\mathcal{ALC}$ [65], which is the smallest propositionally-closed DL, and the light-weight $\mathcal{EL}$ [1].

DLs are knowledge representation formalisms that are specially targeted to encode the terminological knowledge of an application domain. Their main ingredients are *individuals*, *concepts* (that is, classes of individuals), and *roles* providing relationships between individuals. Formally, they are nullary, unary, and binary predicates from first-order logic, respectively. The knowledge of the

---

[1] I am an informal (some will say impolite) Mexican, who insists on using the term *Franz* when referring to Franz Baader. I will do this often from now on. Please bear with me.

domain is represented via a set of *axioms* that restrict the way in which those ingredients can be interpreted.

Let N_I, N_C, and N_R be three mutually disjoint sets of *individual, concept,* and *role names*, respectively. $\mathcal{ALC}$ *concepts* are built following the syntactic rule $C ::= A \mid \neg C \mid C \sqcap C \mid \exists r.C$, where $A \in N_C$ and $r \in N_R$. A *general concept inclusion* (GCI) is an expression of the form $C \sqsubseteq D$, where C and D are concepts. An *assertion* is an expression of the form $C(a)$ (called a *concept assertion*) or $r(a, b)$ (*role assertion*), where $a, b \in N_I$, C is a concept, and $r \in N_R$. Historically, DLs separate the knowledge in *terminological* and *assertional knowledge*. The former describes general relations between the terms, encoded via a *TBox*, which is a finite set of GCIs. The latter refers to the knowledge about the individuals, which is encoded in a finite set of assertions called an *ABox*. Here, we use the general term *axiom* to refer to both GCIs and assertions. An *ontology* is a finite set of axioms; that is, the union of an ABox and a TBox.

The semantics of $\mathcal{ALC}$ is defined in terms of *interpretations*. These are tuples of the form $\mathcal{I} = (\Delta^{\mathcal{I}}, \cdot^{\mathcal{I}})$, where $\Delta^{\mathcal{I}}$ is a non-empty set called the *domain*, and $\cdot^{\mathcal{I}}$ is the *interpretation function* that maps every individual name $a \in N_I$ to an element $a^{\mathcal{I}} \in \Delta^{\mathcal{I}}$, every concept name $A \in N_C$ to a set $A \subseteq \Delta^{\mathcal{I}}$, and every role name $r \in N_R$ to a binary relation $r^{\mathcal{I}} \subseteq \Delta^{\mathcal{I}} \times \Delta^{\mathcal{I}}$. This interpretation function is extended to arbitrary concepts in the usual manner; that is, $(\neg C)^{\mathcal{I}} := \Delta^{\mathcal{I}} \setminus C^{\mathcal{I}}$, $(C \sqcap D)^{\mathcal{I}} := C^{\mathcal{I}} \cap D^{\mathcal{I}}$, and $(\exists r.C)^{\mathcal{I}} := \{x \in \Delta^{\mathcal{I}} \mid \exists y \in C^{\mathcal{I}}.(x, y) \in r^{\mathcal{I}}\}$. The interpretation $\mathcal{I}$ *satisfies* the GCI $C \sqsubseteq D$ iff $C^{\mathcal{I}} \subseteq D^{\mathcal{I}}$; the assertion $C(a)$ iff $a^{\mathcal{I}} \in C^{\mathcal{I}}$; and the assertion $r(a, b)$ iff $(a^{\mathcal{I}}, b^{\mathcal{I}}) \in r^{\mathcal{I}}$. $\mathcal{I}$ is a *model* of the ontology $\mathcal{O}$, denoted by $\mathcal{I} \models \mathcal{O}$, iff it satisfies all the axioms in $\mathcal{O}$.

It is often useful to consider abbreviations of complex concepts. We define $\bot := A \sqcap \neg A$ for an arbitrary $A \in N_C$; $\top := \neg\bot$; $C \sqcup D := \neg(\neg C \sqcap \neg D)$; and $\forall r.C := \neg\exists r.\neg C$. In particular, in Sect. 4.1 dealing with these abbreviations is fundamental for the algorithm.

Once that the knowledge of a domain has been encoded in an ontology, we are interested in *reasoning*; that is, making inferences about this knowledge, which are implicit in the ontology. The most basic reasoning task is to decide *consistency*; i.e., whether a given ontology has at least one model. Other important reasoning problems are: *subsumption* (does $C^{\mathcal{I}} \subseteq D^{\mathcal{I}}$ hold in every model $\mathcal{I}$ of $\mathcal{O}$?); *instance checking* (does $a^{\mathcal{I}} \in A^{\mathcal{I}}$ hold in every model $\mathcal{I}$ of $\mathcal{O}$?); and *classification* (finding all the subsumption relations between concept names appearing in $\mathcal{O}$). It has been shown that consistency, subsumption, and instance checking w.r.t. $\mathcal{ALC}$ ontologies is ExpTime-complete [31,63]. Since the number of concept names appearing in $\mathcal{O}$ is bounded by the size of $\mathcal{O}$, it also follows that the ontology can be classified in exponential time.

The light-weight DL $\mathcal{EL}$ is the sublogic of $\mathcal{ALC}$ that disallows the negation constructor $\neg$, but includes the special concept $\top$ that specifies a tautology; that is $\top^{\mathcal{I}} := \Delta^{\mathcal{I}}$ for all interpretations $\mathcal{I}$ (compare to the abbreviation defined before). All other definitions are analogous as for $\mathcal{ALC}$. Interestingly, every $\mathcal{EL}$ ontology is consistent, and hence the consistency problem becomes trivial;

moreover, subsumption and instance checking are decidable in polynomial time via a so-called completion algorithm that in fact classifies the whole ontology.

3   What Is Axiom Pinpointing?

Although the term "Axiom Pinpointing" was originally coined in the context of description logics, and the main focus in this work is on its development in DLs as well, closely related problems have been also studied—with different names—in other areas such as databases [47], propositional satisfiability [44], or constraint satisfaction problems [48]. Most of the basic ideas can, in fact, be traced to Raymond Reiter [61]. For that reason, we introduce the problem in the most general terms possible, trying to preserve readability.

We consider an abstract *ontology language*, which is composed of a class $\mathcal{A}$ of (well-formed) *axioms*, and a *consequence relation* $\models: 2^{\mathcal{A}} \to \mathcal{A}$.[2] An *ontology* is a finite set $\mathcal{O} \subseteq \mathcal{A}$ of axioms. If $\mathcal{O} \models c$, where $c \in \mathcal{A}$, we say that $\mathcal{O}$ *entails* c, or that c is a *consequence* of $\mathcal{O}$. For reasons that will become clear later, we focus solely on *monotone* ontology languages, which are such that if $\mathcal{O} \subseteq \mathcal{O}'$ and $\mathcal{O} \models c$, then also $\mathcal{O}' \models c$.

Notice that these definitions follow the typical terminology from description logics as seen in the previous section, but they are not restricted exclusively to DLs. For example, the set of propositional clauses $\bigvee_{i=1}^{n} \ell_i$, where each ℓ_i, $1 \le i \le n$ is a literal, forms an ontology language under the standard entailment relation between formulas. In this case, an ontology is a formula in conjunctive normal form (CNF), and one common entailment of interest is whether such an ontology entails the empty clause $\bot := \bigvee_{\ell \in \emptyset} \ell$; that is, whether a formula is unsatisfiable. Clearly, this ontology language is monotone as well.

Historically, most of the work on ontology languages focuses on *reasoning*; that is, on studying and solving the problem of deciding whether $\mathcal{O} \models c$ holds; e.g. deciding subsumption or instance checking in $\mathcal{ALC}$. For the most prominent ontology languages, such as DLs, the computational complexity of this problem is perfectly understood, and efficient methods have been already developed and implemented. The number of available tools for solving these reasoning tasks is too large to enumerate them. However, knowing that a consequence follows from an ontology is only part of the story. Once that this fact has been established, we are left with the issue of answering *why* the consequence holds.

Why do we want to answer why? Well, mainly because the consequence relation is often far from trivial, specially when it depends on the inter-relation of several potentially complex axioms. Hence, it might not be obvious that a given axiom follows from an ontology. In particular, when the ontology is large, surprising or erroneous consequences are bound to appear. Explaining them is important to confirm their correctness or, alternatively, understand the causes of error when they are incorrect. Other applications that are based on this general setting are described in Sect. 7.

---

[2] We use the infix notation for the consequence relation.

In *axiom pinpointing*, we explain consequences by computing so-called *justifications*.[3] Formally, a justification for the entailment $\mathcal{O} \models c$, where $\mathcal{O}$ is an ontology and c is an axioms, is a minimal (w.r.t. set inclusion) subontology $\mathcal{M} \subseteq \mathcal{O}$ that still entails the consequence; more precisely, $\mathcal{M} \models c$ and for every $\mathcal{M}' \subset \mathcal{M}$, $\mathcal{M}' \not\models c$. We emphasise that the minimality is considered here w.r.t. set inclusion. Notice that justifications are not unique. In fact, a single consequence relation $\mathcal{O} \models c$ may allow for exponentially many justifications, measured on the number of axioms in $\mathcal{O}$ [16]. Depending on the situation at hand, one may want to compute one, several, or all these justifications. Axiom pinpointing approaches can be broadly grouped into three categories: *black-box* methods that use unmodified reasoners as an oracle, *glass-box* methods that adapt the reasoning procedure to trace the axioms used, and *gray-box* approaches, which combine the benefits of the other two.

As mentioned already, the basic idea behind the glass-box approach is to modify the reasoning algorithm to keep track of the axioms used throughout the reasoning process, in order to identify the elements of the justification. As we will see later, this process is effective for computing all justifications, but incurs an additional cost, either in terms of complexity, or in the need of a post-processing step. Alternatively, if one is only interested in one justification, the cost of tracing is essentially insignificant, but the result is only an approximation of a justification: it may contain superfluous axioms. Thus, it is often combined with a black-box minimisation step that guarantees that the resulting set is indeed a justification. In the following sections we describe these approaches in greater detail.

4   Finding All Justifications

The black-box method for finding all justifications was introduced and studied in detail by the groups in Maryland and Manchester [37,39,54]. Very briefly, this method uses a sub-procedure that can compute one justification at a time (we will see how to achieve this in the following section) and systematically removes and restores axioms from the ontology, following Reiter's Hitting Set enumeration method, to find new justifications.

4.1   Tableaux-Based Axiom Pinpointing

The history of glass-box methods for axiom pinpointing in DLs started more than 20 years ago, when Baader and Hollunder [5] extended the standard tableau-based algorithm for testing the consistency of an $\mathcal{ALC}$ ABox by a labelling technique that traced the application of the tableau rules, and ultimately the axioms responsible.

---

[3] I rather prefer the name *MinA* coined by Franz Baader as we started our work on this topic. However, despite my best efforts, *justification* has become the *de facto* standard name in DLs. Even I must admit that it is catchier.

In a nutshell, the standard tableaux-based algorithm for deciding consistency tries to build a forest-shaped model of the input ABox, where each node represents an individual of the domain, by decomposing the complex concepts to which each individual is required to belong to smaller pieces, until either an obvious contradiction is observed, or a model is obtained. For instance, the algorithm will decompose the assertion Tall⊓LongHaired(franz) into the two simpler assertions Tall(franz) and LongHaired(franz). Similarly, existential restrictions are solved by introducing new individuals in a tree-like fashion; that is, for decomposing the assertion ∃hasStudent.¬LongHaired(franz), the algorithm introduces a new (anonymous) individual x and the assertions hasStudent(franz, x) and ¬LongHaired(x). To keep all the decomposition steps positive, the algorithm transforms the concepts to negation normal form (NNF), where negations can only occur in front of concept names. Thus, the algorithm needs to handle explicit disjunctions as well. To decompose an assertion like Serious⊔Angry(franz), where we do not know whether Franz is serious or angry, the ABox is duplicated, and one of the alternatives is added to each of the copies to analyse.

Since every ABox has a finite forest-shaped model, it can be shown that this process terminates after finitely many decomposition steps. Ultimately, the tableaux algorithm produces a set of ABoxes 𝔄 that represent the possible alternatives for building a model of the input ABox $\mathcal{A}$. An obvious contradiction (called a *clash*) is observed if the decomposed ABox contains two assertions $A(a), \neg A(a)$. It follows that $\mathcal{A}$ is consistent iff there is an ABox in 𝔄 without any clash; in fact, such clash-free ABox is a representation of a model of $\mathcal{A}$.

The tracing extension proposed in [5] labels each assertion obtained through the decomposition process with a value (formally a propositional variable) that represents the original axioms responsible for it. For example, suppose that the input ABox $\mathcal{A}$ contains the assertion ∃hasStudent.¬LongHaired(franz). The algorithm first provides a unique name for this assertions, say a_1. Then, when the decomposition process generates the two assertions hasStudent(franz, x) and ¬LongHaired(x), it marks both of them with the label a_1 to express that they were caused by that original assertion. If an assertion is caused by more than one original axiom, its label is modified to be the disjunction of the variables associated with those axioms. A clash is caused by two contradictory assertions, each of which is labelled by a disjunction of variables. The conjunction of these labels describes the combinations of axioms required to produce this clash. Each ABox in 𝔄 may have more than one clash, but only one of them is required for inconsistency. Hence, one can define the *pinpointing formula*[4]

$$\varphi_{\mathcal{A}} := \bigwedge_{\mathcal{B} \in 𝔄} \bigvee_{A(a), \neg A(a) \in \mathcal{B}} \mathsf{lab}(A(a)) \wedge \mathsf{lab}(\neg A(a)).$$

This formula expresses all the combinations of axioms that lead to inconsistency, in the following way. If $\mathcal{V}$ is a valuation that satisfies $\phi_{\mathcal{A}}$, then the set of axioms

---

[4] In the original paper [5], this was called a *clash formula*, since it explains the clashes obtained by the algorithm. The name was later changed to pinpointing formula to reflect its more general purpose.

$\{\alpha \in \mathcal{A} \mid \mathsf{lab}(\alpha) \in \mathcal{V}\}$ is an inconsistent sub-ABox of $\mathcal{A}$. In particular, minimal valuations satisfying $\phi_{\mathcal{A}}$ define minimal inconsistent sub-ABoxes. Hence, the pinpointing formula can be seen as a (compact) representation of all the justifications for ABox inconsistency.

This original approach considered only ABoxes, but did not use any terminological knowledge in the form of GCIs. Later on, Schlobach and Cornet [64] extended the tracing method to explain inconsistency and concept unsatisfiability with respect to so-called unfoldable $\mathcal{ALC}$ terminologies, which allow for only a limited use of GCIs as concept definitions. This extension required adapting an additional tableau rule, necessary for handling the concept definitions, but followed the main steps from [5], based on the finite tree model property. This paper, which coined the term *axiom pinpointing*, started a series of extensions including additional constructors or different kinds of axioms [39, 49]. The correctness of these extensions was shown on an individual basis at each paper, despite all of them being based on the same principles: take the original tableau-based algorithm for the logic under consideration, and include a tracing mechanism—based on the execution of the tableau rules—that associates each newly derived assertion with the sets of axioms responsible for its occurrence. It was only when the notion of blocking was considered for handling arbitrary GCIs [41] that new ideas had to be developed to avoid stopping the process too early. As we will see, observing that the classical notion of blocking did not suffice for axiom pinpointing was the first hint of the problems that would later arise for this glass-box idea.

So, what do you do when you observe several instances of a process, and understand the principles behind it? You generalise them, of course! It was at that time that we started our attempts to develop a general notion of pinpointing extensions of tableau algorithms. This required two steps: giving a precise definition of what an abstract tableau algorithm (what we called a *general tableau* at the time) is, and describing how to implement the tracing mechanism on top of it, while guaranteeing correctness from an axiom pinpointing perspective. Fortunately, we could build on previous work by Franz for both steps. Indeed, a general notion of tableau had been defined a few years earlier to study the connections between tableau and automata reasoning methods [3]. Moreover, as mentioned before, the tracing mechanism and its correctness for axiom pinpointing was originally presented in [5]. Putting together both ingredients, after some necessary modifications, serious considerations on the notion and effect of blocking, and extension of the correctness proofs, led to the first general glass-box approach for axiom pinpointing in DLs and other logics [12].

An obvious drawback of these pinpointing extensions, which was almost immediately observed, is that the tracing mechanism requires the main optimisations of the tableau method to be disallowed. More precisely, to ensure efficiency, tableau algorithms stop exploring an ABox in the set of alternatives once a clash has been found. While this is correct for deciding a consequence (e.g., consistency), stopping at this point is bound to ignore some of the potential causes of the consequence, leading to an incomplete pinpointing method.

In reality, this inefficiency hid a larger problem that took some time to be understood and solved effectively.

After Baader and Hollunder proved the correctness of their approach in full detail, the following work took a rather abstract and simplified view on termination of the pinpointing extensions. In essence, most of the work starting from [64] argued that termination of the pinpointing method was a direct consequence of termination of the original tableau algorithm. This argument was convincing enough to make us believe that it should hold in general, but we needed a formal proof of this fact. After struggling for a long time, we ended up finding out that it is not true: there are terminating tableaux whose pinpointing extension does not terminate [12].[5] Fortunately for all the work coming before this counterexample was discovered, we were later able to identify a class of tableau methods where termination is guaranteed [15]. This class strictly contained all the previously studied algorithms in DL. Hence, for them we were still able to guarantee termination.

For the light-weight DL $\mathcal{EL}$, polynomial-time reasoning is achieved through a *completion* algorithm, which is an instance of what later became known as *consequence-based* methods [66,67]. As tableau methods, consequence-based approaches apply extension rules to try to prove an entailment from an ontology. The difference is that, while tableau algorithms attempt to construct a model of a special kind, consequence-based methods try to enumerate the consequences of the ontology that are relevant for the reasoning problem considered. Despite these differences, some consequence-based algorithms—including the completion method for $\mathcal{EL}$—can be seen as simple tableau approaches that are guaranteed to terminate. Thus, we can obtain a pinpointing formula for explaining consequences of $\mathcal{EL}$ ontologies [16,17]. This instance is a perfect example of the cost of adding the tracing mechanism to a tableau algorithm: while the original completion algorithm is guaranteed to terminate in polynomial time [10], its pinpointing extension may require exponentially many rule applications, and hence can only terminate in exponential time.

4.2   Automata-Based Axiom Pinpointing

In addition to tableau-based (and consequence-based) algorithms, automata-based techniques are often used to reason in DLs. Automata-based algorithms are mostly considered in theoretical settings to prove complexity results, but to the best of our knowledge, only one experimental automata-based reasoner exists [23] (and stopped being developed long ago). The main reason for this is that, even though automata's worst-case behaviour is often better than for tableau-based algorithms, their *best-case* behaviour matches the worst case, and for practical ontologies tableau algorithms tend to be more efficient due to their goal-directed nature. However, as mentioned several times already, the tracing mechanism

---

[5] I still remember when I managed to construct the first counterexample just before the deadline for submitting the paper. Imagine a scared first-year PhD student interrupting his supervisor's holidays to tell him the bad news.

implemented for pinpointing reduces the efficiency of tableau methods. It is hence worth analysing the possibility of modifying automata-based algorithms to compute a pinpointing formula as well.

Automata-based reasoning methods for DLs exploit the fact that these logics often allow for well-structured models. For example, as we have seen before, every $\mathcal{ALC}$ satisfiable concept has a tree-shaped model. In this case, we can build an automaton that accepts the tree-shaped models of a given concept C w.r.t. an ontology. It then follows that C is unsatisfiable iff the language accepted by this automaton is empty. At a very high level, the automata-based algorithm for $\mathcal{ALC}$ tries to build a tree-shaped model by labelling the nodes of an infinite tree (which will form the domain of the interpretation) with a set of concepts to which they most belong. These sets of concepts, which form a type, correspond to the states of the automaton. The automaton should label the root node with a set containing the input concept C; the children of each node are then labelled in a way that satisfies the existential and value restrictions appearing in the parent node. Moreover, each type should be consistent with the constraints specified in the TBox. More precisely, if the TBox contains an axiom $C \sqsubseteq D$, then every type that contains C must also contain D. The automaton *accepts* the infinite tree iff such a labelling is possible. Importantly, deciding whether such a labelling exists is polynomial on the number of states of the automaton [29]. It is also important to notice that the automaton accepts *infinite* trees; hence, it does not require any special blocking technique to search for periodicity of a model.

To transform this reasoning algorithm into a pinpointing method, we modify the underlying automata model into a *weighted* automaton [33]. Very briefly, weighted automata generalise the classical notion of automata to provide an *initial weight* to each state (as a generalisation of the set of initial states), and a *weight* to each transition generalising the transition relation. Weighted automata require an algebraic structure called a *semiring* that has a *domain S* and two binary relations $\oplus$ (addition) and $\otimes$ (product), where $\otimes$ distributes over $\oplus$ and a few additional properties hold. Rather than dividing runs into successful and unsuccessful, weighted automata give a weight to every run, which is computed as the product of the weights of all the transitions used, and the weight of the initial state. The *behaviour* of the automaton is a function that maps every input tree T into a semiring value computed as the addition of the weights of all the runs this automaton over T.

For axiom pinpointing, we realise that the class of all propositional formulas over a finite alphabet of variables with the logical disjunction $\vee$ and conjunction $\wedge$ forms a distributive lattice, which is a specific kind of semiring. Thus, automata-based axiom pinpointing uses formulas as weights, and the two logical operators mentioned, to construct the pinpointing formula of a consequence. Recall that the states of the automaton used to decide concept satisfiability enforce that all the GCIs in the TBox are satisfied. Instead, we now allow these types to violate some of these constraints (e.g., even if the TBox contains the axiom $C \sqsubseteq D$, we allow a state that contains C but not D). The weights are used to keep track of exactly which axioms are being violated while labelling the input tree. Thus, the behaviour of this automaton tells us which axioms need to be violated to obtain a model; i.e., a pinpointing formula for unsatisfiability. This was precisely the approach we followed in [13,14].

At this point, we only need to find out how to compute the behaviour of a weighted automaton. Interestingly, despite extensive work made on different weighted automata models, by the time we were considering this issue there was no behaviour computation algorithm, and the computational complexity of this problem was not very well understood. So we had to develop our own techniques.[6] In the end, we developed a technique that extended the ideas behind the emptiness test for unweighted automata [4,71] to track the weights; an alternative approach was independently developed around the same time [32]. Our approach, which only works on distributed lattices [42], runs in polynomial time on the size of the automaton, which matches the complexity of deciding emptiness of unweighted automata.

Thus, we showed that automata-based methods can also provide tight complexity bounds for axiom pinpointing, without worrying about issues like termination. There is, however, a small caveat. In order to guarantee a polynomial-time behaviour computation, the algorithm does not really compute the pinpointing formula, but a compact representation of it built through structure sharing. If the automaton is exponential on the size of the ontology, as is the case for $\mathcal{ALC}$, this representation could be expanded into a formula without any cost in terms of computational complexity. However, for less expressive logics such as $\mathcal{EL}$, this expansion may yield an exponential blow-up. Still, this blow-up may in fact be unavoidable in some of these logics. For example, it has been shown that there exist $\mathcal{EL}$ ontologies and consequences whose smallest pinpointing formula is of super-polynomial length [55]. In particular, this pinpointing formula cannot be generated in polynomial time.

For axiom pinpointing, the advantage of using automata-based methods over the tableau-based approach is that the problem with termination does not show up. Moreover, given that the computational resources needed to compute the behaviour are polynomially bounded on the size of the automaton, automata yield tight complexity bounds for this problem as well. However, it preserves the disadvantage observed for classical reasoning; namely, that its best-case complexity matches the worst-case one. Indeed, the first step of this approach corresponds to the construction of the automaton from the input ontology.

Overall, we have seen two methods for computing a pinpointing formula, and by extension all justifications, for a consequence from an ontology. In both cases, the methods present some issues that make them impractical. In fact, this is not very surprising given the now known complexity results for problems related to axiom pinpointing. As mentioned already, a single consequence may have exponentially many justifications, and hence it is impossible to enumerate them all

---

[6] As a historical remark, an important reason why I ended up working with Franz was because I fell in love with automata theory while I was doing my masters in Dresden. Being the Chair for Automata Theory, it only made sense to ask him for a topic. Little did I know at the time where this would take me.

in polynomial time. What is more interesting, though, is that even if a consequence has polynomially many justifications, there is no guarantee that they can be all found in polynomial time. In particular, any enumeration algorithm will necessarily lead to a longer waiting time between any two successive solutions, and even counting the number of justifications is a hard counting problem [57]. Given these hardness results, it makes sense to focus on the issue of computing only one justification. This is the topic of the next section.

5   Finding One Justification

The black-box approach for finding one justification relies on a straight-forward deletion procedure. The idea is to remove superfluous axioms from the ontology one at a time through a sequence of entailment tests. Starting from the full input ontology $\mathcal{O}$, the method iteratively removes one axiom at a time and checks whether the consequence c follows from the remaining ontology. If it does, the axiom is permanently removed; otherwise, the axiom is re-inserted in the ontology and the process continues. At the end of this process, when every axiom has been tested for deletion, the remaining ontology is guaranteed to be a justification for c [17,38]. Clearly, this process incurs in a linear overhead over reasoning: a standard reasoner is called as many times as there are axioms in the ontology. In terms of complexity, this means that as long as reasoning is at least polynomial, finding one justification is not noticeably harder than just deciding a consequence.

In practice, it has been observed empirically that justifications tend to be small, containing only a few axioms. If the original ontology is large, as in the case of SNOMED CT, the linear overhead may actually make the process unfeasible. Indeed, even if reasoning takes only a millisecond, repeating the process half a million times would require well over 5 min. Thus different optimisations have been proposed to try to prune several axioms at a time [37].

Trying to obtain a glass-box approach for computing one justification, one can adapt the tracing technique extending tableau-based algorithms to focus on *one* cause of derivation only. Recall that the idea behind the tracing technique is to keep track of the axioms used when deriving any new information throughout the execution of the tableau algorithm. To find only a justification, it thus makes sense to apply the same tracing technique, but ignore alternative derivations of the same fact. In other words, rather than preserving a (monotone) Boolean formula describing all the derivation of a given assertion, one just stores a conjunction of propositional variables obtained by the first derivation of this assertion. The method hence preserves one derivation of each assertion. When a clash is found, we can immediately find out the axioms known to cause this clash by conjoining the labels of the two assertions forming it. Likewise, to explain the presence of a clash in all the ABoxes generated by the execution of the algorithm, we simply conjoin the labels the clashes of each of them. Overall, this conjunction yields a set of axioms that is guaranteed to entail the consequence under consideration; e.g., inconsistency of the input ABox.

Interestingly, the problems surrounding the glass-box approach for computing *all* justifications do not play a role in this case. Indeed, the execution of the original tableau-based algorithm does not need to be modified, but only some additional memory is required to preserve the conjunction of labels at each generated assertion. In particular, the fundamental optimisations of the classical tableau methods are preserved: the expansion of an ABox may be stopped once a clash is found in it, without the need to find other possible clashes. Hence, this method runs within the same resource bounds as the original tableau method. Moreover, termination of the method is not affected by this tracing technique, and there is no need to adapt any blocking procedure used, since standard blocking remains correct for this setting. Unfortunately, as was almost immediately noted, the set of axioms generated by this modified algorithm is not necessarily a justification, since it might not be minimal [17]. In fact, it is not difficult to build an example where this approach produces a result with superfluous axioms, potentially caused by the order of rule applications or other dependencies between the assertions generated.

To obtain a justification, this glass-box approach can be combined with the black-box method described before. After finding an approximate justification through the glass-box approach, it can be minimised by deleting the superfluous axioms via the black-box method. Empirical evaluations of this *gray-box* approach show that it behaves well in practice, even for very large ontologies like SNOMED. Indeed, an intensive analysis started by Boontawee Suntisrivaraporn[7] [68] and concluded by Kazakov and Skocovský [40] shows that consequences in this ontology tend to have very small justifications, mostly containing 10 axioms or less. Moreover, the average number of axioms that appear in at least one justification is less than 40 [56]. More interestingly, although the tracing algorithm has no guarantee of finding a justification, it very often does; and even when it does not, it usually gives only one or two superfluous axioms [68]. Thus, the minimisation step might not be needed in some applications.

To date, there is no automata-based method targeted to compute only one justification. Although it is possible to imagine a way to adapt the idea used in tableaux (i.e., rather than preserving formulas, weights are associated with conjunctions of propositional variables), this would require larger changes in terms of the chosen semiring and the use of the operations, and would not yield any real benefit in terms of complexity. Indeed, the automaton would be of the same size, and computing the behaviour requires the same resources, while still not guaranteeing minimality of the set of axioms computed. Hence, automata-based methods are not really meaningful in the context of computing only one justification, and we will not cover them further.

6   Extensions

The original question of axiom pinpointing, as described at the beginning of this chapter, deals exclusively with *whole* axioms. This is a reasonable approach

---

[7] Also known, and herewith referred as *Meng*.

when trying to debug errors in a hand-written ontology, as one can assume that ontology engineers follow modelling guidelines, which limit the variability in the expression of specific piece of knowledge. However, this also makes the results dependent on the representation chosen. For example, {Tall ⊓ Professor(franz)} is logically equivalent to {Tall(franz), Professor(franz)}, but in axiom pinpointing both ontologies are treated differently: the first contains only one axiom, and is thus the only possible justification, while the second has a more fine-grained view where each of the individual axioms (or both together) may serve as a justification. We notice that the choice of the shape of the axioms is not banal. It may affect the underlying modelling language—and by extension the reasoning method chosen—among other things. Consider, for example, the difference between the $\mathcal{ALC}$ GCI $\top \sqsubseteq \neg$LongHaired $\sqcup$ Professor and the logically equivalent $\mathcal{EL}$ GCI LongHaired $\sqsubseteq$ Professor. More importantly, it may also affect the complexity of axiom pinpointing itself. As shown in [57], allowing conjunctions as in the axiom Tall ⊓ Professor(franz) can increase the complexity of axiom pinpointing related tasks, like counting or enumerating justifications.

Depending on the application, the domain, the user, and the shape of the ontology, it may be desirable to produce a finer or a coarser view to a justification. For instance, if the goal is to *repair* an error in an ontology, it makes sense to try to view the ontology in as much detail as possible. Indeed, knowing that a long and (syntactically) complicated axiom is causing an error is less helpful for correcting it than observing more precise pieces, which highlight where the errors occur. As a simple example, knowing that the axiom Tall ⊓ Professor ⊓ ¬LongHaired(franz) causes an error is less informative than knowing that ¬LongHaired(franz) is causing it. On the other hand, if the goal is to *understand* why a consequence follows from an ontology, then a coarser view, where some of the irrelevant details are hidden from the user, may be more informative.

Variations of axiom pinpointing targeting finer or coarser justifications have been proposed throughout the years. Specifically for description logics, coarser justifications—called *lemmata*—were proposed in [35]. The idea in this case is to combine several axioms within a justification into one (simpler) axiom that follows from them and explains their relationship to the remaining ones in the justification. A simple example of this approach is a chain of atomic subsumptions $A_0 \sqsubseteq A_1, A_1 \sqsubseteq A_2, \ldots A_{n-1} \sqsubseteq A_n$ summarised into the lemma $A_0 \sqsubseteq A_n$. If a user is interested in observing the details of the lemma, then it can be expanded to its original form. Of course, the challenge is to summarise more complex combinations of axioms going beyond simple sequences of implications. To understand how this is done, it is worth looking at the original work in detail.

The approach for providing finer justifications originally took the form of so-called precise and laconic justifications [34]. In a nutshell, the idea of these justifications is to *cut* axioms into smaller, but still meaningful, pieces in a way that only the relevant pieces are presented to the user. In our previous example, instead of the (original) axiom Tall ⊓ Professor ⊓ ¬LongHaired(franz), the piece ¬LongHaired(franz) could be used as a laconic justification. Note that in

practice, the pieces of axioms derived for these laconic justifications are generalisations of the original axioms, with the additional property that they tend to
be shorter and easier to read. One can think of taking this idea a step further,
and trying to explain a consequence through the most general variants of the
axioms possible. We will explore a closely related task in the next section.

Another important extension of the original approach to axiom pinpointing
refers to the way axioms are related to each other. Recall that a justification is
a minimal set of axioms (from the original ontology) that entails a given consequence. From this point of view, all axioms are independent in the sense that
their presence or absence in the ontology does not depend on the existence of any
other axiom. However, it is not difficult to find cases where this independence is
not necessary. Without going far, we can consider the completion algorithm for
$\mathcal{EL}$, which requires that the input ontology is written in a special normal form.
Before calling the algorithm proper, the axioms in the ontology are transformed
to this form; for example, an axiom $A \sqsubseteq B \sqcap C$ is replaced by the two axioms
$A \sqsubseteq B$, $A \sqsubseteq C$. Note that this transformation does not affect the logical properties of the ontology, but as mentioned before, it may have an effect on axiom
pinpointing. In this case, one should notice that, as the two latter axioms originate from the same input axiom, they are also bound to appear together. That
is, whenever a justification from the normalised ontology contains $A \sqsubseteq B$, then
it must also contain $A \sqsubseteq C$, and *vice-versa*. Another example is when the axioms
are available to some users only, through an access control mechanism [7,8]. In
this case, all axioms at the same access level should be available simultaneously,
along with those at more public levels.

One way to deal with this dependency between axioms is by means of *contexts*. From a very simplistic point of view, a context is merely a sub-ontology
containing inter-related axioms. Technically, this inter-relation is expressed by
a label associated with each axiom. In a nutshell, axioms that share the same
label should always appear together. More complex relationships can then be
expressed by a variation of the labelling language; e.g., by using propositional
formulas, one can say that an axiom is available when another one is not.

In this context-based scenario, axiom pinpointing corresponds to finding the
contexts from which a consequence can be derived, rather than just finding the
specific axioms within those contexts. Since several axioms may belong to the
same context, this provides a coarser explanation of the causes of the entailment.
Interestingly, all the methods described in Sects. 4 and 5 can be adapted to this
variant of context-based reasoning, by using context labels—rather than the
representation of an axiom—within the tracing process, and by including or
removing whole contexts together in the black-box approach.

7   Applications

Now that we know what is axiom pinpointing, some of its variants, and how to
solve these issues, we will see what they are useful for. Of course, we have already
hinted to various applications, which have motivated the previous descriptions,
but now we try to cover them in larger detail.

The first obvious application is about correcting errors in an ontology. An early motivation for axiom pinpointing—definitely one that caught the interest of Franz—arose from working on the very large ontology SNOMED CT. At some point, it was observed that this ontology entailed the consequence AmputationOfFinger ⊑ AmputationOfHand; i.e., according to SNOMED, every amputation of a finger was also an amputation of a hand (and indeed, also of an arm, suggesting that there was something wrong with part-whole representations). This is a clearly erroneous conclusion that may have extreme consequences if the ontology is used to reason about real-world events.[8] So it was important to find the cause of the error, and correct it adequately. Using the gray-box approach for finding one justification, combined with the hitting-set tree method for finding successive ones, Franz and Meng managed to identify the *six* specific axioms that caused this error [18]. Perhaps more interestingly, they showed that there was a systematic error in the modelling approach used for the construction of the ontology, which was leading to these erroneous consequences. Hence, they proposed to use a slightly more expressive logic than $\mathcal{EL}$, in order to provide a more direct and intuitive mechanism for modelling relations between parts [69]. Since then, this error was erradicated from the ontology.

Staying in the context of correcting the errors in an ontology, recent efforts consider the approach that originated with laconic justifications, but rather than trying to find a justification over generalised axioms from the ontology, they generalise one further step to automatically remove the consequence. This line of research started from the idea of finding a consensus between mutually inconsistent ontologies by different agents [58]. It has then been continued by the research group in Bolzano [70] and, through a different motivation based on privacy, by Franz and his group [9].

Understanding and correcting the causes for an unwanted consequence to follow is a hard and time-consuming task that often requires the involvement of experts to decide which of the potentially exponentially many options to choose. In addition, updates to ontologies are often planned according to a stable calendar; for example, new versions of SNOMED are published twice per year. Hence, one should expect to wait some time before a known error is corrected in an ontology. In the meantime, one should still be able to use the ontology deriving meaningful consequences that avoid the potentially erroneous parts. This is the basic idea underlying *inconsistency tolerant* [19,43], and more generally *error tolerant reasoning* [45]. Essentially, suppose that one knows that an ontology entails an erroneous consequence; the goal is to derive other consequences that would still hold in the absence of this error, and hence would still make sense after the ontology is repaired. To this end, three main semantics have been

---

[8] Imagine someone making an insurance claim after having a finger amputated. If the insurer makes this kind of error, they might end of paying a larger lump for an amputated arm.

defined, based on the notion of a *repair* [43]:[9] *brave* semantics consider consequences that follow from at least one repair; *cautious* semantics require that the conclusion is entailed by all repairs; and the *intersection* semantics, originally proposed for efficiency reasons, uses as a correct ontology the intersection of all the repairs [20]. In DLs, this aspect was originally motivated by the analogous notion in databases. From a similar motivation, currently the idea of tracing the provenance of a consequence in an ontology is gaining interest in the DL world. The difference with axiom pinpointing is that for provenance, the minimality assumption is relaxed; in fact, one is rather interested in finding all the possible ways in which a consequence can be derived [52].

Other important applications use axiom pinpointing as a background step for doing other complex inferences that depend on the combinations of axioms that entail some consequence. These applications are usually, although not always, based on the context-based generalisation described in the previous section. The first one, which we have already mentioned, is about access control. In this scenario, axioms have an access degree that limits the class of users that are able to retrieve them. These access degrees are extended also to the implicit consequences of the ontology in the obvious way: a user can observe a consequence iff they have access to a set of axioms that entail this consequence. Hence, for a given consequence, the problem is now to identify the access levels that can observe it, so that it remains hidden from all others. This becomes a problem of axiom pinpointing at the context level, where each access degree defines one context, and one is not interested in the specific axioms entailing the consequence, but rather their contexts. In [8] this problem is solved through a purely black-box approach using expressive DLs as underlying ontology language.

In order to handle uncertainty about the knowledge in an ontology, probabilistic extensions of DLs have been proposed [46]. A relevant example for this chapter are the DLs with so-called *distribution semantics* originally proposed for probabilistic logic programming [60,62]. In this semantics, every axiom is associated with a probability of being true, and all axioms are assumed to be probabilistically independent. Then, the probability of a consequence is derived from the probabilities of all the combinations of axioms that entail the consequence. The most recent implementation of a reasoner for these logics uses the tableau-based glass-box method to compute a pinpointing formula, which is later fed to a propositional probabilistic reasoner [73]. While this approach seems to work well in empirical evaluations, the assumption of probabilistic independence is too strong to model realistic situations. For that reason, a more general formalism based on contexts was proposed. The idea of these newer probabilistic logics is to model certain knowledge that holds in uncertain contexts. That is, axioms are interrelated via context labels as described before, but the contexts are associated with a probability distribution, which in this case is expressed via a Bayesian network. Hence, these logics are often known as Bayesian logics. The

---

[9] Formally, a repair is a maximal subontology that does not entail the consequence. This is the dual notion of a justification, which is also studied in variations of axiom pinpointing in different fields.

first Bayesian DL was studied, as a variant of another probabilistic extension of DL-Lite [30], as an extension of $\mathcal{EL}$ [26], but it was immediately clear that the underlying ideas could be extended to other ontology languages as well [28]. The reasoning methods proposed for Bayesian $\mathcal{EL}$ included a black-box approach, and reductions to pure Bayesian networks [27] and to probabilistic logic programming with distribution semantics [25]. Later on, a glass-box approach based on a modification of the tracing algorithm for $\mathcal{ALC}$ was considered in [21,22].

To conclude this section, and without going into excessive detail, we note that axiom pinpointing is also an effective sub-procedure for dealing with other extensions of logical reasoning. Examples of this are reasoning about preferences, possibilistic reasoning, and belief revision. More generally, whenever a reasoning problem can be expressed as a sum of products of weights from a semiring, where weights are associated with axioms, the weights of axioms in an ontology entailing a consequence are multiplied, and the results of different derivations are added, axiom pinpointing is a perfect companion to any reasoning method.

8   Conclusions

We have attempted to explain what is axiom pinpointing in the context of Description Logics, as studied by Franz Baader, and his academic successors. As mentioned throughout this work, the idea of axiom pinpointing is not restricted to the DL community, and pops out whenever people study some kind of monotonic entailment relation in detail. Unfortunately, each area chose a different name for this task, thus causing confusions and hindering communication about techniques, successes, and failures.

After finishing my dissertation, I have been trying to collect names and examples of axiom pinpointing *in the wild*, and keep on finding them in many different places. I guess it is true that when all you have is a hammer, then everything looks like a nail. But this work is not about me, but about Franz, who not only proposed this topic to me as I started working with him, but was also the first to propose a solution to a special case, well before the name *axiom pinpointing* was coined by Schlobach and Cornet. Since Franz has a vested interest in DLs, and my work together with him has mainly focused on this area as well, this chapter does not go in much detail about other logical languages. Still, it would be wrong to leave anyone with the impression that the work presented here is exclusive for DLs. Many of the ideas are applicable to, and have often been independently developed for, other ontology languages as well. Famous examples are databases, propositional satisfiability, and constraint satisfaction problems; but there are many more. Even some that I have not yet encountered.

Perhaps more importantly, axiom pinpointing is not dead. Throughout the years, I have tried to leave aside the topic a couple of times, thinking that there cannot possibly exist much more to explore. And each time it came knocking back to my door, under different disguises. It is a fortune to us all that Franz continues exploring these topics as well. I am looking forward to the next years of axiom pinpointing-related research, in DLs and in other fields. And to the results that Franz, and those of us that grew under him, will still be able to contribute.

References

1. Baader, F., Brandt, S., Lutz, C.: Pushing the $\mathcal{EL}$ envelope. In: Kaelbling, L., Saffiotti, A. (eds.) Proceedings of the 19th International Joint Conference on Artificial Intelligence (IJCAI 2005), pp. 364–369. Professional Book Center (2005)
2. Baader, F., Calvanese, D., McGuinness, D., Nardi, D., Patel-Schneider, P. (eds.): The Description Logic Handbook: Theory, Implementation, and Applications, 2nd edn. Cambridge University Press, Cambridge (2007)
3. Baader, F., Hladik, J., Lutz, C., Wolter, F.: From tableaux to automata for description logics. Fundamenta Informaticae **57**(2–4), 247–279 (2003). http://content. iospress.com/articles/fundamenta-informaticae/fi57-2-4-08
4. Baader, F., Hladik, J., Peñaloza, R.: Automata can show PSpace results for description logics. Inf. Comput. **206**(9–10), 1045–1056 (2008). https://doi.org/10.1016/j. ic.2008.03.006
5. Baader, F., Hollunder, B.: Embedding defaults into terminological knowledge representation formalisms. J. Autom. Reason. **14**(1), 149–180 (1995). https://doi.org/ 10.1007/BF00883932
6. Baader, F., Horrocks, I., Lutz, C., Sattler, U.: An Introduction to Description Logic. Cambridge University Press, Cambridge (2017)
7. Baader, F., Knechtel, M., Peñaloza, R.: A generic approach for large-scale ontological reasoning in the presence of access restrictions to the ontology's axioms. In: Bernstein, A., et al. (eds.) ISWC 2009. LNCS, vol. 5823, pp. 49–64. Springer, Heidelberg (2009). https://doi.org/10.1007/978-3-642-04930-9_4
8. Baader, F., Knechtel, M., Peñaloza, R.: Context-dependent views to axioms and consequences of semantic web ontologies. J. Web Semant. **12**, 22–40 (2012). https://doi.org/10.1016/j.websem.2011.11.006
9. Baader, F., Kriegel, F., Nuradiansyah, A., Peñaloza, R.: Making repairs in description logics more gentle. In: Thielscher, M., Toni, F., Wolter, F. (eds.) Proceedings of the Sixteenth International Conference on Principles of Knowledge Representation and Reasoning (KR 2018), pp. 319–328. AAAI Press (2018). https://aaai. org/ocs/index.php/KR/KR18/paper/view/18056
10. Baader, F., Lutz, C., Suntisrivaraporn, B.: CEL—a polynomial-time reasoner for life science ontologies. In: Furbach, U., Shankar, N. (eds.) IJCAR 2006. LNCS (LNAI), vol. 4130, pp. 287–291. Springer, Heidelberg (2006). https://doi.org/10. 1007/11814771_25
11. Baader, F., Lutz, C., Suntisrivaraporn, B.: Efficient reasoning in $\mathcal{EL}^+$. In: Parsia et al. [53]. http://ceur-ws.org/Vol-189/submission_8.pdf
12. Baader, F., Peñaloza, R.: Axiom pinpointing in general tableaux. In: Olivetti, N. (ed.) TABLEAUX 2007. LNCS (LNAI), vol. 4548, pp. 11–27. Springer, Heidelberg (2007). https://doi.org/10.1007/978-3-540-73099-6_4
13. Baader, F., Peñaloza, R.: Automata-based axiom pinpointing. In: Armando, A., Baumgartner, P., Dowek, G. (eds.) IJCAR 2008. LNCS (LNAI), vol. 5195, pp. 226–241. Springer, Heidelberg (2008). https://doi.org/10.1007/978-3-540-71070-7_19

14. Baader, F., Peñaloza, R.: Automata-based axiom pinpointing. J. Autom. Reason. **45**(2), 91–129 (2010). https://doi.org/10.1007/s10817-010-9181-2
15. Baader, F., Peñaloza, R.: Axiom pinpointing in general tableaux. J. Log. Comput. **20**(1), 5–34 (2010). https://doi.org/10.1093/logcom/exn058
16. Baader, F., Peñaloza, R., Suntisrivaraporn, B.: Pinpointing in the description logic $\mathcal{EL}$. In: Calvanese, D., et al. (eds.) Proceedings of the 2007 International Workshop on Description Logics (DL 2007). CEUR Workshop Proceedings, vol. 250. CEUR-WS.org (2007)
17. Baader, F., Peñaloza, R., Suntisrivaraporn, B.: Pinpointing in the description logic $\mathcal{EL}^+$. In: Hertzberg, J., Beetz, M., Englert, R. (eds.) KI 2007. LNCS (LNAI), vol. 4667, pp. 52–67. Springer, Heidelberg (2007). https://doi.org/10.1007/978-3-540-74565-5_7
18. Baader, F., Suntisrivaraporn, B.: Debugging SNOMED CT using axiom pinpointing in the description logic $\mathcal{EL}^+$. In: Cornet, R., Spackman, K.A. (eds.) Proceedings of the Third International Conference on Knowledge Representation in Medicine. CEUR Workshop Proceedings, vol. 410. CEUR-WS.org (2008). http://ceur-ws.org/Vol-410/Paper01.pdf
19. Bertossi, L.E., Hunter, A., Schaub, T. (eds.): Inconsistency Tolerance. LNCS, vol. 3300. Springer, Heidelberg (2005). https://doi.org/10.1007/b104925
20. Bienvenu, M., Rosati, R.: Tractable approximations of consistent query answering for robust ontology-based data access. In: Rossi, F. (ed.) Proceedings of the 23rd International Joint Conference on Artificial Intelligence (IJCAI 2013), pp. 775–781. AAAI Press/IJCAI (2013). http://www.aaai.org/ocs/index.php/IJCAI/IJCAI13/paper/view/6904
21. Botha, L., Meyer, T., Peñaloza, R.: The Bayesian description logic $\mathcal{BALC}$. In: Ortiz and Schneider [51]. http://ceur-ws.org/Vol-2211/paper-09.pdf
22. Botha, L., Meyer, T., Peñaloza, R.: The Bayesian description logic $\mathcal{BALC}$. In: Calimeri, F., Leone, N., Manna, M. (eds.) JELIA 2019. LNCS, vol. 11468, pp. 339–354. Springer, Cham (2019). https://doi.org/10.1007/978-3-030-19570-0_22
23. Calvanese, D., Carbotta, D., Ortiz, M.: A practical automata-based technique for reasoning in expressive description logics. In: Walsh, T. (ed.) Proceedings of the 22nd International Joint Conference on Artificial Intelligence (IJCAI 2011), pp. 798–804. AAAI Press/IJCAI (2011). https://doi.org/10.5591/978-1-57735-516-8/IJCAI11-140
24. Calvanese, D., De Giacomo, G., Lembo, D., Lenzerini, M., Rosati, R.: Tractable reasoning and efficient query answering in description logics: the DL-Lite family. J. Autom. Reason. **39**(3), 385–429 (2007)
25. Ceylan, İ.İ., Mendez, J., Peñaloza, R.: The Bayesian ontology reasoner is BORN! In: Dumontier, M., et al. (eds.) Informal Proceedings of the 4th International Workshop on OWL Reasoner Evaluation (ORE-2015). CEUR Workshop Proceedings, vol. 1387, pp. 8–14. CEUR-WS.org (2015). http://ceur-ws.org/Vol-1387/paper_5.pdf
26. Ceylan, İ.İ., Peñaloza, R.: The Bayesian description logic $\mathcal{BEL}$. In: Demri, S., Kapur, D., Weidenbach, C. (eds.) IJCAR 2014. LNCS (LNAI), vol. 8562, pp. 480–494. Springer, Cham (2014). https://doi.org/10.1007/978-3-319-08587-6_37
27. Ceylan, İ.İ., Peñaloza, R.: Reasoning in the description logic $\mathcal{BEL}$ using Bayesian networks. In: Proceedings of the 2014 AAAI Workshop on Statistical Relational Artificial Intelligence. AAAI Workshops, vol. WS-14-13. AAAI (2014). http://www.aaai.org/ocs/index.php/WS/AAAIW14/paper/view/8765
28. Ceylan, İ.İ., Peñaloza, R.: The Bayesian ontology language $\mathcal{BEL}$. J. Autom. Reason. **58**(1), 67–95 (2017). https://doi.org/10.1007/s10817-016-9386-0

29. Comon, H., et al.: Tree automata techniques and applications (2007). http://www. grappa.univ-lille3.fr/tata. Accessed 12 Oct 2007

30. d'Amato, C., Fanizzi, N., Lukasiewicz, T.: Tractable reasoning with Bayesian description logics. In: Greco, S., Lukasiewicz, T. (eds.) SUM 2008. LNCS (LNAI), vol. 5291, pp. 146–159. Springer, Heidelberg (2008). https://doi.org/10.1007/978-3-540-87993-0_13

31. Donini, F.M., Massacci, F.: Exptime tableaux for $\mathcal{ALC}$. Artif. Intell. **124**(1), 87–138 (2000). https://doi.org/10.1016/S0004-3702(00)00070-9

32. Droste, M., Kuich, W., Rahonis, G.: Multi-valued MSO logics overwords and trees. Fundamenta Informaticae **84**(3–4), 305–327 (2008). http://content.iospress.com/articles/fundamenta-informaticae/fi84-3-4-02

33. Droste, M., Kuich, W., Vogler, H.: Handbook of Weighted Automata, 1st edn. Springer, Heidelberg (2009). https://doi.org/10.1007/978-3-642-01492-5

34. Horridge, M., Parsia, B., Sattler, U.: Laconic and precise justifications in OWL. In: Sheth, A., et al. (eds.) ISWC 2008. LNCS, vol. 5318, pp. 323–338. Springer, Heidelberg (2008). https://doi.org/10.1007/978-3-540-88564-1_21

35. Horridge, M., Parsia, B., Sattler, U.: Lemmas for justifications in OWL. In: Grau, B.C., Horrocks, I., Motik, B., Sattler, U. (eds.) Proceedings of the 22nd International Workshop on Description Logics (DL 2009). CEUR Workshop Proceedings, vol. 477. CEUR-WS.org (2009). http://ceur-ws.org/Vol-477/paper_24.pdf

36. Horrocks, I., Kutz, O., Sattler, U.: The even more irresistible $\mathcal{SROIQ}$. In: Doherty, P., Mylopoulos, J., Welty, C.A. (eds.) Proceedings of the 10th International Conference on Principles of Knowledge Representation and Reasoning (KR 2006), pp. 57–67. AAAI Press (2006)

37. Kalyanpur, A.: Debugging and repair of OWL ontologies. Ph.D. thesis, University of Maryland College Park, USA (2006)

38. Kalyanpur, A., Parsia, B., Sirin, E., Cuenca-Grau, B.: Repairing unsatisfiable concepts in OWL ontologies. In: Sure, Y., Domingue, J. (eds.) ESWC 2006. LNCS, vol. 4011, pp. 170–184. Springer, Heidelberg (2006). https://doi.org/10.1007/11762256_15

39. Kalyanpur, A., Parsia, B., Sirin, E., Hendler, J.A.: Debugging unsatisfiable classes in OWL ontologies. J. Web Semant. **3**(4), 268–293 (2005). https://doi.org/10.1016/j.websem.2005.09.005

40. Kazakov, Y., Skočovský, P.: Enumerating justifications using resolution. In: Galmiche, D., Schulz, S., Sebastiani, R. (eds.) IJCAR 2018. LNCS (LNAI), vol. 10900, pp. 609–626. Springer, Cham (2018). https://doi.org/10.1007/978-3-319-94205-6_40

41. Lee, K., Meyer, T.A., Pan, J.Z., Booth, R.: Computing maximally satisfiable terminologies for the description logic $\mathcal{ALC}$ with cyclic definitions. In: Parsia et al. [53]. http://ceur-ws.org/Vol-189/submission_29.pdf

42. Lehmann, K., Peñaloza, R.: The complexity of computing the behaviour of lattice automata on infinite trees. Theor. Comput. Sci. **534**, 53–68 (2014). https://doi.org/10.1016/j.tcs.2014.02.036

43. Lembo, D., Lenzerini, M., Rosati, R., Ruzzi, M., Savo, D.F.: Inconsistency-tolerant semantics for description logics. In: Hitzler, P., Lukasiewicz, T. (eds.) RR 2010. LNCS, vol. 6333, pp. 103–117. Springer, Heidelberg (2010). https://doi.org/10.1007/978-3-642-15918-3_9

44. Liffiton, M.H., Sakallah, K.A.: Algorithms for computing minimal unsatisfiable subsets of constraints. J. Autom. Reason. **40**(1), 1–33 (2008). https://doi.org/10.1007/s10817-007-9084-z

45. Ludwig, M., Peñaloza, R.: Error-tolerant reasoning in the description logic $\mathcal{EL}$. In: Fermé, E., Leite, J. (eds.) JELIA 2014. LNCS (LNAI), vol. 8761, pp. 107–121. Springer, Cham (2014). https://doi.org/10.1007/978-3-319-11558-0_8

46. Lukasiewicz, T., Straccia, U.: Managing uncertainty and vagueness in description logics for the semantic web. J. Web Semant. 6(4), 291–308 (2008)

47. Meliou, A., Gatterbauer, W., Halpern, J.Y., Koch, C., Moore, K.F., Suciu, D.: Causality in databases. IEEE Data Eng. Bull. 33(3), 59–67 (2010). http://sites.computer.org/debull/A10sept/suciu.pdf

48. Mencía, C., Marques-Silva, J.: Efficient relaxations of over-constrained CSPs. In: Proceedings of the 26th IEEE International Conference on Tools with Artificial Intelligence (ICTAI 2014), pp. 725–732. IEEE Computer Society (2014). https://doi.org/10.1109/ICTAI.2014.113

49. Meyer, T.A., Lee, K., Booth, R., Pan, J.Z.: Finding maximally satisfiable terminologies for the description logic $\mathcal{ALC}$. In: Proceedings of The Twenty-First National Conference on Artificial Intelligence (AAAI 2006), pp. 269–274. AAAI Press (2006). http://www.aaai.org/Library/AAAI/2006/aaai06-043.php

50. Motik, B., Patel-Schneider, P.F., Cuenca Grau, B. (eds.): OWL 2 Web Ontology Language: Direct Semantics. W3C Recommendation, 27 October 2009. http://www.w3.org/TR/owl2-direct-semantics/

51. Ortiz, M., Schneider, T. (eds.): Proceedings of the 31st International Workshop on Description Logics (DL 2018). CEUR Workshop Proceedings, vol. 2211. CEUR-WS.org (2018)

52. Ozaki, A., Peñaloza, R.: Provenance in ontology-based data access. In: Ortiz and Schneider [51]. http://ceur-ws.org/Vol-2211/paper-28.pdf

53. Parsia, B., Sattler, U., Toman, D. (eds.): Proceedings of the 2006 International Workshop on Description Logics (DL 2006). CEUR Workshop Proceedings, vol. 189. CEUR-WS.org (2006)

54. Parsia, B., Sirin, E., Kalyanpur, A.: Debugging OWL ontologies. In: Ellis, A., Hagino, T. (eds.) Proceedings of the 14th International Conference on World Wide Web (WWW 2005), pp. 633–640. ACM (2005). https://doi.org/10.1145/1060745.1060837

55. Peñaloza Nyssen, R.: Axiom pinpointing in description logics and beyond. Ph.D. thesis, Technische Universität Dresden, Germany (2009). http://nbn-resolving.de/urn:nbn:de:bsz:14-qucosa-24743

56. Peñaloza, R., Mencía, C., Ignatiev, A., Marques-Silva, J.: Lean kernels in description logics. In: Blomqvist, E., Maynard, D., Gangemi, A., Hoekstra, R., Hitzler, P., Hartig, O. (eds.) ESWC 2017. LNCS, vol. 10249, pp. 518–533. Springer, Cham (2017). https://doi.org/10.1007/978-3-319-58068-5_32

57. Peñaloza, R., Sertkaya, B.: Understanding the complexity of axiom pinpointing in lightweight description logics. Artif. Intell. 250, 80–104 (2017). https://doi.org/10.1016/j.artint.2017.06.002

58. Porello, D., Troquard, N., Confalonieri, R., Galliani, P., Kutz, O., Peñaloza, R.: Repairing socially aggregated ontologies using axiom weakening. In: An, B., Bazzan, A., Leite, J., Villata, S., van der Torre, L. (eds.) PRIMA 2017. LNCS (LNAI), vol. 10621, pp. 441–449. Springer, Cham (2017). https://doi.org/10.1007/978-3-319-69131-2_26

59. Price, C., Spackman, K.: Snomed clinical terms. Br. J. Healthcare Comput. Inf. Manag. 17(3), 27–31 (2000)

60. Raedt, L.D., Kimmig, A., Toivonen, H.: ProbLog: a probabilistic prolog and its application in link discovery. In: Veloso, M.M. (ed.) Proceedings of the 20th International Joint Conference on Artificial Intelligence (IJCAI 2007), pp. 2462–2467. IJCAI (2007). http://ijcai.org/Proceedings/07/Papers/396.pdf

61. Reiter, R.: A theory of diagnosis from first principles. Artif. Intell. **32**(1), 57–95 (1987). https://doi.org/10.1016/0004-3702(87)90062-2

62. Riguzzi, F., Bellodi, E., Lamma, E., Zese, R.: Probabilistic description logics under the distribution semantics. Semant. Web **6**(5), 477–501 (2015). https://doi.org/10.3233/SW-140154

63. Schild, K.: A correspondence theory for terminological logics: preliminary report. In: Mylopoulos, J., Reiter, R. (eds.) Proceedings of the 12th International Joint Conference on Artificial Intelligence (IJCAI 1991), pp. 466–471. Morgan Kaufmann (1991)

64. Schlobach, S., Cornet, R.: Non-standard reasoning services for the debugging of description logic terminologies. In: Proceedings of the 18th International Joint Conference on Artificial Intelligence (IJCAI 2003), pp. 355–360. Morgan Kaufmann Publishers Inc. (2003)

65. Schmidt-Schauß, M., Smolka, G.: Attributive concept descriptions with complements. J. Artif. Intell. **48**, 1–26 (1991)

66. Simancik, F., Motik, B., Horrocks, I.: Consequence-based and fixed-parameter tractable reasoning in description logics. Artif. Intell. **209**, 29–77 (2014)

67. Simančík, F.: Consequence-based reasoning for ontology classification. Ph.D. thesis, University of Oxford (2013). https://ethos.bl.uk/OrderDetails.do?uin=uk.bl.ethos.581368

68. Suntisrivaraporn, B.: Polynomial time reasoning support for design and maintenance of large-scale biomedical ontologies. Ph.D. thesis, Technische Universität Dresden, Germany (2009). http://hsss.slub-dresden.de/deds-access/hsss.urlmapping.MappingServlet?id=1233830966436-5928

69. Suntisrivaraporn, B., Baader, F., Schulz, S., Spackman, K.: Replacing SEP-triplets in SNOMED CT using tractable description logic operators. In: Bellazzi, R., Abu-Hanna, A., Hunter, J. (eds.) AIME 2007. LNCS (LNAI), vol. 4594, pp. 287–291. Springer, Heidelberg (2007). https://doi.org/10.1007/978-3-540-73599-1_38

70. Troquard, N., Confalonieri, R., Galliani, P., Peñaloza, R., Porello, D., Kutz, O.: Repairing ontologies via axiom weakening. In: McIlraith, S.A., Weinberger, K.Q. (eds.) Proceedings of The Thirty-Second AAAI Conference on Artificial Intelligence (AAAI 2018), pp. 1981–1988. AAAI Press (2018). https://www.aaai.org/ocs/index.php/AAAI/AAAI18/paper/view/17189

71. Vardi, M.Y., Wolper, P.: Automata-theoretic techniques for modal logics of programs. J. Comput. Syst. Sci. **32**(2), 183–221 (1986). https://doi.org/10.1016/0022-0000(86)90026-7

72. Xiao, G., et al.: Ontology-based data access: a survey, pp. 5511–5519. ijcai.org (2018). https://doi.org/10.24963/ijcai.2018/777

73. Zese, R., Bellodi, E., Riguzzi, F., Cota, G., Lamma, E.: Tableau reasoning for description logics and its extension to probabilities. Ann. Math. Artif. Intell. **82**(1–3), 101–130 (2018). https://doi.org/10.1007/s10472-016-9529-3

Asymmetric Unification
and Disunification

Veena Ravishankar[1]([✉]), Kimberly A. Cornell[2], and Paliath Narendran[3]

[1] University of Mary Washington, Fredericksburg, USA
vravisha@umw.edu
[2] The College of Saint Rose, Albany, USA
cornellk@strose.edu
[3] University at Albany-SUNY, Albany, USA
pnarendran@albany.edu

Abstract. We compare two kinds of unification problems: Asymmetric Unification and Disunification, which are variants of Equational Unification. Asymmetric Unification is a type of Equational Unification where the instances of the right-hand sides of the equations are in normal form with respect to the given term rewriting system. In Disunification we solve equations and disequations with respect to an equational theory for the case with free constants. We contrast the time complexities of both and show that the two problems are incomparable: there are theories where one can be solved in polynomial time while the other is NP-hard. This goes both ways. The time complexity also varies based on the termination ordering used in the term rewriting system.

Keywords: Asymmetric unification · Disunification ·
Time complexity analysis

1 Introduction and Motivation

We survey two variants of unification, namely asymmetric unification [16] and disunification [6,14]. Asymmetric unification is a new paradigm comparatively, which requires one side of the equation to be irreducible. Asymmetric unification was introduced by Catherine Meadows [16] for symbolic cryptographic protocol analysis as a state space reduction technique. These reduction techniques are crucial in cryptographic protocol analysis tools such as Maude-NPA to help narrow the exponential search space by identifying infeasible states [16,26]. Disunification [14] like equational unification deals with solving equations, however it also allows disequations. Disequations enable us to use additional constraints such as a variable is not equivalent to a term by our equational theory E (e.g., $x \not\approx_E a$). Disunification has applications in Logic Programming and Artificial Intelligence [12].

The main contribution of this paper is to contrast asymmetric and disunification in terms of their time complexities for different equational theories in the

© Springer Nature Switzerland AG 2019
C. Lutz et al. (Eds.): Baader Festschrift, LNCS 11560, pp. 497–522, 2019.
https://doi.org/10.1007/978-3-030-22102-7_23

case where terms in the input can also have free constant symbols. Complexity analysis has been performed separately on asymmetric unification [10,17] and disunification by Baader and Schulz [6,12], but not much work has been done on contrasting the two paradigms[1]. Initially, it was thought that the two are reducible to one another [17], but our results indicate that they are not, at least where polynomial-time reducibility is concerned[2]. These two variants were thought to be reducible due to an example disunification problem that the authors in [16] were able to simulate using asymmetric unification. They posed the connection between these problems as an open problem. The rewrite rule that the authors added to the term rewriting system R is $f(x,x) \to g(x)$. If the rewrite system R can be extended by adding such a rule, then, in the new system R_1, disunification can be reduced to asymmetric unification. They simulated the disequation $s \neq t$, for terms s and t, by the following asymmetric unification problem, where the downward arrow means irreducible on one side of the equation:

$$\{s \approx_\downarrow^? u, t \approx_\downarrow^? v, w \approx_\downarrow^? f(u,v)\}$$

In Sect. 8 we show that the time complexity of asymmetric unification varies depending on the symbol ordering chosen for the theory. Lastly we conclude and outline our future work.

2   Notations and Preliminaries: Term Rewriting Systems, Equational Unification

We assume the reader is accustomed with the terminologies of term rewriting systems (TRS) [4], equational rewriting [4,8], equational unification [7], and disunification [6,14].

Term Rewriting Systems: A term rewriting system (TRS) [4] is a set of rewrite rules, where a rewrite rule is an identity $l \approx r$ such that l is not a variable and $\mathscr{V}ar(l) \supseteq \mathscr{V}ar(r)$. It is often written or denoted as $l \to r$.

A term is *reducible* by a term rewriting system if and only if a subterm of it is an instance of the left-hand side of a rule. In other words, a term t is *reducible* modulo R if and only if there is a rule $l \to r$ in R, a subterm t' at position p of t, and a substitution σ such that $\sigma(l) = t'$. The term $t[\sigma(r)]_p$ is the result of *reducing* t by $l \to r$ at p. The *reduction relation* $\to_R$ associated with a term rewriting system R is defined as follows: $s \to_R t$ if and only if there exist p in $Pos(s)$ and $l \to r$ in R such that t is the result of reducing s by $l \to r$ at p, i.e., $t = s[\sigma(r)]_p$.

A term is in *normal form* with respect to a term rewriting system if and only if no rule can be applied to it. A term rewriting system is *terminating* if and

---

[1] Symmetric and asymmetric unification were contrasted in [16].

[2] For the theory **ACUN** that we consider in Sect. 6, disunification can be reduced to asymmetric unification.

only if there are no infinite rewrite chains. We write $s \rightarrow^{!}_{R} t$ if $s \rightarrow^{*}_{R} t$ and t is in normal form.

Two terms s and t are said to be *joinable* modulo a term rewriting system R if and only if there exists a term u such that $s \rightarrow^{*}_{R} u$ and $t \rightarrow^{*}_{R} u$, denoted as $s \downarrow t$.

The equational theory $\mathscr{E}(R)$ associated with a term rewriting system R is the set of equations obtained from R by treating every rule as a (bidirectional) equation. Thus the equational congruence $\approx_{\mathscr{E}(R)}$ is the congruence $(\rightarrow_R \cup \leftarrow_R)^{*}$.

A term rewriting system R is said to be *confluent* if and only if the following ("diamond") property holds:

$$\forall t \forall u \forall v \left[(t \rightarrow^{*}_{R} u \wedge t \rightarrow^{*}_{R} v) \Rightarrow \exists w(u \rightarrow^{*}_{R} w \wedge v \rightarrow^{*}_{R} w) \right]$$

R is *convergent* if and only if it is terminating and confluent. In other words, R is *convergent* if and only if it is terminating and, besides, every term has a *unique* normal form.

An equational theory $\approx_E$ is said to be *subterm-collapsing* if and only if there exist terms s, t such that $s \approx_E t$ and t is a proper subterm of s. If the theory has a convergent term rewriting system R, then it is subterm-collapsing if and only if $s \rightarrow^{+}_{R} s|_p$ for some term s and $p \in \mathscr{P}os(s)$. An equational theory is said to be *non-subterm-collapsing* or *simple* [13] if and only if it is not subterm-collapsing.

An equational term rewriting system consists of a set of identities E (which often contains identities such as commutativity and associativity) and a set of rewrite rules R. Two notion of rewriting are defined in the literature: *Class rewriting*, $\rightarrow_{R/E}$, is defined as $\approx_E \circ \rightarrow_R \circ \approx_E$, and *Extended rewriting* modulo E, $\longrightarrow_{R,E}$, is defined as

$$s \longrightarrow_{R,E} t \iff \exists p \in \mathscr{P}os(s) \text{ such that } s|_p \approx_E \sigma(l) \text{ and } t = s[\sigma(r)]_p$$

for some rule $l \rightarrow r$ and substitution σ.

Example 1. *Let* $E = \{(x + y) + z \approx x + (y + z), x + y \approx y + x\}$ *and* $R = \{0 + x \rightarrow x\}$. *Then*

$$(a + 0) + b \longrightarrow_{R,E} a + b$$

since $a + 0$ *matches with* $0 + x$ *modulo* E.

Definition 2.1. *We call* (Σ, E, R) *a* decomposition *of an equational theory* Δ *over a signature* Σ *if* $\Delta = R \uplus E$ *and* R *and* E *satisfy the following conditions:*

1. E *is variable preserving, i.e., for each* $s \approx t$ *in* E *we have* $\mathscr{V}ar(s) = \mathscr{V}ar(t)$.
2. E *has a finitary and complete unification algorithm, i.e., an algorithm that produces a finite complete set of unifiers.*
3. *For each* $l \rightarrow r \in R$ *we have* $\mathscr{V}ar(r) \subseteq \mathscr{V}ar(l)$.
4. R *is confluent and terminating modulo* E, *i.e., the relation* $\rightarrow_{R/E}$ *is confluent and terminating.*
5. $\rightarrow_{R,E}$ *is* E-coherent, *i.e.,* $\forall t_1, t_2, t_3$ *if* $t_1 \rightarrow_{R,E} t_2$ *and* $t_1 \approx_E t_3$ *then* $\exists t_4, t_5$ *such that* $t_2 \rightarrow^{*}_{R,E} t_4$, $t_3 \rightarrow^{+}_{R,E} t_5$, *and* $t_4 \approx_E t_5$.

A set of equations is said to be in *dag-solved form* (or *d-solved form*) if and only if they can be arranged as a list

$$X_1 \approx^? t_1, \ \ldots, \ X_n \approx^? t_n$$

where (a) each left-hand side X_i is a distinct variable, and (b) $\forall 1 \leq i \leq j \leq n$: X_i does not occur in t_j.

Equational Unification: Two terms s and t are unifiable modulo an equational theory E iff there exists a substitution θ such that $\theta(s) \approx_E \theta(t)$. The unification problem modulo equational theory E is the problem of solving a set of equations $\mathscr{S} = \{s_1 \approx_E^? t_1, \ldots, s_n \approx_E^? t_n\}$, whether there exists σ such that $\sigma(s_1) \approx_E \sigma(t_1), \cdots, \sigma(s_n) \approx_E \sigma(t_n)$. This is also referred to as *semantic* unification where equational equivalence [7] or congruence is considered among the terms being unified, rather than syntactic identity. Some of the standard equational theories used are *associativity* and *commutativity*.

A unifier δ is *more general than* another unifer ρ over a set of variables X iff a substitution equivalent to the latter can be obtained from the former by suitably composing it with a third substitution:

$$\delta \preceq_E^X \rho \quad \text{iff} \quad \exists \sigma : \delta \circ \sigma(x) \approx_E \rho(x) \text{ for all } x \in X.$$

A substitution θ is a *normalized* substitution with respect to a term rewrite system R if and only for every x, $\theta(x)$ is in R-normal form. In other words, terms in the range of θ are in normal form. (These are also sometimes referred to as as *irreducible* substitutions.) When R is convergent, one can assume that all unifiers modulo R are normalized substitutions.

3 Asymmetric Unification

Definition 1. *Given a decomposition (Σ, E, R) of an equational theory, a substitution σ is an asymmetric R, E-unifier of a set Γ of asymmetric equations $\{s_1 \approx_\downarrow^? t_1, \ldots, s_n \approx_\downarrow^? t_n\}$ iff for each asymmetric equation $s_i \approx_\downarrow^? t_i$, σ is an $(E \cup R)$-unifier of the equation $s_i \approx^? t_i$, and $\sigma(t_i)$ is in R, E-normal form. In other words, $\sigma(s_i) \rightarrow_{R,E}^! \sigma(t_i)$.*

Note that symmetric unification can be reduced to asymmetric unification: the unification problem $\{s \approx_R^? t\}$ is solvable if and only if the asymmetric problem $\{s \approx_\downarrow^? X, t \approx_\downarrow^? X\}$, where X is a new variable, is solvable. Thus we could also include symmetric equations in a problem instance.

Example 2. *Let $R = \{x + a \rightarrow x\}$ be a rewrite system. An asymmetric unifier θ for $\{u + v \approx_\downarrow^? v + w\}$ modulo this system is $\theta = \{u \mapsto v, w \mapsto v\}$. However, another unifier $\rho = \{u \mapsto a, v \mapsto a, w \mapsto a\}$ is not an asymmetric unifier. But note that $\theta \preceq_E \rho$ over $\{u, v, w\}$: i.e., ρ is an instance of θ, or, alternatively, θ is more general than ρ. This shows that instances of asymmetric unifiers need not be asymmetric unifiers.*

The *ground asymmetric unification problem* is a restricted version of the asymmetric unification problem where the solutions have to be ground substitutions over the signature Σ and a finite set of constants that includes the constants appearing in the problem instance. Following Baader and Schulz [6] we denote ground asymmetric unification problems in the form (Γ, C) where Γ is a set of asymmetric equations and C is a set of constants that includes the constants that appear in Γ. In other words, the terms that appear in Γ are from $T(\Sigma \cup C, V)$ and $\mathcal{V\!R}an(\sigma)$, for any solution σ, must be a subset of $T(\Sigma \cup C)$.

4 Disunification

Disunification deals with solving a set of equations and disequations with respect to a given equational theory.

Definition 2. *For an equational theory E, a disunification problem is a set of equations and disequations* $\mathscr{L} = \{s_1 \approx_E^? t_1, \ldots, s_n \approx_E^? t_n\} \cup \{s_{n+1} \not\approx_E^? t_{n+1}, \ldots, s_{n+m} \not\approx_E^? t_{n+m}\}$.

A solution to this problem is a substitution σ such that:

$$\sigma(s_i) \approx_E \sigma(t_i) \qquad (i = 1, \ldots, n)$$

and

$$\sigma(s_{n+j}) \not\approx_E \sigma(t_{n+j}) \qquad (j = 1, \ldots, m).$$

Example 3. *Given $E = \{x + a \approx x\}$, a disunifier θ for $\{u + v \not\approx_E v + u\}$ is $\theta = \{u \mapsto a, v \mapsto b\}$.*

If $a + x \approx x$ is added to the identities E, then $\theta = \{u \mapsto a, v \mapsto b\}$ is clearly no longer a disunifier modulo this equational theory.

The *ground disunification problem* [6] for an equational theory Δ, denoted as (Γ, C) consists of a set of constants C and a set Γ of equations and disequations over terms from $T(Sig(\Gamma) \cup C, V)$. For any solution σ, $\mathcal{V\!R}an(\sigma) \subset T(\Sigma \cup C)$.

5 A Theory for Which Asymmetric Unification Is in P Whereas Disunification Is NP-Hard

Let R_1 be the following term rewriting system:

$$h(a) \rightarrow f(a, c)$$
$$h(b) \rightarrow f(b, c)$$

We show that asymmetric unifiability modulo this theory can be solved in polynomial time.

Note that reversing the directions of the rules also produces a convergent system, i.e.,

$$f(a,c) \rightarrow h(a)$$
$$f(b,c) \rightarrow h(b)$$

is also terminating and confluent. We assume that the input equations are in standard form, i.e., of one of four kinds: $X \approx^? Y$, $X \approx^? h(Y)$, $X \approx^? f(Y,Z)$ and $X \approx^? d$ where X, Y, Z are variables and d is any constant. Asymmetric equations will have the extra downarrow, e.g., $X \approx^?_{\downarrow} h(Z)$.

Our algorithm transforms an asymmetric unification problem to a set of equations in *dag-solved form* along with *clausal constraints*, where each atom is of the form $(\langle variable \rangle = \langle constant \rangle)$. We use the notation $EQ \parallel \Gamma$, where EQ is set of equations in standard form as mentioned above, and Γ is a set of clausal constraints. Initially Γ is empty.

Lemma 5.1. *(Removing asymmetry) If s is an irreducible term, then $h(s)$ is (also) irreducible iff $s \neq a$ and $s \neq b$.*

Proof. If $s = a$ or $s = b$, then clearly $h(s)$ is reducible. Conversely, if s is irreducible and $h(s)$ is reducible, then s has to be either a (for the first rule to apply) or b (for the second rule).

Hence we first apply the following inference rule (until finished) that gets rid of asymmetry:

$$\frac{\mathcal{EQ} \uplus \{X \approx^?_{\downarrow} h(Y)\} \parallel \Gamma}{\mathcal{EQ} \uplus \{X \approx^? h(Y)\} \parallel \Gamma \cup \{\neg(Y = a)\} \cup \{\neg(Y = b)\}}$$

Lemma 5.2. *(Cancellativity) $h(s) \downarrow_{R_1} h(t)$ iff $s \downarrow_{R_1} t$. Similarly, $f(s_1, s_2) \downarrow_{R_1} f(t_1, t_2)$ iff $s_1 \downarrow_{R_1} t_1$ and $s_2 \downarrow_{R_1} t_2$.*

Proof. The *if* part is straightforward. If s and t are joinable, this implies $h(s)$ and $h(t)$ are joinable modulo R_1.

Only if part: Suppose $h(s)$ is joinable with $h(t)$. Without loss of generality assume s and t are in normal form. If $s = t$ then we are done. Otherwise, if $s \neq t$, since we assumed s and t are in normal forms, $h(s)$ or $h(t)$ must be reducible. If $h(s)$ is reducible, then s has to be either a or b, which reduces $h(s)$ to $f(s,c)$. Then $h(t)$ must also be reducible and joinable with $f(s,c)$. Hence s and t will be equivalent.

The proof of the second part is straightforward. □

Lemma 5.3. *(Root Conflict)*

$$s \rightarrow^! a, t_1 \rightarrow^! a, t_2 \rightarrow^! c$$

$$or$$

$$s \rightarrow^! b, t_1 \rightarrow^! b, t_2 \rightarrow^! c.$$

Proof. The *if* part is straightforward. If s and t_1 reduce to a (resp., b) and t_2 reduces to c, then $h(a)$ reduces to $f(a,c)$ (resp., $f(b,c)$).

Only if part: Suppose $h(s)$ is joinable with $f(t_1, t_2)$ modulo R_1. We can assume wlog that s, t_1, t_2 are in normal forms. Then $h(s)$ must be reducible, i.e., $s = a$ or $s = b$. If $s = a$, then $t_1 = a$ and $t_2 = c$; else if $s = b$, then $t_1 = b$ and $t_2 = c$ (from our rules). □

Now for E-unification, we have the inference rules

(a)
$$\frac{\{X \approx^? V\} \uplus \mathscr{E2} \parallel \Gamma}{\{X \approx^? V\} \cup [V/X](\mathscr{E2}) \parallel [V/X](\Gamma)} \qquad \text{if } X \text{ occurs in } \mathscr{E2} \text{ or } \Gamma$$

(b)
$$\frac{\mathscr{E2} \uplus \{X \approx^? h(Y), X \approx^? h(T)\} \parallel \Gamma}{\mathscr{E2} \cup \{X \approx^? h(Y), T \approx^? Y\} \parallel \Gamma}$$

(c)
$$\frac{\mathscr{E2} \uplus \{X \approx^? f(V,Y), X \approx^? f(W,T)\} \parallel \Gamma}{\mathscr{E2} \cup \{X \approx^? f(V,Y), W \approx^? V, T \approx^? Y\} \parallel \Gamma}$$

(d)
$$\frac{\mathscr{E2} \uplus \{X \approx^? h(Y), X \approx^? f(U,V)\} \parallel \Gamma}{\mathscr{E2} \cup \{U \approx^? Y, V \approx^? c, X \approx^? f(Y,V)\} \parallel \Gamma \cup \{(Y = a) \vee (Y = b)\}}$$

The above inference rules are applied with rule (a) having the highest priority and rule (d) the lowest.

The following are the failure rules, which, of course, have the highest priority.

(F1)
$$\frac{\mathscr{E2} \uplus \{X \approx^? d, X \approx^? f(U,V)\} \parallel \Gamma}{FAIL} \qquad d \in \{a,b,c\}$$

(F2)
$$\frac{\mathscr{E2} \uplus \{X \approx^? d, X \approx^? h(V)\} \parallel \Gamma}{FAIL} \qquad d \in \{a,b,c\}$$

(F3)
$$\frac{\mathscr{E2} \uplus \{X \approx^? c, X \approx^? d\} \parallel \Gamma}{FAIL} \qquad d \in \{a,b\}$$

(F4)
$$\frac{\mathscr{E2} \uplus \{X \approx^? b, X \approx^? a\} \parallel \Gamma}{FAIL}$$

Lemma 5.4. R_1 *is non-subterm-collapsing, i.e., no term is equivalent to a proper subterm of it.*

Proof. Since the rules in R_1 are size increasing, no term can be reduced to a proper subterm of it.   □

Because of the above lemma, we can have an extended occur-check or cycle check [21] as another failure rule.

(F5) $$\frac{\{X_0 \approx^? s_1[X_1], \ldots, X_n \approx^? s_n[X_0]\} \uplus \mathcal{E}\mathcal{Q} \parallel \Gamma}{FAIL}$$

where the X_i's are variables and s_j's are non-variable terms.

Once these inference rules have been exhaustively applied, we are left with a set of equations in *dag-solved form* along with clausal constraints. Thus the set of equations is of the form

$$\{X_1 =^? t_1, \ldots, X_m =^? t_m\}$$

where the variables on the left-hand sides are all distinct (i.e., $X_i \neq X_j$ for $i \neq j$).

Steps for polynomial time solvability of equations and clauses:

1. Add to the list of clauses Γ more clauses derived from the solved form, to generate Γ':
   (a) If $X \approx^? t$ is an equation where t is not a variable or equal to a or b, then we add unit clauses $X \neq a$ and $X \neq b$.
   (b) If $X \approx^? a$ is an equation, then we add unit clauses $X = a$ and $X \neq b$. (Vice versa for $X \approx^? b$.)
2. Check for satisfiability of Γ'.

Soundness of this algorithm follows from the Lemmas 5.1 through 5.4.

As for **termination**, we first observe that none of the inference rules introduce a new variable, i.e., the number of variables never increases. With the first inference rule which removes asymmetry, asymmetric equations are eliminated from $\mathcal{E}\mathcal{Q}$, i.e., the number of asymmetric equations goes down. For the E-unification rules, we can see that in each case either the overall size of equations decreases or some function symbols are lost. In rule (a), we replace X by V and are left with an isolated X, hence the number of unsolved variables goes down [4]. In rules (b) and (d) the number of occurrences of h goes down and in rule (c) the number of occurrences of f goes down.

Note that getting the dag-solved form can be achieved in polynomial time. The clausal constraints are either negative unit clauses of the form $\neg(Y = a)$ or $\neg(Y = b)$ or positive two-literal clauses of the form $(Y = a) \vee (Y = b)$. The solvability of such a system of equations and clauses can be checked in polynomial time because treating each equational atom $Y = a$ as a propositional variable—and similarly with $Y = b$—will result in a 2SAT problem (well-known to be solvable in polynomial time [1]). •

However, disunification modulo R_1 is NP-hard. The proof is by a polynomial-time reduction from the three-satisfiability (3SAT) problem.

Let $U = \{x_1, x_2, \ldots, x_n\}$ be the set of variables, and $B = \{C_1, C_2, \ldots, C_m\}$ be the set of clauses. Each clause C_k, where $1 \leq k \leq m$, has 3 literals.

We construct an instance of a disunification problem from 3SAT. There are 8 different combinations of T and F assignments to the variables in a clause in 3SAT, out of which there is exactly one truth-assignment to the variables in the clause that makes the clause evaluate to false. For the 7 other combinations of T and F assignments to the literals, the clause is rendered true. We represent T by a and F by b. Hence for each clause C_i we create a disequation DEQ_i of the form

$$f(x_p, f(x_q, x_r)) \not\approx_{R_1} f(d_1, f(d_2, d_3))$$

where x_p, x_q, x_r are variables, $d_1, d_2, d_3 \in \{a, b\}$, and (d_1, d_2, d_3) corresponds to the falsifying truth assignment. For example, given a clause $C_k = x_p \vee \overline{x_q} \vee x_r$, we create the corresponding disequation $DEQ_k = f(x_p, f(x_q, x_r)) \not\approx_{R_1} f(b, f(a, b))$.

We also create the equation $h(x_j) \approx_{R_1} f(x_j, c)$ for each variable x_j. These make sure that each x_j is mapped to either a or b.

Thus for B, the instance of disunification constructed is

$$S = h(x_1) \approx f(x_1, c),\ h(x_2) \approx f(x_2, c),\ \ldots,\ h(x_n) \approx f(x_n, c)\}$$
$$\cup$$
$$\{DEQ_1, DEQ_2, \ldots, DEQ_m\}$$

Example 4. *Given $U = \{x_1, x_2, x_3\}$ and $B = \{x_1 \vee \overline{x_2} \vee x_3,\ \overline{x_1} \vee \overline{x_2} \vee x_3\}$, the constructed instance of disunification is*

$$\{h(x_1) \approx f(x_1, c),\ h(x_2) \approx f(x_2, c),\ h(x_3) \approx f(x_3, c),$$
$$f(x_1, f(x_2, x_3)) \not\approx f(b, f(a, b)),$$
$$f(x_1, f(x_2, x_3)) \not\approx f(a, f(a, b))\}$$

We expect that membership in NP would not be hard to show since R_1 is saturated by paramodulation [27]. We have not worked out the details though.

6   A Theory for Which Disunification Is in P Whereas Asymmetric Unification Is NP-Hard

The theory we consider consists of the following term rewriting system R_2:

$$x + x \rightarrow 0$$
$$x + 0 \rightarrow x$$
$$x + (y + x) \rightarrow y$$

and the equational theory AC:

$$(x + y) + z \approx x + (y + z)$$
$$x + y \approx y + x$$

This theory is called **ACUN** because it consists of *associativity, commutativity, unit* and *nilpotence*. This is the theory of the boolean XOR operator. An algorithm for *general* **ACUN** unification is provided by Liu [26] in his Ph.D. dissertation. (See also [16, Sect. 4].)

Disunification modulo this theory can be solved in polynomial time by what is essentially Gaussian Elimination[3] over $\mathbb{Z}_2$. Suppose we have m variables $x_1, x_2, \ldots, x_m$, and n constant symbols $c_1, c_2, \ldots, c_n$, and q such equations and disequations to be unified. We can assume an ordering on the variables and constants $x_1 > x_2 > \ldots > x_m > c_1 > c_2 > \ldots > c_n$. We first pick an equation with leading variable x_1 and eliminate x_1 from all *other* equations and disequations. We continue this process with the next equation consisting of leading variable x_2, followed by an equation containing leading variable x_3 and so on, until no more variables can be eliminated. The problem has a solution if and only if (i) there are no equations that contain only constants, such as $c_3 + c_4 \approx c_5$, and (ii) there are no disequations of the form $0 \not\approx 0$. This way we can solve the disunification problem in polynomial time using Gaussian Elimination over $\mathbb{Z}_2$.

Example 5. *Suppose we have two equations* $x_1 + x_2 + x_3 + c_1 + c_2 \approx^?_{R_2,AC} 0$ *and* $x_1 + x_3 + c_2 + c_3 \approx^?_{R_2,AC} 0$, *and a disequation* $x_2 \not\approx^?_{R_2,AC} 0$.

Eliminating x_1 *from the second equation, results in the equation* $x_2 + c_1 + c_3 \approx_{R_2,AC} 0$. *We can now eliminate* x_2 *from the first equation, resulting in* $x_1 + x_3 + c_2 + c_3 \approx_{R_2,AC} 0$. x_2 *can also be eliminated from the disequation* $x_2 \not\approx_{R_2,AC} 0$, *which gives us* $c_1 + c_3 \not\approx_{R_2,AC} 0$. *Thus the procedure terminates with*

$$x_1 + x_3 + c_2 + c_3 \approx_{R_2,AC} 0$$
$$x_2 + c_1 + c_3 \approx_{R_2,AC} 0$$
$$c_1 + c_3 \not\approx_{R_2,AC} 0$$

Thus we get

$$x_2 \approx_{R_2,AC} c_1 + c_3$$
$$x_1 + x_3 \approx_{R_2,AC} c_2 + c_3$$

and the following substitution is clearly a solution:

$$\{x_1 \mapsto c_2, \ x_2 \mapsto c_1 + c_3, \ x_3 \mapsto c_3\}$$

Example 6. *Suppose we have two disequations* $x + a \not\approx_{R_2,AC} 0$ *and* $x \not\approx_{R_2,AC} 0$.

There is no variable that can be eliminated by the Gaussian elimination technique. The identity substitution is a solution for this problem, but no ground solution exists if a *is the only free constant.*

However, asymmetric unification is NP-hard. The proof is by a polynomial-time reduction from the graph 3-colorability problem. Let $G = (V, E)$ be a graph where $V = \{v_1, v_2, v_3, \ldots, v_n\}$ are the vertices, $E = \{e_1, e_2, e_3, \ldots, e_m\}$

---

[3] Gaussian elimination over $\mathbb{Z}_2$, or GF(2), is discussed in several papers [9,25].

the edges and $C = \{c_1, c_2, c_3\}$ the color set with $n \geq 3$. G is 3-colorable if none of the adjacent vertices $\{v_i, v_j\} \in E$ have the same color assigned from C. We construct an instance of asymmetric unification as follows. We create variables for vertices and edges in G: for each vertex v_i we assign a variable y_i and for each edge e_k we assign a variable z_k. Now for every edge $e_k = \{v_i, v_j\}$ we create an equation $EQ_k = c_1 + c_2 + c_3 \approx_{\downarrow}^? y_i + y_j + z_k$. Note that each z_k appears in only one equation.

Thus for E, the instance of asymmetric unification problem constructed is

$$S = \{EQ_1, EQ_2, \ldots, EQ_m\}$$

If G is 3-colorable, then there is a color assignment $\theta : V \to C$ such that $\theta v_i \neq \theta v_j$ if $e_k = \{v_i, v_j\} \in E$. This can be converted into an asymmetric unifier α for S as follows: We assign the color of v_i, $\theta(v_i)$ to y_i, $\theta(v_j)$ to y_j, and the remaining color to z_k. Thus $\alpha(v_i + v_j + z_k) \approx_{AC} c_1 + c_2 + c_3$ and therefore α is an asymmetric unifier of S. Note that the term $c_1 + c_2 + c_3$ is clearly in normal form modulo the rewrite relation $\longrightarrow_{R_2, AC}$.

Suppose S has an asymmetric unifier β. Note that β cannot map y_i, y_j or z_k to 0 or to a term of the form $u + v$ since $\beta(y_i + y_j + z_k)$ has to be in normal form or irreducible. Hence for each equation EQ_k, it must be that $\beta(y_i), \beta(y_j), \beta(z_k) \in \{c_1, c_2, c_3\}$ and $\beta(y_i) \neq \beta(y_j) \neq \beta(z_k)$. Thus β is a 3-coloring of G.

Example 7. *Given $G = (V, E), V = \{v_1, v_2, v_3, v_4\}$, $E = \{e_1, e_2, e_3, e_4\}$, where $e_1 = \{v_1, v_3\}$, $e_2 = \{v_1, v_2\}$, $e_3 = \{v_2, v_3\}$, $e_4 = \{v_3, v_4\}$ and $C = \{c_1, c_2, c_3\}$, the constructed instance of asymmetric unification is*

$$
\begin{aligned}
EQ_1 &= c_1 + c_2 + c_3 \approx_{\downarrow}^? y_1 + y_3 + z_1 \\
EQ_2 &= c_1 + c_2 + c_3 \approx_{\downarrow}^? y_1 + y_2 + z_2 \\
EQ_3 &= c_1 + c_2 + c_3 \approx_{\downarrow}^? y_2 + y_3 + z_3 \\
EQ_4 &= c_1 + c_2 + c_3 \approx_{\downarrow}^? y_3 + y_4 + z_4.
\end{aligned}
$$

Now suppose the vertices in the graph G are given this color assignment: $\theta = \{v_1 \mapsto c_1, v_2 \mapsto c_2, v_3 \mapsto c_3, v_4 \mapsto c_1\}$. We can create an asymmetric unifier based on this θ by mapping each v_i to $\theta(v_i)$ and, for each edge e_j, mapping z_j to the remaining color from $\{c_1, c_2, c_3\}$ after both its vertices are assigned. For instance, for $e_1 = \{v_1, v_3\}$, since y_1 is mapped to c_1 and y_3 is mapped to c_2, we have to map z_1 to c_3. Similarly for $e_2 = \{v_1, v_2\}$, we map z_2 to c_2 since y_1 is mapped to c_1 and y_2 is mapped to c_3. Thus the asymmetric unifier is

$$\{y_1 \mapsto c_1, \; y_2 \mapsto c_3, \; y_3 \mapsto c_2, \; z_1 \mapsto c_3, \; z_2 \mapsto c_2, \; z_3 \mapsto c_1, \; z_4 \mapsto c_3\}$$

We have not yet looked into whether the problem is in NP, but we expect it to be so.

7 A Theory for Which Ground Disunifiability Is in P Whereas Ground Asymmetric Unification Is NP-Hard

This theory is the same as the one mentioned in previous section, **ACUN**, but with a homomorphism added. It has an AC-convergent term rewriting system, which we call R_3:

$$x + x \rightarrow 0$$
$$x + 0 \rightarrow x$$
$$x + (y + x) \rightarrow y$$
$$h(x + y) \rightarrow h(x) + h(y)$$
$$h(0) \rightarrow 0$$

7.1 Ground Disunification

Ground disunifiability [6] problem refers to checking for ground solutions for a set of disequations and equations. The restriction is that only the set of constants provided in the input, i.e., the equational theory and the equations and disequations, can be used; no new constants can be introduced.

We show that ground disunifiability modulo this theory can be solved in polynomial time, by reducing the problem to that of solving systems of linear equations. This involves finding the Smith Normal Form [19,28,30] which can be done in polynomial time [22,23]. This gives us a general solution to all the variables or unknowns.

Suppose we have m equations in our ground disunifiability problem. We can assume without loss of generality that the disequations are of the form $z \neq 0$. For example, if we have disequations of the form $e_1 \neq e_2$, we introduce a new variable z and set $z = e_1 + e_2$ and $z \neq 0$. Let n be the number of variables or unknowns for which we have to find a solution.

For each constant in our ground disunifiability problem, we follow the approach similar to [20], of forming a set of linear equations and solving them to find ground solutions. (This approach was pioneered by Baader [2,3] and Nutt [5,29] for commutative/monoidal theories, of which **ACUNh** is one.)

We use $h^k x$ to represent the term $h(h(\ldots h(x)\ldots))$ and $H^k = h^{k_1}x + h^{k_2}x + \cdots + h^{k_n}x$ is a polynomial over $\mathbb{Z}_2[h]$.

Let $\{s_1 \approx^?_E t_1, \ldots, s_m \approx^?_E t_m\}$ be the set of equations in a disunifiability problem. Without loss of generality we can assume that each s_i and t_i are of the following forms:

$$s_i = H_{i1}x_1 + H_{i2}x_2 + \ldots + H_{im}x_n, \quad H_{ij} \in \mathbb{Z}_2[h]$$

$$t_i = H'_{i1}c_1 + H'_{i2}c_2 + \ldots + H'_{im}c_l, \quad H'_{ij} \in \mathbb{Z}_2[h]$$

where $\{c_1, \ldots c_l\}$ is the set of constants and $\{x_1, \ldots x_n\}$ is the set of variables.

For each constant c_i, $1 \leq i \leq l$, and each variable x, we create a variable x^{c_i}. We then generate, for each constant c_i, a set of linear equations S^{c_i} of the form $AX =^? B$ with coefficients from the polynomial ring $\mathbb{Z}_2[h]$.

The solutions are found by computing the Smith Normal Form of A. We now outline that procedure[4]:

Note that the dimension of matrix A is $m \times n$ where m is the number of equations and n is the number of unknowns. The dimension of of matrix B is $m \times 1$. Every matrix A, of rank r, is equivalent to a diagonal matrix D, given by

$$D = diag(d_{11}, d_{22}, \ldots d_{rr}, 0, \ldots, 0)$$

Each entry d_{kk} is different from 0 and the entries form a divisibility sequence.

The diagonal matrix D, of size $m \times n$, is the Smith Normal Form (SNF) of matrix A. There exist invertible *matrices* P, of size $m \times m$, and Q, of size $n \times n$ such that

$$D = PAQ \tag{1}$$

and let

$$\overline{D} = diag(d_{11}, d_{22}, \ldots, d_{rr})$$

be the submatrix consisting of the first r rows and the first r columns of D. Suppose $AX = B$. We have, from (1),

$$PAX = PB$$

Since Q is invertible we can write

$$PAQ(Q^{-1}X) = PB$$

Let $C = PB$ and

$$Y = (Q^{-1}X) = \begin{bmatrix} \overline{Y} \\ Z \end{bmatrix}$$

with $\overline{Y}$ being first r rows of the $n \times 1$ matrix Y, and Z the remaining $(n - r)$ rows of Y.

$$C \text{ can be written as } \begin{bmatrix} \overline{C} \\ U \end{bmatrix}$$

with $\overline{C}$ the first r rows of C, and U a matrix of zeros.

---

[4] We follow the notation and procedure similar to Greenwell and Kertzner [19].

Then $DY = PB = C$ translates into

$$\begin{bmatrix} \overline{D} & 0 \\ 0 & 0 \end{bmatrix} \begin{bmatrix} \overline{Y} \\ Z \end{bmatrix} = \begin{bmatrix} \overline{C} \\ U \end{bmatrix}$$

We solve for Y in $DY = C$, by first solving $\overline{D}\,\overline{Y} = \overline{C}$:

$$\begin{bmatrix} d_{11} & & \\ & \ddots & \\ & & d_{rr} \end{bmatrix} \begin{bmatrix} y_1 \\ y_2 \\ y_3 \\ \vdots \\ y_r \end{bmatrix} = \begin{bmatrix} c_1 \\ c_2 \\ c_3 \\ \vdots \\ c_r \end{bmatrix}$$

A solution exists if and only if each d_{ii} divides c_i. If this is the case let $\widehat{y}_i = c_i/d_{ii}$. Now to find a general solution plug in values of Y in $X = QY$:

$$\begin{bmatrix} r & n-r \\ \hline Q_1 & Q_2 \end{bmatrix} \begin{bmatrix} \widehat{y}_1 \\ \widehat{y}_2 \\ \vdots \\ \widehat{y}_r \\ z_{r+1} \\ \vdots \\ z_n \end{bmatrix}$$

First r columns of Q are referred to as Q_1 and remaining $n - r$ columns are referred to as Q_2. To find a particular solution, for any x_j, we take the dot product of the jth row of Q_1 and $(\widehat{y}_1, \ldots, \widehat{y}_r)$.

Similarly, to find a general solution, we take the dot product of ith row of Q_1 with $(\widehat{y}_1, \ldots, \widehat{y}_r)$, plus the dot product of the ith row of Q_2, with a vector $(z_{r+1}, \ldots, z_n)$ consisting of distinct variables.

If we have a disequation of the form $x_i \neq 0$, to check for solvability for x_i, we first check whether the particular solution is 0. If it is not, then we are done. Otherwise, check whether all the values in i^{th} row of Q_2 are identically 0. If it is not, then we have a solution since $z_{r+1}, \ldots, z_n$ can take any arbitrary values. This procedure has to be repeated for all constants.

Example 8. *Consider the disunification problem*

$$\{h(X_1) + h(h(X_2)) + X_2 \approx h(h(a)) + h(a) + a, \; X_1 \not\approx X_2\}$$

This is transformed into a system of two equations

$$hx_1 + (h^2 + 1)x_2 = h^2 + h + 1 \tag{2}$$
$$x_+ x_2 + z = 0 \tag{3}$$

along with the disequation $Z \neq 0$. Thus the matrices are

$$A = \begin{bmatrix} h & h^2+1 & 0 \\ 1 & 1 & 1 \end{bmatrix} \quad \text{and } B = \begin{bmatrix} h^2+h+1 \\ 0 \end{bmatrix}$$

Computing the Smith normal form, we get

$$P = \begin{bmatrix} 0 & 1 \\ 1 & h \end{bmatrix}, \quad Q = \begin{bmatrix} 1 & h & h^2+1 \\ 0 & 1 & h \\ 0 & h+1 & h^2+h+1 \end{bmatrix}, \quad \text{and } D = \begin{bmatrix} 1 & 0 & 0 \\ 0 & 1 & 0 \end{bmatrix}$$

$$C = PB = \begin{bmatrix} 0 \\ h^2+h+1 \end{bmatrix}$$

Since $DY = C$, we get $y_1 = 0$, $y_2 = h^2+h+1$, i.e,

$$Y = \begin{bmatrix} 0 \\ h^2+h+1 \\ z_3 \end{bmatrix}$$

Now

$$QY = \begin{bmatrix} 1 & h & \vline & h^2+1 \\ 0 & 1 & \vline & h \\ 0 & h+1 & \vline & h^2+h+1 \end{bmatrix} \begin{bmatrix} 0 \\ h^2+h+1 \\ z_3 \end{bmatrix},$$

i.e.,

$$Q_1 = \begin{bmatrix} 1 & h \\ 0 & 1 \\ 0 & h+1 \end{bmatrix} \quad \text{and } Q_2 = \begin{bmatrix} h^2+1 \\ h \\ h^2+h+1 \end{bmatrix}$$

The particular solution in this case is

$$x_1 = h^3+h^2+h, \quad x_2 = h^2+h+1$$

which gives us a unifier

$$\{X_1 \mapsto h(h(h(a))) + h(h(a)) + h(a), \quad X_2 \mapsto h(h(a)) + h(a) + a\}$$

7.2   Ground Asymmetric Unification

However, asymmetric unification modulo R_3 is NP-hard. Decidability can be shown by automata-theoretic methods as for Weak Second Order Theory of One successor (WS1S) [11,15]. In WS1S we consider quantification over finite sets of natural numbers, along with one successor function. All equations or formulas are transformed into finite-state automata which accepts the strings that correspond to a model of the formula [24,31].

This automata-based approach is key to showing decidability of WS1S, since the satisfiability of WS1S formulas reduces to the automata intersection-emptiness problem. We follow the same approach here.

For ease of exposition, let us consider the case where there is only one constant a. Thus every ground term can be represented as a set of natural numbers. The homomorphism h is treated as a successor function. Just as in WS1S, the input to the automata are column vectors of bits. The length of each column vector is the number of variables in the problem.

$$\Sigma = \left\{ \begin{pmatrix} 0 \\ 0 \\ \vdots \\ 0 \end{pmatrix}, \ldots, \begin{pmatrix} 1 \\ 1 \\ \vdots \\ 1 \end{pmatrix} \right\}$$

The deterministic finite automata (DFA) are illustrated here The + operator behaves like the *symmetric set difference* operator.

We illustrate how automata are constructed for each equation in standard form. In order to avoid cluttering up the diagrams the dead state has been included only for the first automaton. The missing transitions lead to the dead state by default for the others. Recall that we are considering the case of one constant a. The homomorphism h is treated as successor function.

7.3   P = Q + R

Figure 1: Let P_i, Q_i and R_i denote the ith bits of P, Q and R *respectively*. P_i has a value 1, when either Q_i or R_i has a value 1. We need 3-bit alphabet symbols for this equation. For example, if $R_2 = 0$, $Q_2 = 1$, then $P_2 = 1$. The corresponding alphabet symbol is $\begin{pmatrix} P_2 \\ Q_2 \\ R_2 \end{pmatrix} = \begin{pmatrix} 1 \\ 0 \\ 1 \end{pmatrix}$. Hence, only strings with the alphabet symbols from $\left\{ \begin{pmatrix} 0 \\ 0 \\ 0 \end{pmatrix}, \begin{pmatrix} 0 \\ 1 \\ 1 \end{pmatrix}, \begin{pmatrix} 1 \\ 0 \\ 1 \end{pmatrix}, \begin{pmatrix} 1 \\ 1 \\ 0 \end{pmatrix} \right\}$ are accepted by this automaton. Rest of the input symbols $\begin{pmatrix} 0 \\ 0 \\ 1 \end{pmatrix}, \begin{pmatrix} 1 \\ 1 \\ 1 \end{pmatrix}, \begin{pmatrix} 0 \\ 1 \\ 0 \end{pmatrix}, \begin{pmatrix} 1 \\ 0 \\ 0 \end{pmatrix}$ lead to the dead state D as they violate the XOR property.

Note that the string $\begin{pmatrix} 1 \\ 0 \\ 1 \end{pmatrix}\begin{pmatrix} 1 \\ 1 \\ 0 \end{pmatrix}$ is accepted by automaton. This corresponds to P = a + h(a). Q = h(a) and R = a.

7.4 $P \approx_{\downarrow} Q + R$

Figure 2: To preserve asymmetry on the right-hand side of this equation, Q + R should be irreducible. If either Q or R is empty, or if they have any term in common, then a reduction will occur. For example, if Q = h(a) and R = h(a) + a, there is a reduction, whereas if R = h(a) and Q = a, irreducibility is preserved, since there is no common term and neither one is empty. Since neither Q nor R can be empty, any accepted string should have one occurrence of $\begin{pmatrix} 1 \\ 0 \\ 1 \end{pmatrix}$ and one occurrence of $\begin{pmatrix} 1 \\ 1 \\ 0 \end{pmatrix}$.

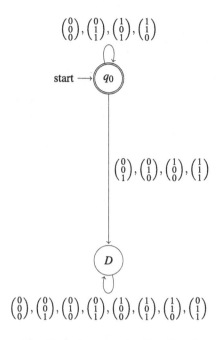

Fig. 1. Automaton for $P = Q + R$

7.5 $X = h(Y)$

Figure 3: We need 2-bit vectors as alphabet symbols since we have two unknowns X and Y. Note again that h acts like the successor function. q_0 is the only accepting state. A state transition occurs with bit vectors $\binom{1}{0}, \binom{0}{1}$. If $Y = 1$ in current state, then $X = 1$ in the next state, hence a transition occurs from q_0 to q_1, and vice versa. The ordering of variables is $\binom{Y}{X}$.

7.6 $X \approx_{\downarrow} h(Y)$

Figure 4: In this equation, $h(Y)$ should be in normal form. So Y cannot be either 0 or of the form $u + v$. Thus Y has to be a string of the form $0^i 1 0^j$ and X then has to be $0^{i+1} 1 0^{j-1}$. Therefore the bit vector $\binom{1}{0}$ has to be succeeded by $\binom{0}{1}$.

7.7   An Example

Let $\{U \approx_{\downarrow} V + Y, \ W = h(V), \ Y \approx_{\downarrow} h(W)\}$ be an asymmetric unification problem. We need 4-bit vectors and 3 automata since we have 4 unknowns in 3 equations, with bit-vectors represented in this ordering of set variables: $\begin{pmatrix} V \\ W \\ Y \\ U \end{pmatrix}$.

We include the $\times$ ("don't-care") symbol in state transitions to indicate that the values can be either 0 or 1. This is essentially to avoid cluttering the diagrams.

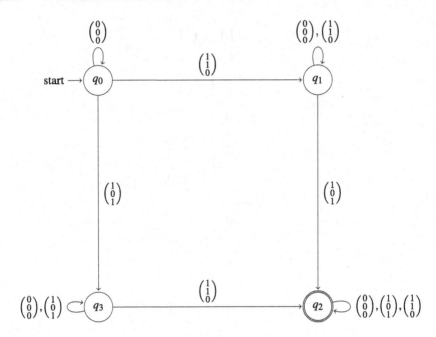

Fig. 2. Automaton for $P \approx_{\downarrow} Q + R$

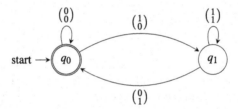

Fig. 3. Automaton for $X = h(Y)$

Note that here this × symbol is a placeholder for the variable W which does not have any significance in this automaton. The automata constructed for this example are indicated in Figs. 5, 6 and 7.

NOTE: As before, the symbol × in the vectors means that the bit value can be either 0 or 1.

The string $\begin{pmatrix} 1 \\ 0 \\ 1 \end{pmatrix}\begin{pmatrix} 0 \\ 1 \\ 0 \end{pmatrix}\begin{pmatrix} 0 \\ 0 \\ 1 \end{pmatrix}\begin{pmatrix} 0 \\ 0 \\ 0 \end{pmatrix}$ is accepted by all the three automata. The corresponding asymmetric unifier is

$$\{V \mapsto a, W \mapsto h(a), Y \mapsto h^2(a), U \mapsto (h^2(a) + a)\}.$$

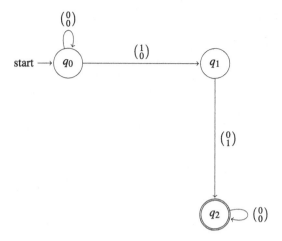

Fig. 4. Automaton for $X \approx_\downarrow h(Y)$

Once we have automata constructed for all the formulas, we take the intersection and check if there exists a string accepted by corresponding automata. If the intersection is not empty, then we have a solution or an asymmetric unifier for set of formulas.

This technique can be extended to the case where we have more than one constant. Suppose we have k constants, say $c_1, \ldots, c_k$. We express each variable X in terms of the constants as follows:

$$X = X^{c_1} + \ldots + X^{c_k}$$

effectively grouping subterms that contain each constant under a new variable. Thus if $X = h^2(c_1) + c_1 + h(c_3)$, then $X^{c_1} = h^2(c_1) + c_1$, $X^{c_2} = 0$, and $X^{c_3} = h(c_3)$. If the variables are $X_1, \ldots, X_m$, then we set

$$X_1 = X_1^{c_1} + \ldots + X_1^{c_k}$$
$$X_2 = X_2^{c_1} + \ldots + X_2^{c_k}$$
$$\vdots$$
$$X_m = X_m^{c_1} + \ldots + X_m^{c_k}$$

For example, if Y and Z are set variables and a, b, c are constants, then we can write $Y = Y^a + Y^b + Y^c$ and $Z = Z^a + Z^b + Z^c$ as our terms with constants. For each original variable, say Z, we refer to Z^{c_1} etc. as its *components* for ease of exposition.

If the equation to be solved is: $X = h(Y)$, with a, b, c as constants, then we create the equations

$$X^a = h(Y^a), X^b = h(Y^b), X^c = h(Y^c).$$

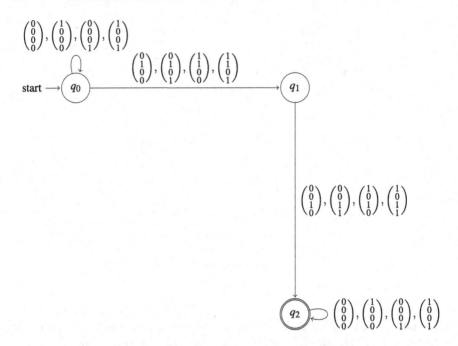

Fig. 5. Automaton for $\mathsf{Y}\approx_\downarrow \mathsf{h(W)}$

However, if the equation is asymmetric, i.e., $\mathsf{X}\approx_\downarrow \mathsf{h(Y)}$, then Y has to be a term of the form $h^i(d)$ where d is either a, b, or c. All components except one have to be 0 and we form the equation $\mathsf{X}^d\approx_\downarrow \mathsf{h(Y}^d)$ since $\mathsf{Y} \neq 0$. The other components for X and Y have to be 0.

Similarly, if the equation to be solved is $\mathsf{X} = \mathsf{W} + \mathsf{Z}$, with a, b, c as constants, we form the equations

$$\mathsf{X}^a = \mathsf{W}^a + \mathsf{Z}^a, \mathsf{X}^b = \mathsf{W}^b + \mathsf{Z}^b \; and \; \mathsf{X}^c = \mathsf{W}^c + \mathsf{Z}^c$$

and solve the equations. If we have an asymmetric equation $\mathsf{X}\approx_\downarrow \mathsf{W} + \mathsf{Z}$, then clearly one of the components of each original variable has to be non-zero; e.g., in

$$\mathsf{W} = \mathsf{W}^a + \mathsf{W}^b + \mathsf{W}^c,$$

all the components cannot be 0 simultaneously. It is ok for W^a and Z^a to be 0 simultaneously, provided either one of W^b or W^c is non-zero *and* one of Z^b or Z^c, is non-zero. For example, $\mathsf{W} = \mathsf{W}^b$ and $\mathsf{Z} = \mathsf{Z}^c$ is fine, i.e, W can be equal to its b-component and Z can be equal to its c-component, respectively, as in the solution $\{W \mapsto h^2(b) + h(b), \; Z \mapsto h(c) + c, \; X \mapsto h^2(b) + h(b) + h(c) + c\}$.

If W^a and Z^a are non-zero, they cannot have anything in common, or otherwise there will be a reduction. In other words, X^a, W^a and Z^a must be solutions of the asymmetric equation $\mathsf{X}^a\approx_\downarrow \mathsf{W}^a + \mathsf{Z}^a$.

Our approach is to design a nondeterministic algorithm. We guess which constant component in each variable has to be 0, i.e., for each variable X and

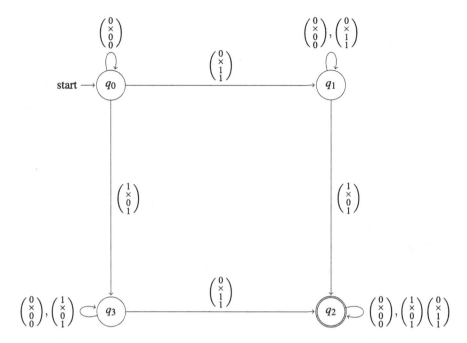

Fig. 6. Automaton for $\mathbf{U} \approx_\downarrow \mathbf{V} + \mathbf{Y}$

each constant a, we "flip a coin" as to whether X^a will be set equal to 0 by the target solution. Now for the case $X \approx_\downarrow W + Z$, we do the following:

for all constants a do:
    if $X^a = W^a = Z^a = 0$ then skip
    else if $W^a = 0$ then set $X^a = Z^a$
      if $Z^a = 0$ then set $X^a = W^a$
      if both W^a and Z^a are non-zero then set $X^a \approx_\downarrow W^a + Z^a$

In the asymmetric case $X \approx_\downarrow h(Y)$, if more than one of the components of Y happens to be non-zero, it is clearly an error. ("The guess didn't work."). Otherwise, i.e., if exactly one of the components is non-zero, we form the asymmetric equation as described above.

Nondeterministic Algorithm when we have more than one constant

1. If there are m variables and k constants, then represent each variable in terms of its k constant components.
2. Guess which constant components have to be 0.
3. Form symmetric and asymmetric equations for each constant.
4. Solve each set of equations by the Deterministic Finite Automata (DFA) construction.

The exact complexity of this problem is open.

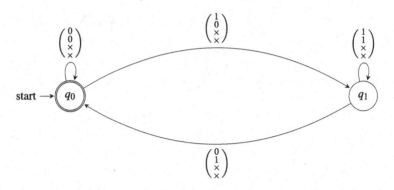

Fig. 7. Automaton for $W = h(V)$

8 A Theory for Which Time Complexity of Asymmetric Unification Varies Based on Ordering of Function Symbols

Let E_4 be the following equational theory:

$$g(a) \approx f(a, a, a)$$
$$g(b) \approx f(b, b, b)$$

Let R_4 denote

$$f(a, a, a) \rightarrow g(a)$$
$$f(b, b, b) \rightarrow g(b)$$

This is clearly terminating, as can be easily shown by the *lexicographic path ordering (lpo)* [4] using the symbol ordering $f > g > a > b$. We show that asymmetric unification modulo the rewriting system R_4 is NP-hard. The proof is by a polynomial-time reduction from the Not-All-Equal Three-Satisfiability (NAE-3SAT) problem [18].

Let $U = \{x_1, x_2, \dots, x_n\}$ be the set of variables, and $C = \{C_1, C_2, \dots, C_m\}$ be the set of clauses. Each clause C_k, has to have at least one *true* literal and at least one *false* literal.

We create an instance of asymmetric unification as follows. We represent T by a and F by b. For each variable x_i we create the equation

$$f(x_i, x_i, x_i) \approx_{R_4^?} g(x_i)$$

These make sure that each x_i is mapped to either a or b. For each clause $C_j = x_p \vee x_q \vee x_r$, we introduce a new variable z_j and create an asymmetric equation EQ_j:

$$z_j \approx_{\downarrow}^? f(x_p, x_q, x_r)$$

Thus for any C, the instance of asymmetric unification problem constructed is

$$\mathscr{S} = \Big\{ f(x_1, x_1, x_1) \approx g(x_1), \ \ldots, \ f(x_n, x_n, x_n) \approx g(x_n) \Big\} \cup \Big\{ EQ_1, EQ_2, \ldots, EQ_m \Big\}$$

If $\mathscr{S}$ has an asymmetric unifier γ, then, x_p, x_q and x_r cannot map to all a's or all b's since these will cause a reduction. Hence for EQ_j, $\gamma(x_p)$, $\gamma(x_q)$ and $\gamma(x_r)$ should take at least one a and at least one b. Thus γ is also a solution for NAE-3SAT.

Suppose C has a satisfying assignment. Then $\{x_p, x_q, x_r\}$ cannot all be T or all F, i.e., $\{x_p, x_q, x_r\}$ needs to have at least one true literal and at least one false literal. Thus if σ is a satisfying assignment, we can convert σ into an asymmetric unifier θ as follows: $\theta(x_p) := \sigma(x_p)$, the value of $\sigma(x_p)$, a or b, is assigned to $\theta(x_p)$. Similarly $\theta(x_q) := \sigma(x_q)$ and $\theta(x_r) := \sigma(x_r)$. Recall that we also introduce a unique variable z_j for each clause C_j in C. Thus if $C_j = \{x_p, x_q, x_r\}$ we can map z_j to $\theta(f(x_p, x_q, x_r))$. Thus θ is an asymmetric unifier of S and $z_j \approx^?_\downarrow f(x_p, x_q, x_r)$. Note that $f(x_p, x_q, x_r)$ is clearly in normal form modulo the rewrite relation $\longrightarrow_{R_4}$, since x_p, x_q, x_r can't all be same.

Example 9. *Given $U = \{x_1, x_2, x_3, x_4\}$ and $C = \{x_1 \vee x_2 \vee x_3, \ x_1 \vee x_2 \vee x_4, \ x_1 \vee x_3 \vee x_4, \ x_2 \vee x_3 \vee x_4\}$ the constructed instance of asymmetric unification $\mathscr{S}$ is*

$$\begin{aligned}
&\{ f(x_1, x_1, x_1) \approx g(x_1), \ f(x_2, x_2, x_2) \approx g(x_2), \ f(x_3, x_3, x_3) \approx g(x_3), \\
&f(x_4, x_4, x_4) \approx g(x_4), \\
&z_1 \ \approx^?_\downarrow \ f(x_1, x_2, x_3), \\
&z_2 \ \approx^?_\downarrow \ f(x_1, x_2, x_4), \\
&z_3 \ \approx^?_\downarrow \ f(x_1, x_3, x_4), \\
&z_4 \ \approx^?_\downarrow \ f(x_2, x_3, x_4) \}
\end{aligned}$$

As in the case of R_1, we believe membership in NP can be shown using the fact that R_4 is saturated by paramodulation [27].

However, if we orient the rules the other way, i.e., when $g > f > a > b$, we can show that asymmetric unifiability modulo this theory can be solved in polynomial time, i.e., when the term rewriting system is

$$g(a) \to f(a, a, a)$$
$$g(b) \to f(b, b, b)$$

Let R_5 denote the above term rewriting system. We assume that the input equations are in standard form, i.e., of one of four kinds: $X \approx^? Y$, $X \approx^? g(Y)$, $X \approx^? f(U, V, W)$ and $X \approx^? d$ where X, Y, U, V, W are variables and d is any constant. Asymmetric equations will have the extra downarrow, e.g., $X \approx^?_\downarrow g(Y)$.

As in Sect. 5, our algorithm transforms an asymmetric unification problem to a set of equations in dag-solved form along with clausal constraints, where each atom is of the form ($\langle variable \rangle = \langle constant \rangle$). We use the notation $EQ \parallel \Gamma$, where EQ is set of equations in standard form as mentioned above, and Γ is a set of clausal constraints. Initially Γ is empty.

We first apply the following inference rule (until finished) that gets rid of asymmetry:

$$\frac{\mathscr{E}\mathscr{Q} \uplus \{X \approx^?_\downarrow g(Y)\} \;\|\; \Gamma}{\mathscr{E}\mathscr{Q} \uplus \{X \approx^? g(Y)\} \;\|\; \Gamma \cup \{\neg(Y = a)\} \cup \{\neg(Y = b)\}}$$

Now for E-unification, we have the inference rules

(a) $\dfrac{\{X \approx^? V\} \uplus \mathscr{E}\mathscr{Q} \;\|\; \Gamma}{\{X \approx^? V\} \cup [V/X](\mathscr{E}\mathscr{Q}) \;\|\; [V/X](\Gamma)}$ if X occurs in $\mathscr{E}\mathscr{Q}$

(b) $\dfrac{\mathscr{E}\mathscr{Q} \uplus \{X \approx^? g(Y),\, X \approx^? g(T)\} \;\|\; \Gamma}{\mathscr{E}\mathscr{Q} \cup \{X \approx^? g(Y),\, T \approx^? Y\} \;\|\; \Gamma}$

(c) $\dfrac{\mathscr{E}\mathscr{Q} \uplus \{X \approx^? f(U_1, V_1, W_1),\, X \approx^? f(U_2, V_2, W_2)\} \;\|\; \Gamma}{\mathscr{E}\mathscr{Q} \cup \{X \approx^? f(U_1, V_1, W_1),\, U_1 \approx^? U_2,\, V_1 \approx^? V_2,\, W_1 \approx^? W_2\} \;\|\; \Gamma}$

(d) $\dfrac{\mathscr{E}\mathscr{Q} \uplus \{X \approx^? g(Y),\, X \approx^? f(U, V, W)\} \;\|\; \Gamma}{\mathscr{E}\mathscr{Q} \cup \{U \approx^? Y, V \approx^? Y, W \approx^? Y, X \approx^? f(Y, Y, Y)\} \;\|\; \Gamma \cup \{(Y = a) \vee (Y = b)\}}$

The inference rules are applied in the descending order of priority from (a), the highest, to (d) the lowest. Occurrence of equations of the form $X \approx^? a$ and $X \approx^? f(U, V, W)$ will make the equations unsolvable. Hence we have failure rules as in Sect. 5. Since the equational theory is non-subterm-collapsing, we have an extended occur-check or cycle check rule here as well:

(Cycle-check) $\dfrac{\{X_0 \approx^? s_1[X_1],\; \ldots,\; X_n \approx^? s_n[X_0]\} \uplus \mathscr{E}\mathscr{Q} \;\|\; \Gamma}{FAIL}$

where the X_i's are variables and s_j's are non-variable terms.

The rest of the algorithm is along the same lines as the one in Sect. 5 for the system R_1. Soundness, termination and polynomial-time complexity can be shown in the same way.

9 Conclusions and Future Work

We have compared and contrasted asymmetric unification and disunification to prove that they are not reducible to each other in terms of time complexity, or, more precisely, the problems are not polynomial-time reducible to one another in general. There is still the issue of designing asymmetric unification algorithms for *specific theories* that come up in protocol analysis. For instance, one such theory is ACUNh which was discussed in Sect. 7 with AC as the background theory. We are working on developing an asymmetric unification algorithm for ACUNh, with ACh as background theory (as identities E).

We are also working on comparing asymmetric unification and disunification in terms of decidability, i.e., whether there are theories where asymmetric unification is decidable and disunification is undecidable, and vice versa. Another topic that has not been explored is the *unification type(s)* of asymmetric unification problems: for instance are there theories for which unification is finitary and asymmetric unification is infinitary (or even nullary)? There are some hurdles to be crossed here since, as has already been pointed out, an instance of an asymmetric unifier need not be an asymmetric unifier.

Acknowledgements. We wish to thank Franz Baader for all his remarkable contributions in this field. We also wish to thank the anonymous reviewers for their detailed comments and suggestions which helped us greatly in improving this paper.

References

1. Aspvall, B., Plass, M.F., Tarjan, R.E.: A linear-time algorithm for testing the truth of certain quantified boolean formulas. Inf. Process. Lett. **8**(3), 121–123 (1979)
2. Baader, F.: Unification in commutative theories. J. Symb. Comput. **8**(5), 479–497 (1989)
3. Baader, F.: Unification in commutative theories, Hilbert's basis theorem, and Gröbner bases. J. ACM **40**(3), 477–503 (1993)
4. Baader, F., Nipkow, T.: Term Rewriting and All That. Cambridge University Press, Cambridge (1999)
5. Baader, F., Nutt, W.: Combination problems for commutative/monoidal theories or how algebra can help in equational unification. Appl. Algebra Eng. Commun. Comput. **7**(4), 309–337 (1996)
6. Baader, F., Schulz, K.U.: Combination techniques and decision problems for disunification. Theor. Comput. Sci. **142**(2), 229–255 (1995)
7. Baader, F., Snyder, W.: Unification theory. In: Handbook of Automated Reasoning, vol. 1, pp. 445–532 (2001)
8. Bachmair, L.: Canonical Equational Proofs. Birkhauser, Boston (1991)
9. Bogdanov, A., Mertens, M.C., Paar, C., Pelzl, J., Rupp, A.: A parallel hardware architecture for fast Gaussian elimination over GF(2). In: 14th IEEE Symposium on Field-Programmable Custom Computing Machines (FCCM 2006), Napa, CA, USA, 24–26 April 2006, Proceedings, pp. 237–248. IEEE Computer Society (2006)
10. Brahmakshatriya, S., Danturi, S., Gero, K.A., Narendran, P.: Unification problems modulo a theory of until. In: Korovin, K., Morawska, B. (eds.) 27th International Workshop on Unification, UNIF 2013, Eindhoven, Netherlands, 26 June 2013. EPiC Series in Computing, vol. 19, pp. 22–29. EasyChair (2013). http://www.easychair.org/publications/?page=723757558
11. Büchi, J.R.: Weak second-order arithmetic and finite automata. Math. Logic Q. **6**(1–6), 66–92 (1960)
12. Buntine, W.L., Bürckert, H.-J.: On solving equations and disequations. J. ACM **41**(4), 591–629 (1994)
13. Bürckert, H.-J., Herold, A., Schmidt-Schauss, M.: On equational theories, unification, and (un)decidability. J. Symb. Comput. **8**(1–2), 3–49 (1989)
14. Comon, H.: Disunification: a survey. In: Lassez, J.-L., Plotkin, G.D. (eds.) Computational Logic - Essays in Honor of Alan Robinson, pp. 322–359. The MIT Press (1991)
15. Elgot, C.C.: Decision problems of finite automata design and related arithmetics. Trans. Am. Math. Soc. **98**(1), 21–51 (1961)
16. Erbatur, S., et al.: Asymmetric unification: a new unification paradigm for cryptographic protocol analysis. In: Bonacina, M.P. (ed.) CADE 2013. LNCS (LNAI), vol. 7898, pp. 231–248. Springer, Heidelberg (2013). https://doi.org/10.1007/978-3-642-38574-2_16

17. Erbatur, S., Kapur, D., Marshall, A.M., Meadows, C., Narendran, P., Ringeissen, C.: On asymmetric unification and the combination problem in disjoint theories. In: Muscholl, A. (ed.) FoSSaCS 2014. LNCS, vol. 8412, pp. 274–288. Springer, Heidelberg (2014). https://doi.org/10.1007/978-3-642-54830-7_18
18. Garey, M., Johnson, D.S.: Computers and Intractability: A Guide to the Theory of NP-Completeness. W. H. Freeman & Co., New York (1979)
19. Greenwell, R.N., Kertzner, S.: Solving linear diophantine matrix equations using the Smith normal form (more or less). Int. J. Pure Appl. Math. 55(1), 49–60 (2009)
20. Guo, Q., Narendran, P., Wolfram, D.A.: Complexity of nilpotent unification and matching problems. Inf. Comput. 162(1–2), 3–23 (2000)
21. Jouannaud, J.-P., Kirchner, C.: Solving equations in abstract algebras: a rule-based survey of unification. In: Computational Logic - Essays in Honor of Alan Robinson, pp. 257–321 (1991)
22. Kaltofen, E., Krishnamoorthy, M.S., Saunders, B.D.: Fast parallel computation of Hermite and Smith forms of polynomial matrices. SIAM J. Algebraic Discrete Methods 8(4), 683–690 (1987)
23. Kannan, R.: Solving systems of linear equations over polynomials. Theor. Comput. Sci. 39, 69–88 (1985)
24. Klaedtke, F., Ruess, H.: Parikh automata and monadic second-order logics with linear cardinality constraints (2002)
25. Koç, Ç.K., Arachchige, S.N.: A fast algorithm for Gaussian elimination over GF(2) and its implementation on the GAPP. J. Parallel Distrib. Comput. 13(1), 118–122 (1991)
26. Liu, Z.: Dealing efficiently with exclusive-OR, abelian groups and homomorphism in cryptographic protocol analysis. Ph.D. thesis, Clarkson University (2012)
27. Lynch, C., Morawska, B.: Basic syntactic mutation. In: Voronkov, A. (ed.) CADE 2002. LNCS (LNAI), vol. 2392, pp. 471–485. Springer, Heidelberg (2002). https://doi.org/10.1007/3-540-45620-1_37
28. MacDuffee, C.C.: The Theory of Matrices. Chelsea Publishing Company, New York (1961)
29. Nutt, W.: Unification in monoidal theories. In: Stickel, M.E. (ed.) CADE 1990. LNCS, vol. 449, pp. 618–632. Springer, Heidelberg (1990). https://doi.org/10.1007/3-540-52885-7_118
30. Rotman, J.J.: Advanced Modern Algebra. Prentice Hall. 1st edn (2002); 2nd printing (2003)
31. Vardi, M.Y., Wilke, T.: Automata: from logics to algorithms. Logic Automata 2, 629–736 (2008)

Building and Combining
Matching Algorithms

Christophe Ringeissen$^{(\boxtimes)}$ (ID)

Université de Lorraine, CNRS, Inria, LORIA, 54000 Nancy, France
`Christophe.Ringeissen@loria.fr`

Abstract. The concept of matching is ubiquitous in declarative programming and in automated reasoning. For instance, it is a key mechanism to run rule-based programs and to simplify clauses generated by theorem provers. A matching problem can be seen as a particular conjunction of equations where each equation has a ground side. We give an overview of techniques that can be applied to build and combine matching algorithms. First, we survey mutation-based techniques as a way to build a generic matching algorithm for a large class of equational theories. Second, combination techniques are introduced to get combined matching algorithms for disjoint unions of theories. Then we show how these combination algorithms can be extended to handle non-disjoint unions of theories sharing only constructors. These extensions are possible if an appropriate notion of normal form is computable.

Keywords: Matching · Unification · Combination of theories · Syntactic theories

1 Introduction

Both unification and matching procedures play a central role in automated reasoning and in various declarative programming paradigms such as functional programming or (constraint) logic programming. In particular, unification is an essential engine in the execution of logic programs. In functional programming, functions are defined by pattern matching. In rule-based programming [18,20,39], matching is needed to apply a rule and so to perform a computation. In automated theorem proving [4,5,14,15,17,41], unification is applied to deduce new facts via expansion inferences, while matching is useful to simplify existing facts via contraction inferences. For the verification of security protocols, dedicated provers [16,28,37] handle protocols specified in a symbolic way. In these reasoning tools, the capabilities of an intruder are modeled using equational theories [1], and the reasoning is supported by decision procedures and solvers modulo equational theories, including equational unification and equational matching.

Unification and matching procedures aim at solving equations in term-generated structures [12,30]. A unification problem is a set of arbitrary equations

© Springer Nature Switzerland AG 2019
C. Lutz et al. (Eds.): Baader Festschrift, LNCS 11560, pp. 523–541, 2019.
https://doi.org/10.1007/978-3-030-22102-7_24

between terms. A matching problem is a unification problem where each equation has a ground side, that is, a ground term possibly built over free constants. Thus, a matching problem is a particular unification problem with free constants. In practice, syntactic unification, as well as syntactic matching, are particularly popular. In that singular case, the underlying equational theory is simply the empty theory and the well-known syntactic unification algorithm computes a most general solution when the input is solvable. More generally, we may consider equational unification and equational matching, where the problems are defined modulo an arbitrary equational theory, such as, for instance, one that defines a function symbol to be associative (A), commutative (C) or associative and commutative (AC). Equational unification and equational matching are undecidable in general. However, specialized techniques have been developed to solve both problems for particular classes of equational theories, many of high practical interest, including for example AC. The successful application of equational rewriting in rule-based programming languages [18,20,39] has demonstrated the interest of developing dedicated equational matching algorithms. Compared to unification, matching can be considered as a simpler problem. Hence A-matching is finitary, that is, the set of solutions of an A-matching problem is finite, whereas A-unification is infinitary. The decision problems for AC-matching and AC-unification are both NP-complete [32] even if for AC-matching the number of solutions is bounded by a single-exponential while a double-exponential [33] is needed to get a bound for AC-unification.

In this paper, we focus on the matching problem. We mainly consider the case of regular theories, that is, theories axiomatized by equalities such that both sides have the same set of variables. Matching in regular theories has a remarkable property: any solution of any matching problem is necessarily ground. This property eases the construction of matching algorithms. We survey two general techniques that allow us to design matching algorithms for a large class of (regular) theories.

First, we focus on mutation techniques that generalize the classical decomposition rule known from the syntactic case [30]. Using a more general mutation rule, it is possible to get a complete unification procedure for theories having the property of being syntactic [34,43]. Unfortunately, the resulting unification procedure only terminates for some particular classes of theories, such as shallow theories [21] or theories saturated by paramodulation [36]. However this unification procedure can be adapted to construct a matching procedure, which is actually terminating for the whole class of finite syntactic theories, as pointed out by Nipkow in the early 1990s [43]. Many permutative theories of practical interest for equational rewriting belong to that class, including A, C and AC [34,43]. For the class of finite syntactic theories, we present a rule-based matching algorithm along the lines of the classical rule-based syntactic unification algorithm.

Second, when a theory is defined as a union of theories, it is quite natural to consider methods that combine the matching algorithms available in the individual theories. In the early 1990s, a first combination method has been proposed by Nipkow for the matching problem in the union of disjoint regular theories [42]. Then the Baader-Schulz combination method has been a seminal contribution for the unification problem in the union of disjoint arbitrary theories [11]. Compared to other combination methods previously developed by Schmidt-Schauss [49] and Boudet [19], the Baader-Schulz method permits us to solve both the unification problem and its related decision problem. Based on an approach à la Baader-Schulz, it is possible to develop new combination methods for the matching problem and its related decision problem, as shown in [46]. In this paper, we survey the existing combined matching methods, by showing how to reconstruct them thanks to the Baader-Schulz combination method and the underlying combination techniques. We also discuss their possible extensions to non-disjoint unions of theories, more precisely, theories sharing free constructors. We show that an approach à la Baader-Schulz can be applied to both the matching problem and the word problem in these non-disjoint unions. For the word problem, this leads to a method very similar to the one proposed by Baader and Tinelli [13] using an approach à la Nelson-Oppen [40].

The paper is organized as follows. Section 2 introduces the main technical concepts and notations. In Sect. 3, we present a mutation-based matching algorithm for a large class of syntactic theories. The combined unification problem is briefly discussed in Sect. 4, where we focus on the Baader-Schulz combination method. The combined word problem is detailed in Sect. 5 while Sect. 6 revisits the combined matching problem. In Sect. 7, we discuss the combined matching problem in the union of non-disjoint theories sharing free constructors. Eventually, Sect. 8 concludes with some final remarks about ongoing and future works.

2   Preliminaries

We use the standard notation of equational unification [12] and term rewriting systems [10]. A signature Σ is a set of function symbols with fixed arity. Given a signature Σ and a (countable) set of variables V, the set of Σ-terms over variables V is denoted by $T(\Sigma, V)$. The set of variables in a term t is denoted by $Var(t)$. A term is *linear* if all its variables occur only once. For any position p in a term t (including the root position ϵ), $t(p)$ is the symbol at position p, $t_{|p}$ is the subterm of t at position p, and $t[u]_p$ is the term t in which $t_{|p}$ is replaced by u. A substitution is an endomorphism of $T(\Sigma, V)$ with only finitely many variables not mapped to themselves. A substitution is denoted by $\sigma = \{x_1 \mapsto t_1, \ldots, x_m \mapsto t_m\}$, where the domain of σ is $Dom(\sigma) = \{x_1, \ldots, x_m\}$. Application of a substitution σ to t is written $t\sigma$.

A term t is *ground* if $Var(t) = \emptyset$. The set of ground Σ-terms is denoted by $T(\Sigma)$. When $\mathcal{C}$ denotes a finite set of constants not occurring in Σ, $\Sigma \cup \mathcal{C}$ is a signature defined as the union of Σ and $\mathcal{C}$, and $T(\Sigma \cup \mathcal{C})$ denotes the set of ground $(\Sigma \cup \mathcal{C})$-terms.

2.1   Equational Theories and Rewrite Systems

A Σ-axiom is a pair of Σ-terms, denoted by $l = r$. Variables in an axiom are implicitly universally quantified. Given a set E of Σ-axioms, the *equational* *theory* $=_E$ presented by E is the closure of E under the laws of reflexivity, symmetry, transitivity, substitutivity and congruence (by a slight abuse of terminology, E is often called an equational theory). Equivalently, $=_E$ can be defined as the reflexive transitive closure $\leftrightarrow_E^*$ of an equational step $\leftrightarrow_E$ defined as follows: $s \leftrightarrow_E t$ if there exist a position p of s, $l = r$ (or $r = l$) in E, and substitution σ such that $s|_p = l\sigma$ and $t = s[r\sigma]_p$. An axiom $l = r$ is *regular* if $Var(l) = Var(r)$. An axiom $l = r$ is *permutative* if l and r have the same multiset of symbols (including function symbols and variables). An axiom $l = r$ is *linear* (resp., *collapse-free*) if l and r are linear (resp. non-variable terms). An equational theory is *regular* (resp., permutative/linear/collapse-free) if all its axioms are regular (resp., permutative/linear/collapse-free). An equational theory E is *finite* if for each term t, there are only finitely many terms s such that $t =_E s$. One can remark that a permutative theory is finite and a finite theory is regular and collapse-free. Well-known theories such as the associativity $A = \{x + (y + z) = (x + y) + z\}$, the commutativity $C = \{x + y = y + x\}$, and the associativity-commutativity $AC = A \cup C$ are permutative. Unification in permutative theories is undecidable in general [50].

A theory E is *syntactic* if it has a finite *resolvent presentation* S, defined as a finite set of axioms S such that each equality $t =_E u$ has an equational proof $t \leftrightarrow_S^* u$ with at most one step $\leftrightarrow_S$ applied at the root position. The theories A, C and AC are syntactic [43]. For C and AC, syntacticness can be shown as a consequence of the fact that any collapse-free theory is syntactic if it admits a unification algorithm [34].

A *term rewrite system* (TRS) is given by a set R of oriented Σ-axioms called rewrite rules and of the form $l \rightarrow r$ such that l, r are Σ-terms, l is not a variable and $Var(r) \subseteq Var(l)$. A term s *rewrites* to a term t w.r.t R, denoted by $s \rightarrow_R t$, if there exist a position p of s, $l \rightarrow r \in R$, and substitution σ such that $s|_p = l\sigma$ and $t = s[r\sigma]_p$. Given an equational theory E, $\longleftrightarrow_{R \cup E}$ denotes the symmetric relation $\leftarrow_R \cup \rightarrow_R \cup =_E$. A TRS R is Church-Rosser modulo E if $\longleftrightarrow_{R \cup E}^*$ is included in $\rightarrow_R^* \circ =_E \circ \leftarrow_R^*$. A reduction ordering $>$ is a well-founded ordering on terms closed under context and substitution. A reduction ordering $>$ is said to be E-compatible if $s > t$ implies $s' > t'$ for any terms s, t, s', t' such that $s' =_E s$ and $t' =_E t$. If $\rightarrow_R$ is included in an E-compatible reduction ordering, then there is no infinite sequence w.r.t $=_E \circ \rightarrow_R \circ =_E$ and according to [31] the following properties are equivalent:

1. R is Church-Rosser modulo E,
2. for any terms t, t', $t \longleftrightarrow_{R \cup E}^* t'$ if and only if $t \downarrow_R \ =_E \ t' \downarrow_R$, where $t \downarrow_R$ (resp., $t' \downarrow_R$) denotes any normal form of t (resp., t') w.r.t $\rightarrow_R$.

A substitution σ is *R-normalized* if, for every variable x in the domain of σ, $x\sigma$ is a normal form w.r.t $\rightarrow_R$.

2.2 Unification and Matching

From now on, we assume a signature Σ and a Σ-theory E such that Σ may include finitely many function symbols not occurring in the axioms of E. These additional function symbols are said to be free in E. A Σ-equation is a pair of Σ-terms denoted by $s =^? t$. Variables in an equation are implicitly existentially quantified. When t is ground, an equation $s =^? t$ is called a match-equation, also denoted by $s \leq^? t$. An E-unification problem is a set of Σ-equations, $\Gamma = \{s_1 =^? t_1, \ldots, s_n =^? t_n\}$, or equivalently a conjunction of Σ-equations. We distinguish the following classes of E-unification problems Γ:

- if there is no free function symbol in Γ, then Γ is an elementary E-unification problem;
- if some free constants (resp., free symbols) occur in Γ, then Γ is an E-unification problem with free constants (resp., a general E-unification problem);
- if Γ is an E-unification problem with free constants (resp., a general E-unification problem) including only ground equations, Γ is an E-word problem (resp., a general E-word problem);
- if Γ is an E-unification problem with free constants (resp., a general E-unification problem) including only match-equations, Γ is an E-matching problem (resp., a general E-matching problem);
- If $\Gamma = \{x_1 =^? t_1, \ldots, x_n =^? t_n\}$ such that $x_1, \ldots, x_n$ are variables occurring only once in Γ, then Γ is called a solved form.

Consider any E-unification problem Γ. The set of variables in Γ is denoted by $Var(\Gamma)$. A solution to Γ, called an E-unifier, is a substitution σ such that $s_i\sigma =_E t_i\sigma$ for all $1 \leq i \leq n$. A substitution σ is more general modulo E than θ on a set of variables V, denoted as $\sigma \leq_E^V \theta$, if there is a substitution τ such that $x\sigma\tau =_E x\theta$ for all $x \in V$. A Complete Set of E-Unifiers of Γ, denoted by $CSU_E(\Gamma)$, is a set of substitutions such that each $\sigma \in CSU_E(\Gamma)$ is an E-unifier of Γ, and for each E-unifier θ of Γ there exists $\sigma \in CSU_E(\Gamma)$ such that $\sigma \leq_E^{Var(\Gamma)} \theta$. A class of E-unification problems is said to be finitary (resp., unitary) if any Γ in that class admits a $CSU_E(\Gamma)$ whose cardinality is finite (resp., at most 1). When E is an empty set of Σ-axioms, E is the empty Σ-theory denoted by $\emptyset$ where $\emptyset$-unification is unitary: a syntactic unification algorithm computes a $CSU_\emptyset(\Gamma)$ whose cardinality is at most 1 for any unification problem Γ.

Two signatures are disjoint if their respective sets of function symbols are disjoint. Two theories are disjoint if their respective signatures are disjoint. Given two disjoint signatures Σ_1 and Σ_2 and any $i = 1, 2$, Σ_i-terms (including the variables) and Σ_i-equations (including the equations between variables) are called i-pure. A term t is said to be Σ_i-rooted if its root symbol is in Σ_i. An alien subterm of a Σ_i-rooted term t is a Σ_j-rooted subterm s ($i \neq j$) such that all superterms of s are Σ_i-rooted. The position of an alien subterm of t is called an alien position of t. The set of alien positions of t is denoted by $APos(t)$. A term with at least one alien subterm is said to be impure.

3   Matching in Syntactic Theories

The interest of syntactic theories is to admit a mutation-based unification procedure that bears similarities with the rule-based unification algorithm known for the empty theory [30]. In addition to the classical decomposition rule, additional mutation rules are needed, one for each equational axiom in a resolvent presentation of a syntactic theory. Unfortunately, this mutation-based unification procedure is not terminating for the class of syntactic theories. However, some important subclasses of syntactic theories actually admit a terminating mutation-based unification procedures, such as *shallow theories* [21], *forward-closed convergent theories* [24], and *equational theories saturated by paramodulation* [36]. When restricting to the matching problem, it is possible to get termination for a large class of theories of practical interest. Actually, Nipkow has shown that the class of finite syntactic theories admits a mutation-based matching algorithm presented as a Prolog-like program in [43]. We give in Fig. 1 a rule-based presentation of this mutation-based matching algorithm. An implementation for the AC case of this rule-based algorithm has been studied in the $UNIF_{AC}$ system developed in Nancy in the early 1990s [2,3]. As shown in [35], this AC-matching algorithm can be easily prototyped using a rule-based programming environment.

Theorem 1. *Consider the* MFS *inference system given in Fig. 1 where* Mutate *is assumed to be applied in a non-deterministic way in addition to* Dec. *The* MFS *inference system provides a mutation-based matching algorithm for any finite syntactic theory admitting a resolvent presentation S.*

Alternatively, there exists a brute force method to get a matching algorithm for finite theories via a reduction to syntactic matching: the finite set of substitutions $\{\sigma \in CSU_\emptyset(s =^? t') \mid t' =_E t\}$ is a $CSU_E(s \leq^? t)$. Compared to this brute force method, the interest of MFS is to show that a slight adaptation of the classical syntactic matching algorithm, i.e., the addition of a single rule, is sufficient to get a matching algorithm for the class of finite syntactic theories. One can notice that MFS can be turned into a decision procedure for the word problem. Moreover, the class of finite syntactic theories being closed by disjoint union [43], MFS can be applied for any union of disjoint finite syntactic theories. To consider more general unions of disjoint theories, we need to rely on combination methods discussed in the rest of the paper.

4   Unification in Unions of Disjoint Theories

There exist several combination methods for the unification problem, in which we find different forms of unification: elementary unification, unification with free constants and unification with free function symbols (also called general unification). Each of these combination methods corresponds to a given class of theories: regular collapse-free theories, regular theories and arbitrary theories. We briefly recall the modularity results that can be derived from these combination methods.

Theorem 2. *The following modularity results are consequences of existing combination methods:*

Mutate

$$\frac{\Gamma \wedge f(s_1, \ldots, s_m) \leq^? g(t_1, \ldots, t_n)}{\Gamma \wedge r_1 \leq^? t_1 \wedge \cdots \wedge r_n \leq^? t_n \wedge s_1 =^? l_1 \wedge \cdots \wedge s_m =^? l_m}$$

where $f(l_1, \ldots, l_m) = g(r_1, \ldots, r_n)$ is a fresh renaming of an axiom in S

Dec

$$\frac{\Gamma \wedge f(s_1, \ldots, s_m) \leq^? f(t_1, \ldots, t_m)}{\Gamma \wedge s_1 \leq^? t_1 \wedge \cdots \wedge s_m \leq^? t_m}$$

Clash

$$\frac{\Gamma \wedge f(s_1, \ldots, s_m) \leq^? g(t_1, \ldots, t_n)}{\bot}$$

where $f \neq g$ and **Mutate** does not apply

Rep

$$\frac{\Gamma \wedge x \leq^? u \wedge t =^? t'}{\Gamma \wedge x \leq^? u \wedge t =^? t'\{x \mapsto u\}} \quad \text{if } x \in Var(t')$$

RemEq

$$\frac{\Gamma \wedge t =^? t'}{\Gamma \wedge t \leq^? t'} \quad \text{if } t' \text{ is ground}$$

Merge

$$\frac{\Gamma \wedge x \leq^? t \wedge x \leq^? t'}{\Gamma \wedge x \leq^? t \wedge t \leq^? t'}$$

Fig. 1. MFS matching algorithm for finite syntactic theories

1. *The class of* **regular collapse-free theories** *admitting an* **elementary unification algorithm** *is closed under disjoint union* [51,53].
2. *The class of* **regular theories** *admitting a* **unification algorithm with free constants** *is closed under disjoint union.*
3. *The class of* **equational theories** *admitting a* **general unification algorithm** *is closed under disjoint union* [11,49].
4. *The class of* **equational theories** *admitting a* **general unification decision procedure** *is closed under disjoint union* [11].

We briefly outline the principles of a combination method for the unification problem in a union of two disjoint theories. First, the input problem is separated into two pure problems. Then, solutions of the pure problems must be carefully combined in order to construct solutions for the input problem. Two cases may appear.

Conflict of theories: The same variable can be instantiated simultaneously in both theories. To solve this conflict, the solution is to select the theory in which the variable is instantiated, meaning that it will be considered as a free constant in the other theory. This transformation of variables into free constants requires an identification of variables to take care of the fact that two variables equally instantiated in one theory must be considered as the same free constant in the other theory. Then, applying unification algorithms with free constants is sufficient to avoid all these conflicts of theories.

Compound cycle: the conjunction of two pure solved forms can be a compound cycle such as $x_1 = t_1[x_2] \land x_2 = t_2[x_1]$ where t_i is i-pure for $i = 1, 2$.

To tackle both the conflicts of theories and the compound cycles, the Baader-Schulz combination method [11] considers a general form of unification called unification with linear constant restriction. It has been shown in [11] that unification with linear constant restriction and general unification are two equivalent notions, leading to the modularity result of the general unification problem given in Theorem 2. In the combination method proposed by Schmidt-Schauss [49], each pure problem is solved in its theory thanks to a unification algorithm with free constants together with a constant elimination algorithm. Actually, constant elimination is useful to break compound cycles [19]. The Schmidt-Schauss method combines unification algorithms while the Baader-Schulz method is also able to combine unification decision procedures. A major application of the Baader-Schulz combination method is to provide a way to show the decidability of general A-unification. The combination techniques developed by Baader and Schulz allow us to reconstruct the combination methods known for regular collapse-free theories and for regular theories:

- for collapse-free theories, a conflict of theories has no solutions,
- for regular theories, a compound cycle has no solutions.

Hence, for regular collapse-free theories, both the conflicts of theories and the compound cycles have no solutions. In the following, we show how to apply the Baader-Schulz combination method and the underlying techniques to build combination methods for two particular unification problems with free constants: the word problem and the matching problem.

5   The Word Problem in Unions of Disjoint Theories

In this section we consider unification problems with free constants where all equations are ground. In other words, we are interested in checking the equality of terms modulo an equational theory, that is, deciding the word problem. The development of a disjoint combination method for the word problem has been considered in [42,49,51] as a first step before investigating more general combination problems. Actually, it was already successfully addressed in [45].

The Baader-Schulz combination method can be applied to reconstruct a combination method dedicated to the word problem, where the word problem is viewed as a particular unification problem with (free) constants for which the theory selection can be simplified and no linear constant restriction is needed. To get a deterministic theory selection, it is useful to normalize the layers related to theories occurring in an impure term.

Definition 1. *An impure term t is in* layer-reduced form *if its alien subterms are in layer-reduced form and if t is not equal to one of its alien subterms. A pure term is in* layer-reduced form *if it is not equal to one of its variables or free constants.*

Example 1. Let $E_1 = \{x + 0 = x, 0 + x = x\}$ and $E_2 = \{g(x, x) = x\}$. The term $g(a, g(a+0, 0+a))$ is not in layer-reduced form but $g(a, g(a+0, 0+a)) =_{E_1 \cup E_2} a$ where a is in layer-reduced form. The term $g(a, b) + g(a, a + 0)$ is not in layer-reduced form but $g(a, b) + g(a, a + 0) =_{E_1 \cup E_2} g(a, b) + a$ where $g(a, b) + a$ is in layer-reduced form.

---

1. **Purify**
   Apply as long as possible the following rule:

   VA
   $$\frac{\exists \bar{v} : \Gamma \wedge s =^? t}{\exists y, \bar{v} : s =^? t[y]_p \wedge y =^? t_{|p}} \quad \text{if } \{s\} \cap \bar{v} = \emptyset, p \in APos(t), y \text{ is a fresh variable}$$

2. **Identify**
   Apply as long as possible the following rule:

   $$\frac{\exists y, y', \bar{v} : \Gamma \wedge y =^? u \wedge y' =^? u'}{\exists y, \bar{v} : \Gamma\{y' \mapsto y\} \wedge y =^? u} \quad \text{if } u =_{E_1 \cup E_2} u'$$

Fig. 2. Abstract algorithm

Lemma 1. *Let s and t be two terms in layer-reduced form. If s and t are free constants, $s =_{E_1 \cup E_2} t$ iff $s = t$. If s is Σ_i-rooted and t is not Σ_i-rooted, then $s \neq_{E_1 \cup E_2} t$. If s and t are both Σ_i-rooted, the* Abstract *algorithm (cf. Fig. 2) applied to $s =^? t$ returns a set of equations including only one i-pure equation $s_i =^? t_i$ between Σ_i-rooted terms such that $s =_{E_1 \cup E_2} t$ iff $s_i =_{E_i} t_i$.*

Lemma 1 provides a recursive combination method for the word problem. A non-recursive version would purify all the alien subterms and would be followed by a variable identification phase to identify variables denoting pure terms that are equal in the related component theory.

Lemma 1 assumes that the input terms are in layer-reduced form. If we have decision procedures for the word problem in the individual theories of the considered union, then the computation of an equivalent term in layer-reduced form is effective by using a bottom-up process which consists in repeatedly checking whether a pure term is equal to one of its variable or free constant. Let us call LRF the algorithm obtained from Abstract (cf. Fig. 2) by adding the steps depicted in Fig. 3 after Identify.

3. **Collapse**
   Apply the following rule:

$$\frac{\exists \bar{v} : \Gamma \wedge x =^? t}{\exists \bar{v} : \Gamma \wedge x =^? c} \quad \text{if} \quad \begin{cases} \{x\} \cap \bar{v} = \emptyset \\ t \text{ is a non-variable } i\text{-pure term} \\ c \text{ is a free constant or a variable in } t \text{ such that } t =_{E_i} c \end{cases}$$

4. **Eliminate**
   Apply as long as possible the following rule:

$$\frac{\exists y, \bar{v} : \Gamma \wedge y =^? u}{\exists \bar{v} : \Gamma\{y \mapsto u\}}$$

Fig. 3. Collapsing for the layer-reduced form computation

Lemma 2. *Let t be any Σ_i-rooted term. Assume all the aliens subterms of t are in layer-reduced form. Given a variable x not occurring in t, the* LRF *algorithm applied to $x =^? t$ returns an equation $x =^? u$ such that $u =_{E_1 \cup E_2} t$ and u is in layer-reduced form.*

By Lemmas 1 and 2 we get the following modularity result:

Theorem 3. *The class of* **equational theories** *admitting a* **decision procedure for the word problem** *is closed by disjoint union.*

6   Matching in Unions of Disjoint Theories

We now study the design of (disjoint) combination methods for the matching problem. To get a dedicated combination method, it is not sufficient to plug matching algorithms into a combination method initially designed for the unification problem. Indeed, the purification phase does not preserve the property of being a matching problem, and so we would have to solve pure equational problems that are not just matching problems. We focus on two simple cases where it is possible to generate equational problems that can be solved thanks to matching algorithms.

6.1   Regular Collapse-Free Theories

In the particular case of regular collapse-free theories, the purification phase can be adapted to introduce only match-equations instead of solved equations. Consider an $E_1 \cup E_2$-matching problem $\{s \leq^? t\}$ where s is impure. Suppose σ is a substitution such that $s\sigma =_{E_1 \cup E_2} t$. When E_1 and E_2 are regular collapse-free, $s\sigma$ and t are rooted in the same theory and any alien subterm of $s\sigma$ is $E_1 \cup E_2$-equal to some alien subterm of t which is necessarily ground. Thus, any alien subterm of s can be unified with some ground alien subterm of t. This leads to a

VA(RCF)

$$\frac{\Gamma \wedge s \leq^? t}{\bigvee_{(p,p') \in P} \Gamma \wedge s[x]_p \leq^? t \wedge s_{|p} \leq^? t_{|p'} \wedge x \leq^? t_{|p'}} \quad \text{if} \quad \begin{cases} P = APos(s) \times APos(t) \\ APos(s) \neq \emptyset \\ s(\epsilon), t(\epsilon) \in \Sigma_i \\ x \text{ is a fresh variable} \end{cases}$$

Conflict

$$\frac{\Gamma \wedge s \leq^? t}{\bot} \quad \text{if } s(\epsilon) \in \Sigma_i, t(\epsilon) \notin \Sigma_i$$

Fig. 4. Purification for regular collapse-free theories

particular purification phase (cf. Fig. 4) producing *left-pure* matching problems that can be handled by the Baader-Schulz combination method. Consequently, it is sufficient to use matching decision procedures.

Theorem 4 ([46]). *The class of* **regular collapse-free theories** *admitting a* **matching decision procedure** *is closed under disjoint union.*

6.2   Regular Theories

Following the approach initiated by Nipkow [42], we present a deterministic combination method described by the inference system given in Fig. 5. Here, we do not care about introducing a pending equation $x =^? s$. In regular theories, this variable x occurring elsewhere in a match-equation will be eventually unified with a ground term. Indeed, solving a match-equation including x generates a conjunction of solved match-equations, in particular a match-equation of the form $x \leq^? t$. Thus, the pending equation $x =^? s$ can be turned into the match-equation $s \leq^? t$.

Theorem 5 ([42]). *The class of* **regular theories** *admitting a* **matching algorithm** *is closed under disjoint union.*

6.3   Arbitrary Theories

The combination of regular theories with linear ones is problematic as shown in the following example borrowed from [42].

Example 2. Consider the two theories $E_1 = \{f(f(x)) = a\}$ and $E_2 = \{g(x,x) = x\} \cup DA$ where

$$DA = \begin{cases} x + (y + z) = (x + y) + z \\ x * (y + z) = x * y + x * z \\ (x + y) * z = x * z + y * z \end{cases}$$

The theory E_1 is a linear theory where the unification with free constants is decidable. The theory E_2 is a union of two disjoint regular theories, each of

---

LeftVA

$$\frac{\Gamma \wedge s \leq^? t}{\Gamma \wedge s[x]_p \leq^? t \wedge x =^? s_{|p}} \quad \text{if } p \in APos(s), x \text{ is a fresh variable}$$

Merge

$$\frac{\Gamma \wedge x \leq^? t \wedge x =^? s}{\Gamma \wedge x \leq^? t \wedge s \leq^? t}$$

Match

$$\frac{\Gamma \wedge s \leq^? t}{\bigvee_{\sigma \in CSU_{E_i}(s \leq^? t)} (\Gamma \wedge \bigwedge_{x \in Dom(\sigma)} x \leq^? x\sigma)} \quad \text{if } s \text{ is a non-variable } i\text{-pure term}$$

Delete

$$\frac{\Gamma \wedge x \leq^? t \wedge x \leq^? t'}{\Gamma \wedge x \leq^? t} \quad \text{if } t =_E t'$$

Fail

$$\frac{\Gamma \wedge x \leq^? t \wedge x \leq^? t'}{\bot} \quad \text{if } t \neq_E t'$$

For any $s \leq^? t$ in the above rules, t is supposed to be in layer-reduced form.

Fig. 5. Matching for the union of regular theories

them admitting a matching algorithm. Thus, the combined method presented in Sect. 6.2 can be applied to get an E_2-matching algorithm. However $E_1 \cup E_2$-matching is undecidable since for any terms s and t built over the signature of DA, the $E_1 \cup E_2$-matching problem $\{f(g(f(s), f(t))) \leq^? a\}$ has a solution iff the DA-unification problem $\{s =^? t\}$ has a solution. Since DA-unification is undecidable, $E_1 \cup E_2$-matching is undecidable while E_1-matching and E_2-matching are both decidable.

In [46], we give a combination method à la Baader-Schulz for the matching problem and its related decision problem. This method is complete for a large class of problems, like matching problems in *partially linear* theories, which are an extension of linear theories including regular collapse-free theories. In the class of partially linear theories, applying matching algorithms is sufficient, the linear constant restriction being superfluous even for the combination of matching decision procedures.

6.4   Matching Versus General Matching

The combination method for the matching problem in regular theories E can be applied to construct a general E-matching algorithm from an E-matching algorithm. This leads to a natural question: is there an equational theory for which matching is decidable while general matching is not? A positive answer to this question is given in [47], by considering a (many-sorted) theory that includes DA (cf. Example 2). This result shows that a combination method for the matching problem cannot exist for arbitrary theories. In the same vein, a similar question arises when comparing unification with free constants and general unification: is there an equational theory for which unification with free constants is decidable while general unification is not? A positive answer to this question is given in [44].

7 Matching in Unions of Non-disjoint Theories

We discuss the problem of designing combination methods for unions of theories sharing constructors [23], by focusing on the word problem and the matching problem. To formalize the notion of constructor, it is convenient to rely on a rewrite system, where a constructor is simply a function symbol not occurring in root positions of left-hand sides. However, not every equational theory can be equivalently presented by a rewrite system. Fortunately, it is always possible to rely on a rewrite system that could be obtained by unfailing completion [11]. Alternatively, this rewrite system and the related constructors can be defined with respect to a reduction ordering over a combined signature used to orient combined ground instances of pure valid equalities. We consider below equational rewrite systems to cope with constructors modulo an equational theory E_0. In the case of absolutely free constructors considered in [23], the theory E_0 is empty.

Definition 2. *Let E_i be a Σ_i-theory for $i = 0, 1, 2$ and $\Sigma = \Sigma_1 \cup \Sigma_2$. The theory $E_1 \cup E_2$ is said to be a* combination of theories sharing constructors modulo E_0 *if $\Sigma_0 = \Sigma_1 \cap \Sigma_2$ and for any arbitrary finite set of variables V viewed as free constants, there exists an E_0-compatible reduction ordering $>$ on the set of ground $(\Sigma \cup V)$-terms $T(\Sigma \cup V)$ satisfying the following two properties for the set R_i $(i = 1, 2)$ of rewrite rules $l\psi \to r\psi$ such that $l\psi > r\psi$; l, r are Σ_i-terms, $l =_{E_i} r$, l is $(\Sigma_i \setminus \Sigma_0)$-rooted; $l\psi$ and $r\psi$ are ground $(\Sigma \cup V)$-terms thanks to a (grounding) substitution ψ:*

1. $\longleftrightarrow^*_{R_i \cup E_0}$ coincides with $=_{E_i}$ on $T(\Sigma \cup V)$,
2. R_i is Church-Rosser modulo E_0 on $T(\Sigma \cup V)$.

Example 3. To satisfy Definition 2, it is sufficient to consider a Σ_0-theory E_0 plus two finite TRSs $\mathcal{R}_1$ and $\mathcal{R}_2$ over respectively Σ_1 and Σ_2 such that

- there is no Σ_0-symbol occurring at the root position of any left-hand side of $\mathcal{R}_1 \cup \mathcal{R}_2$,
- $\mathcal{R}_1 \cup \mathcal{R}_2$ is included in an E_0-compatible reduction ordering,
- $\mathcal{R}_1$ and $\mathcal{R}_2$ are both Church-Rosser modulo E_0.

Then E_0, $E_1 = \mathcal{R}_1 \cup E_0$ and $E_2 = \mathcal{R}_2 \cup E_0$ fulfill Definition 2 using the ordering $>$ provided by the transitive closure of $=_{E_0} \circ \to_{\mathcal{R}_1 \cup \mathcal{R}_2} \circ =_{E_0}$.

From now on, we assume that $\Sigma = \Sigma_1 \cup \Sigma_2$ and $E = E_1 \cup E_2$ is a combination of theories sharing constructors modulo E_0, where R_1 and R_2 denote the TRSs introduced in Definition 2. According to this definition, for any $f \in \Sigma_0$, and any terms $t_1, \ldots, t_m$ in $T(\Sigma \cup V)$, $f(t_1, \ldots, t_m) \downarrow_{R_i} =_{E_0} f(t_1 \downarrow_{R_i}, \ldots, t_m \downarrow_{R_i})$. The combined TRS defined by $R = R_1 \cup R_2$ satisfies the following properties: the rewrite relation $(=_{E_0} \circ \to_R \circ =_{E_0})$ is terminating, $(\longleftrightarrow_R \cup =_{E_0})^*$ coincides with $=_E$ on $T(\Sigma \cup V)$ and R is Church-Rosser modulo E_0 on $T(\Sigma \cup V)$. Thus, for any terms $s, t \in T(\Sigma \cup V)$, $s =_E t$ iff $s \downarrow_R =_{E_0} t \downarrow_R$, and for any $f \in \Sigma_0$, and any terms $t_1, \ldots, t_m$ in $T(\Sigma \cup V)$, $f(t_1, \ldots, t_m) \downarrow_R =_{E_0} f(t_1 \downarrow_R, \ldots, t_m \downarrow_R)$. R-normal forms are useful to define the notion of variable abstraction in a way similar to [11].

Definition 3 (Variable Abstraction). *Let W be a set of variables such that V and W are disjoint. Let $\pi : \{t \downarrow_R \mid t \in T(\Sigma \cup V), t \downarrow_R \notin V\} \longrightarrow W$ be a bijection called a* variable abstraction with range W. *For $i = 1, 2$, the i-abstraction of t is denoted by t^{π_i} and defined as follows:*

- *If $t \in V$, then $t^{\pi_i} = t$.*
- *If t is a Σ_i-rooted term $f(t_1, \ldots, t_n)$, then $t^{\pi_i} = f(t_1^{\pi_i}, \ldots, t_n^{\pi_i})$.*
- *Otherwise, if $t \downarrow_R \notin V$ then $t^{\pi_i} = \pi(t \downarrow_R)$ else $t^{\pi_i} = t \downarrow_R$.*

The notion of variable abstraction is instrumental to state technical lemmas showing that unification and matching procedures known in component theories can be reused without loss of completeness in the combination of theories sharing constructors modulo E_0.

Lemma 3 (Unification). *Consider any $i = 1, 2$, any i-pure terms s and t, and any R-normalized substitution σ. We have that $s\sigma =_E t\sigma$ iff $s\sigma^{\pi_i} =_{E_i} t\sigma^{\pi_i}$.*

In general, R is infinite and so it may be difficult to assume the computability of R-normal forms. In practice, we can rely on a notion of layer-reduced form, just like in the disjoint case. In this non-disjoint setting, a term t is said to be in *layer-reduced* normal form if $t^{\pi_i} =_{E_i} (t \downarrow_R)^{\pi_i}$ for any $i = 1, 2$. Let us assume that for any term, it is possible to compute an E-equal term in layer-reduced form.

Lemma 4 (Word problem). *Consider any $i = 1, 2$ and any terms s and t in layer-reduced form. We have that $s =_E t$ iff $s^{\pi_i} =_{E_i} t^{\pi_i}$.*

Notice that Abstract (cf. Fig. 2) applied to $s =^? t$ computes an i-pure equation which is a renaming of $s^{\pi_i} =^? t^{\pi_i}$ when s and t are Σ_i-rooted terms in layer-reduced form.

Lemma 5 (Matching). *Consider any $i = 1, 2$, any i-pure term s, any term t in layer-reduced form and any R-normalized substitution σ. We have that $s\sigma =_E t$ iff $s\sigma^{\pi_i} =_{E_i} t^{\pi_i}$.*

By Lemma 5 and assuming the computability of layer-reduced forms, the combination methods developed for the matching problem in the union of disjoint theories (cf. Sects. 6.1 and 6.2) can be reused to obtain:

- a combination method for matching decision procedures, deciding in a modular way the matching problem in the combination of regular collapse-free theories sharing constructors (modulo E_0);
- a combination method for matching algorithms, solving in a modular way the matching problem in the combination of regular theories sharing constructors (modulo E_0).

The results presented in this section rely on the use of R-normal forms. The word problem in unions of theories sharing non-absolutely free constructors has been successfully studied in [13], by introducing a computable notion of G-normal forms where G is a particular set of generators called Σ_0-base. In the above setting, a Σ_0-base G corresponds to the set of R-normalized terms that are not Σ_0-rooted. We have not discussed how to compute layer-reduced forms. A possibility is to build them by using normal forms that can computed in component theories, like in [13] for the computation of G-normal forms. The case of absolutely free constructors has been initiated in [23], with some preliminary results for the word problem and the matching problem. Then, a particular form of non-absolutely free constructors has been investigated for a class of theories sharing "inner" constructors, by focusing on the matching problem [48].

More recently, a form of hierarchical combination has been considered in [26]. In that case, the combined theory is given by a term rewrite system R_1 together with an equational Σ_2-theory E_2 such that Σ_2-symbols can occur only below the root positions of right-hand sides of R_1. Thus, under appropriate assumptions on R_1 and E_2, it is possible to design a combination method leading to a $R_1 \cup E_2$-matching algorithm [26]. This procedure uses the combination rules of Fig. 5, the decomposition rules of Fig. 1 for R_1, and applies an E_2-matching algorithm.

8 Conclusion

In this paper, we survey general techniques to build equational matching algorithms for a large class of (combined) theories of practical interest, e.g., in rule-based programming [18,20,39]. Furthermore, we show that the non-disjoint combination of matching procedures can be envisioned when the combined theory admits a computable notion of normal form. The non-disjoint combination of unification procedures remains a challenging problem. There are preliminary results for particular classes of theories such as shallow theories [26] and forward-closed theories [24]. For these particular classes of theories, a mutation-based approach [21] or a variant-based approach [22,29,38] can be successfully applied to solve the (combined) unification problem, but we believe it is always interesting to point out a combination-based alternative when the background theory is a union of "separable" theories. As shown here with the matching problem, some particular decision problems can admit non-disjoint combination methods. In that direction, non-disjoint combination methods have been developed in [27] for two decision problems related to (context) unification and of practical interest in the analysis of security protocols, namely the deduction problem and the indistinguishability problem [1].

Acknowledgments. I am grateful to the reviewers for their insightful remarks.

I would like to thank Hélène Kirchner and Claude Kirchner who gave me the opportunity to start doing research in equational theorem proving, and all my co-authors involved in joint works in connection to this survey:

- Unification in constructor-sharing equational theories [23]: Eric Domenjoud and Francis Klay.
- Satisfiability in constructor-sharing theories [52]: Cesare Tinelli.
- Unification and matching in hierarchical combinations of equational theories [25,26]: Serdar Erbatur, Deepak Kapur, Andrew Marshall and Paliath Narendran.

Last but not least, a special thanks to Franz Baader for his continuous help. The combination papers (co-)authored by Franz Baader are the most influential in my research activity. This collection of papers (see [6–9,11,13] for a non-exhaustive list) is a very precious guide for a journey in combination of theories.

References

1. Abadi, M., Cortier, V.: Deciding knowledge in security protocols under equational theories. Theor. Comput. Sci. **367**(1–2), 2–32 (2006)
2. Adi, M.: Calculs Associatifs-Commutatifs—Etude et réalisation du système *UNIF_AC*. Ph.D. thesis, Université de Nancy 1 (1991)
3. Adi, M., Kirchner, C.: AC-unification race: the system solving approach, implementation and benchmarks. J. Symb. Comput. **14**(1), 51–70 (1992)
4. Armando, A., Bonacina, M.P., Ranise, S., Schulz, S.: New results on rewrite-based satisfiability procedures. ACM Trans. Comput. Log. **10**(1), 4:1–4:51 (2009)
5. Armando, A., Ranise, S., Rusinowitch, M.: A rewriting approach to satisfiability procedures. Inf. Comput. **183**(2), 140–164 (2003)
6. Baader, F.: Combination of compatible reduction orderings that are total on ground terms. In: Winskel, G. (ed.) Proceedings of the Twelfth Annual IEEE Symposium on Logic in Computer Science (LICS 1997), pp. 2–13. IEEE Computer Society Press, Warsaw (1997)
7. Baader, F., Schulz, K.: Combination of constraint solvers for free and quasi-free structures. Theor. Comput. Sci. **192**, 107–161 (1998)
8. Baader, F., Schulz, K.U.: Combination techniques and decision problems for dis-unification. Theor. Comput. Sci. **142**(2), 229–255 (1995)
9. Baader, F., Ghilardi, S., Tinelli, C.: A new combination procedure for the word problem that generalizes fusion decidability results in modal logics. Inf. Comput. **204**(10), 1413–1452 (2006)
10. Baader, F., Nipkow, T.: Term Rewriting and All That. Cambridge University Press, Cambridge (1998)
11. Baader, F., Schulz, K.U.: Unification in the union of disjoint equational theories: combining decision procedures. J. Symb. Comput. **21**(2), 211–243 (1996)
12. Baader, F., Snyder, W.: Unification theory. In: Robinson, J.A., Voronkov, A. (eds.) Handbook of Automated Reasoning (in 2 volumes), pp. 445–532. Elsevier and MIT Press (2001)
13. Baader, F., Tinelli, C.: Deciding the word problem in the union of equational theories. Inf. Comput. **178**(2), 346–390 (2002)

14. Bachmair, L., Ganzinger, H.: Rewrite-based equational theorem proving with selection and simplification. J. Log. Comput. **4**(3), 217–247 (1994)
15. Bachmair, L., Ganzinger, H., Lynch, C., Snyder, W.: Basic paramodulation. Inf. Comput. **121**(2), 172–192 (1995)
16. Blanchet, B.: Modeling and verifying security protocols with the applied Pi calculus and ProVerif. Found. Trends Priv. Secur. **1**(1–2), 1–135 (2016)
17. Bonacina, M.P.: A taxonomy of theorem-proving strategies. In: Wooldridge, M.J., Veloso, M. (eds.) Artificial Intelligence Today. LNCS (LNAI), vol. 1600, pp. 43–84. Springer, Heidelberg (1999). https://doi.org/10.1007/3-540-48317-9_3
18. Borovanský, P., Kirchner, C., Kirchner, H., Moreau, P.: ELAN from a rewriting logic point of view. Theor. Comput. Sci. **285**(2), 155–185 (2002)
19. Boudet, A.: Combining unification algorithms. J. Symb. Comput. **16**(6), 597–626 (1993)
20. Clavel, M., et al. (eds.): All About Maude - A High-Performance Logical Framework. LNCS, vol. 4350. Springer, Heidelberg (2007). https://doi.org/10.1007/978-3-540-71999-1
21. Comon, H., Haberstrau, M., Jouannaud, J.P.: Syntacticness, cycle-syntacticness, and shallow theories. Inf. Comput. **111**(1), 154–191 (1994)
22. Comon-Lundh, H., Delaune, S.: The finite variant property: how to get rid of some algebraic properties. In: Giesl, J. (ed.) RTA 2005. LNCS, vol. 3467, pp. 294–307. Springer, Heidelberg (2005). https://doi.org/10.1007/978-3-540-32033-3_22
23. Domenjoud, E., Klay, F., Ringeissen, C.: Combination techniques for non-disjoint equational theories. In: Bundy, A. (ed.) CADE 1994. LNCS, vol. 814, pp. 267–281. Springer, Heidelberg (1994). https://doi.org/10.1007/3-540-58156-1_19
24. Eeralla, A.K., Erbatur, S., Marshall, A.M., Ringeissen, C.: Rule-based unification in combined theories and the finite variant property. In: Martín-Vide, C., Okhotin, A., Shapira, D. (eds.) LATA 2019. LNCS, vol. 11417, pp. 356–367. Springer, Cham (2019). https://doi.org/10.1007/978-3-030-13435-8_26
25. Erbatur, S., Kapur, D., Marshall, A.M., Narendran, P., Ringeissen, C.: Hierarchical combination. In: Bonacina, M.P. (ed.) CADE 2013. LNCS (LNAI), vol. 7898, pp. 249–266. Springer, Heidelberg (2013). https://doi.org/10.1007/978-3-642-38574-2_17
26. Erbatur, S., Kapur, D., Marshall, A.M., Narendran, P., Ringeissen, C.: Unification and matching in hierarchical combinations of syntactic theories. In: Lutz, C., Ranise, S. (eds.) FroCoS 2015. LNCS (LNAI), vol. 9322, pp. 291–306. Springer, Cham (2015). https://doi.org/10.1007/978-3-319-24246-0_18
27. Erbatur, S., Marshall, A.M., Ringeissen, C.: Notions of knowledge in combinations of theories sharing constructors. In: de Moura, L. (ed.) CADE 2017. LNCS (LNAI), vol. 10395, pp. 60–76. Springer, Cham (2017). https://doi.org/10.1007/978-3-319-63046-5_5
28. Escobar, S., Meadows, C., Meseguer, J.: Maude-NPA: cryptographic protocol analysis modulo equational properties. In: Aldini, A., Barthe, G., Gorrieri, R. (eds.) FOSAD 2007-2009. LNCS, vol. 5705, pp. 1–50. Springer, Heidelberg (2009). https://doi.org/10.1007/978-3-642-03829-7_1
29. Escobar, S., Sasse, R., Meseguer, J.: Folding variant narrowing and optimal variant termination. J. Log. Algebr. Program. **81**(7–8), 898–928 (2012)
30. Jouannaud, J.P., Kirchner, C.: Solving equations in abstract algebras: a rule-based survey of unification. In: Lassez, J.L., Plotkin, G. (eds.) Computational Logic. Essays in honor of Alan Robinson, chap. 8, pp. 257–321. MIT Press, Cambridge (1991)

31. Jouannaud, J.P., Kirchner, H.: Completion of a set of rules modulo a set of equations. SIAM J. Comput. **15**(4), 1155–1194 (1986)
32. Kapur, D., Narendran, P.: Complexity of unification problems with associative-commutative operators. J. Autom. Reason. **9**(2), 261–288 (1992)
33. Kapur, D., Narendran, P.: Double-exponential complexity of computing a complete set of AC-unifiers. In: Proceedings of the Seventh Annual Symposium on Logic in Computer Science (LICS 1992), Santa Cruz, California, USA, 22–25 June 1992, pp. 11–21. IEEE Computer Society (1992)
34. Kirchner, C., Klay, F.: Syntactic theories and unification. In: Proceedings of the Fifth Annual Symposium on Logic in Computer Science (LICS 1990), Philadelphia, Pennsylvania, USA, 4–7 June 1990, pp. 270–277. IEEE Computer Society (1990)
35. Kirchner, C., Ringeissen, C.: Rule-based constraint programming. Fundam. Inform. **34**(3), 225–262 (1998)
36. Lynch, C., Morawska, B.: Basic syntactic mutation. In: Voronkov, A. (ed.) CADE 2002. LNCS (LNAI), vol. 2392, pp. 471–485. Springer, Heidelberg (2002). https://doi.org/10.1007/3-540-45620-1_37
37. Meier, S., Schmidt, B., Cremers, C., Basin, D.: The TAMARIN prover for the symbolic analysis of security protocols. In: Sharygina, N., Veith, H. (eds.) CAV 2013. LNCS, vol. 8044, pp. 696–701. Springer, Heidelberg (2013). https://doi.org/10.1007/978-3-642-39799-8_48
38. Meseguer, J.: Variant-based satisfiability in initial algebras. Sci. Comput. Program. **154**, 3–41 (2018)
39. Moreau, P., Ringeissen, C., Vittek, M.: A pattern matching compiler for multiple target languages. In: Hedin, G. (ed.) CC 2003. LNCS, vol. 2622, pp. 61–76. Springer, Heidelberg (2003). https://doi.org/10.1007/3-540-36579-6_5
40. Nelson, G., Oppen, D.C.: Simplification by cooperating decision procedures. ACM Trans. Program. Lang. Syst. **1**(2), 245–257 (1979)
41. Nieuwenhuis, R., Rubio, A.: Paramodulation-based theorem proving. In: Robinson, J.A., Voronkov, A. (eds.) Handbook of Automated Reasoning (in 2 volumes), pp. 371–443. Elsevier and MIT Press (2001)
42. Nipkow, T.: Combining matching algorithms: the regular case. J. Symb. Comput. **12**(6), 633–654 (1991)
43. Nipkow, T.: Proof transformations for equational theories. In: Proceedings of the Fifth Annual Symposium on Logic in Computer Science (LICS 1990), Philadelphia, Pennsylvania, USA, 4–7 June 1990, pp. 278–288. IEEE Computer Society (1990)
44. Otop, J.: E-unification with constants vs. general E-unification. J. Autom. Reason. **48**(3), 363–390 (2012)
45. Pigozzi, D.: The joint of equational theories. In: Colloquium Mathematicum, pp. 15–25 (1974)
46. Ringeissen, C.: Combining decision algorithms for matching in the union of disjoint equational theories. Inf. Comput. **126**(2), 144–160 (1996)
47. Ringeissen, C.: Matching with free function symbols—a simple extension of matching? In: Middeldorp, A. (ed.) RTA 2001. LNCS, vol. 2051, pp. 276–290. Springer, Heidelberg (2001). https://doi.org/10.1007/3-540-45127-7_21
48. Ringeissen, C.: Matching in a class of combined non-disjoint theories. In: Baader, F. (ed.) CADE 2003. LNCS (LNAI), vol. 2741, pp. 212–227. Springer, Heidelberg (2003). https://doi.org/10.1007/978-3-540-45085-6_17
49. Schmidt-Schauß, M.: Unification in a combination of arbitrary disjoint equational theories. J. Symb. Comput. **8**, 51–99 (1989)
50. Schmidt-Schauß, M.: Unification in permutative equational theories is undecidable. J. Symb. Comput. **8**(4), 415–421 (1989)

51. Tidén, E.: Unification in combinations of collapse-free theories with disjoint sets of function symbols. In: Siekmann, J.H. (ed.) CADE 1986. LNCS, vol. 230, pp. 431–449. Springer, Heidelberg (1986). https://doi.org/10.1007/3-540-16780-3_110
52. Tinelli, C., Ringeissen, C.: Unions of non-disjoint theories and combinations of satisfiability procedures. Theor. Comput. Sci. **290**(1), 291–353 (2003)
53. Yelick, K.A.: Unification in combinations of collapse-free regular theories. J. Symb. Comput. **3**(1/2), 153–181 (1987)

Presburger Concept Cardinality Constraints in Very Expressive Description Logics

Allegro sexagenarioso ma non ritardando

"Johann" Sebastian Rudolph$^{(\boxtimes)}$ (iD)

Computational Logic Group, TU Dresden, Dresden, Germany
sebastian.rudolph@tu-dresden.de

Abstract. Quantitative information plays an increasingly important role in knowledge representation. To this end, many formalisms have been proposed that enrich traditional KR formalisms by counting and some sort of arithmetics. Baader and Ecke (2017) propose an extension of the description logic $\mathcal{ALC}$ by axioms which express correspondences between the cardinalities of concepts by means of Presburger arithmetics. This paper extends their results, enhancing the expressivity of the underlying logic as well as the constraints while preserving complexities. It also widens the scope of investigation from finite models to the classical semantics where infinite models are allowed. We also provide first results on query entailment in such logics. As opposed to prior work, our results are established by polynomially encoding the cardinality constraints in the underlying logic.

Keywords: F♯ Major · B Major

Prelude: Dedication

This work is inspired by Franz. It was him who introduced me to the idea of extending the quantitative capabilities in description logics using Presburger arithmetic constraints. Quite fittingly, this happened during a colloquial discussion at a workshop of our DFG PhD training group QuantLA – *Quantitative Logics and Automata*. Already back then, I was musing that it should be possible to encode certain cardinality constraints in expressive description logics, using some auxiliary vocabulary.

Thankfully, this festschrift provided the most suitable occasion, motivation, and pressure[1] to explore that option thoroughly, arguably with quite decent results. I'd like to dedicate this "composition" to Franz and thank him for bringing up this interesting topic and, generally, for plenty of inspiring discussions, occupational support, and pleasant collaboration.

---

[1] I hereby sincerely apologize to Anni, Carsten, Cesare, Frank, and Uli for pushing the submission deadline.

© Springer Nature Switzerland AG 2019
C. Lutz et al. (Eds.): Baader Festschrift, LNCS 11560, pp. 542–561, 2019.
https://doi.org/10.1007/978-3-030-22102-7_25

1 Introduction: Toward Quantitative Description Logics

Enriching knowledge representation formalisms with features for counting and basic arithmetic regarding domain individuals is a worthwhile endeavor. So far, mainstream description logics provide only very limited support in this respect: *(qualified) number restrictions* allow for enforcing concrete upper and lower bounds on the number of an individual's role neighbors. These limited capabilities fall short of some basic practical knowledge representation requirements, such as expressing statistical information [21]. As an example, assume that on the occasion of a distinguished scientist's 60th birthday, fellow researchers group together to produce a festschrift consisting of distinct contributed papers. The publisher requires the festschrift to have not less than 140 and not more than 800 pages. This could intuitively be expressed by a statement like

$$140 \leq |\mathsf{Page}| \leq 800,$$

assuming Page denotes the class of all the festschrift's pages. Let's say, the editors manage to recruit a total of 73 authors:

$$|\mathsf{Author}| = 73.$$

They assume that the average number of contributors per paper is between 2 and 3:

$$2 \cdot |\mathsf{Paper}| \leq |\mathsf{Author}| \leq 3 \cdot |\mathsf{Paper}|.$$

They impose the condition that each paper must have at least 10 and at most 40 pages:

$$\mathsf{Paper} \sqsubseteq {\geqslant} 10\, \mathsf{OnPage}.\top \sqcap {\leqslant} 40\, \mathsf{OnPage}.\top.$$

They also notice that just one author (the "outlier") contributes to two papers, whereas the others contribute to one.

$$\mathsf{Author} \sqcap \neg\{\mathsf{outlier}\} \sqsubseteq {=}1\, \mathsf{Contributes}.\top \qquad \{\mathsf{outlier}\} \sqsubseteq {=}2\, \mathsf{Contributes}.\top$$

Some background knowledge for our domain needs to be specified: roughly speaking, authors are precisely contributors, papers are precisely "contributees" and they are precisely the things occurring on pages and on nothing but pages.

$$\mathsf{Author} \equiv \exists\mathsf{Contributes}.\top \quad \mathsf{Paper} \equiv \exists\mathsf{Contributes}^-.\top$$
$$\mathsf{Paper} \equiv \exists\mathsf{OnPage}.\top \qquad \mathsf{Page} \equiv \exists\mathsf{OnPage}^-.\top$$

As an important last ingredient, it needs to be specified that no two distinct papers can occur on the same page (i.e., OnPage is inverse functional):

$$\top \sqsubseteq {\leqslant} 1\, \mathsf{OnPage}^-.\top.$$

If a reasoner supporting all the modeling features in this specification existed, the editors could now find out by a satisfiability check if the planned festschrift can be published under the given assumptions (which is the case). Also, by removing the first statement and checking for its entailment instead, they could find out if the publisher's space constraints are guaranteed to be met in view of the given information (which is not the case). Alas, currently, axioms alike the first three statements are not supported by mainstream description logics.

In a line of recent work, Franz and others have addressed the shortcomings in description logics on the quantitative side. For instance, extending results from [12], Franz proposed $\mathcal{ALCSCC}$ [1], an extension of the basic description logic $\mathcal{ALC}$ by constraints expressed in the quantifier-free fragment of Boolean Algebra with Presburger Arithmetic (QFBAPA) [19] over role successors. The described constraints are *local*, as they always refer to an individual under consideration, as opposed to *global* constraints, which range over the full domain and compare cardinalities of concepts. The latter were introduced in [5], giving rise to the notions of $\mathcal{ALC}$ extended cardinality boxes (ECboxes) and – striving for more favorable complexity results – their "light version" $\mathcal{ALC}$ restricted cardinality boxes (RCboxes). As a natural next step, [2] introduced and investigated $\mathcal{ALCSCC}$ ECboxes and RCboxes, enabling both local and global cardinality constraints in a joint formalism. Pushing the envelope further, [3] showed that local and global constraints can be tightly integrated leading to the starkly more expressive logic $\mathcal{ALCSCC}^{++}$, for which ECbox consistency checking is still NExpTime-complete. On the downside, conjunctive query entailment becomes undecidable in this logic. Moreover, as an (albeit massively calculation-enhanced) version of plain $\mathcal{ALC}$, $\mathcal{ALCSCC}^{++}$ is lacking basic modeling features that are normally taken for granted in description logics. Most notably, it does not feature role inverses, which are crucial to draw level with popular logics from other families, such as two-variable logics.

Decidability and complexity results for the logics discussed above were established via the solution of large systems of (in)equalities as well as elaborate constructions and transformations of models. We show that, if we limit our attention to global cardinality constraints, we can use an alternative, reduction-based approach, and expand the existing results simultaneously in three directions:

- *Incorporation of role inverses.* As stated earlier, the existing results are for description logics without the feature of role inverses. In fact, it is notoriously difficult to incorporate this feature into the (in)equality-system-based machinery hitherto used. Interestingly, the method proposed in this paper not only allows for incorporating role inverses, it actually does require their presence in the logic.

- *Relaxation of restrictions on RCboxes.* As mentioned above, RCboxes were introduced as a light version of ECboxes in order to obtain more favorable complexity results. We show that some of the restrictions made can be relaxed without endangering this complexity gain. This actually motivates us to introduce *extended restricted cardinality boxes* (ERCboxes) as low-complexity, high-expressivity middle-ground between ECboxes and RCboxes.

- *Reasoning over finite and arbitrary models.* Previous results confine themselves to a finite-model setting, which is arguably the right choice for practical modeling tasks in concrete scenarios where arithmetic is applied. However, the traditional semantics of description logics allows for infinite models. Hence we extend the scope of our investigations to also include the case of arbitrary models. Next to an appropriate extension of the underlying arithmetic (as described in the beginning of Section 2) this raises some deeper model-theoretic concerns, which we address in Section 5.

2 Ostinato: Preliminaries

Numbers. We recall that $\mathbb{N}$ denotes the set of natural numbers (including 0). Throughout this paper, whenever natural numbers occur in some expression, we assume binary encoding. We let $\mathbb{N}^\infty = \mathbb{N} \cup \{\infty\}$. Basic arithmetic and comparison operations are extended from $\mathbb{N}$ to $\mathbb{N}^\infty$ in the straightforward way, in particular adding anything to ∞ yields ∞, $0 \cdot \infty = 0$, and $n \cdot \infty = \infty$ for every $n \geq 1$. For $n \in \mathbb{N}^\infty$, we let $[n] = \{i \mid i < n\}$, in particular $[\infty] = \mathbb{N}$. For some set S, we let $|S|$ denote the number of elements of S if it is finite and ∞ otherwise.

Description Logics. We give the definition of the extremely expressive description logic $\mathcal{SROIQB}$ which is obtained from the well-known description logic $\mathcal{SROIQ}$ [16] by allowing arbitrary Boolean constructors on simple roles. We assume that the reader is familiar with description logics [4, 6, 26].

The description logics considered in this paper are based on four disjoint sets of *individual names* $\mathbf{N_I}$, *concept names* $\mathbf{N_C}$, *simple role names* $\mathbf{N_R^s}$, and *non-simple role names* $\mathbf{N_R^n}$ (containing the *universal role* $\overline{\top} \in \mathbf{N_R}$). Furthermore, we let $\mathbf{N_R} := \mathbf{N_R^s} \cup \mathbf{N_R^n}$.

Definition 1 (syntax of $\mathcal{SROIQB}$). *A $\mathcal{SROIQB}$ Rbox for $\mathbf{N_R}$ is based on a set $\mathbf{R}$ of atomic roles defined as $\mathbf{R} := \mathbf{N_R} \cup \{R^- \mid R \in \mathbf{N_R}\}$, where we set $\mathrm{Inv}(R) := R^-$ and $\mathrm{Inv}(R^-) := R$ to simplify notation. In turn, we distinguish simple atomic roles $\mathbf{R^s} := \mathbf{N_R^s} \cup \mathrm{Inv}(\mathbf{N_R^s})$ and non-simple roles $\mathbf{R^n} := \mathbf{N_R^n} \cup \mathrm{Inv}(\mathbf{N_R^n})$.*
The set of simple roles $\mathbf{B}$ is defined as follows:

$$\mathbf{B} ::= \mathbf{N_R^s} \mid \neg\mathbf{B} \mid \mathbf{B} \cap \mathbf{B} \mid \mathbf{B} \cup \mathbf{B} \mid \mathbf{B} \setminus \mathbf{B}.$$

Moreover, a simple role will be called safe, *if it does not contain $\neg$.*
A generalized role inclusion axiom (RIA) is a statement of the form $S \sqsubseteq R$ with simple roles S and R, or of the form

$$S_1 \circ \ldots \circ S_n \sqsubseteq R$$

where each S_i is a (simple or non-simple) role, and where R is a non-simple atomic role, none of them being $\overline{\top}$. A set of such RIAs will be called a generalized role hierarchy. A role hierarchy is regular if there is a strict partial order $\prec$ on the non-simple roles $\mathbf{R^n}$ such that

- $S \prec R$ *iff* $\mathrm{Inv}(S) \prec R$, *and*
- *every RIA is of one of the forms*

$$R \circ R \sqsubseteq R \quad R^- \sqsubseteq R \quad S_1 \circ \ldots \circ S_n \sqsubseteq R \quad R \circ S_1 \circ \ldots \circ S_n \sqsubseteq R \quad S_1 \circ \ldots \circ S_n \circ R \sqsubseteq R$$

such that $R \in \mathbf{N_R}$ is a (non-inverse) role name, and $S_i \prec R$ for $i = 1, \ldots, n$ whenever S_i is non-simple.
A $\mathcal{SROIQB}$ Rbox is a regular role hierarchy.[2]

---

[2] The original definition of $\mathcal{SROIQ}$ Rboxes also features explicit axioms expressing role reflexivity, asymmetry, and role disjointness. However, in the presence of (safe) Boolean role constructors, these can be expressed, so we omit them here.

Table 1. Semantics of $\mathcal{SROIQB}$ role and concept constructors for interpretation $\mathcal{I} = (\Delta^{\mathcal{I}}, \cdot^{\mathcal{I}})$.

Name	Syntax	Semantics
inverse role	R^-	$\{(x,y) \in \Delta^{\mathcal{I}} \times \Delta^{\mathcal{I}} \mid (y,x) \in R^{\mathcal{I}}\}$
universal role	$\top$	$\Delta^{\mathcal{I}} \times \Delta^{\mathcal{I}}$
role negation	$\neg S$	$\{(x,y) \in \Delta^{\mathcal{I}} \times \Delta^{\mathcal{I}} \mid (x,y) \notin R^{\mathcal{I}}\}$
role conjunction	$S \cap R$	$S^{\mathcal{I}} \cap R^{\mathcal{I}}$
role disjunction	$S \cup R$	$S^{\mathcal{I}} \cup R^{\mathcal{I}}$
role difference	$S \setminus R$	$S^{\mathcal{I}} \setminus R^{\mathcal{I}}$
top	$\top$	$\Delta^{\mathcal{I}}$
bottom	$\bot$	$\emptyset$
negation	$\neg C$	$\Delta^{\mathcal{I}} \setminus C^{\mathcal{I}}$
conjunction	$C \sqcap D$	$C^{\mathcal{I}} \cap D^{\mathcal{I}}$
disjunction	$C \sqcup D$	$C^{\mathcal{I}} \cup D^{\mathcal{I}}$
nominals	$\{a\}$	$\{a^{\mathcal{I}}\}$
univ. restriction	$\forall R.C$	$\{x \in \Delta^{\mathcal{I}} \mid (x,y) \in R^{\mathcal{I}} \text{ implies } y \in C^{\mathcal{I}}\}$
exist. restriction	$\exists R.C$	$\{x \in \Delta^{\mathcal{I}} \mid \text{ for some } y \in \Delta^{\mathcal{I}}, (x,y) \in R^{\mathcal{I}} \text{ and } y \in C^{\mathcal{I}}\}$
Self concept	$\exists S.\text{Self}$	$\{x \in \Delta^{\mathcal{I}} \mid (x,x) \in S^{\mathcal{I}}\}$
qualified number	$\leqslant n\,S.C$	$\{x \in \Delta^{\mathcal{I}} \mid \lvert\{y \in \Delta^{\mathcal{I}} \mid (x,y) \in S^{\mathcal{I}} \text{ and } y \in C^{\mathcal{I}}\}\rvert \leq n\}$
restriction	$\geqslant n\,S.C$	$\{x \in \Delta^{\mathcal{I}} \mid \lvert\{y \in \Delta^{\mathcal{I}} \mid (x,y) \in S^{\mathcal{I}} \text{ and } y \in C^{\mathcal{I}}\}\rvert \geq n\}$

Given a $\mathcal{SROIQB}$ Rbox $\mathcal{R}$, the set of concept expressions *(short: concepts)* **C** *is inductively defined as follows:*

- $\mathbf{N_C} \subseteq \mathbf{C}, \top \in \mathbf{C}, \bot \in \mathbf{C}$,
- *for $C, D \in \mathbf{C}$ concepts, $R \in \mathbf{B} \cup \mathbf{R}^n$ a (simple or non-simple) role, $S \in \mathbf{B}$ a simple role, $\mathsf{a} \in \mathbf{N_I}$, and $n \in \mathbb{N}$ a non-negative integer, the expressions $\neg C$, $C \sqcap D$, $C \sqcup D$, $\{\mathsf{a}\}$, $\forall R.C$, $\exists R.C$, $\exists S.\text{Self}$, $\leqslant n\,S.C$, and $\geqslant n\,S.C$ are also concepts.*

Throughout this paper, the symbols C, D will be used to denote concepts. A $\mathcal{SROIQB}$ Tbox is a set of general concept inclusion axioms *(GCIs) of the form $C \sqsubseteq D$. We use $C \equiv D$ as a shorthand for $C \sqsubseteq D$, $D \sqsubseteq C$.*

An individual assertion *can have any of the following forms: $C(\mathsf{a})$, $R(\mathsf{a},\mathsf{b})$, $\neg S(\mathsf{a},\mathsf{b})$, $\mathsf{a} \approx \mathsf{b}$, $\mathsf{a} \not\approx \mathsf{b}$, with $\mathsf{a},\mathsf{b} \in \mathbf{N_I}$ individual names, $C \in \mathbf{C}$ a concept, and $R, S \in \mathbf{B} \cup \mathbf{R}^n$ roles with S simple. A $\mathcal{SROIQB}$ Abox is a set of individual assertions.*

A $\mathcal{SROIQB}$ knowledge base $\mathcal{K}$ is a triple $(\mathcal{A}, \mathcal{T}, \mathcal{R})$ where $\mathcal{R}$ is a regular Rbox while $\mathcal{A}$ and $\mathcal{T}$ are an Abox and a Tbox for $\mathcal{R}$, respectively. We use the term axiom *to uniformly refer to any single statement contained in $\mathcal{A}$, $\mathcal{T}$, or $\mathcal{R}$.*

We further provide the semantics of $\mathcal{SROIQB}$ knowledge bases.

Definition 2 (semantics of $\mathcal{SROIQB}$). *An interpretation $\mathcal{I} = (\Delta^{\mathcal{I}}, \cdot^{\mathcal{I}})$ consists of a set $\Delta^{\mathcal{I}}$ called* domain *together with a function $\cdot^{\mathcal{I}}$ mapping individual names to elements of $\Delta^{\mathcal{I}}$, concept names to subsets of $\Delta^{\mathcal{I}}$, and role names to subsets of $\Delta^{\mathcal{I}} \times \Delta^{\mathcal{I}}$.*

The function $\cdot^{\mathcal{I}}$ is inductively extended to roles and concepts as shown in Table 1. An interpretation $\mathcal{I}$ satisfies an axiom φ (written: $\mathcal{I} \models \varphi$) if the respective condition is satisfied:

- $\mathcal{I} \models S \sqsubseteq R$ *if* $S^{\mathcal{I}} \subseteq R^{\mathcal{I}}$,
- $\mathcal{I} \models S_1 \circ \ldots \circ S_n \sqsubseteq R$ *if* $S_1^{\mathcal{I}} \circ \ldots \circ S_n^{\mathcal{I}} \sqsubseteq R^{\mathcal{I}}$ (*$\circ$ being overloaded to denote the standard composition of binary relations here*),
- $\mathcal{I} \models C \sqsubseteq D$ *if* $C^{\mathcal{I}} \subseteq D^{\mathcal{I}}$,
- $\mathcal{I} \models C(\mathsf{a})$ *if* $\mathsf{a}^{\mathcal{I}} \in C^{\mathcal{I}}$,
- $\mathcal{I} \models R(\mathsf{a}, \mathsf{b})$ *if* $(\mathsf{a}^{\mathcal{I}}, \mathsf{b}^{\mathcal{I}}) \in R^{\mathcal{I}}$,
- $\mathcal{I} \models \neg S(\mathsf{a}, \mathsf{b})$ *if* $(\mathsf{a}^{\mathcal{I}}, \mathsf{b}^{\mathcal{I}}) \notin S^{\mathcal{I}}$,
- $\mathcal{I} \models \mathsf{a} \approx \mathsf{b}$ *if* $\mathsf{a}^{\mathcal{I}} = \mathsf{b}^{\mathcal{I}}$,
- $\mathcal{I} \models \mathsf{a} \not\approx \mathsf{b}$ *if* $\mathsf{a}^{\mathcal{I}} \neq \mathsf{b}^{\mathcal{I}}$.

An interpretation $\mathcal{I}$ satisfies *a knowledge base* $\mathcal{K}$ (*we then also say that* $\mathcal{I}$ *is a* model *of* $\mathcal{K}$ *and write* $\mathcal{I} \models \mathcal{K}$) *if it satisfies all axioms of* $\mathcal{K}$. *A knowledge base* $\mathcal{K}$ *is* (finitely) satisfiable *if it has a (finite) model. Two knowledge bases are* equivalent *if they have exactly the same models. They are* (finitely) equisatisfiable *if either both are (finitely) unsatisfiable or both are (finitely) satisfiable.*

The description logic $\mathcal{SHOIQB}$ is obtained from $\mathcal{SROIQB}$ by discarding the universal role $\overline{\top}$ as well as the Self concept and allowing only RIAs of the form $R \sqsubseteq S$ or $R \circ R \sqsubseteq R$. If we also disallow $R \circ R \sqsubseteq R$, we obtain $\mathcal{ALCHOIQB}$. For any of these three logics, replacing $\mathcal{B}$ in the name by b disallows role negation (but preserves role difference) while removing $\mathcal{B}$ entirely also disallows role conjunction, disjunction and difference. Dropping $\mathcal{O}$ from any description logic's name disables nominal concepts $\{o\}$, while dropping $\mathcal{I}$ disables role inverses $\cdot^{-}$, and dropping $\mathcal{H}$ disables RIAs of the form $R \sqsubseteq S$. For any description logic $\mathcal{L}$ that does not feature the Self concept (the universal role $\overline{\top}$), we denote by $\mathcal{L}^{\mathsf{Self}}$ (by $\mathcal{L}_{\overline{\top}}$) the logic with this feature added.

Queries.   In queries, we use *variables* from a countably infinite set $\mathbf{V}$. A Boolean *positive two-way regular path query* (P2RPQ) is a formula $\exists \boldsymbol{x}.\varphi$, where φ is a positive Boolean expression (i.e., one using only $\wedge$ and $\vee$) over atoms of the form $C(t)$ or $T(s,t)$, where s and t are elements of $\boldsymbol{x} \cup \mathbf{N_I}$, C is a concept, and T is a *regular role expression* from $\mathbf{T}$, defined by

$$\mathbf{T} ::= \mathbf{R} \mid \mathbf{T} \cup \mathbf{T} \mid \mathbf{T} \circ \mathbf{T} \mid \mathbf{T}^* \mid id(\mathbf{C}).$$

If q does not use disjunction and all T are simple roles, it is called a *conjunctive query* (CQ). A *variable assignment* π for $\mathcal{I}$ is a mapping $\mathbf{V} \to \Delta^{\mathcal{I}}$. For $x \in \mathbf{V}$, we set $x^{\mathcal{I},\pi} := \pi(x)$; for $\mathsf{c} \in \mathbf{N_I}$, we set $\mathsf{c}^{\mathcal{I},\pi} := \mathsf{c}^{\mathcal{I}}$. $T(s,t)$ evaluates to true under π and $\mathcal{I}$ if $(s^{\mathcal{I},\pi}, t^{\mathcal{I},\pi}) \in T^{\mathcal{I}}$, with $T^{\mathcal{I}}$ obtained as detailed in Table 2. $C(t)$ evaluates to true under π and $\mathcal{I}$ if $t^{\mathcal{I},\pi} \in C^{\mathcal{I}}$. A P2RPQ $q = \exists \boldsymbol{x}.\varphi$ is *satisfied* by $\mathcal{I}$ (written: $\mathcal{I} \models q$) if there is a variable assignment π (called *match*) such that φ evaluates to true under $\mathcal{I}$ and π. A P2RPQ q is *(finitely) entailed* from a KB $\mathcal{K}$ if every (finite) model of $\mathcal{K}$ satisfies q.

3   Subject: Extending Knowledge Bases by Presburger-Style Concept Cardinality Constraints

In this section, we introduce extended cardinality boxes (and several restricted versions thereof) as means for expressing quantitative global knowledge.

Table 2. Semantics of regular role expressions for interpretation $\mathcal{I} = (\Delta^{\mathcal{I}}, \cdot^{\mathcal{I}})$.

Name	Syntax	Semantics
union	$T_1 \cup T_2$	$T_1^{\mathcal{I}} \cup T_2^{\mathcal{I}}$
concatenation	$T_1 \circ T_2$	$T_1^{\mathcal{I}} \circ T_2^{\mathcal{I}}$
Kleene star	T^*	$\bigcup_{i \geq 0}(T^{\mathcal{I}})^i$
concept test	$id(C)$	$\{(x,x) \mid x \in C^{\mathcal{I}}\}$

Definition 3 (concept cardinality constraint, ECbox, RCbox, ERCbox). *A* concept cardinality constraint *(short: constraint)* c *is an expression of the form*

$$n_0 + n_1|\mathsf{A}_1| + \ldots + n_k|\mathsf{A}_k| \leq m_0 + m_1|\mathsf{B}_1| + \ldots + m_\ell|\mathsf{B}_\ell|, \tag{1}$$

where $\mathsf{A}_1, \ldots, \mathsf{A}_k, \mathsf{B}_1, \ldots, \mathsf{B}_\ell$ *are concept names and all* n_i *and* m_i *are natural numbers. A concept cardinality constraint is* restricted *if* $n_0 = m_0 = 0$ *and* semi-restricted *if* $m_0 = 0$. *An* extended cardinality box *(ECbox) is a positive Boolean combination of concept cardinality constraints. A* restricted cardinality box *(RCbox) is a conjunction of restricted cardinality constraints. An* extended restricted cardinality box *(ERCbox) is a positive Boolean combination of semi-restricted cardinality constraints.*
Satisfaction of a concept cardinality constraint c *by an interpretation* $\mathcal{I}$ *(written as* $\mathcal{I} \models$ c*) is verified as follows: every expression* $|\mathsf{A}|$ *is mapped to* $|\mathsf{A}^{\mathcal{I}}|$. *The constraint is evaluated in the straightforward way over* $\mathbb{N}^\infty$. *Satisfaction of constraints is then lifted to satisfaction of ECboxes in the obvious manner.*

Definition 4 (EKB, ERKB, RKB). *For some description logic* $\mathcal{L}$, *an* extended $\mathcal{L}$ knowledge base *($\mathcal{L}$ EKB) is a quadruple* $(\mathcal{A}, \mathcal{T}, \mathcal{R}, \mathcal{E})$ *where* $(\mathcal{A}, \mathcal{T}, \mathcal{R})$ *is an* $\mathcal{L}$ *knowledge base and* $\mathcal{E}$ *is an ECbox. An EKB is a* restricted knowledge base *(RKB) if* $\mathcal{E}$ *is an RCbox. It is an* extended restricted knowledge base *(ERKB) if* $\mathcal{E}$ *is an ERCbox.*

Obviously, RKBs (RCboxes) are properly subsumed by ERKBs (ERCboxes) which in turn are properly subsumed by EKBs (ECboxes). One general insight of this paper is that upper complexity bounds persist when generalizing the previously defined RCboxes to the newly defined, more expressive ERCboxes.

Our syntactic formulation of ECboxes is somewhat more restrictive than that in prior work [5], but we will show that the differences are immaterial. Using our more restricted form allows for a more uniform presentation of our results.

First, the original work allows expressions $|C|$ for arbitrary concept descriptions C. We note that our definition does not restrict expressivity since general Tboxes allow for axioms $\mathsf{A} \equiv C$, so complex concept expressions in cardinality constraints can be replaced by fresh concept names and defined in the Tbox. Resorting to plain concept names in constraints allows us to consider cardinality boxes uniformly independently from the used description logic.

Second, instead of positive weighted sums of concept cardinalities as left and right hand sides, the original work allows for arbitrary functions built from integers z and expressions of the form $|\mathsf{A}|$ using functions $+$ (binary) and $z\cdot$ (unary). It is, however, easy to see that each comparison on such more liberal expressions can be polynomially translated into an equivalent comparison of positive weighted sums.

Third, the original work allows for extended cardinality constraints using other modes of comparison than just "$\leq$": $\alpha = \beta$, $\alpha < \beta$, or n dvd α. However, all these constraints can be rewritten into (combinations of) constraints only using "$\leq$" as follows: $\alpha = \beta$ is replaced by $(\alpha \leq \beta) \wedge (\beta \leq \alpha)$, $\alpha < \beta$ by $\alpha+1 \leq \beta$,[3] and n dvd α can be rewritten into $(n|\mathsf{A}| \leq \alpha) \wedge (\alpha \leq n|\mathsf{A}|)$ for some fresh concept name A.

Fourth, the original work defined ECboxes as arbitrary (not just positive) Boolean combinations of constraints. However, this work only considered finite models. We note that under this assumption, negated constraints can be rewritten into negation-free (combinations of) constraints in the following way: replace $\neg(\alpha \leq \beta)$ by $\beta+1 \leq \alpha$, $\neg(\alpha < \beta)$ by $\beta \leq \alpha$, $\neg(\alpha = \beta)$ by $(\alpha+1 \leq \beta) \vee (\beta+1 \leq \alpha)$, and $\neg(n$ dvd $\alpha)$ by $(n|\mathsf{A}| + |\mathsf{B}| \leq \alpha) \wedge (\alpha \leq n|\mathsf{A}| + |\mathsf{B}|) \wedge (1 \leq |\mathsf{B}|) \wedge (|\mathsf{B}| + 1 \leq n)$ for fresh concept names A and B. Note again, that this rewriting is not equivalent for infinite models.[4]

As observed before, ECboxes allow to express nominal concepts by enforcing that a concept must have cardinality exactly one. However, this is not possible with RCboxes nor ERCboxes.

4   Exposition: Statement of Results

With the notion of ECboxes, RCboxes and ERCboxes in place, we can now formally state results that can be derived from prior work before giving an outlook on the results established in this paper. We first note some results that can be obtained as easy consequences of previous publications.

- Finite satisfiability of Abox-free $\mathcal{SHQ}b$ RKBs is in ExpTime. For $\mathcal{ALCHQ}b$, this is an immediate consequence of earlier work on $\mathcal{ALCSCC}$ RCboxes [2]. Adding transitivity is possible since it can be handled via the classical "box pushing" approach [33, 29].

- Finite satisfiability of $\mathcal{SHOQB}$ EKBs is in NExpTime. This follows from [3] together with the observation that the logic $\mathcal{ALCSCC}^{++}$ considered there allows to express qualified number restrictions, nominals (hence also Aboxes), and arbitrary Boolean role expressions. Again, transitivity can be dealt with via "box pushing".

- Finite CQ entailment over Abox-free $\mathcal{ALCHQ}b$ RKBs is in 2ExpTime, as immediate consequence of the corresponding result for $\mathcal{ALCSCC}$ RCboxes in [3].

At the core of our method is the insight that expressive description logics in and of themselves hold enough expressive means to simulate ECboxes without noteworthy blow-up. We will show that:

---

[3] For this, we have to postulate $\infty < \infty$, which is debatable, but could be justified by the fact that there is an injective, non-surjective mapping between any two countably infinite sets.

[4] In fact, the constraint expression $(1 + |\mathsf{A}| = |\mathsf{A}|) \wedge \neg(|\mathsf{A}| = |\mathsf{B}|)$ would enforce finiteness of the extension of B, which is not axiomatizable in first order logic, neither finitely nor infinitely. For good reasons (see Section 5) we define ECboxes in a way that a first-order axiomatization is still possible.

(A) ERCboxes can be succinctly simulated in any description logic that can express $\mathcal{ALCIQ}_{\top}$ GCIs and

(B) ECboxes can be succinctly simulated in any description logic that can express $\mathcal{ALCOIQ}$ GCIs.

This "simulation", made formally precise in Section 6, is sufficiently authentic for both satisfiability checking and query entailment. Consequently, we are able to significantly strengthen the aforementioned results as follows (further detailed in Section 7):

- Satisfiability and finite satisfiability of $\mathcal{SHIQb}_{\top}^{\mathsf{Self}}$ ERKBs is ExpTime-complete.
- Satisfiability and finite satisfiability of $\mathcal{SHOIQB}^{\mathsf{Self}}$ EKBs is NExpTime-complete.
- Entailment of P2RPQs as well as finite entailment of CQs from $\mathcal{ALCHIQb}_{\top}^{\mathsf{Self}}$ ERKBs are 2ExpTime-complete.
- Entailment of unions of conjunctive queries from $\mathcal{ALCHOIQb}$ EKBs is decidable and coN2ExpTime-hard.

Yet, before going into the details of our translation, we have to take care of a nuisance, arising from counting in the presence of infinity.

5 Interlude: Countability

Dealing with infinity can be tricky [7,9,10,27]. In the general case, allowing for infinite models might require us to account for the presence of several distinct infinite cardinalities. In the realms of first-order logic, the Löwenheim-Skolem Theorem [31] ensures that it suffices to consider models of countable cardinality, in which only one type of infinity can occur.

In our setting, however, expressibility in first-order logic cannot be easily taken for granted. In fact, even the very simple RCbox $(|A| \leq |B|) \wedge (|B| \leq |A|)$, stating that A and B contain the same number of individuals, cannot be expressed using a first-order sentence (as can be shown by an easy argument using Ehrenfeucht-Fraïssé games).

We manage to resolve the issue by showing that ECboxes can be expressed by countable (but possibly infinite) first-order theories, noting that Löwenheim-Skolem still applies in this case.

Lemma 1. *Let $\mathcal{E}$ be an ECbox. Then there exists a countable first-order theory $\Phi_{\mathcal{E}}$ logically equivalent to $\mathcal{E}$.*

Proof. We construct $\Phi_{\mathcal{E}}$ from $\mathcal{E}$. For convenience, we introduce some notation: Given a concept name A and a number $n \in \mathbb{N}$, we let $\boldsymbol{f}(|A| \geq n)$ denote the first-order sentence

$$\exists x_1, \ldots x_n. \bigwedge_{1 \leq i \leq n} A(x_i) \wedge \bigwedge_{1 \leq i < j \leq n} x_i \neq x_j,$$

and we let $\boldsymbol{f}(|A| \leq n)$ denote the first-order sentence

$$\forall x_0, x_1, \ldots x_n. \left(\bigwedge_{0 \leq i \leq n} A(x_i) \right) \to \bigvee_{0 \leq i < j \leq n} x_i = x_j.$$

Note that the first-order sentences have precisely the intended meaning. Now, considering some cardinality constraint $\mathfrak{c}$ of the form

$$n_0 + n_1 |A_1| + \ldots + n_k |A_k| \leq m_0 + m_1 |B_1| + \ldots + m_\ell |B_\ell|,$$

we let $Bad_{\mathfrak{c}}$ denote the (most likely infinite) set of first-order sentences

$$\left\{ \bigwedge_{1 \leq i \leq k} \boldsymbol{f}(|A_i| \geq a_i) \wedge \bigwedge_{1 \leq i \leq \ell} \boldsymbol{f}(|B_i| \leq b_i) \;\middle|\; n_0 + \sum_{1 \leq i \leq k} n_i a_i > m_0 + \sum_{1 \leq i \leq \ell} m_i b_i \right\},$$

and note that $\mathcal{I} \models \mathfrak{c}$ if and only if $\mathcal{I} \not\models \varphi$ holds for all $\varphi \in Bad_{\mathfrak{c}}$. Next, let $\mathfrak{C}$ denote the set of all constraints occurring in $\mathcal{E}$. We let $\mathfrak{Bad}$ consist of all sets $\mathfrak{D} \subseteq \mathfrak{C}$ for which the Boolean expression obtained from $\mathcal{E}$ by replacing all $\mathfrak{c} \in \mathfrak{D}$ with **false** and all $\mathfrak{c} \in \mathfrak{C} \setminus \mathfrak{D}$ with **true** evaluates to **false**. Finally, we let $\Phi_{\mathcal{E}}$ consist of all sentences $\neg(\varphi_1 \wedge \ldots \wedge \varphi_m)$ for which there is some $\{\mathfrak{c}_1, \ldots, \mathfrak{c}_m\} \in \mathfrak{Bad}$ such that $\varphi_i \in Bad_{\mathfrak{c}_i}$ for $1 \leq i \leq m$. ◠

As planned, we can now use this insight to make sure that even in the presence of ECboxes, we can restrict our attention to countable models, as long as the rest of the knowledge base is expressible in first-order logic.

Theorem 2. *Let $\mathcal{L}$ be a description logic such that any $\mathcal{L}$ knowledge base $(\mathcal{A}, \mathcal{T}, \mathcal{R})$ is equivalent to a countable first-order logic theory $\Psi_{(\mathcal{A}, \mathcal{T}, \mathcal{R})}$.*

1. *Every satisfiable $\mathcal{L}$ EKB has a countable model.*
2. *An $\mathcal{L}$ EKB $\mathcal{K}$ entails a P2RPQ q iff q is satisfied by all countable models of $\mathcal{K}$.*

Proof. 1. This is actually a special case of the case below: pick $q = \exists x. \bot(x)$.
2. The "only if" direction is trivial. For the "if" direction, first observe that any Boolean P2RPQ q can be expressed as a possibly infinite disjunction $\bigvee_{q' \in Q_q} q'$ of (finite) Boolean CQs. Let $\mathcal{K} = (\mathcal{A}, \mathcal{T}, \mathcal{R}, \mathcal{E})$. Toward a contradiction, suppose $\mathcal{K} \not\models q$, i.e., there is a model $\mathcal{I}$ (of arbitrary cardinality) such that $\mathcal{I} \models \mathcal{K}$ but $\mathcal{I} \not\models q$. Then, by Lemma 1, we know that $\mathcal{I}$ is a model of the countable first-order theory $\Psi_{(\mathcal{A}, \mathcal{T}, \mathcal{R})} \cup \Phi_{\mathcal{E}} \cup \{\neg q' \mid q' \in Q_q\}$. Now we can apply the Löwenheim-Skolem Theorem downward and obtain that there must be a countable model $\mathcal{J}$ of this theory as well. By construction, $\mathcal{J}$ is a countable model of $\mathcal{K}$ but does not satisfy q, a contradiction. ◠

6 Development: Eliminating Cardinality Boxes

The basic underlying idea of our method is to model satisfaction of cardinality constraints by performing the necessary calculations and comparisons "physically" inside the model, using the domain elements for tallying. However, in the case of finite interpretations, it might happen that evaluating the cardinality constraints produces numbers that are greater than the number of domain elements (note that this danger is material, since expressive description logics allow for enforcing restricted domain sizes). Hence, we somehow have to make sure that our models are allowed to contain enough domain elements.

6.1 Shift: Making Space through Relativization

To this end, we employ a folklore technique called *relativization*, through which a (possibly domain-restricting) knowledge base $\mathcal{K}$ is transformed into a knowledge base $\mathcal{K}^\natural$ that allows for models with arbitrary domain sizes (by means of admitting "silent" or "non-active" domain elements, which do not participate in any relation), but every model of $\mathcal{K}^\natural$ "contains" a model of $\mathcal{K}$ in a formally defined way, so $\mathcal{K}^\natural$ is an authentic replacement of $\mathcal{K}$ when it comes to satisfiability testing or querying. In the context of description logics, similar techniques have been applied in [15, 18].

Definition 5 (relativization). *We let* T_{new} *be a fresh concept name. The function* $\cdot^\natural$ *mapping concepts to concepts is recursively defined as follows:*

$$
\begin{aligned}
A^\natural &= A & \{a\}^\natural &= \{a\} \\
\mathsf{T}^\natural &= \mathsf{T}_{new} & (\forall R.C)^\natural &= \mathsf{T}_{new} \sqcap \forall R.(\neg \mathsf{T}_{new} \sqcup C^\natural) \\
\bot^\natural &= \bot & (\exists R.C)^\natural &= \exists R.C^\natural \\
(\neg C)^\natural &= \mathsf{T}_{new} \sqcap \neg C^\natural & (\geqslant nS.C)^\natural &= \geqslant nS.C^\natural \\
(C_1 \sqcap C_2)^\natural &= C_1^\natural \sqcap C_2^\natural & (\leqslant nS.C)^\natural &= \leqslant nS.C^\natural \\
(C_1 \sqcup C_2)^\natural &= C_1^\natural \sqcup C_2^\natural & (\exists S.\mathsf{Self})^\natural &= \exists S.\mathsf{Self}
\end{aligned}
$$

Given a P2RPQ q, *we let* $q^\natural$ *denote the query obtained by replacing each concept* C *in* q *by* $C^\natural$. *Moreover, we extend* $\cdot^\natural$ *to EKBs* $\mathcal{K} = (\mathcal{A}, \mathcal{T}, \mathcal{R}, \mathcal{E})$ *by letting* $\mathcal{K}^\natural = (\mathcal{A}', \mathcal{T}', \mathcal{R}, \mathcal{E})$, *where* $\mathcal{A}'$ *contains*

- $\mathsf{T}_{new}(a)$ *for every* $a \in \mathbf{N_I}$ *occurring in* $\mathcal{K}$,
- *for every assertion* $C(a)$ *from* $\mathcal{A}$ *the assertion* $C^\natural(a)$, *and*
- *all assertions of the form* $a \approx b$, $a \not\approx b$, $\neg S(a, b)$, *and* $R(a, b)$ *from* $\mathcal{A}$,

while $\mathcal{T}'$ *contains*

- $A \sqsubseteq \mathsf{T}_{new}$ *for every* $A \in \mathbf{N_C}$ *occurring in* $\mathcal{K}$,
- $\exists P.\mathsf{T} \sqsubseteq \mathsf{T}_{new}$ *and* $\mathsf{T} \sqsubseteq \forall P.\mathsf{T}_{new}$ *for every* $P \in \mathbf{N_R} \setminus \{\mathsf{T}\}$ *occurring in* $\mathcal{K}$, *as well as*
- *for every GCI* $C_1 \sqsubseteq C_2$ *from* $\mathcal{T}$ *the GCI* $C_1^\natural \sqsubseteq C_2^\natural$.

It is now not too hard to establish the following lemma, explicating the formerly claimed very close connection between the models of $\mathcal{K}$ and $\mathcal{K}^\natural$.

Lemma 3 (relativization: model synchronicity). *Let* $\mathcal{K} = (\mathcal{A}, \mathcal{T}, \mathcal{R}, \mathcal{E})$ *be an EKB and let* $\mathcal{J} = (\Delta^{\mathcal{J}}, \cdot^{\mathcal{J}})$ *be in interpretation with* $\mathsf{T}_{new}^{\mathcal{J}} \neq \emptyset$. *Then* $\mathcal{J}$ *is a (finite) model of* $\mathcal{K}^\natural$ *if and only if there exists a (finite) set* $\Delta_\blacksquare$ *and a (finite) model* $\mathcal{I} = (\Delta^{\mathcal{I}}, \cdot^{\mathcal{I}})$ *of* $\mathcal{K}$ *such that* $\Delta^{\mathcal{J}} = \Delta^{\mathcal{I}} \cup \Delta_\blacksquare$ *and* $\cdot^{\mathcal{J}} = \cdot^{\mathcal{I}} \cup \{\mathsf{T}_{new} \mapsto \Delta^{\mathcal{I}}\}$.

Proof. (Sketch.) The not immediate cases are a direct consequence of the correspondence $C^{\mathcal{I}} = C^{\natural \mathcal{J}}$ which is proven by induction over the structure of C.   ⌐

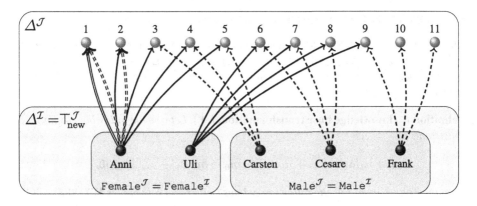

Fig. 1. Illustration of construction enforcing constraints.

6.2 Episode: Illustrative Example

We now know how to ensure that a model can contain enough elements for counting and hence are able to avoid "out of memory errors" in the course of our model-internal computation. Next, we describe in detail how to express concept cardinality constraints and consequently ECboxes polynomially with in-house means of expressive description logics, requiring just some extra vocabulary.

We first describe the core idea behind our modeling by means of an easy example. Assume, we would like to implement the constraint

$$1 + 4 \cdot |\texttt{Female}| \leq 2 + 3 \cdot |\texttt{Male}|$$

on the set of this volume's editors, which, of course, can be simplified by subtracting one on both sides, but we will not do so for the sake of the example. Assume that, by means of relativization, we have already ensured that as many as needed "silent elements" can be present in a model. In order to ensure that the constraint is satisfied, we proceed as follows (aiming at a setting as displayed in Fig. 1, where the silent elements are in the top line while the "proper elements" can be found in the bottom line): we introduce several types of left-hand-side roles (denoted $\mathsf{LHS}_{...}$, depicted by solid arrows in the figure) and right-hand-side roles (denoted $\mathsf{RHS}_{...}$, depicted by dashed arrows in the figure) and we make sure that every individual in $\texttt{Female}$ has (at least) four outgoing (single-line) left-hand-side roles, while every individual in $\texttt{Male}$ has (at most) three outgoing (single-line) right-hand-side roles and any individual not in $\texttt{Male}$ has no such outgoing roles whatsoever. Also, to account for the left- and right-hand-side constant terms, we pick one volunteering domain element, say Anni, as the source of one (double-line) left-hand-side role and of two (double-line) right-hand side roles. Then, we make sure that every domain element may receive at most one left-hand-side role. Under these circumstances, the "$\leq$" condition can be enforced by requiring that any element receiving a left-hand-side role must also be receiving a right-hand-side role. This (somewhat simplified) example will hopefully elucidate the modeling presented in the following.

6.3 Modulation: General Construction and Proof

After explaining the underlying ideas of our construction, we now provide the general definition of the technique of eliminating ECboxes from EKBs yielding plain KBs. We introduce the applied transformation and formally show that it has the announced properties.

Definition 6 (knowledge base transformation, $\mathcal{K}^{\text{tr}}$). *Let c be the following cardinality constraint:*

$$n_0 + n_1|A_1| + \ldots + n_k|A_k| \leq m_0 + m_1|B_1| + \ldots + m_\ell|B_\ell|.$$

Then the Tbox $\mathcal{T}_c$ contains the following axioms (with fresh role names AUX, RHS_c^i, LHS_c^i and – if needed – a fresh individual name o):

$$\top \sqsubseteq \exists \text{AUX}.\left(\top_{\text{new}} \sqcap \geqslant n_0 \text{LHS}_c^0.\top\right) \tag{2}$$

$$A_i \sqsubseteq \geqslant n_i \text{LHS}_c^i.\top \qquad\qquad\qquad for \ \ 1 \leq i \leq k \tag{3}$$

$$\exists \text{LHS}_c^{i\,-}.\top \sqcap \exists \text{LHS}_c^{j\,-}.\top \sqsubseteq \bot \qquad for \ \ 0 \leq i < j \leq k \tag{4}$$

$$\top \sqsubseteq \leqslant 1.\text{LHS}_c^{i\,-}.\top \qquad\qquad\qquad for \ \ 0 \leq i \leq k \tag{5}$$

$$\exists \text{RHS}_c^0.\top \sqsubseteq \bot \qquad\qquad\qquad in \ case \ \ m_0 = 0 \tag{6}$$

$$\exists \text{RHS}_c^0.\top \sqsubseteq \{o\} \qquad\qquad\qquad in \ case \ \ m_0 > 0 \tag{7}$$

$$\exists \text{RHS}_c^i.\top \sqsubseteq B_i \qquad\qquad\qquad for \ \ 1 \leq i \leq \ell \tag{8}$$

$$\top \sqsubseteq \leqslant m_i.\text{RHS}_c^i \qquad\qquad\qquad for \ \ 0 \leq i \leq \ell \tag{9}$$

$$\text{Cstrt}_c \sqcap \bigsqcup_{0 \leq i \leq k} \exists \text{LHS}_c^{i\,-}.\top \sqsubseteq \bigsqcup_{0 \leq i \leq \ell} \exists \text{RHS}_c^{i\,-}.\top \tag{10}$$

Given an ECbox $\mathcal{E}$, let $C_\mathcal{E}$ be the concept expression obtained from $\mathcal{E}$ by replacing every c in $\mathcal{E}$ by Cstrt_c, every $\wedge$ by $\sqcap$, and every $\vee$ by $\sqcup$. Let $\text{Sync}(\text{Cstrt}_c)$ denote $\{\exists \top.\text{Cstrt}_c \sqsubseteq \text{Cstrt}_c\}$ if the underlying description logic supports the universal role $\top$ and $\{\top \sqsubseteq \exists \text{AUX}^-.\{o\}, \exists \text{AUX}.\text{Cstrt}_c \sqsubseteq \forall \text{AUX}.\text{Cstrt}_c\}$ otherwise.
Then, we let

$$\mathcal{T}_\mathcal{E} = \{\top \sqsubseteq C_\mathcal{E}\} \cup \bigcup_{c \ from \ \mathcal{E}} \mathcal{T}_c \cup \text{Sync}(\text{Cstrt}_c). \tag{11}$$

Finally, for an EKB $\mathcal{K} = (\mathcal{A}, \mathcal{T}, \mathcal{R}, \mathcal{E})$ with $\mathcal{K}^{\natural} = (\mathcal{A}', \mathcal{T}', \mathcal{R}, \mathcal{E})$ let $\mathcal{K}^{\text{tr}} = (\mathcal{A}', \mathcal{T}' \cup \mathcal{T}_\mathcal{E}, \mathcal{R}, \emptyset)$ be the corresponding transformed *KB.*

The following observations are immediate from the construction of $\mathcal{K}^{\text{tr}}$.

Lemma 4 (syntactic properties of $\mathcal{K}^{\text{tr}}$). *For any EKB $\mathcal{K}$ in some description logic $\mathcal{L}$:*

1. *$\mathcal{K}^{\text{tr}}$ can be computed from $\mathcal{K}$ in polynomial time.*
2. *If $\mathcal{L}$ subsumes $\mathcal{ALCIQ}_\top$ and $\mathcal{K}$ is an ERKB, then $\mathcal{K}^{\text{tr}}$ is a (plain) $\mathcal{L}$ KB.*
3. *If $\mathcal{L}$ subsumes $\mathcal{ALCOIQ}$, then $\mathcal{K}^{\text{tr}}$ is a (plain) $\mathcal{L}$ KB.*

Next we prove a rather close relationship between the models of $\mathcal{K}$ and $\mathcal{K}^{\text{tr}}$.

Lemma 5 (semantic properties of $\mathcal{K}^{+}$). *For an EKB $\mathcal{K}$ in some description logic $\mathcal{L}$, the following hold:*

1. *For every (finite) model $\mathcal{J} = (\Delta^{\mathcal{J}}, \cdot^{\mathcal{J}})$ of $\mathcal{K}^{+}$, the interpretation $\mathcal{I} = (\Delta^{\mathcal{I}}, \cdot^{\mathcal{I}})$ with $\Delta^{\mathcal{I}} = \mathsf{T}_{\mathrm{new}}^{\mathcal{J}}$ and $\cdot^{\mathcal{I}}$ the appropriate restriction of $\cdot^{\mathcal{J}}$ is a (finite) model of $\mathcal{K}$.*
2. *For every (finite) countable model $\mathcal{I} = (\Delta^{\mathcal{I}}, \cdot^{\mathcal{I}})$ of $\mathcal{K}$ there is a (finite) countable model $\mathcal{J} = (\Delta^{\mathcal{J}}, \cdot^{\mathcal{J}})$ of $\mathcal{K}^{+}$ such that with $\Delta^{\mathcal{I}} = \mathsf{T}_{\mathrm{new}}^{\mathcal{J}}$ and $\cdot^{\mathcal{I}}$ is the appropriate restriction of $\cdot^{\mathcal{J}}$.*

Proof. Let $\mathcal{K} = (\mathcal{A}, \mathcal{T}, \mathcal{R}, \mathcal{E})$. We show the two parts consecutively.

1. To show the first part, assume a (finite) model $\mathcal{J} = (\Delta^{\mathcal{J}}, \cdot^{\mathcal{J}})$ of $\mathcal{K}^{+}$. We now show that $\mathcal{I} = (\Delta^{\mathcal{I}}, \cdot^{\mathcal{I}})$ is a model of $\mathcal{K}$. For all Abox, Tbox, and Rbox axioms, satisfaction follows from Lemma 3. Now consider $\mathcal{E}$. We pick an arbitrary $\delta \in \Delta^{\mathcal{J}}$ and let $\mathfrak{C} = \{\mathfrak{c} \mid \delta \in \mathtt{Cstrt}_{\mathfrak{c}}^{\mathcal{J}}\}$. Due to the axiom $\top \sqsubseteq C_{\mathcal{E}}$, we know that $\mathfrak{C}$ is such that simultaneous satisfaction of all $\mathfrak{c} \in \mathfrak{C}$ implies satisfaction of $\mathcal{E}$. Hence we proceed to prove that this is indeed the case. First note that the axioms $\mathsf{Sync}(\mathtt{Cstrt}_{\mathfrak{c}})$ ensure $\mathtt{Cstrt}_{\mathfrak{c}}^{\mathcal{J}} = \Delta^{\mathcal{J}}$ for every $\mathfrak{c} \in \mathfrak{C}$. Furthermore, for any $\mathfrak{c} \in \mathfrak{C}$ of the form

$$n_0 + n_1|\mathtt{A}_1| + \ldots + n_k|\mathtt{A}_k| \leq m_0 + m_1|\mathtt{B}_1| + \ldots + m_\ell|\mathtt{B}_\ell|,$$

satisfaction of the $\mathfrak{c}$ by $\mathcal{J}$ follows from the following three inequalities:

$$n_0 + n_1|\mathtt{A}_1^{\mathcal{J}}| + \ldots + n_k|\mathtt{A}_k^{\mathcal{J}}| \leq \sum_{i=0}^{k} |\{\delta \mid (\delta', \delta) \in \mathtt{LHS}_{\mathfrak{c}}^{i\,\mathcal{J}}\}|, \qquad (\#)$$

$$\sum_{i=0}^{k} |\{\delta \mid (\delta', \delta) \in \mathtt{LHS}_{\mathfrak{c}}^{i\,\mathcal{J}}\}| \leq \sum_{i=0}^{\ell} |\{\delta \mid (\delta', \delta) \in \mathtt{RHS}_{\mathfrak{c}}^{i\,\mathcal{J}}\}|, \qquad (\%)$$

$$\sum_{i=0}^{\ell} |\{\delta \mid (\delta', \delta) \in \mathtt{RHS}_{\mathfrak{c}}^{i\,\mathcal{J}}\}| \leq m_0 + m_1|\mathtt{B}_1^{\mathcal{J}}| + \ldots + m_\ell|\mathtt{B}_\ell^{\mathcal{J}}|. \qquad (\spadesuit)$$

We will now consecutively show each of these statements.

($\#$) On one hand, the axioms $\top \sqsubseteq \,\leqslant\! 1.\mathtt{LHS}_{\mathfrak{c}}^{i\,-}.\top$ make sure that

$$|\{\delta \mid (\delta', \delta) \in \mathtt{LHS}_{\mathfrak{c}}^{i\,\mathcal{J}}\}| = |\mathtt{LHS}_{\mathfrak{c}}^{i\,\mathcal{J}}|.$$

On the other hand, whenever $1 \leq i \leq k$, the axiom $\mathtt{A}_i \sqsubseteq \,\geqslant\! n_i\mathtt{LHS}_{\mathfrak{c}}^{i}.\top$ ensures

$$|\mathtt{A}_i^{\mathcal{J}}| \cdot n_i \leq |\mathtt{LHS}_{\mathfrak{c}}^{i\,\mathcal{J}}|,$$

while for $i = 0$, the axiom $\top \sqsubseteq \exists \mathtt{AUX}.\!\geqslant\! n_0\mathtt{LHS}_{\mathfrak{c}}^{0}.\top$ enforces that there must be some $\delta \in (\geqslant\! n_0\mathtt{LHS}_{\mathfrak{c}}^{0}.\top)^{\mathcal{J}}$ and consequently

$$n_0 \leq |\mathtt{LHS}_{\mathfrak{c}}^{0\,\mathcal{J}}|.$$

Putting everything together, we obtain

$$n_0 + \sum_{i=1}^{k} n_i|\mathtt{A}_i^{\mathcal{J}}| \leq \sum_{i=0}^{k} |\mathtt{LHS}_{\mathfrak{c}}^{i\,\mathcal{J}}| = \sum_{i=0}^{k} |\{\delta \mid (\delta', \delta) \in \mathtt{LHS}_{\mathfrak{c}}^{i\,\mathcal{J}}\}|$$

as claimed.

($\maltese$) First note that the axioms of the form $\exists \mathrm{LHS}_c^{i^-}.\top \sqcap \exists \mathrm{LHS}_c^{j^-}.\top \sqsubseteq \bot$ ensure

$$|\{\delta \mid (\delta', \delta) \in \mathrm{LHS}_c^{i\,\mathcal{J}}, 0 \leq i \leq k\}| = \sum_{i=0}^{k} |\{\delta \mid (\delta', \delta) \in \mathrm{LHS}_c^{i\,\mathcal{J}}\}|.$$

Also, the axiom $\mathtt{Cstrt}_c \sqcap \bigsqcup_{0 \leq i \leq k} \exists \mathrm{LHS}_c^{i^-}.\top \sqsubseteq \bigsqcup_{0 \leq i \leq \ell} \exists \mathrm{RHS}_c^{i^-}.\top$ enforces

$$\{\delta \mid (\delta', \delta) \in \mathrm{LHS}_c^{i\,\mathcal{J}}, 0 \leq i \leq k\} \subseteq \{\delta \mid (\delta', \delta) \in \mathrm{RHS}_c^{i\,\mathcal{J}}, 0 \leq i \leq \ell\}$$

(remembering that $\mathtt{Cstrt}_c^{\mathcal{J}} = \Delta^{\mathcal{J}}$) and consequently

$$|\{\delta \mid (\delta', \delta) \in \mathrm{LHS}_c^{i\,\mathcal{J}}, 0 \leq i \leq k\}| \leq |\{\delta \mid (\delta', \delta) \in \mathrm{RHS}_c^{i\,\mathcal{J}}, 0 \leq i \leq \ell\}|.$$

It remains to note that

$$|\{\delta \mid (\delta', \delta) \in \mathrm{RHS}_c^{i\,\mathcal{J}}, 0 \leq i \leq \ell\}| \leq \sum_{i=0}^{\ell} |\{\delta \mid (\delta', \delta) \in \mathrm{RHS}_c^{i\,\mathcal{J}}\}|$$

holds unconditionally, so putting the established correspondences together shows our claim.

($\oplus$) First note that, the axiom $\exists \mathrm{RHS}_c^0.\top \sqsubseteq \{o\}$ (or, alternatively, $\exists \mathrm{RHS}_c^0.\top \sqsubseteq \bot$ in case $m_0 = 0$) ensures $\delta = o^{\mathcal{J}}$ whenever we find $(\delta, \delta') \in \mathrm{RHS}_c^{0\,\mathcal{J}}$. But therefrom, using the axiom $\top \sqsubseteq \, \leqslant m_0.\mathrm{RHS}_c^0$, we can derive

$$|\mathrm{RHS}_c^{0\,\mathcal{J}}| \leq m_0.$$

Likewise, for $1 \leq i \leq \ell$, we obtain $\delta' \in B_i^{\mathcal{J}}$ for each $(\delta', \delta) \in \mathrm{RHS}_c^{i\,\mathcal{J}}$ due to the axiom $\exists \mathrm{RHS}_c^i.\top \sqsubseteq B_i$. Yet, for every such δ', at most m_i distinct corresponding δ can exist due to the axiom $\top \sqsubseteq \, \leqslant m_i.\mathrm{RHS}_c^i$ and hence

$$|\mathrm{RHS}_c^{i\,\mathcal{J}}| \leq |B_i^{\mathcal{J}}| \cdot m_i.$$

Moreover, as projecting will never increase the size of a set, we obtain

$$|\{\delta \mid (\delta', \delta) \in \mathrm{RHS}_c^{i\,\mathcal{J}}\}| \leq |\mathrm{RHS}_c^{i\,\mathcal{J}}|.$$

Yet then, combining these statements yields

$$\sum_{i=0}^{\ell} |\{\delta \mid (\delta', \delta) \in \mathrm{RHS}_c^{i\,\mathcal{J}}\}| \leq \sum_{i=0}^{\ell} |\mathrm{RHS}_c^{i\,\mathcal{J}}| \leq m_0 + \sum_{i=1}^{\ell} m_i |B_i^{\mathcal{J}}|$$

as claimed.

2. Let $\mathcal{I} = (\Delta^{\mathcal{I}}, \cdot^{\mathcal{I}})$ be a (finite) countable model of $\mathcal{K}$. We first give a construction for $\mathcal{J} = (\Delta^{\mathcal{J}}, \cdot^{\mathcal{J}})$ and then show modelhood for $\mathcal{K}^{\maltese}$. Let $n_{\max} \in \mathbb{N}^{\infty}$ be the largest value obtained when evaluating all the left and right hand sides of all the constraints in $\mathcal{E}$. Then let $\Delta^{\mathcal{J}} = \Delta^{\mathcal{I}} \cup [n_{\max}]$. Note that $\Delta^{\mathcal{J}}$ is finite, whenever $\Delta^{\mathcal{I}}$ is. We let $\cdot^{\mathcal{J}}$ coincide with $\cdot^{\mathcal{I}}$ for all individual names, concept names and role names from $\mathcal{K}$. It remains to define the fresh auxiliary vocabulary of $\mathcal{K}^{\maltese}$. To this end, let $\mathfrak{C}$ be the set of cardinality constraints occurring in $\mathcal{E}$ which are satisfied in $\mathcal{I}$. Now, pick one δ' from $\Delta^{\mathcal{I}}$ and let

- $o^{\mathcal{J}} = \delta'$,
- $\text{AUX}^{\mathcal{J}} = \Delta^{\mathcal{J}} \times \{\delta'\}$,
- $\top_{\text{new}}^{\mathcal{J}} = \Delta^{\mathcal{I}}$,
- $\text{Cstrt}_{\mathfrak{c}}^{\mathcal{I}} = \Delta^{\mathcal{J}}$ whenever $\mathfrak{c} \in \mathfrak{C}$ and $\text{Cstrt}_{\mathfrak{c}}^{\mathcal{I}} = \emptyset$ otherwise,
- For any $\mathfrak{c}$ of the form $n_0 + n_1|A_1| + \ldots + n_k|A_k| \leq m_0 + m_1|B_1| + \ldots + m_\ell|B_\ell|$, let

$$\mathfrak{L}_{\mathfrak{c}} = \{0\} \times [n_0] \cup \{1\} \times [n_1] \times A_1^{\mathcal{J}} \cup \ldots \cup \{k\} \times [n_k] \times A_k^{\mathcal{J}} \text{ and}$$
$$\mathfrak{R}_{\mathfrak{c}} = \{0\} \times [m_0] \cup \{1\} \times [m_1] \times B_1^{\mathcal{J}} \cup \ldots \cup \{\ell\} \times [m_\ell] \times B_\ell^{\mathcal{J}}.$$

Then we let $\cdot^{\mathfrak{c}\sharp} : \mathfrak{L}_{\mathfrak{c}} \to \left[|\mathfrak{L}_{\mathfrak{c}}|\right]$ and $\cdot^{\mathfrak{c}\flat} : \mathfrak{R}_{\mathfrak{c}} \to \left[|\mathfrak{R}_{\mathfrak{c}}|\right]$ be bijective enumeration functions for $\mathfrak{L}_{\mathfrak{c}}$ and $\mathfrak{R}_{\mathfrak{c}}$. Now we let

- $\text{LHS}_{\mathfrak{c}}^{0\mathcal{J}} = \{(\delta', (0, j)^{\mathfrak{c}\sharp}) \mid j \in [n_0]\}$,
- $\text{LHS}_{\mathfrak{c}}^{i\mathcal{J}} = \{(\delta, (i, j, \delta)^{\mathfrak{c}\sharp}) \mid \delta \in A_i^{\mathcal{J}}, \ j \in [n_i]\}$, for $1 \leq i \leq k$,
- $\text{RHS}_{\mathfrak{c}}^{0\mathcal{J}} = \{(\delta', (0, j)^{\mathfrak{c}\flat}) \mid j \in [m_0]\}$, and
- $\text{RHS}_{\mathfrak{c}}^{i\mathcal{J}} = \{(\delta, (i, j, \delta)^{\mathfrak{c}\flat}) \mid \delta \in B_i^{\mathcal{J}}, \ j \in [m_i]\}$, for $1 \leq i \leq \ell$.

It is now straightforward to check that by construction, $\mathcal{J}$ satisfies all axioms from $\mathcal{T}_{\mathcal{E}}$. As far as the relativized Abox and Tbox axioms and the unchanged Rbox axioms from $\mathcal{K}^{\natural}$ are concerned, their satisfaction follows from Lemma 3. $\qquad \frown$

With these model correspondences in place, we can now establish the results regarding preservation of satisfiability and query entailment as well as their complexities.

Theorem 6 (eliminability of ECboxes). *Let $\mathcal{K}$ be an EKB in some (finitely or at least countably) first-order expressible description logic $\mathcal{L}$. Then the following hold:*

1. *$\mathcal{K}$ and $\mathcal{K}^{\maltese}$ are (finitely) equisatisfiable.*
2. *Given a P2RPQ q, $\mathcal{K}$ (finitely) entails q exactly if $\mathcal{K}^{\maltese}$ (finitely) entails $q^{\natural}$.*
3. *If $\mathcal{L}$ subsumes $\mathcal{ALCIQ}_{\mp}$, then the complexities of (finite) satisfiability and (finite) CQ or P2RPQ entailment for $\mathcal{L}$ ERKBs coincide with those of plain $\mathcal{L}$ KBs.*
4. *If $\mathcal{L}$ subsumes $\mathcal{ALCOIQ}$, then the complexities of (finite) satisfiability and (finite) CQ or P2RPQ entailment for $\mathcal{L}$ EKBs coincide with those of plain $\mathcal{L}$ KBs.*

Proof. 1. On one hand, given a (finite) model of $\mathcal{K}$, Theorem 2 makes sure that we can assume it is countable and thus, Item 2 of Lemma 5 provides us with a (finite) model of $\mathcal{K}^{\maltese}$. On the other hand, given a (finite) model of $\mathcal{K}^{\maltese}$, we can invoke Item 1 of Lemma 5 to obtain a (finite) model of $\mathcal{K}$.

2. We show the equivalent statement that $\mathcal{K}$ does not (finitely) entail q exactly if $\mathcal{K}^{\maltese}$ does not (finitely) entail $q^{\natural}$. Consider a (finite) $\mathcal{I}$ with $\mathcal{I} \models \mathcal{K}$ but $\mathcal{I} \not\models q$. Theorem 2 allows us to assume that $\mathcal{I}$ is countable. Then Item 2 of Lemma 5 ensures that there is a model of $\mathcal{K}^{\maltese}$ which by construction does not satisfy $q^{\natural}$. Vice versa, consider a (finite) $\mathcal{J}$ with $\mathcal{J} \models \mathcal{K}^{\maltese}$ but $\mathcal{J} \not\models q^{\natural}$. Then Item 1 of Lemma 5 provides us with a (finite) model of $\mathcal{K}$ not satisfying q by construction.

3. This follows from the two previous items and Lemma 4, Items 1 and 2.

4. This follows from the two previous items and Lemma 4, Items 1 and 3. $\qquad \frown$

We note that this theorem does not only hold for CQs and P2RPQs, but it easily extends to all query formalisms where non-satisfaction can be expressed via countable first-order theories. Among others, this includes all Datalog queries [28].

7   Recapitulation: Results

Theorem 6 can now be put to use by harvesting a number of findings from known results. We will go through the results announced in Section 4 and discuss their provenance and possible further ramifications.

- Satisfiability and finite satisfiability of $\mathcal{SHIQ}b_{\mp}^{\mathsf{Self}}$ ERKBs is ExpTime-complete. Noting that transitivity can be equisatisfiably removed via box-pushing (along the lines of [33, 29]), yielding $\mathcal{ALCHIQ}b_{\mp}^{\mathsf{Self}}$, ExpTime-completeness for the latter can be obtained via minor extensions of [33] for arbitrary models and [20] for finite models. Both also follow from the corresponding result for $\mathcal{GC}^2$, the guarded two-variable fragment as defined by Pratt-Hartmann [23]. Using these results, the application of standard techniques [11, 17, 29] allow to establish 2ExpTime-completeness for finite and arbitrary satisfiability of $\mathcal{SRIQ}b$ ERKBs.

- Satisfiability and finite satisfiability of $\mathcal{SHOIQB}^{\mathsf{Self}}$ EKBs is NExpTime-complete. Again, as laid out in [29], transitivity can be removed preserving satisfiability (yielding $\mathcal{ALCHOIQB}^{\mathsf{Self}}$), which is just a syntatic variant of $\mathcal{C}^2$, the two-variable fragment of first-order logic, for which the respective complexity results were established by Pratt-Hartmann [22]. Based on these findings, N2ExpTime-completeness of finite and arbitrary satisfiability of $\mathcal{SROIQB}$ EKBs is a rather direct consequence [17, 29].

- P2RPQ entailment as well as finite CQ entailment from $\mathcal{ALCHIQ}b_{\mp}^{\mathsf{Self}}$ ERKBs are 2ExpTime-complete. Note that $\mathcal{ALCHIQ}b_{\mp}^{\mathsf{Self}}$ is a syntactic variant of $\mathcal{GC}^2$, therefore 2ExpTime-completeness of finite entailment of CQs follows from [24], while P2RPQ entailment is a consequence of [8].

- Entailment of unions of CQs from $\mathcal{ALCHOIQ}b$ EKBs is decidable and coN2ExpTime-hard. This is a consequence of the respective results for plain $\mathcal{ALCHOIQ}b$ KBs [27, 14].

8   Coda: Conclusion

Inspired by previous work on quantitative extensions of $\mathcal{ALC}$ driven by Franz [1, 5, 2, 3], we investigated the possibility of extending the expressivity of the underlying logic in the presence of global cardinality constraints. Using a novel idea of simulating the cardinality information via modeling features readily available in mainstream description logics, we were able to show that significant complexity-neutral extensions are possible. Moreover, we laid the formal foundations for adequately dealing with models of infinite domain size.

There are plenty of avenues for future work. We reiterate, that the logics considered here are tailored toward "global counting", whereas "local counting" (that is Presburger constraints over individuals' role successors) is not supported. For example, $\mathcal{ALCSCC}$ [1]

would allow us to express that some course is gender-balanced if and only if it has as many female participants as it has male ones. While this is beyond the capabilities of any of the logics considered here, the presence of inverses and nominals allows us to at least enforce that a concrete given course tcs is indeed gender-balanced, using the following Tbox and ERCbox statements:

$$\text{MalInC} \equiv \text{Male} \sqcap \exists \text{hasParticipant}^-.\{\text{tcs}\} \tag{12}$$

$$\text{FemInC} \equiv \text{Female} \sqcap \exists \text{hasParticipant}^-.\{\text{tcs}\} \tag{13}$$

$$(|\text{MalInC}| \leq |\text{FemInC}|) \wedge (|\text{FemInC}| \leq |\text{MalInC}|) \tag{14}$$

In fact, we can even go one step further and express that tcs is gender-balenced exactly if GendBal(tcs) holds as follows:

$$\text{MalInC} \equiv \text{Male} \sqcap \exists \text{hasParticipant}^-.\{\text{tcs}\} \tag{15}$$

$$\text{FemInC} \equiv \text{Female} \sqcap \exists \text{hasParticipant}^-.\{\text{tcs}\} \tag{16}$$

$$\text{BalTCS} \equiv \text{GendBal} \sqcap \{\text{tcs}\} \tag{17}$$

$$\left(|\text{BalC}| \leq 0 \wedge (|\text{MalInC}|+1 \leq |\text{FemInC}| \vee |\text{FemInC}|+1 \leq |\text{MalInC}|)\right)$$
$$\vee \left(1 \leq |\text{BalC}|\right) \wedge (|\text{MalInC}| \leq |\text{FemInC}| \wedge |\text{FemInC}| \leq |\text{MalInC}|)) \tag{18}$$

A more thorough investigation about which local counting features can be realized by global ones using advanced description logic modeling features is clearly an interesting starting point for future work.

On another note, in the case of reasoning with arbitrary models, it would be very handy from a modeler's perspective to have a way of expressing that a concept may have only finitely many elements. As mentioned before, with such statements, we leave the realms of first-order logic for good. However, for instance, an inspection of Pratt-Hartmann's work on the two-variable fragment of first-order logic with counting strongly suggests that such "finiteness constraints" can be accommodated at no additional complexity cost [22, 25].

Finally, the reduction presented in this paper could potentially turn out to be of practical value, since it allows to express elaborate quantitative information by means of standardized ontology languages, which are supported by existing, highly optimized reasoning engines [13, 30, 32]. This having said, this proposal would only work for reasoning under the classical (i.e., arbitrary-model) semantics and, admittedly, it is also rather questionable if existing reasoners would cope well with large values in qualified number restrictions. Yet, conversely, this work might motivate developers of reasoning engines to come up with better implementations as to support statistical and other quantitative modeling.

Acknowledgements. This work (and, in particular, the conception of ERCboxes) was supported by the European Research Council (ERC) Consolidator Grant 771779 *A Grand Unified Theory of Decidability in Logic-Based Knowledge Representation* (DeciGUT). The author would also like to thank the two anonymous reviewers as well as Andreas Ecke and Bartosz Bednarczyk for their comments.

References

1. Baader, F.: A new description logic with set constraints and cardinality constraints on role successors. In: Dixon, C., Finger, M. (eds.) FroCoS 2017. LNCS (LNAI), vol. 10483, pp. 43–59. Springer, Cham (2017). https://doi.org/10.1007/978-3-319-66167-4_3
2. Baader, F.: Expressive cardinality constraints on $\mathcal{ALCSCC}$ concepts. In: Proceedings of the 34th ACM/SIGAPP Symposium on Applied Computing (SAC 2019). ACM (2019)
3. Baader, F., Bednarczyk, B., Rudolph, S.: Satisfiability checking and conjunctive query answering in description logics with global and local cardinality constraints. In: Proceedings of the 32nd International Workshop on Description Logics (DL 2019). CEUR Workshop Proceedings. CEUR-WS.org (2019, submitted)
4. Baader, F., Calvanese, D., McGuinness, D., Nardi, D., Patel-Schneider, P. (eds.): The Description Logic Handbook: Theory, Implementation, and Applications, 2nd edn. Cambridge University Press, Cambridge (2007)
5. Baader, F., Ecke, A.: Extending the description logic $\mathcal{ALC}$ with more expressive cardinality constraints on concepts. In: 3rd Global Conference on Artificial Intelligence (GCAI 2017). EPiC Series in Computing, vol. 50, pp. 6–19. EasyChair (2017)
6. Baader, F., Horrocks, I., Lutz, C., Sattler, U.: An Introduction to Description Logic. Cambridge University Press, Cambridge (2017)
7. Bach, J.S.: Canon circularis per tonos. In: The Musical Offering, BWV, vol. 1079. Leipzig (1747)
8. Bednarczyk, B., Rudolph, S.: Worst-case optimal querying of very expressive description logics with path expressions and succinct counting. In: Proceedings of the 32nd International Workshop on Description Logics (DL 2019). CEUR Workshop Proceedings, CEUR-WS.org (2019, submitted)
9. Cantor, G.: Beiträge zur Begründung der transfiniten Mengenlehre. Math. Ann. **46**(4), 481–512 (1895)
10. Cantor, G.: Beiträge zur Begründung der transfiniten Mengenlehre. Math. Ann. **49**(2), 207–246 (1897)
11. Demri, S., Nivelle, H.: Deciding regular grammar logics with converse through first-order logic. J. Logic Lang. Inf. **14**(3), 289–329 (2005)
12. Demri, S., Lugiez, D.: Complexity of modal logics with Presburger constraints. J. Appl. Logic **8**(3), 233–252 (2010)
13. Glimm, B., Horrocks, I., Motik, B., Stoilos, G., Wang, Z.: HermiT: an OWL 2 reasoner. J. Autom. Reason. **53**(3), 245–269 (2014)
14. Glimm, B., Kazakov, Y., Lutz, C.: Status QIO: an update. In: Rosati, R., Rudolph, S., Zakharyaschev, M. (eds.) Proceedings of the 24th International Workshop on Description Logics (DL 2011). CEUR Workshop Proceedings, vol. 745. CEUR-WS.org (2011)
15. Glimm, B., Rudolph, S., Völker, J.: Integrated metamodeling and diagnosis in OWL 2. In: Patel-Schneider, P.F., et al. (eds.) ISWC 2010, Part I. LNCS, vol. 6496, pp. 257–272. Springer, Heidelberg (2010). https://doi.org/10.1007/978-3-642-17746-0_17
16. Horrocks, I., Kutz, O., Sattler, U.: The even more irresistible $\mathcal{SROIQ}$. In: Doherty, P., Mylopoulos, J., Welty, C.A. (eds.) Proceedings of the 10th International Conference on Principles of Knowledge Representation and Reasoning (KR 2006), pp. 57–67. AAAI Press (2006)
17. Kazakov, Y.: $\mathcal{RIQ}$ and $\mathcal{SROIQ}$ are harder than $\mathcal{SHOIQ}$. In: Brewka, G., Lang, J. (eds.) Proceedings of the 11th International Conference on Principles of Knowledge Representation and Reasoning (KR 2008), pp. 274–284. AAAI Press (2008)

18. Krötzsch, M., Rudolph, S.: Nominal schemas in description logics: complexities clarified. In: Baral, C., De Giacomo, G., Eiter, T. (eds.) Proceedings of the 14th International Conference on Principles of Knowledge Representation and Reasoning (KR 2014), pp. 308–317. AAAI Press (2014)
19. Kuncak, V., Rinard, M.: Towards efficient satisfiability checking for Boolean algebra with Presburger arithmetic. In: Pfenning, F. (ed.) CADE 2007. LNCS (LNAI), vol. 4603, pp. 215–230. Springer, Heidelberg (2007). https://doi.org/10.1007/978-3-540-73595-3_15
20. Lutz, C., Sattler, U., Tendera, L.: The complexity of finite model reasoning in description logics. Inf. Comput. **199**(1–2), 132–171 (2005)
21. Peñaloza, R., Potyka, N.: Towards statistical reasoning in description logics over finite domains. In: Moral, S., Pivert, O., Sánchez, D., Marín, N. (eds.) SUM 2017. LNCS (LNAI), vol. 10564, pp. 280–294. Springer, Cham (2017). https://doi.org/10.1007/978-3-319-67582-4_20
22. Pratt-Hartmann, I.: Complexity of the two-variable fragment with counting quantifiers. J. Logic Lang. Inf. **14**, 369–395 (2005)
23. Pratt-Hartmann, I.: Complexity of the guarded two-variable fragment with counting quantifiers. J. Log. Comput. **17**(1), 133–155 (2007)
24. Pratt-Hartmann, I.: Data-complexity of the two-variable fragment with counting quantifiers. Inf. Comput. **207**(8), 867–888 (2009)
25. Pratt-Hartmann, I.: Personal communication, 19 March 2019
26. Rudolph, S.: Foundations of description logics. In: Polleres, A., et al. (eds.) Reasoning Web 2011. LNCS, vol. 6848, pp. 76–136. Springer, Heidelberg (2011). https://doi.org/10.1007/978-3-642-23032-5_2
27. Rudolph, S., Glimm, B.: Nominals, inverses, counting, and conjunctive queries or: why infinity is your friend! J. Artif. Intell. Res. **39**, 429–481 (2010)
28. Rudolph, S., Krötzsch, M.: Flag & check: data access with monadically defined queries. In: Hull, R., Fan, W. (eds.) Proceedings of the 32nd Symposium on Principles of Database Systems (PODS 2013), pp. 151–162. ACM (2013)
29. Rudolph, S., Krötzsch, M., Hitzler, P.: Cheap Boolean role constructors for description logics. In: Hölldobler, S., Lutz, C., Wansing, H. (eds.) JELIA 2008. LNCS (LNAI), vol. 5293, pp. 362–374. Springer, Heidelberg (2008). https://doi.org/10.1007/978-3-540-87803-2_30
30. Sirin, E., Parsia, B., Grau, B.C., Kalyanpur, A., Katz, Y.: Pellet: a practical OWL-DL reasoner. J. Web Seman. **5**(2), 51–53 (2007)
31. Skolem, T.: Über einige Grundlagenfragen der Mathematik. Skrifter utgitt av det Norske Videnskaps-Akademi i Oslo, I. Matematisk-naturvidenskabelig Klasse **7**, 1–49 (1929)
32. Steigmiller, A., Liebig, T., Glimm, B.: Konclude: system description. J. Web Seman. **27**, 78–85 (2014)
33. Tobies, S.: Complexity results and practical algorithms for logics in knowledge representation. Ph.D. thesis, RWTH Aachen, Germany (2001)

A Note on Unification, Subsumption and Unification Type

Manfred Schmidt-Schauß[✉]

Department of Computer Science and Mathematics, Goethe-University Frankfurt,
Frankfurt, Germany
schauss@ki.informatik.uni.Jrankfurt.de

Abstract. Various forms of subsumption preorders are used in the literature for comparing unifiers and general solutions of a unification problem
for generality and for defining the unification type. This note presents
some of them and discusses their pros and cons. In particular arguments
against the exist-substitution-based subsumption preorder (ess) are discussed. A proposal for a further partition of unification type nullary is
made. Also some historical notes are included.

Keywords: Unification · Subsumption preorder ·
Most general unifiers · Unification type

1 Introduction

Algorithms for solving equations are of widespread use in Computer Science
and Mathematics. An important issue which accompanies the computation of
solutions of equations is the quest for general solutions or most general solutions.
The design choice should support efficient use of the solutions. As a first example,
consider an equation over the integers: $x^2 \doteq 1$, which has two solutions: $x \mapsto 1$,
and $x \mapsto -1$. The solutions are necessary and non-redundant. As next example,
consider the first-order equation $x \doteq y$, which has several solutions: $\sigma_1 = \{x \mapsto
y\}$, $\sigma_2 = \{y \mapsto x\}$, $\sigma_3 = \{y \mapsto z, x \mapsto z\}$, $\sigma_4 = \{y \mapsto 0, x \mapsto 0\}$, $\sigma_5 =
\{y \mapsto z, x \mapsto z, w \mapsto 0\}$, Here the intuition is that σ_4 is too special, and that
the others look like variations of a single general solution, perhaps with some
redundant information. The subsumption preorder in use nowadays will tell us
that $\sigma_1, \sigma_2, \sigma_3, \sigma_5$ are most general unifiers of the equation and that subsume
each other. The exist-substitution-based subsumption preorder (ess) will only
accept σ_1, σ_2 as most general. This will be detailed below in Sect. 2.

The subsumption preorder is also essential to define for problem classes (and
unification problems) in equational theories their unification type, i.e., whether
at most one most general unifier (unitary), at most finitely many (finitary),
sometimes infinitely many (infinitary) or sometimes no set of minimal unifiers
exist (type nullary).

ⓒ Springer Nature Switzerland AG 2019
C. Lutz et al. (Eds.): Baader Festschrift, LNCS 11560, pp. 562–572, 2019.
https://doi.org/10.1007/978-3-030-22102-7_26

1.1   Some Notes on the (Short) History of Subsumption Preorders

When I was undertaking research for my doctoral thesis in the field of deduction and unification, around 1986, I had trouble with exactly the problem of an acceptable definition of subsumption of unifiers, in particular in a sorted logic. The problem was that the available papers and people all employed the proposal of the ess-subsumption preorder that only accepts σ_1, σ_2, but not the others. There were several problems: the variant of ess subsumption preorder was not transitive, and it did not deal properly with sorted unifiers, for example one like σ_3, and the number of most general unifiers was too high. I had a proposal for a subsumption preorder of unifiers, which has all the necessary nice properties, but depends on the set of variables in the unification problem. Around this time, I also managed to produce a proof that shows that there is a natural equational theory (idempotent semigroups) that has unification type zero (nullary). Coincidentally, Jörg Siekmann, my doctoral-supervisor at this time, got a submission of a paper that shows exactly the same result, by an unknown student of Mathematics, Franz Baader. I looked at the paper, and checked whether it was correct: yes, it was correct. But more surprising for me was that Franz used also exactly the same subsumption preorder that I preferred as a definition. As a side remark, Franz detected an error in my paper, which I could correct later (see [1,15]).

The type zero result (later called nullary) was assessed as important at this time, and the subsumption preorder discussion as minor, however, in retrospect, the latter turns out to have more influence.

The interest of Franz in unification theory, unification type and related properties was always active. He was coauthor of two often cited overviews on unification [8,9]. He undertook deep analyses of the unification properties of commutative theories [2]. A more recent investigation into unification types was presented in the unification workshop 2016 in Warsaw [5], where we exchanged ideas on unification and subsumption preorders. He also put unification to work in description logics, which is still a hot topic, in particular see [4,6].

1.2   Practical Advice

This is another motivation to write down these notes: Submitted papers dealing with unification in a substantial percentage claim that they use the ess-subsumption preorder, however, at other places in their papers the authors implicitly use another subsumption preorder. Fortunately, in most cases there are no effects on the technical correctness of such papers, since the definition is not used in a technical sense. So the opportunity of contributing to the Festschrift is a motivation for me to add this apparently not deep, but as I hope, helpful remarks.

2   Subsumption Preorders and Unification Type

In this section we recall four subsumption preorders. The plan is to discuss pros and cons and summarize some known results.

2.1 The Unification Problem and Several Subsumption Preorders

First we describe a frame for unification problems, where the intention is to cover most forms of unification, but not all variants.

A general one would be to consider existential formula in some logic. Here is our working description:

There is a term-language $L[X]$ with an infinite set X of variables, a ground language $L \subseteq L[X]$ and an equivalence relation $\equiv$ on $L[X]$. There is also a notion of an occurrence of a variable x in a term s. We assume that functions $\phi : X \to L[X]$ define a unique mapping (also called substitution) $\phi : L[X] \to L[X]$, like a replacement of variables by terms. We can compute the domain, codomain, and the set $VRan(.)$ of variables occurring in the codomain of a substitution. There is also an equivalence relation $\equiv$ on substitutions: The following consistencies must hold:

1. $s \equiv t$ implies $s\rho \equiv t\rho$ for all substitutions ρ.
2. $\sigma_1 \equiv \sigma_2$ implies $s\sigma_1 \equiv s\sigma_2$ for all terms s.
3. $x\sigma_1 \equiv x\sigma_2$ for all $x \in Var(s)$ implies $s\sigma_1 \equiv s\sigma_2$ for terms s.

As a generic notion we also consider constrained substitutions $(C[X], \sigma)$, where $C[X]$ is a constraint (from a fixed language). The semantics of $(C[X], \sigma)$ is the set of ground instances that satisfy C; a bit more formally: the set $\{\sigma \circ \rho \mid C[X]\rho$ is valid$\}$. This deviates from the substitution view, and there might be some obscure selection of instances. However, it makes sense in many applications, where C is usually of restricted expressiveness.

- A *unification problem* is given by a set Γ of equations to be solved: $\Gamma = \{s_1 \doteq t_1, \ldots, s_n \doteq t_n\}$.
- A *ground unifier (ground solution)* of Γ is a substitution $\sigma : X \to L$, such that $s_1\sigma \equiv t_1\sigma, \ldots, s_n\sigma \equiv t_n\sigma$ holds in L.
- A *general unifier (solution)* of Γ is a substitution $\phi : X \to L[X]$, such that $s_1\phi \equiv t_1\phi, \ldots, s_n\phi \equiv t_n\phi$ holds in $L[X]$.

In the constrained expression case a general unifier is a substitution where only instances are permitted that satisfy C. Insofar it can be seen as a partially defined instantiation function.

Of course, from an application point of view, a single general unifier covering all solutions/unifiers is preferable over larger sets of general unifiers, or even a large (infinite) set of solutions. The constrained expression method was for example successfully applied to have small general sets in a form of higher-order unification [17].

The following are variants of subsumption preorders on two general solutions ϕ_1, ϕ_2 of a unification problem Γ.

- *ess*-subsumption: (exists substitution subsumption) $\phi_1 \leq_{ess} \phi_2$ (ϕ_1 is ess-more general than ϕ_2): There exists a substitution $\rho : L[X] \to L[X]$, such that $\phi_1 \circ \rho \equiv \phi_2$.

- *vrs*-subsumption: (variable-restricted subsumption) $\phi_1 \leq_{vrs} \phi_2$ (ϕ_1 is vrs-more general than ϕ_2): There exists a substitution $\rho : L[X] \to L[X]$, such that $(x)\phi_1 \circ \rho \equiv (x)\phi_2$ for all variables x occurring in Γ
- *lrs*-subsumption (language-restricted subsumption) (also called exactness preorder [5]): $\phi_1 \leq_{lrs} \phi_2$ (ϕ_1 is lrs-more general than ϕ_2): for all expressions s, t in the language such that $Var(s) \subseteq Var(\Gamma)$ and $Var(t) \subseteq Var(\Gamma)$: $s\phi_1 \equiv t\phi_1 \implies s\phi_2 \equiv t\phi_2$. [10, 11].
- *sem*-subsumption: (semantical subsumption) $\phi_1 \leq_{sem} \phi_2$ (ϕ_1 is sem-more general than ϕ_2): $S_2 \subseteq S_1$ holds, where S_i is the set of all ground solutions represented by ϕ_i for $= 1, 2$, and the comparison is modulo $\equiv$, extended from expressions to substitutions. This means $S_i = \{\sigma \mid dom(\sigma) \subseteq Var(\Gamma), \sigma : L[X] \to L$ is a ground solution of Γ, and there is some $\rho : L[X] \to L$ such that for all $x \in Var(\Gamma)$: $(x)\sigma \equiv (x)(\phi_i \circ \rho)$ $\}$.

Remark 2.1.

1. The vrs-subsumption preorder is defined w.r.t. $Var(\Gamma)$. It could also be generalized to supersets of $Var(\Gamma)$ in order to match other applications of unification like Knuth-Bendix completion. For the purposes of this paper, the unification problem is always implicitly given, hence it is sufficient to restrict vrs-subsumption to $Var(\Gamma)$.
2. The semantical subsumption preorder in the equational theory with defining axiom $f(x, 0) = x$ and where only the symbols $f, 0$ and variables are permitted implies that the identity substitution and the substitution $\sigma_1 = \{x \mapsto 0\}$ subsume each other, since the ground terms consist only of the single element 0 modulo the equation. This is sometimes called the *elementary* case. If further constants are permitted, then these semantical subsumption relations do not longer hold, since the set of ground terms is richer.
   We encode this distinction (with/without constants) in the language $L[X]$, which also fixes the available signature, symbols and terms.

A first insight into the subsumption preorders is their relative strength.

Proposition 2.2. *The subsumption preorders are ordered by subset-relations as follows:*
$$\leq_{ess} \subseteq \leq_{vrs} \subseteq \leq_{lrs} \subseteq \leq_{sem}.$$

Proof. The first subset-relation holds, since in the first case the comparison is on all variables, whereas in the second on less variables. The second subset-relation holds, since if a subsumed substitution solves a problem then also the subsumer. The third subset-relation is again valid since the set in the sem-subsumptions is restricted to ground terms. □

Definition 2.3. *Let Γ be a unification problem, M be a set of unifiers of Γ, $X = Var(\Gamma)$, and let $\leq_s$ be a subsumption preorder. A set M_c of unifiers is $\leq_s$-complete for M, if for every $\sigma \in M$, there is some $\tau \in M_c$, such that $\tau \leq_s \sigma$. We also say the set M_c is minimal, if no subset of M_c is complete for M.*

The unification type of a problem class is the number of minimal (w.r.t. the subsumption preorder $\leq_s$) general unifiers (constrained unifiers) that are necessary to represent sets of unifiers in a given class of unification problems.

- If at most one most general unifier is required for every set of unifiers of every Γ: the problem class is *unitary*.
- If finitely many general unifiers are always sufficient, but the problem class is not unitary: the problem class is *finitary*.
- If in every case there is a minimal infinite set that is complete, and sometimes a minimal infinite complete set of unifiers is necessary: the problem class is *infinitary*.
- If sometimes there is no minimal complete set of unifiers for a unification problem: the problem class is *nullary*.

In order to have a more systematic picture and nicer relationships between the various subsumption preorders, and since it is known that the subsumption type of a problem class may change if the subsumption preorder changes, we will partition the problem class nullary further. A motivation is that we can show a theorem (see Theorem 2.4) on the possible changes of unification type triggered by switching the subsumption preorder. It turns out that using the unitary, finitary, infinitary, and nullary classification, the prediction of the changes of the unification type is rather weak.

Therefore we generalize the notion of a complete set as follows. We represent a set of unifiers M by a set M' that may contain: single unifiers from M, and countably infinite linear strictly descending (w.r.t. the subsumption preorder) chains of unifiers (which must be in M). I.e. the set M' may contain single unifiers and sets of unifiers.
The set M' must cover all unifiers in M w.r.t. the chosen subsumption preorder $\leq_s$ as follows: For every unifier $\sigma \in M$, either there exists a single unifier $\sigma' \in M'$ such that $\sigma' \leq_s \sigma$, or there exists a chain $\{\sigma_1 > \sigma_2 > \ldots\}$ in M', and an i, such that $\sigma_i \leq \sigma$. In this case we say that M' is complete for M.
The further partitioning of nullary is as follows:

- 1-nullary: The problem class is nullary and for all unification problems, the complete set M' of M can always be chosen to be of cardinality 1.
- n-nullary: The problem class is nullary and for all unification problems, the complete set M' of M can be chosen as a finite set, and the problem class is not 1-nullary.
- ∞-nullary: The problem class is nullary, but not n-nullary and not 1-unary: This means there is at least one Γ_1, such that the set of unifiers M_1 needs a complete set M_1' with at least one descending chain; and there is at least one Γ_2 (perhaps different from Γ_1) such that its set of unifiers M_2 needs an infinite complete set M_2' of M_2.

Currently, there are no investigations that show that there are equational theories of one of these unification types. However, it is known that there are nullary equational theories, hence at least one of the classes 1-nullary, n-nullary, and ∞-nullary is not trivial. To determine the exact type of nullary theories within this classification schema is future work.

We define the following transfer-relation $\rightarrow$ between these unification types, where we will later use the reflexive transitive closure $\xrightarrow{*}$. Intuitively, the transfer relation shows the possible outcomes of the operation of adding $\leq$-relationships between unifiers.

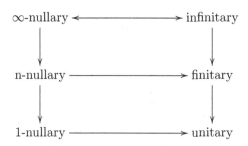

The picture suggests that the worst unification types are ∞-nullary and infinitary, and that n-nullary, finitary are the moderate cases, and 1-nullary and unitary are the good ones.

The relation of the unification type to the subsumption preorders is as follows: if the subsumption preorder increases, then the unification type may decrease along $\xrightarrow{*}$:

Theorem 2.4. *If we vary for a problem class the subsumption preorder spo, then we obtain the following relation to the unification type:*
If $spo_1 \leq spo_2$, then $type_1 \xrightarrow{*} type_2$.

Proof. The argument is simply that if $spo_1 \leq spo_2$, then in a set of unifiers, w.r.t. spo_2, there may be more $\leq$-relations between unifiers in a set, and the picture above shows the possible modifications in the structure of a set of unifiers.

This theorem may lead to a more systematic investigations of the changes of unification type of theories triggered by the change of the subsumption preorder.

3 Varying the Subsumption Preorder: Examples

3.1 First-Order Unification

The Ess-Subsumption Preorder in First-Order Unification. Assume that $L[X]$ consists of the first order terms with set of variables X and that unification is first-order unification [7]. The solutions (i.e., the *unifiers*) of unification problems Γ are substitutions, which are (free) mappings from terms to terms, with a finite domain, usually written as $\{x_1 \mapsto t_1, \ldots, x_n \mapsto t_n\}$. The domain of a substitution σ is the set $\{x \mid \sigma(x) \neq x\}$; the codomain is the set $\{\sigma(x) \mid x \in dom(\sigma)\}$, and the set of variables of the codomain is $VRan(\sigma)$.

Definition 3.1 (The ess-subsumption preorder in first-order unification). *Two unifiers σ, σ' of a first-order unification problem Γ are in the ess-subsumption preorder (σ is more general than σ'), denoted as $\sigma \leq_{ess} \sigma'$, iff there is another substitution ρ, such that $\sigma \circ \rho = \sigma'$, where equality means equality of functions.*

This ess-subsumption was studied for example in [13].

We reconsider the example above, and also add some consequences:

Example 3.2. Let the equation be $x \doteq y$. Some unifiers are

$$
\begin{aligned}
\sigma_1 &= && \{x \mapsto y\} \\
\sigma_2 &= && \{y \mapsto x\} \\
\sigma_3 &= && \{y \mapsto z, x \mapsto z\} \\
\sigma_4 &= && \{y \mapsto 0, x \mapsto 0\} \\
\sigma_5 &= \{y \mapsto z, x \mapsto z, w \mapsto 0\}
\end{aligned}
$$

The substitution σ_1 is a most general unifier. Using the ess-subsumption preorder, we get

$$
\begin{aligned}
\sigma_2 &= \sigma_1 \circ \sigma_2 \\
\sigma_3 &= \sigma_1 \circ \sigma_3 \\
\sigma_4 &= \sigma_1 \circ \sigma_4 \\
\sigma_5 &= \sigma_1 \circ \sigma_5
\end{aligned}
$$

which is nice, and gives some confidence into this formalism. Also σ_2 is a most general unifier.

However, the ess-subsumption preorder is not very flexible: For example σ_3 is like a renamed most general unifier, but according to the subsumption preorder, it is not most general: There is no substitution ρ such that $\sigma_1 = \sigma_3 \circ \rho$. The reason is that $\rho(z)$ must be x or y, however, then $z \mapsto x$ or $z \mapsto y$ would be a component of σ_1, which prevents σ_3 from being most general.

As a general observation we have:

Lemma 3.3. *Let Γ be a first-order unification problem. If the ess-subsumption preorder is used, then for every most general unifier σ the relations $dom(\sigma) \subseteq Var(\Gamma)$ and $VRan(\sigma) \subseteq Var(\Gamma)$ hold.*

The Vrs-Subsumption Preorder in First-Order Unification is defined as follows.

Definition 3.4 (The vrs-subsumption preorder). *Two unifiers σ, σ' of a unification problem Γ are in the vrs-subsumption relation, $\sigma \leq \sigma'$, iff there is another substitution ρ, such that $(x)\sigma \circ \rho = (x)\sigma'$ for all $x \in Var(\Gamma)$.*

Using this preorder, we see that besides σ_1, σ_2, also σ_3 and σ_5 are most general unifiers of $x \doteq y$:

Any unifier $\sigma = \{x \mapsto s, y \mapsto s, \ldots\}$ is covered by σ_3 as follows:

$$(x)\sigma_3 \circ \{z \mapsto s\} = \sigma(x)$$
$$(y)\sigma_3 \circ \{z \mapsto s\} = \sigma(y)$$

This renaming possibility is an (obvious) feature of most general unifiers w.r.t. the vrs-subsumption preorder.

Summary for First-Order Unification The most general unifier property and the unification type for first-order terms is independent of the choice of the subsumption preorder of substitutions.

However, using the ess-subsumption preorder, one has to be aware that most general unifiers can only use (in domain and codomain) variables from the problem Γ. Note that this must also hold for the respective unification algorithms if these are claimed to be correct w.r.t. the ess-subsumption preorder.

3.2   Unification of Terms Modulo Equational Theories

We consider unification of term-equations w.r.t. equational theories.

The basics are that there is a congruence $=_E$, on the set of terms, where $=_E$ is also called equational theory. The task is to solve equations modulo $=_E$ and in case there is a unifier, also to compute a most general unifier or a complete set of unifiers. The general case is that more than one unifier is required to represent all unifiers of a problem. Usually, the vrs-subsumption preorder is employed. However, if the ess-subsumption preorder is chosen, then the situation gets confusing.

Lemma 3.5. *If for some Γ, there is an ess-minimal unifier σ with $dom(\sigma) \subseteq Var(\Gamma)$, and where $Var(\sigma(Var(\Gamma))) \nsubseteq Var(\Gamma)$, then there is an infinite number of incomparable minimal unifiers.*

Proof. The reason is that renamed variants of σ are minimal, but incomparable using the ess-subsumption preorder: Let $Var(\sigma(Var(\Gamma))) \setminus Var(\Gamma) = \{z_1, \ldots, z_n\}$. The intuition is as follows: Another unifier is σ', where $\{z_1, \ldots, z_n\}$ is replaced by $\{z_1', \ldots, z_n'\}$ respectively. Again there is no ρ with $\sigma \circ \rho = \sigma'$ and vice versa: The substitution ρ would introduce components into the substitution that modify z_1, which is not a part of σ'.

Now we exhibit a concrete example of an equational theory for the effect mentioned in the previous lemma.

Proposition 3.6. *There is an equational theory E_0 and a set of equations Γ with $|Var(\Gamma)| = 2$, and a minimal unifier σ of Γ such that $|VRan(\sigma)| = 3 > 2 = |Var(\Gamma)|$. The unification type of Γ w.r.t. ess-subsumption is infinitary, whereas the unification type of Γ w.r.t. vrs-subsumption is unitary.*

Proof. Let the single axiom of theory E_0 be $\{f_1(g(x,y,z)) = f_2(a)$, and let $\Gamma = \{f_1(x) \doteq f_2(y)\}$. Then $\sigma = \{x \mapsto g(x_1, y_1, z_1)), y \mapsto a\}$ is a unifier. Usual reasoning shows that this is a minimal unifier. Since the names of variables are irrelevant, the same holds for $\sigma' = \{x \mapsto g(x_2, y_2, z_2)), y \mapsto a\}$. The question is whether these unifiers are independent from each other.

Using ess-subsumption, an ess-subsumption of σ by σ' requires a substitution ρ with $\sigma' \circ \rho =_E \sigma$. A potential candidate for ρ is $\rho = \{x_2 \mapsto x_1, y_2 \mapsto y_1, z_2 \mapsto z_1\}$. Computing the result of $\sigma' \circ \rho$ results in $\{x \mapsto g(x_1, y_1, z_1)), y \mapsto a, x_2 \mapsto x_1, y_2 \mapsto y_1, z_2 \mapsto z_1\}$, which is different from σ, since there are components mapping x_1, y_2, z_1 to expressions different x_1, y_2, z_1, respectively. Obviously, these components cannot be eliminated, and since the situation is symmetric, the result is that the two unifiers σ, σ' are independent from each other. Hence there is an infinite number of independent minimal unifiers. Indeed Γ requires an infinite number of minimal unifiers.

Now let us use the vrs-subsumption preorder: Then the comparison is only on the variables $Var(\Gamma) = \{x, y\}$. Then the vrs-subsumption relation between σ and σ' holds. □

Another example due to Franz is the theory ACUI (i.e. the equational theory of a binary function symbol that is associative, commutative, idempotent and there is a unit) [3] which is of ess-unification type nullary or infinitary, and which is of unification type unitary for the vrs-subsumption preorder.

3.3 Known Results on the Unification Type W.r.t. the Various Subsumption Preorders

Interesting examples for unification types of equational theories are the commutative theories [2]. These are of unification type unitary or nullary for the vrs-subsumption preorder as well as for the lrs-subsumption preorder. Also the theory ACUIh is of type nullary for vrs-subsumption preorder and lrs-subsumption preorder [5]. Theories where the unification type improves by using the lrs-subsumption preorder are exhibited in [10]: The theories of idempotent semi-groups and distributive lattices are both nullary w.r.t. the vrs-subsumption preorder, and are finitary or unitary, respectively, for the lrs-subsumption preorder.

Here, I can even put a question to Franz: is the theory of commutative monoids either of type unitary or 1-nullary (w.r.t elementary unification) and w.r.t. vrs-subsumption preorder?

3.4 Unification Using Sorts or Types

Consider the following unification example: The equation is $x : S_1 \doteq y : S_2$ w.r.t. a first-orderterm algebra of three sorts: S_1, S_2 and a common subsort S_3. The restriction for substitutions is that the instantiation for every variable x must have the same sort, or lower it. Then a most general unifier of $x : S_1 \doteq y : S_2$ must look like $\sigma = \{x \mapsto z : S_3, y \mapsto z : S_3\}$. The ess-subsumption preorder will result in infinitely many minimal unifiers in the set of all unifiers, hence the unification type of first-order sorted unification would be infinitary. However, if we apply the vrs-subsumption preorder, then one most general unifier is sufficient, and we get that first-order sorted unification is unitary, provided the sort structure is a lower semi-lattice. The unification behavior is then as expected.

3.5   Nominal Unification

I will add some notes on unification that results in singleton sets of minimal unifiers together with constraints, which could be called unitary.. I know that Franz is not fond of this, since it permits cheating. For example, you simply can say the set of equations itself is the constraint representing the solutions, and hence it is unitary. He is right that the unification type for constrainted representations cannot be compared to the unification type if unifiers are explicitly computed substitutions.

Nevertheless, practical need is to opt for small sets of general unifiers (with constraints) which eases further processing. Constraints are a chance to lazily compute unifiers or solutions and already use the partial results in applications.

An interesting example is nominal unification, which can be roughly described as unification of expressions in higher-order languages modulo permitted name changes (i.e. modulo α-equivalence). A nice result is a polynomial-time unification algorithm computing a unique unifier together with a constraint (see [17] for the first algorithm) and [12,14] for quadratic algorithms. Also, there are generalizations of nominal unification that keep the property of being unitary [16].

Acknowledgements. I thank the editors, the anonymous reviewers for their help in making this paper possible, I thank the members of the orthopedics department of the hospital in Heppenheim, and my wife Marlies, for curing me and supporting me during writing.

References

1. Baader, F.: The theory of idempotent semigroups is of unification type zero. J. Autom. Reasoning **2**(3), 283–286 (1986)
2. Baader, F.: Unification in commutative theories. J. Symb. Comput. **8**(5), 479–497 (1989)
3. Baader, F.: Remarks on ACUI, personal communication (2016)
4. Franz, B., Borgwardt, S., Morawska, B.: Extending unification in EL to disunification: the case of dismatching and local disunification. Log. Methods Comput. Sci. **12**(4) (2016)
5. Baader, F., Ludmann, P.: The exact unification type of commutative theories. In: Ghilardi, S., Schmidt-Schauß, M. (eds.) Informal Proceedings of the 30th International Workshop on Unification (UNIF 2016) (2016)
6. Baader, F., Morawska, B.: Unification in the description logic $\mathcal{EL}$. In: Treinen, R. (ed.) RTA 2009. LNCS, vol. 5595, pp. 350–364. Springer, Heidelberg (2009). https://doi.org/10.1007/978-3-642-02348-4_25

7. Baader, F., Nipkow, T.: Term Rewriting and All That. Cambridge University Press, Cambridge (1998)
8. Baader, F., Siekmann, J.H.: Unification theory. In: Gabbay, D.M., Hogger, C.J., Robinson, J.A., Siekmann, J.H. (eds.) Handbook of Logic in Artificial Intelligence and Logic Programming, Deduction Methodologies, vol. 2, pp. 41–126. Oxford University Press (1994)
9. Baader, F., Snyder, W.: Unification theory. In: Robinson, J.A., Voronkov, A. (eds.) Handbook of Automated Reasoning, vol. 2, pp. 445–532. Elsevier and MIT Press (2001)
10. Cabrer, L.M., Metcalfe, G.: From admissibility to a new hierarchy of unification types. In: Kutsia, T., Ringeissen, C. (eds.) Proceedings of the 28th International Workshop on Unification (UNIF 2014) (2014)
11. Cabrer, L.M., Metcalfe, G.: Exact unification and admissibility. Log. Methods Comput. Sci. **11**(3) (2015)
12. Calvès, C., Fernández, M.: A polynomial nominal unification algorithm. Theor. Comput. Sci. **403**(2–3), 285–306 (2008)
13. Eder, E.: Properties of Substitutions and unifications. In: Neumann, B. (ed.) GWAI-83. Informatik-Fachberichte, vol. 76, pp. 197–206. Springer, Heidelberg (1983). https://doi.org/10.1007/978-3-642-69391-5_18
14. Levy, J., Villaret, M.: An efficient nominal unification algorithm. In: Lynch, C. (ed.) Proceedings of the 21st RTA, volume 6 of LIPIcs, pp. 209–226. Schloss Dagstuhl (2010)
15. Schmidt-Schauß, M.: Unification under associativity and idempotence is of type nullary. J. Autom. Reasoning **2**(3), 277–281 (1986)
16. Schmidt-Schauß, M., Sabel, D., Kutz, Y.D.K.: Nominal unification with atom-variables. J. Symb. Comput. **90**, 42–64 (2019)
17. Urban, C., Pitts, A., Gabbay, M.: Nominal unification. In: Baaz, M., Makowsky, J.A. (eds.) CSL 2003. LNCS, vol. 2803, pp. 513–527. Springer, Heidelberg (2003). https://doi.org/10.1007/978-3-540-45220-1_41

15 Years of Consequence-Based Reasoning

David Tena Cucala$^{(\boxtimes)}$, Bernardo Cuenca Grau, and Ian Horrocks

University of Oxford, Oxford OX1 3QD, UK
{david.tena.cucala,bernardo.cuenca.grau,ian.horrocks}@cs.ox.ac.uk

Abstract. Description logics (DLs) are a family of formal languages for knowledge representation with numerous applications. Consequence-based reasoning is a promising approach to DL reasoning which can be traced back to the work of Franz Baader and his group on efficient subsumption algorithms for the $\mathcal{EL}$ family of DLs circa 2004. Consequence-based reasoning combines ideas from hypertableaux and resolution in a way that has proved very effective in practice, and it still remains an active field of research. In this paper, we review the evolution of the field in the last 15 years and discuss the various consequence-based calculi that have been developed for different DLs, from the lightweight $\mathcal{EL}$ to the expressive $\mathcal{SROIQ}$. We thus provide a comprehensive and up-to-date analysis that highlights the common characteristics of these calculi and discusses their implementation.

Keywords: Description Logics · Automated reasoning · Ontologies · Knowledge representation

1 Introduction

Description logics (DLs) are a prominent family of languages for knowledge representation and reasoning with well-understood formal properties [3]. Interest in DLs has been spurred by their applications to the representation of ontologies: for instance, the DL $\mathcal{SROIQ}$ provides the formal underpinning for the Web Ontology Language OWL 2 [46].

A central component of most DL applications is a scalable reasoner, which can be used to discover logical inconsistencies, classify the concepts of an ontology in a subsumption hierarchy, or answer database-style queries over an ontology and a dataset. Two traditional approaches to concept classification (and to DL reasoning more broadly) are tableaux [4] and resolution [9].

Tableau and hyper-tableau calculi underpin many of the existing DL reasoners [17,18,40,41,44]. To check whether a concept subsumption relationship holds, (hyper-)tableau calculi attempt to construct a finite representation of an ontology model disproving the given subsumption. The constructed models can, however, be large—a source of performance issues; this problem is exacerbated in classification tasks due to the large number of subsumptions to be tested.

Another major category of DL reasoning calculi comprises methods based on first-order logic resolution [9]. A common approach to ensure both termination

© Springer Nature Switzerland AG 2019
C. Lutz et al. (Eds.): Baader Festschrift, LNCS 11560, pp. 573–587, 2019.
https://doi.org/10.1007/978-3-030-22102-7_27

and worst-case optimal running time is to parametrise resolution to ensure that the calculus only derives a bounded number of clauses [15,21,22,28,34,37]. This technique has been implemented, for instance, in the KAON2 reasoner [32] for $\mathcal{SHIQ}$. Resolution can also be used to simulate model-building (hyper)tableau techniques [20], including blocking methods which ensure termination [16].

Consequence-based (CB) calculi emerged as a promising approach to DL reasoning combining features of (hyper)tableau and resolution [2,10,24,26]. On the one hand, similarly to resolution, they derive formulae entailed by the ontology (thus avoiding the explicit construction of large models), and they are typically worst-case optimal. On the other hand, clauses are organised into *contexts* arranged as a graph structure reminiscent of that used for model construction in (hyper)tableau; this prevents CB calculi from drawing many unnecessary inferences and yields a nice goal-oriented behaviour. Furthermore, in contrast to both resolution and (hyper)tableau, CB calculi can verify a large number of subsumptions in a single execution, allowing for one-pass classification. Finally, CB calculi are very effective in practice, and systems based on them have shown outstanding performance. Leading reasoners for lightweight DLs such as ELK [27] or Snorocket [31] are based on consequence-based calculi. Furthermore, prototypical implementations of consequence-based calculi for more expressive languages, such as Sequoia [11] or Avalanche [45], show promising results.

The first CB calculi were proposed by Franz Baader, Sebastian Brandt, and Carsten Lutz for the $\mathcal{EL}$ family of DLs [2,12]. They were later extended to more expressive logics like Horn-$\mathcal{SHIQ}$ [24], Horn-$\mathcal{SROIQ}$ [35], and $\mathcal{ALCH}$ [38]. A unifying framework for CB reasoning was developed in [39] for $\mathcal{ALCHI}$, introducing the notion of *contexts* as a mechanism for constraining resolution inferences and making them goal-directed. The framework has been extended to the DLs $\mathcal{ALCHIQ}$, which supports number restrictions and inverse roles [10]; $\mathcal{ALCHOI}$, which supports inverse roles and nominals [42]; $\mathcal{ALCHOQ}$, supporting nominals and number restrictions [23], and finally to $\mathcal{ALCHOIQ}$, which supports all of the aforementioned constructs [43].

This paper reviews the development of the consequence-based approach to DL reasoning since its first appearance fifteen years ago. In Sect. 2 we introduce the core ideas behind consequence-based reasoning, using a simplified version of the original CB calculus in [12]. In Sect. 3 we discuss the evolution of consequence-based calculi in the first decade after the introduction of the original calculus, which focused mostly in lightweight or Horn DLs. In Sect. 4 we discuss the introduction in [39] of a unifying framework for consequence-based reasoning. This piece of work describes an abstract structure which more explicitly captures the defining features of consequence-based calculi. By varying the parameters in this structure, it becomes possible to simulate many previously existing calculi. Finally, in Sect. 5 we discuss recent progress in consequence-based reasoning, including the design of calculi for more expressive DLs, and the introduction of hybrid methods that combine the consequence-based approach with other well-known reasoning techniques such as tableaux or integer linear programming.

We assume familiarity with the basics of Description Logics, and refer the reader to [6] for a comprehensive introduction to DLs.

2 Consequence-Based Reasoning

This section introduces the consequence-based approach to DL reasoning. We start by describing one of the simplest and best known consequence-based calculi: the classification algorithm for $\mathcal{EL}$. We next give a summary of the common features of consequence-based calculi, and use the $\mathcal{EL}$ calculus as an illustrative example of these characteristics.

2.1 The $\mathcal{EL}$ Consequence-Based Calculus

The calculus presented in this section is a restricted form of the classification procedure given in [12]. The original presentation used a notation different from the terminology used in this section, which is due to [24] and has been used in many of the subsequent consequence-based calculi, as well as in a recent textbook on Description Logics [6].

Consider an arbitrary $\mathcal{EL}$ ontology $\mathcal{O}$ in normal form, which is defined as a set of axioms of the form $A \sqsubseteq B$, $A_1 \sqcap A_2 \sqsubseteq B$, $A \sqsubseteq \exists R.B$, or $\exists R.A \sqsubseteq B$, with A, B atomic concepts, and R an atomic role. It is well-known that any $\mathcal{EL}$ ontology can be normalised to this form in polynomial time. The calculus builds a set $\mathcal{S}$ containing inclusions entailed by $\mathcal{O}$, which are called *consequences* of $\mathcal{O}$. Set $\mathcal{S}$ is initialised by using rules IR1 and IR2 from Fig. 1. Next, the algorithm repeatedly applies rules CR1-CR4 to saturate $\mathcal{S}$. These rules use the existing consequences in $\mathcal{S}$ and the axioms of $\mathcal{O}$ to derive further consequences; e.g., rule CR1 uses consequence $A \sqsubseteq B$ and the fact that $B \sqsubseteq C$ is an axiom of $\mathcal{O}$ to conclude that $A \sqsubseteq C$ is also a consequence.

$$\text{IR1} \; \frac{}{A \sqsubseteq A} \qquad \text{IR2} \; \frac{}{A \sqsubseteq \top} \qquad \text{CR1} \; \frac{A \sqsubseteq B \quad B \sqsubseteq C \in \mathcal{O}}{A \sqsubseteq C}$$

$$\text{CR2} \; \frac{A \sqsubseteq B \quad A \sqsubseteq C \quad B \sqcap C \sqsubseteq D \in \mathcal{O}}{A \sqsubseteq D}$$

$$\text{CR3} \; \frac{A \sqsubseteq B \quad B \sqsubseteq \exists R.C \in \mathcal{O}}{A \sqsubseteq \exists R.C} \qquad \text{CR4} \; \frac{A \sqsubseteq \exists R.B \quad B \sqsubseteq C \quad \exists R.C \sqsubseteq D \in \mathcal{O}}{A \sqsubseteq D}$$

Fig. 1. Inference rules for the simplified $\mathcal{EL}$ calculus

The calculus in Fig. 1 ensures that any subsumption of the form $A \sqsubseteq B$ that is logically entailed by the ontology $\mathcal{O}$ will be contained in $\mathcal{S}$ after saturation; hence, the classification of $\mathcal{O}$ can be read directly from $\mathcal{S}$. The resulting algorithm works in worst-case polynomial time in the size of the input ontology $\mathcal{O}$.

2.2   The Defining Characteristics of Consequence-Based Reasoning

Although consequence-based calculi described in the literature may appear rather different from each other at first glance, they all share several defining characteristics. We discuss these common features using the $\mathcal{EL}$ calculus as an example.

Materialisation of Derived Consequences. Similarly to resolution-based approaches, consequence-based calculi proceed by deriving formulae entailed by the input ontology. This approach can have significant practical advantages over (hyper-)tableau calculi, since these construct (a representation of) a model of the input ontology, and such (representations of) models may be very large. We illustrate this with an example: suppose we would like to check whether the subsumption $B_0 \sqsubseteq C$ is entailed by the following $\mathcal{EL}$ ontology:

$$\{B_i \sqsubseteq \exists R.B_{i+1} \mid 0 \le i \le n-1\} \qquad B_n \sqsubseteq C \qquad \exists R.C \sqsubseteq C$$

$$\{B_i \sqsubseteq \exists S.B_{i+1} \mid 0 \le i \le n-1\} \qquad \qquad \exists S.C \sqsubseteq C$$

If we use a tableau algorithm for this task, the size of the generated model will depend on the order in which inference rules are applied. In particular, building the tableau in a breadth-first manner will lead to an exponentially large model (Fig. 2).

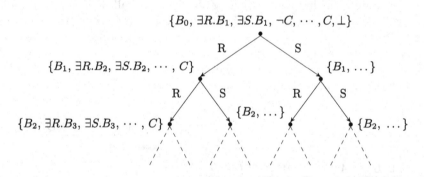

Fig. 2. Model built by a tableau-like procedure. All node labels are derived from the root downwards, except for C, which is derived first at the leaves and then propagated upwards, and $\bot$.

In contrast, the subsumption can be proved using the algorithm of Sect. 2.1 in a linear number of steps. Indeed, initialisation would produce $B_i \sqsubseteq \top$ and $B_i \sqsubseteq B_i$ for each $0 \le i \le n$, together with $C \sqsubseteq \top$ and $C \sqsubseteq C$. Rule CR1 would then produce $B_n \sqsubseteq C$; afterwards, rule CR3 would produce at least one of $B_i \sqsubseteq \exists R.B_{i+1}$ and $B_i \sqsubseteq \exists S.B_{i+1}$ for each $0 \le i \le n$. Finally, CR4 would generate all inferences of the form $B_i \sqsubseteq C$, including our target subsumption.

Locality of Consequences. Another key characteristic of consequence-based calculi is their emphasis on computing "local" consequences only. Unlike methods based on resolution, where many other kinds of entailments can be derived, CB calculi derive only clauses of a very restricted form which constrain only a "central" domain element t and (possibly) domain elements that are directly connected to t via roles (binary predicates). This can be easily seen in the $\mathcal{EL}$ calculus above, where consequences have the same form as ontology axioms, and these are equivalent to first-order clauses universally quantified over a single variable x, which may also introduce an existentially quantified variable y in clauses with a role atom of the form $R(x, y)$ acting as a guard.

Focus on Contexts. The search for local consequences in consequence-based calculi is driven by *contexts* [39], which group similar types of local consequences together. For instance, in the $\mathcal{EL}$ calculus, inference rules always use a premise of the form $A \sqsubseteq \Delta$, where A is an atomic concept, and Δ is a concept. Hence, we can define one context per atomic concept A and group together in the same context all consequences having A in the left-hand-side. This is an advantage with respect to resolution-based methods, for it prevents the derivation of irrelevant clauses. For example, a DL-clause $B_1 \sqcap B_2 \sqsubseteq B_3$, where B_1, B_2, B_3 are atomic concepts, is only used in the $\mathcal{EL}$ calculus if we have derived consequences $A \sqsubseteq B_1$ and $A \sqsubseteq B_2$. In contrast, a resolution method could resolve the same axiom with consequences of the form $A_1 \sqsubseteq B_1$ and $A_2 \sqsubseteq B_2$ even if A_1 and A_2 never appear together in the same context, which suggests that the derived consequence would not be relevant for the taxonomy.

Other Features. The main characteristics described above enable other features that are often regarded as distinctive traits of consequence-based reasoning.

- *Goal-oriented behaviour.* It is often possible to focus only on contexts that are relevant for the given query, as well as contexts that are found to be relevant during the saturation phase. For instance, if we use the $\mathcal{EL}$ calculus to check a single subsumption of the form $A \sqsubseteq B$, we can initialise the algorithm simply with $A \sqsubseteq A$ and $A \sqsubseteq \top$, and then restrict the application of rules IR1 and IR2 only to those atomic concepts that are generated in S during saturation.
- *Reusability of consequences and one-pass classification.* Since consequences derived while answering a query are entailments of the input ontology, they can be re-used in reasoning for any further queries. For instance, in the previous example, any consequences generated while checking whether $B_i \sqsubseteq C$ for some $0 < i < n$ can be reused to determine whether $B_0 \sqsubseteq C$. Therefore, if we initialise the calculus by applying rules IR1 and IR2 to every atomic concept, all subsumptions that follow from an $\mathcal{EL}$ ontology are computed in a single run of the algorithm. Consequences can also be reused when the input ontology $\mathcal{O}$ is extended with new axioms.
- *Parallelisation.* Emphasis on locality makes consequence-based calculi very amenable to parallel implementations, since many inferences in each context

can be done independently of other contexts. Furthermore, if one can predict which contexts will interact little or not at all with each other, the workload can be better divided in order to increase the progress that can be made in parallel.

3   The Early Years: 2004–2013

This section gives a historical perspective of the first 10 years of consequence-based reasoning, which mostly focused on lightweight or Horn DLs.

3.1   The First Consequence-Based Calculus

The first consequence-based calculi were introduced by Franz Baader and colleagues around the year 2004. A first version was presented in [12] for the DL $\mathcal{ELH}$, and this was extended shortly afterwards to handle additional constructs such as the constant for unsatisfiable concepts ($\bot$), concrete domains, and role chains [2]. The $\mathcal{EL}$ family of DLs was starting to receive significant attention at the time: on the one hand, $\mathcal{EL}$ reasoning had been proven to be tractable; on the other hand, the logics in the $\mathcal{EL}$ family had been shown to be expressive enough to capture real-world ontologies. The logic $\mathcal{EL}++$ eventually became the basis for one of the standardised profiles [33] of the ontology language OWL 2.

The first analysis of fragments of $\mathcal{EL}$ used structural subsumption algorithms [5,7], but the new technique introduced in [12] and [2] was radically different. This technique starts by normalising the ontology, following a procedure analogous to that devised in [30], and then it applies a series of inference rules until saturation is reached. The calculus in Sect. 2.1 represents an example of this approach. Although these calculi were not referred to as "consequence-based" at the time, they already displayed the defining features that have been discussed in Sect. 2.2.

The calculus in [2] was implemented in the reasoner CEL [8]; experiments with this system on life-science ontologies showed that efficient reasoning was feasible even for very large ontologies. The elegance and simplicity of this technique was quickly recognised, and it inspired similar methods for more expressive DLs, which we discuss in the following sections.

At the same time, interest was sparked into developing more efficient implementations of consequence-based calculi for the $\mathcal{EL}$ family of DLs. Research in this area has covered topics such as alternative consequence-based calculi for $\mathcal{EL}$ and its variants, efficient strategies for saturation based on tailored data structures and indices, or incremental reasoning. Reasoners such as ELK [27] and Snorocket [31] draw upon this work in order to provide highly efficient, robust, and scalable implementations for lightweight extensions of $\mathcal{EL}$.

3.2   Going Beyond $\mathcal{EL}$: Horn-$\mathcal{SHIQ}$

In 2009, a new consequence-based calculus was introduced in [24] for the DL Horn-$\mathcal{SHIQ}$, which includes expressive constructs not supported in any variant of $\mathcal{EL}$, such as universal restrictions ($\forall R.A$), inverse roles, and number restrictions. The addition of these constructs led to new challenges for the development of consequence-based calculi. In particular, it becomes necessary to allow for the derivation of additional types of consequences in order to ensure completeness. For instance, a consequence like $A \sqsubseteq \exists R.B$ combined with an axiom like $A \sqsubseteq \forall R.C$ entails consequence $A \sqsubseteq \exists R.(B \sqcap C)$, which then becomes relevant for deriving additional atomic subsumptions. The calculus in [24] allows for consequences of the forms $\sqcap_{i=1}^{n} A_i \sqsubseteq B$ and $\sqcap_{i=1}^{n} A_i \sqsubseteq \exists R.(\sqcap_{j=1}^{m} B_j)$, where A_i, B_j are atomic concepts, and R is an atomic role. These inferences may lead to conjunctions that can be as long as the number of atomic concepts in the ontology. This syntax is also useful to represent consequences of axioms involving number restrictions, as seen in the following inference rule from the calculus in [24]:

$$\text{R5} \quad \frac{M \sqsubseteq \exists R_1.N_1 \quad N_1 \sqsubseteq B \quad R_1 \sqsubseteq S \in \mathcal{O}}{M \sqsubseteq \exists R_2.N_2 \quad N_2 \sqsubseteq B \quad R_2 \sqsubseteq S \in \mathcal{O} \quad M \sqsubseteq \,\leq 1S.B}{M \sqsubseteq \exists R_1.(N_1 \sqcap N_2)}$$

In this inference rule, M, N_1, and N_2 are conjunctions of atomic concepts, B is an atomic concept, and R_1, R_2, and S are atomic roles. The calculus shares many desirable properties with the $\mathcal{ELH}$ procedure: it is worst-case optimal (it works in exponential time for a logic that is EXPTIME-complete) and it allows for one-pass classification. Furthermore, it displays a pay-as-you-go behaviour, as it becomes analogous to the $\mathcal{ELH}$ procedure on an $\mathcal{ELH}$ ontology. Such behaviour is very convenient in applications, because the calculus can deal very effectively with ontologies that are "mostly" $\mathcal{ELH}$. This was proved in practice when the calculus was implemented in a reasoner called CB [24], which classified for the first time the full version of the GALEN ontology.

3.3   Reasoning with Nominals: Horn-$\mathcal{SROIQ}$ and $\mathcal{ELHO}$

Although the calculus in [2] for $\mathcal{EL}++$ included nominals, it was later found to be incomplete for handling them [26]. Complete consequence-based calculi for logics involving nominals were later developed for Horn-$\mathcal{SROIQ}$ [35] and $\mathcal{ELHO}$ [26]. In order to ensure completeness, the calculus in [26] had to keep track of "conditional" consequences, which only hold in models where some atomic concepts are non-empty. To see why this may be necessary, observe that an axiom of the form $C \sqsubseteq \{o\}$ in an ontology splits the models of the ontology into two kinds: those in which C is interpreted as the empty set and those where C is interpreted as a singleton; this obviously affects also the subconcepts of C.

Both of the aforementioned calculi introduced additional syntax to keep track of this kind of consequences. The Horn-$\mathcal{SROIQ}$ calculus in [35] introduces a predicate rel which identifies non-empty concepts and inference rules to propagate this predicate when necessary. Similarly, the $\mathcal{ELHO}$ calculus from [26] introduces a new type of consequences which can be written as $G : C \sqsubseteq D$. This represents that inclusion $C \sqsubseteq D$ holds whenever concept G is not empty. Furthermore, the calculus also uses the dependency $G \rightsquigarrow H$ to represent that concept H is non-empty whenever concept G is non-empty. Notice that such consequences are not always "local" in the sense described in Sect. 2.2; indeed, the restriction to local consequences is difficult to marry with the presence of nominals in ontologies. It is often possible, however, to carefully constrain the syntax of consequences so that, in the absence of nominals, local forms are recovered.

Another interesting aspect of the Horn-$\mathcal{SROIQ}$ method is the introduction of a predicate equal to represent equality between individuals, together with inference rules that ensure such relations are congruences. Later calculi for DLs with number restrictions and/or nominals also require the use of equalities, as we show in Sect. 5.1, but instead of axiomatising equality, they use paramodulation-based inference rules [47]. The Horn-$\mathcal{SROIQ}$ calculus introduces also a rule to deal with the "terrible trifecta," a simultaneous interaction between nominals, inverse roles, and functional restrictions. The rule ensures that any two concepts or types related by a role R to the same nominal, and such that the inverse of R is functionally restricted, are satisfied by a single, unique domain element. This is recorded with the help of a special predicate same, which ensures that such a pair behaves, for the purposes of the calculus, like a nominal. Similar strategies have been used for other consequence-based calculi or in other approaches to reasoning for $\mathcal{SHOIQ}$, such as the resolution-based calculus in [28] or the tableau calculus in [19].

Both calculi are worst-case optimal, and the $\mathcal{ELHO}$ algorithm is also pay-as-you-go in the sense that it reduces to the standard $\mathcal{ELH}$ calculus in the absence of nominals.

3.4   Embracing Disjunction: $\mathcal{ALCH}$

The consequence-based calculus presented in [38] for $\mathcal{ALCH}$ was the first to support concept disjunction. The introduction of disjunction leads to difficulties: while previous calculi are such that a canonical model can be built from the saturated set of consequences to disprove any subsumption $A \sqsubseteq B$ absent from this set, such a model may not exist for ontologies with disjunction. The calculus proposed in [38] used an explicit representation of disjunctions which was similar to that used in resolution calculi for fragments of first-order logic. The calculus also introduced ordering, which dramatically reduces the number of inferences to consider [36] and helps to single out a model as "canonical," in case one exists. Some of the more recent calculi have also adopted this approach for dealing with disjunction [10,39].

This calculus, like those for Horn DLs, transforms the input ontology into a normal form, which now allows for conjunctions of concepts of the form A, $\exists R.A$ and $\forall R.A$, and negation in front of atomic concepts. The presence of disjunction

increases the complexity of the representation of consequences, which may now
be of the form:

$$\prod_{i=1}^{n} L_i \sqsubseteq \bigsqcup_{i=1}^{m} A_i \sqcup \exists R. \left(\prod_{i=n+1}^{k} L_i \right),$$

where each expression of the form L_i is an atomic concept or its negation, and
each A_i is an atomic concept. Once again, the calculus is worst-case optimal
and pay-as-you-go, as it simulates the Horn-$\mathcal{ALCH}$ and the $\mathcal{ELH}$ approaches for
ontologies of the corresponding expressivity. One-pass classification is retained.

4   Interlude: A Unifying Framework for CB Reasoning

In 2014, Simančík and colleagues [39] extended the calculus in [38] to $\mathcal{ALCHI}$
while ensuring worst-case optimality, pay-as-you-go behaviour, and one-pass
classification. As part of this work, they introduced a unifying framework
for consequence-based reasoning that explicitly captures many of the aspects
described in Sect. 2.2. This framework describes a graph-based *context structure*, where each node represents a context and the presence of an edge between
contexts indicates that information can be transferred between those contexts.
The set $\mathcal{S}$ of consequences derived by the calculus is split into sets $\mathcal{S}(v)$ associated to each context v. Furthermore, each context is associated to a particular
set of concepts, called the *core* of the context; consequences in $\mathcal{S}(v)$ are relevant
for all elements of a model that satisfy the core. Inference rules are applied to
individual contexts, or to pairs of contexts that share an edge. Edges between
contexts v and w are defined in cases where each element satisfying the core of
v may have an R-filler that satisfies the core of w for some role R; each edge
is labelled by the existential concept generating this connection. Figure 3 shows
how the example of Sect. 2.2 could look like in this framework.

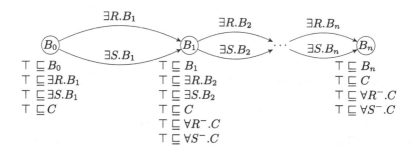

Fig. 3. Context structure for the example in Sect. 2.2. Cores of contexts are written
inside their respective nodes, and below each node all the consequences derived for that
context are listed. The first three inclusions in each of these sets are derived first and
left-to-right; the inclusions involving C are derived afterwards and right-to-left.

Contexts in the example have only atomic concepts as cores; however, conjunctions of concepts are also allowed to represent cores. This may be relevant

whenever consequences of the form $\top \sqsubseteq \exists R.A$ and $\top \sqsubseteq \forall R.B$ appear in the same context; in that case, one may wish to draw an edge to a context with core $A \sqcap B$. However, the calculus also makes it possible to draw an edge to a context with core A. Furthermore, if the context with the chosen core does not exist, the calculus may create it fresh. The policy for deciding how to draw edges between contexts and whether to introduce new contexts or re-use those already in the context structure is a parameter of the calculus called the *expansion strategy*. By suitably choosing this parameter, the framework can be used to simulate some of the calculi from previous sections. For instance, by allowing only cores with atomic concepts, the calculus described in Sect. 2.1 can be simulated; in contrast, to simulate the Horn-$\mathcal{SHIQ}$ calculus in [24], we need a strategy such that, for every relevant conjunction of atomic concepts M, a new context v_M is introduced with core M.

The calculus for $\mathcal{ALCHI}$ discussed in this section does not support nominals, and therefore it cannot simulate the calculi described in Sect. 3.3. However, one can find correspondences between constructs introduced in Sect. 3.3 to deal with nominals, and properties of a context structure created according to a suitable expansion strategy. For instance, assume an expansion strategy that uses a successor context v_A with core A whenever $\top \sqsubseteq \exists R.A$ appears in a context w (and otherwise it re-uses a context $v_\perp$ with empty core); then, the existence of a path from a context u to a context v implies that $\mathsf{core}_u \leadsto \mathsf{core}_v$, where $\leadsto$ is the reachability relation in the $\mathcal{ELHO}$ calculus in Sect. 3.3. Similarly, any context clause $\Gamma \sqsubseteq \Delta$ in a context v reachable from context u can be seen as a conditional consequence $\mathsf{core}_u \colon \Gamma \sqsubseteq \Delta$ from the $\mathcal{ELHO}$ calculus. These correspondences are exploited by some of the calculi for DLs with nominals discussed in Sect. 5.1.

5   Recent Developments: 2014–2019

This section discusses recent developments in consequence-based reasoning, including the extension of calculi to expressive DLs and the hybridisation of this approach with other well-known reasoning techniques.

5.1   Calculi for Expressive DLs

The framework described in Sect. 4 has been recently modified and extended to the DLs $\mathcal{SRIQ}$ [10] and $\mathcal{SROIQ}$ [43]. One of the most noteworthy aspects of these calculi is their use of first-order logic syntax with equality to represent both ontology axioms and derived consequences. This choice is motivated by the need of representing consequences stemming from the interactions between number restrictions, inverse roles, and/or nominals. Equalities participate in inferences that implement well-known paramodulation rules for equality reasoning [47].

These calculi are still pay-as-you-go and retain the one-pass classification property. In addition, the $\mathcal{SRIQ}$ calculus is also worst-case optimal, while the $\mathcal{SROIQ}$ calculus terminates within worst-case optimal bounds except when ontologies feature a simultaneous interaction of disjunction with nominals, inverse roles, and "at-most" quantifiers. In practice, these occurrences are very rare.

The $\mathcal{SRIQ}$ calculus has been implemented in the reasoner Sequoia [11], which has proved to be competitive with state-of-the-art reasoners despite being an early prototype. The main outstanding practical issues in systems such as Sequoia are the following:

- The application of inference rules in the system can lead to the generation of clauses involving a large number of disjuncts. Resolution of such clauses with others can lead to combinatorial explosions due to repetitive inferences. For instance, suppose a clause of the form $\top \to A_1(x) \vee \cdots \vee A_n(x)$ is generated in a context, and suppose the ontology contains clause $A_i(x) \to B_i(x) \vee C_i(x)$ for each $1 \leq i \leq n$. If the ordering between literals in the context makes atoms of the form $B_i(x)$ and $C_j(x)$ smaller than atoms of the form $A_k(x)$, the calculus will derive 2^n clauses in the context. Most of these clauses are typically irrelevant since they do not participate in the derivation of target subsumptions.
- The system performs poorly in the presence of number restrictions involving elements of the model with many different successors by a particular role. For instance, if clauses $\{\top \to R(x, f_i(x)) \mid 1 \leq i \leq n\}$ are derived in a context, and the ontology contains a number restriction enforcing that no element may have more than m successors by R, then the calculus will derive $\binom{n}{m+1}$ different clauses. Furthermore, each of these clauses will have no less than $\binom{m+1}{2}$ disjuncts in the head; these long disjunctions may, in turn, exacerbate the issue discussed above.

5.2 Hybrid Methods

The consequence-based approach has been successfully combined with other well-known reasoning methods that can help overcome some of the aforementioned practical limitations. It has been suggested that the problem of generating long disjunctions may be addressed by means of an algebraic reasoner [28]. There currently exist consequence-based calculi that follow this strategy. The calculus in [45] for the DL $\mathcal{ELQ}$ incorporates calls to an external Integer Linear Programming component that is able to find solutions to algebraic constraint satisfaction problems based on the numeric restrictions appearing in derived consequences. This approach has been extended to the DL $\mathcal{SHOQ}$ in [23]. From a theoretical perspective, it remains unclear whether these calculi are worst-case optimal or whether they show pay-as-you-go behaviour. However, these systems have been implemented into the reasoner Avalanche [45], which shows promising results in ontologies with large number restrictions.

A different kind of hybridisation is proposed in [41], which describes the reasoner Konclude. This system works very efficiently in a wide range of expressive DLs. In contrast to all consequence-based calculi discussed so far, Konclude is based on a tableau calculus, and therefore it attempts to answer queries by building a model. However, Konclude uses an incomplete version of a consequence-based calculus in order to generate as many (sound) consequences of the ontology as possible. These consequences are then used to aid in the construction of the tableau, and the result is a system that is highly competitive even in lightweight description logics such as $\mathcal{EL}$.

Another method for building a hybrid of tableau and consequence-based calculus is presented in [25] for the description logic $\mathcal{ALCHI}$. This calculus is once again based in deriving ontology entailments. However, it deals with disjunction in a way that is very different from the method described in Sect. 3.4. Instead of freely using resolution-like inference rules to generate long disjunctions, this algorithm makes non-deterministic choices during the saturation procedure which are reminiscent of those used by tableau-based algorithms. This calculus can explore alternative branches and backtrack when necessary, unlike previously discussed consequence-based calculi, which are purely deterministic. This prevents the problem of generating long disjunctions. The calculus is also worst-case optimal and enjoys pay-as-you-go behaviour. A prototype of this calculus has been implemented, and the evaluation shows promising results.

The reasoner MORe [1] provides yet another way of combining consequence-based reasoning with other techniques. The algorithm presented in [1] addresses the problem of ontology classification by decomposing an input ontology into modules, and then classifying each module using a reasoner most suited for the language used in that module. MORe has shown it can effectively classify ontologies using a consequence-based reasoner on $\mathcal{ELH}$ modules, and a reasoner such as HermiT [17], which is based on a tableau calculus, on the remaining modules. This approach is particularly useful for ontologies which have most axioms in a lightweight DL and a few axioms in an expressive logic.

6   Conclusion and Future Directions

Consequence-based reasoning has been a very active area of research for the last 15 years, and progress on this field shows no signs of slowing down. There is still no consequence-based calculi covering all expressive features of OWL 2 DL, as there is yet little progress in the area of consequence-based reasoning in expressive DLs with concrete domains. Further research is also needed in the area of optimisations and implementation techniques for consequence-based reasoning, especially for those calculi for the more expressive DLs. Furthermore, it is yet unclear whether hybrid approaches with algebraic reasoning and non-determinism can be effectively implemented for the whole of $\mathcal{SROIQ}$.

The application of consequence-based calculi to problems other than subsumption or classification (such as conjunctive query answering) remains also a fairly unexplored topic. A solid basis for this line of research is provided by "combined" approaches [13,14,29] to query answering that start with a materialisation phase similar to saturation in consequence-based reasoning, which is then followed by query-rewriting techniques.

For all these reasons, we think that consequence-based reasoning will continue being a dynamic area of research, one which holds promise for delivering the next generation of robust, efficient, and scalable reasoners for Description Logics.

References

1. Armas Romero, A., Kaminski, M., Cuenca Grau, B., Horrocks, I.: Module extraction in expressive ontology languages via datalog reasoning. J. Artif. Intell. Res. **55**, 499–564 (2016)
2. Baader, F., Brandt, S., Lutz, C.: Pushing the $\mathcal{EL}$ envelope. In: Proceedings of the 19th International Joint Conference on Artificial Intelligence, pp. 364–369. Morgan Kaufmann Publishers, Edinburgh (2005)
3. Baader, F., Calvanese, D., McGuinness, D., Nardi, D., Patel-Schneider, P.F. (eds.): The Description Logic Handbook: Theory, Implementation and Applications. Cambridge University Press, Cambridge (2003)
4. Baader, F., Sattler, U.: An overview of tableau algorithms for description logics. Studia Logica **69**, 5–40 (2001)
5. Baader, F.: Terminological cycles in a description logic with existential restrictions. In: Proceedings of the Eighteenth International Joint Conference on Artificial Intelligence, pp. 325–330. Morgan Kaufmann Publishers, Acapulco (2003)
6. Baader, F., Horrocks, I., Lutz, C., Sattler, U.: An Introduction to Description Logic. Cambridge University Press, Cambridge (2017)
7. Baader, F., Küsters, R., Molitor, R.: Computing least common subsumers in description logics with existential restrictions. In: Proceedings of the Sixteenth International Joint Conference on Artificial Intelligence, pp. 96–103 (1999)
8. Baader, F., Lutz, C., Suntisrivaraporn, B.: CEL—a polynomial-time reasoner for life science ontologies. In: Furbach, U., Shankar, N. (eds.) IJCAR 2006. LNCS (LNAI), vol. 4130, pp. 287–291. Springer, Heidelberg (2006). https://doi.org/10.1007/11814771_25
9. Bachmair, L., Ganzinger, H.: Resolution theorem proving. In: Robinson, A., Voronkov, A. (eds.) Handbook of Automated Reasoning, pp. 19–99. Elsevier Science, London (2001)
10. Bate, A., Motik, B., Cuenca Grau, B., Simancik, F., Horrocks, I.: Extending consequence-based reasoning to SRIQ. In: Principles of Knowledge Representation and Reasoning: Proceedings of the Fifteenth International Conference, pp. 187–196. AAAI Press, Cape Town (2016)
11. Bate, A., Motik, B., Cuenca Grau, B., Tena Cucala, D., Simancik, F., Horrocks, I.: Consequence-based reasoning for description logics with disjunctions and number restrictions. J. Artif. Intell. Res. **63**, 625–690 (2018)
12. Brandt, S.: Polynomial time reasoning in a description Logic with existential restrictions, GCI axioms, and—what else? In: Proceedings of the 16th European Conference on Artificial Intelligence, Valencia, Spain, pp. 298–302 (2004)
13. Carral, D., Dragoste, I., Krötzsch, M.: The combined approach to query answering in horn-ALCHOIQ. In: Principles of Knowledge Representation and Reasoning: Proceedings of the Sixteenth International Conference, pp. 339–348. AAAI Press, Tempe (2018)
14. Feier, C., Carral, D., Stefanoni, G., Grau, B.C., Horrocks, I.: The combined approach to query answering beyond the OWL 2 profiles. In: Proceedings of the Twenty-Fourth International Joint Conference on Artificial Intelligence, pp. 2971–2977. AAAI Press, Buenos Aires (2015)
15. Ganzinger, H., de Nivelle, H.: A superposition decision procedure for the guarded fragment with equality. In: Proceedings of the 14th IEEE Symposium on Logic in Computer Science, pp. 295–305. IEEE Computer Society, Trento (1999)

16. Georgieva, L., Hustadt, U., Schmidt, R.A.: Hyperresolution for guarded formulae. J. Symb. Comput. **36**(1–2), 163–192 (2003)
17. Glimm, B., Horrocks, I., Motik, B., Stoilos, G., Wang, Z.: HermiT: an OWL 2 reasoner. J. Autom. Reason. **53**(3), 245–269 (2014)
18. Haarslev, V., Hidde, K., Möller, R., Wessel, M.: The RacerPro knowledge representation and reasoning system. Semant. Web **3**(3), 267–277 (2012)
19. Horrocks, I., Sattler, U.: A tableaux decision procedure for $\mathcal{SHOIQ}$. In: Proceedings of the Nineteenth International Joint Conference on Artificial Intelligence, Edinburgh, UK, pp. 448–453 (2005)
20. Hustadt, U., Schmidt, R.A.: Issues of decidability for description logics in the framework of resolution. In: Caferra, R., Salzer, G. (eds.) FTP 1998. LNCS (LNAI), vol. 1761, pp. 191–205. Springer, Heidelberg (2000). https://doi.org/10.1007/3-540-46508-1_13
21. Hustadt, U., Motik, B., Sattler, U.: Deciding expressive description logics in the framework of resolution. Inf. Comput. **206**(5), 579–601 (2008)
22. Hustadt, U., Schmidt, R.A.: Using resolution for testing modal satisfiability and building models. J. Autom. Reason. **28**(2), 205–232 (2002)
23. Karahroodi, N.Z., Haarslev, V.: A consequence-based algebraic calculus for $\mathcal{SHOQ}$. In: Proceedings of the 30th International Workshop on Description Logics. CEUR Workshop Proceedings, Montpellier, France, vol. 1879 (2017)
24. Kazakov, Y.: Consequence-driven reasoning for horn $\mathcal{SHIQ}$ ontologies. In: Proceedings of the 21st International Joint Conference on Artificial Intelligence, Pasadena, CA, USA, pp. 2040–2045 (2009)
25. Kazakov, Y., Klinov, P.: Bridging the gap between tableau and consequence-based reasoning. In: Informal Proceedings of the 27th International Workshop on Description Logics, Vienna, Austria, pp. 579–590 (2014)
26. Kazakov, Y., Krötzsch, M., Simančík, F.: Practical reasoning with nominals in the EL family of description logics. In: Proceedings of the Thirteenth International Conference on Principles of Knowledge Representation and Reasoning (2012)
27. Kazakov, Y., Krötzsch, M., Simančík, F.: The incredible ELK—from polynomial procedures to efficient reasoning with $\mathcal{EL}$ ontologies. J. Autom. Reason. **53**(1), 1–61 (2014)
28. Kazakov, Y., Motik, B.: A resolution-based decision procedure for $\mathcal{SHOIQ}$. J. Autom. Reason. **40**(2–3), 89–116 (2008)
29. Kontchakov, R., Lutz, C., Toman, D., Wolter, F., Zakharyaschev, M.: The combined approach to query answering in DL-Lite. In: Principles of Knowledge Representation and Reasoning: Proceedings of the Twelfth International Conference. AAAI Press, Toronto (2010)
30. Lutz, C.: Complexity of terminological reasoning revisited. In: Ganzinger, H., McAllester, D., Voronkov, A. (eds.) LPAR 1999. LNCS (LNAI), vol. 1705, pp. 181–200. Springer, Heidelberg (1999). https://doi.org/10.1007/3-540-48242-3_12
31. Metke-Jimenez, A., Lawley, M.: Snorocket 2.0: concrete domains and concurrent classification. In: Informal Proceedings of the 2nd International Workshop on OWL Reasoner Evaluation, Ulm, Germany, vol. 1015, pp. 32–38 (2013)
32. Motik, B.: KAON2 - scalable reasoning over ontologies with large data sets. ERCIM News **2008**(72), 19–20 (2008)
33. Motik, B., Cuenca Grau, B., Horrocks, I., Wu, Z., Fokoue, A., Lutz, C.: OWL 2 Web Ontology Language Profiles. http://www.w3.org/TR/owl2-profiles/
34. de Nivelle, H., Schmidt, R.A., Hustadt, U.: Resolution-based methods for modal logics. Log. J. IGPL **8**(3), 265–292 (2000)

35. Ortiz, M., Rudolph, S., Simkus, M.: Worst-case optimal reasoning for the horn-DL fragments of OWL 1 and 2. In: Proceedings of the Twelfth International Conference on the Principles of Knowledge Representation and Reasoning, Toronto, Canada. AAAI Press (2010)
36. Robinson, A., Voronkov, A. (eds.): Handbook of Automated Reasoning. Elsevier, Amsterdam (2001)
37. Schmidt, R.A., Hustadt, U.: First-order resolution methods for modal logics. In: Voronkov, A., Weidenbach, C. (eds.) Programming Logics. LNCS, vol. 7797, pp. 345–391. Springer, Heidelberg (2013). https://doi.org/10.1007/978-3-642-37651-1_15
38. Simančík, F., Kazakov, Y., Horrocks, I.: Consequence-based reasoning beyond Horn ontologies. In: Proceedings of the 22nd International Joint Conference on Artificial Intelligence, pp. 1093–1098. IJCAI/AAAI, Barcelona (2011)
39. Simančík, F., Motik, B., Horrocks, I.: Consequence-based and fixed-parameter tractable reasoning in description logics. Artif. Intell. **209**, 29–77 (2014)
40. Sirin, E., Parsia, B., Cuenca Grau, B., Kalyanpur, A., Katz, Y.: Pellet: a practical OWL-DL reasoner. J. Web Semant. **5**(2), 51–53 (2007)
41. Steigmiller, A., Liebig, T., Glimm, B.: Konclude: system description. J. Web Semant. **27**(1), 78–85 (2014)
42. Tena Cucala, D., Cuenca Grau, B., Horrocks, I.: Consequence-based reasoning for description logics with disjunction, inverse roles, and nominals. In: Proceedings of the 30th International Workshop on Description Logics. CEUR-WS.org, Montpelier (2017)
43. Tena Cucala, D., Cuenca Grau, B., Horrocks, I.: Consequence-based reasoning for description logics with disjunction, inverse roles, number restrictions, and nominals. In: Proceedings of the Twenty-Seventh International Joint Conference on Artificial Intelligence, pp. 1970–1976. ijcai.org, Stockholm (2018)
44. Tsarkov, D., Horrocks, I.: FaCT++ description logic reasoner: system description. In: Furbach, U., Shankar, N. (eds.) IJCAR 2006. LNCS (LNAI), vol. 4130, pp. 292–297. Springer, Heidelberg (2006). https://doi.org/10.1007/11814771_26
45. Vlasenko, J., Daryalal, M., Haarslev, V., Jaumard, B.: A saturation-based algebraic reasoner for ELQ. In: Proceedings of the 5th Workshop on Practical Aspects of Automated Reasoning co-located with International Joint Conference on Automated Reasoning, Coimbra, Portugal, pp. 110–124 (2016)
46. W3C OWL Working Group: OWL 2 Web Ontology Language Overview. http://www.w3.org/TR/owl2-overview/
47. Wos, L., Robinson, G.: Paramodulation and theorem-proving in first-order theories with equality. In: Michie, D., Meltzer, B. (eds.) Proceedings of the 4th Annual Machine Intelligence Workshop. Edinburgh University Press, Edinburgh (1969)

Maximum Entropy Calculations for the Probabilistic Description Logic $\mathcal{ALC}^{\mathsf{ME}}$

Marco Wilhelm$^{(\boxtimes)}$ and Gabriele Kern-Isberner

Department of Computer Science, TU Dortmund, Dortmund, Germany
`marco.wilhelm@tu-dortmund.de`

Abstract. The probabilistic Description Logic $\mathcal{ALC}^{\mathsf{ME}}$ extends the classical Description Logic $\mathcal{ALC}$ with probabilistic conditionals of the form $(D|C)[p]$ stating that "D follows from C with probability p." Conditionals are interpreted based on the aggregating semantics where probabilities are understood as degrees of belief. For reasoning with probabilistic conditional knowledge bases in $\mathcal{ALC}^{\mathsf{ME}}$, the principle of maximum entropy provides a valuable methodology following the idea of completing missing information in a most cautious way. In this paper, we give a guidance on calculating approximations of the maximum entropy distribution for $\mathcal{ALC}^{\mathsf{ME}}$-knowledge bases. For this, we discuss the benefits of solving the dual maximum entropy optimization problem instead of the primal problem. In particular, we show how representations of approximations of the maximum entropy distribution can be derived from the dual problem in time polynomial in the size of the underlying domain. The domain is *the* crucial quantity in practical applications. For a compact representation of the objective function of the dual maximum entropy optimization problem, we apply the principle of typed model counting.

Keywords: Probabilistic Description Logics · Aggregating semantics · Principle of maximum entropy · Polynomial-time optimization · Typed model counting

1 Introduction

Description Logics (*$\mathcal{DL}$s*) [1] constitute a well-investigated family of logic-based knowledge representation languages. In Description Logics it is possible to represent *terminological* (i.e., conceptual) knowledge which can then be used to state *factual* knowledge about single individuals and objects. In many application domains, however, knowledge is not always certain which motivates the development of extensions of Description Logics that deal with uncertainty.

One of the most powerful frameworks for uncertain reasoning is probability theory, as it combines qualitative as well as quantitative aspects of uncertainty. However, reasoning based on probabilities is problematic if the available information is incomplete what is usually the case. For example, the popular Bayesian network approach [6,18] does not work in such cases. Here, we focus on the principle of maximum entropy [15] which yields a unique probability distribution

© Springer Nature Switzerland AG 2019
C. Lutz et al. (Eds.): Baader Festschrift, LNCS 11560, pp. 588–609, 2019.
https://doi.org/10.1007/978-3-030-22102-7_28

that satisfies prescribed conditional probabilities and enriches the incomplete information in a most cautious way [9, 22]. Therewith, it constitutes a most appropriate form of commonsense probabilistic reasoning [16].

In this paper, we investigate maximum entropy calculations for the probabilistic Description Logic $\mathcal{ALC}^{ME}$ [25]. The logic $\mathcal{ALC}^{ME}$ extends the prototypical Description Logic $\mathcal{ALC}$ with the possibility of representing probabilistic conditional statements of the form "if C holds, then D follows with probability p" with probabilities understood as degrees of belief based on the *aggregating semantics* [11]. The aggregating semantics generalizes the statistical interpretation of conditional probabilities by combining it with subjective probabilities based on probability distributions over possible worlds. Therewith, it mimics statistical probabilities from a subjective point of view. The core idea of the aggregating semantics for Description Logics is to define the probability of a concept C by the sum of the probabilities of all possible worlds, here classical $\mathcal{DL}$-interpretations, weighted by the number of individuals that satisfy the concept in the respective possible world, i.e., $\mathcal{P}(C) = \sum_{\mathcal{I}} |C^{\mathcal{I}}| \cdot \mathcal{P}(\mathcal{I})$. This interpretation is in contrast to other approaches for probabilistic Description Logics which handle either subjective [12] or statistical probabilities [19], or are essentially classical terminologies over probabilistic databases [2].

The models of $\mathcal{ALC}^{ME}$-knowledge bases are probability distributions over a set of $\mathcal{DL}$-interpretations that serve as possible worlds. We assume that all these $\mathcal{DL}$-interpretations are defined with respect to the same *fixed finite domain* Δ in order to ensure that they have the same scope and that the counts of individuals considered in the aggregating semantics are well-defined. In many application domains, the size k of the domain Δ is large. At the same time, it crucially influences the costs of maximum entropy calculations, as the size of the sample space, i.e. the number of possible worlds, is exponential in k. In [25] we have shown that drawing inferences in $\mathcal{ALC}^{ME}$ (without assertions) is possible in time polynomial in k once an approximation of the maximum entropy distribution is given. Approximations are necessary since there is no closed form of the maximum entropy distribution in general. The complexity results in [25] are based on sophisticated strategies on consolidating and counting possible worlds based on the notions of conditional structures of possible worlds [10] and of types in Description Logics [20, 21].

The main contributions of the current paper are the following: We extend the probabilistic Description Logic $\mathcal{ALC}^{ME}$ with assertions in an ABox, which have been left out in [25]. We discuss *typed model counting* [24] as a superordinated framework for the counting strategies presented in [25]. The benefit of typed model counting is that it can deal with assertional knowledge which is not directly possible with the typed-based approach in [25]. And, mainly, we investigate polynomial time algorithms for calculating approximations of the maximum entropy distribution. Therewith, we close the gap of calculating the maximum entropy distribution when drawing inferences in $\mathcal{ALC}^{ME}$ which has been left for future work in [25].

Table 1. Compounded concepts and their semantics.

Concept	Syntax	Semantics
Top concept	$\top$	$\Delta^{\mathcal{I}}$
Bottom concept	$\bot$	$\emptyset$
Negation	$\neg C$	$\Delta^{\mathcal{I}} \backslash C^{\mathcal{I}}$
Conjunction	$C \sqcap D$	$C^{\mathcal{I}} \cap D^{\mathcal{I}}$
Disjunction	$C \sqcup D$	$C^{\mathcal{I}} \cup D^{\mathcal{I}}$
Existential restriction	$\exists r.C$	$\{x \in \Delta^{\mathcal{I}} \mid \exists y \in \Delta^{\mathcal{I}} : (x,y) \in r^{\mathcal{I}} \wedge y \in C^{\mathcal{I}}\}$
Universal restriction	$\forall r.C$	$\{x \in \Delta^{\mathcal{I}} \mid \forall y \in \Delta^{\mathcal{I}} : (x,y) \in r^{\mathcal{I}} \Rightarrow y \in C^{\mathcal{I}}\}$

The rest of the paper is organized as follows: In Sect. 2, we recall the syntax and the semantics of the Description Logic $\mathcal{ALC}^{\mathsf{ME}}$ and extend it with assertions. After that, we discuss polynomial time algorithms for calculating approximations of the maximum entropy distribution for $\mathcal{ALC}^{\mathsf{ME}}$-knowledge bases (Sect. 3). Finally, we elaborate on typed model counting for $\mathcal{ALC}^{\mathsf{ME}}$ and highlight its importance for the aforementioned maximum entropy calculations in Sect. 4 before we conclude.

2   The Description Logic $\mathcal{ALC}^{\mathsf{ME}}$

Let $\mathcal{N}_I$, $\mathcal{N}_C$, and $\mathcal{N}_R$ be disjoint sets of individual, concept and role names, respectively. A *concept* is either a concept name or of the form

$$\top, \quad \bot, \quad \neg C, \quad C \sqcup D, \quad C \sqcap D, \quad \exists r.C, \quad \forall r.C,$$

where C and D are concepts and r is a role name. An *interpretation* $\mathcal{I} = (\Delta^{\mathcal{I}}, \cdot^{\mathcal{I}})$ is a tuple of a non-empty set $\Delta^{\mathcal{I}}$ called *domain* and an *interpretation function* $\cdot^{\mathcal{I}}$ that maps every individual name $a \in \mathcal{N}_I$ to an element $a^{\mathcal{I}} \in \Delta^{\mathcal{I}}$, every concept name $C \in \mathcal{N}_C$ to a subset $C^{\mathcal{I}} \subseteq \Delta^{\mathcal{I}}$, and every role name $r \in \mathcal{N}_R$ to a binary relation $r^{\mathcal{I}} \subseteq \Delta^{\mathcal{I}} \times \Delta^{\mathcal{I}}$. The interpretation of compounded concepts is recursively defined as shown in Table 1.

A *general concept inclusion* is a statement of the form $(C \sqsubseteq D)$ where C and D are concepts. An interpretation $\mathcal{I}$ *satisfies* a general concept inclusion $(C \sqsubseteq D)$, written $\mathcal{I} \models (C \sqsubseteq D)$, iff $C^{\mathcal{I}} \subseteq D^{\mathcal{I}}$. Therewith, general concept inclusions allow one to express terminological knowledge like "every individual that has property C also has property D." On the contrary, assertional knowledge is represented in $\mathcal{ALC}^{\mathsf{ME}}$ by statements of the form $A(a)$ or $r(a,b)$ where a and b are individual names, A is a constant name, and r is a role name. While $A(a)$ states that "a has property A," $r(a,b)$ stands for "a is related to b via r." Formally, an *assertion* $A(a)$ holds in an interpretation $\mathcal{I}$, written $\mathcal{I} \models A(a)$, iff $a^{\mathcal{I}} \in A^{\mathcal{I}}$. Analogously, $\mathcal{I} \models r(a,b)$ iff $(a^{\mathcal{I}}, b^{\mathcal{I}}) \in r^{\mathcal{I}}$. General concept inclusions and assertions express strict knowledge that is certainly true.

In many application domains like medicine, however, it is necessary to formalize uncertain knowledge. For example, the heart of most humans is located on the left-hand side of their thorax. However, there are a few people with their heart on the right-hand side. In order to deal with this kind of uncertain knowledge, we equip the language $\mathcal{ALC}^{\mathsf{ME}}$ with *(probabilistic) conditionals*. A conditional $(D|C)[p]$, where C and D are concepts and $p \in (0,1)$ is a probability, is a statement of the form "if C holds, then D follows with probability p." Here, the probability p is understood as a reasoner's *belief* in D in the presence of C rather than a statistical probability. Accordingly, the conditional

$$(\mathsf{HasHeartOnTheirLeft}|\mathsf{Patient})[0.98]$$

could mean that a doctor believes that her patients typically have their heart on the left-hand side, here with probability 0.98.

The semantics of conditionals is formally given by the so-called *aggregating semantics* [23]. Before we discuss the aggregating semantics in detail, we define knowledge bases and make some remarks.

Definition 1 (Knowledge Base). *Let $\mathcal{T}, \mathcal{A}$, and $\mathcal{C}$ be finite sets of general concept inclusions, assertions, and of conditionals, respectively. Then, the tuple $\mathcal{R} = (\mathcal{T}, \mathcal{A}, \mathcal{C})$ is called a* knowledge base.

Remarks

- Disallowing probabilities $p = 0$ and $p = 1$ in conditionals does not limit the expressivity of the language as it turns out that a conditional of the form $(D|C)[p]$ is semantically equivalent to the general concept inclusion $(C \sqsubseteq D)$ iff $p = 1$ and to $(C \sqsubseteq \neg D)$ iff $p = 0$ (cf. Definition 2). Hence, the restriction $p \in (0,1)$ just means a clear separation of strict terminological knowledge in form of general concept inclusions and of uncertain beliefs in form of conditionals.
- Without loss of generality, we assume that concepts C in existential restrictions $(\exists r.C)$ and in universal restrictions $(\forall r.C)$ are concept names. If not, introduce a fresh concept name A, replace C by A and add the general concept inclusions $(A \sqsubseteq C)$ and $(C \sqsubseteq A)$ to the knowledge base.
- When investigating computability, we assume that input probabilities, i.e. probabilities in a knowledge base $\mathcal{R} = (\mathcal{T}, \mathcal{A}, \mathcal{C})$, are rational numbers. We refer to the conditionals in $\mathcal{C}$ with $(D_i|C_i)[s_i/t_i]$ for $i = 1, \ldots, n$ where $n = |\mathcal{C}|$ and s_i, t_i are natural numbers satisfying $0 < s_i < t_i$.
- The language $\mathcal{ALC}^{\mathsf{ME}}$ as defined in this paper does not allow for uncertain assertions. This is a design decision following the idea that assertional knowledge is (usually) unambiguously true or false and can be verified or falsified by observation. This design decision makes formal arguments a bit simpler but it is not a necessary precondition for our further analysis.

The formal semantics of conditionals and of knowledge bases is based on probability distributions over possible worlds. A possible world is a formal description of the possible state of the real world according to the reasoner's knowledge. Here, classical $\mathcal{DL}$-interpretations serve as possible worlds. We make the following prerequisites:

(A1) $\mathcal{N}_I$, $\mathcal{N}_C$, and $\mathcal{N}_R$ are finite sets.
(A2) $\Delta^{\mathcal{I}} = \mathcal{N}_I =: \Delta$ and $a^{\mathcal{I}} = a$ for all $a \in \mathcal{N}_I$ and for all interpretations $\mathcal{I}$.

The second assumption (A2) is known as the *unique name assumption*. Both assumptions (A1) and (A2) together imply a fixed finite domain Δ for all interpretations $\mathcal{I}$, which ensures that there are only finitely many interpretations all of which have the same scope. Therewith, all interpretations are comparable and entail a well-defined probability space. We call the set $\mathfrak{I}$ consisting of all interpretations that satisfy the assumptions (A1) and (A2) the *set of possible worlds*. A probability distribution $\mathcal{P}$ on the set of possible worlds can then be seen as a reasoner's epistemic state, and the probability of a single interpretation $\mathcal{I}$ is the degree of the reasoner's belief in the fact that $\mathcal{I}$ describes the real world properly. In this context, a knowledge base is a set of constraints on the probability distributions that may serve as the reasoner's epistemic state:

Definition 2 (Aggregating Semantics). *Let $\mathcal{R} = (\mathcal{T}, \mathcal{A}, \mathcal{C})$ be a knowledge base, and let $\mathfrak{I}$ be the set of possible worlds. A probability distribution $\mathcal{P} \colon \mathfrak{I} \to [0, 1]$ is a* model *of $\mathcal{R}$ iff both*

1. $\mathcal{P}(\mathcal{I}) = 0$ *for every $\mathcal{I} \in \mathfrak{I}$ with $\mathcal{I} \not\models f$ for any $f \in \mathcal{T} \cup \mathcal{A}$,*
2. $\mathcal{P} \models (D|C)[p]$ *for every conditional $(D|C)[p] \in \mathcal{C}$, i.e.*

$$\frac{\sum_{\mathcal{I} \in \mathfrak{I}} |C^{\mathcal{I}} \cap D^{\mathcal{I}}| \cdot \mathcal{P}(\mathcal{I})}{\sum_{\mathcal{I} \in \mathfrak{I}} |C^{\mathcal{I}}| \cdot \mathcal{P}(\mathcal{I})} = p. \tag{1}$$

The first condition (1.) of Definition 2 states that all the facts in $\mathcal{R}$, either general concept inclusions or assertions, have to be true in interpretations with non-zero probability. Conversely, the probabilities of the interpretations that satisfy all general concept inclusions and all assertions have to sum up to 1, hence

$$\sum_{\substack{\mathcal{I} \in \mathfrak{I} \\ \mathcal{I} \models \mathcal{T}, \mathcal{I} \models \mathcal{A}}} \mathcal{P}(\mathcal{I}) = 1.$$

We denote the set of the interpretations that satisfy all general concept inclusions and all assertions in $\mathcal{R}$ by $\mathfrak{I}_{\mathcal{R}}$, i.e.

$$\mathfrak{I}_{\mathcal{R}} = \{\mathcal{I} \in \mathfrak{I} \mid \mathcal{I} \models \mathcal{T} \text{ and } \mathcal{I} \models \mathcal{A}\}.$$

Equation (1), which is the aggregating semantics for probabilistic conditionals, captures the definition of conditional probabilities by weighting probabilities $\mathcal{P}(\mathcal{I})$ with the number of individuals for which the conditional $(D|C)[p]$ is *applicable* ($|C^{\mathcal{I}}|$) respectively *verified* ($|C^{\mathcal{I}} \cap D^{\mathcal{I}}|$) in $\mathcal{I}$. Hence, the aggregating semantics mimics statistical probabilities from a subjective point of view, and probabilities can be understood as degrees of belief in accordance with type 2 probabilities in the classification of Halpern [8]. If $\mathcal{P}$ is the Dirac distribution which is the probability distribution that assigns the probability 1 to a single interpretation $\mathcal{I}$ and which again means that the reasoner is certain that $\mathcal{I}$ is the real world, then the aggregating semantics means counting relative frequencies within $\mathcal{I}$. To the contrary, if $\mathcal{P}$ is the uniform distribution on $\mathfrak{I}_\mathcal{R}$ which means that the agent is maximally unconfident with her beliefs, then the aggregating semantics means counting relative frequencies spread over all interpretations. Finally, if $|\Delta| = 1$, the aggregating semantics boils down to computing conditional probabilities. Note that Eq. (1) implicitly states that the conditional $(D|C)[p]$ has to be applicable in at least one interpretation in $\mathfrak{I}_\mathcal{R}$.

A knowledge base with at least one model is called *consistent*. Note that consistency depends on the size of the underlying domain Δ. Consistent knowledge bases typically have infinitely many models (even if the domain and hence the sample space is finite). For reasoning tasks it is gainful to select a certain one among them, as reasoning based on the whole set of models leads to monotonic and often uninformative inferences. Any selected model $\mathcal{P}$ yields the non-monotonic inference relation

$$\mathcal{R} \models_\mathcal{P} (D|C)[p] \qquad \text{iff} \qquad \mathcal{P} \models (D|C)[p] \tag{2}$$

for conditionals $(D|C)[p]$. For factual knowledge f, whether $f = (C \sqsubseteq D)$ is a general concept inclusion or an assertion $f = C(a)$ or $f = r(a,b)$, respectively, one has $\mathcal{R} \models_\mathcal{P} f$ iff $\sum_{\mathcal{I} \models f} \mathcal{P}(\mathcal{I}) = 1$. Obviously, all general concept inclusions, assertions, and conditionals in $\mathcal{R}$ can be inferred from $\mathcal{R}$ w.r.t. the relation $\models_\mathcal{P}$.

From a commonsense point of view, the *maximum entropy distribution* $\mathcal{P}_\mathcal{R}^{\mathsf{ME}}$ is the model of $\mathcal{R}$ which fits best to the model selection task. The maximum entropy distribution is the unique distribution among all the models of $\mathcal{R}$ which has maximum entropy. From an information theoretical point of view, it adds as less information as possible to $\mathcal{R}$. Hence, the benefit of the maximum entropy distribution is that it assigns a concrete probability p to a query conditional $(D|C)$ (and not a whole interval of conceivable probabilities) while being as cautious as possible. The distinct probability p can then be seen as a most expected value of observing D under the presence of C.

Definition 3 (Maximum Entropy Distribution). *Let $\mathcal{R}$ be a consistent knowledge base, and let $\mathfrak{P}$ be the set of all probability distributions over $\mathfrak{I}$. The probability distribution*

$$\mathcal{P}_\mathcal{R}^{\mathsf{ME}} = \arg\max_{\substack{\mathcal{P} \in \mathfrak{P} \\ \mathcal{P} \models \mathcal{R}}} - \sum_{\mathcal{I} \in \mathfrak{I}} \mathcal{P}(\mathcal{I}) \cdot \log \mathcal{P}(\mathcal{I}) \tag{3}$$

is called the maximum entropy distribution *of $\mathcal{R}$. In Eq. (3), the convention $0 \cdot \log 0 = 0$ applies.*

Since the maximum entropy distribution $\mathcal{P}_\mathcal{R}^{\mathsf{ME}}$ is the solution of a nonlinear optimization problem, there is no closed form of $\mathcal{P}_\mathcal{R}^{\mathsf{ME}}$ in general. Hence, $\mathcal{P}_\mathcal{R}^{\mathsf{ME}}$ has to be calculated approximatively. We will investigate this approximation process in the next section.

3   Calculating the Maximum Entropy Distribution $\mathcal{P}_{\mathcal{R}}^{\mathsf{ME}}$

Let Δ be a fixed finite domain and let $\mathcal{R} = (\mathcal{T}, \mathcal{A}, \mathcal{C})$ be a consistent knowledge base. Computing the maximum entropy distribution $\mathcal{P}_{\mathcal{R}}^{\mathsf{ME}}$ efficiently is a non-trivial but considerable problem. Here, the domain size $k := |\Delta|$ is *the* crucial quantity, since it is typically large in application domains, and the size of $\mathcal{P}_{\mathcal{R}}^{\mathsf{ME}}$, more precisely the underlying sample space, depends exponentially on k. Therefore, we are interested in algorithms that approximate $\mathcal{P}_{\mathcal{R}}^{\mathsf{ME}}$ in time polynomial in k.

Recall our convention $\mathcal{C} = \{(D_1|C_1)[s_1/t_1], \ldots, (D_n|C_n)[s_n/t_n]\}$. Since it holds that $\mathcal{P}_{\mathcal{R}}^{\mathsf{ME}}(\mathcal{I}) = 0$ for all $\mathcal{I} \in \mathfrak{I} \backslash \mathfrak{I}_{\mathcal{R}}$ due to Definition 2, it is sufficient to maximize the entropy w.r.t. the interpretations in $\mathfrak{I}_{\mathcal{R}}$ when calculating $\mathcal{P}_{\mathcal{R}}^{\mathsf{ME}}$. More precisely, let $\mathcal{P}^*$ be the solution of the optimization problem (cf. Eq. (3))

$$maximize \qquad -\sum_{\mathcal{I} \in \mathfrak{I}_{\mathcal{R}}} \mathcal{P}(\mathcal{I}) \cdot \log \mathcal{P}(\mathcal{I}) \qquad (\mathsf{ME}_{\mathcal{R}})$$

$$subject\ to \qquad \sum_{\mathcal{I} \in \mathfrak{I}_{\mathcal{R}}} \mathcal{P}(\mathcal{I}) = 1$$

$$\sum_{\mathcal{I} \in \mathfrak{I}_{\mathcal{R}}} \left(t_i \cdot |C_i^{\mathcal{I}} \cap D_i^{\mathcal{I}}| - s_i \cdot |C_i^{\mathcal{I}}| \right) \cdot \mathcal{P}(\mathcal{I}) = 0 \qquad i = 1, \ldots, n$$

$$\mathcal{P}(\mathcal{I}) \in \mathbb{R}_{\geq 0} \qquad \forall \mathcal{I} \in \mathfrak{I}_{\mathcal{R}}.$$

Then,

$$\mathcal{P}_{\mathcal{R}}^{\mathsf{ME}}(\mathcal{I}) = \begin{cases} \mathcal{P}^*(\mathcal{I}), & \mathcal{I} \in \mathfrak{I}_{\mathcal{R}} \\ 0, & \text{otherwise} \end{cases}. \qquad (4)$$

For the rest of the paper, we assume that $(\mathsf{ME}_{\mathcal{R}})$ has a feasible point in its relative interior, i.e., we assume that there is a *positive* probability distribution $\mathcal{P}$ which satisfies the constraints of $(\mathsf{ME}_{\mathcal{R}})$. In the most general sense, this condition is known as *Slater's condition* (cf. [3]). If Slater's condition holds, the solution of $(\mathsf{ME}_{\mathcal{R}})$ lies in the relative interior of $(\mathsf{ME}_{\mathcal{R}})$, too. Again, this is a well-known result that is called *Paris' open-mindedness principle* in the field of knowledge representation and reasoning [17]. As a consequence of this principle, it holds that $0 < \mathcal{P}_{\mathcal{R}}^{\mathsf{ME}}(\mathcal{I}) < 1$ for all $\mathcal{I} \in \mathfrak{I}_{\mathcal{R}}$.

Actually, the adherence of Slater's condition is a restriction for the knowledge base $\mathcal{R}$ and means that strict knowledge has to be formalized by general concept inclusions or assertions and must not be the implicit outcome of conditionals. We illustrate this by means of an example.

Example 1. Let A and B be concept names, and let $\mathcal{R}_1 = (\mathcal{T}_1, \mathcal{A}_1, \mathcal{C}_1)$ be a knowledge base with $\mathcal{T}_1 = \emptyset$, $\mathcal{A}_1 = \emptyset$, and

$$\mathcal{C}_1 = \{(A \sqcap B|\top)[1/3], \quad (A \sqcap \neg B|\top)[1/3], \quad (\neg A \sqcap B|\top)[1/3]\}.$$

Obviously, the conditional $(\neg A \sqcap \neg B|\top)[0]$ can be inferred from $\mathcal{R}_1$. Hence, $\mathcal{P}_{\mathcal{R}_1}^{\mathsf{ME}}(\mathcal{I}) = 0$ for all interpretations $\mathcal{I}$ with $(\neg A \sqcap \neg B)^{\mathcal{I}} \neq \emptyset$. Since $\mathcal{I} \in \mathfrak{I}_{\mathcal{R}_1}$

for these interpretations $\mathcal{I}$, Slater's condition does not hold. To overcome this problem, one can make the implicit knowledge $(\neg A \sqcap \neg B | \top)[0]$ explicit and add the general concept inclusion $(\top \sqsubseteq A \sqcup B)$ to $\mathcal{T}_1$ such that one obtains the knowledge base $\mathcal{R}_1' = (\mathcal{T}_1', \mathcal{A}_1, \mathcal{C}_1)$ with $\mathcal{T}_1' = \{\top \sqsubseteq A \sqcup B\}$ which is semantically equivalent to $\mathcal{R}_1$. Now, $\mathcal{I} \in \mathfrak{I} \backslash \mathfrak{I}_{\mathcal{R}_1'}$ for interpretations $\mathcal{I}$ with $(\neg A \sqcap \neg B)^{\mathcal{I}} \neq \emptyset$. As required, $\mathcal{P}_{\mathcal{R}_1'}^{\mathsf{ME}}(\mathcal{I}) = 0$ holds for these interpretations due to Condition (1.) of Definition 2, but the interpretations are excluded from the optimization problem $(\mathsf{ME}_{\mathcal{R}_1'})$. As a consequence, Slater's condition is recovered for $\mathcal{R}_1'$.   ∎

Knowledge bases that entail the adherence of Slater's condition are called *p-consistent* [7]. As shown in Example 1, Slater's condition can be established by prescient knowledge engineering (or, mathematically, by linear programming). For p-consistent knowledge bases, we may reduce the domain of $(\mathsf{ME}_{\mathcal{R}})$ to the positive reals:

$$\textit{maximize} \qquad -\sum_{\mathcal{I} \in \mathfrak{I}_F} \mathcal{P}(\mathcal{I}) \cdot \log \mathcal{P}(\mathcal{I}) \qquad (\mathsf{ME}_{\mathcal{R}}^+)$$

$$\textit{subject to} \qquad \sum_{\mathcal{I} \in \mathfrak{I}_F} \mathcal{P}(\mathcal{I}) = 1$$

$$\sum_{\mathcal{I} \in \mathfrak{I}_F} \left(t_i \cdot |C_i^{\mathcal{I}} \cap D_i^{\mathcal{I}}| - s_i \cdot |C_i^{\mathcal{I}}| \right) \cdot \mathcal{P}(\mathcal{I}) = 0 \qquad i = 1, \ldots, n$$

$$\mathcal{P}(\mathcal{I}) \in \mathbb{R}_{>0} \qquad \forall \mathcal{I} \in \mathfrak{I}_{\mathcal{R}}.$$

According to [3], the optimization problem $(\mathsf{ME}_{\mathcal{R}}^+)$ is convex and can be transformed into an equivalent self-concordant optimization problem in standard form. Basically, one has to make the domain constraints $\mathcal{P}(\mathcal{I}) \in \mathbb{R}_{>0}$ explicit by adding the constraints $\mathcal{P}(\mathcal{I}) > 0$ to the optimization problem. We do not discuss the theory of self-concordant problems here but note that optimization problems of that form can be solved up to any fixed precision in polynomial time by so-called *interior point methods* [14]. However, there are two reasons why solving $(\mathsf{ME}_{\mathcal{R}}^+)$ in this way is problematic:

- The number of constraints in $(\mathsf{ME}_{\mathcal{R}}^+)$ is exponential in k since $|\mathfrak{I}_{\mathcal{R}}|$ is exponential in k, and hence, solving $(\mathsf{ME}_{\mathcal{R}}^+)$ is exponential in k, too, even for the efficient interior point methods.
- The complexity results for the interior point methods hold modulo an oracle which returns the values of the objective function of the problem, which is $-\sum_{\mathcal{I} \in \mathfrak{I}_{\mathcal{R}}} \mathcal{P}(\mathcal{I}) \cdot \log \mathcal{P}(\mathcal{I})$ in this case, and the gradient thereof, at any feasible point [14]. However, the exact evaluation of the objective function of $(\mathsf{ME}_{\mathcal{R}}^+)$ is possible in real number arithmetic only, and hence, the complexity results hold only over the reals, too.

A better idea is to investigate the unconstrained optimization problem dual to (ME$_\mathcal{R}^+$) which is, following the method of Lagrange multipliers (cf. [3]),

$$maximize \qquad -\log\left(\sum_{\mathcal{I}\in\mathfrak{I}_\mathcal{R}}\exp\left(-\sum_{i=1}^{n}g_i(\mathcal{I})\cdot\nu_i\right)\right) \qquad \text{(ME}_\mathcal{R}^d\text{)}$$

$$\nu\in\mathbb{R}^n,$$

in which we abbreviated

$$g_i(\mathcal{I}) = t_i\cdot|C_i^\mathcal{I}\cap D_i^\mathcal{I}| - s_i\cdot|C_i^\mathcal{I}|$$

for all $\mathcal{I}\in\mathfrak{I}_\mathcal{R}$ and $i = 1,\ldots,n$. The optimization problem (ME$_\mathcal{R}^d$) is convex, and, due to the strong duality between (ME$_\mathcal{R}^+$) and (ME$_\mathcal{R}^d$), has a unique solution $\nu^*\in\mathbb{R}^n$ from which the solution of the primal problem (ME$_\mathcal{R}$) can be derived by

$$\mathcal{P}^*(\mathcal{I}) = \frac{\exp(-\sum_{i=1}^{n}g_i(\mathcal{I})\cdot\nu_i^*)}{\sum_{\mathcal{I}'\in\mathfrak{I}_\mathcal{R}}\exp(-\sum_{i=1}^{n}g_i(\mathcal{I}')\cdot\nu_i^*)}, \qquad \mathcal{I}\in\mathfrak{I}_\mathcal{R}.$$

The benefit of (ME$_\mathcal{R}^d$) is that the length of the solution vector is independent of the domain size. Unfortunately, the objective function of (ME$_\mathcal{R}^d$) is also evaluable exactly only over the reals. To overcome this obstacle, we substitute

$$\alpha_i = \exp(-\nu_i), \qquad i = 1,\ldots,n, \tag{5}$$

and observe the unconstrained optimization problem

$$minimize \qquad \sum_{\mathcal{I}\in\mathfrak{I}_\mathcal{R}}\prod_{i=1}^{n}\alpha_i^{g_i(\mathcal{I})} \qquad \text{(ME}_\mathcal{R}^\alpha\text{)}$$

$$\alpha\in\mathbb{R}_{>0}^n.$$

equivalent to (ME$_\mathcal{R}^d$). The optimization problem (ME$_\mathcal{R}^\alpha$) eventually combines a vast number of beneficial properties:

- Since (5) is a bijection between $\mathbb{R}$ and $\mathbb{R}_{>0}$, the problem (ME$_\mathcal{R}^\alpha$) has a unique solution $\alpha^*\in\mathbb{R}_{>0}^n$ which satisfies

$$\mathcal{P}_\mathcal{R}^{\mathsf{ME}}(\mathcal{I}) = \frac{\prod_{i=1}^{n}(\alpha_i^*)^{g_i(\mathcal{I})}}{\sum_{\mathcal{I}'\in\mathfrak{I}_\mathcal{R}}\prod_{i=1}^{n}(\alpha_i^*)^{g_i(\mathcal{I}')}}, \qquad \mathcal{I}\in\mathfrak{I}_\mathcal{R}. \tag{6}$$

- The length of the solution vector of (ME$_\mathcal{R}^\alpha$) is independent of the domain size k and, therewith, a very compact representation of $\mathcal{P}_\mathcal{R}^{\mathsf{ME}}$.
- The objective function $\phi_\mathcal{R}(\alpha) = \sum_{\mathcal{I}\in\mathfrak{I}_\mathcal{R}}\prod_{i=1}^{n}\alpha_i^{g_i(\mathcal{I})}$ can be computed *exactly* at any rational point $\alpha\in\mathbb{Q}_{>0}^n$. Furthermore, the techniques used in [25] guarantee that these computations can be performed in time polynomial in k for $\mathcal{ALC}^{\mathsf{ME}}$-knowledge bases without assertions. In Sect. 4 we will discuss a method with which it is possible to handle assertions, too.

– Once a (rational) approximation $\beta \in \mathbb{Q}_{>0}^n$ of α^* is computed, an approxima-
tion $\mathcal{P}_{\mathcal{R}}^{\beta}$ of the maximum entropy distribution $\mathcal{P}_{\mathcal{R}}^{\mathsf{ME}}$ can be computed *exactly*
via (cf. Eqs. (4) and (6))

$$\mathcal{P}_{\mathcal{R}}^{\beta}(\mathcal{I}) = \begin{cases} \dfrac{\prod_{i=1}^{n} \beta_i^{g_i(\mathcal{I})}}{\sum_{\mathcal{I}' \in \mathfrak{I}_{\mathcal{R}}} \prod_{i=1}^{n} \beta_i^{g_i(\mathcal{I}')}}, & \mathcal{I} \in \mathfrak{I}_{\mathcal{R}} \\ 0, & \text{otherwise} \end{cases} .$$

The drawback of the Substitution (5) is that ($\mathsf{ME}_{\mathcal{R}}^{\alpha}$) is no longer convex
in general. Hence, complexity results for convex optimization problems do not
apply in contrast to ($\mathsf{ME}_{\mathcal{R}}^{d}$). However, provided that upper and lower bounds for
the optimal solution α^* are known, the following theorem applies.

Theorem 1 *(cf. [13]). Let $f, g : \mathbb{R}^n \to \mathbb{R}$ be Lipschitz continuous functions,
i.e., there is $K \in \mathbb{R}_{>0}$ with $\|f(\boldsymbol{x}) - f(\boldsymbol{y})\| \leq K \cdot \|\boldsymbol{x} - \boldsymbol{y}\|$ for all $\boldsymbol{x}, \boldsymbol{y} \in \mathbb{R}^n$ (the
same for g), and let $\lambda \in \mathbb{R}$. If the optimization problem*

$$\begin{array}{lll} \textit{minimize} & f(x) & (\mathsf{OptLip}) \\ \textit{subject to} & g(x) \leq 0 & \\ & |x_i| \leq \lambda & i = 1, \ldots, n \\ & \boldsymbol{x} \in \mathbb{R}^n & \end{array}$$

*has a solution, then (OptLip) has an additive polynomial time approximation
scheme (PTAS) modulo an oracle which returns function evaluations of f and g.*

For a proof and for technical details of Theorem 1, especially for precise
complexity bounds, please see [13]. In order to apply Theorem 1 to our dual
maximum entropy optimization problem ($\mathsf{ME}_{\mathcal{R}}^{\alpha}$), we set

$$f(\boldsymbol{x}) = \sum_{\mathcal{I} \in \mathfrak{I}_{\mathcal{R}}} \prod_{i=1}^{n} x_i^{g_i(\mathcal{I})}, \qquad g(\boldsymbol{x}) = l - x_i, \qquad \lambda = u,$$

where $0 < l < u$ are real numbers. If $l \leq \alpha_i^* \leq u$ for all $i = 1, \ldots, n$, then f
is obviously Lipschitz continuous on $[l, u]^n$ due to the compactness of $[l, u]^n$, g
is Lipschitz continuous in any case, and ($\mathsf{ME}_{\mathcal{R}}^{\alpha}$) is of the form (OptLip). Hence,
there is a **PTAS** modulo oracle for ($\mathsf{ME}_{\mathcal{R}}^{\alpha}$) provided that the bounds l and u
are known. Unfortunately, calculating l and u in general is a non-trivial task.
However, we show how these bounds can be calculated in a concrete example.

Example 2. Let A, B, and C be concept names, and let $k = |\Delta|$ be the size
of an arbitrary finite domain Δ with $k > 0$. We consider the knowledge base
$\mathcal{R}_2 = (\mathcal{T}_2, \mathcal{A}_2, \mathcal{C}_2)$ with $\mathcal{T}_2 = \emptyset$, $\mathcal{A}_2 = \emptyset$, and

$$\mathcal{C}_2 = \{(C|A)[s_1/t_1], \quad (C|B)[s_2/t_2]\}.$$

Even for this simple knowledge base, the solution $\boldsymbol{\alpha^*} = (\alpha_1^*, \alpha_2^*) \in \mathbb{R}_{>0}^2$ of the
optimization problem ($\mathsf{ME}_{\mathcal{R}_2}^{\alpha}$) has no closed form expression. Actually, this holds

for any domain size k and has already been noticed for the case $k = 1$ in [10]. The knowledge base $\mathcal{R}_2$ is of interest because it is part of the *antecedent conjunction problem*:

For which probability $q \in [0, 1]$ does $\mathcal{R}_2 \models^{\text{ME}}_{\mathcal{R}_2} (C|A \sqcap B)[q]$ hold?

An answer to that question would reveal clearly how maximum entropy inference combines evidences.

Here, we calculate lower and upper bounds for the solution $\boldsymbol{\alpha}$: The objective function $\phi_{\mathcal{R}_2}(\boldsymbol{\alpha})$ of $(\text{ME}^{\alpha}_{\mathcal{R}_2})$ is

$$\left(\alpha_1^{t_1-s_1}\alpha_2^{t_2-s_2} + \alpha_1^{-s_1}\alpha_2^{-s_2} + \alpha_1^{t_1-s_1} + \alpha_2^{t_2-s_2} + \alpha_1^{-s_1} + \alpha_2^{-s_2} + 2\right)^k =: f(\boldsymbol{\alpha}^*)^k. \quad (7)$$

We will give an explanation for this representation of $\phi(\boldsymbol{\alpha})$ in the next section (cf. Example (3)). It holds that the gradient of $\phi(\boldsymbol{\alpha})$ vanishes in the minimum $\boldsymbol{\alpha}^*$. Hence,

$$0 = \frac{\partial}{\partial \alpha_1}\phi(\boldsymbol{\alpha}^*) = \frac{\partial}{\partial \alpha_1}f(\boldsymbol{\alpha}^*)^k$$

$$= \left((t_1 - s_1) \cdot (\alpha_1^*)^{t_1-s_1-1} \cdot (\alpha_2^*)^{t_2-s_2} + (-s_1) \cdot (\alpha_1^*)^{-s_1-1} \cdot (\alpha_2^*)^{-s_2}\right.$$

$$\left. + (t_1 - s_1) \cdot (\alpha_1^*)^{t_1-s_1-1} + (-s_1) \cdot (\alpha_1^*)^{-s_1-1}\right) \cdot k \cdot f(\boldsymbol{\alpha}^*)^{k-1}$$

Since $\alpha_1^*, \alpha_2^* > 0$, the second factor $k \cdot f(\boldsymbol{\alpha}^*)^{k-1}$ of the right-hand side cannot be zero, and the first factor must be zero. We multiply both sides with $\frac{(\alpha_1^*)^{s_1+1} \cdot (\alpha_2^*)^{s_2}}{k \cdot f(\boldsymbol{\alpha}^*)^{k-1}}$ and get

$$0 = (t_1 - s_1) \cdot (\alpha_1^*)^{t_1} \cdot (\alpha_2^*)^{t_2} - s_1 + (t_1 - s_1) \cdot (\alpha_1^*)^{t_1} \cdot (\alpha_2^*)^{s_2} - s_1 \cdot (\alpha_2^*)^{s_2}.$$

Analogously, we have

$$0 = (t_2 - s_2) \cdot (\alpha_2^*)^{t_2} \cdot (\alpha_1^*)^{t_1} - s_2 + (t_2 - s_2) \cdot (\alpha_2^*)^{t_2} \cdot (\alpha_1^*)^{s_1} - s_2 \cdot (\alpha_1^*)^{s_1}.$$

The first of these two equations can be solved for α_1^* and the resulting expression can be plugged into the second (and vice versa). We get (analogously for the case the other way around)

$$0 = (t_2 - s_2) \cdot \left(\frac{s_1 \cdot (1 + (\alpha_2^*)^{s_2})}{(t_1 - s_1) \cdot ((\alpha_2^*)^{t_2} + (\alpha_2^*)^{s_2})} + \left(\frac{s_1 \cdot (1 + (\alpha_2^*)^{s_2})}{(t_1 - s_1) \cdot ((\alpha_2^*)^{t_2} + (\alpha_2^*)^{s_2})}\right)^{s_1/t_1}\right)$$

$$\cdot (\alpha_2^*)^{t_2} - s_2 \cdot \left(1 + \left(\frac{s_1 \cdot (1 + (\alpha_2^*)^{s_2}}{(t_1 - s_1) \cdot ((\alpha_2^*)^{t_2} + (\alpha_2^*)^{s_2})}\right)^{s_1/t_1}\right).$$

In case of $\alpha_2^* < 1$, we estimate

$$0 \leq (t_2 - s_2) \cdot \left(\frac{s_1}{t_1 - s_1} + \left(\frac{s_1}{t_1 - s_1}\right)^{s_1/t_1}\right) \cdot \frac{1 + (\alpha_2^*)^{s_2}}{(\alpha_2^*)^{t_2} + (\alpha_2^*)^{s_2}} \cdot (\alpha_2^*)^{t_2}$$

$$- s_2 \cdot \left(1 + \left(\frac{s_1}{t_1 - s_1}\right)^{s_1/t_1}\right).$$

With

$$\frac{1+(\alpha_2^*)^{s_2}}{(\alpha_2^*)^{t_2}+(\alpha_2^*)^{s_2}} \cdot (\alpha_2^*)^{t_2} = \frac{(\alpha_2^*)^{-s_2}+1}{(\alpha_2^*)^{t_2-s_2}+1} \cdot (\alpha_2^*)^{t_2} \le 2 \cdot (\alpha_2^*)^{t_2-s_2},$$

it further follows that

$$l_2 := \left(\frac{s_2 \cdot \left(1 + \left(\frac{s_1}{t_1-s_1}\right)^{s_1/t_1}\right)}{2 \cdot (t_2 - s_2) \cdot \left(\frac{s_1}{t_1-s_1} + \left(\frac{s_1}{t_1-s_1}\right)^{s_1/t_1}\right)} \right)^{1/t_2-s_2} \le \alpha_2^*.$$

Otherwise, in case of $\alpha_2^* \ge 1$, we estimate

$$0 \ge (t_2 - s_2) \cdot \left(\frac{s_1}{t_1 - s_1} + \left(\frac{s_1}{t_1 - s_1}\right)^{s_1/t_1}\right) \cdot \frac{1+(\alpha_2^*)^{s_2}}{(\alpha_2^*)^{t_2}+(\alpha_2^*)^{s_2}} \cdot (\alpha_2^*)^{t_2}$$
$$- s_2 \cdot \left(1 + \left(\frac{s_1}{t_1 - s_1}\right)^{s_1/t_1}\right).$$

With

$$\frac{1+(\alpha_2^*)^{s_2}}{(\alpha_2^*)^{t_2}+(\alpha_2^*)^{s_2}} \cdot (\alpha_2^*)^{t_2} = \frac{(\alpha_2^*)^{-s_2}+1}{(\alpha_2^*)^{t_2-s_2}+1} \cdot (\alpha_2^*)^{t_2} \ge \frac{1}{2} \cdot (\alpha_2^*)^{s_2},$$

it follows that

$$u_2 := \left(\frac{2 \cdot s_2 \cdot \left(1 + \left(\frac{s_1}{t_1-s_1}\right)^{s_1/t_1}\right)}{(t_2 - s_2) \cdot \left(\frac{s_1}{t_1-s_1} + \left(\frac{s_1}{t_1-s_1}\right)^{s_1/t_1}\right)} \right)^{1/s_2} \ge \alpha_2^*.$$

Putting both estimations together, we have

$$\min\{1, l_2\} \le \alpha_2^* \le \max\{1, u_2\},$$

where l_2 and u_2 depend on the input probabilities only. Analogously, it holds that

$$\min\{1, l_1\} \le \alpha_1^* \le \max\{1, u_1\},$$

where

$$l_1 := \left(\frac{s_1 \cdot \left(1 + \left(\frac{s_2}{t_2-s_2}\right)^{s_2/t_2}\right)}{2 \cdot (t_1 - s_1) \cdot \left(\frac{s_2}{t_2-s_2} + \left(\frac{s_2}{t_2-s_2}\right)^{s_2/t_2}\right)} \right)^{1/t_1-s_1},$$

$$u_1 := \left(\frac{2 \cdot s_1 \cdot \left(1 + \left(\frac{s_2}{t_2-s_2}\right)^{s_2/t_2}\right)}{(t_1 - s_1) \cdot \left(\frac{s_2}{t_2-s_2} + \left(\frac{s_2}{t_2-s_2}\right)^{s_2/t_2}\right)} \right)^{1/s_1}.$$

For example, if $\frac{s_1}{t_1} = \frac{3}{4}$ and $\frac{s_2}{t_2} = \frac{2}{5}$, one has $1 \le \alpha_1^* \le 1,96$ and $0,59 \le \alpha_2^* \le 1$. These estimations can now be used as the starting points for solving ($\mathsf{ME}_{\mathcal{R}}^\alpha$) according to Theorem 1. ∎

In the next section we focus on the oracle that is needed for Theorem 1. For this, we make use of the principle of typed model counting [24].

4   Typed Model Counting for $\mathcal{ALC}^{\mathsf{ME}}$

Algorithms that solve the optimization problem $(\mathsf{ME}_{\mathcal{R}}^{\alpha})$ in time polynomial in k require the evaluation of the objective function

$$\phi_{\mathcal{R}}(\boldsymbol{\alpha}) = \sum_{\mathcal{I} \in \mathfrak{I}_{\mathcal{R}}} \prod_{i=1}^{n} \alpha_i^{g_i(\mathcal{I})}$$

in time polynomial in k. It is not obvious that this is possible since $\phi(\boldsymbol{\alpha})$ is a sum over all interpretations in $\mathfrak{I}_{\mathcal{R}}$ and $|\mathfrak{I}_{\mathcal{R}}|$ is exponential in k. Hence, a more compressed, i.e. factorized representation of $\phi(\boldsymbol{\alpha})$ is needed. In particular, it is not efficient to set up the sum by calculating $g_i(\mathcal{I})$, $i = 1, \ldots, n$, for the single interpretations independently. Instead, we take an individual based perspective: We characterize individuals by the concepts and role memberships they satisfy, and we determine which combinations of individuals can occur in an interpretation in $\mathfrak{I}_{\mathcal{R}}$. By exploiting combinatorial arguments in order to determine these combinations, it is possible to simplify computations. As interpretations are considered jointly in this approach, the number of interpretations becomes simply a parameter in many expressions.

Formally, setting up $\phi(\boldsymbol{\alpha})$ is a weighted model counting problem where the interpretations in $\mathfrak{I}_{\mathcal{R}}$ are the models and $\prod_{i=1}^{n} \alpha_i^{g_i(\mathcal{I})}$ for $\mathcal{I} \in \mathfrak{I}_{\mathcal{R}}$ are the weights. Let $\mathcal{R} = (\mathcal{T}, \mathcal{A}, \mathcal{C})$ be a p-consistent knowledge base, let $\mathcal{I} \in \mathfrak{I}_{\mathcal{R}}$ be a fixed interpretation, and let $a \in \mathcal{N}_I$ be an individual. While a general concept inclusion $(C \sqsubseteq D)$ can be satisfied $(a \in C^{\mathcal{I}} \cap D^{\mathcal{I}})$ or not by the individual a, a conditional $\mathfrak{c} = (D|C)[p]$ leads to a three-valued interpretation w.r.t. a:

- The individual a *verifies* $\mathfrak{c}$ iff $a \in C^{\mathcal{I}} \cap D^{\mathcal{I}}$,
- a *falsifies* $\mathfrak{c}$ iff $a \in C^{\mathcal{I}} \backslash D^{\mathcal{I}}$,
- and $\mathfrak{c}$ does not *apply* to a iff $a \notin C^{\mathcal{I}}$.

In order to set up $\phi_{\mathcal{R}}(\boldsymbol{\alpha})$, more precisely, in order to set up $g_i(\mathcal{I})$ for $i = 1, \ldots, n$ and $\mathcal{I} \in \mathfrak{I}_{\mathcal{R}}$, it is necessary to record these evaluations of the conditionals in $\mathcal{C}$ for all individuals in $\mathcal{N}_I$. Hence, determining the weights of the weighted model counting problem of setting up $\phi_{\mathcal{R}}(\boldsymbol{\alpha})$ requires a fine-grained evaluation of the interpretations in $\mathfrak{I}_{\mathcal{R}}$. In the following, we carry out this evaluation for all interpretations in one step. For this, we make use of *typed model counting* [24] which allows us to determine the weights and to perform weighted model counting simultaneously. The basic idea of typed model counting is to include algebraic elements symbolizing verification $(\mathbf{v_i})$, falsification $(\mathbf{f_i})$, and non-applicability $(\mathbf{1})$ directly into formulas. When counting the models of the formulas, the algebraic elements are collected and constitute the weight of the respective model.

Table 2. Canonical mapping π from Description Logic to first-order logic.

$\pi_x(A) = A(x)$	A is a concept name
$\pi_x(\top) = \top$	top concept
$\pi_x(\bot) = \bot$	bottom concept
$\pi_x(\neg C) = \neg \pi_x(C)$	negation
$\pi_x(C \sqcap D) = \pi_x(C) \wedge \pi_x(D)$	conjunction
$\pi_x(C \sqcup D) = \pi_x(C) \vee \pi_x(D)$	disjunction
$\pi_x(\exists r.C) = \bigvee_{y \in \Delta}(R(x,y) \wedge \pi_y(C))$	existential restriction
$\pi_x(\forall r.C) = \bigwedge_{y \in \Delta}(R(x,y) \Rightarrow \pi_y(C))$	universal restriction
$\pi(C \sqsubseteq D) = \forall x \in \Delta (\pi_x(C) \Rightarrow \pi_x(D))$	general concept inclusion
$\pi(A(a)) = A(a)$	assertions
$\pi(r(a,b)) = r(a,b)$	

The procedure of applying typed model counting in order to determine $\phi(\alpha)$ is as follows: First, one converts every assertion and every general concept inclusion in $\mathcal{R}$ into a first-order sentence (i.e., a first-order formula without free variables). This is canonically done by the mapping $\pi(= \pi(\mathcal{R}))$ (cf. Table 2). Note that every such mapping π induces a first-order signature Σ_π that consists of exactly those predicates that are introduced by π. Further, the set of constants shall be Δ. Therewith, the first-order interpretations w.r.t. Σ_π are also fixed (by their evaluation of the ground atoms built upon Σ_π), and they are in a one-to-one correspondence to the $\mathcal{DL}$-interpretations.

The sentences $\pi(f)$ for all $f \in \mathcal{T} \cup \mathcal{A}$ build a theory $\Pi(\mathcal{R})$ and the models of this theory correspond to the interpretations in $\mathfrak{I}_\mathcal{R}$. So far, counting the models of $\Pi(\mathcal{R})$ means calculating $|\mathfrak{I}_\mathcal{R}|$.

Further, the conditionals $(D_i|C_i)[s_i/t_i]$, $i = 1, \ldots, n$, are translated into *structured formulas*, more precisely into structured sentences, by

$$\pi((D_i|C_i)[s_i/t_i]) = \bigwedge_{x \in \Delta}\left(\mathbf{v_i} \circ \pi_x(C_i \sqcap D_i) \vee \mathbf{f_i} \circ \pi_x(C_i \sqcap \neg D_i) \vee \mathbf{1} \circ \pi_x(\neg C_i)\right), \quad (8)$$

where $\mathbf{v_i}$, $\mathbf{f_i}$, and $\mathbf{1}$ are the algebraic elements mentioned above. The idea behind (8) is as follows: Consider a fixed interpretation $\mathcal{I}$. If an individual (i.e., a constant) a verifies the conditional $(D_i|C_i)[s_i/t_i]$, then $\pi_a(C_i \sqcap D_i)$ is true (and $\pi_a(C_i \sqcap \neg D_i)$ as well as $\pi_a(\neg C_i)$ are false), and the algebraic element $\mathbf{v_i}$ is stored. Otherwise, if a falsifies or does not apply to the conditional, then $\pi_a(C_i \sqcap \neg D_i)$ or $\pi_a(\neg C_i)$ are true, respectively, and $\mathbf{f_i}$ or $\mathbf{1}$ are stored. By the outer conjunction in (8) this is done for all individuals in the domain. As a result, one gets a factor $\mathbf{v_i}$ for every individual that verifies the conditional, and a factor $\mathbf{f_i}$ for every individual that falsifies the conditional. If one does this for all conditionals in $\mathcal{C}$ and commutatively concatenates the algebraic elements thus obtained, one gets the so-called *conditional structure* of $\mathcal{I}$ (cf. [10]):

$$\sigma_{\mathcal{R}}(\mathcal{I}) = \prod_{i=1}^{n} \mathbf{v_i}^{|C_i^{\mathcal{I}} \cap D_i^{\mathcal{I}}|} \mathbf{f_i}^{|C_i^{\mathcal{I}} \setminus D_i^{\mathcal{I}}|}.$$

Hence, the conditional structure $\sigma_{\mathcal{R}}(\mathcal{I})$ is a compact representation of how often the conditionals in $\mathcal{C}$ are verified and falsified in $\mathcal{I}$. Formally, conditional structures are elements of a free Abelian group with identity element $\mathbf{1}$. We extend this group to a commutative semiring $\mathcal{S}$ by allowing to add conditional structures and by introducing the zero element $\mathbf{0}$. Elements in $\mathcal{S}$ are called *structural element* as they are used to structure formulas here.

Obviously, the objective function $\phi_{\mathcal{R}}(\boldsymbol{\alpha})$ highly relates to conditional structures: Let $\rho_{\mathcal{R}}(X)$ be the mapping which substitutes every occurrence of $\mathbf{v_i}$ in X with $\alpha_i^{t_i - s_i}$, every occurrence of $\mathbf{f_i}$ with $\alpha_i^{-s_i}$, and every occurence of $\mathbf{1}$ with 1. Then,

$$\phi_{\mathcal{R}}(\boldsymbol{\alpha}) = \sum_{\mathcal{I} \models \Pi(\mathcal{R})} \rho_{\mathcal{R}}(\sigma_{\mathcal{R}}(\mathcal{I})).$$

The formal semantics of structured sentences is defined as follows: Let $\mathcal{I}$ be a classical first-order interpretation which maps every first order sentence to $\mathbf{0}$ or $\mathbf{1}$. Then, $\mathcal{I}$ is extended to structured sentences by

- $\mathcal{I}(A \wedge B) = \mathcal{I}(A) \cdot \mathcal{I}(B)$,
- $\mathcal{I}(A \vee B) = \begin{cases} \mathcal{I}(A), & \mathcal{I}(B) = \mathbf{0} \\ \mathcal{I}(B), & \mathcal{I}(A) = \mathbf{0} \\ \mathcal{I}(A) \cdot \mathcal{I}(B) & \text{otherwise} \end{cases}$
- $\mathcal{I}(\mathbf{s} \circ A) = \mathbf{s} \cdot \mathcal{I}(A)$,
- $\mathcal{I}(\bigvee_{x \in \mathcal{D}} A) = \mathcal{I}(\bigvee_{a \in \mathcal{D}} A[x/a])$,
- $\mathcal{I}(\bigwedge_{x \in \mathcal{D}} A) = \mathcal{I}(\bigwedge_{a \in \mathcal{D}} A[x/a])$,

where A and B are structured sentences, $\mathbf{s}$ is a structural element, $\mathcal{D} \subseteq \Delta$, and $[x/a]$ is the substitution of variable x by constant a. This definition of structured interpretations coincides with first-order interpretations in the sense that they evaluate classical first-order sentences in the same way. Note that we do not consider structured sentences with negation in the scope of algebraic elements, as this is not well-defined.

With the help of structured sentences and structured interpretations, we can now formally define setting up $\phi(\boldsymbol{\alpha})$ as a weighted model counting problem. Let $\mathfrak{I}_{\mathcal{R}}^{S}$ be the set of all structured interpretations w.r.t. Σ_{π}. Then, counting the typed models of $\mathcal{R}$ means calculating

$$\mathsf{TMC}(\pi(\mathcal{R})) = \sum_{\mathcal{I} \in \mathfrak{I}_{\mathcal{R}}^{S}} \mathcal{I}(\pi(\mathcal{R})),$$

where

$$\pi(\mathcal{R}) = \bigwedge_{f \in \mathcal{T} \cup \mathcal{A}} \pi(f) \wedge \bigwedge_{\mathfrak{c} \in \mathcal{C}} \pi(\mathfrak{c}).$$

Theorem 2 *(cf. [24]). Let $\mathcal{R} = (\mathcal{T}, \mathcal{A}, \mathcal{C})$ be a p-consistent knowledge base. Then,*

$$\phi_{\mathcal{R}}(\alpha) = \rho_{\mathcal{R}}(\mathsf{TMC}(\mathcal{R})).$$

Before we illustrate typed model counting by means of examples, we discuss some basic compilation strategies that allow one to perform typed model counting more efficiently. These strategies were adopted from first-order model counting (cf., e.g., [4]) and are discussed in [24] in more depth.

1. **Literal Conditioning:** Let $A\langle a/\top \rangle$ be the structured sentence A in which every occurrence of the constant a is substituted by $\top$, and let $A\langle a/\bot \rangle$ be defined analogously. Then,

$$A \equiv a \wedge A\langle a/\top \rangle \vee \neg a \wedge A\langle a/\bot \rangle, \tag{9}$$

where $\equiv$ is the logical equivalence between structured sentences, i.e., $A \equiv B$ iff $\mathcal{I}(A) = \mathcal{I}(B)$ for all structured interpretations $\mathcal{I}$.

2. **Decomposable Conjunction:** Let A and B be structured sentences such that A and B do not share any ground atoms. Then,

$$\mathsf{TMC}(A \wedge B) = \mathsf{TMC}(A) \cdot \mathsf{TMC}(B).$$

For example, the conjunctions in (9) are decomposable.

3. **Smooth Deterministic Disjunction:** Let A and B be structured sentences that mention the same ground atoms. If A and B are mutually exclusive, i.e., $\mathcal{I}(A) \cdot \mathcal{I}(B) = \mathbf{0}$ for all structured interpretations $\mathcal{I}$, then

$$\mathsf{TMC}(A \vee B) = \mathsf{TMC}(A) + \mathsf{TMC}(B).$$

For example, the disjunction in (9) is smooth deterministic.

Corollary 1. *Let A be a structured sentence, and let a be a constant. Then,*

$$\mathsf{TMC}(A) = \mathsf{TMC}(A\langle a/\top \rangle) + \mathsf{TMC}(A\langle a/\bot \rangle).$$

Proof. This corollary is a direct consequence of the fact that the disjunction in (9) is smooth deterministic. Obviously, at most one of the disjuncts in (9) can be satisfied by a given interpretation since either a or $\neg a$ is true.   □

We now discuss some examples. For all the example knowledge bases $\mathcal{R}$, the objective function $\phi(\mathcal{R})$ can be evaluated in time polynomial in k.

Example 3. Recall $\mathcal{R}_2$ from Example 2. Then,

$$\pi(\mathcal{R}_2) \equiv \Big(\bigwedge_{x \in \Delta} \mathbf{v_1} \circ \pi_x(A \sqcap C) \vee \mathbf{f_1} \circ \pi_x(A \sqcap \neg C) \vee 1 \circ \pi_x(\neg A) \Big)$$

$$\wedge \Big(\bigwedge_{x \in \Delta} \mathbf{v_2} \circ \pi_x(B \sqcap C) \vee \mathbf{f_2} \circ \pi_x(B \sqcap \neg C) \vee 1 \circ \pi_x(\neg B) \Big)$$

$$\equiv \Big(\bigwedge_{x \in \Delta} \mathbf{v_1} \circ A(x) \wedge C(x) \vee \mathbf{f_1} \circ A(x) \wedge \neg C(x) \vee 1 \circ \neg A(x) \Big)$$

$$\wedge \Big(\bigwedge_{x \in \Delta} \mathbf{v_2} \circ B(x) \wedge C(x) \vee \mathbf{f_2} \circ B(x) \wedge \neg C(x) \vee 1 \circ \neg B(x) \Big)$$

$$\equiv \bigwedge_{x \in \Delta} \Big(\big(\mathbf{v_1} \circ A(x) \wedge C(x) \vee \mathbf{f_1} \circ A(x) \wedge \neg C(x) \vee 1 \circ \neg A(x) \big)$$

$$\wedge \big(\mathbf{v_2} \circ B(x) \wedge C(x) \vee \mathbf{f_2} \circ B(x) \wedge \neg C(x) \vee 1 \circ \neg B(x) \big) \Big)$$

$$\equiv \bigwedge_{x \in \Delta} \Big(C(x) \wedge \big(\mathbf{v_1} \circ A(x) \vee 1 \circ \neg A(x) \big) \wedge \big(\mathbf{v_2} \circ B(x) \vee 1 \circ \neg B(x) \big) \Big)$$

$$\vee \Big(\neg C(x) \wedge \big(\mathbf{f_1} \circ A(x) \vee 1 \circ \neg A(x) \big) \wedge \big(\mathbf{f_2} \circ B(x) \vee 1 \circ \neg B(x) \big) \Big)$$

All conjunctions in the last expression are decomposable and all disjunctions are smooth deterministic. Hence,

$$\mathsf{TMC}(\mathcal{R}_2) = \sum_{x \in \Delta} \mathsf{TMC}\Big(\big(C(x) \wedge \big(\mathbf{v_1} \circ A(x) \vee 1 \circ \neg A(x) \big)$$

$$\wedge \big(\mathbf{v_2} \circ B(x) \vee 1 \circ \neg B(x) \big) \big)$$

$$\vee \big(\neg C(x) \wedge \big(\mathbf{f_1} \circ A(x) \vee 1 \circ \neg A(x) \big) \wedge \big(\mathbf{f_2} \circ B(x) \vee 1 \circ \neg B(x) \big) \big) \Big)$$

$$= \sum_{x \in \Delta} \Big(1 \cdot (\mathbf{v_1} + 1) \cdot (\mathbf{v_2} + 1) + 1 \cdot (\mathbf{f_1} + 1) \cdot (\mathbf{f_2} + 1) \Big)$$

$$= \sum_{x \in \Delta} \Big(\mathbf{v_1 v_2} + \mathbf{f_1 f_2} + \mathbf{v_1} + \mathbf{v_2} + \mathbf{f_1} + \mathbf{f_2} + 2 \Big)$$

$$= \Big(\mathbf{v_1 v_2} + \mathbf{f_1 f_2} + \mathbf{v_1} + \mathbf{v_2} + \mathbf{f_1} + \mathbf{f_2} + 2 \Big)^k$$

If one applies $\rho_{\mathcal{R}_2}$ to the last expression, one obtains the objective function $\phi_{\mathcal{R}_2}(\boldsymbol{\alpha})$ (cf. Eq. (7)). In $\phi_{\mathcal{R}_2}(\boldsymbol{\alpha})$, the domain size k occurs only as a parameter. ∎

Compiling structured sentences into sentences that mention decomposable disjunctions and smooth deterministic conjunctions only is a powerful tool when counting typed models. In the absence of existential/universal restrictions and of assertions, the objective function $\phi_\mathcal{R}(\alpha)$ of any p-consistent knowledge base $\mathcal{R}$ can be calculated similar to Example 3. The basic idea is to exploit interchangeability of constants. It remains to show how assertions and existential/universal restrictions can be handled.

If assertions occur, one first applies literal conditioning w.r.t. those ground atoms that mention the named individuals. Doing so, the resulting formula is symmetric in the remaining constants and one can proceed as before.

Example 4. Let A and B be constant names, let a be a individual name, and let $\mathcal{R}_3 = (\mathcal{T}_3, \mathcal{A}_3, \mathcal{C}_3)$ be a p-consistent knowledge base with $\mathcal{T}_3 = \emptyset, \mathcal{A}_3 = \{A(a)\}$, and

$$\mathcal{C}_3 = \{(B|A)[s_1/t_1]\},$$

where $s_1, t_1 \in \mathbb{N}$ with $s_1 < t_1$. Then,

$$\pi(\mathcal{R}_3) \equiv \pi(A(a)) \wedge \bigwedge_{x \in \Delta} \left(\mathbf{v_1} \circ \pi_x(A \sqcap B) \vee \mathbf{f_1} \circ \pi_x(A \sqcap \neg B) \vee \mathbf{1} \circ \pi_x(\neg A) \right)$$

$$\equiv A(a) \wedge \bigwedge_{x \in \Delta} \left(\mathbf{v_1} \circ A(x) \wedge B(x) \vee \mathbf{f_1} \circ A(x) \wedge \neg B(x) \vee \mathbf{1} \circ \neg A(x) \right)$$

$$\equiv A(a) \wedge (\mathbf{v_1} \circ B(x) \vee \mathbf{f_1} \circ \neg B(x)) \wedge$$

$$\bigwedge_{x \in \Delta \setminus a} \left(\mathbf{v_1} \circ A(x) \wedge B(x) \vee \mathbf{f_1} \circ A(x) \wedge \neg B(x) \vee \mathbf{1} \circ \neg A(x) \right)$$

$$\equiv A(a) \wedge \left(\mathbf{v_1} \circ B(x) \vee \mathbf{f_1} \circ \neg B(x) \right) \wedge$$

$$\bigwedge_{x \in \Delta \setminus a} \left(A(x) \wedge \left(\mathbf{v_1} \circ B(x) \vee \mathbf{f_1} \circ \neg B(x) \right) \vee \neg A(x) \wedge \left(B(x) \vee \neg B(x) \right) \right)$$

Hence,

$$\mathsf{TMC}(\mathcal{R}_3) = (\mathbf{v_1} + \mathbf{f_1}) \cdot \left(\mathbf{v_1} + \mathbf{f_1} + 2 \right)^{k-1}.$$

∎

Since the number of named individuals is independent of k, evaluating $\phi_\mathcal{R}(\alpha)$ for knowledge bases $\mathcal{R}$ that contain assertions is still polynomial in k.

In order to deal with existential and universal restrictions we make use of the principle of skolemization (in analogy to [5]).

Example 5. Let A and B be a concept names, let r be a role name, and let $\mathcal{R}_4 = (\mathcal{T}_4, \mathcal{A}_4, \mathcal{C}_4)$ be a p-consistent knowledge base with $\mathcal{T}_4 = \emptyset, \mathcal{A}_4 = \emptyset$, and

$$\mathcal{C}_4 = \{(\forall r.B|A)[s_1/t_1]\}.$$

Then,

$$\pi(\mathcal{R}_4) \equiv \bigwedge_{x \in \Delta} \left(\mathbf{v_1} \circ \pi_x(A \sqcap \forall r.B) \vee \mathbf{f_1} \circ \pi_x(A \sqcap \neg \forall r.B) \vee \mathbf{1} \circ \pi_x(\neg A) \right).$$

The difficulty when counting the models of $\pi(\mathcal{R}_4)$ is the concealed existential restriction $\neg \forall r.B$, which holds w.r.t. a fixed constant a in all models except for those in which $a^\mathcal{I} \in (\forall r.B)^\mathcal{I}$. The idea of skolemization in weighted model counting is to count all models and to subtract the models in which $a^\mathcal{I} \in (\forall r.B)^\mathcal{I}$ holds by assigning them a negative weight. Here, we write the negative weight directly into the formula in form of the structural element $-\mathbf{1}$. For this, we introduce two fresh predicates $S/1$ and $Z/1$, replace $\pi_x(\neg \forall r.B)$ by $Z(x)$ and, analogously, replace $\pi_x(\forall r.B)$ by $\neg Z(x)$. In order to fix the model counts, we finally have to add

$$\bigwedge_{x \in \Delta} \left(Z(x) \wedge \left(S(x) \vee -\mathbf{1} \circ \neg S(x) \wedge \pi_x(\forall r.B) \right) \vee \neg Z(x) \wedge S(x) \wedge \pi_x(\forall r.B) \right)$$

to $\pi(\mathcal{R}_4)$. We get

$$\pi(\mathcal{R}_4) \equiv \bigwedge_{x \in \Delta} \left(\mathbf{v_1} \circ A(x) \wedge \neg Z(x) \vee \mathbf{f_1} \circ A(x) \wedge Z(x) \vee \mathbf{1} \circ \neg A(x) \right) \wedge$$
$$\bigwedge_{x \in \Delta} \left(Z(x) \wedge \left(S(x) \vee -\mathbf{1} \circ \neg S(x) \wedge \bigwedge_{y \in \Delta} (\neg r(x,y) \vee B(y)) \right) \right.$$
$$\left. \vee \neg Z(x) \wedge S(x) \wedge \bigwedge_{y \in \Delta} (\neg r(x,y) \vee B(y)) \right)$$

Now we split the domain Δ into two parts: $\mathcal{B}$ and $\mathcal{B}^c = \Delta \backslash \mathcal{B}$. $\mathcal{B}$ shall contain those individuals that satisfy the concept B, and $\mathcal{B}^c$ shall contain those that do not. Obviously $\mathcal{B}$ depends implicitly on an interpretation $\mathcal{I}$.

It is,

$$\phi(\mathcal{R}_4) \equiv \bigvee_{\mathcal{B} \subseteq \Delta} \Big(\bigwedge_{x \in \mathcal{B}} B(x) \wedge \bigwedge_{x \in \mathcal{B}^c} \neg B(x) \wedge$$

$$\bigwedge_{x \in \Delta} \Big(\mathbf{v_1} \circ A(x) \wedge \neg Z(x) \vee \mathbf{f_1} \circ A(x) \wedge Z(x) \vee \mathbf{1} \circ \neg A(x) \Big) \wedge$$

$$\Big(Z(x) \wedge \Big(S(x) \vee -\mathbf{1} \circ \neg S(x) \wedge \bigwedge_{y \in \mathcal{B}^c} \neg r(x,y) \Big)$$

$$\vee \neg Z(x) \wedge S(x) \wedge \bigwedge_{y \in \mathcal{B}^c} \neg r(x,y) \Big) \Big)$$

$$\equiv \bigvee_{\mathcal{B} \subseteq \Delta} \Big(\bigwedge_{x \in \mathcal{B}} B(x) \wedge \bigwedge_{x \in \mathcal{B}^c} \neg B(x) \wedge$$

$$\bigwedge_{x \in \Delta} \Big(Z(x) \wedge (\mathbf{f_1} \circ A(x) \vee \mathbf{1} \circ \neg A(x))$$

$$\wedge (S(x) \vee -\mathbf{1} \circ \neg S(x) \wedge \bigwedge_{y \in \mathcal{B}^c} \neg r(x,y))$$

$$\vee \neg Z(x) \wedge (\mathbf{v_1} \circ A(x) \vee \mathbf{1} \circ \neg A(x)) \wedge S(x) \wedge \bigwedge_{y \in \mathcal{B}^c} \neg r(x,y) \Big) \Big)$$

$$\equiv \bigvee_{\mathcal{B} \subseteq \Delta} \Big(\bigwedge_{x \in \mathcal{B}} B(x) \wedge \bigwedge_{x \in \mathcal{B}^c} \neg B(x) \wedge \bigwedge_{x \in \Delta} \wedge \bigwedge_{y \in \mathcal{B}} (\neg r(x,y) \vee r(x,y)) \wedge$$

$$\Big(Z(x) \wedge (\mathbf{f_1} \circ A(x) \vee \mathbf{1} \circ \neg A(x)) \wedge (S(x) \wedge \bigwedge_{y \in \mathcal{B}^c} (\neg r(x,y) \vee r(x,y))$$

$$\vee -\mathbf{1} \circ \neg S(x) \wedge \bigwedge_{y \in \mathcal{B}^c} \neg r(x,y))$$

$$\vee \neg Z(x) \wedge (\mathbf{v_1} \circ A(x) \vee \mathbf{1} \circ \neg A(x)) \wedge S(x) \wedge \bigwedge_{y \in \mathcal{B}^c} \neg r(x,y) \Big) \Big)$$

In the last expression all conjunctions are decomposable and all disjunctions are smooth deterministic. In order to count the typed models of $\pi(\mathcal{R}_4)$, it is necessary to compute the numbers of possible subsets $\mathcal{B} \subseteq \Delta$. Obviously, $|\mathcal{B}|$ can vary from 0 to k, and there are $\binom{k}{m}$-many subsets of Δ of size m. Hence,

$$\mathsf{TMC}(\mathcal{R}_4) = \sum_{m=0}^{k} \binom{k}{m} \Big(2^m \cdot (\mathbf{f_1} + 1) \cdot (2^{k-m} - 1 + (\mathbf{v_1} + 1)) \Big)^k.$$

Here, the domain size k is not simply a parameter, but evaluating the expression $\phi(\mathcal{R}_4) = \rho_{\mathcal{R}}(\mathsf{TMC}(\mathcal{R}_4))$ is still possible in time polynomial in k. If there are more than one existential/universal restriction in $\mathcal{R}$ one gets nested sums, but evaluation is still possible in time polynomial in k. Only the degree of the polynomial increases. ∎

5   Conclusion and Future Work

The Description Logic $\mathcal{ALC}^{ME}$ is a probabilistic extension of the well-known Description Logic $\mathcal{ALC}$ which allows for probabilistic conditional statements of the form "if concept C holds, then concept D follows with probability p." Probabilities are understood as degrees of beliefs and a reasoner's belief state is established by the principle of maximum entropy based on the aggregating semantics. In [25] we showed that drawing inferences in $\mathcal{ALC}^{ME}$ (without assertions) is possible in time polynomial in the size k of the domain of discourse, provided that (an approximation of) the maximum entropy distribution is given.

In this paper, we proved that approximations of the maximum entropy distribution can be calculated in time polynomial in k, too, with only little restrictions. Further, we discussed typed model counting as a helpful framework for maximum entropy calculations based on $\mathcal{ALC}^{ME}$-knowledge bases, even in the presence of assertional knowledge.

In future work, we want to apply our methods to more expressive Description Logics, and we want to investigate how robust maximum entropy calculations are against changes in the domain size k.

Acknowledgements. This work was supported by the German Research Foundation (DFG) within the Research Unit FOR 1513 "Hybrid Reasoning for Intelligent Systems".

References

1. Baader, F., Calvanese, D., McGuinness, D.L., Nardi, D., Patel-Schneider, P.F. (eds.): The Description Logic Handbook: Theory, Implementation, and Applications. Cambridge University Press, Cambridge (2003)
2. Baader, F., Koopmann, P., Turhan, A.-Y.: Using ontologies to query probabilistic numerical data. In: Dixon, C., Finger, M. (eds.) FroCoS 2017. LNCS (LNAI), vol. 10483, pp. 77–94. Springer, Cham (2017). https://doi.org/10.1007/978-3-319-66167-4_5
3. Boyd, S., Vandenberghe, L.: Convex Optimization. Cambridge University Press, Cambridge (2004)
4. Van den Broeck, G., Taghipour, N., Meert, W., Davis, J., De Raedt, L.: Lifted probabilistic inference by first-order knowledge compilation. In: Proceedings of the 22th International Joint Conference on Artificial Intelligence (IJCAI), pp. 2178–2185. AAAI Press (2011)
5. den Broeck, G.V., Meert, W., Darwiche, A.: Skolemization for weighted first-order model counting. In: Proceedings of the Fourteenth International Conference on Principles of Knowledge Representation and Reasoning, KR 2014, Vienna, Austria, 20–24 July 2014 (2014)
6. Cowell, R., Dawid, A., Lauritzen, S., Spiegelhalter, D.: Probabilistic Networks and Expert Systems. Springer, New York (1999). https://doi.org/10.1007/b97670
7. Finthammer, M.: Concepts and algorithms for computing maximum entropy distributions for knowledge bases with relational probabilistic conditionals. Ph.D. thesis, University of Hagen (2016)
8. Halpern, J.Y.: An analysis of first-order logics of probability. Artif. Intell. **46**(3), 311–350 (1990)

9. Jaynes, E.: Papers on Probability, Statistics and Statistical Physics. D. Reidel Publishing Company, Dordrecht (1983)
10. Kern-Isberner, G. (ed.): Conditionals in Nonmonotonic Reasoning and Belief Revision. LNCS (LNAI), vol. 2087. Springer, Heidelberg (2001). https://doi.org/10.1007/3-540-44600-1
11. Kern-Isberner, G., Thimm, M.: Novel semantical approaches to relational probabilistic conditionals. In: Proceedings of the 12th International Conference on the Principles of Knowledge Representation and Reasoning (KR), pp. 382–392. AAAI Press (2010)
12. Lutz, C., Schröder, L.: Probabilistic description logics for subjective uncertainty. In: Proceedings of the 12th International Conference on Principles of Knowledge Representation and Reasoning (KR), pp. 393–403. AAAI Press (2010)
13. Mintz, Y., Aswani, A.: Polynomial-time approximation for nonconvex optimization problems with an L1-constraint. In: 56th IEEE Annual Conference on Decision and Control, CDC 2017, Melbourne, Australia, 12–15 December 2017, pp. 682–687 (2017)
14. Nemirovskii, A.: Interior point polynomial time methods in convex programming (1996). Lecture script
15. Paris, J., Vencovská, A.: A note on the inevitability of maximum entropy. Int. J. Approx. Reason. 4(3), 183–223 (1990)
16. Paris, J.B.: Common sense and maximum entropy. Synthese 117(1), 75–93 (1999)
17. Paris, J.B.: The Uncertain Reasoner's Companion: A Mathematical Perspective. Cambridge University Press, Cambridge (2006)
18. Pearl, J.: Probabilistic Reasoning in Intelligent Systems. Morgan Kaufmann, San Francisco (1988)
19. Peñaloza, R., Potyka, N.: Towards statistical reasoning in description logics over finite domains. In: Moral, S., Pivert, O., Sánchez, D., Marín, N. (eds.) SUM 2017. LNCS (LNAI), vol. 10564, pp. 280–294. Springer, Cham (2017). https://doi.org/10.1007/978-3-319-67582-4_20
20. Pratt, V.R.: Models of program logics. In: Proceedings of the 20th Annual Symposium on Foundations of Computer Science (FOCS), pp. 115–122. IEEE Computer Society (1979)
21. Rudolph, S., Krötzsch, M., Hitzler, P.: Type-elimination-based reasoning for the description logic SHIQbs using decision diagrams and disjunctive datalog. Log. Methods Comput. Sci. 8(1), 1–38 (2012)
22. Shore, J., Johnson, R.: Axiomatic derivation of the principle of maximum entropy and the principle of minimum cross-entropy. IEEE Trans. Inf. Theory 26(1), 26–37 (1980)
23. Thimm, M., Kern-Isberner, G.: On probabilistic inference in relational conditional logics. Log. J. IGPL 20(5), 872–908 (2012)
24. Wilhelm, M., Finthammer, M., Kern-Isberner, G., Beierle, C.: First-order typed model counting for probabilistic conditional reasoning at maximum entropy. In: Moral, S., Pivert, O., Sánchez, D., Marín, N. (eds.) SUM 2017. LNCS (LNAI), vol. 10564, pp. 266–279. Springer, Cham (2017). https://doi.org/10.1007/978-3-319-67582-4_19
25. Wilhelm, M., Kern-Isberner, G., Ecke, A., Baader, F.: Counting strategies for the probabilistic description logic $\mathcal{ALC}^{ME}$ under the principle of maximum entropy. In: Calimeri, F., Leone, N., Manna, M. (eds.) JELIA 2019. LNCS, vol. 11468, pp. 434–449. Springer, Cham (2019). https://doi.org/10.1007/978-3-030-19570-0_28

Automating Automated Reasoning
The Case of Two Generic Automated Reasoning Tools

Yoni Zohar[1], Dmitry Tishkovsky[2], Renate A. Schmidt[2(✉)], and Anna Zamansky[3]

[1] Computer Science Department, Stanford University, Stanford, USA
[2] School of Computer Science, University of Manchester, Manchester, UK
[3] Information Systems Department, University of Haifa, Haifa, Israel

Abstract. The vision of automated support for the investigation of logics, proposed decades ago, has been implemented in many forms, producing numerous tools that analyze various logical properties (e.g., cut-elimination, semantics, and more). However, full 'automation of automated reasoning' in the sense of automatic generation of efficient provers has remained a 'holy grail' of the field. Creating a generic prover which can efficiently reason in a given logic is challenging, as each logic may be based on a different language, and involve different inference rules, that require different implementation considerations to achieve efficiency, or even tractability. Two recently introduced generic automated provers apply different approaches to tackle this challenge. MetTeL, based on the formalism of tableaux, automatically generates a prover for a given tableau calculus, by implementing generic proof-search procedures with optimizations applicable to many tableau calculi. Gen2sat, based on the formalism of sequent calculi, shifts the burden of search to the realm of off-the-shelf SAT solvers by applying a uniform reduction of derivability in sequent calculi to SAT. This paper examines these two generic provers, focusing in particular on criteria relevant for comparing their performance and usability. To this end, we evaluate the performance of the tools, and describe the results of a preliminary empirical study where user experiences of expert logicians using the two tools are compared.

1 Introduction

The idea of automated support for the investigation of logics has been envisioned more than twenty years ago by Ohlbach [49], who wrote: 'not every designer of an application program, which needs logic in some of its components, is a logician and can develop the optimal logic for his purposes, neither can he hire a trained logician to do this for him. In this situation we could either resign and live with non-optimal solutions, or we could try to give more or less automated support and guidance for developing new logics'. Ohlbach's vision has been successfully applied to the paradigm of 'logic engineering', a term coined by Areces [3] to refer to approaches that systematically investigate and construct new logical formalisms with specific desired properties (such as decidability, expressive power,

© Springer Nature Switzerland AG 2019
C. Lutz et al. (Eds.): Baader Festschrift, LNCS 11560, pp. 610–638, 2019.
https://doi.org/10.1007/978-3-030-22102-7_29

and effective reasoning methods), for a particular need or application. Many tools that implement automated approaches for the investigation of large families of proof systems have been introduced, including the linear logic-based framework for specifying and reasoning about proof systems of [44,47,48], the reformulation of paraconsistent and substructural logics in terms of analytic calculi in [13,14], and the automatic production of sequent calculi for many-valued logics [8]. Generic tools for correspondence theory include [16,20,50], that compute frame conditions for modal axioms, from which it is then possible to obtain corresponding tableau rules using the tableau synthesis method of [54,60]. Related forgetting tools [39,66] compute uniform interpolants and give users the ability to decompose logical theories and ontologies.

While the mentioned tools offer useful automated support for studying logics and building logic-based systems, they are not prover generators. Unlike classical logic, which has efficient provers that make use of state-of-the-art SAT technologies, there has been insufficient work to create provers for the wide variety of non-classical logics investigated in the literature that can enable their easy integration in applications. This calls for *generic* provers, as well as tools for *automated generation* of such provers.

The naive approach for generation of a prover for a given logic is associating a basic proof-search algorithm for a given calculus, without considerations for reducing the search space for this particular logic. This, however, yields impractical, non-efficient provers. Implementing an efficient prover from scratch is a significant investment of time, and requires relevant expertise and experience of the developer. Thus, realization of *efficient* provers has long remained the 'holy grail' of automated support in line with Ohlbach [49].

There are many tools that approach this problem by focusing on a specific *family* of logics, that have a shared syntax and structure of inference rules. Examples of such tools include the Logic WorkBench [30], the Tableau Workbench [1], LoTREC [23], focusing on modal-like logics, and COOL [27], focusing on modal and hybrid logics.

Two recently developed provers take different approaches to achieve considerable genericity. The first approach is implemented in MetTeL [61–63] (available at [35]). MetTeL is a powerful platform for automatically generating provers from the definition of tableau calculi of very general forms. It achieves efficiency by using strong, general heuristics and optimizations that are broad enough to apply to a wide variety of inference rules on the one hand, and are efficient and non-trivial on the other hand. Such generic techniques, when identified, can enhance any generated proof search algorithm, making it less naive and more practical. MetTeL differs from the above mentioned tools mainly by being completely logic and language independent, and the language and inference rules are completely defined by the user.

The second approach is implemented in Gen2sat [68] (available in [67]). Gen2sat is a platform which provides a method for deciding the derivability of a sequent in a given sequent calculus, via a uniform polynomial reduction to the classical satisfiability problem. Looking for specific heuristics for a given calculus is bypassed in Gen2sat, by shifting the actual search to the realm of off-the-shelf SAT solvers. Using (classical) SAT-solving for non-classical logics was also employed, e.g., in [11,26,33,40] for various modal (and description) logics, where the non-modal part was fixed to be classical.

While there are several papers discussing the theoretical aspects and implementation details of MetTeL and Gen2sat (developed by the second and first authors) [41,54,62,68], this paper is concerned with comparing these two tools with respect to criteria relating to their performance and usability.

To this end we carry out a performance analysis of the tools, as well as a preliminary empirical study of usability, with five expert logicians providing user feedback on both tools. While the former form of evaluation is rather standard in the automated reasoning community, empirical studies with real users are scarce. We discuss the insights received from our study participants which will be instrumental for improving the tools.

The paper is structured as follows. Sections 2 and 3 describe the approaches taken in the development of MetTeL and Gen2sat and provide a short overview of each. Section 4 provides a comparison of the performance of the tools on a collection of benchmarks. Section 5 discusses the tools from the users' perspective, and presents the results of a preliminary empirical study on their usability. Section 6 concludes with a summary and a discussion of several directions for further research.

2   Generic Automated Reasoning with Tableau

In this section we describe the prover generator MetTeL, aimed at supporting researchers and practitioners who use *tableau calculi* for the specification of logics. MetTeL automatically generates and compiles Java code of a tableau prover from specifications of the syntax of a logic and a set of tableau rules for the logic. The specification language of MetTeL is designed to be as simple as possible for the user on the one hand, and as expressive as the traditional notation used in logic and automated reasoning textbooks, on the other hand.

2.1   Tableau Synthesis

Of all the different forms of tableau calculi, semantic tableau calculi [7,21,58] are widely used and widely taught in logic and computer science courses, because the rules of inference are easily explained and understood, and deductions are carried out in a completely goal-directed way. In explicit semantic tableau approaches the application of the inference rules is order independent (because these approaches are proof confluent), which avoids the overhead and complication associated with handling don't know non-determinism of non-invertible rules in direct methods [1] (see also the discussion in [32]). Because semantic tableau approaches construct and return (counter-)models, they are suitable for error finding, which is useful for ontology development, theory creation and applications such as multi-agent systems.

MetTeL is an outcome of research on the systematic development of explicit semantic tableau systems and automated generation of tableau provers [54,60–63]. In line with the aims of logic engineering, the vision of this research is to allow the steps of developing a tableau calculus and prover to be automated as much as possible. The idea is that from the semantic definition of a logic a sound, complete and often terminating calculus is generated that can then be input into MetTeL, which will produce a prover for the logic. In the endeavour of finding systematic general ways of generating elegant, natural tableau systems, a lot of research went into finding ways to ensure the systems produce smaller proofs and have smaller search space, both for the theoretical tableau synthesis framework and the provers generated by MetTeL. The research involved generalising clever backtracking techniques such as backjumping and dynamic backtracking to the tableau synthesis framework so that they can be combined naturally with new systems. The research also involved finding new, more powerful ways to do blocking in order to ensure termination of tableau derivations for decidable logics. Refinement techniques were developed to devise more effective deduction calculi and improve the way that deductions are performed in tableau systems.

Having devised a calculus for a logic, the next step is the implementation of a prover for it. To avoid the burden of developing a prover from scratch, or extend and adapt an existing prover, the MetTeL system was developed to automatically generate code of fully-functioning stand-alone provers specialised for the user's application. MetTeL takes as input a high-level specification of a logic, or a theory, together with a set of deduction rules and then generates a prover for this logic and calculus. Together with the tableau synthesis framework, this provides a systematic and nearly fully automated methodology for obtaining tableau provers for a logic.

The rule specification language of MetTeL is based on a powerful many-sorted first-order language designed to be as general as possible. The language extends the meta-language of the tableau synthesis framework which enables calculi obtained in the tableau synthesis framework to be implemented with little effort in MetTeL. Concrete case studies undertaken with MetTeL include:

- Labelled, semantic tableau calculi for standard modal logics K, KT, $S4$ [60].
- Labelled, semantic tableau calculi for propositional intuitionistic logic [54].
- Tableau calculi for hybrid modal logic with counting quantifiers [37,65].
- Internalized tableau calculi for hybrid logics and description logics such as $\mathcal{SO}$, $\mathcal{ALCO}$ and $\mathcal{SHOI}$ [36,54]. Various specialisations of the blocking mechanism were defined and evaluated, and simulation of the standard blocking techniques was shown. This work has evaluated the use of flexibly generated refined rules for ontology TBox axioms to reduce the search space and improve performance.
- A terminating tableau calculus for the description logic $\mathcal{ALBO}$id allowing compound role expressions which gives it the same expressive power as the two-variable fragment of first-order logic [55]. MetTeL was used to implement a tableau decision procedure for this logic.
- Linear temporal logic with Boolean constraints [19]. A method of dealing with fixpoints in the linear time temporal logic was developed and tested.

- Interrogative-epistemic logics for reasoning about questions and queries of multiple agents [45].
- Logics and algebras of hypergraphs with relevance to image processing [53,59]. MetTeL played an important role in the introduction and investigation of a novel bi-intuitionistic modal logic, called BISKT, and a related modal tense logic and the development of tableau decision procedures for these.
- The extension $K_m(\neg)$ of the basic multi-modal logic K_m with relational negation, the modal logic of 'some', 'all' and 'only' [31], is used to illustrate the atomic rule refinement techniques investigated [60].
- Unlabelled deduction calculi for Boolean logic and three-valued Łukasiewicz logic (which we consider in Sect. 4), and a calculus for simple equational reasoning about lists (given in Fig. 1) [62].

These applications have shown it is easy to generate provers for a wide variety of logics, including new logics. They have also shown the approach is especially useful for systematic comparisons of different sets of tableaux rules for a specific logic, different strategies, and techniques. This is useful for research purposes but also in teaching and learning environments.

At present, MetTeL does not accommodate languages with first-order quantifiers directly, although the syntax specification language of MetTeL has enough expressive power to represent languages of first-order theories with a finite number of logical operators, predicate symbols and functional symbols.

2.2   MetTeL Features and Usage

First, an input file containing a definition of the syntax used in the tableau rules and the definition of the tableau rules themselves needs to be prepared. Figure 1 shows the contents of an input file defining a simple 'non-logical' example of a syntax and tableau calculus for describing and comparing lists. The line `specification lists` defines `lists` to be the name of the user-defined logical language. The `syntax lists` block consists of the declaration of the sorts and definitions of logical operators. Here, three sorts are declared: `formula`, `element` and `list`. For the sort `element`, no operators are defined, which means that all `element` formulas are atomic. There are two operators for the sort `list`: a nullary operator `empty` (to be used for the empty list) and a binary operator `composite` (used to inductively define non-empty lists). The next two lines are the formation rules for formulas of sort `formula`, namely inequalities between elements and lists.

The `tableau lists` block defines the tableau rules of the calculus. In the tableau rule specification language of MetTeL, the premises and conclusions of a rule are separated by / and each rule is terminated by $;. Branching rules can have two or more sets of conclusions which are separated by $|. Premises and conclusions are formulas in the user-defined logical language specified in the previous block. As is illustrated, the rules can be annotated with priority values, that determine the order by which the rules are applied. The default priority value of any rule with unspecified priority is 0. The `tableau lists`

```
specification lists;
syntax lists{
     sort formula, element, list;
     list empty = '<>' | composite = '<' element list '>';
     formula elementInequality = '[' element '!=' element ']';
     formula listInequality =  '{' list '!=' list '}';
}
tableau lists{
     [a != a] / priority 0 $;
     {L != L} / priority 0 $;
     {L0 != L1} / {L1 != L0} priority 1 $;
     {<a L0> != <b L1>} / [a != b] $| {L0 != L1} priority 2 $;
}
```

Fig. 1. Input to MetTeL: A specification of syntax and tableau rules.

```
Input

{<a (<b L>)> != <a (<b L>)>}

Output

Unsatisfiable.
Contradiction:
[({(<a (<b L>)>) != (<a (<b L>)>)})]
```

```
Input

{<a (<b L0>)> != <a (<b L1>)>}

Output

Satisfiable.
Model:
[({(<a (<b L0>)>) != (<a (<b L1>)>)}),
({(<b L0>) != (<b L1>)}), ({L0 != L1})]
```

Fig. 2. An unsatisfiable instance   **Fig. 3.** A satisfiable instance

block includes two closure rules in which the right hand sides of / are empty, reflecting that inequality is irreflexive, as well as additional rules for handling inequalities.

Having prepared an input file named lists.s (say) MetTeL can be run from the command line using: java -jar mettel2.jar -i lists.s. The generated prover lists.jar can be run from the command line using: java -jar lists.jar.

The generated provers return the answers Satisfiable or Unsatisfiable. If the answer is Unsatisfiable and the prover is able to extract the input formulas needed for deriving the contradiction, they are printed. If the answer is Satisfiable then all the formulas within the completed open branch are output as a model. For efficiency reasons MetTeL does not output proofs. Although proofs are useful for the user, the overhead of outputting proofs is high because tableau proofs for unsatisfiable problems may be very big. Additionally, the backtracking techniques and the destructive nature of the rewriting for equality reasoning and blocking (described below), make the problem of generating a human-readable proof harder. For some instances, however, MetTeL is able to output the assumptions that are used to show unsatisfiability, and so some information about such proofs is recovered.

Figures 2 and 3 present satisfiable and unsatisfiable runs of the generated prover for the lists example. The list [a,b,L] is identical to itself, and thus the inequality in Fig. 2 cannot be satisfied. On the other hand, the lists [a,b,L0] and [a,b,L1] differ from one another, and so the inequality in Fig. 3 is satisfiable.

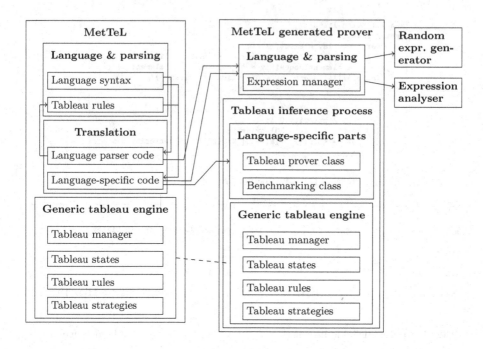

Fig. 4. Architecture of MetTeL and MetTeL generated provers

The online-version [35] of MetTeL consist of several screens and includes predefined specifications for several tableau calculi, some of which are mentioned in Sect. 2.1.

2.3 Under the MetTeL Hood and Features of the Generated Provers

Along with easy use and efficiency of the generated provers, the goals and objectives of the implementation of MetTeL included modularity of generated code and a hierarchy of public JAVA classes and interfaces that can be extended and integrated with other systems.

A top-level view of the architecture of MetTeL and MetTeL generated provers is given in Fig. 4. The user-defined syntax for formulas is parsed using the ANTLR parser generator, and is internally represented as an abstract syntax tree (AST). All generated Java classes for formula representation implement the basic `MettelExpression` interface. At runtime, the creation of formula objects is done according to the factory design pattern, via the interface `MettelObjectFactory`. The two most important methods that formula classes implement are: (i) a method that returns a substitution that matches the current object with the formula object supplied as a parameter; and (ii) a method that returns an instance of the current formula with respect to a given substitution.

Every tableau rule is applied within a tableau state. A sequence of formulas from the set of active formulas associated with the tableau state is selected and the formulas in the sequence are matched with the premises of the chosen rule. All selected formulas are deleted from the set of active formulas associated with the rule. If the selected formulas in the sequence match the premises of the rule, the resulting substitution object is passed to the conclusions of the rule. The final result of a rule application is a set of branches, which are sets of formulas obtained by applying the substitution to the conclusions of the rule.

Important concerns in the creation of MetTeL were efficiency of the generated provers and providing decision procedures via generic blocking. MetTeL includes two particular built-in optimizations for reducing the search space in tableau derivations. The first is *dynamic backtracking* [25], that avoids repeating the same rule applications in parallel branches. The second is *conflict directed backjumping* [22,51], that derives conflict sets of formulas from a derivation, thus causing branches with the same conflict sets to be discarded. Usually, these optimizations are designed for a particular tableau procedure with a fixed syntax, while in MetTeL, they are both implemented in a logic-independent way.

To achieve termination for semantic tableau approaches some form of blocking is usually necessary. Because of its generality and independence from the logic or the tableau calculus, blocking in MetTeL generated provers use an equality-based approach from the tableau synthesis framework [55]. The forms of blocking available include unrestricted blocking, which is the strongest form of blocking, and predecessor blocking. They can be incorporated through inference rules added to the calculus. In a semantic tables calculus for hybrid modal logic the shape of unrestricted blocking is

```
@s P  @t Q  / [s=t] $| (not([s=t]))  priority 9 $;
```

and predecessor blocking may look like this:

```
R(s,t) / [s=t] $| (not([s=t]))        priority 9 $;
```

These rules use in-built equality to merge terms s and t in the right branch which is selected first. This either leads to a model, or it does not, in which case s and t cannot be equal. While in the second case the rule is only applied if s is a predecessor of t in the R relation, in the first case the premises are not really constraining, meaning the rule is potentially applied for all terms in a derivation, for P and Q are matched with any modal formulas. We should note that R is also matched with any relational formula, but here we are assuming relational formulas can only be atomic as would be specified in the input file. Unrestricted blocking rule can be used to achieve termination for logics with the finite model property and finitely satisfiable formulas [54,55].

To realise blocking, the generated provers support equational rewriting of terms with respect to congruence relations defined in the language specification. In particular, if the definition of a rule involves equality as above then (ordered) rewriting is triggered. Rewriting allows derivations to be simplified on the fly and the search space to be reduced: for example when [f(i,P)=i] exhaustive rewriting reduces the term $f(f(f(i,P),P),P)$ to i. Refinements of equality reasoning and equality-based blocking in semantic tableau-like approaches have been studied in [9,36,53,56].

The default search strategy in the derivation process of the core tableau engine of MetTeL is depth-first left-to-right search, which is implemented as a `MettelSimpleLIFOBranchSelectionStrategy` request to the `MettelSimpleTableauManager`. Breadth-first search is implemented as a `MettelSimpleFIFOBranchSelectionStrategy` request and can be used by introducing a small modification in the generated Java code. Users can also implement their own search strategy and pass it to `MettelSimpleTableauManager`.

The rule selection strategy can be controlled by specifying priority values for the rules in the tableau calculus specification. The rule selection algorithm checks the applicability of rules and returns a rule that can be applied to formulas on the current branch according to the rule priority values. First, the algorithm selects a group of rules with the same priority value. Selection within a group with higher priority value is made only if no rules with smaller priority values are applicable. Second, rules with the same priority values are checked for applicability sequentially. To ensure fair treatment of rules within the same priority group all rules within the group are checked for applicability an equal number of times.

3   Generic Automated Reasoning with Sequent Calculi

In this section we describe Gen2sat, which, like MetTeL is a generic tool written in Java. In contrast to MetTeL, Gen2sat aims to support researchers and practitioners who use *sequent calculi* for the specification of logics. Sequent calculi, introduced in [24], are a prominent proof-theoretic framework, suitable for a wide variety of different logics (see, e.g., [6,64]). Unlike the usual method of *proof search* that is common in decision procedures for sequent calculi [18], Gen2sat employs a *uniform* reduction to SAT [41]. Shifting the intricacies of implementation and heuristic considerations to the realm of off-the-shelf SAT solvers, the tool is lightweight and focuses solely on the transformation of derivability to a SAT instance. As such, it also has the potential to serve as a tool that can enhance learning and research of concepts related to proof theory and semantics of non-classical logics, in particular those of sequent calculi.

3.1   Analytic Pure Sequent Calculi

We start by precisely defining the family of calculi for which Gen2sat is applicable. An inference rule is called *pure* if it does not enforce any limitations on the context formulas (following [5], the adjective *pure* stands for this requirement). For example, the right introduction rule of implication in classical logic $\frac{\Gamma,\varphi\Rightarrow\psi,\Delta}{\Gamma\Rightarrow\varphi\supset\psi,\Delta}$ is pure, as it can be applied with any Γ and Δ. However, in intuitionistic logic, the corresponding rule is $\frac{\Gamma,\varphi\Rightarrow\psi}{\Gamma\Rightarrow\varphi\supset\psi}$ (in other words, Δ must be empty).

Thus the latter rule is impure. A sequent calculus is called *pure* if it includes all the standard structural rules:[1] weakening, identity and cut; and all its inference rules are pure.

For a finite set $\circledcirc$ of unary connectives, we say that a formula φ is a $\circledcirc$-*subformula* of a formula ψ if either φ is a subformula of ψ, or $\varphi = \circ\psi'$ for some $\circ \in \circledcirc$ and proper subformula ψ' of ψ. A pure calculus is $\circledcirc$-*analytic* if whenever a sequent s is derivable in it, s can be derived using only formulas from $sub^{\circledcirc}(s)$, the set of $\circledcirc$-subformulas of s. We call a calculus *analytic* if it is $\circledcirc$-analytic for some set $\circledcirc$. Note that $\emptyset$-analyticity amounts to the usual subformula property. Many well-known logics can be represented by analytic pure sequent calculi, including three and four-valued logics, various paraconsistent logics, and extensions of primal infon logic ([41] presents several examples).

Gen2sat is capable of handling impure rules of the form $(*i)\ \dfrac{\Gamma \Rightarrow \Delta}{*\Gamma \Rightarrow *\Delta}$ for *Next*-operators. $(*i)$ is the usual rule for *Next* in LTL (see, e.g., [34]). It is also used as $\square$ (and $\lozenge$) in the modal logic $KD!$ of functional Kripke frames (also known as KF and $KDalt1$). In primal infon logic [17] *Next* operators play the role of quotations.

3.2   Gen2sat Features and Usage

From the command line, Gen2sat is called by: `java -jar gen2sat.jar <path>`, where `path` points to a property file with the following fields:

Connectives. A comma separated list of connectives, each specified by its symbol and arity, separated by a colon.

Next operators. A comma separated list of the symbols for the next operators.

Rules. Each rule is specified in a separate line that starts with `rule:`. The rule itself has two parts separated by /: the premises, which is a semicolon separated list of sequents, and the conclusion, which is a sequent.

Analyticity. For the usual subformula property this field is left empty. For other forms of analyticity, it contains a comma separated list of unary connectives.

Input sequent. The sequent whose derivability should be decided.

If the sequent is unprovable, Gen2sat outputs a countermodel. If it is provable, a full proof is unobtainable, due to the semantic approach Gen2sat undertakes. However, in case the sequent is provable, Gen2sat is able to recover a sub-calculus in which the sequent is already provable, that is, a subset of rules that suffice to prove the sequent.

Figures 5 and 6 present examples for the usage of Gen2sat. In Fig. 5, the input contains a sequent calculus for the Dolev-Yao intruder model [15]. The connectives E and P correspond to encryption and pairing. The sequent is provable, meaning that given two messages m_1 and m_2 that are paired and encrypted twice with k, the intruder can discover m_1 if it knows k. In Fig. 6, the input file

---

[1] Here sequents are taken to be pairs of *sets* of formulas, and therefore exchange and contraction are built in.

```
Input file

connectives: P:2, E:2
rule: =>a; =>b / =>aPb
rule: a=> / aPb=>
rule: b=> / aPb=>
rule: =>a; =>b / =>aEb
rule: =>b; a=> / aEb=>
analyticity:        •
inputSequent: (((m1 P m2 ) E k) E k),k=>m1



Output

provable
There's a proof that uses only these rules:
[=>b; a=> / a E b=>, a=> / a P b=>]
```

```
Input file

connectives: AND:2,OR:2,IMPLIES:2,TOP:0
nextOperators: q1 said, q2 said, q3 said
rule: =>p1; =>p2 / =>p1 AND p2
rule: p1,p2=> / p1 AND p2=>
rule: =>p1,p2 / =>p1 OR p2
rule: =>p2 / =>p1 IMPLIES p2
rule: =>p1; p2=> / p1 IMPLIES p2=>
rule: / => TOP
analyticity:
inputSequent: =>q1 said (p IMPLIES p)


Output

unprovable
Countermodel:
q1said p=false, q1said(p IMPLIES p)=false
```

Fig. 5. A provable instance   **Fig. 6.** An unprovable instance

contains a sequent calculus for primal infon logic, where the implication connective is not reflexive, and hence the input sequent is unprovable. Note that the rules for the next operators are fixed, and therefore they are not included in the input file. Both calculi are ∅-analytic, and hence the analyticity field is left empty. In general, whenever the calculus is ⊚-analytic, this field lists the elements of ⊚.

Gen2sat also includes an online version, in which the user fills a form that corresponds to the input file of the command line version. When all the information is filled, the user clicks the 'submit' button, and gets both an abbreviated result and a detailed result. The web-based version includes predefined forms for some propositional logics (e.g. classical logic, primal infon logic and more). In addition, it allows the user to import sequent calculi from *Paralyzer*.[2]

3.3   Under the Gen2sat Hood

The core of Gen2sat is a reduction to SAT, thus it leaves the 'hard work' and heuristic considerations of optimizations to state of the art SAT solvers, allowing the user to focus solely on the *logical* considerations.

The theoretical background on which Gen2sat is based can be found in [41]. Below are the relevant results from that paper.

In order to decide derivability in sequent calculi, Gen2sat adopts a *semantic* view of them. Thus, two-valued valuations functions (bivaluations), normally defined over formulas, are extended to sequents in the following, natural way: $v(\Gamma \Rightarrow \Delta) = 1$ if $v(\varphi) = 0$ for some $\varphi \in \Gamma$ or $v(\psi) = 1$ for some $\psi \in \Delta$. This extended semantics gives way for a semantic interpretation of pure rules:

---

[2] Paralyzer is a tool that transforms Hilbert calculi of a certain general form into equivalent analytic sequent calculi. It was described in [12] and can be found at http://www.logic.at/people/lara/paralyzer.html.

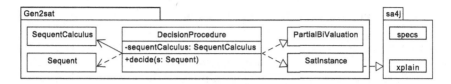

Fig. 7. A partial class diagram of Gen2sat

Definition 1. Let **G** be a pure sequent calculus. A **G**-*legal bivaluation* is a function v from some set of formulas to $\{0, 1\}$ that respects each rule of **G**, that is, for every instance of a rule, if v assigns 1 to all premises, it also assigns 1 to the conclusion.

Example 1. If **G** is taken to be the Dolev-Yao calculus from Fig. 5, then one of the conditions for being **G**-legal is that if $v(\Rightarrow a) = 1$ and $v(\Rightarrow b) = 1$, then also $v(\Rightarrow aEb) = 1$.

Theorem 1 [41]. *Let $\odot$ be a set of unary connectives, **G** a $\odot$-analytic pure sequent calculus, and s a sequent. s is provable in **G** if and only if there is no **G**-legal bivaluation v with domain $sub^{\odot}(s)$ such that $v(s) = 0$.*

Thus, given a $\odot$-analytic calculus **G** and a sequent s as its input, Gen2sat does not search for a proof. Instead, it searches for a countermodel of the sequent, by encoding in a SAT instance the following properties of the countermodel: (i) assigning 0 to s; and (ii) being **G**-legal with domain $sub^{\odot}(s)$. The addition of *Next*-operators requires some adaptations of the above reduction, that are described in [41].

Gen2sat is implemented in Java and uses sat4j [42] as its underlying SAT solver. Since this approach is based on a 'one-shot' reduction to SAT, no changes are needed in the SAT solver itself. In particular, sat4j can be easily replaced by other available solvers. Figure 7 includes a partial class diagram of Gen2sat, that shows the main modules of the tool. The two main modules of sat4j that are used are **specs**, which provides the solver itself, and **xplain**, which searches for an unsatisfiable core. The main class of Gen2sat is **DecisionProcedure**, that is instantiated with a specific **SequentCalculus**. Its main method **decide** checks whether the input sequent is provable. Given a **Sequent** s, **decide** generates a **SatInstance** stating that s has a countermodel, by applying the rules of the calculus on the relevant formulas, as described above. **SatInstance** is the only class that uses sat4j directly, and thus it is the only class that will change if another SAT solver is used.

For satisfiable instances, the **specs** module returns a satisfying assignment, which is directly translated to a countermodel in the form of a **PartialBivaluation**. For unsatisfiable instances, the **xplain** module generates a subset of clauses that is itself unsatisfiable. Tracking back to the rules that induced these clauses, it is possible to recover a smaller sequent calculus in which s is already provable. For this purpose, a multi-map is maintained, that saves for each clause of the SAT instance the set of sequent rules that induced it. Note however, that the smaller calculus need not be analytic, and then the correctness, that relies on Theorem 1 might fail. Nevertheless, correctness is preserved in this case, as the 'if' part of Theorem 1 holds even for non-analytic calculi. Thus, although Gen2sat does not provide a proof of the sequent, it does provide useful information about the rules that were used in it.

4   Performance Evaluation

We describe an evaluation performed on Gen2sat and MetTeL. The goal of this evaluation is two-fold: first, it shows that both tools are usable in practice. Second, it sheds some light on the effect that the internal differences between the tools and their underlying approaches (described in earlier sections) have on actual benchmarks.

As a case study, we consider Łukasiewicz three-valued logic, denoted by $Ł_3$ [43]. This logic employs three truth values: t, f, and i, representing 'true', 'false', and 'undetermined', respectively, and is defined using the following three-valued truth tables:

$\wedge$	t	f	i		$\supset$	t	f	i		$\vee$	t	f	i		p	$\neg p$
t	t	f	i		t	t	f	i		t	t	t	t		t	f
f	f	f	f		f	t	t	t		f	t	f	i		f	t
i	i	f	i		i	t	i	t		i	t	i	i		i	i

Valid formulas in $Ł_3$ are the formulas that are always assigned the value t. Its implication-free fragment is identical to Kleene's three-valued logic [38]. As a consequence, it does not have implication-free valid formulas. $Ł_3$ is decidable, like every propositional logic that is defined using a finite-valued logical matrix.

We start by describing the different implementations of this logic in both tools. This is followed by a description of the problems (formulas) that were tested. Then, we provide the actual results of this case study, and discuss the various differences between the tools.

4.1   Calculi

The paper [29] presents a tableau calculus for $Ł_3$ (henceforth denoted $\mathcal{T}$), which is available in the online version of MetTeL. The paper [6] presents a sequent calculus for this logic (henceforth denoted $\mathcal{S}$). As it is $\{\neg\}$-analytic and pure, it can be implemented easily in Gen2sat. The most straightforward comparison would be between MetTeL's implementation of the first calculus and Gen2sat's implementation of the second calculus. Since our goal is to compare the underlying automated reasoning *approaches* rather than specific *calculi*, and in order to avoid the comparison be obscured by differences in the calculi, it is important to

```
specification Lukasiewicz;
syntax Lukasiewicz{
     sort valuation;
     sort formula;
     valuation true = 'T' formula;
     valuation unknown = 'U' formula;
     valuation false = 'F' formula;
     formula true = 'true';
     formula false = 'false';
     formula negation = '~' formula;
     formula conjunction = formula '&' formula;
     formula disjunction = formula '|' formula;
     formula implication = formula '->' formula;
}
tableau Lukasiewicz{
     T P   F P /  priority 0 $;
     T P   U P /  priority 0 $;
     U P   F P /  priority 0 $;
     U P   F P /  priority 0 $;
     T ~P / F P priority 1 $;
     U ~P / U P priority 1 $;
     F ~P / T P priority 1 $;
     T (P & Q) / T P   T Q priority 1 $;
     F (P & Q) /   F P $| F Q priority 2 $;
     U (P & Q) /   T P  U Q $| U P  T Q $| U P   U Q   priority 3 $;
     T (P | Q) / T P $| T Q priority 2 $;
     F (P | Q) /   F P   F Q priority 1 $;
     U (P | Q) /   F P  U Q $| U P  F Q $| U P   U Q   priority 3 $;
     F (P -> Q) /  T P   F Q priority 1 $;
     U (P -> Q) / U P   F Q $|  T P  U Q  priority 2 $;
     T (P -> Q) / T Q $|  F P $| U P  U Q  priority 3 $;
     T false / priority 0 $;
     U false / priority 0 $;
     U true / priority 0 $;
     F true / priority 0 $;
}
```

Fig. 8. Definition of $\mathcal{T}$ in MetTeL

evaluate both frameworks on the same calculus. For this purpose, we have translated the sequent calculus $\mathcal{S}$ to a tableau calculus (henceforth denoted $\mathcal{ST}$).[3] To summarize, we have considered three implementations of L$_3$:

$\mathcal{T}$ the tableau calculus from [29], implemented in MetTeL, specified in Fig. 8.
$\mathcal{S}$ the sequent calculus from [6], implemented in Gen2sat, specified in Fig. 9.
$\mathcal{ST}$ a translation of $\mathcal{S}$ as a tableau calculus, implemented in MetTeL, specified in Fig. 10.

The calculus $\mathcal{T}$ is *three-valued* (corresponding to the three values of L$_3$). This means that in order to check the validity of a given formula φ, one needs to apply $\mathcal{T}$ both on $F : \varphi$ and on $U : \varphi$. Only if both turn out to be unsatisfiable, then the formula is valid. Obviously, once one of them is found satisfiable, there is no need to check the second. In contrast, the calculus $\mathcal{S}$ is *two-valued*, and thus checking the validity of a formula φ amounts to applying the calculus once on the sequent $\Rightarrow \varphi$.

---

[3] Note that a translation of $\mathcal{T}$ to a sequent calculus is less obvious, as this is a three-sided calculus, where Gen2sat employs ordinary two-sided sequents.

```
name: S
displayName: L3
connectives: &:2,|:2,->:2,!:1
rule: =>p1; =>p2 / => p1 & p2
rule: p1,p2=> / p1 & p2 =>
rule: =>p1,p2 / => p1 | p2
rule: p1=>; p2=> / p1 | p2 =>
rule: a=> / !! a=>
rule: =>a / => !! a
rule: !A, !B=> / !(A | B)=>
rule: =>!A; =>!B / => !(A | B)
rule: !A=>; !B=> / !(A & B)=>
rule: =>!A, !B / => !(A & B)
rule: /! A, A=>
rule: ! A => ; B =>; => A,! B / A -> B=>
rule: A=>B; ! B=>! A / => A -> B
rule: A, ! B=> / ! (A -> B)=>
rule: =>A; =>!B / => ! (A -> B)
analyticity: !
details: false
```

Fig. 9. Definition of $\mathcal{S}$ in Gen2sat

In Gen2sat, the ability to provide a sub-calculus in which a given sequent is provable is expensive, as it relies on finding unsatisfiable cores. Thus, for this evaluation we have compiled a non-verbose version of the tool, that does not provide this information.

Overall, the five implementations we consider are:

$\mathcal{S}_m$ the implementation of $\mathcal{S}$ in the non-verbose version of Gen2sat.
$\mathcal{S}$ the implementation of $\mathcal{S}$ in the usual (slower) version of Gen2sat.
$\mathcal{ST}$ the implementation of $\mathcal{ST}$ in MetTeL.
$\mathcal{T}$-F the implementation of $\mathcal{T}$ in MetTeL, applied on inputs of the form $F : \varphi$.
$\mathcal{T}$-U the implementation of $\mathcal{T}$ in MetTeL, applied on inputs of the form $U : \varphi$.

These implementations allow two interesting types of comparisons: the first is comparing different implementations in the same tool: the first two for Gen2sat, and the last three for MetTeL. The second is to compare the two tools, which is best achieved by comparing either $\mathcal{S}$ or $\mathcal{S}_m$ against $\mathcal{ST}$.

4.2   Benchmarks

As benchmark problems we used the four problem classes from [52]:

(1) $(A^n \vee B^n) \supset (A \vee B)^n$ (2) $(A \vee B)^n \supset (A^n \vee B^n)$

(3) $(n \cdot (A \wedge B)) \supset ((n \cdot A) \wedge (n \cdot B))$ (4) $((n \cdot A) \wedge (n \cdot B)) \supset (n \cdot (A \wedge B))$

where $A^0 = \top$, $A^{n+1} = A \odot A^n$, $0 \cdot A = \bot$, $(n+1) \cdot A = A \oplus (n \cdot A)$, $A \odot B = \neg(\neg A \oplus \neg B)$ and $A \oplus B = \neg A \supset B$. We only considered the language $\{\wedge, \vee, \supset, \neg\}$, and so we defined $\top$ as $p \supset p$ and $\bot$ as $\neg\top$. We produced formulas for $0 \le n \le 300$ of intervals of 5.

```
specification ST;
syntax ST{
    sort valuation;
    sort formula;
    valuation true = 'T' formula;
    valuation false = 'F' formula;
    formula negation = '!' formula;
    formula conjunction = formula '&' formula;
    formula disjunction = formula '|' formula;
    formula implication = formula '->' formula;
}
tableau ST{
    T P   F P / priority 0 $;
    T (P & Q) / T P   T Q priority 1 $;
    F (P & Q) /   F P $| F Q priority 2 $;
    T (P | Q) / T P $| T Q priority 2 $;
    F (P | Q) /   F P   F Q priority 1 $;
    F (!(!(P))) / F P priority 1 $;
    T (!(!(P))) / T P priority 1 $;
    F !(P) / T P priority 1 $;
    T (!(P | Q)) / T !P   T !Q priority 1 $;
    F (!(P | Q)) / F !P $| F !Q priority 2 $;
    T (!(P & Q)) / T !P $| T !Q priority 2 $;
    F (!(P & Q)) / F !P   F !Q priority 1 $;
    T (P->Q) / F P   F !Q $| F P   T !P $| T Q   F !Q $| T Q   T !P priority 3 $;
    F (P->Q) / T P   F Q   F !P   $| T !Q   F Q   F !P priority 2 $;
    T !(P->Q) / T P   T !Q priority 1 $;
    F !(P->Q) / F P $| F !Q priority 2 $;
}
```

Fig. 10. Definition of $\mathcal{ST}$ in MetTeL

These problems were designed to test provers for infinite-valued Łukasiewicz logic, and are all valid in it, as well as in L_3. Non-valid formulas were obtained by adding a negation. In [52], problems of the first and third class are considered easy, while problems of the second and forth class are considered hard. There are several explanations to this classification in [52] (e.g., hard problems require cuts and branching proofs), that are backed by experimental results of several implementations of calculi for infinite-valued Łukasiewicz logic.

4.3   Results

The experiments were made on a dedicated Linux machine with four dual-core 2.53 Ghz AMD Opteron 285 processors and 8 GB RAM. The Java heap limit was 4 GB. Figure 11 exhibits the main results. A timeout of 10000 ms was imposed on all problems, and anything higher appears in these figures as '11000'. The benchmarks themselves are available online.[4]

Figure 11 presents running times. In every problem class, both Gen2sat implementations of $\mathcal{S}$ performed better than the MetTeL implementations of $\mathcal{ST}$ and $\mathcal{T}$.

Notably, there was a big difference between the performances of the verbose and non-verbose versions of Gen2sat ($\mathcal{S}$ and $\mathcal{S}_m$, respectively), but only on *provable instances*. The reason is that on such instances, the largest amount of

---

[4] https://github.com/yoni206/gen2satvsmettel.

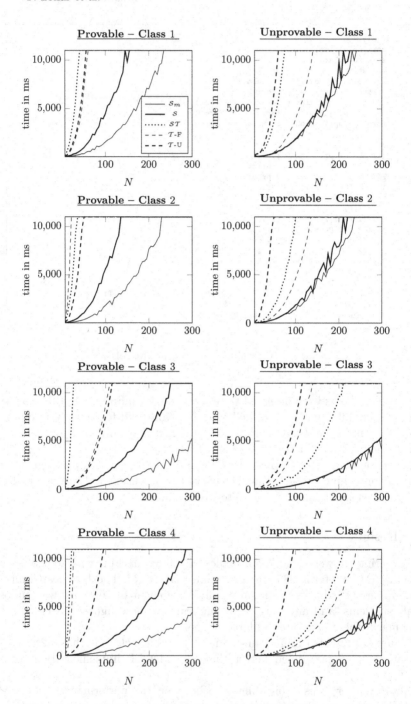

Fig. 11. Running times on provable and unprovable instances of classes 1–4 from Rothenberg's problems. N is the size of the Rothenberg problem. See top left graph for the legend.

computation time is spent on calls to the xplain module of sat4j, which is used only for $\mathcal{S}$ in order to produce unsatisfiable cores. On unprovable problems, for which this module was never called, the difference between the two versions of Gen2sat was negligible.

Comparing the different implementations of MetTeL between themselves, we did not get consistent results. Focusing on $\mathcal{T}$ however, we did see that problems of the form $U : \varphi$ are processed slower than problems of the form $F : \varphi$, whenever φ was not valid. In all these formulas, it was possible to assign F to the Rothenberg formula, but not U. This is not surprising, as the rules for U in $\mathcal{T}$ involve three-way branching, that significantly increases the search space for MetTeL. When φ was valid, however, F-problems and U-problems either performed similarly, or U-problems were processed faster. Thus, when using the prover generated by MetTeL for $\mathcal{T}$, it is better to first use it with an F-label and only if it was not satisfiable, run it again with U.

On the other hand, almost all the rules in $\mathcal{ST}$ have one premise, which explains the better performance of this calculus over $\mathcal{T}$. Moreover, few fine grained priority values improved the performance for this calculus. For example, raising the priority value of T (P->Q) from 3 to 4, and that of F (P->Q) from 2 to 3 resulted in some improvement in running times.

Both MetTeL and Gen2sat performed better on unprovable problems than on provable ones. An exception is the non-verbose implementation $\mathcal{S}_m$, whose performance was the same on provable and unprovable problems.

Figure 12 shows that Rothenberg's original classification [52] of hard vs. easy problems does not hold for the provers MetTeL and Gen2sat generated for Łukasiewicz three-valued logic. In $\mathcal{S}$, $\mathcal{S}_m$ and $\mathcal{T}$-U, we have that classes 3 and 4 were easier than classes 1 and 2. In $\mathcal{ST}$, the exact opposite was observed. Only in $\mathcal{T}$-F, the Rothenberg classification survived, and classes 1 and 3 were easier than classes 2 and 4.

The fact that the original classification did not survive the transition from infinite-valued Łukasiewicz logic to the three-valued one, is not surprising. First, these are two different logics, and second, the calculi for them are much simpler than the calculi for the infinite-valued version. For example, the sequent calculus that we consider here is cut-free, while only hyper-sequent calculi that are cut-free are known for the infinite case.

In the three-valued case, we however uncovered a different classification, according to which classes 1 and 2 are harder than classes 3 and 4 (this was the case for four out of five implementations of the calculi for L_3). This is consistent with the fact that the problems of classes 3 and 4 are less complex than those of 1 and 2. At least in Gen2sat, where the complexity of the input has a big effect on the parsing stage, this is to be expected.

5   Usability Evaluation

In this section we complement the performance evaluation of Gen2sat and MetTeL with an evaluation of another important aspect of provers: *usabil-*

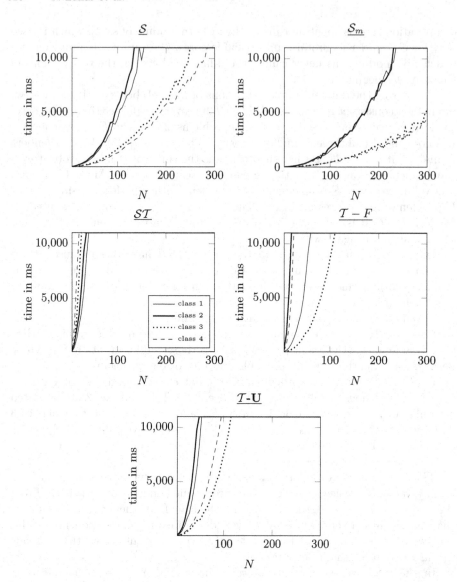

Fig. 12. Running times of provable instances of all classes from Rothenberg's problems. N is the size of the Rothenberg problem.

ity. Usability of software systems [46] is measured by evaluation of their user-interfaces. Standard approaches for evaluating usability require some form of user involvement: user testing, focus groups and other types of feedback collection from users. Studies of usability in the field of automated reasoning are scarce, and have mainly been carried out in the context of interactive theorem proving (see, e.g., [2,4,10]).

According to [28], understanding who the users are and what are their tasks is key in designing for usability. Some design and usability-related decisions obviously had to be made while developing Gen2sat and MetTeL. For example, using the online version of MetTeL, the user can download a standalone prover and then use it locally, and also edit its source code. When using the online version of Gen2sat, in contrast, the sequent calculus in question can be changed throughout the working session of the user. Obviously, different users have different preferences regarding such issues, that are reflected in the data presented below. While it is probably impossible to cater for every taste of users, prover developers need to be very clear about their intended audience.

In what follows we describe a preliminary usability study we carried out, aiming to better understand the impact of the different approaches taken in Gen2sat and MetTeL on their usability. Our participants were five expert logicians, carefully selected according to the following criteria: (i) published research in the fields of proof theory and/or automated reasoning, and (ii) familiar with both sequent and tableau formalisms. The participants received detailed instructions introducing the tools and asking them to perform various tasks using them (the instructions are given in Fig. 13). They then answered several questions concerning their experiences using the tools. Below we provide a summary and some quotes from their answers to open-ended questions, as well as some further discussion on these results.

5.1   Results

Both tools received good reception from the users, that found them potentially useful and convenient to use. Remarkably, even in our very focused group of logicians working in proof theory and automated reasoning, we got a range of different responses to the features of the tools, that can be used in future developments of the tools considered here, and also of new tools being developed.

The average satisfaction score (on a scale of 1–5) for MetTeL was 3.8, while for Gen2sat it was 3.6. MetTeL was indeed pointed out as a more user-friendly tool:

- *It was easy to understand how to define the calculus looking at the predefined systems.*
- *MetTeL is user-friendly when it comes to specifying the calculus, and the fact that we can download a prover for the system is a big plus.*
- *I find the GUI of MetTeL more well-polished.*

The main issues with Gen2sat were related to lack of documentation and customizability:

- *The system worked perfectly after I realized how to specify the proof system in a suitable way. I think that some instruction about the format of rules would be very helpful.*
- *At first appearance, the system seems a bit less customizable by the practitioner.*

In what follows you will investigate the logic P1 [57]. Its sequent calculus is obtained from the calculus for classical logic by replacing the rule ($\neg \Rightarrow$) with the following four rules:

$$\frac{\Gamma \Rightarrow \neg A, \Delta}{\Gamma, \neg\neg A \Rightarrow \Delta} \qquad \frac{\Gamma \Rightarrow A \wedge B, \Delta}{\Gamma, \neg(A \wedge B) \Rightarrow \Delta} \qquad \frac{\Gamma \Rightarrow A \vee B, \Delta}{\Gamma, \neg(A \vee B) \Rightarrow \Delta} \qquad \frac{\Gamma \Rightarrow A \supset B, \Delta}{\Gamma, \neg(A \supset B) \Rightarrow \Delta}$$

Its tableau calculus is obtained from the calculus for classical logic by replacing the rule ($T\neg$) by the following rules:

$$\frac{T : \neg\neg A}{F : \neg A} \qquad \frac{T : \neg(A \wedge B)}{F : A \wedge B} \qquad \frac{T : \neg(A \vee B)}{F : A \vee B} \qquad \frac{T : \neg(A \supset B)}{F : A \supset B}$$

Step 1. *Use Gen2sat to check the validity of the formula $X = (A \wedge \neg A) \supset (B \wedge \neg B)$ in P1. Please describe your experience in a few sentences. Relate to difficulty of the task, user-friendliness of the tool, comprehensibility of the result and other usability aspects.*

Step 2. *Use MetTeL to check the validity of the same formula X in P1. Please describe your experience in a few sentences. Relate to difficulty of the task, user-friendliness of the tool, comprehensibility of the result and other usability aspects.*

Step 3. *You will now check the validity of the formula $Y = ((A \wedge A) \wedge (\neg(A \wedge A))) \supset (B \wedge \neg B)$ in P1 using MetTeL. Please describe your experience in a few sentences. Relate to difficulty of the task, user-friendliness of the tool, comprehensibility of the result and other usability aspects.*

Step 4. *You will now check the validity of the same formula Y in P1 using Gen2sat. Please describe your experience in a few sentences. Relate to difficulty of the task, user-friendliness of the tool, comprehensibility of the result and other usability aspects.*

Concluding Questions:

1. On a scale of 1–5 (where 1 is the least satisfied, and 5 the most satisfied), how satisfied are you with MetTeL? Why did you choose this score?
2. On a scale of 1–5 (where 1 is the least satisfied, and 5 the most satisfied), how satisfied are you with Gen2sat? Why did you choose this score?
3. Which of the two tools do you think would be more helpful for you in a study of a specific logic (which has both a sequent and a tableaux representation)? Why?
4. Which of the two tools do you think would be more helpful for you in a study of a family of logics (which have both a sequent and a tableaux representation)? Why?
5. In your own research, do you envision that Gen2sat and/or MetTeL could be utilized? If so, describe how your research could benefit from these tools, how you would utilize them, and which of them would be preferable. If your answer is negative, please explain why.
6. Can you offer any further feedback on Gen2sat, MetTeL and the difference between them? You can refer to pros and cons of each tool, relying on your experience with them.

Fig. 13. User questionnaire

- *Specifying the system is a bit tricky, not having contexts for sequent calculi seem strange.*
- *Some features were not fine-tuned for best user experience.*

The preferences of the participants with respect to the two tools were mixed:

- *Gen2Sat would be more helpful to me. Since I am more familiar with sequent calculus, I would prefer it over MetTeL.*
- *I would prefer MetTeL if the family of logics were more naturally defined by the models of the logics, but I would prefer Gen2Sat in case the class were more naturally defined as a class of axiomatic systems.*
- *It depends on the (language of the) logic, I guess. If there is no need to play around with fancy syntaxes, or if a sufficiently similar example is already available to be modified, I would guess that Gen2Sat is a bit easier to work with, at first. I think both are equally good to investigate a family of logics (sharing syntax).*

Most participants found the tools potentially useful for tasks performed by logician users:

- *My work could definitely profit from the use of the provers. It is often useful to have a way to find out whether a formula is a theorem, for example to find examples or counterexample easily.*
- *I think that having a prover that is easy to use and available online can surely help in playing with a new logic system, in order to get a feeling of what is and what is not provable by it.*
- *I think they could be utilized if I was tweaking around with known logics, like LK or LJ, and trying to understand what happens if this or that connective is changed... playing around with them to figure out provable and non-provable sequents may turn out to be useful.*
- *It is often helpful to check derivability of some statements, and doing it by hand is tedious.*

Two participants explicitly referred to the fact that the tools do not provide full proofs:

- *If the tool could actually give the proof (in some form) so that the user (or another tool) can check it, it would be great.*
- *Having proofs exhibited in some form would be helpful.*

Three participants pointed out that the tools are limited in their ability to reason about meta-logical properties:

- *As I am usually most interested in meta-results concerning such systems, though, these tools would probably not be extremely useful.*
- *I am usually doing this to investigate the meta-properties of a calculus, which is something both tools lack.*
- *I think both tools could be extended to check the consistency of the rules input by the user.*

Other interesting comments included:

- *It is not obvious how the steps done in specifying a proof system in prover A maps into a step in prover B.*
- *The use of 'priority' in MetTeL's third step would seem to allow one to easily define a proof strategy, which might be advantageous in some situations.*

5.2   Discussion

The participants acknowledged the potential both tools have in logical research. Indeed, it is useful to have an automated tool to mechanically check the validity of certain formulas. Also, when studying the effect each inference rule in a given system has, both tools can be of great help, as they allow for an easy specification of calculi.

Two participants, however, noted that despite their usefulness in testing formulas in a given logic, both Gen2sat and MetTeL lack the ability to reason on the meta-logical level, and assist in proving properties *about* the investigated proof systems. We note that some of these abilities can be recovered using the tools, perhaps with the aid of other related tools. For example, in order to check whether a sequent calculus is consistent, it often suffices to check for derivability of the empty sequent. This can be done in Gen2sat. Moreover, many logics include in their language a formula from which all formulas follow (e.g., $\perp$). Checking for derivability of this formula can be done in both tools.

Some participants noted that presenting the actual proofs of provable formulas would be very helpful, while neither of the tools provides this information. This issue can be attributed in part for the genericity of the tools, as well as efficiency considerations. In Gen2sat, the proofs are inherently unobtainable, as they are lost in the translation to a SAT-instance, that goes through a *semantic* representation of the sequent calculus. The correctness of this translation was shown in [41] by usual non-constructive completeness arguments, from which one cannot extract proofs. In MetTeL, sophisticated algorithms and heuristics are employed in the search process, that sacrifice the possibility to output a tableau proof for the sake of efficiency.

It is interesting to note that three participants scored satisfaction from Gen2sat higher due to their personal preference of sequent calculi over tableaux (while one participant noted these are equivalent due to their duality). This raises the question of how generic provers should address user preferences of representations, and whether customization and personalization can be increased. Due to the bias towards sequent calculi, we plan to address this issue in a follow-up study by recruiting participants who have a preference towards tableaux.

The user interface and documentation of Gen2sat should be improved according to the feedback above. For example, several participants found it odd that they are not expected (and actually, are expected not) to provide the tool with any structural rules, as Gen2sat automatically enforces their inclusion in the background. Such hidden assumptions should be clearly stated.

General usability remarks on both tools, such as the ability to download a prover with MetTeL, versus the ability to change the calculus instantly in Gen2sat, can be easily addressed in each tool. One has to take into deeper consideration, however, the identity of the users for each tool, in order to decide which of these changes should be made.

6   Conclusion and Future Work

In this paper we compared two generic provers, MetTeL and Gen2sat with respect to their performance and usability. Both tools aim at providing auto-mated support to researchers and practical users of non-classical logics, but take completely different approaches to achieve this goal. In this paper we scrutinized the impact the chosen approaches have on the performance and usability of the respective tools.

Our performance evaluation was performed on several implementations of Łukasiewicz three-valued logic in both tools. The results are encouraging: both tools performed well, despite the fact that this particular logic has not been investigated with either tools before. A future research direction in this respect is to make the performance comparison between the tools wider, to include more logics and more problems. Such comparisons may shed some light on the strengths and weaknesses of each tool, and possibly yield a classification of logic problems according to the tools that are best for each.

Some insights worth further exploring arise from considering the usability of the two tools. While MetTeL got a better usability score mainly due to its higher level of customizability, a polished user-interface and well-developed doc-umentation, a preference towards Gen2sat was expressed mainly due to its sim-plicity and also its use of the sequent formalism, which some participants found more intuitive and familiar. This indicates a need to take user preference into consideration when developing generic automated reasoning tools, and perhaps considering providing the possibility to work with the formalism of the users' choice when applicable.

While performance analysis is a standard approach for evaluating provers and other automated reasoning tools, few empirical usability studies have been undertaken in this domain (to the best of our knowledge, none of them were in the realm of non-classical logics). We have found the feedback received from our participants helpful in improving the user interface and documentation in both tools and intend to expand the usability studies of the tools to a wider range of participants. We also hope that this paper has further demonstrated the potential of such studies in the field of automated reasoning in general, and for generic provers in particular. The wide variety of feedback that we got from our small sample of users stresses the significant value systematic user studies may have for the development of new provers. It is our hope that this paper will start a discourse towards a more user-centric development of automated reasoning tools.

A desired capability that is currently missing in both tools is *proof produc-tion*. The size of tableau-style proofs that are searched for in MetTeL makes it difficult to store and produce them in an efficient manner. The issue is more fundamental in Gen2sat, whose first step is to translate the derivability problem to a semantic variant, which is in turn translated into a SAT-instance. An avenue for future work for both tools is overcoming these difficulties. Techniques for minimizing and concisely storing tableau proofs could be considered for MetTeL, while for Gen2sat, unsatisfiability proofs from SAT-solvers could be utilized in order to certify solution for the translated semantic variant of the original proof-theoretical problem.

Finally, a further research task that seems beneficial both from a performance and usability point of view is to consider a combination between the tools. For example, both provers can run in parallel for a given problem, thus providing the faster performance between the two for each problem separately. Also, exchanging information between the provers in runtime can be useful, both for performance, and for providing the user with additional meaningful output. Combining the tools into one suite could also help logicians and logic students to use their preferred formalism for defining logic on the one hand, and get a better understanding on the connection between these formalisms on the other hand.

Acknowledgments. We thank Francesco Genco, Yotam Feldman, Roman Kuznets, Giselle Reis, João Marcos, and Bruno Woltzenlogel Paleo for providing valuable feedback on both tools. e also thank Mohammad Khodadadi for useful discussions and setting up the MetTeL website. he research of the first and fourth authors was supported by The Israel Science Foundation (grant no. 817-15). The research of the second and third authors was supported by UK EPSRC research grant EP/H043748/1.

Last but not least, we extend our best wishes to Franz Baader on the occasion of his 60th birthday. It is an immense privilege to have been asked to contribute to this volume.

References

1. Abate, P., Goré, R.: The tableau workbench. Electron. Notes Theor. Comput. Sci. **231**, 55–67 (2009)
2. Aitken, S., Melham, T.: An analysis of errors in interactive proof attempts. Interact. Comput. **12**(6), 565–586 (2000)
3. Areces, C.E.: Logic engineering: the case of description and hybrid logics. Ph.D. thesis, University of Amsterdam (2000)
4. Asperti, A., Coen, C.S.: Some considerations on the usability of interactive provers. In: Autexier, S., et al. (eds.) CICM 2010. LNCS (LNAI), vol. 6167, pp. 147–156. Springer, Heidelberg (2010). https://doi.org/10.1007/978-3-642-14128-7_13
5. Avron, A.: Simple consequence relations. Inf. Comput. **92**(1), 105–139 (1991)
6. Avron, A.: Classical Gentzen-type methods in propositional many-valued logics. In: Fitting, M., Orłowska, E. (eds.) Beyond Two: Theory and Applications of Multiple-Valued Logic. Studies in Fuzziness and Soft Computing, vol. 114, pp. 117–155. Physica-Verlag, Heidelberg (2003). https://doi.org/10.1007/978-3-7908-1769-0_5
7. Baader, F., Sattler, U.: An overview of tableau algorithms for description logics. Stud. Logica **69**, 5–40 (2001)

8. Baaz, M., Fermüller, C.G., Salzer, G., Zach, R.: MUltlog 1.0: towards an expert system for many-valued logics. In: McRobbie, M.A., Slaney, J.K. (eds.) CADE 1996. LNCS, vol. 1104, pp. 226–230. Springer, Heidelberg (1996). https://doi.org/10.1007/3-540-61511-3_84

9. Baumgartner, P., Schmidt, R.A.: Blocking and other enhancements for bottom-up model generation methods. J. Autom. Reason. 1–27 (2019). https://doi.org/10.1007/s10817-019-09515-1

10. Beckert, B., Grebing, S., Böhl, F.: A usability evaluation of interactive theorem provers using focus groups. In: Canal, C., Idani, A. (eds.) SEFM 2014. LNCS, vol. 8938, pp. 3–19. Springer, Cham (2015). https://doi.org/10.1007/978-3-319-15201-1_1

11. Caridroit, T., Lagniez, J.M., Le Berre, D., de Lima, T., Montmirail, V.: A SAT-based approach for solving the modal logic S5-satisfiability problem. In: Singh, S.P., Markovitch, S. (eds.) Thirty-First AAAI Conference on Artificial Intelligence, pp. 3864–3870. AAAI Press (2017)

12. Ciabattoni, A., Lahav, O., Spendier, L., Zamansky, A.: Automated support for the investigation of paraconsistent and other logics. In: Artemov, S., Nerode, A. (eds.) LFCS 2013. LNCS, vol. 7734, pp. 119–133. Springer, Heidelberg (2013). https://doi.org/10.1007/978-3-642-35722-0_9

13. Ciabattoni, A., Lahav, O., Spendier, L., Zamansky, A.: Taming paraconsistent (and other) logics: an algorithmic approach. ACM Trans. Comput. Logic 16(1), 5:1–5:23 (2014)

14. Ciabattoni, A., Spendier, L.: Tools for the investigation of substructural and paraconsistent logics. In: Fermé, E., Leite, J. (eds.) JELIA 2014. LNCS (LNAI), vol. 8761, pp. 18–32. Springer, Cham (2014). https://doi.org/10.1007/978-3-319-11558-0_2

15. Comon-Lundh, H., Shmatikov, V.: Intruder deductions, constraint solving and insecurity decision in presence of exclusive or. In: Logic in Computer Science (LICS 2003), pp. 271–280. IEEE Computer Society (2003)

16. Conradie, W., Goranko, V., Vakarelov, D.: Algorithmic correspondence and completeness in modal logic I: the core algorithm SQEMA. Log. Methods Comput. Sci. 2, 1–5 (2006)

17. Cotrini, C., Gurevich, Y.: Basic primal infon logic. J. Logic Comput. 26(1), 117 (2016)

18. Degtyarev, A., Voronkov, A.: The inverse method. In: Robinson, J.A., Voronkov, A. (eds.) Handbook of Automated Reasoning, pp. 179–272. MIT Press (2001)

19. Dixon, C., Konev, B., Schmidt, R.A., Tishkovsky, D.: Labelled tableaux for temporal logic with cardinality constraints. In: Proceedings of the 14th International Symposium on Symbolic and Numeric Algorithms for Scientific Computing (SYNASC 2012), pp. 111–118. IEEE Computer Society (2012)

20. Doherty, P., Łukaszewicz, W., Szałas, A.: Computing circumscription revisited: a reduction algorithm. J. Autom. Reason. 18(3), 297–336 (1997)

21. Fitting, M.: Tableau methods of proof for modal logics. Notre Dame J. Formal Logic 13(2), 237–247 (1972)

22. Gaschnig., J.: Performance measurement and analysis of certain search algorithms. Ph.D. thesis, Carnegie-Mellon University (1979)

23. Gasquet, O., Herzig, A., Longin, D., Sahade, M.: LoTREC: logical tableaux research engineering companion. In: Beckert, B. (ed.) TABLEAUX 2005. LNCS (LNAI), vol. 3702, pp. 318–322. Springer, Heidelberg (2005). https://doi.org/10.1007/11554554_25

24. Gentzen, G.: Untersuchungen über das logische Schliessen. Mathematische Zeitschrift **39**, 176–210 (1934)
25. Ginsberg, M.L., McAllester, D.A.: GSAT and dynamic backtracking. In: Doyle, J., Sandewall, E., Torasso, P. (eds.) Principles of Knowledge Representation and Reasoning (KR 1994), pp. 226–237. Morgan Kaufmann (1994)
26. Giunchiglia, E., Tacchella, A., Giunchiglia, F.: SAT-based decision procedures for classical modal logics. J. Autom. Reason. **28**(2), 143–171 (2002)
27. Gorín, D., Pattinson, D., Schröder, L., Widmann, F., Wißmann, T.: COOL – a generic reasoner for coalgebraic hybrid logics (system description). In: Demri, S., Kapur, D., Weidenbach, C. (eds.) IJCAR 2014. LNCS (LNAI), vol. 8562, pp. 396–402. Springer, Cham (2014). https://doi.org/10.1007/978-3-319-08587-6_31
28. Gould, J.D., Lewis, C.: Designing for usability: key principles and what designers think. Commun. ACM **28**(3), 300–311 (1985)
29. Hähnle, R.: Tableaux and related methods. In: Robinson, J.A., Voronkov, A. (eds.) Handbook of Automated Reasoning, pp. 100–178. Elsevier and MIT Press (2001)
30. Heuerding, A., Jäger, G., Schwendimann, S., Seyfried, M.: The logics workbench LWB: a snapshot. Euromath Bull. **2**(1), 177–186 (1996)
31. Humberstone, I.L.: The modal logic of 'all and only'. Notre Dame J. Formal Logic **28**(2), 177–188 (1987)
32. Hustadt, U., Schmidt, R.A.: Simplification and backjumping in modal tableau. In: de Swart, H. (ed.) TABLEAUX 1998. LNCS (LNAI), vol. 1397, pp. 187–201. Springer, Heidelberg (1998). https://doi.org/10.1007/3-540-69778-0_22
33. Kaminski, M., Tebbi, T.: InKreSAT: modal reasoning via incremental reduction to SAT. In: Bonacina, M.P. (ed.) CADE 2013. LNCS (LNAI), vol. 7898, pp. 436–442. Springer, Heidelberg (2013). https://doi.org/10.1007/978-3-642-38574-2_31
34. Kawai, H.: Sequential calculus for a first order infinitary temporal logic. Math. Logic Q. **33**(5), 423–432 (1987)
35. Khodadadi, M., Schmidt, R.A., Tishkovsky, D.: MetTeL. http://www.mettel-prover.org
36. Khodadadi, M., Schmidt, R.A., Tishkovsky, D.: A refined tableau calculus with controlled blocking for the description logic $\mathcal{SHOI}$. In: Galmiche, D., Larchey-Wendling, D. (eds.) TABLEAUX 2013. LNCS (LNAI), vol. 8123, pp. 188–202. Springer, Heidelberg (2013). https://doi.org/10.1007/978-3-642-40537-2_17
37. Khodadadi, M., Schmidt, R.A., Tishkovsky, D., Zawidzki, M.: Terminating tableau calculi for modal logic K with global counting operators (2012). http://www.mettel-prover.org/papers/KEn12.pdf
38. Kleene, S.C.: Introduction to Metamathematics. Van Nostrand, New York (1950)
39. Koopmann, P., Schmidt, R.A.: LETHE: saturation-based reasoning for non-standard reasoning tasks. In: Dumontier, M., et al. (eds.) OWL Reasoner Evaluation (ORE-2015), CEUR Workshop Proceedings, vol. 1387, pp. 23–30 (2015)
40. Lagniez, J.M., Le Berre, D., de Lima, T., Montmirail, V.: On checking Kripke models for modal logic K. In: Fontaine, P., Schulz, S., Urban, J. (eds.) Practical Aspects of Automated Reasoning (PAAR 2016), CEUR Workshop Proceedings, vol. 1635, pp. 69–81 (2016)
41. Lahav, O., Zohar, Y.: SAT-based decision procedure for analytic pure sequent calculi. In: Demri, S., Kapur, D., Weidenbach, C. (eds.) IJCAR 2014. LNCS (LNAI), vol. 8562, pp. 76–90. Springer, Cham (2014). https://doi.org/10.1007/978-3-319-08587-6_6
42. Le Berre, D., Parrain, A.: The SAT4J library, release 2.2. J. Satisf. Boolean Model. Comput. **7**, 59–64 (2010)

43. Lukasiewicz, J., Tarski, A.: Investigations into the sentential calculus. Borkowski **12**, 131–152 (1956)
44. Miller, D., Pimentel, E.: A formal framework for specifying sequent calculus proof systems. Theor. Comput. Sci. **474**, 98–116 (2013)
45. Minica, S., Khodadadi, M., Schmidt, R.A., Tishkovsky, D.: Synthesising and implementing tableau calculi for interrogative epistemic logics. In: Fontaine, P., Schmidt, R.A., Schulz, S. (eds.) Practical Aspects of Automated Reasoning (PAAR-2012). EPiC Series in Computing, vol. 21, pp. 109–123. EasyChair (2012)
46. Nielsen, J.: Usability inspection methods. In: Plaisant, C. (ed.) Conference on Human Factors in Computing Systems (CHI 1994), pp. 413–414. ACM (1994)
47. Nigam, V., Pimentel, E., Reis, G.: An extended framework for specifying and reasoning about proof systems. J. Logic Comput. **26**, 539–576 (2014)
48. Nigam, V., Reis, G., Lima, L.: Quati: an automated tool for proving permutation lemmas. In: Demri, S., Kapur, D., Weidenbach, C. (eds.) IJCAR 2014. LNCS (LNAI), vol. 8562, pp. 255–261. Springer, Cham (2014). https://doi.org/10.1007/978-3-319-08587-6_18
49. Ohlbach, H.J.: Computer support for the development and investigation of logics. Logic J. IGPL **4**(1), 109–127 (1996)
50. Ohlbach, H.J.: SCAN—elimination of predicate quantifiers. In: McRobbie, M.A., Slaney, J.K. (eds.) CADE 1996. LNCS, vol. 1104, pp. 161–165. Springer, Heidelberg (1996). https://doi.org/10.1007/3-540-61511-3_77
51. Prosser, P.: Hybrid algorithms for the constraint satisfaction problem. Comput. Intell. **9**, 268–299 (1993)
52. Rothenberg, R.: A class of theorems in Lukasiewicz logic for benchmarking automated theorem provers. In: TABLEAUX, Automated Reasoning with Analytic Tableaux and Related Methods, Position Papers, vol. 7, pp. 101–111 (2007)
53. Schmidt, R.A., Stell, J.G., Rydeheard, D.: Axiomatic and tableau-based reasoning for Kt(H, R). In: Goré, R., Kooi, B., Kurucz, A. (eds.) Advances in Modal Logic, vol. 10, pp. 478–497. College Publications (2014)
54. Schmidt, R.A., Tishkovsky, D.: Automated synthesis of tableau calculi. Log. Methods Comput. Sci. **7**(2), 1–32 (2011)
55. Schmidt, R.A., Tishkovsky, D.: Using tableau to decide description logics with full role negation and identity. ACM Trans. Comput. Logic **15**(1), 7:1–7:31 (2014)
56. Schmidt, R.A., Waldmann, U.: Modal tableau systems with blocking and congruence closure. In: De Nivelle, H. (ed.) TABLEAUX 2015. LNCS (LNAI), vol. 9323, pp. 38–53. Springer, Cham (2015). https://doi.org/10.1007/978-3-319-24312-2_4
57. Sette, A.M.: On the propositional calculus P1. Math. Japonicae **18**(13), 173–180 (1973)
58. Smullyan, R.M.: First Order Logic. Springer, Berlin (1971)
59. Stell, J.G., Schmidt, R.A., Rydeheard, D.E.: A bi-intuitionistic modal logic: foundations and automation. J. Log. Algebraic Methods Program. **85**(4), 500–519 (2016)
60. Tishkovsky, D., Schmidt, R.A.: Rule refinement for semantic tableau calculi. In: Schmidt, R.A., Nalon, C. (eds.) TABLEAUX 2017. LNCS (LNAI), vol. 10501, pp. 228–244. Springer, Cham (2017). https://doi.org/10.1007/978-3-319-66902-1_14
61. Tishkovsky, D., Schmidt, R.A., Khodadadi, M.: *MetTeL*: a tableau prover with logic-independent inference engine. In: Brünnler, K., Metcalfe, G. (eds.) TABLEAUX 2011. LNCS (LNAI), vol. 6793, pp. 242–247. Springer, Heidelberg (2011). https://doi.org/10.1007/978-3-642-22119-4_19

62. Tishkovsky, D., Schmidt, R.A., Khodadadi, M.: MetTeL2: towards a tableau prover generation platform. In: Fontaine, P., Schmidt, R.A., Schulz, S. (eds.) Practical Aspects of Automated Reasoning (PAAR-2012). EPiC Series in Computing, vol. 21, pp. 149–162. EasyChair (2012)
63. Tishkovsky, D., Schmidt, R.A., Khodadadi, M.: The tableau prover generator MetTeL2. In: del Cerro, L.F., Herzig, A., Mengin, J. (eds.) JELIA 2012. LNCS (LNAI), vol. 7519, pp. 492–495. Springer, Heidelberg (2012). https://doi.org/10.1007/978-3-642-33353-8_41
64. Wansing, H.: Sequent systems for modal logics. In: Gabbay, D.M., Guenthner, F. (eds.) Handbook of Philosophical Logic, vol. 8, pp. 61–145. Springer, Dordrecht (2002)
65. Zawidzki, M.: Deductive systems and decidability problem for hybrid logics. Ph.D. thesis, Faculty of Philosophy and History, University of Lodz (2013)
66. Zhao, Y., Schmidt, R.A.: Forgetting concept and role symbols in $\mathcal{ALCOIH}\mu^+(\nabla, \sqcap)$-ontologies. In: Kambhampati, S. (ed.) International Joint Conference on Artificial Intelligence (IJCAI 2016), pp. 1345–1352. AAAI Press/IJCAI (2016)
67. Zohar, Y.: Gen2sat. http://www.cs.tau.ac.il/research/yoni.zohar/gen2sat.html
68. Zohar, Y., Zamansky, A.: Gen2sat: an automated tool for deciding derivability in analytic pure sequent calculi. In: Olivetti, N., Tiwari, A. (eds.) IJCAR 2016. LNCS (LNAI), vol. 9706, pp. 487–495. Springer, Cham (2016). https://doi.org/10.1007/978-3-319-40229-1_33

On Bounded-Memory Stream Data Processing with Description Logics

Özgür Lütfü Özçep$^{(\boxtimes)}$ and Ralf Möller

Institute of Information Systems (IFIS), University of Lübeck, Lübeck, Germany
{oezcep,moeller}@ifis.uni-luebeck.de

Abstract. Various research groups of the description logic community, in particular the group of Franz Baader, have been involved in recent efforts on temporalizing or streamifying ontology-mediated query answering (OMQA). As a result, various temporal and streamified extensions of query languages for description logics with different expressivity were investigated. For practically useful implementations of OMQA systems over temporal and streaming data, efficient algorithms for answering continuous queries are indispensable. But, depending on the expressivity of the query and ontology language, finding an efficient algorithm may not always be possible. Hence, the aim should be to provide criteria for easily checking whether an efficient algorithm exists at all and, possibly, to describe such an algorithm for a given query. In particular, for stream data it is important to find simple criteria that help deciding whether a given OMQA query can be answered with sub-linear space w.r.t. the length of a growing stream prefix. An important special case dealt with under the term "bounded memory" is that of testing for constant space. This paper discusses known syntactical criteria for bounded-memory processing of SQL queries over relational data streams and describes how these criteria from the database community can be lifted to criteria of bounded-memory query answering in the streamified OMQA setting. For illustration purposes, a syntactic criterion for bounded-memory processing of queries formulated in a fragment of the stream-temporal query language STARQL is given.

Keywords: Streams · Bounded memory ·
Ontology-mediated query answering · Ontology-based data access

1 Introduction

Ontology-mediated query answering (OMQA) [8] is a paradigm for accessing data via declarative queries whose intended sets of answers is constrained by an ontology. Usually, the ontology is represented in a formal logic such as a description logic (DL). Though OMQA has been of interest both for researchers as well as users from industry, a real benefit for the latter heavily depends on the possibility to handle temporal and streaming data. So, various research groups of the DL community, in particular the group of Franz Baader (see, e.g., [6,10,27]),

© Springer Nature Switzerland AG 2019
C. Lutz et al. (Eds.): Baader Festschrift, LNCS 11560, pp. 639–660, 2019.
https://doi.org/10.1007/978-3-030-22102-7_30

have been involved in recent efforts on temporalizing or streamifying classical OMQA. As a result, various temporal and streamified extensions of query languages for description logics with different expressivity were investigated.

For practically useful implementations of OMQA systems over temporal and streaming data, efficient algorithms for answering continuous queries are indispensable. But, depending on the expressivity of the query and ontology languages, finding an efficient algorithm may not always be possible. Hence, the aim should be to provide criteria for easily checking whether an efficient algorithm exists at all and, possibly, to describe such an algorithm for a given query. For those queries that are provably bounded-memory computable, one knows that there exists an algorithm using only constant space [3]. That means, if one had a (preferably simple) criterion for testing whether a given query is bounded-memory computable and, moreover, if one had a constructive procedure to generate a memory-bounded algorithm producing exactly the answers of the original query (over all streams), then one would make a considerably big step towards performant stream processing.

A special sub-scenario for performant query answering over streams is to provide simple criteria that help deciding whether a given OMQA query can be answered with sub-linear space w.r.t. the length of the growing stream prefix. An even more special (but important) case dealt with under the term "bounded memory" [13] is that of testing for computability in constant space. We note that, in particular in research on low-level data stream processing for sensor networks [1], there is an equal interest in considering other sub-linear space constraints such as (poly)logarithmic space. In this paper, we focus on constant space requirements, however.

Usually, when considering bounded-memory computability one is interested in what we call here *bounded-memory computability w.r.t. the input*: It denotes the constraint that at most constant space (in the length of the input) is required to store the relevant information of the ever growing input stream(s). Following the approach of [3] we are also going to consider what we call *bounded-memory computability w.r.t. the output*: This notion denotes the constraint of constant space required to memorize the required information of the output produced so far in order to compute new output correctly. This kind of constraint is required in particular for a processing model where the output in each time point consists only of the delta w.r.t. the output written in earlier time points. Such an output model is implemented, e.g., with the so-called IStream operator ("I" for "inserted") in the relational stream query language CQL [4].

As bounded-memory computability is motivated by implementability and performance, it has a flavour of a low-level issue that should be handled when considering the implementation of an OMQA system. However, it would be an asset to have criteria for bounded-memory computability at the ontology level, i.e., to have criteria deciding whether a given query w.r.t. a stream of abox assertions, an ontology and, possibly, integrity constraints can be computed in constant space w.r.t. the length of the stream of abox axioms. The reason is that ontology axioms or integrity constraints may have effects on bounded-memory computability, either positively or negatively. For example, if the ontology allows to formulate rigidity assumptions, then bounded-memory computability may not hold anymore [12]. On the other hand, functional integrity constraints over the whole stream may lead to a bound on an otherwise unbounded set of possible values, thereby ensuring bounded-memory computability.

Of course, if one considers ontology-mediated query answering in the strict sense, namely so-called ontology-based data access (OBDA), there is an obvious alternative approach for testing bounded-memory computability. In OBDA, a query is rewritten w.r.t. the tbox and then unfolded w.r.t. some mappings into an SQL query, so that answers to the original query can be calculated by answering a streamified SQL query over a backend data stream management system. For streamified OBDA this means that one can reduce the bounded-memory test of a query on the ontology level to a bounded-memory test of the transformed query over the backend data stream and then use the known bounded-memory criteria for queries on relational data streams. In this paper, we do not deal with the aspects of mapping and unfolding. Instead we consider how to lift the known criteria [3] for relational stream queries to ontological queries. For this we consider the case of lightweight description logics as representation languages for ontologies so that perfect rewriting of queries according to the OBDA paradigm is possible.

Quite a common scenario of stream processing (in particular for model checking of infinite words [7]) is that the queries on a stream have to be answered over the whole growing prefixes of a stream. Using the window metaphor, this corresponds to applying a window whose right end slides whereas its left end is set to a constant, i.e., to a fixed time point. Of course, the question arises whether the problem of ensuring bounded-memory computability is not solved by using a *finite* sliding window over the data stream. If one considers only row-based windows, i.e., windows where the width-parameter denotes the number of elements that make up its content (see [4]), then bounded-memory computability is always guaranteed by definition of the semantics of row-based windows. But sometimes one cannot easily decide on the appropriate width of the window that is required to capture relevant information on the prefixes of the streams. And even if it would be possible, the necessary size of the window could still be too big so that in optimizing algorithms one could benefit from the use of less memory.

For example, a naively implemented query that requires a quadratic number of comparisons such as a query asking for the monotonic increase of temperature value sensors, may be implementable more efficiently with a data structure storing a state with relevant data that are updated during stream processing. In general, any optimized algorithm would have to rely on some appropriate state data structure. The data structure for states we are going to consider stores values in registers and allows manipulating them with basic arithmetical operators. In low-level stream processing scenarios, where the queries (such as top-k) are required to be answered only approximately, the state data structures are called *sketches*, *summaries* or *synopses*, as these data structures really give some approximate summary of the stream prefixes [1,14].

In this paper we discuss known syntactical criteria for bounded-memory processing of SQL queries over relational data streams [3] and describe how and to what extent these criteria from the database community can be lifted to criteria of bounded-memory query answering in the streamified OMQA setting. For illustration purposes we consider a syntactic criterion of bounded-memory computability applied to a fragment of STARQL [18,19,21,23,24] which was developed as a general query framework for accessing temporal and streaming data in the OMQA paradigm.

2   The STARQL Framework

STARQL is a stream query language framework for OMQA scenarios with temporal and streaming data. As such, it is part of recent efforts of streamifying and temporalizing OMQA [5,6,9–11,15,23,26,28] with, amongst others, contributions by Franz Baader and members of his group. We are referring to STARQL as a framework, because it describes a whole class of query languages which differ regarding the expressivity of the DL used for the tbox and regarding the embedded query languages used to query the individual intra-window aboxes constructed in the sequencing operation (see below).

2.1   Example

The following example for an information need in an agent scenario illustrates the main constructors of STARQL. A rational agent has different sensors, in particular different temperatures attached to different components. The agent receives both, high-level messages and low-level measurement messages, from a single input stream Sin. The agent has stored in a tbox some background knowledge on the sensors. In particular, the tbox contains an axiom stating that all temperature sensors are sensors and that all type-X temperature sensors are temperature sensors. Factual knowledge on the sensors is stored in a (static) abox. For example, the abox may contain assertions *type-X-temperature-Sensor(tcc125)*, *attachedTo(tcc125,c1)*, *locatedAt(c1,rear)* stating that there is a temperature sensor of type X named *tcc125* that is attached to some component *c1* at the rear. There is no explicit statement that *tcc125* is a temperature sensor, this can be derived only with the axioms of the tbox.

The agent has to recognize whether the sensed temperature is critical. Due to some heuristics, a critical state is identified with the following pattern: In the last 5 min there was a monotonic increase on some interval followed by an alert message. As we assume that temperature values have been pre-processed via a smoothing operation, monotonic increase is not prevented from appearing quite often. The agent is expected to output every 10 s all temperature sensors showing this pattern and to mark them as critical. A STARQL formalization of the information need the agent is going to satisfy is given in the listing of Fig. 1.

```
 1 CREATE STREAM Sout AS
 2 CONSTRUCT   GRAPH NOW { ?s a :inCriticalState }
 3 FROM    Sin[NOW-5min, NOW]->10s
 4         <http://www.ifis.uni-luebeck.de/abox>
 5         <http://www.ifis.uni-luebeck.de/tbox>
 6 USING PULSE AS START = 0s, FREQUENCY = 10s
 7 WHERE { ?s a :TempSens }
 8 SEQUENCE BY StdSeq
 9 HAVING
10 EXISTS i1, i2, i3:
11 0 < i1 AND i2 < MAX AND plus(i2,1,i3) AND i1 < i2
12 GRAPH i3 { ?s :message ?m  . ?m a :AlertMessage } AND
13 FORALL i, j, ?x,?y:
14    IF  i1 <= i  AND i <  j  AND j <= i2  AND
15       GRAPH i { ?s :val ?x }  AND GRAPH j { ?s :val ?y }
16    THEN ?x <= ?y
```

Fig. 1. Example STARQL query

The CONSTRUCT operator (line 2) fixes the format of the output stream. Here, as well as in the HAVING clause (see below), STARQL uses the named-graph notation of the W3C recommended RDF[1] query language SPARQL[2] for specifying a basic graph pattern (BGP) and attaching a time expression. The output stream contains expressions of the form

$$\text{GRAPH NOW } \{ \text{ ?s a :inCriticalState } \}$$

where NOW is instantiated by time points and ?s by constants fulfilling the required conditions as specified in the following lines of the query. The evolvement of the time NOW is specified in the pulse declaration (line 6).

The resources to which the query refers are specified using the keyword FROM (line 3–5). Following this keyword one can refer to one or more streams (by names or further stream expressions) and to URIs to a tbox and an abox, which are understood as static knowledge bases. In this example, only one stream is referenced, the stream named S_{in}. In this case, the stream consists, first, of timestamped triples matching the BGPs of the form

$$\text{GRAPH t1 } \{ \text{ ?s :val ?y } \}$$

stating that ?s has value ?y at time t1. In logical notation, these subgraphs would be written as timestamped abox assertions of the form $val(?s, ?y)\langle t1 \rangle$. Secondly, the input stream may contain timestamped triples matching BGPs of the form

---

[1] https://www.w3.org/RDF/.
[2] https://www.w3.org/TR/rdf-sparql-query/.

GRAPH t2 { ?m a :AlertMessage }

stating that at time point t2 an alert message arrived. In DL-notation this would be expressed as: $AlertMessage(?m)\langle t2 \rangle$. The window operator attached to the input stream, [NOW-5 min, NOW]->10 s, is meant to give snapshots of the stream with the slide of 10 s (update frequency) and range of 5 min (all stream elements within last 5 min).

For both types of BGPs the number of possible triples in a stream are unbounded: in the first case this is due to the attribute val with its range being real numbers, an infinite (even dense and continuous) domain to represent possible measurement values. In the second case, this is due to the possibly infinite number of messages that are generated. We think of messages being produced by a controller that generates message IDs from a (discrete but still) infinite domain.

The WHERE clause (line 7) specifies the sensors ?s that are relevant for the information need, namely temperature sensors. Already here it becomes clear that the agent has to incorporate his background knowledge from the tbox: in order to get all temperature sensors ?s it also has to find all type-X sensors. The WHERE clause is evaluated only against the static abox. The stream-temporal conditions are specified in HAVING clause.

For every binding of ?s, the query evaluates conditions that are specified in the HAVING clause (lines 9–16). A sequencing method (here StdSeq) maps an input stream to a sequence of aboxes, annotated by states[3] i, j, according to a grouping criterion. Note that the index variables for states $i1, i2$ are not prefixed by a question mark, as is done for the other variables. This is to indicate the different types of variables. Index variables need to be bound by a quantifier, they are not allowed as answer variables. The built-in sequencing method StdSeq is called *standard sequencing*. It puts all stream elements with the same timestamp into the same mini abox. Note that abox sequencing gives the user the flexibility of defining its own abox sequence—whereas most of the other approaches of temporalized and streamified OMQA, such as [6] already presuppose a sequence of aboxes. This flexibility, on the other hand, means a burden for classical OBDA where queries have to be transformed to queries over the backend. But fortunately for simple sequencing strategies (and possibly for others) such as standard sequencing one can get rid of the additional sequencing layer by reducing the state indexes to the timestamps of the triples in the stream (see [24]).

Testing for conditions at a state is done with the SPARQL sub-graph mechanism. So, e.g.,

GRAPH i3 {?s :message ?m . ?m a :AlertMessage}   (line 12)

asks whether ?s showed an alert message at a state annotated by the variable i3. State i3 is further determined as the successor of the end state i2 in the

---

[3] Note that we prefer to use the term "state" instead of the temporally connotated "stage", because we allow in principle sequencing methods that are not temporal, e.g., sequencing by clustering.

$$
\begin{aligned}
\textit{starqlQuery} &\longrightarrow [\textit{prefix}]\ \textit{createExp} \\
\textit{createExp} &\longrightarrow \texttt{CREATE STREAM}\ \textit{sName}\ \texttt{AS} \\
&\qquad \textit{constrExp} \\
\textit{pulseExp} &\longrightarrow \texttt{PULSE AS} \\
&\qquad \texttt{START = } \textit{start}, \\
&\qquad \texttt{FREQUENCY = } \textit{freq} \\
\textit{constrExp} &\longrightarrow \texttt{CONSTRUCT}\ \textit{cHead}(x, y) \\
&\qquad \texttt{FROM}\ \textit{listWStrExp} \\
&\qquad \underline{URI - To - abox}, \\
&\qquad \underline{URI - To - tbox} \\
&\qquad [\texttt{USING}\ \textit{pulseExp}] \\
&\qquad [\texttt{WHERE}\ \textit{whereCl}(x)] \\
&\qquad \texttt{SEQUENCE BY}\ \textit{seqMeth} \\
&\qquad \texttt{HAVING}\ \textit{safeHCl}(x, y) \\
\textit{cHead}(x, y) &\longrightarrow \texttt{GRAPH}\ \textit{timeExp}\ \textit{triple}(x, y) \\
&\qquad \{\ .\ \textit{cHead}(x, y)\} \\
\textit{listWStrExp} &\longrightarrow (\textit{sName} \mid \textit{constrExp})\ \textit{winExp} \\
&\qquad [,\ \textit{listWStrExp}] \\
\textit{winExp} &\longrightarrow [\textit{timeExp}_1, \textit{timeExp}_2]\text{-}{>}sl \\
\textit{timeExp} &\longrightarrow \texttt{NOW} \mid \texttt{NOW - } \textit{constant} \mid \textit{constant}
\end{aligned}
$$

$$
\begin{aligned}
\textit{whereCl}(x) &\longrightarrow \boxed{\textit{ECL}(x)} \\
\textit{seqMeth} &\longrightarrow \texttt{StdSeq} \mid \textit{seqMeth}(\sim) \\
\textit{term}(i) &\longrightarrow i \\
\textit{term}() &\longrightarrow \texttt{MAX} \mid 0 \mid 1 \\
\textit{arAt}(i_1, i_2) &\longrightarrow \textit{term}_1(i_1)\ \textit{op}\ \textit{term}_2(i_2) \\
&\qquad (\textit{op} \in \{\texttt{<,<=, =, >, >=}\}) \\
\textit{arAt}(i_1, i_2, i_3) &\longrightarrow \texttt{plus}(\textit{term}_1(i_1), \\
&\qquad \textit{term}_2(i_2), \\
&\qquad \textit{term}_3(i_3)) \\
\textit{stateAt}(x, i) &\longrightarrow \texttt{GRAPH}\ i\ \boxed{\textit{ECL}}\ (x) \\
\textit{atom}(x) &\longrightarrow \textit{arAt}(x) \mid \textit{stateAt}(x) \\
\textit{hCl}(x) &\longrightarrow \textit{atom}(x) \mid \textit{hCl}(x)\ \texttt{OR}\ \textit{hCl}(x) \\
\textit{hCl}(x, y) &\longrightarrow \textit{hCl}(x)\ \texttt{AND}\ \textit{hCl}(y) \\
\textit{hCl}(x) &\longrightarrow \textit{hCl}(x)\ \texttt{AND FORALL}\ y \\
&\qquad \texttt{IF}\ \textit{hCl}(x, y)\ \texttt{THEN}\ \textit{hCl}(x, y) \\
\textit{hCl}(x, z) &\longrightarrow \texttt{EXISTS}\ y\ \textit{hCl}(x, y)\ \texttt{AND} \\
&\qquad \textit{hCl}(z, y) \\
\textit{safeHCl}(x) &\longrightarrow \textit{hCl}(x) \\
&\qquad (x\ \text{contains no}\ i\ \text{variable})
\end{aligned}
$$

Fig. 2. Syntax for STARQL(OL, ECL) template.

interval [i1, i2] (line 11). Over the interval [i1, i2] the usual monotonicity condition (FORALL condition, lines 13–15) is expressed using a first-order logic pattern. Note that a naive implementation of this condition would store all values received so and make a quadratic number of comparisons over them.

As in the case of the WHERE clause, also for the evaluation of the HAVING clause the background knowledge (static tbox and static abox) must be incorporated in order to guarantee a complete set of answers. For example, the tbox may contain a taxonomy of different types of messages, in particular different sub-types of alert messages. If only instances of these subtypes are mentioned in the abox, then their super-types have to be inferred by the agent.

2.2 Syntax

The example in the previous subsection illustrated the syntax and the intended semantics of STARQL. For the sake of completeness we recapitulate here the grammar that captures the syntax of STARQL. This grammar leads to a sub-fragment of the original STARQL language [22]. In particular, the HAVING clauses are less expressive than the original ones. Further we leave out aggregation constructors and macro definitions. For a full description see [22,23]. For the full STARQL language, the bounded-memory results of this paper may not hold anymore.

The grammar (Fig. 2) is denoted STARQL (OL, ECL) and it contains parameters that have to be specified in its instantiations: the ontology language OL and the embedded condition language ECL. OL constrains the languages of the aboxes and the tboxes that are referred to in the grammar (underlined in Fig. 2). ECL is a query language referring to the signature of the ontology language. STARQL uses ECL conditions in its WHERE and HAVING clauses. The adequate instantiation of STARQL(OL, ECL) may vary depending on the requirements of the use case.

We are not going to discuss the whole grammar but only make some comments on the most interesting part, which is the set of rules for the specification of HAVING clauses (abbreviated hCl in the grammar). In the full STARQL grammar (see [23]) HAVING clauses are allowed to use arbitrary first-order logic constructors, in particular all boolean connectors, and also exists- as well as forall-quantifiers. As STARQL allows infinite domains (such as the real numbers in order to specify, say, temperature values) queries using FOL constructors have to be used with care in order to give safe queries, i.e., queries that output only finite sets of bindings. A query such as $\phi(y) = \neg val(tcc125, y)$ for example is not safe as it would require outputting all of the infinitely many ys not being values of $tcc125$.

This problem is known since the beginning of classical DB theory and it has been handled by describing syntactical rules guaranteeing safeness. A similar approach for handling safeness, but relying on adornments, is described in [23]. The grammar presented here has no adornments but still reflects safety conditions. For example, the boolean connector for disjunction (or) is allowed to be applied only for disjunctions with the same set of open variables. Furthermore, the existential and the forall quantifiers are allowed to quantify only over variables which are guarded. Hence, an exists quantifier over x is allowed only if x is bounded by a safe hCL clause appearing as conjunction in the scope of the exists quantifier. And universally bounded variables are allowed only if they are guarded in with the antecedent of an implication in the scope of the for-all quantifier.

We further note that the grammar allows also unbounded windows, that is, windows of the form [*constant*, NOW] where the left interval point is fixed, so that the window content is going to contain the whole prefixes beginning with the start time point of the query (set to *constant*).

2.3 Semantics

The explication of the semantics for STARQL queries rests on the semantics of the instantiations of the parameter values OL and ECL. The only presumption we make is that the OL and ECL have to fulfill the following condition: There must be a notion of a certain answer of an ECL w.r.t. an ontology. The motivation for such a layered—or as we call it here: separated—definition of the semantics is a strict separation of the semantics provided by the embedded condition languages ECL and the semantics for the stream query language on top of it. Hence the separated semantics has a plug-in-flavor, allowing users to embed any preferred ECL without repeatedly redefining the semantics of the whole query language.

For ease of exposition we assume that the query specifies only one output sub-graph pattern and that there is exactly one static abox $\mathcal{A}_{st}$ and one tbox $\mathcal{T}$. Similar to the approach of [9], the tbox is assumed to be non-temporal in the sense that there are no special temporal or stream constructors. We give a denotational specification $[\![S_{out}]\!]$ of S_{out} recursively by defining the denotations of the components. We will refer to the notion of a temporal abox within this denotation semantics and also later on. A *temporal abox* or *intra-window abox* is a finite set of timestamped abox axioms $ax\langle t\rangle$, with $t \in T$. We call structures of the form $\langle (\mathcal{A}_i)_{i\in[n]}, \mathcal{T}\rangle$ consisting of a finite sequence of aboxes and a pure tbox a *sequenced ontology (SO)*. The index i of the abox $\mathcal{A}_i$ is called its *state index*. So assume that the following query template is given.

$$S_{out} = \texttt{CONSTRUCT GRAPH}\ timeExpCons\ \Theta(\boldsymbol{x},\boldsymbol{y})$$
$$\texttt{FROM}\ S_1\ winExp_1, \ldots, S_m\ winExp_m, \mathcal{A}_{st}, \mathcal{T}$$
$$\texttt{WHERE}\ \psi(\boldsymbol{x})\ \texttt{SEQUENCE BY}\ seqMeth\ \texttt{HAVING}\ \phi(\boldsymbol{x},\boldsymbol{y})$$

Windowing. Let $[\![S_i]\!]$ for $i \in [m]$ be the streams of timestamped abox assertions. The denotation of the windowed stream $ws_i = S_i\ [timeExp_1^i, timeExp_2^i]\texttt{->}sl_i$ is defined by specifying a function F^{winExp_i} s.t.: $[\![ws_i]\!] = F^{winExp_i}([\![S_i]\!])$.

$[\![ws_i]\!]$ is a stream with timestamps from the set $T' \subseteq T$, where $T' = (t_j)_{j\in\mathbb{N}}$ is fixed by the pulse declaration with t_0 being the starting time point of the pulse. The domain of the resulting stream consists of temporal aboxes.

Assume that $\lambda t.g_1^i(t) = [\![timeExp_1^i]\!]$ and $\lambda t.g_2^i(t) = [\![timeExp_2^i]\!]$ are the unary functions of time denoted by the time expressions in the window. For example, if $timeExp_2^i$ is $\texttt{NOW - 5}$, then the function g_2^i is just the function $\lambda t.(t-5)$. We assume that for all t $[\![timeExp_1^i]\!](t) \leq [\![timeExp_w^i]\!](t)$, as otherwise the window would not denote a proper interval. We have to define for every t_j the temporal abox $\tilde{\mathcal{A}}_{t_j}^i \in [\![ws_i]\!]$. If $t_j < sl - 1$, then $\tilde{\mathcal{A}}_{t_j}^i = \emptyset$. Otherwise set first $t_{start}^i = \lfloor t_j/sl \rfloor \times sl$ and $t_{end}^i = max\{t_{start} - (g_2^i(t_j) - g_1^i(t_i)), 0\}$, and define on that basis $\tilde{\mathcal{A}}_{t_j}^i = \{ax\langle t\rangle \mid ax\langle t\rangle \in [\![S]\!]$ and $t_{end}^i \leq t \leq t_{start}^i\}$. Now, the denotations of all windowed streams are joined w.r.t. the timestamps in T': $js([\![ws_1]\!], \ldots, [\![ws_m]\!]) := \{(\bigcup_{i\in[m]} \tilde{\mathcal{A}}_t^i)\langle t\rangle \mid t \in T'$ and $\tilde{\mathcal{A}}_t^i\langle t\rangle \in [\![ws_i]\!]\}$.

Sequencing. The stream $js([\![ws_1]\!], \ldots, [\![ws_m]\!])$ is processed according to the sequencing method specified in the query. The output stream has timestamps from T'. The stream domain now consists of finite sequences of pure aboxes.

The sequencing methods used in STARQL refer to an equivalence relation $\sim$ to specify which assertions go into the same intra-window abox. The relation $\sim$ is required to respect the time ordering, i.e., it has to be a congruence over T. The equivalence classes are referred to as states and are denoted by variables i, j etc.

Let $\tilde{\mathcal{A}}_t\langle t\rangle$ be the temporal abox of $js([\![ws_1]\!], \ldots, [\![ws_m]\!])$ at t. Let $T'' = \{t_1, \ldots, t_l\}$ be the time points occurring in $\tilde{\mathcal{A}}_t$ and let k' be the number of

648   Ö. L. Özçep and R. Möller

equivalence classes generated by the time points in T''. Then define the sequence at t as $(\mathcal{A}_0, \ldots, \mathcal{A}_{k'})$ where for every $i \in [k']$ the abox $\mathcal{A}_i$ is $\mathcal{A}_i = \{ax\langle t' \rangle \mid ax\langle t' \rangle \in \tilde{\mathcal{A}}_t$ and t' in i^{th} equiv. class$\}$. The standard sequencing method StdSeq is just the one using the identity $=$ as equivalence relation. Let $F^{seqMeth}$ be the function realizing the sequencing.

WHERE Clause. In the WHERE clause only $\mathcal{A}_{st}$ and $\mathcal{T}$ are relevant for the answers. So, purely static conditions (e.g. asking for sensor types as in the example above) are evaluated only on $\mathcal{A}_{st} \cup \mathcal{T}$. The result are bindings $a_{wh} \in cert(\psi(\boldsymbol{x}), \langle \mathcal{A}_{st}, \mathcal{T} \rangle)$. This set of bindings is applied to the HAVING clause $\phi(\boldsymbol{x}, \boldsymbol{y})$.

HAVING Clause. STARQL's semantics for the HAVING clauses relies on the certain-answer semantics of the embedded ECL conditions.

The semantics of $\phi(\boldsymbol{a}_{wh}, \boldsymbol{y})$, i.e., the set of certain answers containing bindings for $\boldsymbol{y}$, is defined for every binding $\boldsymbol{a}_{wh}$ from the evaluation of the WHERE clause. The semantics depends on t. Assume that the sequence of aboxes at t is $seq = (\mathcal{A}_0, \ldots, \mathcal{A}_k)$. We define the set of *separation-based certain answers*, denoted: $cert_{sep}(\phi(\boldsymbol{a}_{wh}, \boldsymbol{y}), \langle \mathcal{A}_i \cup \mathcal{A}_{st}, \mathcal{T} \rangle)$.

If for any i the pure ontology $\langle \mathcal{A}_i \cup \mathcal{A}_{st}, \mathcal{T} \rangle$ is inconsistent, then we set $cert_{sep} = \text{NIL}$, where NIL is a new constant not contained in the signature. In the other case, the bindings are defined as follows. For t one constructs a sorted first-order logic structure $\mathfrak{I}_t$: the domain of $\mathfrak{I}_t$ consists of the index set $\{0, \ldots, k\}$ as well as the set of all individual constants of the signature. For every state atom $stateAt$ GRAPH i $ECL(\boldsymbol{z})$ in $\phi(\boldsymbol{a}_{wh}, \boldsymbol{y})$ with free variables $\boldsymbol{z}$ having length l, say, introduce an $(l+1)$-ary symbol R and replace GRAPH i $ECL(\boldsymbol{z})$ by $R(\boldsymbol{z}, i)$. The denotation of R in $\mathfrak{I}_t$ is then defined as the set of certain answers of the embedded condition $ECL(\boldsymbol{z})$ w.r.t. the ith abox $\mathcal{A}_i$: $R^{\mathfrak{I}_t} = \{(\boldsymbol{b}, i) \mid \boldsymbol{b} \in cert(ECL(\boldsymbol{z}), \langle \mathcal{A}_i \cup \mathcal{A}_{st}, \mathcal{T} \rangle)\}$. Constants denote themselves in $\mathfrak{I}_t$. This fixes a structure $\mathfrak{I}_t$ with finite denotations of its relation symbols. The evaluation of the HAVING clause is then nothing more than evaluating the FOL formula (after substitutions) on the structure $\mathfrak{I}_t$.

Let $F^{\phi(a_{wh}, y)}$ be the function that maps a stream of abox sequences to the set of bindings $(\boldsymbol{b}, t)$ where $\boldsymbol{b}$ is the binding for $\boldsymbol{y}$ in $\phi(\boldsymbol{a}_{wh}, \boldsymbol{y})$ at time point t.

Summing up, the following denotational decomposition results:

$$[\![S_{out}]\!] = \{\text{GRAPH } [\![timeExpCons]\!] \, \Theta(\boldsymbol{a}_{wh}, \boldsymbol{b}) \mid \boldsymbol{a}_{wh} \in cert(\psi(\boldsymbol{x}), \mathcal{A}_{st} \cup \mathcal{T}) \text{ and}$$
$$(\boldsymbol{b}, t) \in F^{\phi(a_{wh}, y)}\left(F^{seqMeth}(js(F^{winExp_1}([\![S_1]\!]), \ldots, F^{winExp_m}([\![S_m]\!])))\right)\}$$

Regarding the following considerations on bounded-memory processing we note two points: First, the output is controlled by the pulse. At each evolving time point the whole set of elements with timestamps falling into the current time interval of the window is considered for the calculation of the output. This means that there may be more than one RDF tuple to be processed at each time point. We assume that at each time point the set of RDF tuples to be processed is bounded by a constant, otherwise the stream system could eventually fall behind the pulse. (Of course, it may also fall behind the pulse without the assumption on boundedness by a constant.) But even under this restrictive assumption, bounded-memory computability is an issue due to the non-bounded number of triples in the ever growing prefixes of the input streams. Hence a systematic consideration is in order.

Further we note that the semantics is defined such that at every time point the whole set of bindings that make the WHERE clause and the HAVING clause true is returned, and not the delta of new bindings. That is, the semantics of STARQL follows the idea of the RStream operator of the relational stream query language CQL [4] and not that of the IStream operator.

2.4   Properties of STARQL

Non-reified Approach. A relevant question from the representational point of view is how to represent events and, in particular, time in the query language. For STARQL, the decision was to use a non-reified approach, where time is handled as an annotation for sentences whose evaluation depends on the associated time. As illustrated by the agent example above, the abox assertions (RDF triples in SPARQL speak) are tagged with timestamps. This method is similar to adding an extra time argument for concept and roles as in [5]. The non-reified approach allows for representing time-dependent facts such as the fact that some sensor showed some value at a given time point. This time point is relevant for the window semantics in STARQL.

As the reified approach is more conservative and does not require to change the semantics (the time attribute is treated as an ordinary attribute), a natural question is why STARQL follows the non-reified strategy. The main reason is that time requires a special treatment as it has specific constraints for reasoning. For example, in the measurement scenario one would like to express the constraint that, at every time point, a sensor shows at most one value. This can be done with a classical DL-Lite axiom by stating $(func\ val) \in \mathcal{T}$. Note that under such a constraint it is necessary that the window semantics preserves the timestamps, as is indeed the case for the STARQL window semantics. Otherwise two timestamped stream elements of the form $val(tcc125, 92°)\langle 3\,s\rangle$ and of the form $val(tcc125, 95°)\langle 5\,s\rangle$ would lead to an inconsistency.

On the other hand, if one follows the reified approach such a time-dependent constraint is not expressible in a DL: One would have to formulate that there are no two measurements with the same associated sensor and same timestamp but different values. As DLs are concept oriented, they are not suited to expressing non-tree-shaped constraints with tbox axioms.

Homogeneous Interface. For the syntax and the semantics of STARQL queries the exact resource of the input stream is not relevant: It may be a stream of elements arriving in real-time via a TCP port, but equally it can be a simulated stream of data produced by reading out a text file or a temporal database.

In the former case, one can speak of (genuine) stream querying, whereas in the latter case we use the term *history querying*. So STARQL offers the same interface to real-time queries (as required, for example, in monitoring scenarios) and history queries (as required, e.g., for reactive diagnostics). And, indeed such a homogenous interface to two different modes of querying has proved useful for real industrial use cases, in particular, for the turbine-diagnostics use case of SIEMENS in the context of the OPTIQUE project [17,19,20].

Separation Between Static and Temporal Conditions. As illustrated in the example above, STARQL allows to separate the conditions expressed in an information need into conditions that concern only the static part of the background knowledge (tbox $\mathcal{T}$ and static abox $\mathcal{A}_{st}$) and into conditions which require both, the static part and the streams. The former can be queried in the WHERE clause, the latter in the HAVING clause. For the semantics of HAVING clauses we also incorporated the tbox and the static abox (which is always added to each abox in the sliding window). And indeed, this reference, at first sight, is not eliminable. The reference to the tbox can be eliminated by just rewriting the HAVING clause into a new HAVING clause using the standard perfect rewriting technique. Still, the theoretical question remains whether it is possible to push all references to the static abox (all occurrences of concept and role symbols that appear in the static abox) into the WHERE clause, so that the HAVING clause can be evaluated only on the streams and the bindings resulting from the evaluation of the WHERE clause. In other terms, is the HAVING clause separable in a pure static part and a part containing only role and concept symbols not part of the static abox? This is an open problem.

Even if separability in the sense above holds, in terms of feasible implementation, the reference to a large static abox remains a challenging problem. As far as we know, this problem has not been solved satisfactorily by any of the current temporal and streamified OMQA systems.

OBDA Rewritability. STARQL queries with standard sequencing can be rewritten into queries over backend data stream management system. This is possible because the two layers in STARQL, the semantics of the outer temporal FOL template and the semantics of the embedded ECL queries, are separated. This is similar to the temporal conjunctive queries (TCQs) of [9]. For the details of rewritability and a comparison of STARQL with the query languages of TCQ we refer the reader to [24].

An Alternative Operational Semantics. The window semantics defined above is denotational and mimics the window operator definitions for CQL [4], which is one of the first relational data stream query languages. From the implementation point of view, an operational semantics is more helpful—at least it gives a different perspective on the intended semantics of windows. Furthermore, the operational view also sheds light on why the window definition was chosen exactly the way as stated above. For the details of the operational semantics we refer the reader to [18].

3 A Criterion for Bounded-Memory Computability for SQL Queries over Streams

We have seen that in STARQL, queries can refer to streams that may contain infinitely many different RDF triples. Moreover, we saw that naive implementations of queries such as the linear-space, quadratic-time implementation of monotonicity may lead to non-efficient query processing. Hence, finding good criteria for bounded-memory processing of STARQL queries is a real issue. In order to find such criteria, in the following, we consider criteria known to hold for SQL queries over relational data streams.

The early work of Arasu and colleagues [3] gives syntactic criteria for bounded memory computability of queries in the SPJ (select-project-join fragment) of SQL and also for an extension of SPJ with aggregation operators. In each case they consider both natural set semantics and multi-set (alias: bag) semantics. Moreover they describe an algorithm that in case of bounded-memory computability constructs a corresponding bounded-memory stream algorithm. Though SQL, per se, does not provide stream specific operators as in specific stream query languages (such as CQL [4]), the results are still fundamental enough in order to be adaptable to genuine stream query languages.

The underlying computation model for the bounded-memory results is described only informally in [4]. It is a register machine model extended to handle infinite input streams. Such a computation model can be formally described by streaming abstract state machines [16].

3.1 Query Language and the Query Model

We assume that the user is familiar with the SPJ-Fragment of SQL. We just restate some SPJ queries from [3] in order to illustrate the memory-boundedness criterion.

Assume that you have two homogeneous data streams, one containing tuples of the form $S(A, B, C)$ with a ternary relation S and a stream containing tuples of the form $T(D, E)$. All attributes (here A, B, C, D, E) are assumed to range over the integers. The queries are constructed using a projection operator $\Pi \in \{\pi, \dot{\pi}\}$, where π is the duplicate eliminating projection operator and $\dot{\pi}$ is the duplicate-preserving operator. The selection operator σ is restricted to conjunctions of atoms of the form $X = Y$ and $X > Y$, where X, Y are either attributes or integer constants. The join is a full join with the cartesian product $\times$. An example query which is evaluated with multi-set semantics, i.e., Duplicate Preserving, is the following query:

$$Q_3^{DP} = \dot{\pi}_A(\sigma_{(A=D \wedge A>10 \wedge D<20)}(S \times T))$$

The query asks for all values A (with duplicates) with $20 > A > 10$ such that there are tuples of the form $S(A, \cdot, \cdot)$ and $T(A, \cdot)$. In logical notation, this is the conjunctive query

$$\exists B, C, D, E.S(A, B, C) \wedge T(D, E) \wedge A = D \wedge A > 10 \wedge D < 20$$

What is the process model for evaluating this query, when S and T do not stand for static tables but streams?

The query is answered over one inhomogeneous big stream of tuples. The stream is inhomogeneous in the sense that tuples belonging to different relations may arrive (in case of the above query: tuples from S and tuples from T.) This is usually the case in the area of complex event processing (see, e.g., [2]). The idea is that the big stream is the result of merging—or *interleaving* as Arasu and colleagues call merging—many homogeneous streams (i.e. streams where every tuple belongs to exactly to one relation, here: the two homogeneous streams associated with S and T). Interleaving means that an arbitrary sequence of tuples is fixed which consists of tuples from the referenced homogeneous stream.[4] For example, if $S = \langle S(1, 1, 1), S(2, 2, 2), S(3, 3, 3), \ldots \rangle$ and $T = \langle T(1, 1), T(2, 2), T(3, 3), \ldots \rangle$, the following big stream is a possible interleaving

$$BS_1 = \langle S(1, 1, 1), S(2, 2, 2), S(3, 3, 3), T(1, 1), T(2, 2), T(3, 3), \ldots \rangle$$

Another is

$$BS_2 = \langle S(1, 1, 1), T(1, 1), S(2, 2, 2), T(2, 2), S(3, 3, 3), T(2, 2), \ldots \rangle$$

and so on. In many modern stream query languages following a pipeline architecture, these kinds of interleavings are not completely outsourced to a system but are controlled with the query language using cascading of stream queries. Such a control is given also in STARQL. The criteria of memory-boundedness mentioned below are to be understood to hold for all (!) possible interleavings.

Now, how is a query such as Q_3^{DP} evaluated? Every time t a new tuple in the big stream BS arrives it is stored in an ordinary SQL DB containing all tuples arrived so far. The query is evaluated on the accumulated DB with the last tuple. So, one has a notion of an output of a query Q at time t over the big input stream BS, $ans(Q, t, BS)$, which is defined as $Q^{DB(BS^{\leq t})}$ that is the answer of the query Q on the accumulated DB from the t-prefix of the stream BS. The output at every time t is a set (or multi-set). This definition of the output stream corresponds to the IStream semantics of CQL [4].

Now one could associate with a query and a big stream BS the stream of answers $(ans(Q, t, BS)_{t \in \mathbb{N}})$. But actually, the authors of [3] associate an output stream with a query over the input big stream in a different way as they want to have a stream of tuples again. So they consider a stream of elements produced so far and consider the multi-set-union over this prefix as the intended answer of

---

[4] We note that there is no fairness assumption for the interleavings in [3].

the query. As the authors consider only monotonic queries they assume that the answer stream can be given by reference to the answers produced so far multi-unioned with the answer produced at the current time stamp. To formalize this, let us assume that a query Q maps an input stream BS_{in} into an output stream $Q(BS_{in}) = BS_{out}$. Then, one demands that for every arrival time point t one has $ans(Q, BS_{in}^{\leq t}) = \uplus BS_{out}^{\leq t}$, where $\uplus$ is defined for a sequence of elements $(s_i)_{i \leq t}$ as the multi-set of elements by multi-union of all the $\{s_i\}$.

3.2   Criterion for SPJ Queries

The following table gives some example queries and states which of them are bounded-memory computable. Duplicate preserving queries have a DP superscript, duplicate eliminating ones have a DE superscript. As before, we assume two homogenous streams, with elements of the form $S(A, B, C)$ and the other with elements of the form $T(D, E)$.

Acronym	Query	Memory-Bounded?
Q_1^{DP}	$= \dot{\pi}_A(\sigma_{(A>10)}(S))$	yes
Q_1^{DE}	$= \pi_A(\sigma_{(A>10)}(S))$	no
Q_3^{DP}	$= \dot{\pi}_A(\sigma_{(A=D \wedge A>10 \wedge D<20)}(S \times T))$	yes
Q_3^{DE}	$= \pi_A(\sigma_{(A=D \wedge A>10 \wedge D<20)}(S \times T))$	yes
Q_4^{DP}	$= \dot{\pi}_A(\sigma_{(B<D \wedge A=10)}(S \times T))$	no
Q_4^{DE}	$= \pi_A(\sigma_{(B<D \wedge A=10)}(S \times T))$	yes

The first query Q_1^{DP} is memory bounded as it acts as a simple filter: there is no join condition and the query answering system does not have to eliminate duplicates. This is different for the query Q_1^{DE} which is the same as the first except for using duplicate elimination: As A is not bounded from above, the system would have to store any $A > 10$ arrived in S so far in order not to output them a second time.

Both queries Q_3^{DP}, Q_3^{DE} are memory-bounded. The algorithm in the duplicate-preserving case has synopses for S and T, resp. Both synopses consist of registers for all integer values v in the range $[11, 19]$. A register for value v in the S-synopsis stores the number of S-tuples having $A = v$. Similarly the register for value v in the T-synopsis counts all T-tuples arrived so far with value $D = v$. Now, assume for example that the next element in the big stream is an S-tuple with $A = v$. If v is not in the interval $[11, 19]$, it is ignored. Otherwise one considers the number of T-tuples in the v-register of the T-synopsis. This number of v tuples is put onto the output stream. The duplicate-eliminating case is similar but one stores just boolean values in the registers instead of number counts.

In case of the pair of queries Q_4^{DP}, Q_4^{DE} the duplicate-preserving one is not bounded-memory computable, whereas the duplicate eliminating query is. Regarding the latter one constructs a synopsis for S where the minimum value of attribute B among all tuples of S with $A = 10$ having arrived so far are stored, and one has a T-synopsis, in which the maximum value of attribute D among all tuples of T so far is stored. For the duplicate-preserving query Q_4^{DP} it is not enough to know whether a stream joins with a past stream but one has to count the number—and these numbers are not bounded.

We state the syntactic criterion for the duplicate-eliminating case only as we are going to consider the set-semantics for STARQL only. The syntactic criterion is formulated for SPJ-queries that have a special form. These queries are called locally totally ordered queries, for short, *LTO queries*. As every query is equivalent to a union of LTO queries, a query is memory-bounded iff all its LTOs are (Theorem 5.3 in [3]). An *LTO query* Q is a query in which for every stream S referenced in Q the union of attributes in S and all constants occurring in Q are totally ordered.

Let P be a selection predicate, i.e., a conjunction of atoms of the form $X > Y, or X = Y$. Let P^+ denote the set of all atoms entailed by P that contain only symbols of P.

An attribute A is called *lower-bounded* (resp. *upper-bounded*) if there exists an atom $A > k \in P^+$ (resp. $A < k \in P^+$), or an atom $A = k \in P^+$ for some constant k. A is *bounded* if it is both upper-bounded and lower-bounded. Two elements e and d (variables or constants) are called equivalent w.r.t. a set of predicates iff $e = d \in P^+$. Then $\mid E \mid_{eq}$ denotes the number of equivalence classes into which a set of elements E is partitioned according to the equivalence relation above.

Now consider a stream S_i referenced in a query Q. $MaxRef(S_i)$ is defined as the set of all lower bounded but not upper-bounded attributes A of S_i such that A appears in a non-redundant inequality join $(S_j.B < S_i.A)$, $i \neq j$, in P^+. The definition of $MinRef(Si)$ is the dual of the definition of $MaxRef(S_i)$. With this definition the following characterization can be proved.

Theorem 1. ([3] , Theorem 5.10). *Let* $Q = \pi_L(\sigma_P(S_1 \times \cdots \times S_n))$ *be an LTO query.* Q *is bounded-memory computable iff:*

C1: *Every attribute in the list* L *of projected attributes is bounded.*
C2: *For every equality join predicate* $(S_i.A = S_j.B)$, $i \neq j$, $S_i.A$ *and* $S_j.B$ *are bounded.*
C3: $\mid MaxRef(S_i) \mid_{eq} + \mid MinRef(S_i) \mid_{eq} \leq 1$ *for* $i \in \{1, \dots, n\}$.

4   Lifting the Criteria of Bounded-Memory Criteria Bounded-Memory Computability to OMQA

As STARQL can be used with unbounded windows, in STARQL, we face problems similar to the ones for the model of SQL stream processing described in the previous section. Even if one were to consider finite windows, considerations on bounded-memory stream processing could give insights into optimization means. In the previous section the relational tuples were defined over the integers, a discrete, infinite domain. If one considers also dense domains, then one has a similar if not a more difficult problem of ensuring bounded-memory computability. But even here one can sometimes guarantee bounded-memory computability as the following monotonicity example (a variant of the example from the beginning) suggests.

Consider the simple monotonicity query in Fig. 3, asking every second whether the temperature in sensor s0 increased monotonically up to the current time point.

```
1 ...
2 {FROM SMsmt [0, NOW]->1s }
3 ...
4 {HAVING FORALL i < j IN SEQ1,?x,?y:
5         IF {s0 val ?x}<i>  AND {s0  val ?y}<j>
6         THEN ?x <= ?y  }
```

Fig. 3. Simple monotonicity query

A straight-forward implementation of this query is to construct from scratch sequences of [0, NOW]-windows and test on these the monotonicity condition by iterating trough all possible state-pairs (i, j). But this results in a test of quadratic order (w.r.t. time). It is not hard to see that first one can find a complete and correct algorithm that is not quadratic in time and uses constant space only (w.r.t. the uniform cost measure in register machines): it just stores the maximal temperature value for the last time point and compares it with the values arriving at the current time point. Of course, this optimization is possible only if it can be guaranteed that the input streams are not out of sync, i.e., if tuples with earlier time stamps than the current time point are excluded. In such an asynchronous case one would have to store all possible measurement values.

The results of [3] can be adapted to formulate criteria for bounded-memory computability on STARQL queries. For this we consider the following variant of STARQL, which, syntactically, is a simple fragment, called CQ-fragment of STARQL and denoted STARQL^{CQ}, but which, semantically, differs in applying the IStream semantics and not the RStream semantics.

Definition 1. *Syntactically STARQL^{CQ} is defined as that fragment of STARQL adhering to the following constraints:*

1. *The FROM clause refers only to streams that are streams of abox axioms (or RDF tuples).*
2. *All sliding windows are unbounded windows (with the same slide which is identical to the pulse).*
3. *The WHERE clause is allowed to be any reasonable query language allowing a certain answer semantics.*
4. *The sequencing strategy is that of standard sequencing.*

5. The `HAVING` fragment is restricted to consist of conjunctive queries of the form

$$EXISTS\ y_1, y_2, \ldots, y_n A(x_1, \ldots, x_m, y_1, \ldots, y_n)$$

where the x_i are non-state variables which may occur as free variables in the `WHERE` clause or the x_i are state variables, and the y_i are state variables or non-state variables bounded by the `EXISTS` quantifier. The expression after the quantifiers $A(x_1, \ldots, x_m, y_1, \ldots, y_n)$ is a conjunction of atoms in which only variables $x_1, \ldots, x_n, y_1, \ldots, y_n$ may occur and which have one of the following forms.
- `GRAPH` i $r(x, y)$ where
  r is a role symbol, x is a constant or a variable in $\{x_1, \ldots, x_m, y_1, \ldots, y_n\}$
- `GRAPH` i $C(x)$ where
  C is an atomic concept symbol, x is a constant or a variable in the set $\{x_1, \ldots, x_m, y_1, \ldots, y_n\}$
- x op y where
  $op \in \{<, >, =\}$ and x is a constant or variable in $\{x_1, \ldots, x_m, y_1, \ldots, y_n\}$

Semantically, $STARQL^{CQ}$ uses the IStream semantics.

Please note that in the first item of the definition we exclude the reference to streams that are constructed with other STARQL queries. Otherwise we would have to consider criteria for composed queries. We do not exclude that this is possible, but the adaptation would be rather awkward.

Now we get the following adapted version of the syntactic criterion.

Proposition 1. Let Q be an LTO query in the $STARQL^{CQ}$ fragment. We make the following assumptions

1. The streams are interleaved in a synchronized way, i.e., the time points of tuples adheres to the arrival ordering.
2. At every time point only a finite number of elements bounded by some constant can arrive.
3. The tbox is empty. (But see Corollary 1 where this assumption is dropped.)

Then, Q is memory-bounded iff it fulfills the following constraints.

C1: Every variable appearing in the `HAVING` clause is either bounded by `EXISTS` or is a-bounded or occurs free in the `WHERE` clause.
C2: For every non-state variable that occurs in two atoms or in an identity atom: if it is not a state-variable, then it is bounded or occurs free in the `WHERE` clause.
C3: $|\ MaxRef(S_i)\ |_{eq} + |\ MinRef(S_i)\ |_{eq} \leq 1$ for $i \in \{1, \ldots, n\}$.

Please note that we have the following differences w.r.t. the criterion of [3]: First of all, there is a `WHERE` clause. Evaluating the variables occurring in a `WHERE` clause does not pose a problem regarding bounded-memory computability as these variables refer to the static database (which is finite). So all variables bounded by the WHERE clause can be considered to be bounded in the sense of Arasu.

The STARQL query language allows composing queries, i.e., a STARQL query may refer to streams produced by other STARQL queries. This composition leads to constraints regarding interleaving which are not captured by the criteria of Arasu (hence we considered only non-cascaded queries).

In the case of STARQL and STARQL^{CQ} the time flow is not restricted to a discrete domain. In the case of non-discrete time domains it is important to know that the streams are synchronized. Otherwise, as we mentioned above in the monotonicity example, one would have to store measurement values for all time points of tuples having arrived so far as one cannot exclude the case that values for time points in between may arrive.

As the proposition shows, the impossibility of bounded-memory processing may also be due to non-bounded memory computability w.r.t. the output. This was one reason why, in the original definition of STARQL, we decided to use the RStream semantics for STARQL. The other reason was that for non-monotonic queries and aggregation queries one cannot rely on IStream semantics. On the other hand, the use of an RStream semantics means that due to the possibility of unbounded sets of answers at every time point, the system may fall behind the pulse requirements. So, we have here a classical opposition of time and space constraints.

In the proposition we assume that the tbox is empty. In the STARQL framework the tbox is assumed to be atemporal. But even then one cannot exclude that a tbox axiom may lead to the loss of bounded-memory computability, although the query w.r.t. the empty tbox is bounded-memory computable. For example, a simple role inclusion axiom may lead to a self join in the query, which is handled via other syntactic criteria. It is an interesting open question to find criteria on the tbox that preserve the correctness and completeness of the syntactical criteria for bounded-memory computability.

On the other hand, if we consider OBDA, which allows for perfect rewriting of queries, then we can apply the criterion of Proposition 1 to each completion of each conjunctive query in the rewritten query (which is a union of CQs, for short a UCQ).

Corollary 1. *Let Q be a query in the STARQL^{CQ} fragment w.r.t. some DL allowing for perfect rewritability. Let Q_{rew} be the rewritten UCQ. We assume that each prefix of the abox stream is consistent with the ontology.*

If each of the LTO queries to each CQ in Q_{rew} fulfills the conditions mentioned in Proposition 1, then and only then, Q is bounded-memory computable.

Note that we assume consistency of the abox-stream prefixes with the tbox. Otherwise one would have to test for inconsistency. This test can be reduced to first-order logic queries but the resulting queries are not bounded-memory computable. Consider, e.g., a negative inclusion $A \sqsubseteq \neg B$ which would lead to an unbounded query $\exists A(x) \land B(x)$. Similar considerations follow for functional constraints.

On the other hand, when considering such axioms to hold only on the extensional part of the ontology, i.e., if we consider integrity constraints on the abox streams, then it is possible to gain bounded-memory computability. In the monotonicity example above, we already discussed a similar form of integrity constraint, though it was not formulated as a DL axiom. A systematic study of the consequences of integrity constraints for bounded memory computability is left for future work.

5   Conclusion

The study of bounded-memory computability of OMQA queries can profit from corresponding results on bounded-memory processing over relational data streams. The adaptations are not trivial in the presence of tboxes—for languages in which the tbox cannot be compiled away. We presented a syntactical criterion for the information processing paradigm of OBDA in which the test of bounded-memory processing w.r.t. a non-empty tbox could be reduced to a test with an empty tbox. The question of how the syntactical criterion of [3] can be adapted to the general OMQA case, in particular for tboxes in which temporal constructors are allowed [5], is an open problem. We guess that due to the syntacticality of the criterion, the adaptation is not going to be obvious. Hence, as a further future research topic we think of an equivalent semantic criterion for bounded-memory processing using the framework of dynamic complexity [25].

References

1. Aggarwal, C.C. (ed.): Data Streams: Models and Algorithms. Advances in Database Systems, vol. 31. Springer, Heidelberg (2007). https://doi.org/10.1007/978-0-387-47534-9
2. Agrawal, J., Diao, Y., Gyllstrom, D., Immerman, N.: Efficient pattern matching over event streams. In: Proceedings of the 2008 ACM SIGMOD International Conference on Management of Data, pp. 147–160, SIGMOD 2008. ACM, New York (2008)
3. Arasu, A., Babcock, B., Babu, S., McAlister, J., Widom, J.: Characterizing memory requirements for queries over continuous data streams. ACM Trans. Database Syst. **29**(1), 162–194 (2004)
4. Arasu, A., Babu, S., Widom, J.: The CQL continuous query language: semantic foundations and query execution. VLDB J. **15**, 121–142 (2006)
5. Artale, A., Kontchakov, R., Wolter, F., Zakharyaschev, M.: Temporal description logic for ontology-based data access. In: Proceedings of the Twenty-Third International Joint Conference on Artificial Intelligence, IJCAI 2013, pp. 711–717. AAAI Press (2013). http://dl.acm.org/citation.cfm?id=2540128.2540232
6. Baader, F., Borgwardt, S., Lippmann, M.: Temporalizing ontology-based data access. In: Bonacina, M.P. (ed.) CADE 2013. LNCS (LNAI), vol. 7898, pp. 330–344. Springer, Heidelberg (2013). https://doi.org/10.1007/978-3-642-38574-2_23
7. Bauer, A., Küster, J.-C., Vegliach, G.: From propositional to first-order monitoring. In: Legay, A., Bensalem, S. (eds.) RV 2013. LNCS, vol. 8174, pp. 59–75. Springer, Heidelberg (2013). https://doi.org/10.1007/978-3-642-40787-1_4

8. Bienvenu, M.: Ontology-mediated query answering: harnessing knowledge to get more from data. In: Kambhampati, S. (ed.) Proceedings of the Twenty-Fifth International Joint Conference on Artificial Intelligence, IJCAI 2016, New York, NY, USA, 9–15 July 2016, pp. 4058–4061. IJCAI/AAAI Press (2016). http://www.ijcai. org/Abstract/16/600

9. Borgwardt, S., Lippmann, M., Thost, V.: Temporal query answering in the description logic *DL-Lite*. In: Fontaine, P., Ringeissen, C., Schmidt, R.A. (eds.) FroCoS 2013. LNCS (LNAI), vol. 8152, pp. 165–180. Springer, Heidelberg (2013). https:// doi.org/10.1007/978-3-642-40885-4_11

10. Borgwardt, S., Lippmann, M., Thost, V.: Temporalizing rewritable query languages over knowledge bases. J. Web Semant. **33**, 50–70 (2015). https://doi.org/ 10.1016/j.websem.2014.11.007. http://www.sciencedirect.com/science/article/pii/ S157082681400119X

11. Calbimonte, J.P., Jeung, H., Corcho, O., Aberer, K.: Enabling query technologies for the semantic sensor web. Int. J. Semant. Web Inf. Syst. **8**(1), 43–63 (2012). https://doi.org/10.4018/jswis.2012010103

12. Chomicki, J.: Efficient checking of temporal integrity constraints using bounded history encoding. ACM Trans. Database Syst. **20**(2), 149–186 (1995)

13. Chomicki, J., Toman, D.: Temporal databases. In: Handbook of Temporal Reasoning in Artificial Intelligence, vol. 1, pp. 429–467. Elsevier (2005)

14. Cormode, G.: The continuous distributed monitoring model. SIGMOD Rec. **42**(1), 5–14 (2013)

15. Della Valle, E., Ceri, S., Barbieri, D., Braga, D., Campi, A.: A first step towards stream reasoning. In: Domingue, J., Fensel, D., Traverso, P. (eds.) Future Internet - FIS 2008. Lecture Notes in Computer Science, vol. 5468, pp. 72–81. Springer, Heidelberg (2009)

16. Gurevich, Y., Leinders, D., Van den Bussche, J.: A theory of stream queries. In: Arenas, M., Schwartzbach, M.I. (eds.) DBPL 2007. LNCS, vol. 4797, pp. 153–168. Springer, Heidelberg (2007). https://doi.org/10.1007/978-3-540-75987-4_11

17. Kharlamov, E., et al.: Towards analytics aware ontology based access to static and streaming data. In: Groth, P., et al. (eds.) ISWC 2016. LNCS, vol. 9982, pp. 344–362. Springer, Cham (2016). https://doi.org/10.1007/978-3-319-46547-0_31

18. Kharlamov, E., et al.: An ontology-mediated analytics-aware approach to support monitoring and diagnostics of static and streaming data. J. Web Seman. (2018, in print)

19. Kharlamov, E., et al.: Semantic access to streaming and static data at Siemens. Web Semant.: Sci. Serv. Agents World Wide Web **44**, 54–74 (2017). https://doi. org/10.1016/j.websem.2017.02.001

20. Kharlamov, E., et al.: How semantic technologies can enhance data access at siemens energy. In: Mika, P., et al. (eds.) ISWC 2014. LNCS, vol. 8796, pp. 601–619. Springer, Cham (2014). https://doi.org/10.1007/978-3-319-11964-9_38

21. Özçep, Ö.L., Möller, R.: Ontology based data access on temporal and streaming data. In: Koubarakis, M., et al. (eds.) Reasoning Web 2014. LNCS, vol. 8714, pp. 279–312. Springer, Cham (2014). https://doi.org/10.1007/978-3-319-10587-1_7

22. Özçep, Ö.L., Möller, R., Neuenstadt, C., Zheleznyakov, D., Kharlamov, E.: Deliverable D5.1 - a semantics for temporal and stream-based query answering in an OBDA context. Deliverable FP7-318338, EU, October 2013

23. Özçep, Ö.L., Möller, R., Neuenstadt, C.: A stream-temporal query language for ontology based data access. In: Lutz, C., Thielscher, M. (eds.) KI 2014. LNCS (LNAI), vol. 8736, pp. 183–194. Springer, Cham (2014). https://doi.org/10.1007/ 978-3-319-11206-0_18

24. Özçep, Ö.L., Möller, R., Neuenstadt, C.: Stream-query compilation with ontologies. In: Pfahringer, B., Renz, J. (eds.) AI 2015. LNCS (LNAI), vol. 9457, pp. 457–463. Springer, Cham (2015). https://doi.org/10.1007/978-3-319-26350-2_40

25. Patnaik, S., Immerman, N.: Dyn-FO: a parallel, dynamic complexity class. J. Comput. Syst. Sci. **55**(2), 199–209 (1997)

26. Le-Phuoc, D., Dao-Tran, M., Xavier Parreira, J., Hauswirth, M.: A native and adaptive approach for unified processing of linked streams and linked data. In: Aroyo, L., et al. (eds.) ISWC 2011. LNCS, vol. 7031, pp. 370–388. Springer, Heidelberg (2011). https://doi.org/10.1007/978-3-642-25073-6_24

27. Thost, V.: Using ontology-based data access to enable context recognition in the presence of incomplete information. Ph.D. thesis, TU Dresden (2017)

28. Turhan, A., Zenker, E.: Towards temporal fuzzy query answering on stream-based data. In: Nicklas, D., Özçep, Ö.L. (eds.) Proceedings of the 1st Workshop on High-Level Declarative Stream Processing Co-located with the 38th German AI Conference (KI 2015), CEUR Workshop Proceedings, Dresden, Germany, 22 September 2015, vol. 1447, pp. 56–69. CEUR-WS.org (2015). http://ceur-ws.org/Vol-1447/paper5.pdf

Author Index

Printed in the United States
By Bookmasters